CRC HANDBOOK OF

LIQUID-LIQUID EQUILIBRIUM DATA *of* POLYMER SOLUTIONS

CRC HANDBOOK OF

Liquid-Liquid Equilibrium Data of Polymer Solutions

Christian Wohlfarth

CRC Press is an imprint of the
Taylor & Francis Group, an informa business

CRC Press
Taylor & Francis Group
6000 Broken Sound Parkway NW, Suite 300
Boca Raton, FL 33487-2742

First issued in paperback 2021

ISBN 13: 978-1-03-209967-5 (pbk)
ISBN 13: 978-1-4200-6798-9 (hbk)

Publisher's Note
The publisher has gone to great lengths to ensure the quality of this reprint but points out that some imperfections in the original copies may be apparent.

Library of Congress Cataloging-in-Publication Data

Wohlfarth, C.
CRC handbook of liquid-liquid equilibrium data of polymer solutions / Christian Wohlfarth.
p. cm.
Includes bibliographical references and index.
ISBN 978-1-4200-6798-9 (hardback : alk. paper)
1. Polymer solutions. 2. Liquid-liquid equilibrium. I. Title.

QD381.9.S65W6324 2007
547'.70454--dc22 2007035787

Visit the Taylor & Francis Web site at
http://www.taylorandfrancis.com

and the CRC Press Web site at
http://www.crcpress.com

PREFACE

Knowledge of thermodynamic data of polymer solutions is a necessity for industrial and laboratory processes. Furthermore, such data serve as essential tools for understanding the physical behavior of polymer solutions, for studying intermolecular interactions, and for gaining insights into the molecular nature of mixtures. They also provide the necessary basis for any developments of theoretical thermodynamic models. Scientists and engineers in academic and industrial research need such data and will benefit from a careful collection of existing data. The new *CRC Handbook of Liquid-Liquid Equilibrium Data of Polymer Solutions* builds a completely new and reliable collection of LLE data for polymer solutions from the original literature. It will be a very useful completion to the *CRC Handbook of Thermodynamic Data of Copolymer Solutions*, the *CRC Handbook of Thermodynamic Data of Aqueous Polymer Solutions*, and the *CRC Handbook of Thermodynamic Data of Polymer Solutions at Elevated Pressures.*

The *Handbook* is divided into five chapters: (1) Introduction, (2) Liquid-Liquid Equilibrium Data of Binary Polymer Solutions, (3) Liquid-Liquid Equilibrium Data of Ternary Polymer Solutions, (4) Liquid-Liquid Equilibrium Data of Quaternary Polymer Solutions, and (5) References (for all chapters). Finally, appendices quickly route the user to the desired data sets. Thus, the book covers all the necessary systems for researchers and engineers who work in this field.

In comparison with low-molecular systems, the amount of data for polymer solutions is still rather small. About 1000 literature sources were perused for the purpose of this handbook, including some dissertations and diploma papers. About 1210 data sets, i.e., 810 binary systems, 325 ternary systems (among them 110 systems composed of one polymer and two solvents and 215 systems composed of one solvent and two polymers), and 25 quaternary (or higher) systems are reported. Additionally, tables of systems are provided where results were published only in graphical form in the original literature to lead the reader to further sources. Data are included only if numerical values were published or authors provided their numerical results by personal communication (and I wish to thank all those who did so). No digitized data have been included in this data collection. The *Handbook* is the first complete overview about this subject in the world's literature. The closing day for the data collection was December 31, 2006. However, the user who is in need of new additional data sets is kindly invited to ask for new information beyond this book via e-mail at wohlfarth@chemie.uni-halle.de. Additionally, the author will be grateful to all users who call his attention to mistakes and make suggestions for improvements.

The *CRC Handbook of Liquid-Liquid Equilibrium Data of Polymer Solutions* will be useful to researchers, specialists, and engineers working in the fields of polymer science, physical chemistry, chemical engineering, material science, biological science and technology, and those developing computerized predictive packages. The *Handbook* should also be of use as a data source to Ph.D. students and faculty in chemistry, physics, chemical engineering, biotechnology, and materials science departments at universities.

Christian Wohlfarth
Halle, March 2007

About the Author

Christian Wohlfarth is Associate Professor for Physical Chemistry at Martin Luther University Halle-Wittenberg, Germany. He earned his degree in chemistry in 1974 and wrote his Ph.D. thesis on investigations on the second dielectric virial coefficient and the intermolecular pair potential in 1977, both at Carl Schorlemmer Technical University Merseburg. In 1985, he wrote his habilitation thesis, *Phase Equilibria in Systems with Polymers and Copolymers*, at Technical University Merseburg.

Since then, Dr. Wohlfarth's main research has been related to polymer systems. Currently, his research topics are molecular thermodynamics, continuous thermodynamics, phase equilibria in polymer mixtures and solutions, polymers in supercritical fluids, PVT behavior and equations of state, and sorption properties of polymers, about which he has published approximately 100 original papers. He has written the books: *Vapor-Liquid Equilibria of Binary Polymer Solutions*, *CRC Handbook of Thermodynamic Data of Copolymer Solutions*, *CRC Handbook of Thermodynamic Data of Aqueous Polymer Solutions*, *CRC Handbook of Thermodynamic Data of Polymer Solutions at Elevated Pressures*, and *CRC Handbook of Enthalpy Data of Polymer-Solvent Systems*.

He is working on the evaluation, correlation, and calculation of thermophysical properties of pure compounds and binary mixtures resulting in six volumes of the *Landolt-Börnstein New Series*. He is a respected contributor to the *CRC Handbook of Chemistry and Physics*.

CONTENTS

APPENDICES

1. INTRODUCTION

1.1. Objectives of the handbook

Knowledge of thermodynamic data of polymer solutions is a necessity for industrial and laboratory processes. Furthermore, such data serve as essential tools for understanding the physical behavior of polymer solutions, for studying intermolecular interactions, and for gaining insights into the molecular nature of mixtures. They also provide the necessary basis for any developments of theoretical thermodynamic models. Scientists and engineers in academic and industrial research need such data and will benefit from a careful collection of existing data. However, the database for polymer solutions is still modest in comparison with the enormous amount of data for low-molecular mixtures, and the specialized database for polymer solutions is even smaller. On the other hand, especially liquid-liquid demixing in polymer solutions is a quite common phenomenon, and liquid-liquid equilibrium data are needed for optimizing applications, e.g., separation operations of complex mixtures in the synthesis of polymers, recovery of polymer wastes, precipitation, fractionation and purification of polymers, polymer membrane formation and utilization, or polymers in green chemistry processes.

Basic information on polymers can be found in the *Polymer Handbook* (1999BRA), and there is also a chapter on properties of polymers and polymer solutions in the *CRC Handbook of Chemistry and Physics* (2006LID1). Two data books containing a small number of liquid-liquid equilibrium (LLE) data for polymer solutions appeared in the early 1990s (1992WEN, 1993DAN), but most data for polymer solutions have been compiled during the last two decades. More recent data books were written by the author of this *Handbook* containing a certain number of LLE data for copolymer solutions (2001WOH), aqueous polymer solutions (2004WOH), and polymer solutions at elevated pressures (2005WOH). But, much more experimental LLE data of polymer solutions exist today. Thus, the intention of the *Handbook* is to fill this gap and to provide scientists and engineers with an up-to-date compilation from the literature of the available LLE data of polymer solutions. However, LLE data of electrolyte solutions or solutions with supercritical fluids are not included in this *Handbook*. Please find such data in 2004WOH and 2005WOH. The *Handbook* does also not present theories and models for polymer solution thermodynamics. Other publications (1971YAM, 1990FUJ, 1990KAM, 1999KLE, 1999PRA, and 2001KON) can serve as starting points for investigating those issues.

The data within this book are divided into four chapters:

- Cloud-point and/or coexistence data of binary polymer solutions together with a LCST/UCST-data table
- Cloud-point and/or coexistence data of ternary polymer solutions subdivided into systems composed of two solvents and one polymer or two polymers and one solvent
- Cloud-point and/or coexistence data of quaternary polymer solutions (including additionally systems with more than four mixing partners)
- References (for all data together)

Data from investigations applying to more than one chapter are divided and appear in the relevant chapters. Data are included only if numerical values were published or authors provided their results by personal communication (and I wish to thank all those who did so). No digitized data have been included in this data collection. However, every data chapter is completed by a table that includes systems and references for data published only in graphical form as phase diagrams or related figures.

1.2. Experimental methods involved

Analytical methods

Analytical methods involve the determination of the composition of the coexisting phases. This can be done by taking samples from each phase and analyzing them outside the equilibrium cell or by using physicochemical methods of analysis inside the equilibrium cell. If one needs the determination of more information than the total polymer composition, i.e., if one wants to characterize the polymer with respect to molar mass (distribution) or chemical composition (distribution), the sampling technique is unavoidable. However, withdrawing a large sample from an equilibrium cell disturbs the phase equilibrium significantly. Small samples can be withdrawn using capillaries or special sampling devices. Often such sampling devices are directly coupled to analytical equipment. The equilibration time is the most important point for polymer solutions. Due to their (high) viscosity and slow diffusion, equilibration will often need many hours or even days. Sometimes, phases are recirculated to reduce equilibration and/or sampling problems. Before analyzing the compositions of the coexisting phases, the mixture is usually given some time for a clear phase separation. The use of physicochemical methods of analysis inside the equilibrium cell, e.g., by a spectrometer, avoids the problems related to sampling. On the other hand, time-consuming calibrations can be necessary.

Analytical methods can be classified as isothermal methods, isobaric-isothermal methods, and isobaric methods. Using the *isothermal mode*, an equilibrium cell is charged with the system of interest, the mixture is heated to the desired temperature, and this temperature is then kept constant. Thus, the isothermal mode can be applied if a pressure is adjusted in the heterogeneous region above or below the desired equilibrium value depending on how the equilibrium will change pressure. The pressure can be readjusted by adding or withdrawing of material or by changing the volume of the cell if necessary. *Isobaric-isothermal methods* are often also called dynamic methods. Dynamic methods are difficult in their application to highly viscous media like concentrated polymer solutions. Therefore, they were usually not applied to measure liquid-liquid equilibria in polymer solutions, but demixing in supercritical solvents can be investigated (with more sophisticated equipment). *Isobaric methods* are applied for all demixing experiments with polymer solutions at ordinary pressure. Analytical methods need relatively simple and inexpensive laboratory equipment. If carried out carefully, they can produce reliable results. The main advantages of the analytical methods are that systems with more than two components can be studied, several isotherms or isobars can be studied with one filling, and the coexistence data are determined directly. The main disadvantage is that the method is not suitable near critical states or for systems where the phases do not separate well.

Synthetic methods

In synthetic methods, a mixture of known composition is prepared and the phase equilibrium is observed subsequently in an equilibrium cell (the problem of analyzing fluid mixtures is replaced by the problem of "synthesizing" them). After known amounts of the components have been placed into an equilibrium cell, pressure and temperature are adjusted so that *the mixture is homogeneous*. Then temperature or pressure is varied until formation of a new phase is observed. This is the common way to observe cloud points in demixing polymer systems. No sampling is necessary. Therefore, the experimental equipment is often relatively simple and inexpensive. For multicomponent systems, experiments with synthetic methods yield less information than with analytical methods, because the tie lines cannot be determined without additional experiments. This is especially true for polymer solutions where fractionation accompanies demixing. The appearance of a new phase is either detected visually or by monitoring physical properties. Visually, the beginning of turbidity in the system or the meniscus in a view cell can be observed. Otherwise, light scattering is the common method to detect the formation of the new phase. Both visual and non-visual synthetic methods are widely used in investigations on liquid-liquid equilibria in polymer systems. The problem of isorefractive systems (where the coexisting phases have approximately the same refractive index) does not belong to polymer solutions where the usually strong concentration dependence of their refractive index prevents such a behavior. The changes of other physical properties like viscosity, ultrasonic absorption, thermal expansion, dielectric constant, heat capacity, or UV and IR absorption are also applied as non-visual synthetic methods for polymer solutions. The synthetic method is particularly suitable for measurements near critical states.

Some problems related to systems with polymers

Details of special experimental equipment can be found in the original papers compiled for this book and will not be presented here. But, some problems should be summarized that have to be obeyed and solved during the experiment. The polymer solution is often of an amount of some cm^3 and may contain about 1g of polymer or even more. Therefore, the equilibration of prepared solutions can be difficult and equilibration is usually very time consuming (liquid oligomers do not need so much time, of course). Additionally, mixing and demixing kinetics can be different, which leads to differences between cloud-point data measured for example in heating or cooling modes. Increasing viscosity makes the preparation of concentrated solutions more and more difficult with further increasing the amount of polymer. Solutions above 50-60 wt% can hardly be prepared (depending on the solvent/polymer pair under investigation). All impurities in the pure solvents have to be eliminated. Polymers and solvents must keep dry. Sometimes, inhibitors and antioxidants are added to polymers. They may especially influence the position of the demixing equilibrium. The thermal stability of polymers must be obeyed, otherwise depolymerization or formation of networks by chemical processes can change the sample (and its molar mass averages) during the experiment, which will cause large errors in liquid-liquid demixing.

Furthermore, to understand the results of LLE experiments in polymer solutions, one has to take into account the strong influence of polymer distribution functions on LLE, because fractionation occurs during demixing (1968KON, 1972KON, 2001KON). Fractionation takes place with respect to molar mass distribution as well as to chemical distribution if copolymers are involved. Fractionation during demixing leads to some special effects by which the LLE phase behavior differs from that of an ordinary, strictly binary mixture, because a common polymer solution is a multicomponent system.

Cloud-point curves are measured instead of binodals; and per each individual feed concentration of the mixture, *two parts of a coexistence curve* occur below (for upper critical solution temperature, UCST, behavior) or above the cloud-point curve (for lower critical solution temperature, LCST, behavior), i.e., produce an infinite number of coexistence data. Distribution functions of the feed polymer belong only to cloud-point data. On the other hand, each pair of coexistence points is characterized by two new and different distribution functions in each coexisting phase. The critical concentration is the only feed concentration where both parts of the coexistence curve meet each other on the cloud-point curve at the critical point that belongs to the feed polymer distribution function. The threshold point (maximum or minimum corresponding to either UCST or LCST behavior) temperature (or pressure) is not equal to the critical point, since the critical point is to be found at a shoulder of the cloud-point curve. Details were discussed by Koningsveld (1968KON, 1972KON). Thus, LLE data have to be specified in the tables as cloud-point or coexistence data, and coexistence data make sense only if the feed concentration is given. This is not always the case, however.

Special methods are necessary to measure the *critical point*. Only for solutions of *monodisperse polymers*, the critical point is the maximum (or minimum) of the binodal. Binodals of polymer solutions can be rather broad and flat. Then, the exact position of the critical point can be obtained by the method of the rectilinear diameter:

$$\frac{(\varphi_B^{\,I} - \varphi_B^{\,II})}{2} - \varphi_{B,crit} \propto (1 - \frac{T}{T_{crit}})^{1-\alpha} \qquad (1)$$

where:

$\varphi_B^{\,I}$	volume fraction of the polymer B in coexisting phase I
$\varphi_B^{\,II}$	volume fraction of the polymer B in coexisting phase II
$\varphi_{B,\,crit}$	volume fraction of the polymer B at the critical point
T	(measuring) temperature
T_{crit}	critical demixing temperature
α	critical exponent

For solutions of *polydisperse polymers*, such a procedure cannot be used because the critical concentration must be known in advance to measure its corresponding coexistence curve. Two different methods were developed to solve this problem: the *phase-volume-ratio method* (1968KON) where one uses the fact that this ratio is exactly equal to one only at the critical point, and the *coexistence concentration plot* (1969WOL) where an isoplethal diagram of values of $\varphi_B^{\,I}$ and $\varphi_B^{\,II}$ vs. the feed concentration, φ_{0B}, gives the critical point as the intersection of cloud-point curve and shadow curve. Details are not discussed here.

Treating polymer solutions with distribution functions by continuous thermodynamics and procedures to measure and calculate liquid-liquid equilibria of such systems is reviewed in (1989RAE) and (1990RA1).

1.3. Guide to the data tables

Characterization of the polymers

Polymers vary by a number of characterization variables. The molar mass and their distribution function are the most important variables. However, tacticity, sequence distribution, branching, and end groups determine their thermodynamic behavior in solutions too. For copolymers, the chemical distribution and the average chemical composition are also to be given. Unfortunately, much less information is provided with respect to the polymers or copolymers that were applied in most of the thermodynamic investigations in the original literature. In many cases, the samples are characterized only by one or two molar mass averages and some additional information (e.g., T_g, T_m, ρ_B, or how and where they were synthesized). Sometimes even this information is missed.

The molar mass averages are defined as follows:

number average M_n

$$M_n = \frac{\sum_i n_{B_i} M_{B_i}}{\sum_i n_{B_i}} = \frac{\sum_i w_{B_i}}{\sum_i w_{B_i} / M_{B_i}} \tag{2}$$

mass average M_w

$$M_w = \frac{\sum_i n_{B_i} M_{B_i}^2}{\sum_i n_{B_i} M_{B_i}} = \frac{\sum_i w_{B_i} M_{B_i}}{\sum_i w_{B_i}} \tag{3}$$

z-average M_z

$$M_z = \frac{\sum_i n_{B_i} M_{B_i}^3}{\sum_i n_{B_i} M_{B_i}^2} = \frac{\sum_i w_{B_i} M_{B_i}^2}{\sum_i w_{B_i} M_{B_i}} \tag{4}$$

viscosity average M_η

$$M_\eta = \left(\frac{\sum_i w_{B_i} M_{B_i}^a}{\sum_i w_{B_i}} \right)^{1/a} \tag{5}$$

where:

a	exponent in the viscosity-molar mass relationship
M_{Bi}	relative molar mass of the polymer species B_i
n_{Bi}	amount of substance of polymer species B_i
w_{Bi}	mass fraction of polymer species B_i

Measures for the polymer concentration

The following concentration measures are used in the tables of this *Handbook* (where B always denotes the main polymer, A denotes the solvent, and in ternary systems C denotes the third component):

mass/volume concentration

$$c_A = m_A/V \qquad c_B = m_B/V \tag{6}$$

mass fraction

$$w_A = m_A/\Sigma\, m_i \qquad w_B = m_B/\Sigma\, m_i \tag{7}$$

mole fraction

$$x_A = n_A/\Sigma\, n_i \qquad x_B = n_B/\Sigma\, n_i \qquad \text{with } n_i = m_i/M_i \text{ and } M_B = M_n \tag{8}$$

volume fraction

$$\varphi_A = (m_A/\rho_A)/\Sigma\,(m_i/\rho_i) \qquad \varphi_B = (m_B/\rho_B)/\Sigma\,(m_i/\rho_i) \tag{9}$$

segment fraction

$$\psi_A = x_A r_A/\Sigma\, x_i r_i \qquad \psi_B = x_B r_B/\Sigma\, x_i r \tag{10}$$

base mole fraction

$$z_A = x_A r_A/\Sigma\, x_i r_i \qquad z_B = x_B r_B/\Sigma\, x_i r_i \text{ with } r_B = M_B/M_0 \text{ and } r_A = 1 \tag{11}$$

where:

c_A (mass/volume) concentration of solvent A
c_B (mass/volume) concentration of polymer B
m_A mass of solvent A
m_B mass of polymer B
M_A relative molar mass of the solvent A
M_B relative molar mass of the polymer B
M_n number-average relative molar mass of the polymer B
M_0 molar mass of a basic unit of the polymer B
n_A amount of substance of solvent A
n_B amount of substance of polymer B
r_A segment number of the solvent A, usually $r_A = 1$
r_B segment number of the polymer B
V volume of the liquid solution at temperature T
w_A mass fraction of solvent A
w_B mass fraction of polymer B
x_A mole fraction of solvent A
x_B mole fraction of polymer B
z_A base mole fraction of solvent A
z_B base mole fraction of polymer B

φ_A	volume fraction of solvent A
φ_B	volume fraction of polymer B
ρ_A	density of solvent A
ρ_B	density of polymer B
ψ_A	segment fraction of solvent A
ψ_B	segment fraction of polymer B

For high-molecular polymers, a mole fraction is not an appropriate unit to characterize composition. However, for oligomeric products with rather low molar masses, mole fractions were sometimes used. In the common case of a distribution function for the molar mass, $M_B = M_n$ is to be chosen. Mass fraction and volume fraction can be considered as special cases of segment fractions depending on the way by which the segment size is actually determined: $r_i/r_A = M_i/M_A$ or $r_i/r_A = V_i/V_A = (M_i/\rho_i)/(M_A/\rho_A)$, respectively. Classical segment fractions are calculated by applying $r_i/r_A = V_i^{\mathrm{vdW}}/V_A^{\mathrm{vdW}}$ ratios where hard-core van der Waals volumes, V_i^{vdW}, are taken into account. Their special values depend on the chosen equation of state or simply some group contribution schemes, e.g., (1968BON, 1990KRE) and have to be specified. Volume fractions imply a temperature dependence and, as they are defined in equation (9), neglect excess volumes of mixing and, very often, the densities of the polymer in the state of the solution are not known correctly. However, volume fractions can be calculated without the exact knowledge of the polymer molar mass (or its averages). Base mole fractions are sometimes applied for polymer systems in earlier literature. The value for M_0 is the molar mass of a basic unit of the polymer. Sometimes it is chosen arbitrarily, however, and has to be specified.

Tables of experimental data

The data tables in each chapter are provided there in order of the names of the polymers. In this data book, mostly source-based polymer names are applied. These names are more common in use, and they are usually given in the original sources too. For copolymers, their names were usually built by the two names of the co-monomers which are connected by *-co-*, or more specifically by *-alt-* for alternating copolymers, by *-b-* for block copolymers, by *-g-* for graft copolymers, or *-stat-* for statistical copolymers. Structure-based names, for which details about their nomenclature can be found in the *Polymer Handbook* (1999BRA), are chosen in some single cases only. CAS index names for polymers are not applied here. Finally, a list of the polymers in Appendix 1 utilizes the names as given in the chapters of this book.

Within types of polymers the individual samples are ordered by their increasing average molar mass, and, when necessary, systems are ordered by increasing temperature. In ternary systems, ordering is additionally made subsequently according to the name of the third component in the system. Each data set begins with the lines for the solution components, e.g., in binary systems

Polymer (B):	**poly(dimethylsiloxane)**		**1998SCH**
Characterization:	M_n/g.mol^{-1} = 29500, M_w/g.mol^{-1} = 44500, PDMS AK1000, Wacker GmbH, Germany		
Solvent (A):	**anisole**	**C_7H_8O**	**100-66-3**

where the polymer sample is given in the first line together with the reference. The second line provides then the characterization available for the polymer sample. The following line gives the solvent's chemical name, molecular formula, and CAS registry number.

In ternary and quaternary systems, the following lines are either for a second solvent or a second polymer, e.g., in ternary systems with two solvents

Polymer (B):	**polyethersulfone**		**1999BAR**
Characterization:	M_w/g.mol^{-1} = 49000, ρ_B (298 K) = 1.37 g/cm^3, Tg/K = 498, Ultrason E 6020 P, BASF AG, Ludwigshafen, Germany		
Solvent (A):	***N,N*-dimethylformamide**	C_3H_7NO	**68-12-2**
Solvent (C):	**ethanol**	C_2H_6O	**64-17-5**

or, e.g., in ternary systems with a second polymer

Polymer (B):	**polystyrene**		**2000BEH**
Characterization:	M_n/g.mol^{-1} = 93000, M_w/g.mol^{-1} = 101400, M_z/g.mol^{-1} = 111900 BASF AG, Germany		
Solvent (A):	**cyclohexane**	C_6H_{12}	**110-82-7**
Polymer (C):	**polyethylene**		
Characterization:	M_n/g.mol^{-1} = 13000, M_w/g.mol^{-1} = 89000, M_z/g.mol^{-1} = 600000, LDPE, Stamylan, DSM, Geleen, The Netherlands		

or, e.g., in quaternary (or higher) systems like

Polymer (B):	**polyetherimide**		**1989ROE**
Characterization:	M_w/g.mol^{-1} = 32800, ρ= 1.27 g/cm^3, Ultem 1000		
Solvent (A):	**water**	H_2O	**7732-18-5**
Solvent (C):	**1-methyl-2-pyrrolidinone**	C_5H_9NO	**872-50-4**
Polymer (D):	**poly(1-vinyl-2-pyrrolidinone)**		
Characterization:	M_w/g.mol^{-1} = 423000, ρ= 1.22 g/cm^3		

There are some exceptions from this type of presentation within the tables for the UCST and LCST data. These tables are prepared in the forms as chosen in 2001WOH.

The originally measured data for each single system are sometimes listed together with some comment lines if necessary. The data are usually given as published, but temperatures are always given in K. Pressures are sometimes recalculated into kPa or MPa.

Final day for including data into this *Handbook* was December, 31, 2006.

1.4. List of symbols

a	exponent in the viscosity-molar mass relationship
a_A	activity of solvent A
c_A	(mass/volume) concentration of solvent A
c_B	(mass/volume) concentration of polymer B
m_A	mass of solvent A
m_B	mass of polymer B
M	relative molar mass
M_A	molar mass of the solvent A
M_B	molar mass of the polymer B
M_n	number-average relative molar mass of the polymer B
M_w	mass-average relative molar mass of the polymer B
M_η	viscosity-average relative molar mass of the polymer B
M_z	z-average relative molar mass of the polymer B
M_0	molar mass of a basic unit of the polymer B
MI	melting index of the polymer B
n_A	amount of substance of solvent A
n_B	amount of substance of polymer B
P	pressure
P_0	standard pressure (= 0.101325 MPa)
P_{crit}	critical pressure
R	gas constant
r_A	segment number of the solvent A, usually $r_A = 1$
r_B	segment number of the polymer B
T	(measuring) temperature
T_{crit}	critical temperature
T_g	glass transition temperature
T_m	melting transition temperature
V, V_{spez}	volume or specific volume at temperature T
V_0	reference volume
V^E	excess volume at temperature T
V^{vdW}	hard-core van der Waals volume
w_A	mass fraction of solvent A
w_B	mass fraction of polymer B
$w_{B,\ crit}$	mass fraction of the polymer B at the critical point
x_A	mole fraction of solvent A
x_B	mole fraction of polymer B
z_A	base mole fraction of solvent A
z_B	base mole fraction of polymer B

α	critical exponent
φ_A	volume fraction of solvent A
φ_B	volume fraction of polymer B
$\varphi_{B,\ crit}$	volume fraction of the polymer B at the critical point
ρ	density (of the mixture) at temperature *T*
ρ_A	density of solvent A at temperature *T*
ρ_B	density of polymer B at temperature *T*
ψ_A	segment fraction of solvent A
ψ_B	segment fraction of polymer B

2. LIQUID-LIQUID EQUILIBRIUM DATA OF BINARY POLYMER SOLUTIONS

2.1. Cloud-point and/or coexistence data

Polymer (B): **decamethyltetrasiloxane** **1997MCL**
Characterization: M/g.mol^{-1} = 310.69
Solvent (A): **tetradecafluorohexane** C_6F_{14} **355-42-0**

Type of data: cloud points (UCST-behavior)

x_A	0.174	0.294	0.390	0.486	0.594	0.697	0.800	0.899
φ_A	0.104	0.187	0.261	0.343	0.447	0.560	0.689	0.831
T/K	295.15	316.85	325.05	329.95	332.55	332.95	331.15	321.55

Type of data: critical point (UCST-behavior)

$\varphi_{A,\,crit}$ = 0.552 T_{crit}/K = 332.59

Polymer (B): **ethyl(hydroxyethyl)cellulose** **2000PE1**
Characterization: M_w/g.mol^{-1} = 200000, EHEC 230G, Akzo Nobel, Stenungssund, Sweden
Solvent (A): **water** H_2O **7732-18-5**

Type of data: cloud points (LCST-behavior)

w_B 0.01 T/K 338.15

Polymer (B): **ethyl(hydroxyethyl)cellulose** **1995THU**
Characterization: see table
Solvent (A): **water** H_2O **7732-18-5**

Type of data: cloud points (LCST-behavior)

w_B 0.010 was kept constant (D.S. is the degree of substitution)

D.S. (ethyl)	1.5	1.7	1.4	1.9	0.8	1.5	1.4	1.01
D.S. (hydroxyethyl)	0.7	1.1	0.9	1.3	0.8	1.7	1.7-1.8	1.42
Mol% hydroxyethyl	0.32	0.39	0.39	0.41	0.50	0.53	0.55-0.56	0.58
T/K	303.15	307.15	310.15	307.15	335.15	310.15	318.15	331.15

D.S. (ethyl)	1.07	1.1-1.2	1.0	1.12	0.9	0.84	0.75	0.6-0.7
D.S. (hydroxyethyl)	1.55	1.7	1.8	2.24	2.1	1.98	2.1	1.9
Mol% hydroxyethyl	0.59	0.59-0.61	0.64	0.67	0.70	0.70	0.74	0.72-0.75
T/K	330.15	331.15	328.15	336.15	343.15	340.15	343.15	338.15

Polymer (B): **hepta(ethylene glycol) monotetradecyl ether** **1998KUB**
Characterization: M/g.mol^{-1} = 690
Solvent (A): **water** **H_2O** **7732-18-5**

Type of data: cloud points (LCST-behavior)

w_B 0.0122 T/K 332.68

Polymer (B): **hexamethyldisiloxane** **1997MCL**
Characterization: M/g.mol^{-1} = 162.38
Solvent (A): **tetradecafluorohexane** **C_6F_{14}** **355-42-0**

Type of data: cloud points (UCST-behavior)

x_A	0.204	0.213	0.231	0.328	0.352	0.449	0.465	0.498	0.498
φ_A	0.196	0.204	0.222	0.317	0.340	0.435	0.452	0.485	0.485
T/K	288.85	289.55	290.85	295.45	296.05	296.85	296.95	296.95	296.95
x_A	0.501	0.541	0.543	0.546	0.548	0.627	0.732	0.856	0.875
φ_A	0.488	0.528	0.530	0.533	0.535	0.614	0.722	0.849	0.869
T/K	296.95	296.95	296.95	296.95	296.95	296.35	293.15	281.75	278.05

Type of data: critical point (UCST-behavior)

$\varphi_{A, crit}$ = 0.503 T_{crit}/K = 296.95

Polymer (B): **hydroxypropylcellulose** **1971KAG**
Characterization: see table
Solvent (A): **water** **H_2O** **7732-18-5**

Type of data: cloud points (LCST-behavior)

w_B	0.0051	0.0067	0.0082
T/K	318.45	324.45	331.25
M_w/(g/mol)	75000	150000	300000

Polymer (B): **hydroxypropylcellulose** **1990SCH**
Characterization: see table
Solvent (A): **water** **H_2O** **7732-18-5**

Type of data: cloud points (LCST-behavior)

c_B/(g/l)	4.0	4.0	4.0
T/K	319.15	316.05	315.05
M_w/(g/mol)	100000	300000	1000000

Polymer (B): **methylcellulose** **2000PE1**
Characterization: M_w/g.mol^{-1} = 14000, Sigma Chemical Co., Inc., St. Louis, MO
Solvent (A): **water** **H_2O** **7732-18-5**

Type of data: cloud points (LCST-behavior)

w_B 0.01 T/K 323.15

Polymer (B): **methylcellulose** **1972KAG**
Characterization: M_η/g.mol^{-1} = 70000, degree of substitution 26.5–32.0 wt%, Research Institute of Textiles, Yokohama, Japan
Solvent (A): **water** H_2O **7732-18-5**

Type of data: cloud points (LCST-behavior)

φ_B	0.00199	0.00238	0.00265	0.00376	0.00462	0.00546	0.00546	0.00655	0.00684
T/K	339.35	337.85	334.05	330.55	328.55	328.55	330.15	327.35	326.65

φ_B	0.00783	0.00815	0.00892	0.01042	0.01150
T/K	324.55	325.05	324.75	324.05	324.35

Type of data: critical point (LCST-behavior)

$\varphi_{B, crit}$ = 0.00892 T_{crit}/K = 324.75

Polymer (B): **methylcellulose** **1972KAG**
Characterization: M_η/g.mol^{-1} = 150000, degree of substitution 26.5–32.0 wt%, Research Institute of Textiles, Yokohama, Japan
Solvent (A): **water** H_2O **7732-18-5**

Type of data: cloud points (LCST-behavior)

φ_B	0.00016	0.00071	0.00317	0.00386	0.00461	0.00536	0.00614	0.00764	0.01033
T/K	336.35	333.85	329.55	328.55	328.75	327.45	327.35	326.85	352.45

Type of data: critical point (LCST-behavior)

$\varphi_{B, crit}$ = 0.00756 T_{crit}/K = 326.35

Polymer (B): **methylcellulose** **1972KAG**
Characterization: M_η/g.mol^{-1} = 300000, degree of substitution 26.5–32.0 wt%, Research Institute of Textiles, Yokohama, Japan
Solvent (A): **water** H_2O **7732-18-5**

Type of data: cloud points (LCST-behavior)

φ_B	0.00016	0.00078	0.00264	0.00383	0.00463	0.00538	0.00708	0.00882	0.00997
T/K	335.65	333.15	331.25	328.75	328.95	328.75	329.45	331.85	348.45

Type of data: critical point (LCST-behavior)

$\varphi_{B, crit}$ = 0.00539 T_{crit}/K = 328.75

Polymer (B): **octamethyltrisiloxane** **1997MCL**
Characterization: M/g.mol^{-1} = 236.53
Solvent (A): **tetradecafluorohexane** C_6F_{14} **355-42-0**

Type of data: cloud points (UCST-behavior)

continued

continued

x_A	0.147	0.282	0.406	0.476	0.522	0.587	0.626	0.700	0.768
φ_A	0.107	0.215	0.323	0.388	0.433	0.498	0.539	0.620	0.698
T/K	280.65	304.85	313.05	314.15	314.80	315.25	315.35	314.65	313.15

x_A	0.876	0.935
φ_A	0.831	0.909
T/K	303.95	290.35

Type of data: critical point (UCST-behavior)

$\varphi_{A,\ crit} = 0.539$ $T_{crit}/K = 315.35$

Polymer (B): **penta(ethylene glycol) monoheptyl ether** **1992SAS**
Characterization: M_n/g.mol^{-1} = 340, surfactant C_7E_5, Bachem, purity > 98 wt%
Solvent (A): **n-dodecane** **$C_{12}H_{26}$** **112-40-3**

Type of data: cloud points

w_B	0.0499	0.0499	0.0499	0.0499	0.0499	0.0499	0.0499	0.0857	0.0857
T/K	273.71	275.64	277.35	279.18	281.00	282.71	284.51	277.92	279.31
P/MPa	9.80	24.00	36.80	51.00	65.20	79.60	95.00	13.20	23.40

w_B	0.0857	0.0857	0.0857	0.0857	0.0857	0.0999	0.0999	0.0999	0.0999
T/K	280.42	281.97	283.65	285.78	287.59	280.05	282.03	284.17	286.20
P/MPa	31.20	43.40	55.45	69.05	85.25	14.65	28.35	43.95	58.75

w_B	0.0999	0.0999	0.1680	0.1680	0.1680	0.1680	0.1680	0.1680	0.1680
T/K	288.20	290.08	279.57	281.63	283.57	285.66	287.58	289.57	291.52
P/MPa	73.45	88.05	7.95	22.95	37.35	52.95	67.55	82.85	98.35

w_B	0.2037	0.2037	0.2037	0.2037	0.2037	0.2037	0.2037	0.3010	0.3010
T/K	279.68	281.66	283.54	285.57	287.55	289.51	291.48	279.45	281.56
P/MPa	8.22	22.45	36.35	51.45	66.35	81.55	97.05	11.95	27.45

w_B	0.3010	0.3010	0.3010	0.3010	0.4013	0.4013	0.4013	0.4013	0.4013
T/K	283.48	285.51	287.51	289.47	277.39	279.19	281.06	283.75	285.38
P/MPa	41.75	57.05	72.55	88.75	7.55	21.55	35.75	55.55	68.25

w_B	0.4013	0.4013	0.4681	0.4681	0.4681	0.4681	0.4681	0.4681	0.6025
T/K	287.14	289.05	276.54	278.43	280.29	282.47	285.76	288.09	273.38
P/MPa	82.25	97.45	8.45	22.35	36.55	53.10	78.45	97.05	6.15

w_B	0.6025	0.6025	0.6025	0.6025	0.6025	0.6025
T/K	274.50	276.36	278.23	280.03	281.50	283.66
P/MPa	14.35	29.15	43.45	58.25	70.65	88.95

Polymer (B): **penta(ethylene glycol) monoheptyl ether** **1992SAS**
Characterization: M_n/g.mol^{-1} = 340, surfactant C_7E_5, Bachem, purity > 98 wt%
Solvent (A): **water** **H_2O** **7732-18-5**

Type of data: cloud points

continued

continued

w_B	0.0100	0.0100	0.0100	0.0100	0.0100	0.0302	0.0302	0.0302	0.0302
T/K	341.74	342.82	343.92	345.93	348.12	341.34	342.28	343.02	345.04
P/MPa	6.35	15.34	25.65	46.20	71.60	8.35	15.45	21.45	39.35
w_B	0.0302	0.0302	0.0302	0.0799	0.0799	0.0799	0.0799	0.0799	0.0799
T/K	346.98	347.88	349.98	339.94	340.96	341.91	343.15	344.14	345.06
P/MPa	59.25	70.15	96.15	1.88	9.85	17.55	28.65	38.15	47.45
w_B	0.0799	0.0799	0.0987	0.0987	0.0987	0.0987	0.0987	0.0987	0.0987
T/K	347.03	348.93	340.88	341.93	342.97	345.08	346.89	347.77	348.75
P/MPa	70.25	95.95	7.95	16.55	25.75	46.15	66.75	77.65	91.15
w_B	0.1595	0.1595	0.1595	0.1595	0.1595	0.1595	0.1595	0.1862	0.1862
T/K	340.81	341.84	342.85	344.87	346.94	348.05	349.40	340.56	342.65
P/MPa	6.65	15.05	23.85	43.15	66.15	80.05	99.15	2.83	19.65
w_B	0.1862	0.1862	0.1862	0.2488	0.2488	0.2488	0.2488	0.2488	0.2488
T/K	344.78	346.67	348.65	340.79	341.91	342.89	344.06	345.84	347.81
P/MPa	39.45	59.75	84.75	1.03	9.75	17.95	28.55	46.25	68.65
w_B	0.2488	0.2488	0.3489	0.3489	0.3489	0.3489	0.3489	0.3489	0.3489
T/K	348.80	349.79	343.87	345.05	345.92	347.86	349.80	350.75	351.88
P/MPa	81.45	95.35	10.55	20.65	28.55	47.95	70.55	83.05	99.35
w_B	0.4006	0.4006	0.4006	0.4006	0.4006	0.4006	0.4006	0.5029	0.5029
T/K	345.04	346.19	347.36	347.88	350.71	351.73	353.05	349.77	350.70
P/MPa	5.95	15.45	25.35	30.05	58.45	70.35	87.45	7.15	14.45
w_B	0.5029	0.5029	0.5029	0.5029	0.5029	0.5029			
T/K	351.74	352.61	354.53	356.43	357.44	358.37			
P/MPa	23.15	30.85	49.35	69.55	81.55	93.15			

Polymer (B): **poly(acrylic acid)** **1996SAF**
Characterization: M_w/g.mol^{-1} = 120000
Solvent (A): **tetrahydrofuran** C_4H_8O **109-99-9**

Type of data: cloud points, precipitation threshold (LCST-behavior)

w_B 0.027 T/K 268.3

Polymer (B): **poly[bis(2,3-dimethoxypropanoxy)phosphazene]** **1996ALL**
Characterization: M_n/g.mol^{-1} = 1070000, M_w/g.mol^{-1} = 1500000,
T_g/K = 192.2, synthesized in the laboratory
Solvent (A): **water** H_2O **7732-18-5**

Type of data: cloud points (LCST-behavior)

w_B 0.05 T/K 317.15

Polymer (B): **poly[bis(2-(2'-methoxyethoxy)ethoxy)-phosphazene]** **1996ALL**
Characterization: M_n/g.mol^{-1} = 667000, M_w/g.mol^{-1} = 1000000, T_g/K = 189.2, synthesized in the laboratory
Solvent (A): **water** **H_2O** **7732-18-5**

Type of data: cloud points (LCST-behavior)

w_B	0.05	T/K	338.15

Polymer (B): **poly[bis(2,3-bis(2-methoxyethoxy)propanoxy)-phosphazene]** **1996ALL**
Characterization: M_n/g.mol^{-1} = 714000, M_w/g.mol^{-1} = 1000000, T_g/K = 192.2, synthesized in the laboratory
Solvent (A): **water** **H_2O** **7732-18-5**

Type of data: cloud points (LCST-behavior)

w_B	0.05	T/K	311.15

Polymer (B): **poly[bis(2,3-bis(2-(2'-methoxyethoxy)ethoxy)-propanoxy)phosphazene]** **1996ALL**
Characterization: M_n/g.mol^{-1} = 1420000, M_w/g.mol^{-1} = 1700000, T_g/K = 192.2, synthesized in the laboratory
Solvent (A): **water** **H_2O** **7732-18-5**

Type of data: cloud points (LCST-behavior)

w_B	0.05	T/K	322.65

Polymer (B): **poly[bis(2,3-bis(2-(2'-(2''-dimethoxyethoxy)-ethoxy)ethoxy)propanoxy)phosphazene]** **1996ALL**
Characterization: M_n/g.mol^{-1} = 857000, M_w/g.mol^{-1} = 1200000, T_g/K = 192.2, synthesized in the laboratory
Solvent (A): **water** **H_2O** **7732-18-5**

Type of data: cloud points (LCST-behavior)

w_B	0.05	T/K	334.65

Polymer (B): **poly(1-butene)** **2005KOZ**
Characterization: M_n/g.mol^{-1} = 18700, M_w/g.mol^{-1} = 35100, isotactic, T_m/K = 376.83 (Form I)
Solvent (A): **2-butanol** **$C_4H_{10}O$** **78-92-2**

Type of data: cloud points

x_B	0.0002	0.0002	0.0003	0.0005	0.0006
T/K	360.0	360.4	361.5	360.6	360.0

Polymer (B): **poly(1-butene)** **2005KOZ**
Characterization: M_n/g.mol^{-1} = 16800, M_w/g.mol^{-1} = 30700, isotactic, T_m/K = 383.93 (Form I)
Solvent (A): **1-octanol** **$C_8H_{18}O$** **111-87-5**

Type of data: cloud points

x_B	0.0002	0.0003	0.0004	0.0006
T/K	359.5	360.7	361.3	362.0

Polymer (B): **poly(1-butene)** **2005KOZ**
Characterization: M_n/g.mol^{-1} = 18700, M_w/g.mol^{-1} = 35100, isotactic, T_m/K = 376.83 (Form I)
Solvent (A): **1-octanol** **$C_8H_{18}O$** **111-87-5**

Type of data: cloud points

x_B	0.0004	0.0005	0.0006
T/K	360.9	361.5	361.5

Polymer (B): **poly(butyl methacrylate)** **1984SAN, 1986SAN**
Characterization: M_n/g.mol^{-1} = 278000, M_w/g.mol^{-1} = 470000, Roehm GmbH, Darmstadt, Germany
Solvent (A): **ethanol** **C_2H_6O** **64-17-5**

Type of data: cloud points

w_B	0.075	was kept constant				
T/K	315.35	314.15	313.15	312.15	311.15	310.15
P/bar	1	105	198	400	625	1054

Polymer (B): **polycarbonate-tetrabromobisphenol-A** **1993BE1**
Characterization: synthesized in the laboratory
Solvent (A): **bis(2-ethoxyethyl) ether** **$C_8H_{18}O_3$** **112-36-7**

Type of data: cloud points (LCST-behavior)

w_B	0.1057	0.1545	0.2058	0.2551	0.3062	0.4026	0.5070	0.6006	0.7041
T/K	298.55	299.10	299.90	301.65	303.40	308.20	317.55	326.75	346.40

Polymer (B): **poly(*N*-cyclopropylacrylamide)** **1998KUR**
Characterization: synthesized in the laboratory
Solvent (A): **water** **H_2O** **7732-18-5**

Type of data: cloud points (LCST-behavior)

c_B/(g/l) 1.0 T/K 320.25

Polymer (B): **poly(*N*-cyclopropylacrylamide-*co*-vinylferrocene)** **1998KUR**
Characterization: synthesized in the laboratory
Solvent (A): **water** **H_2O** **7732-18-5**

continued

continued

Type of data: cloud points (LCST-behavior)

c_B/(g/l)	1.0	T/K	314.85	for a copolymer of 1.0 mol% vinylferrocene
c_B/(g/l)	1.0	T/K	296.95	for a copolymer of 3.0 mol% vinylferrocene

Polymer (B): **poly(decyl methacrylate)** **1983HER**
Characterization: see table
Solvent (A): **1-butanol** $C_4H_{10}O$ **71-36-3**

Type of data: cloud points, precipitation threshold (UCST-behavior)

M_n/ g mol^{-1}	M_w/ g mol^{-1}	w_B	T/ K
220000	252000	0.139	302.80
390000	468000	0.116	304.85
564000	728000	0.082	305.95

Polymer (B): **poly(decyl methacrylate)** **1983HER**
Characterization: see table
Solvent (A): **1-pentanol** $C_5H_{12}O$ **71-41-0**

Type of data: cloud points, precipitation threshold (UCST-behavior)

M_n/ g mol^{-1}	M_w/ g mol^{-1}	w_B	T/ K
220000	252000	0.141	276.40
390000	468000	0.118	278.40
564000	728000	0.090	278.85

Polymer (B): **poly(*N*,*N*-diethylacrylamide)** **2002MA2**
Characterization: M_w/g.mol^{-1} = 19000, synthesized in the laboratory
Solvent (A): **deuterium oxide** D_2O **7789-20-0**

Type of data: cloud points (LCST-behavior)

w_B	0.10	0.20	0.30	0.40	0.50	5.55	0.60	0.70
T/K	306.45	307.45	308.65	311.05	312.85	313.35	314.75	326.05

Polymer (B): **poly(*N*,*N*-diethylacrylamide)** **2002MA2**
Characterization: M_w/g.mol^{-1} = 19000, synthesized in the laboratory
Solvent (A): **water** H_2O **7732-18-5**

continued

continued

Type of data: cloud points (LCST-behavior)

w_B	0.001	0.005	0.05	0.10	0.20
T/K	307.45	306.55	305.15	305.35	306.05

Polymer (B): **poly(*N,N*-diethylacrylamide)** **1997KUR**
Characterization: synthesized in the laboratory
Solvent (A): **water** H_2O **7732-18-5**

Type of data: cloud points (LCST-behavior)

c_B/(g/l) 1.0 T/K 304.65

Polymer (B): **poly(*N,N*-diethylacrylamide)** **2001CAI, 2001GAN**
Characterization: M_w/g.mol^{-1} = 412000
Solvent (A): **water** H_2O **7732-18-5**

Type of data: cloud points (LCST-behavior)

w_B 0.005 T/K 303.65

Polymer (B): **poly(*N,N*-diethylacrylamide-*co*-acrylic acid)** **2001CAI**
Characterization: M_w/g.mol^{-1} = 319000, 5.98 mol% acrylic acid
Solvent (A): **water** H_2O **7732-18-5**

Type of data: cloud points (LCST-behavior)

w_B 0.005 T/K 305.05
w_B 0.005 T/K 301.25 (in a solution of 0.05 M NaCl)

Polymer (B): **poly(*N,N*-diethylacrylamide-*co*-acrylic acid)** **2001CAI**
Characterization: M_w/g.mol^{-1} = 306000, 13.22 mol% acrylic acid
Solvent (A): **water** H_2O **7732-18-5**

Type of data: cloud points (LCST-behavior)

w_B 0.005 T/K 304.15

Polymer (B): **poly(*N,N*-diethylacrylamide-*co*-acrylic acid)** **2001CAI**
Characterization: M_w/g.mol^{-1} = 308000, 20.64 mol% acrylic acid
Solvent (A): **water** H_2O **7732-18-5**

Type of data: cloud points (LCST-behavior)

w_B 0.005 T/K 300.15

Polymer (B): **poly(*N,N*-diethylacrylamide-*co*-vinylferrocene)** **1997KUR**
Characterization: synthesized in the laboratory
Solvent (A): **water** H_2O **7732-18-5**

continued

continued

Type of data: cloud points (LCST-behavior)

c_B/(g/l)	1.0	T/K	300.35	for a copolymer of 1.0 mol% vinylferrocene
c_B/(g/l)	1.0	T/K	293.55	for a copolymer of 3.0 mol% vinylferrocene

Polymer (B): **poly(*N*,*N*-dimethylacrylamide-*co*-allyl methacrylate)** **2005YI2**
Characterization: M_n/g.mol^{-1} = 9200, M_w/M_n = 1.9, 14.0 mol% allyl methacrylate
Solvent (A): **water** H_2O **7732-18-5**

Type of data: cloud points (LCST-behavior)

w_B 0.005 T/K 345.15

Polymer (B): **poly(*N*,*N*-dimethylacrylamide-*co*-allyl methacrylate)** **2005YI2**
Characterization: M_n/g.mol^{-1} = 10000, M_w/M_n = 2.2, 19.0 mol% allyl methacrylate
Solvent (A): **water** H_2O **7732-18-5**

Type of data: cloud points (LCST-behavior)

w_B 0.005 T/K 327.15

Polymer (B): **poly(*N*,*N*-dimethylacrylamide-*co*-allyl methacrylate)** **2005YI2**
Characterization: M_n/g.mol^{-1} = 12000, M_w/M_n = 2.3, 21.0 mol% allyl methacrylate
Solvent (A): **water** H_2O **7732-18-5**

Type of data: cloud points (LCST-behavior)

w_B 0.005 T/K 313.75

Polymer (B): **poly(*N*,*N*-dimethylacrylamide-*co*-allyl methacrylate)** **2005YI2**
Characterization: M_n/g.mol^{-1} = 13000, M_w/M_n = 2.2, 23.0 mol% allyl methacrylate
Solvent (A): **water** H_2O **7732-18-5**

Type of data: cloud points (LCST-behavior)

w_B 0.005 T/K 302.55

Polymer (B): **poly(*N*,*N*-dimethylacrylamide-*co*-allyl methacrylate)** **2005YI2**
Characterization: M_n/g.mol^{-1} = 13000, M_w/M_n = 2.3, 28.0 mol% allyl methacrylate
Solvent (A): **water** H_2O **7732-18-5**

Type of data: cloud points (LCST-behavior)

w_B 0.005 T/K 296.25

Polymer (B): **poly(*N,N*-dimethylacrylamide-*co*-allyl methacrylate)** **2005YI2**
Characterization: M_n/g.mol^{-1} = 12000, M_w/M_n = 2.1, 30.0 mol% allyl methacrylate
Solvent (A): **water** **H_2O** **7732-18-5**

Type of data: cloud points (LCST-behavior)

w_B 0.005 T/K 288.85

Polymer (B): **poly(*N,N*-dimethylacrylamide-*co*-*N*-phenyl-acrylamide)** **2005YI1**
Characterization: M_n/g.mol^{-1} = 9700, M_w/M_n = 1.09,
12.0 mol% *N*-phenylacrylamide, synthesized in the laboratory
Solvent (A): **water** **H_2O** **7732-18-5**

Type of data: cloud points (LCST-behavior)

w_B 0.01 T/K 343.25

Polymer (B): **poly(*N,N*-dimethylacrylamide-*co*-*N*-phenyl-acrylamide)** **2005YI1**
Characterization: M_n/g.mol^{-1} = 2000, M_w/M_n = 1.08,
15.9 mol% *N*-phenylacrylamide, synthesized in the laboratory
Solvent (A): **water** **H_2O** **7732-18-5**

Type of data: cloud points (LCST-behavior)

w_B 0.01 T/K 312.45

Polymer (B): **poly(*N,N*-dimethylacrylamide-*co*-*N*-phenyl-acrylamide)** **2005YI1**
Characterization: M_n/g.mol^{-1} = 10200, M_w/M_n = 1.07,
15.9 mol% *N*-phenylacrylamide, synthesized in the laboratory
Solvent (A): **water** **H_2O** **7732-18-5**

Type of data: cloud points (LCST-behavior)

w_B 0.01 T/K 316.65

Polymer (B): **poly(*N,N*-dimethylacrylamide-*co*-*N*-phenyl-acrylamide)** **2005YI1**
Characterization: M_n/g.mol^{-1} = 3500, M_w/M_n = 1.06,
16.2 mol% *N*-phenylacrylamide, synthesized in the laboratory
Solvent (A): **water** **H_2O** **7732-18-5**

Type of data: cloud points (LCST-behavior)

w_B 0.01 T/K 313.35

Polymer (B): **poly(*N,N*-dimethylacrylamide-*co*-*N*-phenyl-acrylamide)** **2005YI1**
Characterization: M_n/g.mol^{-1} = 4700, M_w/M_n = 1.05,
16.3 mol% *N*-phenylacrylamide, synthesized in the laboratory
Solvent (A): **water** H_2O **7732-18-5**

Type of data: cloud points (LCST-behavior)

w_B	0.01	T/K	312.25

Polymer (B): **poly(*N,N*-dimethylacrylamide-*co*-*N*-phenyl-acrylamide)** **2005YI1**
Characterization: M_n/g.mol^{-1} = 4600, M_w/M_n = 1.07,
21.7 mol% *N*-phenylacrylamide, synthesized in the laboratory
Solvent (A): **water** H_2O **7732-18-5**

Type of data: cloud points (LCST-behavior)

w_B	0.01	T/K	290.35

Polymer (B): **poly(*N,N*-dimethylacrylamide-*co*-*N*-phenyl-acrylamide)** **2005YI1**
Characterization: M_n/g.mol^{-1} = 8600, M_w/M_n = 1.07,
21.7 mol% *N*-phenylacrylamide, synthesized in the laboratory
Solvent (A): **water** H_2O **7732-18-5**

Type of data: cloud points (LCST-behavior)

w_B	0.01	T/K	293.25

Polymer (B): **poly(*N,N*-dimethylacrylamide-*co*-*N*-phenyl-acrylamide)** **2005YI1**
Characterization: M_n/g.mol^{-1} = 3200, M_w/M_n = 1.07,
21.8 mol% *N*-phenylacrylamide, synthesized in the laboratory
Solvent (A): **water** H_2O **7732-18-5**

Type of data: cloud points (LCST-behavior)

w_B	0.01	T/K	289.45

Polymer (B): **poly(*N,N*-dimethylacrylamide-*co*-*N*-phenyl-acrylamide)** **2005YI1**
Characterization: M_n/g.mol^{-1} = 10600, M_w/M_n = 1.07,
22.0 mol% *N*-phenylacrylamide, synthesized in the laboratory
Solvent (A): **water** H_2O **7732-18-5**

Type of data: cloud points (LCST-behavior)

w_B	0.01	T/K	293.25

Polymer (B): **poly(dimethylsiloxane)** **1998SCH**
Characterization: M_n/g.mol^{-1} = 29500, M_w/g.mol^{-1} = 44500, PDMS AK1000, Wacker GmbH, Germany
Solvent (A): **anisole** C_7H_8O **100-66-3**

Type of data: cloud points (UCST-behavior)

w_B	0.629	0.595	0.502	0.402	0.230	0.220	0.124	0.058	0.031
T/K	306.65	313.95	329.25	339.75	352.75	353.25	356.95	359.65	359.55

w_B	0.013
T/K	357.95

Polymer (B): **poly(dimethylsiloxane)** **1972ZE1**
Characterization: M_η/g.mol^{-1} = 626000, Dow-Corning
Solvent (A): **n-butane** C_4H_{10} **106-97-8**

Type of data: cloud points

T/K = 392.95 w_B = 0.0369 $(dT/dP)_{P=1}$/K.bar^{-1} = +0.92

Polymer (B): **poly(dimethylsiloxane)** **1987BAR**
Characterization: M_n/g.mol^{-1} = 6330, M_w/g.mol^{-1} = 7410
Solvent (A): **2,2-dimethylpropane** C_5H_{12} **463-82-1**

Type of data: cloud points (LCST-behavior)

w_B	0.0175	0.0295	0.0440	0.0679	0.1115
T/K	433.25	433.15	433.35	433.55	433.75

Polymer (B): **poly(dimethylsiloxane) (cyclic)** **1987BAR**
Characterization: M_n/g.mol^{-1} = 9810, M_w/g.mol^{-1} = 10300
Solvent (A): **2,2-dimethylpropane** C_5H_{12} **463-82-1**

Type of data: cloud points (LCST-behavior)

w_B	0.0102	0.0189	0.0312	0.1112
T/K	433.65	433.25	433.15	433.15

Polymer (B): **poly(dimethylsiloxane)** **1987BAR**
Characterization: M_n/g.mol^{-1} = 10060, M_w/g.mol^{-1} = 11570
Solvent (A): **2,2-dimethylpropane** C_5H_{12} **463-82-1**

Type of data: cloud points (LCST-behavior)

w_B	0.0046	0.0165	0.0340	0.0632	0.1098	0.1263	0.1594
T/K	432.45	431.15	430.85	430.95	431.15	431.85	432.55

Polymer (B): **poly(dimethylsiloxane) (cyclic)** **1987BAR**
Characterization: M_n/g.mol^{-1} = 14330, M_w/g.mol^{-1} = 14620
Solvent (A): **2,2-dimethylpropane** C_5H_{12} **463-82-1**

continued

continued

Type of data: cloud points (LCST-behavior)

w_B	0.0128	0.0347	0.0489	0.0533	0.1031	0.1577
T/K	432.15	431.05	430.45	430.35	430.75	431.15

Polymer (B): **poly(dimethylsiloxane)** **1987BAR**
Characterization: M_n/g.mol^{-1} = 14750, M_w/g.mol^{-1} = 16370
Solvent (A): **2,2-dimethylpropane** C_5H_{12} **463-82-1**

Type of data: cloud points (LCST-behavior)

w_B	0.0197	0.0251	0.0453	0.0547	0.0804	0.0923
T/K	428.55	429.25	427.95	428.25	428.25	428.45

Polymer (B): **poly(dimethylsiloxane)** **1987BAR**
Characterization: M_n/g.mol^{-1} = 18240, M_w/g.mol^{-1} = 18970
Solvent (A): **2,2-dimethylpropane** C_5H_{12} **463-82-1**

Type of data: cloud points (LCST-behavior)

w_B	0.0232	0.0255	0.0502	0.0673	0.0996	0.1356
T/K	428.25	428.15	427.25	427.15	427.35	427.75

Polymer (B): **poly(dimethylsiloxane) (cyclic)** **1987BAR**
Characterization: M_n/g.mol^{-1} = 18680, M_w/g.mol^{-1} = 19800
Solvent (A): **2,2-dimethylpropane** C_5H_{12} **463-82-1**

Type of data: cloud points (LCST-behavior)

w_B	0.0025	0.0365	0.0446	0.1017
T/K	428.85	427.65	427.95	428.25

Polymer (B): **poly(dimethylsiloxane)** **1987BAR**
Characterization: M_n/g.mol^{-1} = 21420, M_w/g.mol^{-1} = 22920
Solvent (A): **2,2-dimethylpropane** C_5H_{12} **463-82-1**

Type of data: cloud points (LCST-behavior)

w_B	0.0113	0.0196	0.0524	0.1296	0.1427	0.1725
T/K	427.15	426.65	426.05	426.25	426.55	426.95

Polymer (B): **poly(dimethylsiloxane)** **1987BAR**
Characterization: M_n/g.mol^{-1} = 30510, M_w/g.mol^{-1} = 31120
Solvent (A): **2,2-dimethylpropane** C_5H_{12} **463-82-1**

Type of data: cloud points (LCST-behavior)

w_B	0.0080	0.0264	0.0388	0.0743
T/K	425.95	424.25	424.05	424.05

Polymer (B): **poly(dimethylsiloxane)** **1972ZE1**
Characterization: M_η/g.mol^{-1} = 1200, 10 cSt viscosity, Dow-Corning
Solvent (A): **ethane** C_2H_6 **74-84-0**

Type of data: cloud points

T/K = 280.65 w_B = 0.366 $(dT/dP)_{P=1}$/K.bar^{-1} = +0.92

Polymer (B): **poly(dimethylsiloxane)** **1972ZE1**
Characterization: M_η/g.mol^{-1} = 3200, 50 cSt viscosity, Dow-Corning
Solvent (A): **ethane** C_2H_6 **74-84-0**

Type of data: cloud points

T/K = 273.15 w_B = 0.262 $(dT/dP)_{P=1}$/K.bar^{-1} = +0.90

Polymer (B): **poly(dimethylsiloxane)** **1972ZE1**
Characterization: M_η/g.mol^{-1} = 14200, 292 cSt viscosity, Dow-Corning
Solvent (A): **ethane** C_2H_6 **74-84-0**

Type of data: cloud points

T/K = 272.15 w_B = 0.0364 $(dT/dP)_{P=1}$/K.bar^{-1} = +0.89

Polymer (B): **poly(dimethylsiloxane)** **1972ZE1**
Characterization: M_η/g.mol^{-1} = 626000, Dow-Corning
Solvent (A): **ethane** C_2H_6 **74-84-0**

Type of data: cloud points

T/K = 259.65 w_B = 0.0651 $(dT/dP)_{P=1}$/K.bar^{-1} = +0.85

Polymer (B): **poly(dimethylsiloxane)** **2002SCH**
Characterization: M_n/g.mol^{-1} = 50000, M_w/g.mol^{-1} = 74000,
ρ_B (298.15 K) = 0.97 g/cm^3, Wacker GmbH, Germany
Solvent (A): **ethanol** C_2H_6O **64-17-5**

Type of data: cloud points (UCST-behavior)

w_B	0.8881	0.8650	0.8377	0.8132
T/K	303.15	313.15	323.15	333.15

Polymer (B): **poly(dimethylsiloxane)** **1998SCH**
Characterization: M_n/g.mol^{-1} = 10700, M_w/g.mol^{-1} = 24500, PDMS AK350,
Wacker GmbH, Germany
Solvent (A): **ethoxybenzene** $C_8H_{10}O$ **103-73-1**

Type of data: cloud points (UCST-behavior)

w_B	0.49	0.40	0.30	0.25	0.22	0.20	0.14	0.12	0.09
T/K	303.55	308.45	314.25	316.05	317.05	317.65	319.95	321.45	320.65

w_B	0.06	0.03	0.01
T/K	321.75	321.35	318.05

Polymer (B): **poly(dimethylsiloxane)** **1972ZE1**
Characterization: M_η/g.mol^{-1} = 203000, Dow-Corning
Solvent (A): **propane** C_3H_8 **74-98-6**

Type of data: cloud points

T/K = 340.15 w_B = 0.0400 $(dT/dP)_{P=1}$/K.bar^{-1} = +0.92

Polymer (B): **poly(dimethylsiloxane)** **1972ZE1**
Characterization: M_η/g.mol^{-1} = 626000, Dow-Corning
Solvent (A): **propane** C_3H_8 **74-98-6**

Type of data: cloud points

T/K = 337.75 w_B = 0.0402 $(dT/dP)_{P=1}$/K.bar^{-1} = +0.91

Polymer (B): **poly(dimethylsiloxane)** **1998SCH**
Characterization: M_n/g.mol^{-1} = 29500, M_w/g.mol^{-1} = 44500, PDMS AK1000, Wacker GmbH, Germany
Solvent (A): **2-propanone** C_3H_6O **67-64-1**

Type of data: cloud points (UCST-behavior)

w_B	0.66	0.62	0.55	0.50	0.45	0.40	0.35	0.26	0.21
T/K	299.25	304.15	309.55	312.65	317.15	319.65	321.45	323.15	321.55

w_B	0.15	0.11	0.10	0.07	0.03	0.01
T/K	324.75	325.55	328.05	329.05	328.85	326.55

Polymer (B): **poly(dimethylsiloxane)** **1987BAR**
Characterization: M_n/g.mol^{-1} = 6330, M_w/g.mol^{-1} = 7410
Solvent (A): **tetramethylsilane** $C_4H_{12}Si$ **75-76-3**

Type of data: cloud points (LCST-behavior)

w_B	0.0651	0.0732	0.1148
T/K	449.15	449.15	449.85

Polymer (B): **poly(dimethylsiloxane) (cyclic)** **1987BAR**
Characterization: M_n/g.mol^{-1} = 9810, M_w/g.mol^{-1} = 10300
Solvent (A): **tetramethylsilane** $C_4H_{12}Si$ **75-76-3**

Type of data: cloud points (LCST-behavior)

w_B	0.0291	0.0294	0.0455	0.0639	0.0996	0.1110
T/K	448.15	448.15	448.05	448.05	447.95	448.15

Polymer (B): **poly(dimethylsiloxane)** **1987BAR**
Characterization: M_n/g.mol^{-1} = 10060, M_w/g.mol^{-1} = 11570
Solvent (A): **tetramethylsilane** $C_4H_{12}Si$ **75-76-3**

continued

continued

Type of data: cloud points (LCST-behavior)

w_B	0.0142	0.0490	0.0862	0.1037	0.1420	0.1655
T/K	447.55	446.15	446.15	446.35	447.15	447.95

Polymer (B): **poly(dimethylsiloxane) (cyclic)** **1987BAR**
Characterization: M_n/g.mol^{-1} = 14330, M_w/g.mol^{-1} = 14620
Solvent (A): **tetramethylsilane** $C_4H_{12}Si$ **75-76-3**

Type of data: cloud points (LCST-behavior)

w_B	0.0368	0.0501	0.0810	0.1043	0.1286
T/K	445.95	445.65	445.55	445.55	445.75

Polymer (B): **poly(dimethylsiloxane)** **1987BAR**
Characterization: M_n/g.mol^{-1} = 14750, M_w/g.mol^{-1} = 16370
Solvent (A): **tetramethylsilane** $C_4H_{12}Si$ **75-76-3**

Type of data: cloud points (LCST-behavior)

w_B	0.0151	0.0371	0.0603	0.0822	0.1250
T/K	444.15	443.15	443.15	443.15	443.75

Polymer (B): **poly(dimethylsiloxane)** **1987BAR**
Characterization: M_n/g.mol^{-1} = 18240, M_w/g.mol^{-1} = 18970
Solvent (A): **tetramethylsilane** $C_4H_{12}Si$ **75-76-3**

Type of data: cloud points (LCST-behavior)

w_B	0.0210	0.0484	0.0767	0.1036	0.1404	0.1820
T/K	442.65	441.75	441.55	442.15	442.65	443.35

Polymer (B): **poly(dimethylsiloxane) (cyclic)** **1987BAR**
Characterization: M_n/g.mol^{-1} = 18680, M_w/g.mol^{-1} = 19800
Solvent (A): **tetramethylsilane** $C_4H_{12}Si$ **75-76-3**

Type of data: cloud points (LCST-behavior)

w_B	0.0189	0.0346	0.0776	0.0877	0.1581
T/K	444.65	442.95	443.15	443.25	444.05

Polymer (B): **poly(dimethylsiloxane)** **1987BAR**
Characterization: M_n/g.mol^{-1} = 21420, M_w/g.mol^{-1} = 22920
Solvent (A): **tetramethylsilane** $C_4H_{12}Si$ **75-76-3**

Type of data: cloud points (LCST-behavior)

w_B	0.0050	0.0107	0.0235	0.0425	0.0744	0.1075	0.1370	0.1720
T/K	443.15	442.15	441.55	440.65	440.15	440.95	441.15	441.85

Polymer (B): **poly(dimethylsiloxane)** **1987BAR**
Characterization: M_n/g.mol^{-1} = 30510, M_w/g.mol^{-1} = 31120
Solvent (A): **tetramethylsilane** **$C_4H_{12}Si$** **75-76-3**

Type of data: cloud points (LCST-behavior)

w_B	0.0111	0.0232	0.0424	0.0628	0.1660	0.2060
T/K	440.45	439.35	438.95	439.05	440.05	440.05

Polymer (B): **poly(dimethylsiloxane-*co*-methylphenylsiloxane)** **1998SCH**
Characterization: M_n/g.mol^{-1} = 9100, M_w/g.mol^{-1} = 41200,
15 wt% methylphenylsiloxane, Dow Corning Corp., Midland
Solvent (A): **anisole** **C_7H_8O** **100-66-3**

Type of data: cloud points (UCST-behavior)

w_B	0.010	0.030	0.050	0.100	0.150	0.200	0.226	0.239	0.240
T/K	309.85	309.75	308.65	304.05	299.35	291.55	293.15	291.75	291.45

w_B	0.250	0.300	0.351
T/K	291.15	287.85	284.35

Type of data: critical point (UCST-behavior)

$w_{B, crit}$ 0.240 T_{crit}/K 291.45

Type of data: coexistence data (tie lines)

The total feed concentration of the copolymer in the homogeneous system is: w_B = 0.250

Demixing temperature	w_B		Fractionation during demixing			
	Sol phase	Gel phase	Sol phase		Gel phase	
T/K			M_n/g mol^{-1}	M_w/g mol^{-1}	M_n/g mol^{-1}	M_w/g mol^{-1}
290.65	0.195	0.298	8300	25200	9400	42600
290.29	0.204	0.282	9000	27200	10900	41200
289.38	0.165	0.327	7800	16600	5200	40100
288.27	0.165	0.328	4700	21800	5500	35000
287.00	0.189	0.345	8300	19600	12100	43000
285.96	0.137	0.357	4400	18600	5700	39000
284.83	0.115	0.391	4100	13700	5500	39500
283.36	0.128	0.399	7300	15600	11800	40600

Comments: apparent M_w and M_n values were determined via polystyrene standards.

Polymer (B): **poly(dimethylsiloxane-*co*-methylphenylsiloxane)** **1998SCH**
Characterization: M_n/g.mol^{-1} = 9100, M_w/g.mol^{-1} = 41200,
15 wt% methylphenylsiloxane, Dow Corning Corp., Midland
Solvent (A): **2-propanone** **C_3H_6O** **67-64-1**

Type of data: cloud points (UCST-behavior)

w_B	0.010	0.044	0.064	0.080	0.098	0.152	0.198	0.248	0.298
T/K	296.75	298.15	297.25	296.35	295.15	291.55	288.45	285.25	282.65

w_B	0.310
T/K	282.45

Type of data: critical point (UCST-behavior)

$w_{B, crit}$ 0.310 T_{crit}/K 282.45

Type of data: coexistence data (tie lines)

The total feed concentration of the copolymer in the homogeneous system is: w_B = 0.320

Demixing temperature	w_B		Fractionation during demixing			
	Sol phase	Gel phase	Sol phase		Gel phase	
T/ K			M_n/ g mol^{-1}	M_w/ g mol^{-1}	M_n/ g mol^{-1}	M_w/ g mol^{-1}
281.00	0.229	0.413	8700	22400	13600	52400
280.05	0.186	0.453	8100	18100	13600	51100
279.35	0.191	0.409	9000	22600	13200	53600
278.05	0.183	0.473				
277.25	0.179	0.460	7800	18600	14200	52600
276.05	0.167	0.465	7700	16000	13100	50900

Comments: apparent M_w and M_n values were determined via polystyrene standards.

Polymer (B): **polyester (hyperbranched, aliphatic)** **2003SE1**
Characterization: M_n/g.mol^{-1} = 2830, M_w/g.mol^{-1} = 5100,
hydroxyl functional hyperbranched polyesters produced from polyalcohol cores and hydroxy acids, 64 OH groups per macromolecule, Boltorn H40, Perstorp Specialty Chemicals AB, Perstorp, Sweden
Solvent (A): **water** **H_2O** **7732-18-5**

Type of data: coexistence data (liquid-liquid-vapor three phase equilibrium)

w_B	0.062	0.082	0.100	0.130	0.170	0.201	0.300	0.350	0.398
T/K	364.4	378.1	412.2	414.4	406.0	394.1	381.0 (*)	371.8 (*)	370.0 (*)

w_B	0.300	0.350	0.398
T/K	377.5 (**)	358.8 (**)	331.6 (**)

(*) phase transition LLV to LV and (**) LV to LLV.

Polymer (B):	**poly[2-(2-ethoxy)ethoxyethyl vinyl ether]**	**2005MA2**
Characterization:	M_n/g.mol^{-1} = 21900, M_w/g.mol^{-1} = 29100, M_z/g.mol^{-1} = 37900,	
Solvent (A):	**water** H_2O	**7732-18-5**

Type of data: cloud points (LCST-behavior)

c_B/(g/cm^3)	0.005086	0.010461	0.020509	0.030677	0.039252	0.051767	0.073117	0.109069
T/K	314.59	314.31	314.08	313.95	313.87	313.84	313.76	313.74
c_B/(g/cm^3)	0.152433	0.189095	0.234671	0.274014	0.306242	0.339149	0.376634	
T/K	313.69	313.69	313.62	313.59	313.56	313.52	313.47	

Type of data: coexistence data (tie lines, LCST-behavior)

Comments: The total feed concentration of the polymer is c_B/(g/cm^3) = 0.155. The corresponding cloud point is at 313.69 K.

T/K	314.15	313.95
c_B/(g/cm^3) (sol phase)	0.064894	0.094846
c_B/(g/cm^3) (gel phase)	0.313561	0.298225

Polymer (B):	**poly[2-(2-ethoxy)ethoxyethyl vinyl ether]**	**2005MA2**
Characterization:	M_n/g.mol^{-1} = 30400, M_w/g.mol^{-1} = 38300, M_z/g.mol^{-1} = 49800,	
Solvent (A):	**water** H_2O	**7732-18-5**

Type of data: cloud points (LCST-behavior)

c_B/(g/cm^3)	0.010562	0.019268	0.029420	0.043136	0.059266	0.105665	0.114900	0.130871
T/K	313.87	313.77	313.69	313.66	313.56	313.52	313.59	313.62
c_B/(g/cm^3)	0.144014	0.187642	0.254007	0.316415	0.362050	0.404309	0.442353	0.483084
T/K	313.63	313.58	313.60	313.69	313.70	313.69	313.70	313.65
c_B/(g/cm^3)	0.531654	0.534207						
T/K	313.58	313.55						

Type of data: coexistence data (tie lines, LCST-behavior)

Comments: The total feed concentration of the polymer is c_B/(g/cm^3) = 0.182. The corresponding cloud point is at 313.6 K.

T/K	314.05	313.85	313.65
c_B/(g/cm^3) (sol phase)	0.052796	0.061726	0.088480
c_B/(g/cm^3) (gel phase)	0.324503	0.304274	0.247577

Polymer (B):	**poly(*N*-ethylacrylamide)**	**1997KUR**
Characterization:	synthesized in the laboratory	
Solvent (A):	**water** H_2O	**7732-18-5**

Type of data: cloud points (LCST-behavior)

c_B/(g/l) 1.0 *T*/K 317.85

Polymer (B): **poly(*N*-ethylacrylamide-*co*-vinylferrocene)** **1997KUR**
Characterization: synthesized in the laboratory
Solvent (A): **water** H_2O **7732-18-5**

Type of data: cloud points (LCST-behavior)

c_B/(g/l)	1.0	T/K	335.65	for a copolymer of 1.0 mol% vinylferrocene
c_B/(g/l)	1.0	T/K	317.85	for a copolymer of 3.0 mol% vinylferrocene

Polymer (B): **polyethylene** **2000BEH**
Characterization: M_n/g.mol^{-1} = 13000, M_w/g.mol^{-1} = 89000, M_z/g.mol^{-1} = 600000, LDPE, DSM Stamylan, DSM, Geleen, The Netherlands
Solvent (A): **cyclohexane** C_6H_{12} **110-82-7**

Type of data: cloud points

w_B	0.0766	0.0766	0.0766	0.0766	0.0766	0.1083	0.1083	0.1083
T/K	514.49	517.75	521.00	522.25	526.45	515.38	520.75	526.19
P/bar	26.4	32.9	39.7	41.8	48.6	25.60	30.43	38.43

Polymer (B): **polyethylene** **2002HOR**
Characterization: M_n/g.mol^{-1} = 13000, M_w/g.mol^{-1} = 89000, M_z/g.mol^{-1} = 600000, LDPE, DSM Stamylan, DSM, Geleen, The Netherlands
Solvent (A): **cyclohexane** C_6H_{12} **110-82-7**

Type of data: cloud points

w_B 0.103 was kept constant

T/K	554.06	560.69	564.14	570.03	574.22	578.33	584.85	589.48	592.73
P/bar	62	68	73	78	82	86	91	95	98

T/K	598.74	603.65	610.30	614.57	619.14	628.86
P/bar	103	107	112	115	118	124

Polymer (B): **polyethylene** **1999BEY**
Characterization: M_n/g.mol^{-1} = 20100, M_w/g.mol^{-1} = 106000
Solvent (A): **cyclopentane** C_5H_{10} **287-92-3**

Type of data: cloud points (LCST-behavior)

w_B	0.06	0.06	0.06	0.06	0.06	0.06	0.06	0.06	0.06
T/K	482.95	483.35	492.45	492.95	501.65	502.55	513.65	526.05	528.75
P/bar	47.4	48.5	61.7	62.3	74.5	75.4	90.4	106.4	110.4

Polymer (B): **polyethylene** **1999BEY**
Characterization: M_n/g.mol^{-1} = 20100, M_w/g.mol^{-1} = 106000
Solvent (A): **cyclopentene** C_5H_8 **142-29-0**

Type of data: cloud points (LCST-behavior)

w_B	0.06	0.06	0.06	0.06	0.06	0.06	0.06
T/K	482.85	483.95	488.55	489.15	498.05	498.75	510.35
P/bar	39.4	41.0	47.8	48.6	62.3	63.1	79.6

Polymer (B): **polyethylene** **1991OPS**
Characterization: M_n/g.mol^{-1} = 7900, M_w/g.mol^{-1} = 92000, M_z/g.mol^{-1} = 730000, linear, Marlex-type, Phillips Petroleum Co.
Solvent (A): **1-decanol** $C_{10}H_{22}O$ **112-30-1**

Type of data: critical point (UCST-behavior)

$w_{B,\ crit}$ 0.154 T_{crit}/K 409.45

Polymer (B): **polyethylene** **1979KLE**
Characterization: M_n/g.mol^{-1} = 8000, M_w/g.mol^{-1} = 177000, M_z/g.mol^{-1} = 990000, linear, Marlex-type, Phillips Petroleum Co.
Solvent (A): **diphenyl ether** $C_{12}H_{10}O$ **101-84-8**

Type of data: coexistence data (UCST-behavior)

w_B (total)	0.010	0.010	0.010	0.030	0.030	0.030	0.050	0.050
T/K	423.15	418.15	413.15	420.15	418.15	416.15	418.15	413.25
w_B (sol phase)	0.0085	0.0074	0.0091	0.0254	0.0245	0.0228	0.0422	0.0264
w_B (gel phase)	0.1860	0.0788	0.1330	0.0419	0.0609	0.0800		0.1702
w_B (total)	0.050	0.060	0.060	0.060	0.070	0.070	0.070	0.100
T/K	408.15	416.15	413.25	408.15	413.25	408.15	405.15	408.15
w_B (sol phase)	0.0250	0.0423	0.0398	0.0294	0.0436	0.0243	0.0284	0.0380
w_B (gel phase)	0.1074	0.0819	0.1045	0.1363	0.1053	0.1287	0.1368	0.1000
w_B (total)	0.100	0.100						
T/K	405.15	401.15						
w_B (sol phase)	0.0355	0.0284						
w_B (gel phase)	0.1396	0.1641						

Type of data: critical point (UCST-behavior)

$\varphi_{B,\ crit}$ 0.076 T_{crit}/K 417.8

Polymer (B): **polyethylene** **1979KLE, 1980KL2**
Characterization: see table, linear, HDPE
Solvent (A): **diphenyl ether** $C_{12}H_{10}O$ **101-84-8**

Type of data: critical point (UCST-behavior)

M_n/ g mol^{-1}	M_w/ g mol^{-1}	M_z/ g mol^{-1}	$\varphi_{B,\ crit}$	T_{crit}/ K
19000	36000	126000	0.116	406.6
7900	89000	730000	0.106	411.2
15000	27500	99000	0.099	405.0
12000	150000	900000	0.082	416.2
8600	55000	300000	0.097	410.1
34000	150000	286000	0.0556	420.6
92000	140000	270000	0.050	421.9
200000	680000		0.030	427.8

Polymer (B): **polyethylene** **1979KLE**
Characterization: M_n/g.mol^{-1} = 11000, M_w/g.mol^{-1} = 160000, M_z/g.mol^{-1} = 1800000, branched, 2.34 CH_3-groups/100 C
Solvent (A): **diphenyl ether** $C_{12}H_{10}O$ **101-84-8**

Type of data: spinodal points (UCST-behavior)

φ_B	0.0773	0.0844	0.0852	0.0896	0.0932	0.0944	0.0959	0.0988	0.1043
T/K	404.0	403.8	404.5	403.6	403.8	404.0	403.8	403.4	402.6
φ_B	0.1049	0.1063	0.1069	0.1081	0.1098	0.1146	0.1218	0.1261	0.1282
T/K	402.8	403.2	402.2	402.2	402.0	401.6	400.6	399.8	399.8

Type of data: critical point (UCST-behavior)

$\varphi_{B,\text{crit}}$ 0.113 T_{crit}/K 402.4

Polymer (B): **polyethylene** **1979KLE**
Characterization: M_n/g.mol^{-1} = 14000, M_w/g.mol^{-1} = 70000, M_z/g.mol^{-1} = 550000, branched, 2.21 CH_3-groups/100 C
Solvent (A): **diphenyl ether** $C_{12}H_{10}O$ **101-84-8**

Type of data: spinodal points (UCST-behavior)

φ_B	0.0691	0.0785	0.0848	0.0892	0.0933	0.0965	0.0992	0.1001	0.1020
T/K	405.4	404.9	403.7	403.4	403.6	403.2	403.4	403.3	402.8
φ_B	0.1067	0.1132	0.1152	0.1179	0.1211	0.1241	0.1247	0.1256	0.1274
T/K	402.9	402.4	401.5	401.4	401.8	401.6	401.2	401.4	400.6
φ_B	0.1307	0.1314	0.1589						
T/K	400.1	399.7	400.2						

Type of data: critical point (UCST-behavior)

$\varphi_{B,\text{crit}}$ 0.100 T_{crit}/K 403.6

Polymer (B): **polyethylene** **1979KLE, 1980KL2**
Characterization: M_n/g.mol^{-1} = 23000, M_w/g.mol^{-1} = 247000, M_z/g.mol^{-1} = 2000000, branched, 2.36 CH_3-groups/100 C
Solvent (A): **diphenyl ether** $C_{12}H_{10}O$ **101-84-8**

Type of data: spinodal points (UCST-behavior)

φ_B	0.0506	0.0575	0.0591	0.0606	0.0702	0.0719	0.0731	0.0733	0.0748
T/K	409.0	410.4	410.2	409.6	409.2	409.4	409.2	408.8	408.2
φ_B	0.0760	0.0781	0.0806	0.0828	0.0847	0.0888	0.0918	0.0971	0.1032
T/K	409.0	408.8	408.5	408.6	408.5	408.5	408.0	407.0	406.2

Type of data: critical point (UCST-behavior)

$\varphi_{B,\text{crit}}$ 0.083 T_{crit}/K 409.0

continued

continued

Type of data: coexistence data (UCST-behavior)

w_B (total)	0.03994	0.03979	0.04013	0.04074	0.03995	0.03977	0.03994	0.05990
T/K	405.15	406.15	407.15	408.15	409.15	410.15	411.15	405.15
w_B (sol phase)	0.0220	0.0214	0.0250	0.0263	0.0295	0.0288	0.0329	0.0297
w_B (gel phase)	0.1366	0.1187	0.1166	0.1234	0.1033	0.0823	0.0807	0.1265
w_B (total)	0.05986	0.05983	0.05970	0.05990	0.07009	0.07010	0.06985	0.06945
T/K	406.15	407.15	408.15	409.15	405.15	406.15	407.15	408.15
w_B (sol phase)	0.0341	0.0298	0.0390	0.0396	0.0348	0.0373	0.0389	0.0463
w_B (gel phase)	0.1148	0.1142	0.1043	0.0968	0.1072	0.1072	0.1074	0.0922

Polymer (B): **polyethylene** **1979KLE**
Characterization: M_n/g.mol^{-1} = 34000, M_w/g.mol^{-1} = 230000, M_z/g.mol^{-1} = 1650000, branched, 2.30 CH_3-groups/100 C
Solvent (A): **diphenyl ether** $C_{12}H_{10}O$ **101-84-8**

Type of data: spinodal points (UCST-behavior)

φ_B	0.0287	0.0333	0.0359	0.0360	0.0373	0.0391	0.0396	0.0418	0.0435
T/K	413.3	412.8	413.0	411.2	410.7	411.8	411.6	411.8	410.8
φ_B	0.0464	0.0506	0.0532	0.0595	0.0615	0.0633	0.0724	0.0765	0.0800
T/K	411.0	411.8	412.0	411.8	411.6	411.0	410.0	409.6	409.5
φ_B	0.0831	0.0872	0.0961						
T/K	410.0	409.0	409.4						

Type of data: critical point (UCST-behavior)

$\varphi_{B,\,crit}$ 0.068 T_{crit}/K 411.3

Polymer (B): **polyethylene** **1979KLE, 1980KL2**
Characterization: see table, branched
Solvent (A): **diphenyl ether** $C_{12}H_{10}O$ **101-84-8**

Type of data: critical point (UCST-behavior)

M_n/ g mol^{-1}	M_w/ g mol^{-1}	M_z/ g mol^{-1}	CH_3-groups/ 100 C	$\varphi_{B,\,crit}$	T_{crit}/ K
8500	70000	660000	2.48	0.188	387.0
18000	45000	490000	2.29	0.115	399.0
64000	345000	2900000	2.33	0.061	413.1
65000	420000	2900000	2.23	0.050	415.3
17000	274000		1.80	0.084	409.2
38000	640000		2.41	0.0695	409.8
19000	229000		1.20	0.084	410.2

continued

continued

M_n/ g mol^{-1}	M_w/ g mol^{-1}	M_z/ g mol^{-1}	CH_3-groups/ 100 C	$\varphi_{B,\,crit}$	T_{crit}/ K
24000	470000		0.75	0.086	410.6
27000	420000		1.30	0.085	409.9
29000	219000		1.46	0.075	412.3
7000	54000		5.18	0.139	385.4
8400	32000		5.26	0.133	384.7
19000	84000		1.42	0.089	408.1
30000	525000		2.21	0.074	411.4
25000	385000		2.31	0.084	408.4
35000	375000		2.18	0.067	412.6
29000	165000		2.25	0.074	410.0
31000	800000		2.45	0.075	410.9
24000	123000		2.83	0.121	396.7
25000	600000		1.63	0.084	408.8

Polymer (B): **polyethylene** **1991OPS**
Characterization: M_n/g.mol^{-1} = 7900, M_w/g.mol^{-1} = 92000, M_z/g.mol^{-1} = 730000, linear, Marlex-type, Phillips Petroleum Co.
Solvent (A): **1-dodecanol** **$C_{12}H_{26}O$** **112-53-8**

Type of data: critical point (UCST-behavior)

$w_{B,\,crit}$ 0.148 T_{crit}/K 394.65

Polymer (B): **polyethylene** **1991OPS**
Characterization: M_n/g.mol^{-1} = 8000, M_w/g.mol^{-1} = 177000, M_z/g.mol^{-1} = 990000, linear, Marlex-type, Phillips Petroleum Co.
Solvent (A): **n-heptane** **C_7H_{16}** **142-82-5**

Type of data: critical point (LCST-behavior)

$w_{B,\,crit}$ 0.0893 T_{crit}/K 468.15

Polymer (B): **polyethylene** **1991OPS**
Characterization: M_n/g.mol^{-1} = 15000, M_w/g.mol^{-1} = 27500, M_z/g.mol^{-1} = 99000, linear, Marlex-type, Phillips Petroleum Co.
Solvent (A): **n-heptane** **C_7H_{16}** **142-82-5**

Type of data: critical point (LCST-behavior)

$w_{B,\,crit}$ 0.0996 T_{crit}/K 480.25

Polymer (B): **polyethylene** **2002JOU**
Characterization: M_w/g.mol^{-1} = 125000, Aldrich Chem. Co., Inc., Milwaukee, WI
Solvent (A): **n-heptane** C_7H_{16} **142-82-5**

Type of data: cloud points

w_B	0.024	was kept constant			
T/K	463.15	473.15	483.15	493.15	503.15
P/MPa	2.0	3.2	4.6	5.7	7.1

Polymer (B): **polyethylene** **2004SCH**
Characterization: M_n/g.mol^{-1} = 6280, M_w/g.mol^{-1} = 6500, linear, completely hydrogenated polybutadiene, Polymer Source, Inc., Canada
Solvent (A): **n-hexane** C_6H_{14} **110-54-3**

Type of data: cloud points

w_B	0.018	0.018	0.018	0.018	0.018	0.018	0.0355	0.0355	0.0355
T/K	470.15	479.15	488.15	498.15	505.15	515.15	464.15	473.65	480.15
P/bar	30	41	48	54	60	69	19	31	38
w_B	0.0355	0.0355	0.0355	0.0355	0.051	0.051	0.051	0.051	0.0564
T/K	488.15	497.15	507.15	519.15	472.15	489.15	499.15	508.15	482.15
P/bar	49	58	67	76	28	45	55	66	68
w_B	0.0564	0.0564	0.0564	0.0564	0.065	0.065	0.083	0.083	0.083
T/K	492.15	502.15	514.15	522.15	487.15	496.15	479.15	488.15	497.15
P/bar	75	81	87	97	35	44	28	37	48
w_B	0.083	0.083	0.085	0.085	0.085	0.085	0.085	0.085	0.090
T/K	507.65	516.15	466.15	474.15	483.15	492.15	500.15	510.15	494.15
P/bar	56	63	34	43	52	57	63	72	67
w_B	0.090	0.090	0.090	0.103	0.103	0.103	0.103	0.155	0.155
T/K	504.15	514.15	524.15	489.15	496.15	506.15	515.15	502.15	509.15
P/bar	78	88	100	45	58	66	75	43	45
w_B	0.155	0.279	0.279	0.279					
T/K	521.15	518.15	527.65	536.65					
P/bar	57	38	50	54					

Type of data: critical points

$\varphi_{B,\,crit}$	0.138	0.138	0.138
T/K	451	463	476
P/bar	20	40	60

Polymer (B): **polyethylene** **1990KEN**
Characterization: M_n/g.mol^{-1} = 8000, M_w/g.mol^{-1} = 177000, M_z/g.mol^{-1} = 1000000 HDPE, DSM, Geleen, The Netherlands
Solvent (A): **n-hexane** C_6H_{14} **110-54-3**

continued

continued

Type of data: cloud points

w_B	0.0053	0.0053	0.0053	0.0053	0.0053	0.0055	0.0055	0.0055	0.0055
T/K	412.50	419.93	422.43	427.40	434.77	412.60	415.07	417.59	420.03
P/bar	16.3	29.0	32.2	40.2	51.0	23.1	27.1	31.1	34.6
w_B	0.0055	0.0055	0.0055	0.0095	0.0095	0.0095	0.0095	0.0095	0.0095
T/K	422.50	427.49	432.47	411.76	414.28	416.83	419.29	421.80	421.78
P/bar	38.1	45.6	52.6	18.2	21.2	26.1	30.8	33.1	33.7
w_B	0.0095	0.0095	0.0095	0.0199	0.0199	0.0199	0.0199	0.0199	0.0199
T/K	426.80	429.78	436.70	412.52	417.54	422.51	424.93	427.46	432.36
P/bar	40.8	45.1	55.2	15.2	23.4	30.9	35.0	38.3	46.0
w_B	0.0199	0.0199	0.0298	0.0298	0.0298	0.0298	0.0298	0.0298	0.0298
T/K	437.33	442.29	412.49	417.46	419.93	422.45	424.93	427.41	429.88
P/bar	52.7	59.8	11.3	19.3	23.4	27.2	31.2	35.2	38.6
w_B	0.0298	0.0298	0.0298	0.0423	0.0423	0.0423	0.0423	0.0423	0.0423
T/K	432.39	437.39	442.36	412.52	417.51	419.98	422.48	424.95	427.46
P/bar	42.6	49.8	57.2	7.6	16.1	19.5	23.7	27.8	32.1
w_B	0.0423	0.0423	0.0423	0.0423	0.0498	0.0498	0.0498	0.0498	0.0498
T/K	429.91	432.44	434.90	437.39	417.54	419.95	422.51	425.01	427.48
P/bar	35.2	38.9	42.7	46.3	13.8	17.3	21.8	25.6	29.2
w_B	0.0498	0.0612	0.0612	0.0612	0.0612	0.0612	0.0612	0.0612	0.0612
T/K	432.44	417.45	419.93	422.43	424.90	427.40	427.38	429.86	432.36
P/bar	36.8	10.4	14.5	18.4	22.2	26.2	26.4	30.1	34.0
w_B	0.0612	0.0612	0.0666	0.0666	0.0666	0.0666	0.0666	0.0666	0.0666
T/K	437.33	439.82	421.63	423.11	424.64	426.59	428.09	429.54	431.55
P/bar	41.0	44.6	14.6	17.0	19.2	22.1	24.4	26.5	29.8
w_B	0.0666	0.0666	0.0666	0.0820	0.0820	0.0820	0.0820	0.0820	0.0954
T/K	436.51	439.47	441.43	421.75	423.72	425.69	427.67	429.62	422.52
P/bar	37.1	41.4	44.3	11.3	14.2	17.2	20.4	23.2	9.8
w_B	0.0954	0.0954	0.0954	0.0954	0.0954	0.0954	0.0954	0.0954	0.0954
T/K	424.82	427.30	429.78	432.29	433.79	434.77	437.25	439.82	442.31
P/bar	13.3	17.3	21.1	24.9	27.3	28.7	32.4	36.3	39.9
w_B	0.0954	0.0954	0.0954	0.1313	0.1313	0.1313	0.1313	0.1313	0.1313
T/K	444.80	447.29	452.26	429.15	432.45	434.11	437.41	439.03	446.55
P/bar	43.5	47.0	54.3	10.3	15.5	18.2	23.1	25.4	36.6
w_B	0.1313	0.1313							
T/K	451.46	457.15							
P/bar	43.7	52.0							

continued

continued

Type of data: coexistence data (liquid-liquid-vapor three phase equilibrium)

w_B	0.0055	0.0053	0.0095	0.0199	0.0298	0.0423	0.0498	0.0612	0.0666
T/K	401.45	405.75	404.35	407.05	409.55	411.85	413.75	415.55	417.35
P/bar	4.8	5.5	5.3	5.8	6.3	6.8	7.1	7.4	7.7

w_B	0.0820	0.0954	0.1313
T/K	419.85	421.55	428.15
P/bar	8.0	8.2	8.5

Polymer (B): **polyethylene** **2004CHE**

Characterization: M_n/g.mol^{-1} = 14400, M_w/g.mol^{-1} = 15500, completely hydrogenated polybutadiene, Scientific Polymer Products, Ontario, NY

Solvent (A): **n-hexane** C_6H_{14} **110-54-3**

Type of data: cloud points

w_B	0.0322	0.0322	0.0322	0.0824	0.0824	0.0824	0.0906	0.0906	0.0906
T/K	453.17	463.11	473.15	453.07	463.11	473.12	453.11	463.17	473.09
P/MPa	1.7	3.0	4.3	2.1	3.4	4.8	2.0	3.4	4.7

w_B	0.0984	0.0984	0.0984	0.1187	0.1187	0.1187	0.1379	0.1379	0.1379
T/K	453.18	463.15	473.22	453.21	463.10	473.18	453.17	463.15	473.20
P/MPa	2.0	3.3	4.7	1.9	3.2	4.5	1.8	3.2	4.5

w_B	0.0906	0.0906	0.0906	
T/K	453.13	463.14	473.12	
P/MPa	1.5	1.8	2.0	(three VLLE data points)

Polymer (B): **polyethylene** **1991OPS**

Characterization: M_n/g.mol^{-1} = 15000, M_w/g.mol^{-1} = 27500, M_z/g.mol^{-1} = 99000, linear, Marlex-type, Phillips Petroleum Co.

Solvent (A): **n-hexane** C_6H_{14} **110-54-3**

Type of data: critical point (LCST-behavior)

$w_{B,crit}$ 0.100 T_{crit}/K 442.75

Polymer (B): **polyethylene** **2000BEH**

Characterization: M_n/g.mol^{-1} = 22500, M_w/g.mol^{-1} = 58000, HDPE, metallocene product, BASF AG, Germany

Solvent (A): **n-hexane** C_6H_{14} **110-54-3**

Type of data: cloud points

w_B	0.0497	0.0497	0.0497	0.0497	0.0497	0.0497	0.0497	0.0497	0.0497
T/K	430.52	438.64	446.86	455.16	463.47	471.86	479.83	488.05	496.17
P/bar	15.5	28.4	40.5	52.3	63.9	74.1	83.8	93.9	102.4

continued

continued

w_B	0.0497	0.1037	0.1037	0.1037	0.1037	0.1037	0.1037	0.1037	0.1037
T/K	503.21	430.94	437.29	445.42	453.63	462.15	470.24	478.48	486.85
P/bar	110.0	10.1	20.0	31.0	42.1	53.9	64.9	75.3	84.9
w_B	0.1037	0.1037	0.1508	0.1508	0.1508	0.1508	0.1508	0.1508	0.1508
T/K	495.09	503.07	434.00	439.84	446.61	446.86	454.32	461.36	469.38
P/bar	94.6	103.1	9.5	16.9	26.6	27.0	37.3	46.7	55.5
w_B	0.1508	0.1508	0.1508	0.1508	0.1508				
T/K	475.84	482.86	490.05	497.37	503.34				
P/bar	62.3	73.6	81.8	90.1	96.9				

Polymer (B): **polyethylene** **2004CHE**
Characterization: M_n/g.mol^{-1} = 23300, M_w/g.mol^{-1} = 60400, M_z/g.mol^{-1} = 100700
metallocene LLDPE, unspecified comonomer, industrial source
Solvent (A): **n-hexane** **C_6H_{14}** **110-54-3**

Type of data: cloud points

w_B	0.0049	0.0049	0.0049	0.0049	0.0148	0.0148	0.0148	0.0148	0.0364
T/K	443.13	453.13	463.18	473.22	443.25	453.15	463.14	473.05	443.49
P/MPa	2.5	3.9	5.2	6.6	2.6	4.1	5.5	6.6	2.6
w_B	0.0364	0.0364	0.0364	0.0822	0.0822	0.0822	0.0888	0.0888	0.0888
T/K	453.54	463.58	473.55	433.15	453.10	473.14	443.34	453.36	463.37
P/MPa	4.0	5.4	6.6	1.6	4.5	7.2	2.1	3.4	4.9
w_B	0.0888	0.1083	0.1083	0.1083	0.1083	0.1535	0.1535	0.1535	0.1535
T/K	473.41	443.42	453.38	463.42	473.45	443.03	453.04	463.05	473.01
P/MPa	6.3	1.9	3.2	4.7	6.1	1.4	2.8	4.1	5.4
w_B	0.1952	0.1952	0.1952						
T/K	452.98	463.09	473.21						
P/MPa	2.4	3.8	5.1						
w_B	0.0822	0.0822	0.0822						
T/K	433.18	453.13	473.13						
P/MPa	1.1	1.5	1.9	(three VLLE data points)					

Polymer (B): **polyethylene** **2000BEH**
Characterization: M_n/g.mol^{-1} = 20000, M_w/g.mol^{-1} = 210000,
HDPE, BASF AG, Germany
Solvent (A): **n-hexane** **C_6H_{14}** **110-54-3**

Type of data: cloud points

w_B	0.0488	0.0488	0.0488	0.0488	0.0488	0.0488	0.0488	0.0488	0.0488
T/K	436.15	440.77	446.20	452.04	456.57	461.96	467.33	472.64	478.75
P/bar	42.3	47.1	54.2	61.2	66.8	74.5	81.4	88.4	96.4

continued

continued

w_B	0.0488	0.0488	0.0488	0.0488	0.1097	0.1097	0.1097	0.1097	0.1097
T/K	484.21	487.39	495.09	504.53	423.75	430.13	435.31	441.82	448.99
P/bar	103.0	107.1	114.3	126.2	21.6	29.9	35.1	43.8	51.7
w_B	0.1097	0.1097	0.1097	0.1097	0.1097	0.1097	0.1097	0.1097	0.1097
T/K	455.25	462.02	462.15	469.19	476.63	483.67	490.59	496.95	503.85
P/bar	59.3	67.2	67.4	75.5	83.7	94.4	94.4	105.7	113.5

Polymer (B): **polyethylene** **2000BEH**
Characterization: M_n/g.mol^{-1} = 20000, M_w/g.mol^{-1} = 585000, HDPE, BASF AG, Germany
Solvent (A): **n-hexane** C_6H_{14} **110-54-3**

Type of data: cloud points

w_B	0.0477	0.0477	0.0477	0.0477	0.0477	0.0477	0.0477	0.0477	0.0477
T/K	423.48	429.84	436.07	443.68	451.38	459.90	467.99	475.57	483.40
P/bar	25.6	34.1	43.3	52.4	62.5	74.2	84.9	94.8	104.2
w_B	0.0477	0.0477	0.0477	0.0994	0.0994	0.0994	0.0994	0.0994	0.0994
T/K	491.11	498.15	504.39	422.96	431.18	438.64	447.27	454.83	462.95
P/bar	111.7	120.3	127.0	8.8	19.8	31.0	40.6	51.6	62.5
w_B	0.0994	0.0994	0.0994	0.0994	0.0994	0.1482	0.1482	0.1482	0.1482
T/K	470.39	478.90	487.54	495.36	502.94	438.88	439.03	446.08	452.85
P/bar	72.6	83.7	94.5	103.3	111.5	12.7	13.0	24.9	36.7
w_B	0.1482	0.1482	0.1482	0.1482	0.1482	0.1482	0.1482		
T/K	460.29	467.60	474.64	481.93	488.98	496.56	503.61		
P/bar	47.6	58.9	68.2	77.7	86.0	94.5	102.4		

Polymer (B): **polyethylene** **2006NAG**
Characterization: M_n/g.mol^{-1} = 43700, M_w/g.mol^{-1} = 52000, M_z/g.mol^{-1} = 59000, 2.05 ethyl branches per 100 backbone C atoms, hydrogenated polybutadiene PBD 50000, DSM, The Netherlands
Solvent (A): **n-hexane** C_6H_{14} **110-54-3**

Type of data: cloud points

w_B	0.0005	0.0005	0.0005	0.0005	0.0005	0.0005	0.0005	0.0005	0.0005
T/K	450.65	455.60	460.52	465.47	470.26	475.12	480.14	485.02	490.13
P/MPa	1.706	2.386	3.051	3.746	4.336	4.911	5.611	6.136	6.806
w_B	0.0005	0.0024	0.0024	0.0024	0.0024	0.0024	0.0024	0.0024	0.0024
T/K	495.03	440.92	445.88	450.92	455.86	460.76	466.02	470.62	475.56
P/MPa	7.281	1.198	1.898	2.668	3.408	4.048	4.703	5.333	5.948
w_B	0.0024	0.0024	0.0024	0.0024	0.0049	0.0049	0.0049	0.0049	0.0049
T/K	480.46	485.21	490.07	494.94	440.48	445.50	450.36	455.27	460.43
P/MPa	6.498	7.068	7.643	8.223	1.572	2.322	3.017	3.662	4.402

continued

continued

w_B	0.0049	0.0049	0.0049	0.0049	0.0049	0.0049	0.0049	0.0078	0.0078
T/K	465.40	470.32	475.12	480.08	485.17	490.14	494.91	435.55	440.60
P/MPa	5.052	5.652	6.242	6.867	7.437	8.017	8.547	1.099	1.874

w_B	0.0078	0.0078	0.0078	0.0078	0.0078	0.0078	0.0078	0.0078	0.0078
T/K	445.56	450.47	455.40	460.33	465.27	470.19	475.15	480.05	484.91
P/MPa	2.599	3.274	3.964	4.719	5.269	5.899	6.494	7.099	7.699

w_B	0.0078	0.0078	0.0100	0.0100	0.0100	0.0100	0.0100	0.0100	0.0100
T/K	490.19	495.08	435.95	440.86	445.93	450.76	455.68	460.64	465.65
P/MPa	8.274	8.809	1.223	1.973	2.738	3.453	4.103	4.798	5.428

w_B	0.0100	0.0100	0.0100	0.0100	0.0100	0.0100	0.0223	0.0223	0.0223
T/K	470.57	475.52	480.43	485.28	490.18	495.20	435.86	440.94	445.88
P/MPa	6.023	6.648	7.228	7.788	8.353	8.913	1.563	2.348	3.048

w_B	0.0223	0.0223	0.0223	0.0223	0.0223	0.0223	0.0223	0.0223	0.0223
T/K	450.77	455.64	460.68	465.64	470.62	475.53	480.52	485.32	490.24
P/MPa	3.748	4.408	5.098	5.763	6.388	6.988	7.588	8.168	8.738

w_B	0.0223	0.0452	0.0452	0.0452	0.0452	0.0452	0.0452	0.0452	0.0452
T/K	495.24	435.79	440.73	445.66	450.62	455.59	460.51	465.56	470.53
P/MPa	9.298	1.750	2.490	3.200	3.900	4.590	5.260	5.920	6.560

w_B	0.0452	0.0452	0.0452	0.0452	0.0452	0.0585	0.0585	0.0585	0.0585
T/K	475.45	480.39	485.32	490.13	494.97	435.70	440.62	445.52	450.36
P/MPa	7.165	7.765	8.345	8.905	9.445	2.024	2.719	3.394	4.119

w_B	0.0585	0.0585	0.0585	0.0585	0.0585	0.0585	0.0585	0.0585	0.0606
T/K	460.25	465.20	470.10	475.03	480.03	485.07	490.13	495.02	435.90
P/MPa	5.449	6.084	6.674	7.334	7.969	8.569	9.124	9.699	1.964

w_B	0.0606	0.0606	0.0606	0.0606	0.0606	0.0606	0.0606	0.0606	0.0606
T/K	440.70	445.68	450.67	455.37	460.64	465.45	470.40	475.34	480.24
P/MPa	2.614	3.369	4.099	4.724	5.459	6.094	6.724	7.344	7.944

w_B	0.0606	0.0606	0.0606	0.06242	0.06242	0.06242	0.06242	0.06242	0.06242
T/K	485.23	490.26	495.15	435.72	440.65	445.61	450.46	455.52	460.47
P/MPa	8.519	9.084	9.619	1.894	2.624	3.384	4.119	4.789	5.494

w_B	0.06242	0.06242	0.06242	0.06242	0.06242	0.06242	0.06242	0.065445	0.065445
T/K	465.71	470.67	475.56	480.50	485.43	490.46	495.72	435.89	440.81
P/MPa	6.144	6.734	7.374	7.994	8.559	9.134	9.709	1.924	2.749

w_B	0.065445	0.065445	0.065445	0.065445	0.065445	0.065445	0.065445	0.065445	0.065445
T/K	445.79	450.61	455.61	460.50	465.30	470.25	475.23	480.14	484.88
P/MPa	3.399	4.069	4.774	5.474	6.044	6.724	7.319	7.899	8.469

w_B	0.065445	0.065445	0.06690	0.06690	0.06690	0.06690	0.06690	0.06690	0.06690
T/K	489.85	494.89	435.93	440.80	445.73	450.66	455.55	460.52	465.40
P/MPa	9.069	9.624	1.969	2.594	3.349	4.034	4.684	5.394	5.994

w_B	0.06690	0.06690	0.06690	0.06690	0.06690	0.06690	0.0706	0.0706	0.0706
T/K	470.31	475.21	480.28	485.25	490.00	495.00	436.14	441.05	445.86
P/MPa	6.644	7.244	7.869	8.454	9.014	9.589	1.909	2.684	3.374

continued

continued

w_B	0.0706	0.0706	0.0706	0.0706	0.0706	0.0706	0.0706	0.0706	0.0706
T/K	450.805	455.70	460.81	465.77	470.61	475.57	480.38	485.36	490.26
P/MPa	4.084	4.794	5.434	6.029	6.709	7.324	7.899	8.469	9.029
w_B	0.0706	0.0734	0.0734	0.0734	0.0734	0.0734	0.0734	0.0734	0.0734
T/K	495.323	430.71	435.74	440.62	445.60	450.52	455.46	460.35	465.27
P/MPa	9.629	0.973	1.748	2.503	3.223	3.918	4.603	5.263	5.913
w_B	0.0734	0.0734	0.0734	0.0734	0.0734	0.0734	0.0826	0.0826	0.0826
T/K	470.15	475.21	480.11	485.29	490.16	495.18	436.21	441.16	446.06
P/MPa	6.538	7.188	7.768	8.443	9.018	9.543	1.814	2.564	3.279
w_B	0.0826	0.0826	0.0826	0.0826	0.0826	0.0826	0.0826	0.0826	0.0826
T/K	450.95	455.85	460.80	465.70	470.83	475.71	480.57	485.49	490.59
P/MPa	3.989	4.664	5.329	5.969	6.624	7.229	7.834	8.414	8.989
w_B	0.0826	0.0923	0.0923	0.0923	0.0923	0.0923	0.0923	0.0923	0.0923
T/K	495.31	431.30	436.40	441.11	446.17	451.05	456.00	458.55	460.95
P/MPa	9.529	0.985	1.790	2.475	3.210	3.905	4.590	4.940	5.255
w_B	0.0923	0.0923	0.0923	0.0923	0.0923	0.0923	0.0923	0.0923	0.1534
T/K	463.34	465.91	473.36	475.73	482.45	487.53	491.14	495.73	435.805
P/MPa	5.575	5.905	6.900	7.175	7.990	8.585	8.980	9.485	1.363
w_B	0.1534	0.1534	0.1534	0.1534	0.1534	0.1534	0.1534	0.1534	0.1534
T/K	440.700	445.574	450.492	455.459	460.303	465.374	470.264	475.266	480.228
P/MPa	2.103	2.798	3.513	4.233	4.898	5.563	6.213	6.843	7.438
w_B	0.1534	0.1534	0.1534	0.19463	0.19463	0.19463	0.19463	0.19463	0.19463
T/K	485.118	489.966	495.032	435.888	440.916	446.062	451.113	455.543	460.414
P/MPa	8.028	8.598	9.178	0.989	1.624	2.449	3.149	3.774	4.424
w_B	0.19463	0.19463	0.19463	0.19463	0.19463	0.19463	0.19463	0.2435	0.2435
T/K	465.513	470.451	475.400	480.519	485.453	490.548	495.514	445.479	450.554
P/MPa	5.144	5.799	6.419	7.074	7.649	8.244	8.849	1.644	2.374
w_B	0.2435	0.2435	0.2435	0.2435	0.2435	0.2435	0.2435	0.2435	0.2435
T/K	455.445	460.460	465.535	470.421	475.462	480.029	484.864	489.887	494.885
P/MPa	3.094	3.849	4.499	5.104	5.774	6.199	6.894	7.499	8.099
w_B	0.3031	0.3031	0.3031	0.3031	0.3031	0.3031	0.3031	0.3031	0.3031
T/K	451.011	455.705	460.899	465.863	470.829	475.762	480.675	485.650	490.540
P/MPa	1.597	2.297	3.067	3.847	4.547	4.897	5.642	6.247	6.847

Type of data: coexistence data (liquid-liquid-vapor three phase equilibrium)

w_B	0.0005	0.0005	0.0005	0.0005	0.0005	0.0024	0.0024	0.0024	0.0024
T/K	450.58	460.52	470.32	480.30	490.07	440.94	450.82	460.74	470.75
P/MPa	1.251	1.481	1.736	2.026	2.341	1.068	1.278	1.603	1.758
w_B	0.0024	0.0024	0.0049	0.0049	0.0049	0.0049	0.0049	0.0049	0.0078
T/K	480.45	490.02	440.62	450.62	460.54	470.46	480.31	490.20	440.61
P/MPa	2.053	2.363	1.087	1.292	1.517	1.762	2.052	2.367	1.081

continued

continued

w_B	0.0078	0.0078	0.0078	0.0078	0.0078	0.0223	0.0223	0.0223	0.0223
T/K	450.62	460.54	470.46	480.32	490.20	440.970	450.959	460.669	470.571
P/MPa	1.320	1.519	1.782	2.057	2.371	1.068	1.273	1.538	1.753
w_B	0.0223	0.0223	0.0452	0.0452	0.0452	0.0452	0.0452	0.0452	0.0452
T/K	480.380	490.182	431.016	440.959	450.757	460.601	470.530	480.378	490.215
P/MPa	2.128	2.353	0.899	1.079	1.274	1.499	1.759	2.049	2.364
w_B	0.0923	0.0923	0.0923	0.0923	0.0923	0.0923	0.0923	0.1946	0.1946
T/K	431.31	441.38	451.33	460.96	470.67	480.46	490.32	441.080	450.976
P/MPa	0.890	1.075	1.275	1.490	1.740	2.030	2.355	1.098	1.288
w_B	0.1946	0.1946	0.1946	0.1946	0.2435	0.2435	0.2435	0.3031	0.3031
T/K	460.856	470.696	480.547	491.112	445.488	450.430	460.276	460.793	470.728
P/MPa	1.513	1.773	2.058	2.418	1.164	1.269	1.494	1.521	1.748
w_B	0.3031								
T/K	480.391								
P/MPa	2.087								

Polymer (B): **polyethylene** **2001TOR**
Characterization: M_n/g.mol^{-1} = 43000, M_w/g.mol^{-1} = 105000, M_z/g.mol^{-1} = 190000
HDPE, DSM, Geleen, The Netherlands
Solvent (A): **n-hexane** C_6H_{14} **110-54-3**

Type of data: cloud points

w_B	0.2128	0.2128	0.2128	0.2128	0.2128	0.2128	0.2128	0.2128	0.2128
T/K	446.2	457.4	467.6	478.0	457.4	467.6	478.0	488.2	498.4
P/MPa	1.05	1.33	1.60	1.88	2.29	3.78	5.16	6.45	7.65
w_B	0.2128	0.2128	0.1382	0.1382	0.1382	0.1382	0.1382	0.1382	0.1382
T/K	508.5	519.0	436.5	446.9	456.2	467.5	446.9	456.2	467.5
P/MPa	8.79	9.87	0.92	1.06	1.26	1.52	1.91	3.25	4.81
w_B	0.1382	0.1382	0.1382	0.1382	0.1382	0.0813	0.0813	0.0813	0.0813
T/K	477.7	488.1	498.4	508.5	519.0	431.3	441.2	451.4	461.9
P/MPa	6.13	7.44	8.67	9.80	10.89	0.77	0.93	1.14	1.38
w_B	0.0813	0.0813	0.0813	0.0813	0.0813	0.0813	0.0813	0.0813	0.0582
T/K	441.2	451.4	461.9	472.0	482.5	492.9	502.9	513.0	441.2
P/MPa	1.48	3.00	4.50	5.86	7.16	8.45	9.59	10.60	1.63
w_B	0.0582	0.0582	0.0582	0.0582	0.0221	0.0221	0.0221	0.0221	0.0221
T/K	451.5	461.9	472.2	482.5	464.0	474.4	482.7	492.6	502.9
P/MPa	3.15	4.63	6.01	7.29	5.11	6.49	7.53	8.73	9.90
w_B	0.0221	0.0138	0.0138	0.0138	0.0138	0.0138	0.0138	0.0138	0.0138
T/K	513.3	423.4	433.6	443.4	453.6	462.9	473.4	483.5	493.8
P/MPa	10.98	0.68	0.78	2.15	3.64	4.92	6.30	7.57	8.76
w_B	0.0138	0.0138							
T/K	503.	513.4							
P/MPa	9.82	10.86							

Polymer (B): **polyethylene** **2004SCH**
Characterization: M_n/g.mol^{-1} = 61000, M_w/g.mol^{-1} = 67100, linear, completely hydrogenated polybutadiene, Polymer Source, Inc., Canada
Solvent (A): **n-hexane** **C_6H_{14}** **110-54-3**

Type of data: cloud points

w_B	0.0207	0.0207	0.0207	0.0207	0.0207	0.0207	0.0374	0.0374	0.0374
T/K	436.15	446.15	453.15	463.15	481.15	490.15	423.15	436.65	446.15
P/bar	15	27	37	52	72	94	22	30	46
w_B	0.0374	0.0374	0.0374	0.0374	0.050	0.050	0.069	0.069	0.069
T/K	454.15	463.15	471.15	478.15	428.15	437.15	436.15	454.65	464.15
P/bar	59	71	81	91	22	35	22	49	58
w_B	0.069	0.106	0.106	0.106	0.106	0.106	0.106	0.118	0.118
T/K	473.15	448.65	458.65	466.15	474.15	481.65	493.15	443.55	451.35
P/bar	76	42	54	60	70	80	91	23	34
w_B	0.118	0.118	0.118	0.118	0.285	0.285	0.285	0.285	0.285
T/K	469.65	478.15	487.15	496.15	479.15	488.15	498.15	507.65	517.15
P/bar	63	71	77	92	54	69	83	93	101

Type of data: critical points

$\varphi_{B, crit}$	0.026	0.025	0.024
T/K	429.0	444.0	461.0
P/bar	20	40	60

Polymer (B): **polyethylene** **2004CHE**
Characterization: M_n/g.mol^{-1} = 82000, M_w/g.mol^{-1} = 108000, completely hydrogenated polybutadiene, Scientific Polymer Products, Ontario, NY
Solvent (A): **n-hexane** **C_6H_{14}** **110-54-3**

Type of data: cloud points

w_B	0.0076	0.0076	0.0076	0.0076	0.0186	0.0186	0.0186	0.0186	0.0310
T/K	443.67	453.14	463.18	473.12	443.15	453.16	463.15	473.15	443.15
P/MPa	3.8	5.3	6.6	7.9	3.9	5.4	6.7	8.0	4.2
w_B	0.0310	0.0310	0.0310	0.0536	0.0536	0.0536	0.0536	0.0829	0.0829
T/K	453.16	463.16	473.16	443.17	453.15	463.17	473.15	433.13	453.13
P/MPa	5.6	6.9	8.2	4.0	5.4	6.7	8.0	2.4	5.3
w_B	0.0829	0.0886	0.0886	0.0886	0.0886	0.1026	0.1026	0.1026	0.1026
T/K	473.13	443.24	453.16	463.16	473.17	443.08	453.15	463.16	473.19
P/MPa	8.0	3.8	5.3	6.6	7.9	3.8	5.1	6.5	7.8
w_B	0.1310	0.1310	0.1310	0.1310					
T/K	443.14	453.14	463.18	473.15					
P/MPa	3.6	5.0	6.4	7.7					

w_B	0.0829	0.0829	0.0829	
T/K	433.13	453.13	473.12	
P/MPa	1.1	1.5	2.0	(three VLLE data points)

Polymer (B): **polyethylene** **2004SCH**
Characterization: M_n/g.mol^{-1} = 322000, M_w/g.mol^{-1} = 383000, linear, completely hydrogenated polybutadiene, Polymer Source, Inc., Canada
Solvent (A): **n-hexane** C_6H_{14} **110-54-3**

Type of data: cloud points

w_B	0.00422	0.00422	0.00422	0.00422	0.00422	0.00422	0.00422	0.00422	0.006
T/K	437.65	447.35	456.15	465.65	475.05	484.75	494.55	503.75	434.15
P/bar	50	62	75	87	99	110	119.5	129.5	41

w_B	0.006	0.006	0.006	0.006	0.006	0.006	0.0175	0.0175	0.0175
T/K	442.15	452.15	461.15	469.15	478.15	487.15	453.15	461.15	469.15
P/bar	54	66	79	90	101	111	69	81	91

w_B	0.0175	0.0175	0.0175	0.0428	0.0428	0.0428	0.0428	0.0428	0.0428
T/K	478.15	488.15	496.65	430.15	437.15	445.15	454.15	462.15	473.15
P/bar	104	115	125	29	43	57	69	83	94

w_B	0.0428	0.0428	0.060	0.060	0.060	0.060	0.060	0.060	0.060
T/K	482.15	492.15	429.85	437.55	448.15	456.15	464.85	474.65	483.65
P/bar	107	117	34	48	63	74	85	96	107

w_B	0.060	0.0838	0.0838	0.0838	0.0838	0.0838	0.0838	0.0838	0.127
T/K	491.95	444.15	452.65	460.15	468.15	477.15	486.15	495.15	442.15
P/bar	115	46	58	69	80	93	101	114	32

w_B	0.127	0.127	0.127	0.127	0.127	0.127	0.237	0.237	0.237
T/K	453.15	461.15	472.15	479.15	489.15	497.15	467.65	482.85	494.45
P/bar	43	56	74	86	98	112	47	66	83

Type of data: critical points

$\varphi_{B, crit}$	0.012	0.012	0.013
T/K	416.2	431.4	446.3
P/bar	20	40	60

Polymer (B): **polyethylene** **1991OPS**
Characterization: M_n/g.mol^{-1} = 7900, M_w/g.mol^{-1} = 92000, M_z/g.mol^{-1} = 730000, linear, Marlex-type, Phillips Petroleum Co.
Solvent (A): **1-hexanol** $C_6H_{14}O$ **111-27-3**

Type of data: critical point (UCST-behavior)

$w_{B, crit}$ 0.156 T_{crit}/K 458.15

Polymer (B): **polyethylene** **2002CH3**
Characterization: M_n/g.mol^{-1} = 680, wax
Solvent (A): **octamethylcyclotetrasiloxane** $C_8H_{24}O_4Si_4$ **556-67-2**

Type of data: cloud points (UCST-behavior)

c_B/(g/l)	150	200
T/K	357.85	358.65

Polymer (B): **polyethylene** **2002CH3**
Characterization: M_n/g.mol^{-1} = 1138, M_w/g.mol^{-1} = 1263, wax
Solvent (A): **octamethylcyclotetrasiloxane** $C_8H_{24}O_4Si_4$ **556-67-2**

Type of data: cloud points (UCST-behavior)

c_B/(g/l)	20	50	100	150	200	300
T/K	406.15	402.15	397.65	393.15	388.15	385.15

Polymer (B): **polyethylene** **1991OPS**
Characterization: M_n/g.mol^{-1} = 15000, M_w/g.mol^{-1} = 27500, M_z/g.mol^{-1} = 99000, linear, Marlex-type, Phillips Petroleum Co.
Solvent (A): **n-octane** C_8H_{18} **111-65-9**

Type of data: critical point (LCST-behavior)

$w_{B,\text{ crit}}$ 0.0802 T_{crit}/K 519.15

Polymer (B): **polyethylene** **1991OPS**
Characterization: M_n/g.mol^{-1} = 8000, M_w/g.mol^{-1} = 177000, M_z/g.mol^{-1} = 990000, linear, Marlex-type, Phillips Petroleum Co.
Solvent (A): **n-octane** C_8H_{18} **111-65-9**

Type of data: critical point (LCST-behavior)

$w_{B,\text{ crit}}$ 0.0783 T_{crit}/K 506.05

Polymer (B): **polyethylene** **1991OPS**
Characterization: M_n/g.mol^{-1} = 7900, M_w/g.mol^{-1} = 92000, M_z/g.mol^{-1} = 730000, linear, Marlex-type, Phillips Petroleum Co.
Solvent (A): **1-octanol** $C_8H_{18}O$ **111-87-5**

Type of data: critical point (UCST-behavior)

$w_{B,\text{ crit}}$ 0.156 T_{crit}/K 426.65

Polymer (B): **polyethylene** **2001TOR**
Characterization: M_n/g.mol^{-1} = 43000, M_w/g.mol^{-1} = 105000, M_z/g.mol^{-1} = 190000 HDPE, DSM, Geleen, The Netherlands
Solvent (A): **1-octene** C_8H_{16} **111-66-0**

Type of data: cloud points

w_B	0.1791	0.1791	0.1791	0.1791	0.1791	0.1791	0.1791	0.0897	0.0897
T/K	503.3	513.4	524.3	533.9	524.3	533.9	544.8	503.2	512.8
P/MPa	0.97	1.14	1.35	1.57	1.38	2.30	3.30	0.94	1.13

w_B	0.0897	0.0897	0.0897	0.0897	0.0897	0.0439	0.0439	0.0439	0.0439
T/K	523.8	535.0	523.8	534.9	544.3	503.3	513.4	524.5	533.4
P/MPa	1.35	1.60	1.93	3.07	3.95	0.95	1.11	1.34	1.54

continued

continued

w_B	0.0439	0.0439	0.0439	0.0198	0.0198	0.0198	0.0198	0.0198	0.0198
T/K	524.4	533.4	544.1	503.3	513.3	523.9	533.5	523.9	533.5
P/MPa	2.19	3.09	4.10	0.95	1.14	1.35	1.58	2.22	3.16

w_B	0.0198
T/K	544.5
P/MPa	4.16

Polymer (B): **polyethylene** **2002YEO**
Characterization: M_n/g.mol^{-1} = 11300, M_w/g.mol^{-1} = 41700, Pressure Chemical Company, Pittsburgh, PA
Solvent (A): **n-pentane** C_5H_{12} **109-66-0**

Type of data: cloud points

T/K = 403.15

w_B	0.003	0.010	0.015	0.020	0.030	0.035	0.037	0.039	0.040
P/bar	97.2	113.1	112.4	109.6	108.9	111.7	109.6	111.7	111.7

w_B	0.043	0.045	0.050	0.060	0.070
P/bar	108.2	109.6	108.9	102.0	101.3

critical concentration: $w_{B, crit}$ = 0.0395

T/K = 423.15

w_B	0.003	0.010	0.015	0.020	0.030	0.035	0.037	0.039	0.040
P/bar	124.8	137.9	137.9	133.7	135.1	137.2	135.1	136.5	136.5

w_B	0.043	0.045	0.050	0.060	0.070
P/bar	131.7	133.7	131.0	127.5	127.5

critical concentration: $w_{B, crit}$ = 0.0395

T/K = 443.15

w_B	0.003	0.010	0.015	0.020	0.030	0.035	0.037	0.039	0.040
P/bar	153.7	160.0	160.6	157.9	159.3	158.6	157.2	156.5	158.6

w_B	0.043	0.045	0.050	0.060	0.070
P/bar	155.1	156.5	153.7	150.3	149.6

critical concentration: $w_{B, crit}$ = 0.0395

Polymer (B): **polyethylene** **2002YEO**
Characterization: M_n/g.mol^{-1} = 13900, M_w/g.mol^{-1} = 27700, Pressure Chemical Company, Pittsburgh, PA
Solvent (A): **n-pentane** C_5H_{12} **109-66-0**

Type of data: cloud points

continued

continued

T/K = 403.15

w_B	0.003	0.020	0.040	0.060	0.062	0.065	0.068	0.070	0.075
P/bar	46.2	78.6	70.3	66.2	66.2	66.9	64.1	63.4	62.7

w_B	0.080	0.100	0.110
P/bar	62.0	47.5	42.7

critical concentration: $w_{B, crit}$ = 0.069

T/K = 423.15

w_B	0.003	0.020	0.040	0.060	0.062	0.065	0.068	0.070	0.075
P/bar	79.3	108.9	100.6	93.1	92.4	92.4	90.3	89.6	88.9

w_B	0.080	0.100	0.110
P/bar	88.2	75.8	70.3

critical concentration: $w_{B, crit}$ = 0.069

T/K = 443.15

w_B	0.003	0.020	0.040	0.060	0.062	0.065	0.068	0.070	0.075
P/bar	115.8	130.3	126.2	120.0	119.3	117.9	117.2	115.8	115.1

w_B	0.080	0.100	0.110
P/bar	113.1	102.7	95.8

critical concentration: $w_{B, crit}$ = 0.069

Polymer (B): **poly(ethylene-*co*-vinyl acetate)** **1999BEY**
Characterization: M_n/g.mol^{-1} = 74000, M_w/g.mol^{-1} = 285000, 70.0 wt% vinyl acetate, Scientific Polymer Products, Inc., Ontario, NY
Solvent (A): **cyclopentane** C_5H_{10} **287-92-3**

Type of data: cloud points (LCST-behavior)

w_B	0.06	0.06	0.06	0.06	0.06	0.06
T/K	462.55	480.05	494.45	507.05	518.35	530.35
P/bar	34.6	60.7	80.9	97.4	113.3	125.3

Polymer (B): **poly(ethylene-*co*-vinyl acetate)** **1999BEY**
Characterization: M_n/g.mol^{-1} = 74000, M_w/g.mol^{-1} = 285000, 70.0 wt% vinyl acetate, Scientific Polymer Products, Inc., Ontario, NY
Solvent (A): **cyclopentene** C_5H_8 **142-29-0**

Type of data: cloud points (LCST-behavior)

w_B	0.06	0.06	0.06	0.06	0.06	0.06	0.06	0.06
T/K	475.75	475.95	485.25	486.35	496.75	497.95	505.65	507.15
P/bar	37.3	37.6	51.9	53.4	68.4	70.1	80.7	83.2

Polymer (B): **poly(ethylene-*co*-vinyl acetate)** **1990VAN**
Characterization: see table
Solvent (A): **diphenyl ether** $C_{12}H_{10}O$ **101-84-8**

Type of data: critical point (UCST-behavior)

M_n/ g mol^{-1}	M_w/ g mol^{-1}	M_z/ g mol^{-1}	wt% vinyl acetate	$w_{B, crit}$	T_{crit}/ K
52000	465000	2280000	2.3	0.05115	404.15
47000	280000	1300000	4.0	0.0608	392.45
57000	290000	1100000	4.1	0.0585	394.25
34000	460000	3100000	7.1	0.0072	378.15
53000	350000	1900000	9.5	0.0071	367.25
55000	490000	2400000	9.7	0.0625	370.75
66000	300000	1100000	12.1	0.0595	360.35

Polymer (B): **poly(ethylene-*co*-vinyl acetate)** **1986RAE**
Characterization: M_n/g.mol^{-1} = 14800, M_w/g.mol^{-1} = 41500, M_z/g.mol^{-1} = 79200, 42.6 mol% vinyl acetate, ρ_B (293.15 K) = 1.241 g/cm^3
Solvent (A): **methyl acetate** $C_3H_6O_2$ **79-20-9**

Type of data: cloud points (UCST-behavior)

φ_B	0.0051	0.0117	0.0152	0.0300	0.0453	0.0579	0.0740	0.0817	0.0843
T/K	311.1	314.9	315.9	314.9	313.9	312.4	310.3	309.3	309.0

φ_B	0.0875	0.0934	0.0992	0.1019	0.1031	0.1052	0.1225	0.1278	0.1303
T/K	308.7	308.0	307.5	307.2	307.0	306.8	305.7	305.3	305.7

φ_B	0.1500	0.1650	0.1947	0.2570
T/K	304.4	303.7	302.2	296.3

Type of data: critical point (UCST-behavior)

$\varphi_{B, crit}$ 0.1031 T_{crit}/K 307.0

Polymer (B): **poly(ethylene glycol)** **1991BAE**
Characterization: M_n/g.mol^{-1} = 2100, M_w/g.mol^{-1} = 3350, Union Carbide Corp., New York, NY
Solvent (A): **water** H_2O **7732-18-5**

Type of data: cloud points (closed loop miscibility gap)

w_B	0.049	0.070	0.100	0.150	0.200	0.297	0.399	0.443
T/K	513.65	518.05	518.55	519.55	517.75	512.95	501.65	496.35

w_B	0.049	0.070	0.100	0.150	0.200	0.297	0.399	0.443
T/K	430.75	429.35	428.85	428.65	429.75	431.25	438.05	445.45

Polymer (B): **poly(ethylene glycol)** **1993SAR**
Characterization: M_n/g.mol^{-1} = 2160, M_w/g.mol^{-1} = 2530, M_z/g.mol^{-1} = 2985, Aldrich Chem. Co., Inc., Milwaukee, WI
Solvent (A): **water** **H_2O** **7732-18-5**

Type of data: cloud points (closed loop miscibility gap)

w_B	0.0507	0.0638	0.0920	0.1805	0.2989	0.3788	0.4269
T/K	473.85	483.85	499.55	503.05	500.85	495.45	487.35

w_B	0.0507	0.0638	0.0920	0.1805	0.2989	0.3788	0.4269
T/K	451.15	443.65	432.45	431.85	434.15	433.25	444.05

Polymer (B): **poly(ethylene glycol)** **1993SAR**
Characterization: M_n/g.mol^{-1} = 3970, M_w/g.mol^{-1} = 4490, M_z/g.mol^{-1} = 4980, Aldrich Chem. Co., Inc., Milwaukee, WI
Solvent (A): **water** **H_2O** **7732-18-5**

Type of data: cloud points (closed loop miscibility gap)

w_B	0.0252	0.0487	0.1040	0.1906	0.2957	0.3636	0.4546
T/K	501.45	520.55	529.95	525.65	527.45	527.15	509.35

w_B	0.0252	0.0487	0.1040	0.1906	0.2957	0.3636	0.4546
T/K	431.15	421.65	415.45	417.75	415.45	419.95	431.85

Polymer (B): **poly(ethylene glycol)** **1991BAE**
Characterization: M_n/g.mol^{-1} = 5000, M_w/g.mol^{-1} = 8000
Solvent (A): **water** **H_2O** **7732-18-5**

Type of data: cloud points (closed loop miscibility gap)

w_B	0.012	0.020	0.030	0.049	0.100	0.300	0.400	0.550
T/K	548.65	552.15	554.05	555.05	555.65	552.05	544.35	503.15

w_B	0.012	0.020	0.030	0.049	0.100	0.199	0.300	0.400	0.550
T/K	405.35	402.05	399.85	397.95	397.75	397.95	401.75	409.25	439.75

Polymer (B): **poly(ethylene glycol)** **1996FIS, 2001FIS**
Characterization: M_n/g.mol^{-1} = 6096, M_w/g.mol^{-1} = 6197, M_z/g.mol^{-1} = 6293, Merck KGaA, Darmstadt, Germany
Solvent (A): **water** **H_2O** **7732-18-5**

Type of data: cloud points (LCST-behavior)

φ_B	0.0969	0.0840	0.0807	0.0771	0.0706	0.0659	0.0616	0.0583	0.0547
T/K	405.00	405.10	405.30	405.50	405.70	406.00	406.30	406.60	407.00

φ_B	0.0535	0.0502	0.0475	0.0461	0.0442	0.0418	0.0393	0.0392	0.0376
T/K	407.00	407.40	407.80	408.00	408.20	408.50	408.85	409.00	409.50

continued

continued

φ_B	0.0335	0.0316	0.0312	0.0282	0.0269	0.0254	0.0252	0.0230	0.0219
T/K	410.00	410.50	411.00	411.50	412.00	412.45	412.85	413.40	414.00
φ_B	0.0208	0.0198	0.0187	0.0187	0.0181	0.0152	0.0170	0.0157	
T/K	414.50	415.00	415.50	416.00	416.45	416.65	417.00	417.90	
φ_B	0.2031	0.2264	0.2209	0.2331	0.2368	0.2457	0.2538	0.2608	0.2690
T/K	405.00	405.10	405.30	405.50	405.70	406.00	406.30	406.60	407.00
φ_B	0.2751	0.2778	0.2846	0.2910	0.2921	0.2974	0.3111	0.3048	0.3094
T/K	407.00	407.40	407.80	408.00	408.20	408.50	408.85	409.00	409.50
φ_B	0.3181	0.3253	0.3299	0.3373	0.3488	0.3490	0.3527	0.3591	0.3650
T/K	410.00	410.50	411.00	411.50	412.00	412.45	412.85	413.40	414.00
φ_B	0.3699	0.3738	0.3792	0.3824	0.3859	0.4173	0.3907	0.3986	
T/K	414.50	415.00	415.50	416.00	416.45	416.65	417.00	417.90	

Polymer (B): **poly(ethylene glycol)** **1993SAR**
Characterization: M_n/g.mol^{-1} = 6740, M_w/g.mol^{-1} = 8420, M_z/g.mol^{-1} = 9935, Aldrich Chem. Co., Inc., Milwaukee, WI
Solvent (A): **water** **H_2O** **7732-18-5**

Type of data: cloud points (closed loop miscibility gap)

w_B	0.0242	0.0511	0.1008	0.1607	0.2314	0.3921	0.4992
T/K	553.35	556.55	561.95	557.85	561.35	547.85	533.35
w_B	0.0242	0.0511	0.1008	0.1607	0.2314	0.3921	0.4992
T/K	397.85	396.15	394.05	395.05	396.85	403.75	421.25

Polymer (B): **poly(ethylene glycol)** **2001FIS**
Characterization: M_n/g.mol^{-1} = 10457, M_w/g.mol^{-1} = 11615, M_z/g.mol^{-1} = 12375, Merck KGaA, Darmstadt, Germany
Solvent (A): **water** **H_2O** **7732-18-5**

Type of data: cloud points (LCST-behavior)

φ_B	0.0489	0.0477	0.0484	0.0480	0.0429	0.0242	0.0213	0.0241	0.0241
T/K	393.95	394.15	394.25	394.35	395.55	406.70	407.65	409.60	410.65
φ_B	0.2236	0.2280	0.2246	0.2239	0.2301	0.3411	0.3635	0.3931	0.4023
T/K	393.95	394.15	394.25	394.35	395.55	406.70	407.65	409.60	410.65

Polymer (B): **poly(ethylene glycol)** **2001FIS**
Characterization: M_n/g.mol^{-1} = 12000, Fluka AG, Buchs, Switzerland
Solvent (A): **water** **H_2O** **7732-18-5**

Type of data: cloud points (LCST-behavior)

continued

continued

φ_B	0.0278	0.0255	0.0230	0.0194	0.0202	0.0194	0.0182
T/K	391.14	391.98	393.14	394.97	397.03	398.02	398.97
φ_B	0.2045	0.2322	0.2446	0.2732	0.2917	0.3039	0.3145
T/K	391.14	391.98	393.14	394.97	397.03	398.02	398.97

Polymer (B): **poly(ethylene glycol)** **1994BOR, 2001FIS**
Characterization: M_n/g.mol^{-1} = 12161, M_w/g.mol^{-1} = 15481, M_z/g.mol^{-1} = 18004, trimodal distribution, Merck KGaA, Darmstadt, Germany
Solvent (A): **water** **H_2O** **7732-18-5**

Type of data: cloud points (LCST-behavior)

φ_B	0.0547	0.0469	0.0435	0.0416	0.0493	0.0315	0.0221	0.0225	0.0136
T/K	388.25	388.55	388.75	389.05	389.25	389.75	390.75	391.25	397.75
φ_B	0.1929	0.1879	0.1916	0.1900	0.1944	0.2095	0.2201	0.2443	0.3208
T/K	388.25	388.55	388.75	389.05	389.25	389.75	390.75	391.25	397.75
φ_B	0.0405	0.0353	0.0321	0.0306	0.0295	0.0285	0.0274	0.0254	0.0241
T/K	388.55	388.65	388.75	388.85	388.95	389.15	389.35	389.55	389.75
φ_B	0.0217	0.0229	0.0173	0.0196	0.0194	0.0190	0.0185	0.0182	0.0174
T/K	389.90	390.00	390.15	390.35	390.55	390.75	390.95	391.15	391.35
φ_B	0.2046	0.1949	0.1884	0.2257	0.1982	0.1945	0.2008	0.2069	0.2080
T/K	388.55	388.65	388.75	388.85	388.95	389.15	389.35	389.55	389.75
φ_B	0.2128	0.2172	0.2158	0.2198	0.2236	0.2275	0.2315	0.2345	0.2387
T/K	389.90	390.00	390.15	390.35	390.55	390.75	390.95	391.15	391.35

Polymer (B): **poly(ethylene glycol)** **1991BAE**
Characterization: M_n/g.mol^{-1} = 12500, M_w/g.mol^{-1} = 15000, Polysciences, Inc., Warrington, PA
Solvent (A): **water** **H_2O** **7732-18-5**

Type of data: cloud points (closed loop miscibility gap)

w_B	0.002	0.005	0.007	0.010	0.021	0.050	0.071	0.101	0.202
T/K	543.15	550.75	553.35	557.75	560.65	568.25	569.25	569.05	568.85
w_B	0.294	0.403	0.501	0.002	0.005	0.007	0.010	0.021	0.031
T/K	564.95	555.65	539.25	408.75	403.15	401.55	398.95	396.35	394.85
w_B	0.050	0.071	0.101	0.202	0.294	0.403	0.501		
T/K	393.35	392.95	392.65	393.35	397.35	406.05	419.45		

Polymer (B): **poly(ethylene glycol)** **2001FIS**
Characterization: M_n/g.mol^{-1} = 33500, Hoechst AG, Frankfurt, Germany
Solvent (A): **water** **H_2O** **7732-18-5**

continued

continued

Type of data: cloud points (LCST-behavior)

φ_B	0.0316	0.0233	0.0174	0.0161	0.0140	0.0129	0.0115	0.0110	0.0090
T/K	382.00	382.24	382.35	382.75	383.55	383.67	384.04	384.51	384.83
φ_B	0.0084	0.0078	0.0065	0.0067	0.0054	0.0050	0.0040	0.0037	0.0025
T/K	385.27	386.44	387.47	387.94	388.95	389.97	391.93	392.87	397.86
φ_B	0.1200	0.1252	0.1372	0.1495	0.1579	0.1629	0.1697	0.1786	0.1871
T/K	382.00	382.24	382.35	382.75	383.55	383.67	384.04	384.51	384.83
φ_B	0.1937	0.2131	0.2288	0.2332	0.2500	0.2633	0.2883	0.3000	0.3518
T/K	385.27	386.44	387.47	387.94	388.95	389.97	391.93	392.87	397.86

Polymer (B): **poly(ethylene glycol)** **2001FIS**
Characterization: M_n/g.mol^{-1} = 40850, M_w/g.mol^{-1} = 151000,
WSR-10, Union Carbide Chemicals and Plastics
Solvent (A): **water** **H_2O** **7732-18-5**

Type of data: cloud points (LCST-behavior)

φ_B	0.0076	0.0060	0.0050	0.0041	0.1675	0.1655	0.1997	0.2463	0.2659
T/K	379.95	380.85	385.95	387.95	379.95	380.85	380.95	385.95	387.95

Polymer (B): **poly(ethylene glycol)** **1991BAE**
Characterization: M_n/g.mol^{-1} = 50000, M_w/g.mol^{-1} = 100000,
Scientific Polymer Products, Inc., Ontario, NY
Solvent (A): **water** **H_2O** **7732-18-5**

Type of data: cloud points (LCST-behavior)

w_B	0.005	0.007	0.010	0.030	0.050	0.070	0.100
T/K	376.85	374.65	375.55	375.75	377.45	378.45	379.55

Polymer (B): **poly(ethylene oxide)** **2000SPI**
Characterization: M_n/g.mol^{-1} = 3350, Fluka AG, Buchs, Switzerland
Solvent (A): **dichloromethane** **CH_2Cl_2** **75-09-2**

Type of data: cloud points (LCST-behavior)

w_B	0.017	0.050	0.061	0.099	0.203	0.334	0.381
T/K	193.15	205.15	203.15	208.15	218.15	238.15	235.15

Polymer (B): **poly(ethylene oxide)** **2000SPI**
Characterization: M_n/g.mol^{-1} = 200, Sigma Chemical Co., Inc., St. Louis, MO
Solvent (A): **methanol** **CH_4O** **67-56-1**

Type of data: cloud points (LCST-behavior)

w_B	0.044	0.084	0.140	0.185
T/K	207.15	211.15	214.15	216.15

Polymer (B): **poly(ethylene oxide)** **2000SPI**
Characterization: M_n/g.mol^{-1} = 3350, Fluka AG, Buchs, Switzerland
Solvent (A): **methanol** **CH_4O** **67-56-1**

Type of data: cloud points (LCST-behavior)

w_B	0.010	0.050	0.100	0.150
T/K	273.65	277.15	279.15	280.15

Polymer (B): **poly(ethylene oxide)** **2000SPI**
Characterization: M_n/g.mol^{-1} = 10000, Sigma Chemical Co., Inc., St. Louis, MO
Solvent (A): **methanol** **CH_4O** **67-56-1**

Type of data: cloud points (LCST-behavior)

w_B	0.010	0.044	0.085	0.184	0.407
T/K	243.05	254.95	259.75	265.85	267.05

Polymer (B): **poly(ethylene oxide)** **2000SPI**
Characterization: M_n/g.mol^{-1} = 3350, Fluka AG, Buchs, Switzerland
Solvent (A): **trichloromethane** **$CHCl_3$** **67-66-3**

Type of data: cloud points (LCST-behavior)

w_B	0.009	0.047	0.094	0.140	0.280	0.379
T/K	222.15	234.15	245.15	253.15	261.15	268.85

Polymer (B): **poly(ethylene oxide)** **1992COO**
Characterization: M_n/g.mol^{-1} = 14700, M_w/g.mol^{-1} = 19700,
Pressure Chemical Company, Pittsburgh, PA
Solvent (A): **water** **H_2O** **7732-18-5**

Type of data: cloud points

c_B/(g/dl)	1.0	20.0
T/K	295.65	295.65
P/MPa	557	539

Comments: The complete phase behavior is given in the original source in graphs.

Polymer (B): **poly(ethylene oxide)** **1992COO**
Characterization: M_n/g.mol^{-1} = 248000, M_w/g.mol^{-1} = 270000,
American Polymer Standards
Solvent (A): **water** **H_2O** **7732-18-5**

Type of data: cloud points

c_B/(g/dl)	0.3	1.0
T/K	295.65	295.65
P/MPa	430	435

Comments: The complete phase behavior is given in the original source in graphs.

Polymer (B): **poly(ethylene oxide-*b*-propylene fumarate-*b*-ethylene oxide) dimethyl ether** **2002BEH**
Characterization: M_n/g.mol^{-1} = 2730, about 40 mol% ethylene oxide in 1.9 blocks, M_n/g.mol^{-1} = 570 PEG monomethyl ether block, M_n/g.mol^{-1} = 1660 poly(propylene fumarate) block
Solvent (A): **water** **H_2O** **7732-18-5**

Type of data: cloud points (LCST-behavior)

w_B	0.05	0.15	0.25
T/K	313.15	318.15	318.15

Polymer (B): **poly(ethylene oxide-*b*-propylene fumarate-*b*-ethylene oxide) dimethyl ether** **2002BEH**
Characterization: M_n/g.mol^{-1} = 3120, about 50 mol% ethylene oxide in 1.8 blocks, M_n/g.mol^{-1} = 800 PEG monomethyl ether block, M_n/g.mol^{-1} = 1660 poly(propylene fumarate) block
Solvent (A): **water** **H_2O** **7732-18-5**

Type of data: cloud points (LCST-behavior)

w_B	0.05	0.15	0.25
T/K	328.15	333.15	333.15

Polymer (B): **poly(ethylene oxide-*co*-propylene oxide)** **2000PE2**
Characterization: M_n/g.mol^{-1} = 3400, 20.0 mol% ethylene oxide, Shearwater Polymers, Huntsville, AL
Solvent (A): **water** **H_2O** **7732-18-5**

Type of data: cloud points (LCST-behavior)

w_B 0.10 T/K 303.15

Polymer (B): **poly(ethylene oxide-*co*-propylene oxide)** **2000PE1**
Characterization: M_n/g.mol^{-1} = 3000, 27.0 mol% ethylene oxide, Shearwater Polymers, Huntsville, AL
Solvent (A): **water** **H_2O** **7732-18-5**

Type of data: cloud points (LCST-behavior)

w_B 0.01 T/K 309.15

Polymer (B): **poly(ethylene oxide-*co*-propylene oxide)** **2000PE2**
Characterization: M_n/g.mol^{-1} = 5400, 30.0 mol% ethylene oxide, Shearwater Polymers, Huntsville, AL
Solvent (A): **water** **H_2O** **7732-18-5**

Type of data: cloud points (LCST-behavior)

w_B 0.10 T/K 313.15

Polymer (B): **poly(ethylene oxide-*co*-propylene oxide)** **1997LIM**
Characterization: M_n/g.mol^{-1} = 3000-3500, 33.3 mol% ethylene oxide, Zhejiang Univ. Chem. Factory, PR China
Solvent (A): **water** **H_2O** **7732-18-5**

Type of data: cloud points (LCST-behavior)

w_B	0.10	T/K	340.15

Polymer (B): **poly(ethylene oxide-*co*-propylene oxide)** **2000PE1**
Characterization: M_n/g.mol^{-1} = 5000, 38.5 mol% ethylene oxide, Shearwater Polymers, Huntsville, AL
Solvent (A): **water** **H_2O** **7732-18-5**

Type of data: cloud points (LCST-behavior)

w_B	0.01	T/K	309.15

Polymer (B): **poly(ethylene oxide-*co*-propylene oxide)** **2000LIW**
Characterization: M_n/g.mol^{-1} = 2340, M_w/g.mol^{-1} = 2480, 50.0 mol% ethylene oxide
Solvent (A): **water** **H_2O** **7732-18-5**

Type of data: cloud points (LCST-behavior)

w_B	0.2000
T/K	338.25

Type of data: coexistence data (tie lines, LCST-behavior)

Comments: The total feed concentration of the polymer is w_B = 0.2000.

T/K	341.35	345.25	350.65	357.95	364.65	373.15
w_B (top phase)	0.1870	0.1087	0.0710	0.0424	0.0315	0.0216
w_B (bottom phase)	0.6124	0.6592	0.6942	0.7173	0.7528	0.7654

Polymer (B): **poly(ethylene oxide-*co*-propylene oxide)** **1997LIM**
Characterization: M_n/g.mol^{-1} = 3000-3500, 50.0 mol% ethylene oxide, Zhejiang Univ. Chem. Factory, PR China
Solvent (A): **water** **H_2O** **7732-18-5**

Type of data: cloud points (LCST-behavior)

w_B	0.10	T/K	343.15

Polymer (B): **poly(ethylene oxide-*co*-propylene oxide)** **2000LIW**
Characterization: M_n/g.mol^{-1} = 3640, M_w/g.mol^{-1} = 4040, 50.0 mol% ethylene oxide
Solvent (A): **water** **H_2O** **7732-18-5**

continued

continued

Type of data: cloud points (LCST-behavior)

w_B	0.2000
T/K	324.85

Type of data: coexistence data (tie lines, LCST-behavior)

Comments: The total feed concentration of the polymer is w_B = 0.2000.

T/K	328.85	332.55	339.05	343.30	347.20	352.15	363.35	373.15
w_B (top phase)	0.1876	0.1275	0.0530	0.0320	0.0210	0.0140	0.0079	0.0084
w_B (bottom phase)	0.6151	0.6531	0.6992	0.7394	0.7685	0.7881	0.8315	0.8461

Polymer (B): **poly(ethylene oxide-*co*-propylene oxide)** **2000PE2**
Characterization: M_n/g.mol^{-1} = 3900, 50.0 mol% ethylene oxide, Breox PAG 50A 1000, Int. Speciality Chemicals, Southampton, UK
Solvent (A): **water** **H_2O** **7732-18-5**

Type of data: cloud points (LCST-behavior)

w_B 0.10 T/K 323.15

Polymer (B): **poly(ethylene oxide-*co*-propylene oxide)** **2000PE1**
Characterization: M_n/g.mol^{-1} = 3000, 58.8 mol% ethylene oxide, Shearwater Polymers, Huntsville, AL
Solvent (A): **water** **H_2O** **7732-18-5**

Type of data: cloud points (LCST-behavior)

w_B 0.01 T/K 326.65

Polymer (B): **poly(ethylene oxide-*co*-propylene oxide)** **1997LIM**
Characterization: M_n/g.mol^{-1} = 3000-3500, 66.7 mol% ethylene oxide, Zhejiang Univ. Chem. Factory, PR China
Solvent (A): **water** **H_2O** **7732-18-5**

Type of data: cloud points (LCST-behavior)

w_B 0.10 T/K 359.15

Polymer (B): **poly(ethylene oxide-*b*-propylene oxide-*b*-ethylene oxide)** **2000LAM**
Characterization: M_w/g.mol^{-1} = 4400, 30 mol% ethylene oxide, ρ(298.15 K) = 1.036 g/cm^3
Solvent (A): **water** **H_2O** **7732-18-5**

Type of data: cloud points (LCST-behavior)

w_B 0.05 T/K 286.65

Polymer (B): **poly(ethylene oxide-*b*-propylene oxide-*b*-ethylene oxide)** **2005ZH2**
Characterization: M_n/g.mol^{-1} = 4950, 30 wt% ethylene oxide, EO17PO60EO17
Solvent (A): **water** **H_2O** **7732-18-5**

Type of data: cloud points (LCST-behavior)

w_B 0.10 T/K 325.15

Polymer (B): **poly(ethylene oxide-*b*-propylene oxide-*b*-ethylene oxide)** **2005ZH2**
Characterization: M_n/g.mol^{-1} = 2200, 40 wt% ethylene oxide, EO10PO23EO10
Solvent (A): **water** **H_2O** **7732-18-5**

Type of data: cloud points (LCST-behavior)

w_B 0.10 T/K 346.15

Polymer (B): **poly(ethylene oxide-*b*-propylene oxide-*b*-ethylene oxide)** **2005ZH2**
Characterization: M_n/g.mol^{-1} = 2900, 40 wt% ethylene oxide, EO13PO30EO13
Solvent (A): **water** **H_2O** **7732-18-5**

Type of data: cloud points (LCST-behavior)

w_B 0.10 T/K 333.15

Polymer (B): **poly(ethylene oxide-*b*-propylene oxide-*b*-ethylene oxide)** **2005ZH2**
Characterization: M_n/g.mol^{-1} = 5900, 40 wt% ethylene oxide, EO27PO61EO27
Solvent (A): **water** **H_2O** **7732-18-5**

Type of data: cloud points (LCST-behavior)

w_B 0.10 T/K 351.15

Polymer (B): **poly(ethylene oxide-*b*-propylene oxide-*b*-ethylene oxide)** **2005ZH2**
Characterization: M_n/g.mol^{-1} = 4600, 50 wt% ethylene oxide, EO26PO40EO26
Solvent (A): **water** **H_2O** **7732-18-5**

Type of data: cloud points (LCST-behavior)

w_B 0.10 T/K 359.15

Polymer (B): **poly(ethylene oxide-*b*-propylene oxide-*b*-ethylene oxide)** **2005ZH2**
Characterization: M_n/g.mol^{-1} = 6500, 50 wt% ethylene oxide, EO37PO56EO37
Solvent (A): **water** **H_2O** **7732-18-5**

continued

continued

Type of data: cloud points (LCST-behavior)

w_B	0.10	T/K	367.15

Polymer (B): **poly(ethylene oxide-*b*-propylene oxide-*b*-ethylene oxide)** **1995KIM**
Characterization: M_n/g.mol^{-1} = 2500, 80 wt% ethylene oxide, Pluronic L62, BASF Wyandotte Corp., Parsippany, NJ
Solvent (A): **water** **H_2O** **7732-18-5**

Type of data: cloud points (LCST-behavior)

w_B	0.005	0.010	0.025	0.050	0.075	0.100
T/K	308.15	305.15	301.15	299.15	297.15	296.15

Polymer (B): **poly(ethylethylene)** **1979KLE, 1980KL2**
Characterization: M_n/g.mol^{-1} = 48000, M_w/g.mol^{-1} = 52000, hydrogenated polybutadiene, prepared in the laboratory
Solvent (A): **diphenyl ether** **$C_{12}H_{10}O$** **101-84-8**

Type of data: critical point (UCST-behavior)

$\varphi_{B,\,crit}$	0.0592	T_{crit}/K	411.2

Polymer (B): **polyglycerol** **2002SEI**
Characterization: M_n/g.mol^{-1} = 6500, M_w/g.mol^{-1} = 13650, hyperbranched, acetylated, synthesized in the laboratory
Solvent (A): **water** **H_2O** **7732-18-5**

Type of data: coexistence data (tie lines)

T/K = 295.15

phase I	w_A	0.090	w_B	0.910
phase II	w_A	0.997	w_B	0.003

Polymer (B): **poly(2-hydroxyethyl methacrylate)** **1969DUS**
Characterization: M_η/g.mol^{-1} = 77400, fractionated in the laboratory
Solvent (A): **1-butanol** **$C_4H_{10}O$** **71-36-3**

Type of data: cloud points (UCST-behavior)

w_B	0.001	0.005	0.010	0.015	0.025	0.035	0.050	0.070
T/K	324.65	331.35	333.95	335.55	336.75	337.15	337.25	337.25

Polymer (B): **poly(2-hydroxyethyl methacrylate)** **1969DUS**
Characterization: M_η/g.mol^{-1} = 133000, fractionated in the laboratory
Solvent (A): **1-butanol** **$C_4H_{10}O$** **71-36-3**

continued

continued

Type of data: cloud points (UCST-behavior)

w_B	0.001	0.005	0.010	0.015	0.025	0.035	0.050
T/K	332.15	336.15	337.85	338.75	339.75	339.75	339.65

Polymer (B): **poly(2-hydroxyethyl methacrylate)** **1969DUS**
Characterization: M_η/g.mol^{-1} = 175000, fractionated in the laboratory
Solvent (A): **1-butanol** $C_4H_{10}O$ **71-36-3**

Type of data: cloud points (UCST-behavior)

w_B	0.001	0.005	0.010	0.015	0.020	0.030	0.040	0.050
T/K	336.05	339.45	340.55	341.35	341.75	342.05	342.05	342.05

Polymer (B): **poly(2-hydroxyethyl methacrylate)** **1969DUS**
Characterization: M_η/g.mol^{-1} = 233600, fractionated in the laboratory
Solvent (A): **1-butanol** $C_4H_{10}O$ **71-36-3**

Type of data: cloud points (UCST-behavior)

w_B	0.001	0.005	0.010	0.015	0.025	0.035	0.050
T/K	338.35	341.25	342.25	342.95	343.35	343.35	343.35

Polymer (B): **poly(2-hydroxyethyl methacrylate)** **1969DUS**
Characterization: M_η/g.mol^{-1} = 352200, fractionated in the laboratory
Solvent (A): **1-butanol** $C_4H_{10}O$ **71-36-3**

Type of data: cloud points (UCST-behavior)

w_B	0.001	0.010	0.015	0.020	0.025	0.035	0.050
T/K	341.85	344.55	344.65	344.75	344.85	344.85	344.75

Polymer (B): **poly(2-hydroxyethyl methacrylate)** **1969DUS**
Characterization: M_η/g.mol^{-1} = 557800, fractionated in the laboratory
Solvent (A): **1-butanol** $C_4H_{10}O$ **71-36-3**

Type of data: cloud points (UCST-behavior)

w_B	0.001	0.005	0.010	0.015	0.025	0.035	0.050
T/K	344.65	345.65	346.15	346.25	346.35	346.25	346.15

Polymer (B): **poly(2-hydroxyethyl methacrylate)** **1969DUS**
Characterization: M_η/g.mol^{-1} = 233600, fractionated in the laboratory
Solvent (A): **2-butanol** $C_4H_{10}O$ **78-92-2**

Type of data: cloud points (UCST-behavior)

w_B 0.010 T/K 286.65

Polymer (B): **poly(2-hydroxyethyl methacrylate)** **1969DUS**
Characterization: M_η/g.mol^{-1} = 352200, fractionated in the laboratory
Solvent (A): **2-butanol** $C_4H_{10}O$ **78-92-2**

Type of data: cloud points (UCST-behavior)

w_B 0.010 T/K 287.15

Polymer (B): **poly(2-hydroxyethyl methacrylate)** **1969DUS**
Characterization: M_η/g.mol^{-1} = 233600, fractionated in the laboratory
Solvent (A): **2-methyl-1-propanol** $C_4H_{10}O$ **78-83-1**

Type of data: cloud points (UCST-behavior)

w_B 0.010 T/K 341.95

Polymer (B): **poly(2-hydroxyethyl methacrylate)** **1969DUS**
Characterization: M_η/g.mol^{-1} = 352200, fractionated in the laboratory
Solvent (A): **2-methyl-1-propanol** $C_4H_{10}O$ **78-83-1**

Type of data: cloud points (UCST-behavior)

w_B 0.010 T/K 344.55

Polymer (B): **poly(2-hydroxyethyl methacrylate)** **1969DUS**
Characterization: M_η/g.mol^{-1} = 77400, fractionated in the laboratory
Solvent (A): **1,2,3-propanetriol** $C_3H_8O_3$ **56-81-5**

Type of data: cloud points (UCST-behavior)

w_B 0.010 T/K 344.95

Polymer (B): **poly(2-hydroxyethyl methacrylate)** **1969DUS**
Characterization: M_η/g.mol^{-1} = 133000, fractionated in the laboratory
Solvent (A): **1,2,3-propanetriol** $C_3H_8O_3$ **56-81-5**

Type of data: cloud points (UCST-behavior)

w_B 0.010 T/K 349.15

Polymer (B): **poly(2-hydroxyethyl methacrylate)** **1969DUS**
Characterization: M_η/g.mol^{-1} = 233600, fractionated in the laboratory
Solvent (A): **1,2,3-propanetriol** $C_3H_8O_3$ **56-81-5**

Type of data: cloud points (UCST-behavior)

w_B 0.010 T/K 355.55

Polymer (B): **poly(2-hydroxyethyl methacrylate)** **1969DUS**
Characterization: M_η/g.mol^{-1} = 557800, fractionated in the laboratory
Solvent (A): **1,2,3-propanetriol** $C_3H_8O_3$ **56-81-5**

continued

continued

Type of data: cloud points (UCST-behavior)

w_B	0.010	T/K	356.75

Polymer (B): **poly(2-hydroxyethyl methacrylate)** **1969DUS**
Characterization: M_η/g.mol^{-1} = 77400, fractionated in the laboratory
Solvent (A): **1-propanol** C_3H_8O **71-23-8**

Type of data: cloud points (UCST-behavior)

w_B	0.001	0.005	0.010	0.020	0.050	0.070	0.100
T/K	303.35	307.45	308.95	310.45	311.15	310.85	310.15

Polymer (B): **poly(2-hydroxyethyl methacrylate)** **1969DUS**
Characterization: M_η/g.mol^{-1} = 133000, fractionated in the laboratory
Solvent (A): **1-propanol** C_3H_8O **71-23-8**

Type of data: cloud points (UCST-behavior)

w_B	0.001	0.005	0.010	0.020	0.050	0.070
T/K	307.35	310.85	311.85	312.45	313.15	312.85

Polymer (B): **poly(2-hydroxyethyl methacrylate)** **1969DUS**
Characterization: M_η/g.mol^{-1} = 175000, fractionated in the laboratory
Solvent (A): **1-propanol** C_3H_8O **71-23-8**

Type of data: cloud points (UCST-behavior)

w_B	0.001	0.005	0.010	0.020	0.050	0.070
T/K	310.95	313.35	314.55	314.65	314.85	314.95

Polymer (B): **poly(2-hydroxyethyl methacrylate)** **1969DUS**
Characterization: M_η/g.mol^{-1} = 233600, fractionated in the laboratory
Solvent (A): **1-propanol** C_3H_8O **71-23-8**

Type of data: cloud points (UCST-behavior)

w_B	0.001	0.005	0.010	0.020	0.050
T/K	312.45	314.95	315.45	315.35	315.35

Polymer (B): **poly(2-hydroxyethyl methacrylate)** **1969DUS**
Characterization: M_η/g.mol^{-1} = 352200, fractionated in the laboratory
Solvent (A): **1-propanol** C_3H_8O **71-23-8**

Type of data: cloud points (UCST-behavior)

w_B	0.001	0.005	0.010	0.020	0.050
T/K	315.75	317.15	318.15	318.35	318.15

Polymer (B): **poly(2-hydroxyethyl methacrylate)** **1969DUS**
Characterization: M_η/g.mol^{-1} = 557800, fractionated in the laboratory
Solvent (A): **1-propanol** C_3H_8O **71-23-8**

Type of data: cloud points (UCST-behavior)

w_B	0.001	0.005	0.010	0.020	0.050
T/K	316.75	318.65	319.05	319.65	319.55

Polymer (B): **polyimide** **2001KI1**
Characterization: M_n/g.mol^{-1} = 46000, M_w/g.mol^{-1} = 80000,
Matrimid 5218, Ciba-Geigy Co., Summit, NJ
Solvent (A): **γ-butyrolactone** $C_4H_6O_2$ **96-48-0**

Type of data: cloud points

w_B 0.04 T/K 298.15

Polymer (B): **polyimide** **2001KI1**
Characterization: M_n/g.mol^{-1} = 46000, M_w/g.mol^{-1} = 80000,
Matrimid 5218, Ciba-Geigy Co., Summit, NJ
Solvent (A): ***N,N*-dimethylformamide** C_3H_7NO **68-12-2**

Type of data: cloud points

w_B 0.130 T/K 298.15

Polymer (B): **polyimide** **2001KI1**
Characterization: M_n/g.mol^{-1} = 46000, M_w/g.mol^{-1} = 80000,
Matrimid 5218, Ciba-Geigy Co., Summit, NJ
Solvent (A): **dimethylsulfoxide** C_2H_6OS **67-68-5**

Type of data: cloud points

w_B 0.02 T/K 298.15

Polymer (B): **polyimide** **2001KI1**
Characterization: M_n/g.mol^{-1} = 46000, M_w/g.mol^{-1} = 80000,
Matrimid 5218, Ciba-Geigy Co., Summit, NJ
Solvent (A): **1-methyl-2-pyrrolidinone** C_5H_9NO **872-50-4**

Type of data: cloud points

w_B 0.200 T/K 298.15

Polymer (B): **polyisobutylene** **1972ZE1**
Characterization: M_η/g.mol^{-1} = 6000, M_w/M_n = 1.14 by GPC
Solvent (A): **n-butane** C_4H_{10} **106-97-8**

Type of data: cloud points

continued

continued

T/K = 344.35 w_B = 0.0079 $(dT/dP)_{P=1}$/K.bar^{-1} = +0.47
T/K = 337.85 w_B = 0.0272 $(dT/dP)_{P=1}$/K.bar^{-1} = +0.50
T/K = 321.85 w_B = 0.0466 $(dT/dP)_{P=1}$/K.bar^{-1} = +0.58

Polymer (B): **polyisobutylene** **1972ZE1**
Characterization: M_η/g.mol^{-1} = 703000, M_w/M_n = 1.14 by GPC
Solvent (A): **n-butane** C_4H_{10} **106-97-8**

Type of data: cloud points

T/K = 264.75 w_B = 0.0114 $(dT/dP)_{P=1}$/K.bar^{-1} = +0.37

Polymer (B): **polyisobutylene** **1972ZE1**
Characterization: M_η/g.mol^{-1} = 1660000, M_w/M_n = 1.14 by GPC
Solvent (A): **n-butane** C_4H_{10} **106-97-8**

Type of data: cloud points

T/K = 253.85 w_B = 0.0061 $(dT/dP)_{P=1}$/K.bar^{-1} = +0.37
T/K = 253.85 w_B = 0.0112 $(dT/dP)_{P=1}$/K.bar^{-1} = +0.37
T/K = 253.85 w_B = 0.0466 $(dT/dP)_{P=1}$/K.bar^{-1} = +0.37

Polymer (B): **polyisobutylene** **2002JOU**
Characterization: M_n/g.mol^{-1} = 600000, M_w/g.mol^{-1} = 1000000,
Aldrich Chem. Co., Inc., Milwaukee, WI
Solvent (A): **n-heptane** C_7H_{16} **142-82-5**

Type of data: cloud points

w_B	0.025	was kept constant				
T/K	443.15	453.15	463.95	473.25	483.15	493.15
P/MPa	1.3	2.5	3.9	5.4	6.7	8.3

Polymer (B): **polyisobutylene** **1972ZE1**
Characterization: M_η/g.mol^{-1} = 1660000, M_w/M_n = 1.14 by GPC
Solvent (A): **n-hexane** C_6H_{14} **110-54-3**

Type of data: cloud points

T/K = 409.65 w_B = 0.0204 $(dT/dP)_{P=1}$/K.bar^{-1} = +0.61

Polymer (B): **polyisobutylene** **1972ZE1**
Characterization: M_η/g.mol^{-1} = 6000, M_w/M_n = 1.14 by GPC
Solvent (A): **2-methylbutane** C_6H_{12} **78-78-4**

Type of data: cloud points

T/K = 387.85 w_B = 0.0235 $(dT/dP)_{P=1}$/K.bar^{-1} = +0.55

Polymer (B): **polyisobutylene** **1972ZE1**
Characterization: M_η/g.mol^{-1} = 703000, M_w/M_n = 1.14 by GPC
Solvent (A): **2-methylbutane** C_6H_{12} **78-78-4**

Type of data: cloud points

T/K = 330.55 w_B = 0.0220 $(dT/dP)_{P=1}$/K.bar^{-1} = +0.45

Polymer (B): **polyisobutylene** **1972ZE1**
Characterization: M_η/g.mol^{-1} = 1660000, M_w/M_n = 1.14 by GPC
Solvent (A): **2-methylbutane** C_6H_{12} **78-78-4**

Type of data: cloud points

T/K = 323.35 w_B = 0.0220 $(dT/dP)_{P=1}$/K.bar^{-1} = +0.44

Polymer (B): **polyisobutylene** **1972ZE1**
Characterization: M_η/g.mol^{-1} = 6000, M_w/M_n = 1.14 by GPC
Solvent (A): **n-pentane** C_5H_{12} **109-66-0**

Type of data: cloud points

T/K = 403.55 w_B = 0.0226 $(dT/dP)_{P=1}$/K.bar^{-1} = +0.57

Polymer (B): **polyisobutylene** **1972ZE1**
Characterization: M_η/g.mol^{-1} = 703000, M_w/M_n = 1.14 by GPC
Solvent (A): **n-pentane** C_5H_{12} **109-66-0**

Type of data: cloud points

T/K = 353.55 w_B = 0.0247 $(dT/dP)_{P=1}$/K.bar^{-1} = +0.42

Polymer (B): **polyisobutylene** **1972ZE1**
Characterization: M_η/g.mol^{-1} = 1660000, M_w/M_n = 1.14 by GPC
Solvent (A): **n-pentane** C_5H_{12} **109-66-0**

Type of data: cloud points

T/K = 347.35 w_B = 0.0243 $(dT/dP)_{P=1}$/K.bar^{-1} = +0.45

Polymer (B): **polyisobutylene** **1962BAK**
Characterization: M_η/g.mol^{-1} = 2250000, ρ_B (298 K) = 0.914 g/cm^3,
Solvent (A): **n-pentane** C_5H_{12} **109-66-0**

Type of data: cloud points (LCST-behavior)

w_B	0.000953	0.00178	0.00209	0.00273	0.00530	0.00783	0.0125	0.0195	0.0202
T/K	355.85	352.75	351.75	351.15	350.50	350.03	349.95	349.50	349.65
w_B	0.0282	0.0341	0.0495	0.0754	0.0830	0.0982	0.128	0.162	0.171
T/K	349.33	349.15	348.55	347.05	346.35	345.65	344.25	343.85	343.75
w_B	0.179	0.201	0.225	0.246	0.254				
T/K	343.55	343.35	343.45	343.25	343.25				

Polymer (B): **polyisobutylene** **1972ZE1**
Characterization: M_η/g.mol^{-1} = 2200, M_w/M_n = 1.14 by GPC
Solvent (A): **propane** C_3H_8 **74-98-6**

Type of data: cloud points

T/K = 291.55 w_B = 0.0135 $(dT/dP)_{P=1}$/K.bar^{-1} = +0.33
T/K = 279.85 w_B = 0.0374 $(dT/dP)_{P=1}$/K.bar^{-1} = +0.48

Polymer (B): **1,4-*cis*-polyisoprene** **1989BOH**
Characterization: M_η/g.mol^{-1} = 780000, Sp2 Company, Ontario, NY
Solvent (A): **2,5-dimethylhexane** C_8H_{18} **592-13-2**

Type of data: cloud points, precipitation threshold (LCST-behavior)

φ_B 0.02 T/K 474.15

Polymer (B): **1,4-*trans*-polyisoprene** **1989BOH**
Characterization: M_η/g.mol^{-1} = 780000, Sp2 Company, Ontario, NY
Solvent (A): **2,5-dimethylhexane** C_8H_{18} **592-13-2**

Type of data: cloud points, precipitation threshold (LCST-behavior)

φ_B 0.02 T/K 451.15

Polymer (B): **1,4-*cis*-polyisoprene** **1989BOH**
Characterization: M_η/g.mol^{-1} = 780000, Sp2 Company, Ontario, NY
Solvent (A): **3,4-dimethylhexane** C_8H_{18} **583-48-2**

Type of data: cloud points, precipitation threshold (LCST-behavior)

φ_B 0.02 T/K 520.15

Polymer (B): **1,4-*trans*-polyisoprene** **1989BOH**
Characterization: M_η/g.mol^{-1} = 780000, Sp2 Company, Ontario, NY
Solvent (A): **3,4-dimethylhexane** C_8H_{18} **583-48-2**

Type of data: cloud points, precipitation threshold (LCST-behavior)

φ_B 0.02 T/K 521.15

Polymer (B): **1,4-*cis*-polyisoprene** **1989BOH**
Characterization: M_η/g.mol^{-1} = 780000, Sp2 Company, Ontario, NY
Solvent (A): **2,2-dimethylpentane** C_7H_{16} **590-35-2**

Type of data: cloud points, precipitation threshold (LCST-behavior)

φ_B 0.02 T/K 445.15

Polymer (B): **1,4-*trans*-polyisoprene** **1989BOH**
Characterization: M_η/g.mol^{-1} = 780000, Sp2 Company, Ontario, NY
Solvent (A): **2,2-dimethylpentane** **C_7H_{16}** **590-35-2**

Type of data: cloud points, precipitation threshold (LCST-behavior)

φ_B 0.02 T/K 405.15

Polymer (B): **1,4-*cis*-polyisoprene** **1989BOH**
Characterization: M_η/g.mol^{-1} = 780000, Sp2 Company, Ontario, NY
Solvent (A): **2,3-dimethylpentane** **C_7H_{16}** **565-59-3**

Type of data: cloud points, precipitation threshold (LCST-behavior)

φ_B 0.02 T/K 484.15

Polymer (B): **1,4-*trans*-polyisoprene** **1989BOH**
Characterization: M_η/g.mol^{-1} = 780000, Sp2 Company, Ontario, NY
Solvent (A): **2,3-dimethylpentane** **C_7H_{16}** **565-59-3**

Type of data: cloud points, precipitation threshold (LCST-behavior)

φ_B 0.02 T/K 460.15

Polymer (B): **1,4-*cis*-polyisoprene** **1989BOH**
Characterization: M_η/g.mol^{-1} = 780000, Sp2 Company, Ontario, NY
Solvent (A): **2,4-dimethylpentane** **C_7H_{16}** **108-08-7**

Type of data: cloud points, precipitation threshold (LCST-behavior)

φ_B 0.02 T/K 442.15

Polymer (B): **1,4-*trans*-polyisoprene** **1989BOH**
Characterization: M_η/g.mol^{-1} = 780000, Sp2 Company, Ontario, NY
Solvent (A): **2,4-dimethylpentane** **C_7H_{16}** **108-08-7**

Type of data: cloud points, precipitation threshold (LCST-behavior)

φ_B 0.02 T/K 404.15

Polymer (B): **1,4-*cis*-polyisoprene** **1989BOH**
Characterization: M_η/g.mol^{-1} = 780000, Sp2 Company, Ontario, NY
Solvent (A): **3-ethylpentane** **C_7H_{16}** **617-78-7**

Type of data: cloud points, precipitation threshold (LCST-behavior)

φ_B 0.02 T/K 483.15

Polymer (B): **1,4-*trans*-polyisoprene** **1989BOH**
Characterization: M_η/g.mol^{-1} = 780000, Sp2 Company, Ontario, NY
Solvent (A): **3-ethylpentane** **C_7H_{16}** **617-78-7**

continued

continued

Type of data: cloud points, precipitation threshold (LCST-behavior)

φ_B 0.02 T/K 473.15

Polymer (B): **1,4-*cis*-polyisoprene** **1989BOH**
Characterization: M_η/g.mol^{-1} = 780000, Sp2 Company, Ontario, NY
Solvent (A): **n-heptane** **C_7H_{16}** **142-82-5**

Type of data: cloud points, precipitation threshold (LCST-behavior)

φ_B	0.008	0.015	0.022	0.030
T/K	491.15	488.15	488.15	493.15

Polymer (B): **1,4-*trans*-polyisoprene** **1989BOH**
Characterization: M_η/g.mol^{-1} = 780000, Sp2 Company, Ontario, NY
Solvent (A): **n-heptane** **C_7H_{16}** **142-82-5**

Type of data: cloud points, precipitation threshold (LCST-behavior)

φ_B	0.007	0.010	0.022	0.037
T/K	469.15	471.15	467.15	469.15

Polymer (B): **1,4-*cis*-polyisoprene** **1989BOH**
Characterization: M_η/g.mol^{-1} = 780000, Sp2 Company, Ontario, NY
Solvent (A): **n-hexane** **C_6H_{14}** **110-54-3**

Type of data: cloud points, precipitation threshold (LCST-behavior)

φ_B 0.02 T/K 434.15

Polymer (B): **1,4-*trans*-polyisoprene** **1989BOH**
Characterization: M_η/g.mol^{-1} = 780000, Sp2 Company, Ontario, NY
Solvent (A): **n-hexane** **C_6H_{14}** **110-54-3**

Type of data: cloud points, precipitation threshold (LCST-behavior)

φ_B 0.02 T/K 407.15

Polymer (B): **1,4-*cis*-polyisoprene** **1989BOH**
Characterization: M_η/g.mol^{-1} = 780000, Sp2 Company, Ontario, NY
Solvent (A): **n-nonane** **C_9H_{20}** **111-84-2**

Type of data: cloud points, precipitation threshold (LCST-behavior)

φ_B 0.02 T/K 541.15

Polymer (B): **1,4-*trans*-polyisoprene** **1989BOH**
Characterization: M_η/g.mol^{-1} = 780000, Sp2 Company, Ontario, NY
Solvent (A): **n-nonane** **C_9H_{20}** **111-84-2**

continued

continued

Type of data: cloud points, precipitation threshold (LCST-behavior)

φ_B 0.02 T/K 540.15

Polymer (B): **1,4-*cis*-polyisoprene** **1989BOH**
Characterization: M_η/g.mol^{-1} = 780000, Sp2 Company, Ontario, NY
Solvent (A): **n-octane** C_8H_{18} **111-65-9**

Type of data: cloud points, precipitation threshold (LCST-behavior)

φ_B 0.02 T/K 509.15

Polymer (B): **1,4-*trans*-polyisoprene** **1989BOH**
Characterization: M_η/g.mol^{-1} = 780000, Sp2 Company, Ontario, NY
Solvent (A): **n-octane** C_8H_{18} **111-65-9**

Type of data: cloud points, precipitation threshold (LCST-behavior)

φ_B 0.02 T/K 503.15

Polymer (B): **1,4-*cis*-polyisoprene** **1989BOH**
Characterization: M_η/g.mol^{-1} = 780000, Sp2 Company, Ontario, NY
Solvent (A): **2,2,4,4-tetramethylpentane** C_9H_{20} **1070-87-7**

Type of data: cloud points, precipitation threshold (LCST-behavior)

φ_B 0.02 T/K 518.15

Polymer (B): **1,4-*trans*-polyisoprene** **1989BOH**
Characterization: M_η/g.mol^{-1} = 780000, Sp2 Company, Ontario, NY
Solvent (A): **2,2,4,4-tetramethylpentane** C_9H_{20} **1070-87-7**

Type of data: cloud points, precipitation threshold (LCST-behavior)

φ_B 0.02 T/K 519.15

Polymer (B): **1,4-*cis*-polyisoprene** **1989BOH**
Characterization: M_η/g.mol^{-1} = 780000, Sp2 Company, Ontario, NY
Solvent (A): **2,3,4-trimethylhexane** C_9H_{20} **1070-87-7**

Type of data: cloud points, precipitation threshold (LCST-behavior)

φ_B 0.02 T/K 548.15

Polymer (B): **1,4-*trans*-polyisoprene** **1989BOH**
Characterization: M_η/g.mol^{-1} = 780000, Sp2 Company, Ontario, NY
Solvent (A): **2,3,4-trimethylhexane** C_9H_{20} **1070-87-7**

Type of data: cloud points, precipitation threshold (LCST-behavior)

φ_B 0.02 T/K 548.15

Polymer (B): **1,4-*cis*-polyisoprene** **1989BOH**
Characterization: M_η/g.mol^{-1} = 780000, Sp2 Company, Ontario, NY
Solvent (A): **2,2,4-trimethylpentane C_8H_{18}** **540-84-1**

Type of data: cloud points, precipitation threshold (LCST-behavior)

φ_B	0.02	T/K	471.15

Polymer (B): **1,4-*trans*-polyisoprene** **1989BOH**
Characterization: M_η/g.mol^{-1} = 780000, Sp2 Company, Ontario, NY
Solvent (A): **2,2,4-trimethylpentane C_8H_{18}** **540-84-1**

Type of data: cloud points, precipitation threshold (LCST-behavior)

φ_B	0.02	T/K	441.15

Polymer (B): **poly(*N*-isopropylacrylamide)** **2004SHI**
Characterization: synthesized in the laboratory
Solvent (A): **deuterium oxide D_2O** **7789-20-0**

Type of data: cloud points

c_B/mol.l^{-1} = 0.690 $T/\text{K} = 308.45 - 5.99\ 10^{-4}\ (P/\text{MPa} - 48.2)^2$

Polymer (B): **poly(*N*-isopropylacrylamide)** **2000AFR**
Characterization: M_n/g.mol^{-1} = 2200, M_w/g.mol^{-1} = 10000, M_z/g.mol^{-1} = 20500, synthesized in the laboratory
Solvent (A): **water H_2O** **7732-18-5**

Type of data: cloud points (LCST-behavior)

w_B	0.0009	0.0010	0.0025	0.0049	0.0075	0.0100	0.0249	0.0499	0.1002
T/K	306.98	307.17	306.09	306.40	305.91	305.44	304.70	303.97	303.21
w_B	0.2003	0.3017	0.3986	0.5001					
T/K	301.97	301.01	300.55	300.32					

Type of data: demixing temperatures (LCST-behavior, measured by DSC)

w_B	0.0100	0.0249	0.0499	0.1002	0.1495	0.2003	0.2462	0.3017	0.3505
T/K	306.20	305.63	304.86	304.03	302.70	302.50	302.23	301.36	300.99
w_B	0.4000	0.4501	0.4942	0.5575	0.5984	0.6499	0.6992		
T/K	300.89	300.89	300.55	300.80	300.85	302.23	302.86		

Polymer (B): **poly(*N*-isopropylacrylamide)** **1990SCH**
Characterization: see table, synthesized in the laboratory
Solvent (A): **water** H_2O **7732-18-5**

Type of data: cloud points (LCST-behavior)

M_n/ g mol^{-1}	M_w/ g mol^{-1}	c_B g/l	T/ K
5400	14000	0.40	307.45
11000	76000	0.40	307.25
49000	160000	0.40	307.05
73000	400000	0.40	306.35
160000	440000	0.40	306.35
146000	530000	0.40	305.85

Polymer (B): **poly(*N*-isopropylacrylamide)** **2005XIA**
Characterization: see table, synthesized in the laboratory
Solvent (A): **water** H_2O **7732-18-5**

Type of data: cloud points (LCST-behavior)

M_n/ g.mol^{-1}	M_w/M_n	w_B	T/ K
2800	1.07	0.01	317.15
5000	1.15	0.01	312.05
6500	1.09	0.01	309.45
6700	1.16	0.01	309.55
10900	1.11	0.01	308.65
15700	1.13	0.01	307.75
26500	1.16	0.01	306.45
28900	2.00	0.01	304.35

Polymer (B): **poly(*N*-isopropylacrylamide)** **2005RAY**
Characterization: see table, synthesized in the laboratory
Solvent (A): **water** H_2O **7732-18-5**

Type of data: cloud points (LCST-behavior)

continued

continued

M_n/ g.mol^{-1}	M_w/M_n	Tacticity % meso diads	w_B	T/ K
34000	1.20	45	0.01	304.25
37500	1.20	47	0.01	303.65
37200	1.20	49	0.01	303.25
39900	1.21	51	0.01	302.65
36900	1.23	53	0.01	301.65
35200	1.25	57	0.01	300.85
39800	1.21	62	0.01	297.85
39300	1.24	66	0.01	290.15

Polymer (B): **poly(*N*-isopropylacrylamide)** **1992INO**
Characterization: M_n/g.mol^{-1} = 31700, M_w/g.mol^{-1} = 76000, synthesized in the laboratory
Solvent (A): **water** **H_2O** **7732-18-5**

Type of data: cloud points (LCST-behavior)

w_B	0.014	0.034
T/K	304.35	304.85

Polymer (B): **poly(*N*-isopropylacrylamide)** **2000AFR**
Characterization: M_n/g.mol^{-1} = 36000, M_w/g.mol^{-1} = 53000, M_z/g.mol^{-1} = 69000, synthesized in the laboratory
Solvent (A): **water** **H_2O** **7732-18-5**

Type of data: demixing temperatures (LCST-behavior, measured by DSC)

w_B	0.0490	0.0998	0.1490	0.1999	0.2508	0.3006	0.3575	0.4496	0.4917
T/K	303.50	302.60	301.40	301.40	301.16	301.07	300.85	299.86	299.72

w_B	0.5500	0.6008	0.6449	0.7072	0.7571	0.8016	0.8406	0.9089
T/K	301.01	301.22	301.22	302.87	304.82	311.12	316.82	338.62

Polymer (B): **poly(*N*-isopropylacrylamide)** **2002KUN**
Characterization: M_n/g.mol^{-1} = 49000, M_w/g.mol^{-1} = 88200, synthesized in the laboratory
Solvent (A): **water** **H_2O** **7732-18-5**

Type of data: cloud points (LCST-behavior, determined by DSC)

w_B	0.00001	0.00005	0.00010	0.00020	0.00040
T/K	310.15	309.55	309.55	309.45	309.35

Polymer (B): **poly(*N*-isopropylacrylamide)** **2001DJO**
Characterization: M_n/g.mol^{-1} = 70000, M_w/g.mol^{-1} = 224000
Solvent (A): **water** **H_2O** **7732-18-5**

Type of data: cloud points (LCST-behavior)

w_B	0.01	T/K	305.95

Polymer (B): **poly(*N*-isopropylacrylamide)** **2000AFR**
Characterization: M_n/g.mol^{-1} = 83300, M_w/g.mol^{-1} = 124000, M_z/g.mol^{-1} = 181000
synthesized in the laboratory
Solvent (A): **water** **H_2O** **7732-18-5**

Type of data: demixing temperatures (LCST-behavior, measured by DSC)

w_B	0.0025	0.0049	0.0075	0.0099	0.0149	0.0198	0.0250	0.0299	0.0349
T/K	304.10	303.83	303.73	303.63	303.63	303.36	303.26	303.16	302.96
w_B	0.0399	0.0453	0.0497	0.0749	0.1007	0.1503	0.1998	0.2487	0.3002
T/K	302.80	302.70	302.40	302.40	302.13	301.93	301.06	300.90	300.70
w_B	0.3500	0.4096	0.4523	0.5161	0.5499	0.6338	0.7571		
T/K	300.60	300.23	300.23	299.93	300.23	302.50	304.82		

Polymer (B): **poly(*N*-isopropylacrylamide)** **2002RE1**
Characterization: M_n/g.mol^{-1} = 70000, M_w/g.mol^{-1} = 144000,
synthesized in the laboratory by radical polymerization
Solvent (A): **water** **H_2O** **7732-18-5**

Type of data: cloud points (LCST-behavior)

w_B	0.0035	0.0035	0.0035	0.0035	0.0035	0.0035	0.0035	0.0035	0.0307
T/K	313.0	313.7	314.2	314.9	315.1	315.6	315.8	315.4	310.2
P/bar	100.0	202.0	354.5	507.5	558.5	609.5	660.5	711.5	20.5
w_B	0.0307	0.0307	0.0307	0.0307	0.0307	0.0307	0.0307	0.0307	0.0307
T/K	311.2	311.7	312.5	313.1	313.2	313.4	313.4	313.3	313.0
P/bar	148.5	262.5	365.0	426.0	477.0	507.5	538.5	558.5	609.5

Polymer (B): **poly(*N*-isopropylacrylamide)** **2002RE1**
Characterization: M_n/g.mol^{-1} = 120000, M_w/g.mol^{-1} = 258000,
synthesized in the laboratory by radical polymerization
Solvent (A): **water** **H_2O** **7732-18-5**

Type of data: cloud points (LCST-behavior)

w_B	0.0039	0.0039	0.0039	0.0039	0.0039	0.0039	0.0039	0.0039	0.0295
T/K	307.2	307.5	308.1	308.4	308.4	308.4	308.2	308.0	306.6
P/bar	20.55	50.00	148.50	252.50	354.50	426.00	507.50	609.50	20.55
w_B	0.0295	0.0295	0.0295	0.0295	0.0295	0.0295	0.0295		
T/K	307.0	307.7	308.1	308.1	308.0	307.9	307.9		
P/bar	50.00	148.50	283.50	354.50	426.00	507.50	609.50		

Polymer (B): **poly(*N*-isopropylacrylamide)** **2001GOM**
Characterization: M_n/g.mol^{-1} = 301500, M_w/g.mol^{-1} = 615000,
synthesized in the laboratory by radical polymerization
Solvent (A): **water** **H_2O** **7732-18-5**

Type of data: cloud points (LCST-behavior)

w_B	0.01030	0.02011	0.02502	0.03574	0.03701	0.04167	0.04167	0.04167	0.04167
T/K	307.20	306.74	306.65	306.46	305.59	305.50	305.60	305.70	305.73
P/MPa	0.1	0.1	0.1	0.1	0.1	0.1	1.0	2.5	3.0

w_B	0.04167	0.04167	0.04167	0.04167	0.04167	0.05646	0.06402	0.06774	0.07670
T/K	305.86	305.90	308.12	308.27	308.32	306.10	306.25	306.40	306.35
P/MPa	4.0	5.0	20.0	30.0	40.0	0.1	0.1	0.1	0.1

w_B	0.08839	0.1057	0.1146	0.1474	0.1757
T/K	306.46	305.89	306.20	305.84	305.72
P/MPa	0.1	0.1	0.1	0.1	0.1

Type of data: spinodal points (LCST-behavior)

w_B	0.02011	0.03574	0.04167	0.04167	0.04167	0.04167	0.04167	0.04167	0.04529
T/K	321.6	312.1	309.2	308.5	309.2	308.2	307.6	307.4	310.7
P/MPa	0.10	0.10	0.10	1.0	2.5	3.0	4.0	5.0	0.10

w_B	0.05646	0.06402	0.06774	0.07670	0.08839	0.1146	0.1474	0.1757
T/K	308.9	307.8	308.7	309.1	308.7	309.1	308.7	310.9
P/MPa	0.10	0.10	0.10	0.10	0.10	0.10	0.10	0.10

Polymer (B): **poly(*N*-isopropylacrylamide)** **2002RE1**
Characterization: M_n/g.mol^{-1} = 301500, M_w/g.mol^{-1} = 615000,
synthesized in the laboratory by radical polymerization
Solvent (A): **water** **H_2O** **7732-18-5**

Type of data: cloud points (LCST-behavior)

w_B	0.0103	0.0201	0.0250	0.0357	0.0370	0.0417	0.0417	0.0417	0.0417
T/K	307.2	306.7	306.6	306.5	305.6	305.5	305.6	305.7	305.7
P/bar	1.0	1.0	1.0	1.0	1.0	1.0	10.0	25.0	30.0

w_B	0.0417	0.0417	0.0417	0.0417	0.0417	0.0417	0.0453	0.0565	0.0640
T/K	305.9	305.9	306.7	308.0	308.0	308.0	306.3	306.1	306.2
P/bar	40.0	50.0	100.0	200.0	300.0	400.0	1.0	1.0	1.0

w_B	0.0677	0.0767	0.0884	0.1057	0.1146	0.1474	0.1757
T/K	306.4	306.3	306.5	305.9	306.2	305.8	305.7
P/bar	1.0	1.0	1.0	1.0	1.0	1.0	1.0

Polymer (B): **poly(*N*-isopropylacrylamide)** **1990OTA**
Characterization: M_n/g.mol^{-1} = 43750, M_w/g.mol^{-1} = 2100000,
synthesized in the laboratory
Solvent (A): **water** **H_2O** **7732-18-5**

continued

continued

Type of data: cloud points (LCST-behavior)

w_B	0.059	0.044	0.030	0.015	0.0074	0.0037
T/K	303.05	305.45	306.15	306.95	307.45	308.25

Polymer (B): **poly(*N*-isopropylacrylamide)** **2004SHI**
Characterization: synthesized in the laboratory
Solvent (A): **water** H_2O **7732-18-5**

Type of data: cloud points

c_B/(mol/l) 0.690 $T/\text{K} = 306.75 - 5.59\ 10^{-4}\ (P/\text{MPa} - 51.7)^2$

Polymer (B): **poly(*N*-isopropylacrylamide)** **2000PE1**
Characterization: synthesized in the laboratory
Solvent (A): **water** H_2O **7732-18-5**

Type of data: cloud points (LCST-behavior)

w_B 0.01 T/K 307.15

Polymer (B): **poly(*N*-isopropylacrylamide)** **1998KUR**
Characterization: synthesized in the laboratory
Solvent (A): **water** H_2O **7732-18-5**

Type of data: cloud points (LCST-behavior)

c_B/(g/l) 1.0 T/K 306.45

Polymer (B): **poly(*N*-isopropylacrylamide-*co*-acrylamide)** **2006SHE**
Characterization: 2.2 mol% acrylamide, synthesized in the laboratory
Solvent (A): **water** H_2O **7732-18-5**

Type of data: cloud points (LCST-behavior)

c_B/(g/cm^3) 0.005 T/K 307.55

Polymer (B): **poly(*N*-isopropylacrylamide-*co*-acrylamide)** **2006SHE**
Characterization: 4.5 mol% acrylamide, synthesized in the laboratory
Solvent (A): **water** H_2O **7732-18-5**

Type of data: cloud points (LCST-behavior)

c_B/(g/cm^3) 0.005 T/K 309.45

Polymer (B): **poly(*N*-isopropylacrylamide-*co*-acrylamide)** **2006SHE**
Characterization: 6.6 mol% acrylamide, synthesized in the laboratory
Solvent (A): **water** H_2O **7732-18-5**

Type of data: cloud points (LCST-behavior)

c_B/(g/cm^3) 0.005 T/K 311.15

Polymer (B): **poly(*N*-isopropylacrylamide-*co*-acrylamide)** **2006SHE**
Characterization: M_η/g.mol^{-1} = 15600, 7.4 mol% acrylamide, synthesized in the laboratory
Solvent (A): **water** H_2O **7732-18-5**

Type of data: cloud points (LCST-behavior)

c_B/(g/cm^3) 0.005 *T*/K 311.35

Polymer (B): **poly(*N*-isopropylacrylamide-*co*-acrylamide)** **2006SHE**
Characterization: M_η/g.mol^{-1} = 16900, 8.1 mol% acrylamide, synthesized in the laboratory
Solvent (A): **water** H_2O **7732-18-5**

Type of data: cloud points (LCST-behavior)

c_B/(g/cm^3) 0.005 *T*/K 312.15

Polymer (B): **poly(*N*-isopropylacrylamide-*co*-acrylamide)** **2006SHE**
Characterization: M_η/g.mol^{-1} = 9700, 14.4 mol% acrylamide, synthesized in the laboratory
Solvent (A): **water** H_2O **7732-18-5**

Type of data: cloud points (LCST-behavior)

c_B/(g/cm^3) 0.005 *T*/K 316.85

Polymer (B): **poly(*N*-isopropylacrylamide-*co*-acrylic acid)** **2000YAM**
Characterization: M_n/g.mol^{-1} = 330000, M_w/g.mol^{-1} = 924000
5.4 mol% acrylic acid, synthesized in the laboratory
Solvent (A): **water** H_2O **7732-18-5**

Type of data: cloud points (LCST-behavior)

w_B	0.001	0.001	0.001	0.001	0.001
pH	2.0	3.0	4.0	5.0	6.0
T/K	307.35	307.45	309.35	311.15	312.55

Comments: The pH of the solution was adjusted by adding the appropriate amount of NaOH solution.

Polymer (B): **poly(*N*-isopropylacrylamide-*co*-acrylic acid)** **1999JON**
Characterization: 32 mol% acrylic acid, synthesized in the laboratory
Solvent (A): **water** H_2O **7732-18-5**

Type of data: cloud points (LCST-behavior)

w_B	0.005	0.005	0.005	0.005	0.005	0.005	0.005
pH	1.0	1.5	2.0	2.5	3.0	3.5	4.0
T/K	<273.15	284.15	287.15	291.15	295.15	301.15	316.15

Polymer (B): **poly(*N*-isopropylacrylamide-*co*-acrylic acid)** **2000YAM**
Characterization: M_n/g.mol^{-1} = 340000, M_w/g.mol^{-1} = 1020000
37.9 mol% acrylic acid, synthesized in the laboratory
Solvent (A): **water** H_2O **7732-18-5**

Type of data: cloud points (LCST-behavior)

w_B	0.001	pH	5.0	T/K	315.25

Comments: The pH of the solution was adjusted by adding the appropriate amount of NaOH solution.

Polymer (B): **poly(*N*-isopropylacrylamide-*co*-acrylic acid)** **1999JON**
Characterization: 54 mol% acrylic acid, synthesized in the laboratory
Solvent (A): **water** H_2O **7732-18-5**

Type of data: cloud points (LCST-behavior)

w_B	0.005	0.005	0.005	0.005	0.005
pH	1.0	1.5	2.0	2.5	3.0
T/K	<273.15	283.15	288.15	299.15	318.15

Polymer (B): **poly(*N*-isopropylacrylamide-*co*-acrylic acid)** **2000YAM**
Characterization: M_n/g.mol^{-1} = 94000, M_w/g.mol^{-1} = 169200
75.0 mol% acrylic acid, synthesized in the laboratory
Solvent (A): **water** H_2O **7732-18-5**

Type of data: cloud points (LCST-behavior)

w_B	0.001	pH	5.0	T/K	320.25

Comments: The pH of the solution was adjusted by adding the appropriate amount of NaOH solution.

Polymer (B): **poly(*N*-isopropylacrylamide-*co*-acrylic acid)** **1999JON**
Characterization: 80 mol% acrylic acid, synthesized in the laboratory
Solvent (A): **water** H_2O **7732-18-5**

Type of data: cloud points (LCST-behavior)

w_B	0.005	0.005	0.005	0.005	0.005
pH	1.0	1.5	2.0	2.5	3.0
T/K	<273.15	276.15	289.15	299.15	324.15

Polymer (B): **poly(*N*-isopropylacrylamide-*co*-1-deoxy-1-methacryl-amido-D-glucitol)** **2002RE1**
Characterization: M_n/g.mol^{-1} = 78000, M_w/g.mol^{-1} = 170000, 12.9 mol% glucitol
synthesized in the laboratory by radical polymerization
Solvent (A): **water** H_2O **7732-18-5**

continued

continued

Type of data: cloud points (LCST-behavior)

w_B	0.00500	0.01060	0.02062	0.03010	0.04010	0.05260	0.05440	0.06080	0.07080
T/K	320.1	316.1	314.3	313.6	313.7	312.7	312.8	313.1	313.1
w_B	0.07790	0.08530	0.09720	0.18900					
T/K	313.0	312.8	312.4	310.6					

Polymer (B): **poly(*N*-isopropylacrylamide-*co*-1-deoxy-1-methacryl-amido-D-glucitol)** **2002RE1**

Characterization: M_n/g.mol^{-1} = 28600, M_w/g.mol^{-1} = 56000, 13.3 mol% glucitol synthesized in the laboratory by radical polymerization

Solvent (A): **water** **H_2O** **7732-18-5**

Type of data: cloud points (LCST-behavior)

w_B	0.0102	0.0174	0.0233	0.0263	0.0397	0.0550	0.0102	0.0102	0.0102
T/K	413.2	384.8	368.5	358.1	353.0	366.0	428.6	450.3	465.0
P/bar	1.00	1.00	1.00	1.00	1.00	1.00	2.90	3.87	4.48
w_B	0.0174	0.0174	0.0233	0.0233	0.0233	0.0233	0.0233	0.0397	0.0397
T/K	401.6	432.5	391.3	395.0	419.0	433.5	453.0	377.8	400.0
P/bar	2.27	3.87	2.10	2.27	3.58	5.71	8.57	2.57	3.65
w_B	0.0550	0.0550	0.0550	0.0550	0.0550	0.0550	0.0550	0.0550	
T/K	373.0	388.5	392.2	403.8	411.5	421.6	440.2	451.6	
P/bar	1.62	2.26	2.45	2.90	3.58	4.70	7.43	8.84	

Polymer (B): **poly(*N*-isopropylacrylamide-*co*-1-deoxy-1-methacryl-amido-D-glucitol)** **2002RE1**

Characterization: M_n/g.mol^{-1} = 51600, M_w/g.mol^{-1} = 110000, 13.7 mol% glucitol synthesized in the laboratory by radical polymerization

Solvent (A): **water** **H_2O** **7732-18-5**

Type of data: cloud points (LCST-behavior)

w_B	0.01001	0.03914	0.0400	0.08810	0.0070	0.0070	0.0070	0.0070	0.0070
T/K	321.1	315.6	313.7	315.1	320.9	321.8	322.7	323.4	323.8
P/bar	1.0	1.0	1.0	1.0	20.0	100.0	200.0	300.0	500.0
w_B	0.0070	0.0230	0.0230	0.0230	0.0230	0.0230	0.0230		
T/K	324.0	317.6	319.0	320.3	320.7	321.9	322.4		
P/bar	600.0	20.0	200.0	300.0	400.0	600.0	700.0		

Polymer (B): **poly(*N*-isopropylacrylamide-*co*-1-deoxy-1-methacryl-amido-D-glucitol)** **2002RE1**

Characterization: M_n/g.mol^{-1} = 145000, M_w/g.mol^{-1} = 432000, 14.0 mol% glucitol synthesized in the laboratory by radical polymerization

Solvent (A): **water** **H_2O** **7732-18-5**

continued

continued

Type of data: cloud points (LCST-behavior)

w_B	0.0050	0.0100	0.0305	0.0432	0.0530	0.0983	0.1103	0.1294	0.0038
T/K	313.1	311.3	310.1	309.8	309.3	308.6	308.0	307.3	311.7
P/bar	1.0	1.0	1.0	1.0	1.0	1.0	1.0	1.0	3.5
w_B	0.0038	0.0038	0.0038	0.0038	0.0038	0.0038	0.0038	0.0038	0.0432
T/K	312.0	312.4	313.3	313.6	313.7	314.3	314.3	314.0	310.0
P/bar	62.0	110.0	200.0	300.0	400.0	500.0	600.0	700.0	10.0
w_B	0.0432	0.0432	0.0432	0.0432	0.0432	0.0432	0.0432	0.0432	0.0432
T/K	310.1	310.3	310.5	310.8	311.1	311.4	311.8	311.8	312.0
P/bar	40.0	70.0	100.0	150.0	200.0	250.0	300.0	350.0	400.0
w_B	0.0432	0.0432	0.0432	0.0432					
T/K	312.1	312.3	312.3	312.0					
P/bar	450.0	550.0	600.0	700.0					

Polymer (B): **poly(*N*-isopropylacrylamide-*co*-*N,N*-dimethylacrylamide)** **2006SHE**
Characterization: 10.7 mol% *N,N*-dimethylacrylamide
Solvent (A): **water** H_2O **7732-18-5**

Type of data: cloud points (LCST-behavior)

c_B/(g/cm^3) 0.005 T/K 310.05

Polymer (B): **poly(*N*-isopropylacrylamide-*co*-*N,N*-dimethylacrylamide)** **2003BAR**
Characterization: 13.0 mol% *N,N*-dimethylacrylamide
Solvent (A): **water** H_2O **7732-18-5**

Type of data: cloud points (LCST-behavior)

w_B 0.01 T/K 309.15

Polymer (B): **poly(*N*-isopropylacrylamide-*co*-*N,N*-dimethylacrylamide)** **2006SHE**
Characterization: 14.7 mol% *N,N*-dimethylacrylamide
Solvent (A): **water** H_2O **7732-18-5**

Type of data: cloud points (LCST-behavior)

c_B/(g/cm^3) 0.005 T/K 311.25

Polymer (B): **poly(*N*-isopropylacrylamide-*co*-*N,N*-dimethylacrylamide)** **2006SHE**
Characterization: 17.3 mol% *N,N*-dimethylacrylamide
Solvent (A): **water** H_2O **7732-18-5**

continued

continued

Type of data: cloud points (LCST-behavior)

c_B/(g/cm^3) 0.005 T/K 312.95

Polymer (B): **poly(*N*-isopropylacrylamide-*co*-*N,N*-dimethylacrylamide)** **2006SHE**
Characterization: 21.2 mol% *N,N*-dimethylacrylamide
Solvent (A): **water** H_2O **7732-18-5**

Type of data: cloud points (LCST-behavior)

c_B/(g/cm^3) 0.005 T/K 314.25

Polymer (B): **poly(*N*-isopropylacrylamide-*co*-*N,N*-dimethylacrylamide)** **2003BAR**
Characterization: 27.0 mol% *N,N*-dimethylacrylamide
Solvent (A): **water** H_2O **7732-18-5**

Type of data: cloud points (LCST-behavior)

w_B 0.01 T/K 312.15

Polymer (B): **poly(*N*-isopropylacrylamide-*co*-*N,N*-dimethylacrylamide)** **2003BAR**
Characterization: 30.0 mol% *N,N*-dimethylacrylamide
Solvent (A): **water** H_2O **7732-18-5**

Type of data: cloud points (LCST-behavior)

w_B 0.01 T/K 315.15

Polymer (B): **poly(*N*-isopropylacrylamide-*co*-*N,N*-dimethylacrylamide)** **2006SHE**
Characterization: 31.4 mol% *N,N*-dimethylacrylamide
Solvent (A): **water** H_2O **7732-18-5**

Type of data: cloud points (LCST-behavior)

c_B/(g/cm^3) 0.005 T/K 319.15

Polymer (B): **poly(*N*-isopropylacrylamide-*co*-*N,N*-dimethylacrylamide)** **2003BAR**
Characterization: 50.0 mol% *N,N*-dimethylacrylamide
Solvent (A): **water** H_2O **7732-18-5**

Type of data: cloud points (LCST-behavior)

w_B 0.01 T/K 323.15

Polymer (B): **poly(*N*-isopropylacrylamide-*co*-*N,N*-dimethylacrylamide)** **2003BAR**
Characterization: 60.0 mol% *N,N*-dimethylacrylamide
Solvent (A): **water** **H_2O** **7732-18-5**

Type of data: cloud points (LCST-behavior)

w_B	0.01	T/K	336.15

Polymer (B): **poly(*N*-isopropylacrylamide-*co*-*N,N*-dimethylacrylamide)** **2003BAR**
Characterization: 66.0 mol% *N,N*-dimethylacrylamide
Solvent (A): **water** **H_2O** **7732-18-5**

Type of data: cloud points (LCST-behavior)

w_B	0.01	T/K	345.15

Polymer (B): **poly(*N*-isopropylacrylamide-*co*-*N*-glycineacrylamide)** **2000PRI**
Characterization: M_n/g.mol^{-1} = 30000, M_w/g.mol^{-1} = 77000, 20 mol% N-glycine-acrylamide, synthesized in the laboratory
Solvent (A): **water** **H_2O** **7732-18-5**

Type of data: cloud points (LCST-behavior)

c_B/(g/l)	1.0	1.0	1.0	1.0
pH	2.77	3.83	4.25	5.10
T/K	303.55	305.85	307.55	311.45

Comments: pH values were prepared by buffer solutions from 0.1 M citric acid and 0.1 M sodium hydroxide at constant ionic strength of 0.1 M NaCl.

Polymer (B): **poly(*N*-isopropylacrylamide-*co*-2-hydroxyethyl methacrylate)** **2006SHE**
Characterization: 9.4 mol% 2-hydroxyethyl methacrylate
Solvent (A): **water** **H_2O** **7732-18-5**

Type of data: cloud points (LCST-behavior)

c_B/(g/cm^3)	0.005	T/K	303.45

Polymer (B): **poly(*N*-isopropylacrylamide-*co*-2-hydroxyethyl methacrylate)** **2006SHE**
Characterization: 17.0 mol% 2-hydroxyethyl methacrylate
Solvent (A): **water** **H_2O** **7732-18-5**

Type of data: cloud points (LCST-behavior)

c_B/(g/cm^3)	0.005	T/K	299.75

Polymer (B): **poly(*N*-isopropylacrylamide-*co*-2-hydroxyethyl methacrylate)** **2006SHE**
Characterization: 25.8 mol% 2-hydroxyethyl methacrylate
Solvent (A): **water** **H_2O** **7732-18-5**

Type of data: cloud points (LCST-behavior)

c_B/(g/cm^3)	0.005	T/K	294.55

Polymer (B): **poly(*N*-isopropylacrylamide-*co*-2-hydroxyethyl methacrylate)** **2006SHE**
Characterization: 34.9 mol% 2-hydroxyethyl methacrylate
Solvent (A): **water** **H_2O** **7732-18-5**

Type of data: cloud points (LCST-behavior)

c_B/(g/cm^3)	0.005	T/K	290.15

Polymer (B): **poly(*N*-isopropylacrylamide-*co*-*N*-isopropylmethacrylamide)** **2001DJO**
Characterization: M_n/g.mol^{-1} = 55300, M_w/g.mol^{-1} = 177000, 10.56 mol% *N*-isopropylmethacrylamide, synthesized in the laboratory
Solvent (A): **water** **H_2O** **7732-18-5**

Type of data: cloud points (LCST-behavior)

w_B	0.01	T/K	307.15

Polymer (B): **poly(*N*-isopropylacrylamide-*co*-*N*-isopropylmethacrylamide)** **2001DJO**
Characterization: M_n/g.mol^{-1} = 42800, M_w/g.mol^{-1} = 137000, 20.30 mol% *N*-isopropylmethacrylamide, synthesized in the laboratory
Solvent (A): **water** **H_2O** **7732-18-5**

Type of data: cloud points (LCST-behavior)

w_B	0.01	T/K	308.45

Polymer (B): **poly(*N*-isopropylacrylamide-*co*-*N*-isopropylmethacrylamide)** **2001DJO**
Characterization: M_n/g.mol^{-1} = 28800, M_w/g.mol^{-1} = 92000, 30.00 mol% *N*-isopropylmethacrylamide, synthesized in the laboratory
Solvent (A): **water** **H_2O** **7732-18-5**

Type of data: cloud points (LCST-behavior)

w_B	0.01	T/K	309.75

Polymer (B): **poly(*N*-isopropylacrylamide-*co*-*N*-isopropylmethacrylamide)** **2001DJO**
Characterization: M_n/g.mol^{-1} = 23100, M_w/g.mol^{-1} = 74000, 39.99 mol% *N*-isopropylmethacrylamide, synthesized in the laboratory
Solvent (A): **water** **H_2O** **7732-18-5**

Type of data: cloud points (LCST-behavior)

w_B 0.01 T/K 311.05

Polymer (B): **poly(*N*-isopropylacrylamide-*co*-*N*-isopropylmethacrylamide)** **2001DJO**
Characterization: M_n/g.mol^{-1} = 18800, M_w/g.mol^{-1} = 60000, 50.22 mol% *N*-isopropylmethacrylamide, synthesized in the laboratory
Solvent (A): **water** **H_2O** **7732-18-5**

Type of data: cloud points (LCST-behavior)

w_B 0.01 T/K 312.95

Polymer (B): **poly(*N*-isopropylacrylamide-*co*-*N*-isopropylmethacrylamide)** **2001DJO**
Characterization: M_n/g.mol^{-1} = 23100, M_w/g.mol^{-1} = 74000, 59.89 mol% *N*-isopropylmethacrylamide, synthesized in the laboratory
Solvent (A): **water** **H_2O** **7732-18-5**

Type of data: cloud points (LCST-behavior)

w_B 0.01 T/K 314.65

Polymer (B): **poly(*N*-isopropylacrylamide-*co*-*N*-isopropylmethacrylamide)** **2001DJO**
Characterization: M_n/g.mol^{-1} = 17500, M_w/g.mol^{-1} = 56000, 69.86 mol% *N*-isopropylmethacrylamide, synthesized in the laboratory
Solvent (A): **water** **H_2O** **7732-18-5**

Type of data: cloud points (LCST-behavior)

w_B 0.01 T/K 315.55

Polymer (B): **poly(*N*-isopropylacrylamide-*co*-*N*-isopropylmethacrylamide)** **2001DJO**
Characterization: M_n/g.mol^{-1} = 16600, M_w/g.mol^{-1} = 53000, 79.81 mol% *N*-isopropylmethacrylamide, synthesized in the laboratory
Solvent (A): **water** **H_2O** **7732-18-5**

Type of data: cloud points (LCST-behavior)

w_B 0.01 T/K 317.35

Polymer (B): **poly(*N*-isopropylacrylamide-*co*-*N*-isopropylmethacrylamide)** **2001DJO**
Characterization: M_n/g.mol^{-1} = 14700, M_w/g.mol^{-1} = 47000, 89.99 mol% *N*-isopropylmethacrylamide, synthesized in the laboratory
Solvent (A): **water** **H_2O** **7732-18-5**

Type of data: cloud points (LCST-behavior)

w_B	0.01	T/K	318.75

Polymer (B): **poly(*N*-isopropylacrylamide-*co*-itaconic acid)** **1999ERB**
Characterization: 9.8 mol% itaconic acid, synthesized in the laboratory
Solvent (A): **water** **H_2O** **7732-18-5**

Type of data: cloud points (LCST-behavior)

w_B	0.005	0.005	0.005	0.005	0.005	0.005
pH	2.01	2.30	3.09	3.32	4.08	5.03
T/K	307.55	308.15	309.25	310.15	312.65	332.55

Polymer (B): **poly(*N*-isopropylacrylamide-*co*-itaconic acid)** **1999ERB**
Characterization: 23.0 mol% itaconic acid, synthesized in the laboratory
Solvent (A): **water** **H_2O** **7732-18-5**

Type of data: cloud points (LCST-behavior)

w_B	0.005	0.005	0.005
pH	1.00	2.02	3.14
T/K	310.35	310.65	315.55

Polymer (B): **poly(*N*-isopropylacrylamide-*co*-8-methacryloyloxyoctanoic acid)** **2004SAL**
Characterization: M_w/g.mol^{-1} = 354000, 7.1 mol% 8-methacryloyloxy-octanoic acid, synthesized in the laboratory
Solvent (A): **water** **H_2O** **7732-18-5**

Type of data: cloud points (LCST-behavior)

c_B/(mg/ml)	5.0	5.0	5.0	5.0	5.0
pH	*	7	8	9	10
T/K	290.95	299.65	307.65	310.75	315.05

Comments: pH values were prepared by buffer solutions at constant ionic strength of 0.1 M.
* is deionized water. Cloud points were detected by using modulated DSC.

Polymer (B): **poly(*N*-isopropylacrylamide-*co*-8-methacryloyloxyoctanoic acid)** **2004SAL**
Characterization: M_w/g.mol^{-1} = 318000, 12.4 mol% 8-methacryloyloxy-octanoic acid, synthesized in the laboratory
Solvent (A): **water** **H_2O** **7732-18-5**

continued

continued

Type of data: cloud points (LCST-behavior)

c_B/(mg/ml)	5.0	5.0
pH	7	8
T/K	294.25	313.35

Comments: pH values were prepared by buffer solutions at constant ionic strength of 0.1 M. Cloud points were detected by using modulated DSC.

Polymer (B): **poly(*N*-isopropylacrylamide-*co*-8-methacryloyloxyoctanoic acid)** **2004SAL**
Characterization: M_w/g.mol^{-1} = 275000, 17.4 mol% 8-methacryloyloxy-octanoic acid, synthesized in the laboratory
Solvent (A): **water** **H_2O** **7732-18-5**

Type of data: cloud points (LCST-behavior)

c_B/(mg/ml)	5.0
pH	8
T/K	314.75

Comments: pH values were prepared by buffer solutions at constant ionic strength of 0.1 M. Cloud points were detected by using modulated DSC.

Polymer (B): **poly(*N*-isopropylacrylamide-*co*-5-methacryloyloxypentanoic acid)** **2004SAL**
Characterization: M_w/g.mol^{-1} = 291000, 6.7 mol% 5-methacryloyloxy-pentanoic acid, synthesized in the laboratory
Solvent (A): **water** **H_2O** **7732-18-5**

Type of data: cloud points (LCST-behavior)

c_B/(mg/ml)	5.0	5.0	5.0	5.0	5.0
pH	*	7	8	9	10
T/K	299.45	305.45	307.55	308.15	306.65

Comments: pH values were prepared by buffer solutions at constant ionic strength of 0.1 M. * is deionized water. Cloud points were detected by using modulated DSC.

Polymer (B): **poly(*N*-isopropylacrylamide-*co*-5-methacryloyloxypentanoic acid)** **2004SAL**
Characterization: M_w/g.mol^{-1} = 340000, 10.8 mol% 5-methacryloyloxy-pentanoic acid, synthesized in the laboratory
Solvent (A): **water** **H_2O** **7732-18-5**

Type of data: cloud points (LCST-behavior)

c_B/(mg/ml)	5.0	5.0	5.0	5.0
pH	*	7	8	9
T/K	283.25	307.65	310.25	315.15

continued

continued

Comments: pH values were prepared by buffer solutions at constant ionic strength of 0.1 M.
* is deionized water. Cloud points were detected by using modulated DSC.

Polymer (B): **poly(*N*-isopropylacrylamide-*co*-5-methacryloyloxypentanoic acid)** **2004SAL**
Characterization: M_w/g.mol^{-1} = 241000, 17.9 mol% 5-methacryloyloxypentanoic acid, synthesized in the laboratory
Solvent (A): **water** **H_2O** **7732-18-5**

Type of data: cloud points (LCST-behavior)

c_B/(mg/ml)	5.0	5.0
pH	7	8
T/K	307.55	316.95

Comments: pH values were prepared by buffer solutions at constant ionic strength of 0.1 M.
Cloud points were detected by using modulated DSC.

Polymer (B): **poly(*N*-isopropylacrylamide-*co*-5-methacryloyloxypentanoic acid)** **2004SAL**
Characterization: M_w/g.mol^{-1} = 310000, 23.7 mol% 5-methacryloyloxypentanoic acid, synthesized in the laboratory
Solvent (A): **water** **H_2O** **7732-18-5**

Type of data: cloud points (LCST-behavior)

c_B/(mg/ml)	5.0	5.0
pH	7	8
T/K	304.55	325.95

Comments: pH values were prepared by buffer solutions at constant ionic strength of 0.1 M.
Cloud points were detected by using modulated DSC.

Polymer (B): **poly(*N*-isopropylacrylamide-*co*-11-methacryloyloxyundecanoic acid)** **2004SAL**
Characterization: M_w/g.mol^{-1} = 287000, 6.5 mol% 11-methacryloyloxyundecanoic acid, synthesized in the laboratory
Solvent (A): **water** **H_2O** **7732-18-5**

Type of data: cloud points (LCST-behavior)

c_B/(mg/ml)	5.0	5.0	5.0	5.0	5.0
pH	*	7	8	9	10
T/K	279.15	290.25	306.65	301.65	305.85

Comments: pH values were prepared by buffer solutions at constant ionic strength of 0.1 M.
* is deionized water. Cloud points were detected by using modulated DSC.

Polymer (B): **poly(*N*-isopropylacrylamide-*co*-11-methacryloyloxyundecanoic acid)** **2004SAL**
Characterization: M_w/g.mol^{-1} = 289000, 11.2 mol% 11-methacryloyloxy-undecanoic acid, synthesized in the laboratory
Solvent (A): **water** **H_2O** **7732-18-5**

Type of data: cloud points (LCST-behavior)

c_B/(mg/ml)	5.0	5.0
pH	8	9
T/K	306.35	308.85

Comments: pH values were prepared by buffer solutions at constant ionic strength of 0.1 M.
Cloud points were detected by using modulated DSC.

Polymer (B): **poly(*N*-isopropylacrylamide-*co*-11-methacryloyloxyundecanoic acid)** **2004SAL**
Characterization: M_w/g.mol^{-1} = 231000, 17.2 mol% 11-methacryloyloxy-undecanoic acid, synthesized in the laboratory
Solvent (A): **water** **H_2O** **7732-18-5**

Type of data: cloud points (LCST-behavior)

c_B/(mg/ml)	5.0
pH	8
T/K	306.85

Comments: pH values were prepared by buffer solutions at constant ionic strength of 0.1 M.
Cloud points were detected by using modulated DSC.

Polymer (B): **poly(*N*-isopropylacrylamide-*co*-4-pentenoic acid)** **1995CHE**
Characterization: M_w/g.mol^{-1} = 240000, 0.7 mol% 4-pentenoic acid
Solvent (A): **water** **H_2O** **7732-18-5**

Type of data: cloud points (LCST-behavior)

c_B/(mg/ml)	2.0	2.0
pH	4.0	7.5
T/K	303.05	306.15

Comments: pH values were prepared by phosphate buffer solutions.

Polymer (B): **poly(*N*-isopropylacrylamide-*co*-4-pentenoic acid)** **1995CHE**
Characterization: M_w/g.mol^{-1} = 280000, 1.3 mol% 4-pentenoic acid
Solvent (A): **water** **H_2O** **7732-18-5**

Type of data: cloud points (LCST-behavior)

c_B/(mg/ml)	2.0	2.0
pH	4.0	7.5
T/K	302.85	307.45

Comments: pH values were prepared by phosphate buffer solutions.

Polymer (B): **poly(*N*-isopropylacrylamide-*co*-4-pentenoic acid)** **1995CHE**
Characterization: M_w/g.mol^{-1} = 370000, 2.3 mol% 4-pentenoic acid
Solvent (A): **water** **H_2O** **7732-18-5**

Type of data: cloud points (LCST-behavior)

c_B/(mg/ml)	2.0	2.0
pH	4.0	7.5
T/K	302.05	309.65

Comments: pH values were prepared by phosphate buffer solutions.

Polymer (B): **poly(*N*-isopropylacrylamide-*co*-4-pentenoic acid)** **1995CHE**
Characterization: M_w/g.mol^{-1} = 400000, 6.5 mol% 4-pentenoic acid
Solvent (A): **water** **H_2O** **7732-18-5**

Type of data: cloud points (LCST-behavior)

c_B/(mg/ml)	2.0
pH	4.0
T/K	300.45

Comments: pH values were prepared by phosphate buffer solutions.

Polymer (B): **poly(*N*-isopropylacrylamide-*co*-4-pentenoic acid)** **1995CHE**
Characterization: M_w/g.mol^{-1} = 300000, 10.8 mol% 4-pentenoic acid
Solvent (A): **water** **H_2O** **7732-18-5**

Type of data: cloud points (LCST-behavior)

c_B/(mg/ml)	2.0
pH	4.0
T/K	297.45

Comments: pH values were prepared by phosphate buffer solutions.

Polymer (B): **poly(*N*-isopropylacrylamide-*co*-4-pentenoic acid)** **1995CHE**
Characterization: M_w/g.mol^{-1} = 220000, 21.7 mol% 4-pentenoic acid
Solvent (A): **water** **H_2O** **7732-18-5**

Type of data: cloud points (LCST-behavior)

c_B/(mg/ml)	2.0
pH	4.0
T/K	292.35

Comments: pH values were prepared by phosphate buffer solutions.

Polymer (B): **poly(*N*-isopropylacrylamide-*co*-4-pentenoic acid)** **1995CHE**
Characterization: M_w/g.mol^{-1} = 190000, 28.6 mol% 4-pentenoic acid
Solvent (A): **water** **H_2O** **7732-18-5**

continued

continued

Type of data: cloud points (LCST-behavior)

c_B/(mg/ml)	2.0
pH	4.0
T/K	277.15

Comments: pH values were prepared by phosphate buffer solutions.

Polymer (B): **poly(*N*-isopropylacrylamide-*co*-p-vinylphenylboronic acid)** **2005CIM**
Characterization: 3.2 mol% p-vinylphenylboronic acid
Solvent (A): **water** H_2O **7732-18-5**

Type of data: cloud points (LCST-behavior)

pH	4.0	5.0	7.4
T/K	299.75	300.05	300.95

Polymer (B): **poly(*N*-isopropylacrylamide-*co*-p-vinylphenylboronic acid)** **2005CIM**
Characterization: 7.3 mol% p-vinylphenylboronic acid
Solvent (A): **water** H_2O **7732-18-5**

Type of data: cloud points (LCST-behavior)

pH	4.0	5.0	7.4
T/K	299.05	299.35	300.65

Polymer (B): **poly(*N*-isopropylacrylamide-*co*-p-vinylphenylboronic acid)** **2005CIM**
Characterization: 31.1 mol% p-vinylphenylboronic acid
Solvent (A): **water** H_2O **7732-18-5**

Type of data: cloud points (LCST-behavior)

pH	4.0	5.0	7.4
T/K	298.75	298.95	299.45

Polymer (B): **poly(*N*-isopropylacrylamide-*co*-vinylferrocene)** **1998KUR**
Characterization: synthesized in the laboratory
Solvent (A): **water** H_2O **7732-18-5**

Type of data: cloud points (LCST-behavior)

c_B/(g/l)	1.0	T/K	304.45	for a copolymer of 1.0 mol% vinylferrocene
c_B/(g/l)	1.0	T/K	299.15	for a copolymer of 3.0 mol% vinylferrocene

Polymer (B): **poly(*N*-isopropylacrylamide-*co*-1-vinylimidazole)** **2000PE1**
Characterization: synthesized in the laboratory
Solvent (A): **water** H_2O **7732-18-5**

continued

continued

Type of data: cloud points (LCST-behavior)

w_B	0.01	T/K	309.65	for a copolymer of 4.76 mol% 1-vinylimidazole
w_B	0.01	T/K	318.15	for a copolymer of 33.33 mol% 1-vinylimidazole
w_B	0.01	T/K	329.15	for a copolymer of 50.00 mol% 1-vinylimidazole

Polymer (B): **poly(*N*-isopropylmethacrylamide)** **2001DJO**
Characterization: M_n/g.mol^{-1} = 6250, M_w/g.mol^{-1} = 20000
Solvent (A): **water** H_2O **7732-18-5**

Type of data: cloud points (LCST-behavior)

w_B 0.01 T/K 319.95

Polymer (B): **poly(methoxydiethylene glycol methacrylate)** **2004KIT**
Characterization: M_n/g.mol^{-1} = 21600, M_w/g.mol^{-1} = 31900
Solvent (A): **water** H_2O **7732-18-5**

Type of data: cloud points (LCST-behavior)

c_B/(g/l) 1.0 T/K 298.45

Polymer (B): **poly(methoxydiethylene glycol methacrylate-*co*-dodecyl methacrylate)** **2004KIT**
Characterization: M_n/g.mol^{-1} = 32100, M_w/g.mol^{-1} = 52600, 9.43 mol% methoxydiethylene glycol methacrylate, molar ratio = 1/9.61, synthesized in the laboratory
Solvent (A): **water** H_2O **7732-18-5**

Type of data: cloud points (LCST-behavior)

c_B/(g/l) 1.0 T/K 292.65

Polymer (B): **poly(methoxydiethylene glycol methacrylate-*co*-methoxyoligoethylene glycol methacrylate)** **2004KIT**
Characterization: M_n/g.mol^{-1} = 46600, M_w/g.mol^{-1} = 116000, 2.75 mol% methoxyoligoethylene glycol(9 EO units) methacrylate, molar ratio = 1/35.4, synthesized in the laboratory
Solvent (A): **water** H_2O **7732-18-5**

Type of data: cloud points (LCST-behavior)

c_B/(g/l) 1.0 T/K 300.85

Polymer (B): **poly(methoxydiethylene glycol methacrylate-*co*-methoxyoligoethylene glycol methacrylate)** **2004KIT**

continued

continued

Characterization: M_n/g.mol^{-1} = 35200, M_w/g.mol^{-1} = 67200, 4.12 mol% methoxyoligoethylene glycol(9 EO units) methacrylate, molar ratio = 1/23.3, synthesized in the laboratory

Solvent (A): water H_2O 7732-18-5

Type of data: cloud points (LCST-behavior)

c_B/(g/l) 1.0 *T*/K 302.85

Polymer (B): poly(methoxydiethylene glycol methacrylate-*co*-methoxyoligoethylene glycol methacrylate) 2004KIT

Characterization: M_n/g.mol^{-1} = 13100, M_w/g.mol^{-1} = 20100, 9.69 mol% methoxyoligoethylene glycol(9 EO units) methacrylate, molar ratio = 1/9.32, synthesized in the laboratory

Solvent (A): water H_2O 7732-18-5

Type of data: cloud points (LCST-behavior)

c_B/(g/l) 1.0 *T*/K 312.45

Polymer (B): poly(methoxydiethylene glycol methacrylate-*co*-methoxyoligoethylene glycol methacrylate) 2004KIT

Characterization: M_n/g.mol^{-1} = 39700, M_w/g.mol^{-1} = 76300, 1.83 mol% methoxyoligoethylene glycol(23 EO units) methacrylate, molar ratio = 1/53.5, synthesized in the laboratory

Solvent (A): water H_2O 7732-18-5

Type of data: cloud points (LCST-behavior)

c_B/(g/l) 1.0 *T*/K 302.75

Polymer (B): poly(methoxydiethylene glycol methacrylate-*co*-methoxyoligoethylene glycol methacrylate) 2004KIT

Characterization: M_n/g.mol^{-1} = 18000, M_w/g.mol^{-1} = 34200, 4.40 mol% methoxyoligoethylene glycol(23 EO units) methacrylate, molar ratio = 1/21.7, synthesized in the laboratory

Solvent (A): water H_2O 7732-18-5

Type of data: cloud points (LCST-behavior)

c_B/(g/l) 1.0 *T*/K 309.15

Polymer (B): poly(methoxydiethylene glycol methacrylate-*co*-methoxyoligoethylene glycol methacrylate) 2004KIT

Characterization: M_n/g.mol^{-1} = 33200, M_w/g.mol^{-1} = 56800, 9.43 mol% methoxyoligoethylene glycol(23 EO units) methacrylate, molar ratio = 1/9.61, synthesized in the laboratory

continued

continued

Solvent (A): **water** **H_2O** **7732-18-5**

Type of data: cloud points (LCST-behavior)

c_B/(g/l) 1.0 T/K 334.65

Polymer (B): **poly(methyl methacrylate)** **1962FOX**
Characterization: M_η/g.mol^{-1} = 127000, fractionated in the laboratory
Solvent (A): **acetonitrile** **C_2H_3N** **75-05-8**

Type of data: cloud points, precipitation threshold (UCST-behavior)

φ_B 0.090 T/K 267.15

Polymer (B): **poly(methyl methacrylate)** **1962FOX**
Characterization: M_η/g.mol^{-1} = 147000, fractionated in the laboratory
Solvent (A): **acetonitrile** **C_2H_3N** **75-05-8**

Type of data: cloud points, precipitation threshold (UCST-behavior)

φ_B 0.060 T/K 271.15

Polymer (B): **poly(methyl methacrylate)** **1962FOX**
Characterization: M_η/g.mol^{-1} = 590000, fractionated in the laboratory
Solvent (A): **acetonitrile** **C_2H_3N** **75-05-8**

Type of data: cloud points, precipitation threshold (UCST-behavior)

φ_B 0.050 T/K 294.15

Polymer (B): **poly(methyl methacrylate)** **1962FOX**
Characterization: M_η/g.mol^{-1} = 970000, fractionated in the laboratory
Solvent (A): **acetonitrile** **C_2H_3N** **75-05-8**

Type of data: cloud points, precipitation threshold (UCST-behavior)

φ_B 0.050 T/K 303.15

Polymer (B): **poly(methyl methacrylate)** **1950JE2**
Characterization: M_η/g.mol^{-1} = 50000, fractionated in the laboratory
Solvent (A): **1-butanol** **$C_4H_{10}O$** **71-36-3**

Type of data: cloud points, precipitation threshold (UCST-behavior)

z_B 0.050 T/K 353.25

Polymer (B): **poly(methyl methacrylate)** **1950JE2**
Characterization: M_η/g.mol^{-1} = 90000, fractionated in the laboratory
Solvent (A): **1-butanol** **$C_4H_{10}O$** **71-36-3**

Type of data: cloud points, precipitation threshold (UCST-behavior)

z_B 0.034 T/K 353.85

Polymer (B): **poly(methyl methacrylate)** **1950JE2**
Characterization: M_η/g.mol^{-1} = 180000, fractionated in the laboratory
Solvent (A): **1-butanol** **$C_4H_{10}O$** **71-36-3**

Type of data: cloud points, precipitation threshold (UCST-behavior)

z_B 0.029 T/K 355.05

Polymer (B): **poly(methyl methacrylate)** **1950JE2**
Characterization: M_η/g.mol^{-1} = 400000, fractionated in the laboratory
Solvent (A): **1-butanol** **$C_4H_{10}O$** **71-36-3**

Type of data: cloud points, precipitation threshold (UCST-behavior)

z_B 0.021 T/K 356.65

Polymer (B): **poly(methyl methacrylate)** **2006GAR**
Characterization: M_n/g.mol^{-1} = 66000, M_w/g.mol^{-1} = 68000
Solvent (A): **cyclohexanol** **$C_6H_{12}O$** **108-93-0**

Type of data: cloud points (UCST-behavior)

w_B	0.0014	0.0045	0.0073	0.0106	0.0170	0.0211	0.0511	0.0982	0.1298
T/K	344.48	344.89	345.39	344.02	345.12	345.20	346.15	346.69	346.30

w_B	0.1329
T/K	346.59

Polymer (B): **poly(methyl methacrylate)** **2006GAR**
Characterization: M_n/g.mol^{-1} = 264000, M_w/g.mol^{-1} = 280000
Solvent (A): **cyclohexanol** **$C_6H_{12}O$** **108-93-0**

Type of data: cloud points (UCST-behavior)

w_B	0.0014	0.0026	0.0068	0.0138	0.0234	0.0523	0.0788	0.1005	0.1102
T/K	347.39	347.42	348.72	349.02	349.06	350.85	350.12	350.25	350.38

w_B	0.1163
T/K	350.73

Polymer (B): **poly(methyl methacrylate)** **2006GAR**
Characterization: M_n/g.mol^{-1} = 936000, M_w/g.mol^{-1} = 992000
Solvent (A): **cyclohexanol** **$C_6H_{12}O$** **108-93-0**

Type of data: cloud points (UCST-behavior)

w_B	0.0015	0.0044	0.0100	0.0103	0.0190	0.0475	0.0860	0.0968	0.1475
T/K	348.36	350.05	349.38	350.58	350.40	350.51	349.42	350.24	350.18

w_B	0.1690
T/K	349.49

Polymer (B): **poly(methyl methacrylate)** **1962FOX**
Characterization: M_η/g.mol^{-1} = 970000, fractionated in the laboratory
Solvent (A): **2,2-dimethyl-3-pentanone** $C_7H_{14}O$ **564-04-5**

Type of data: cloud points, precipitation threshold (UCST-behavior)

φ_B 0.0445 T/K 301.55

Polymer (B): **poly(methyl methacrylate)** **1962FOX**
Characterization: M_η/g.mol^{-1} = 127000, fractionated in the laboratory
Solvent (A): **2,4-dimethyl-3-pentanone** $C_7H_{14}O$ **565-80-0**

Type of data: cloud points, precipitation threshold (UCST-behavior)

φ_B 0.018 T/K 280.15

Polymer (B): **poly(methyl methacrylate)** **1962FOX**
Characterization: M_η/g.mol^{-1} = 308000, fractionated in the laboratory
Solvent (A): **2,4-dimethyl-3-pentanone** $C_7H_{14}O$ **565-80-0**

Type of data: cloud points, precipitation threshold (UCST-behavior)

φ_B 0.010 T/K 299.85

Polymer (B): **poly(methyl methacrylate)** **1962FOX**
Characterization: M_η/g.mol^{-1} = 970000, fractionated in the laboratory
Solvent (A): **2,4-dimethyl-3-pentanone** $C_7H_{14}O$ **565-80-0**

Type of data: cloud points, precipitation threshold (UCST-behavior)

φ_B 0.010 T/K 308.85

Polymer (B): **poly(methyl methacrylate)** **1998CHA**
Characterization: M_n/g.mol^{-1} = 200000, M_w/g.mol^{-1} = 264000
Solvent (A): **2-ethoxyethanol** $C_4H_{10}O_2$ **110-80-5**

Type of data: cloud points, precipitation threshold (UCST-behavior)

φ_B 0.085 T/K 312.15

Polymer (B): **poly(methyl methacrylate)** **1962FOX**
Characterization: M_η/g.mol^{-1} = 127000, fractionated in the laboratory
Solvent (A): **2-ethylbutanal** $C_6H_{12}O$ **97-96-1**

Type of data: cloud points, precipitation threshold (UCST-behavior)

φ_B unknown T/K 264.65

Polymer (B): **poly(methyl methacrylate)** **1962FOX**
Characterization: M_η/g.mol^{-1} = 590000, fractionated in the laboratory
Solvent (A): **2-ethylbutanal** $C_6H_{12}O$ **97-96-1**

continued

continued

Type of data: cloud points, precipitation threshold (UCST-behavior)

φ_B 0.042 T/K 280.85

Polymer (B): **poly(methyl methacrylate)** **1962FOX**
Characterization: M_η/g.mol^{-1} = 970000, fractionated in the laboratory
Solvent (A): **2-ethylbutanal** **$C_6H_{12}O$** **97-96-1**

Type of data: cloud points, precipitation threshold (UCST-behavior)

φ_B unknown T/K 284.85

Polymer (B): **poly(methyl methacrylate)** **1962FOX**
Characterization: M_η/g.mol^{-1} = 15900, fractionated in the laboratory
Solvent (A): **4-heptanone** **$C_7H_{14}O$** **123-19-3**

Type of data: cloud points, precipitation threshold (UCST-behavior)

φ_B 0.22 T/K 246.15

Polymer (B): **poly(methyl methacrylate)** **1962FOX**
Characterization: M_η/g.mol^{-1} = 66400, fractionated in the laboratory
Solvent (A): **4-heptanone** **$C_7H_{14}O$** **123-19-3**

Type of data: cloud points, precipitation threshold (UCST-behavior)

φ_B 0.065 T/K 276.35

Polymer (B): **poly(methyl methacrylate)** **1962FOX**
Characterization: M_η/g.mol^{-1} = 147000, fractionated in the laboratory
Solvent (A): **4-heptanone** **$C_7H_{14}O$** **123-19-3**

Type of data: cloud points, precipitation threshold (UCST-behavior)

φ_B 0.050 T/K 286.15

Polymer (B): **poly(methyl methacrylate)** **1962FOX**
Characterization: M_η/g.mol^{-1} = 308000, fractionated in the laboratory
Solvent (A): **4-heptanone** **$C_7H_{14}O$** **123-19-3**

Type of data: cloud points, precipitation threshold (UCST-behavior)

φ_B 0.048 T/K 292.55

Polymer (B): **poly(methyl methacrylate)** **1962FOX**
Characterization: M_η/g.mol^{-1} = 970000, fractionated in the laboratory
Solvent (A): **4-heptanone** **$C_7H_{14}O$** **123-19-3**

Type of data: cloud points, precipitation threshold (UCST-behavior)

φ_B 0.030 T/K 299.95

Polymer (B): **poly(methyl methacrylate)** **1950JE2**
Characterization: M_η/g.mol^{-1} = 50000, fractionated in the laboratory
Solvent (A): **1-methyl-4-isopropylbenzene** $C_{10}H_{14}$ **99-87-6**

Type of data: cloud points, precipitation threshold (UCST-behavior)

z_B 0.075 T/K 400.15

Polymer (B): **poly(methyl methacrylate)** **1950JE2**
Characterization: M_η/g.mol^{-1} = 90000, fractionated in the laboratory
Solvent (A): **1-methyl-4-isopropylbenzene** $C_{10}H_{14}$ **99-87-6**

Type of data: cloud points, precipitation threshold (UCST-behavior)

z_B 0.064 T/K 402.45

Polymer (B): **poly(methyl methacrylate)** **1950JE2**
Characterization: M_η/g.mol^{-1} = 180000, fractionated in the laboratory
Solvent (A): **1-methyl-4-isopropylbenzene** $C_{10}H_{14}$ **99-87-6**

Type of data: cloud points, precipitation threshold (UCST-behavior)

z_B 0.060 T/K 406.75

Polymer (B): **poly(methyl methacrylate)** **1950JE2**
Characterization: M_η/g.mol^{-1} = 400000, fractionated in the laboratory
Solvent (A): **1-methyl-4-isopropylbenzene** $C_{10}H_{14}$ **99-87-6**

Type of data: cloud points, precipitation threshold (UCST-behavior)

z_B 0.050 T/K 413.65

Polymer (B): **poly(methyl methacrylate)** **1962FOX**
Characterization: M_η/g.mol^{-1} = 55000, fractionated in the laboratory
Solvent (A): **2-octanone** $C_8H_{16}O$ **111-13-7**

Type of data: cloud points, precipitation threshold (UCST-behavior)

φ_B 0.067 T/K 281.15

Polymer (B): **poly(methyl methacrylate)** **1962FOX**
Characterization: M_η/g.mol^{-1} = 135000, fractionated in the laboratory
Solvent (A): **2-octanone** $C_8H_{16}O$ **111-13-7**

Type of data: cloud points, precipitation threshold (UCST-behavior)

φ_B 0.057 T/K 299.15

Polymer (B): **poly(methyl methacrylate)** **1962FOX**
Characterization: M_η/g.mol^{-1} = 970000, fractionated in the laboratory
Solvent (A): **2-octanone** $C_8H_{16}O$ **111-13-7**

continued

continued

Type of data: cloud points, precipitation threshold (UCST-behavior)

φ_B 0.030 T/K 317.65

Polymer (B): **poly(methyl methacrylate)** **1962FOX**
Characterization: M_η/g.mol^{-1} = 1400000, fractionated in the laboratory
Solvent (A): **2-octanone** **$C_8H_{16}O$** **111-13-7**

Type of data: cloud points, precipitation threshold (UCST-behavior)

φ_B 0.026 T/K 321.15

Polymer (B): **poly(methyl methacrylate)** **1992XIA**
Characterization: M_n/g.mol^{-1} = 19830, M_w/g.mol^{-1} = 21600,
Pressure Chemical Company, Pittsburgh, PA
Solvent (A): **3-octanone** **$C_8H_{16}O$** **106-68-3**

Type of data: critical point (UCST-behavior)

$\varphi_{B,\,crit}$ 0.175 T_{crit}/K 285.675

Polymer (B): **poly(methyl methacrylate)** **1996XIA**
Characterization: M_n/g.mol^{-1} = 24230, M_w/g.mol^{-1} = 26900,
Pressure Chemical Company, Pittsburgh, PA
Solvent (A): **3-octanone** **$C_8H_{16}O$** **106-68-3**

Type of data: coexistence data (tie lines) (UCST-behavior)

T/K	293.584	293.551	293.525	293.455	293.339	293.185	293.032	292.852
φ_B (sol phase)	0.1264	0.1217	0.1164	0.1124	0.1053	0.0968	0.0919	0.0862
φ_B (gel phase)	0.1743	0.1853	0.1831	0.1957	0.2057	0.2159	0.2235	0.2327

T/K	292.671	292.442	292.118	291.718	291.176	290.168	287.450
φ_B (sol phase)	0.0812	0.0765	0.0704	0.0651	0.0596	0.0504	0.0360
φ_B (gel phase)	0.2402	0.2470	0.2565	0.2676	0.2816	0.3008	0.3440

Type of data: critical point (UCST-behavior)

$\varphi_{B,\,crit}$ 0.156 T_{crit}/K 293.613

Polymer (B): **poly(methyl methacrylate)** **1996XIA**
Characterization: M_n/g.mol^{-1} = 46300, M_w/g.mol^{-1} = 48600,
Pressure Chemical Company, Pittsburgh, PA
Solvent (A): **3-octanone** **$C_8H_{16}O$** **106-68-3**

Type of data: coexistence data (tie lines) (UCST-behavior)

T/K	307.823	307.773	307.663	307.566	307.438	307.254	307.057	306.779
φ_B (sol phase)	0.0929	0.0871	0.0771	0.0742	0.0691	0.0632	0.0587	0.0531
φ_B (gel phase)	0.1466	0.1561	0.1663	0.1760	0.1807	0.1884	0.1959	0.2068

continued

continued

T/K	306.376	305.811	305.123	304.229	302.748
φ_B (sol phase)	0.0470	0.0402	0.0351	0.0294	0.0237
φ_B (gel phase)	0.2162	0.2294	0.2438	0.2589	0.2820

Type of data: critical point (UCST-behavior)

$\varphi_{B, crit}$ 0.125 T_{crit}/K 307.870

Polymer (B): **poly(methyl methacrylate)** **1992XIA**
Characterization: M_n/g.mol^{-1} = 57030, M_w/g.mol^{-1} = 58700,
Polymer Laboratories, Inc., Amherst, MA
Solvent (A): **3-octanone** $C_8H_{16}O$ **106-68-3**

Type of data: critical point (UCST-behavior)

$\varphi_{B, crit}$ 0.120 T_{crit}/K 308.63

Polymer (B): **poly(methyl methacrylate)** **1996XIA**
Characterization: M_n/g.mol^{-1} = 91350, M_w/g.mol^{-1} = 95000,
Pressure Chemical Company, Pittsburgh, PA
Solvent (A): **3-octanone** $C_8H_{16}O$ **106-68-3**

Type of data: coexistence data (tie lines) (UCST-behavior)

T/K	316.850	316.835	316.782	316.685	316.660	316.529	316.350	316.144
φ_B (sol phase)	0.0757	0.0738	0.0664	0.0595	0.0577	0.0516	0.0464	0.0419
φ_B (gel phase)	0.1171	0.1206	0.1262	0.1338	0.1364	0.1391	0.1517	0.1583

T/K	315.894	315.597	315.102	314.293	313.272	311.876
φ_B (sol phase)	0.0385	0.0379	0.0323	0.0260	0.0202	0.0156
φ_B (gel phase)	0.1682	0.1777	0.1894	0.2047	0.2214	0.2415

Type of data: critical point (UCST-behavior)

$\varphi_{B, crit}$ 0.098 T_{crit}/K 316.883

Polymer (B): **poly(methyl methacrylate)** **1992XIA**
Characterization: M_n/g.mol^{-1} = 118200, M_w/g.mol^{-1} = 125300,
Polymer Laboratories, Inc., Amherst, MA
Solvent (A): **3-octanone** $C_8H_{16}O$ **106-68-3**

Type of data: critical point (UCST-behavior)

$\varphi_{B, crit}$ 0.091 T_{crit}/K 319.045

Polymer (B): **poly(methyl methacrylate)** **1962FOX**
Characterization: M_η/g.mol^{-1} = 127000, fractionated in the laboratory
Solvent (A): **3-octanone** $C_8H_{16}O$ **106-68-3**

Type of data: cloud points, precipitation threshold (UCST-behavior)

φ_B 0.050 T/K 315.15

Polymer (B): **poly(methyl methacrylate)** **1996XIA**
Characterization: M_n/g.mol^{-1} = 137500, M_w/g.mol^{-1} = 143000,
Polymer Laboratories, Inc., Amherst, MA
Solvent (A): **3-octanone** **$C_8H_{16}O$** **106-68-3**

Type of data: coexistence data (tie lines) (UCST-behavior)

T/K	323.098	323.018	322.914	322.826	322.656	322.491	322.281	321.954
φ_B (sol phase)	0.0634	0.0532	0.0464	0.0426	0.0385	0.0349	0.0305	0.0265
φ_B (gel phase)	0.0982	0.1082	0.1166	0.1210	0.1301	0.1349	0.1436	0.1510

T/K	321.786	321.284	320.438	319.327	318.053
φ_B (sol phase)	0.0239	0.0209	0.0164	0.0119	0.0079
φ_B (gel phase)	0.1552	0.1659	0.1837	0.2005	0.2183

Type of data: critical point (UCST-behavior)

$\varphi_{B, crit}$ 0.084 T_{crit}/K 323.125

Polymer (B): **poly(methyl methacrylate)** **1962FOX**
Characterization: M_η/g.mol^{-1} = 147000, fractionated in the laboratory
Solvent (A): **3-octanone** **$C_8H_{16}O$** **106-68-3**

Type of data: cloud points, precipitation threshold (UCST-behavior)

φ_B 0.055 T/K 319.65

Polymer (B): **poly(methyl methacrylate)** **1992XIA**
Characterization: M_n/g.mol^{-1} = 217200, M_w/g.mol^{-1} = 225900,
Polymer Laboratories, Inc., Amherst, MA
Solvent (A): **3-octanone** **$C_8H_{16}O$** **106-68-3**

Type of data: critical point (UCST-behavior)

$\varphi_{B, crit}$ 0.074 T_{crit}/K 324.447

Polymer (B): **poly(methyl methacrylate)** **1996XIA**
Characterization: M_n/g.mol^{-1} = 218300, M_w/g.mol^{-1} = 227000,
Polymer Laboratories, Inc., Amherst, MA
Solvent (A): **3-octanone** **$C_8H_{16}O$** **106-68-3**

Type of data: coexistence data (tie lines) (UCST-behavior)

T/K	326.268	326.229	326.129	325.984	325.813	325.616	325.370	325.056
φ_B (sol phase)	0.0464	0.0446	0.0391	0.0338	0.0296	0.0264	0.0218	0.0189
φ_B (gel phase)	0.0888	0.0929	0.1000	0.1065	0.1129	0.1203	0.1266	0.1345

T/K	324.477	323.680	322.623	321.779
φ_B (sol phase)	0.0150	0.0116	0.0085	0.0057
φ_B (gel phase)	0.1468	0.1618	0.1781	0.1908

Type of data: critical point (UCST-behavior)

$\varphi_{B, crit}$ 0.074 T_{crit}/K 326.325

Polymer (B): **poly(methyl methacrylate)** **1962FOX**
Characterization: M_η/g.mol^{-1} = 308000, fractionated in the laboratory
Solvent (A): **3-octanone** **$C_8H_{16}O$** **106-68-3**

Type of data: cloud points, precipitation threshold (UCST-behavior)

φ_B 0.040 *T*/K 325.65

Polymer (B): **poly(methyl methacrylate)** **1992XIA**
Characterization: M_n/g.mol^{-1} = 572400, M_w/g.mol^{-1} = 595300,
Polymer Laboratories, Inc., Amherst, MA
Solvent (A): **3-octanone** **$C_8H_{16}O$** **106-68-3**

Type of data: critical point (UCST-behavior)

$\varphi_{B,\,crit}$ >0.0473 T_{crit}/K 329.88

Polymer (B): **poly(methyl methacrylate)** **1996XIA**
Characterization: M_n/g.mol^{-1} = 573000, M_w/g.mol^{-1} = 596000,
Pressure Chemical Company, Pittsburgh, PA
Solvent (A): **3-octanone** **$C_8H_{16}O$** **106-68-3**

Type of data: coexistence data (tie lines) (UCST-behavior)

T/K	333.364	333.345	333.274	333.146	332.992	332.796	332.620	332.225
φ_B (sol phase)	0.0290	0.0273	0.0245	0.0204	0.0189	0.0149	0.0121	0.0086
φ_B (gel phase)	0.0667	0.0684	0.0722	0.0783	0.0860	0.0913	0.0957	0.1030

T/K	331.724	330.941	329.744
φ_B (sol phase)	0.0062	0.0053	0.0028
φ_B (gel phase)	0.1142	0.1285	0.1489

Type of data: critical point (UCST-behavior)

$\varphi_{B,\,crit}$ 0.050 T_{crit}/K 333.459

Polymer (B): **poly(methyl methacrylate)** **1962FOX**
Characterization: M_η/g.mol^{-1} = 860000, fractionated in the laboratory
Solvent (A): **3-octanone** **$C_8H_{16}O$** **106-68-3**

Type of data: cloud points, precipitation threshold (UCST-behavior)

φ_B 0.030 *T*/K 333.15

Polymer (B): **poly(methyl methacrylate)** **1962FOX**
Characterization: M_η/g.mol^{-1} = 970000, fractionated in the laboratory
Solvent (A): **3-octanone** **$C_8H_{16}O$** **106-68-3**

Type of data: cloud points, precipitation threshold (UCST-behavior)

φ_B 0.023 *T*/K 334.15

Polymer (B): **poly(methyl methacrylate)** **1950JE2**
Characterization: M_η/g.mol^{-1} = 50000, fractionated in the laboratory
Solvent (A): **1-propanol** **C_3H_8O** **71-23-8**

Type of data: cloud points, precipitation threshold (UCST-behavior)

z_B	0.057	T/K	349.95

Polymer (B): **poly(methyl methacrylate)** **1950JE2**
Characterization: M_η/g.mol^{-1} = 90000, fractionated in the laboratory
Solvent (A): **1-propanol** **C_3H_8O** **71-23-8**

Type of data: cloud points, precipitation threshold (UCST-behavior)

z_B	0.051	T/K	350.65

Polymer (B): **poly(methyl methacrylate)** **1950JE2**
Characterization: M_η/g.mol^{-1} = 180000, fractionated in the laboratory
Solvent (A): **1-propanol** **C_3H_8O** **71-23-8**

Type of data: cloud points, precipitation threshold (UCST-behavior)

z_B	0.040	T/K	352.05

Polymer (B): **poly(methyl methacrylate)** **1950JE2**
Characterization: M_η/g.mol^{-1} = 400000, fractionated in the laboratory
Solvent (A): **1-propanol** **C_3H_8O** **71-23-8**

Type of data: cloud points, precipitation threshold (UCST-behavior)

z_B	0.031	T/K	355.25

Polymer (B): **poly(methyl methacrylate)** **1998CHA**
Characterization: M_n/g.mol^{-1} = 200000, M_w/g.mol^{-1} = 264000
Solvent (A): **tetra(ethylene glycol)** **$C_8H_{18}O_5$** **112-60-7**

Type of data: cloud points, precipitation threshold (UCST-behavior)

φ_B	0.095	T/K	390.15

Polymer (B): **poly(methyl methacrylate)** **1950JE2**
Characterization: M_η/g.mol^{-1} = 400000, fractionated in the laboratory
Solvent (A): **toluene** **C_7H_8** **108-88-3**

Type of data: cloud points, precipitation threshold (UCST-behavior)

z_B	0.065	T/K	225.35

Polymer (B): **poly(methyl methacrylate)** **1950JE2**
Characterization: M_η/g.mol^{-1} = 50000, fractionated in the laboratory
Solvent (A): **trichloromethane** **$CHCl_3$** **67-66-3**

continued

continued

Type of data: cloud points, precipitation threshold (UCST-behavior)

z_B	0.135	T/K	231.15

Polymer (B): **poly(methyl methacrylate)** **1950JE2**
Characterization: M_η/g.mol^{-1} = 180000, fractionated in the laboratory
Solvent (A): **trichloromethane** **$CHCl_3$** **67-66-3**

Type of data: cloud points, precipitation threshold (UCST-behavior)

z_B	0.130	T/K	232.65

Polymer (B): **poly(methyl methacrylate)** **1950JE2**
Characterization: M_η/g.mol^{-1} = 400000, fractionated in the laboratory
Solvent (A): **trichloromethane** **$CHCl_3$** **67-66-3**

Type of data: cloud points, precipitation threshold (UCST-behavior)

z_B	0.120	T/K	233.65

Polymer (B): **poly(methyl methacrylate)** **1998CHA**
Characterization: M_n/g.mol^{-1} = 200000, M_w/g.mol^{-1} = 264000
Solvent (A): **tri(ethylene glycol)** **$C_6H_{14}O_4$** **112-27-6**

Type of data: cloud points, precipitation threshold (UCST-behavior)

φ_B	0.090	T/K	407.15

Polymer (B): **poly(α-methylstyrene)** **1995PFO**
Characterization: M_n/g.mol^{-1} = 58500, M_w/g.mol^{-1} = 61400,
Pressure Chemical Company, Pittsburgh, PA
Solvent (A): **butyl acetate** **$C_6H_{12}O_2$** **123-86-4**

Type of data: cloud points (UCST-behavior)

w_B	0.0200	0.0300	0.0400	0.0610	0.0800	0.1020	0.1310	0.1560	0.1890
T/K	241.75	250.55	256.85	259.85	260.95	261.95	262.05	261.75	260.55
w_B	0.2200								
T/K	258.75								

Type of data: cloud points (LCST-behavior)

w_B	0.0050	0.0070	0.0100	0.0200	0.0300	0.0400	0.0610	0.0800	0.1020
T/K	482.65	479.65	474.05	470.45	468.85	466.25	463.05	460.95	459.25
w_B	0.1310	0.1560	0.1890	0.2200					
T/K	457.15	457.35	459.25	461.45					

Polymer (B): **poly(α-methylstyrene)** **1995PFO**
Characterization: M_n/g.mol^{-1} = 58500, M_w/g.mol^{-1} = 61400, Pressure Chemical Company, Pittsburgh, PA
Solvent (A): **cyclohexane** **C_6H_{12}** **110-82-7**

Type of data: cloud points (UCST-behavior)

w_B	0.0603	0.0898	0.1196
T/K	280.85	281.25	281.55

Type of data: cloud points (LCST-behavior)

w_B	0.0603	0.0898	0.1196
T/K	490.05	492.65	490.85

Polymer (B): **poly(α-methylstyrene)** **1991LEE**
Characterization: M_n/g.mol^{-1} = 99100, M_w/g.mol^{-1} = 113000, M_z/g.mol^{-1} = 126600, synthesized and fractionated in the laboratory
Solvent (A): **cyclohexane** **C_6H_{12}** **110-82-7**

Type of data: cloud points (UCST-behavior)

φ_B	0.0709	T/K	293.75	(threshold point)
φ_B	0.0865	T/K	293.55	(critical point)

Polymer (B): **poly(α-methylstyrene)** **1991LEE**
Characterization: M_n/g.mol^{-1} = 137500, M_w/g.mol^{-1} = 165000, M_z/g.mol^{-1} = 193000, synthesized and fractionated in the laboratory
Solvent (A): **cyclohexane** **C_6H_{12}** **110-82-7**

Type of data: cloud points (UCST-behavior)

φ_B	0.0647	T/K	295.65	(threshold point)
φ_B	0.0760	T/K	295.55	(critical point)

Polymer (B): **poly(α-methylstyrene)** **1991LEE**
Characterization: M_n/g.mol^{-1} = 178700, M_w/g.mol^{-1} = 227000, M_z/g.mol^{-1} = 279200, synthesized and fractionated in the laboratory
Solvent (A): **cyclohexane** **C_6H_{12}** **110-82-7**

Type of data: cloud points (UCST-behavior)

φ_B	0.0558	T/K	296.65	(threshold point)
φ_B	0.0679	T/K	296.45	(critical point)

Polymer (B): **poly(α-methylstyrene)** **1991LEE**
Characterization: M_n/g.mol^{-1} = 129000, M_w/g.mol^{-1} = 257000, M_z/g.mol^{-1} = 483000, synthesized and fractionated in the laboratory
Solvent (A): **cyclohexane** **C_6H_{12}** **110-82-7**

continued

continued

Type of data: cloud points (UCST-behavior)

φ_B	0.0487	T/K	298.15	(threshold point)
φ_B	0.0830	T/K	295.35	(critical point)

Polymer (B): **poly(α-methylstyrene)** **1995PFO**
Characterization: M_w/g.mol^{-1} = 289200,
Pressure Chemical Company, Pittsburgh, PA
Solvent (A): **cyclohexane** C_6H_{12} **110-82-7**

Type of data: cloud points (UCST-behavior)

w_B	0.0010	0.0050	0.0097	0.0299	0.0491	0.0685	0.0904	0.1100
T/K	285.05	288.05	290.95	291.95	292.05	291.65	290.35	291.25

Type of data: cloud points (LCST-behavior)

w_B	0.0010	0.0050	0.0097	0.0299	0.0491	0.0685	0.0904	0.1100
T/K	486.15	480.25	475.35	473.15	474.55	473.45	472.95	474.85

Polymer (B): **poly(α-methylstyrene)** **1991LEE**
Characterization: M_n/g.mol^{-1} = 219700, M_w/g.mol^{-1} = 290000, M_z/g.mol^{-1} = 359600, synthesized and fractionated in the laboratory
Solvent (A): **cyclohexane** C_6H_{12} **110-82-7**

Type of data: cloud points (UCST-behavior)

φ_B	0.0532	T/K	297.95	(threshold point)
φ_B	0.0634	T/K	297.75	(critical point)

Polymer (B): **poly(α-methylstyrene)** **1991LEE**
Characterization: M_n/g.mol^{-1} = 273500, M_w/g.mol^{-1} = 424000, M_z/g.mol^{-1} = 610600, synthesized and fractionated in the laboratory
Solvent (A): **cyclohexane** C_6H_{12} **110-82-7**

Type of data: cloud points (UCST-behavior)

φ_B	0.0435	T/K	298.65	(threshold point)
φ_B	0.0588	T/K	298.25	(critical point)

Polymer (B): **poly(α-methylstyrene)** **1991LEE**
Characterization: M_n/g.mol^{-1} = 349000, M_w/g.mol^{-1} = 604000, M_z/g.mol^{-1} = 966000, synthesized and fractionated in the laboratory
Solvent (A): **cyclohexane** C_6H_{12} **110-82-7**

Type of data: cloud points (UCST-behavior)

φ_B	0.0400	T/K	301.25	(threshold point)
φ_B	0.0550	T/K	300.55	(critical point)

Polymer (B): **poly(α-methylstyrene)** **1995PFO**
Characterization: M_n/g.mol^{-1} = 26000, M_w/g.mol^{-1} = 31200,
Pressure Chemical Company, Pittsburgh, PA
Solvent (A): **cyclopentane** **C_5H_{10}** **287-92-3**

Type of data: cloud points (UCST-behavior)

w_B	0.0098	0.0384	0.0474	0.0660	0.0886	0.1038	0.1211	0.1469	0.1601
T/K	268.45	273.05	274.25	275.25	275.95	276.55	276.65	276.25	276.45

Type of data: cloud points (LCST-behavior)

w_B	0.0384	0.0474	0.0660	0.0886	0.1038	0.1211	0.1469	0.1601
T/K	440.45	439.25	437.25	435.95	436.25	435.95	436.25	436.65

Polymer (B): **poly(α-methylstyrene)** **1995PFO**
Characterization: M_n/g.mol^{-1} = 58500, M_w/g.mol^{-1} = 61400,
Pressure Chemical Company, Pittsburgh, PA
Solvent (A): **cyclopentane** **C_5H_{10}** **287-92-3**

Type of data: cloud points (UCST-behavior)

w_B	0.0566	0.0850	0.1141
T/K	287.95	288.95	289.45

Type of data: cloud points (LCST-behavior)

w_B	0.0566	0.0850	0.1141
T/K	424.95	424.75	424.75

Polymer (B): **poly(α-methylstyrene)** **1995PFO**
Characterization: M_n/g.mol^{-1} = 58500, M_w/g.mol^{-1} = 61400,
Pressure Chemical Company, Pittsburgh, PA
Solvent (A): ***trans*-decahydronaphthalene** **$C_{10}H_{18}$** **493-02-7**

Type of data: cloud points (UCST-behavior)

w_B	0.0799	0.1084	0.1399
T/K	260.25	260.65	260.85

Polymer (B): **poly(α-methylstyrene)** **1995PFO**
Characterization: M_w/g.mol^{-1} = 289200,
Pressure Chemical Company, Pittsburgh, PA
Solvent (A): ***trans*-decahydronaphthalene** **$C_{10}H_{18}$** **493-02-7**

Type of data: cloud points (UCST-behavior)

w_B	0.0050	0.0099	0.0246	0.0403	0.0550	0.0692	0.0831	0.1003	0.1150
T/K	267.65	269.95	271.95	272.65	272.85	272.95	272.95	273.05	272.95

Polymer (B): **poly(α-methylstyrene)** **1995PFO**
Characterization: M_n/g.mol^{-1} = 26000, M_w/g.mol^{-1} = 31200,
Pressure Chemical Company, Pittsburgh, PA
Solvent (A): **hexyl acetate** $C_8H_{16}O_2$ **142-92-7**

Type of data: cloud points (UCST-behavior)

w_B	0.0894	0.1200	0.1479
T/K	239.05	238.85	238.15

Type of data: cloud points (LCST-behavior)

w_B	0.0894	0.1200	0.1479
T/K	532.25	532.85	532.55

Polymer (B): **poly(α-methylstyrene)** **1995PFO**
Characterization: M_n/g.mol^{-1} = 69500, M_w/g.mol^{-1} = 76500,
Pressure Chemical Company, Pittsburgh, PA
Solvent (A): **hexyl acetate** $C_8H_{16}O_2$ **142-92-7**

Type of data: cloud points (UCST-behavior)

w_B	0.0100	0.0301	0.0500	0.0702	0.0886	0.1071	0.1239	0.1499	0.1990
T/K	262.35	278.75	282.65	284.55	284.45	285.05	284.45	284.45	284.75

Type of data: cloud points (LCST-behavior)

w_B	0.0100	0.0301	0.0500	0.0702	0.0886	0.1071	0.1239	0.1499	0.1990
T/K	526.25	515.45	513.35	510.25	509.75	508.85	509.15	508.15	508.85

Polymer (B): **poly(α-methylstyrene)** **1999PRU**
Characterization: M_n/g.mol^{-1} = 19700, M_w/g.mol^{-1} = 20100,
Polymer Source, Inc., Quebec, Canada
Solvent (A): **methylcyclohexane** C_7H_{14} **108-87-2**

Type of data: cloud points (UCST-behavior)

w_B	0.065	0.100	0.111	0.122	0.148	0.180	0.199	0.221	0.252
T/K	302.85	304.18	305.50	305.67	305.801	305.776	306.71	305.61	305.44
w_B	0.275	0.287	0.301						
T/K	305.19	302.00	301.78						

Polymer (B): **poly(α-methylstyrene)** **1999PRU**
Characterization: M_n/g.mol^{-1} = 32300, M_w/g.mol^{-1} = 32900,
Polymer Source, Inc., Quebec, Canada
Solvent (A): **methylcyclohexane** C_7H_{14} **108-87-2**

Type of data: cloud points (UCST-behavior)

w_B	0.005	0.010	0.014	0.019	0.023	0.047	0.0477	0.0909	0.092
T/K	<289	303	307	309	313.3	314.0	313	317	317.6
w_B	0.1299	0.130	0.131	0.166	0.1661	0.167	0.197	0.2000	0.230
T/K	317	317.9	317	318.6	317	317	318.4	317	317.5

Polymer (B): **poly(α-methylstyrene)** **2001PEN**
Characterization: M_n/g.mol^{-1} = 32300, M_w/g.mol^{-1} = 32900,
Polymer Source, Inc., Quebec, Canada
Solvent (A): **methylcyclohexane** C_7H_{14} **108-87-2**

Type of data: critical point (UCST-behavior)

$w_{B,crit}$ 0.171 T_{crit}/K 317.72

Polymer (B): **poly(α-methylstyrene)** **1999PRU**
Characterization: M_n/g.mol^{-1} = 72000, M_w/g.mol^{-1} = 75600,
Polymer Source, Inc., Quebec, Canada
Solvent (A): **methylcyclohexane** C_7H_{14} **108-87-2**

Type of data: cloud points (UCST-behavior)

w_B	0.0059	0.0105	0.0147	0.0195	0.0243	0.0481	0.0505	0.0907	0.0918
T/K	315	321	327	327	325.9	329	328.5	329	329.2

w_B	0.1307	0.1308	0.1647	0.1978	0.2304
T/K	328.5	328	327.8	326.6	325

Polymer (B): **poly(α-methylstyrene)** **1995PFO**
Characterization: M_n/g.mol^{-1} = 25000, M_w/g.mol^{-1} = 29000,
Pressure Chemical Company, Pittsburgh, PA
Solvent (A): **pentyl acetate** $C_7H_{14}O_2$ **628-63-7**

Type of data: cloud points (UCST-behavior)

w_B	0.0290	0.0400	0.0500	0.0690	0.0810	0.0890	0.1110	0.1290	0.1510
T/K	240.05	242.55	245.95	250.35	249.65	251.05	251.55	251.55	250.75

w_B	0.1690	0.2110	0.2400
T/K	249.65	248.55	246.65

Type of data: cloud points (LCST-behavior)

w_B	0.0150	0.0200	0.0310	0.0400	0.0500	0.0690	0.0800	0.1100	0.1410
T/K	521.25	514.15	511.85	508.75	506.75	506.45	504.55	503.05	501.55

w_B	0.1700	0.2110	0.2400
T/K	500.65	500.75	502.15

Polymer (B): **poly(α-methylstyrene)** **1995PFO**
Characterization: M_n/g.mol^{-1} = 58500, M_w/g.mol^{-1} = 61400,
Pressure Chemical Company, Pittsburgh, PA
Solvent (A): **pentyl acetate** $C_7H_{14}O_2$ **628-63-7**

Type of data: cloud points (UCST-behavior)

w_B	0.0090	0.0100	0.0300	0.0500	0.0700	0.0940	0.1200	0.1500	0.2000
T/K	258.75	263.35	276.55	281.85	285.05	286.05	287.05	285.95	285.25

continued

continued

w_B	0.2300
T/K	281.85

Type of data: cloud points (LCST-behavior)

w_B	0.0090	0.0100	0.0300	0.0500	0.0700	0.0940	0.1200	0.1500	0.2000
T/K	509.65	500.65	493.05	488.95	488.35	485.15	484.65	485.15	485.65

w_B	0.2300
T/K	486.15

Polymer (B): **polyol (hyperbranched, 2nd generation)** **1999JAN**
Characterization: M_n/g.mol^{-1} = 1215, M_w/g.mol^{-1} = 1750,
Aldrich Chem. Co., Inc., Milwaukee, WI
Solvent (A): **water** H_2O **7732-18-5**

Type of data: cloud points (LCST-behavior)

w_B	0.050	0.050	0.053	0.060	0.088	0.100	0.151	0.229	0.249
T/K	321.6	322.4	324.6	327.9	329.4	330.2	333.0	336.2	338.1

w_B	0.276	0.290	0.299	0.327	0.375
T/K	341.6	341.4	341.5	341.6	341.0

Polymer (B): **polyol (hyperbranched, 3rd generation)** **1999JAN**
Characterization: M_n/g.mol^{-1} = 2780, M_w/g.mol^{-1} = 3600,
Aldrich Chem. Co., Inc., Milwaukee, WI
Solvent (A): **water** H_2O **7732-18-5**

Type of data: cloud points (LCST-behavior)

w_B	0.020	0.100	0.150	0.191	0.250	0.302	0.399	0.413	0.344
T/K	386.5	396.7	395.4	394.2	378.5	366.5	330.1	331.3	334.4

Polymer (B): **polyol (hyperbranched, 4th generation)** **1999JAN**
Characterization: M_n/g.mol^{-1} = 6190, M_w/g.mol^{-1} = 7300,
Aldrich Chem. Co., Inc., Milwaukee, WI
Solvent (A): **water** H_2O **7732-18-5**

Type of data: cloud points (LCST-behavior)

w_B	0.020	0.050	0.100	0.150	0.200	0.250	0.298	0.349
T/K	406.3	436.3	430.1	425.3	421.9	413.5	312.4	331.7

Polymer (B): **polypropylene** **2002MA3**
Characterization: M_η/g.mol^{-1} = 2500000, isotactic, ultrahigh molecular weight,
Mitsui Chemical Inc., Japan
Solvent (A): ***cis*-decahydronaphthalene** $C_{10}H_{18}$ **493-01-6**

Type of data: spinodal points (UCST-behavior)

w_B	0.009	0.010	0.011	0.012
T/K	344.55	348.55	348.95	349.85

Polymer (B): **poly(propylene glycol)** **1972LIR**
Characterization: M_n/g.mol^{-1} = 500
Solvent (A): **n-hexane** C_6H_{14} **110-54-3**

Type of data: cloud points, precipitation threshold (UCST-behavior)

w_B	0.37	T/K	317.15

Polymer (B): **poly(propylene glycol)** **1977VSH**
Characterization: M_n/g.mol^{-1} = 1000, fractionated in the laboratory
Solvent (A): **n-hexane** C_6H_{14} **110-54-3**

Type of data: cloud points (UCST-behavior)

w_B	0.20	T/K	293.15	(threshold point)
w_B	0.36	T/K	288.15	(critical point)

Polymer (B): **poly(propylene glycol)** **1977VSH**
Characterization: M_n/g.mol^{-1} = 1800, fractionated in the laboratory
Solvent (A): **n-hexane** C_6H_{14} **110-54-3**

Type of data: cloud points (UCST-behavior)

w_B	0.19	T/K	258.15	(threshold point)
w_B	0.30	T/K	257.15	(critical point)

Polymer (B): **poly(propylene glycol)** **1977VSH**
Characterization: M_n/g.mol^{-1} = 3100, fractionated in the laboratory
Solvent (A): **n-hexane** C_6H_{14} **110-54-3**

Type of data: cloud points (UCST-behavior)

w_B	0.18	T/K	262.15	(threshold point)
w_B	0.22	T/K	257.15	(critical point)

Polymer (B): **poly(propylene glycol)** **1977VSH**
Characterization: M_n/g.mol^{-1} = 6600, fractionated in the laboratory
Solvent (A): **n-hexane** C_6H_{14} **110-54-3**

Type of data: cloud points (UCST-behavior)

w_B	0.125	T/K	261.15	(threshold point)
w_B	0.28	T/K	258.15	(critical point)

Polymer (B): **poly(propylene oxide)** **1993MAL**
Characterization: M_n/g.mol^{-1} = 400, Fluka AG, Buchs, Switzerland
Solvent (A): **water** H_2O **7732-18-5**

Type of data: cloud points (LCST behavior)

T/K = 326.15

w_A	0.822	0.800	0.747	0.672	0.600	0.529	0.448	0.374
w_B	0.178	0.200	0.253	0.328	0.400	0.471	0.552	0.626

Polymer (B): **polystyrene** **1997DES**
Characterization: M_n/g.mol^{-1} = 2300, M_w/g.mol^{-1} = 2500, Pressure Chemical Company, Pittsburgh, PA
Solvent (A): **acetaldehyde** C_2H_4O **75-07-0**

Type of data: cloud points

w_B	0.1090	0.1090	0.1090	0.1090	0.1090	0.1090	0.1090	0.1090	0.1090
T/K	276.17	276.18	276.17	276.67	276.67	277.17	277.18	277.18	277.67
P/MPa	2.91	2.92	2.92	2.50	2.51	2.02	2.02	2.05	1.55
w_B	0.1090	0.1090	0.1090	0.1090	0.1090	0.1090	0.1090	0.1090	0.1374
T/K	277.67	277.67	278.47	278.48	278.67	278.66	278.97	278.97	281.66
P/MPa	1.57	1.58	0.79	0.74	0.61	0.62	0.34	0.32	2.69
w_B	0.1374	0.1374	0.1374	0.1374	0.1374	0.1374	0.1374	0.1374	0.1374
T/K	281.66	281.66	282.16	282.16	282.16	282.16	282.66	282.66	282.66
P/MPa	2.65	2.67	2.27	2.28	2.26	2.27	1.83	1.81	1.81
w_B	0.1374	0.1374	0.1374	0.1374	0.1374	0.1374	0.1374	0.1374	0.1709
T/K	283.16	283.16	283.16	283.97	283.97	284.16	284.16	284.32	284.16
P/MPa	1.36	1.38	1.37	0.60	0.61	0.46	0.46	0.31	2.96
w_B	0.1709	0.1709	0.1709	0.1709	0.1709	0.1709	0.1709	0.1709	0.1709
T/K	284.16	284.16	284.67	284.67	284.67	285.17	285.16	285.17	285.67
P/MPa	2.96	2.97	2.42	2.41	2.42	1.89	1.89	1.91	1.39
w_B	0.1709	0.1709	0.1709	0.1709	0.1709	0.1709	0.1709	0.1709	0.1709
T/K	285.67	285.87	285.86	286.26	286.26	286.48	286.47	286.66	286.66
P/MPa	1.38	1.14	1.16	0.73	0.74	0.54	0.54	0.34	0.35
w_B	0.1766	0.1766	0.1766	0.1766	0.1766	0.1766	0.1766	0.1766	0.1766
T/K	290.07	290.07	290.07	291.17	291.17	291.38	291.38	289.67	289.67
P/MPa	1.51	1.54	1.50	0.74	0.73	0.65	0.65	1.75	1.72
w_B	0.1766	0.1766	0.1766	0.1766	0.1766	0.1986	0.1986	0.1986	0.1986
T/K	289.46	289.46	288.76	288.76	288.67	286.67	286.66	287.17	287.17
P/MPa	1.93	1.90	2.41	2.43	2.68	3.10	3.10	2.65	2.67
w_B	0.1986	0.1986	0.1986	0.1986	0.1986	0.1986	0.1986	0.1986	0.1986
T/K	287.66	287.66	287.66	288.17	288.17	288.67	288.67	288.97	288.97
P/MPa	2.25	2.25	2.25	1.82	1.83	1.39	1.41	1.19	1.16
w_B	0.1986	0.1986	0.1986	0.1986	0.2076	0.2076	0.2076	0.2076	0.2076
T/K	289.97	289.97	290.27	290.27	281.66	282.16	282.67	283.16	283.16
P/MPa	0.46	0.46	0.22	0.23	2.93	2.55	2.18	1.80	1.75
w_B	0.2076	0.2225	0.2225	0.2225	0.2225	0.2225	0.2225	0.2225	0.2225
T/K	284.68	290.67	290.68	291.17	291.18	291.66	291.66	292.17	292.17
P/MPa	0.43	2.96	3.02	2.64	2.64	2.24	2.23	1.90	1.88
w_B	0.2225	0.2225	0.2225	0.2225	0.2225	0.2225	0.2225	0.2225	0.2329
T/K	292.67	292.67	293.16	293.16	294.16	294.16	294.67	294.68	286.17
P/MPa	1.54	1.56	1.23	1.24	0.58	0.59	0.29	0.29	2.82

continued

continued

w_B	0.2329	0.2329	0.2329	0.2329	0.2329	0.2329	0.2329	0.2329	0.2329
T/K	286.17	286.66	287.17	287.66	287.66	287.67	287.66	288.17	288.17
P/MPa	2.83	2.38	1.99	1.65	1.63	1.63	1.62	1.29	1.28
w_B	0.2329	0.2329	0.2329	0.2329	0.2329	0.2329	0.2631	0.2631	0.2631
T/K	288.17	289.16	289.16	289.57	289.57	289.77	285.68	286.17	286.17
P/MPa	1.26	0.62	0.62	0.30	0.30	0.17	3.08	2.70	2.71
w_B	0.2631	0.2631	0.2631	0.2631	0.2631	0.2631	0.2631	0.2631	0.2631
T/K	286.66	286.66	286.66	287.17	287.17	287.67	287.67	288.17	288.17
P/MPa	2.33	2.31	2.31	1.95	1.94	1.53	1.52	1.15	1.13
w_B	0.2631	0.2631	0.2631	0.2631	0.3363	0.3363	0.3363	0.3363	0.3363
T/K	288.67	288.67	289.16	289.16	284.17	284.17	284.68	284.68	285.17
P/MPa	0.77	0.78	0.43	0.43	2.97	2.96	2.52	2.52	2.11
w_B	0.3363	0.3363	0.3363	0.3363	0.3363	0.3363	0.3363	0.3363	0.3363
T/K	285.17	285.68	285.68	286.17	286.17	286.66	286.66	287.17	287.17
P/MPa	2.10	1.61	1.61	1.27	1.28	0.87	0.88	0.49	0.50

Polymer (B): **polystyrene** **1997DES**
Characterization: M_n/g.mol^{-1} = 3100, M_w/g.mol^{-1} = 3250,
Pressure Chemical Company, Pittsburgh, PA
Solvent (A): **acetaldehyde** C_2H_4O **75-07-0**

Type of data: cloud points

w_B	0.0736	0.0736	0.0736	0.0736	0.0736	0.0736	0.0736	0.0736	0.0736
T/K	286.16	286.65	287.16	287.65	288.66	289.66	290.15	290.67	290.86
P/MPa	4.32	3.91	3.51	3.10	2.33	1.64	1.30	0.96	0.84
w_B	0.0736	0.0736	0.0736	0.0892	0.0892	0.0892	0.0892	0.0892	0.0892
T/K	291.16	291.46	291.76	294.68	295.17	295.67	296.67	296.67	297.16
P/MPa	0.67	0.46	0.26	3.25	3.01	2.72	2.19	2.20	1.87
w_B	0.0892	0.0892	0.0892	0.0892	0.0892	0.1058	0.1058	0.1058	0.1058
T/K	297.65	298.16	299.15	299.56	299.76	297.25	297.05	296.96	296.85
P/MPa	1.60	1.35	0.83	0.62	0.52	0.15	0.26	0.31	0.36
w_B	0.1058	0.1058	0.1058	0.1058	0.1058	0.1058	0.1058	0.1058	0.1058
T/K	296.76	296.66	296.56	296.45	296.35	295.17	294.16	293.16	292.17
P/MPa	0.41	0.47	0.53	0.59	0.63	1.27	1.86	2.47	3.05
w_B	0.1058	0.1058	0.1184	0.1184	0.1184	0.1184	0.1184	0.1184	0.1184
T/K	291.66	291.15	295.66	296.17	296.66	297.66	298.16	299.15	299.66
P/MPa	3.40	3.78	3.34	3.02	2.74	2.19	1.92	1.40	1.17
w_B	0.1184	0.1184	0.1184	0.1184	0.1184	0.1184	0.1415	0.1415	0.1415
T/K	300.65	301.15	301.34	301.56	301.75	301.95	309.14	309.14	309.65
P/MPa	0.65	0.48	0.40	0.29	0.24	0.16	4.18	4.20	3.95

continued

continued

w_B	0.1415	0.1415	0.1415	0.1415	0.1415	0.1415	0.1415	0.1415	0.1415
T/K	310.14	312.15	314.14	317.14	318.65	320.63	321.63	322.63	323.63
P/MPa	3.71	3.00	2.38	1.60	1.21	0.81	0.60	0.44	0.29

w_B	0.1415	0.1561	0.1561	0.1561	0.1561	0.1561	0.1561	0.1561	0.1561
T/K	324.14	315.64	316.64	317.64	318.65	320.63	323.12	326.14	327.63
P/MPa	0.23	4.03	3.72	3.38	3.10	2.57	2.00	1.43	1.21

w_B	0.1561	0.1561	0.1561	0.1561	0.1561	0.1654	0.1654	0.1654	0.1654
T/K	330.62	331.62	333.11	334.12	335.12	314.64	315.64	316.64	317.65
P/MPa	0.80	0.70	0.57	0.46	0.41	3.80	3.45	3.20	2.93

w_B	0.1654	0.1654	0.1654	0.1654	0.1654	0.1654	0.1654	0.1654	0.1654
T/K	318.65	320.63	322.63	325.14	328.63	329.62	330.63	331.63	333.12
P/MPa	2.62	2.13	1.72	1.28	0.80	0.70	0.57	0.48	0.35

w_B	0.1654	0.1654	0.1676	0.1676	0.1676	0.1676	0.1676	0.1676	0.1676
T/K	334.22	334.71	289.85	290.67	291.16	292.17	293.16	294.16	294.37
P/MPa	0.25	0.23	4.08	3.52	3.19	2.55	1.96	1.40	1.26

w_B	0.1676	0.1676	0.1676	0.1676	0.1676	0.1676	0.1676	0.1702	0.1702
T/K	295.46	295.56	295.75	295.96	296.17	296.35	296.56	275.17	275.66
P/MPa	0.69	0.64	0.53	0.45	0.32	0.23	0.13	4.11	3.43

w_B	0.1702	0.1702	0.1702	0.1702	0.1702	0.1702	0.1702	0.1702	0.1702
T/K	276.17	277.17	277.66	277.76	278.26	278.47	278.66	278.76	278.87
P/MPa	2.90	1.89	1.39	1.30	0.84	0.65	0.50	0.40	0.32

w_B	0.1702	0.1786	0.1786	0.1786	0.1786	0.1786	0.1786	0.1786	0.1786
T/K	278.96	270.17	270.68	271.17	271.67	272.67	273.06	273.67	273.76
P/MPa	0.25	4.56	3.95	3.33	2.74	1.66	1.24	0.65	0.55

w_B	0.1786	0.1786	0.1786	0.1932	0.1932	0.1932	0.1932	0.1932	0.1932
T/K	273.87	273.97	274.06	282.15	283.15	284.15	284.67	285.16	285.67
P/MPa	0.42	0.32	0.21	4.11	3.27	2.45	2.06	1.68	1.32

w_B	0.1932	0.1932	0.1932	0.1932	0.1932	0.1932	0.2069	0.2069	0.2069
T/K	286.46	286.66	287.07	287.16	287.35	287.46	312.64	313.13	314.13
P/MPa	0.72	0.60	0.33	0.27	0.14	0.08	4.35	4.18	3.84

w_B	0.2069	0.2069	0.2069	0.2069	0.2069	0.2069	0.2069	0.2069	0.2069
T/K	315.64	317.13	319.14	321.13	323.13	325.64	326.64	327.32	327.93
P/MPa	3.26	2.81	2.21	1.71	1.26	0.78	0.60	0.49	0.40

w_B	0.2069	0.2253	0.2253	0.2253	0.2253	0.2253	0.2253	0.2253	0.2253
T/K	328.43	316.91	317.64	318.65	320.63	323.13	325.64	328.13	329.63
P/MPa	0.33	4.12	3.89	3.58	3.02	2.41	1.88	1.46	1.20

w_B	0.2253	0.2253	0.2253	0.2253	0.2253	0.2253	0.2253	0.2461	0.2461
T/K	333.12	334.11	335.61	337.62	339.10	340.10	341.10	316.14	317.14
P/MPa	0.78	0.67	0.55	0.39	0.30	0.22	0.17	3.89	3.58

continued

continued

w_B	0.2461	0.2461	0.2461	0.2461	0.2461	0.2461	0.2461	0.2461	0.2461
T/K	318.13	320.14	322.13	325.14	328.13	331.62	333.12	334.12	335.12
P/MPa	3.30	2.76	2.29	1.70	1.23	0.79	0.64	0.55	0.45
w_B	0.2461	0.3196	0.3196	0.3196	0.3196	0.3196	0.3196	0.3196	0.3196
T/K	337.12	316.64	317.14	318.13	320.13	322.13	324.14	327.13	330.12
P/MPa	0.34	4.55	4.35	4.02	3.43	2.90	2.44	1.85	1.37
w_B	0.3196	0.3196	0.3196	0.3196	0.3196	0.3196			
T/K	333.62	336.12	338.12	340.10	341.10	342.10			
P/MPa	0.92	0.69	0.54	0.39	0.32	0.27			

Polymer (B): **polystyrene** **1997DES, 1997REB**
Characterization: M_n/g.mol^{-1} = 4000, M_w/g.mol^{-1} = 4140,
Pressure Chemical Company, Pittsburgh, PA
Solvent (A): **acetaldehyde** **C_2H_4O** **75-07-0**

Type of data: cloud points

w_B	0.0563	0.0563	0.0563	0.0563	0.0563	0.0563	0.0563	0.0563	0.0563
T/K	268.07	268.65	269.16	270.19	270.95	271.55	272.07	272.57	428.05
P/MPa	4.94	4.34	3.68	2.33	1.60	0.93	0.53	0.10	5.29
w_B	0.0563	0.0563	0.0703	0.0703	0.0703	0.0703	0.0703	0.0703	0.0703
T/K	427.05	426.23	293.61	295.21	298.19	300.63	302.21	302.79	304.31
P/MPa	4.37	3.41	5.58	4.80	3.19	2.11	1.59	1.26	0.80
w_B	0.0703	0.0703	0.0703	0.0703	0.0985	0.0985	0.0985	0.0985	0.0985
T/K	304.59	305.19	305.47	306.02	323.61	326.12	328.65	332.09	337.08
P/MPa	0.71	0.55	0.40	0.28	4.22	3.65	3.18	2.70	2.14
w_B	0.0985	0.0985	0.0985	0.0985	0.0985	0.0985	0.0985	0.0985	0.0985
T/K	342.07	348.10	353.13	356.60	361.59	363.61	368.08	371.12	375.11
P/MPa	1.89	1.78	1.81	1.92	2.01	2.08	2.46	2.64	3.07
w_B	0.0985	0.0985	0.1115	0.1115	0.1115	0.1115	0.1115	0.1115	0.1115
T/K	380.13	385.12	281.66	282.62	283.73	284.70	285.47	286.57	286.72
P/MPa	3.58	4.04	4.39	3.63	2.81	2.04	1.50	0.68	0.56
w_B	0.1115	0.1115	0.1115	0.1115	0.1115	0.1115	0.1115	0.1253	0.1253
T/K	287.02	287.20	413.23	412.69	412.22	411.75	411.11	278.68	279.57
P/MPa	0.41	0.27	4.27	3.64	3.11	2.54	2.09	5.64	4.77
w_B	0.1253	0.1253	0.1253	0.1253	0.1253	0.1253	0.1253	0.1253	0.1253
T/K	280.74	281.75	282.75	283.68	284.14	284.45	284.74	285.01	285.19
P/MPa	3.75	2.91	2.08	1.35	0.99	0.74	0.53	0.29	0.24
w_B	0.1253	0.1253	0.1253	0.1253	0.1253	0.1419	0.1419	0.1419	0.1419
T/K	414.25	413.66	413.54	413.22	412.84	280.17	281.19	282.17	283.17
P/MPa	4.05	3.44	3.00	2.53	2.18	5.80	4.83	3.87	3.12

continued

continued

w_B	0.1419	0.1419	0.1419	0.1419	0.1419	0.1419	0.1419	0.1419	0.1419
T/K	284.18	285.07	285.66	286.27	286.61	286.76	287.19	415.13	414.13
P/MPa	2.35	1.59	1.24	0.89	0.63	0.48	0.22	4.39	3.79
w_B	0.1419	0.1419	0.1419	0.1630	0.1630	0.1630	0.1630	0.1630	0.1630
T/K	413.15	412.56	412.19	306.08	307.09	309.15	311.05	313.14	315.11
P/MPa	3.14	2.61	2.21	4.78	4.37	3.65	2.98	2.40	1.87
w_B	0.1630	0.1630	0.1630	0.1630	0.1630	0.1630	0.1630	0.1630	0.1630
T/K	317.14	319.62	320.63	322.11	323.53	324.20	389.62	389.12	387.11
P/MPa	1.37	0.87	0.70	0.51	0.26	0.17	4.88	4.58	4.01
w_B	0.1630	0.1630	0.1630	0.1630	0.1630	0.1700	0.1700	0.1700	0.1700
T/K	385.11	383.10	380.16	377.14	376.08	305.02	306.15	308.10	310.09
P/MPa	3.51	2.97	2.26	1.53	1.19	4.70	4.29	3.49	2.80
w_B	0.1700	0.1700	0.1700	0.1700	0.1700	0.1700	0.1700	0.1700	0.1700
T/K	312.03	313.95	315.22	317.02	317.99	319.10	319.93	320.64	391.13
P/MPa	2.13	1.56	1.34	0.89	0.69	0.50	0.34	0.20	4.85
w_B	0.1700	0.1700	0.1700	0.1700	0.1700	0.1700	0.1700	0.1700	0.1731
T/K	390.20	388.18	386.15	384.19	382.15	380.18	378.10	376.09	304.18
P/MPa	4.48	3.98	3.49	3.05	2.61	2.11	1.66	1.23	4.74
w_B	0.1731	0.1731	0.1731	0.1731	0.1731	0.1731	0.1731	0.1731	0.1731
T/K	305.13	307.02	308.99	311.12	313.02	314.13	316.57	317.06	318.07
P/MPa	4.27	3.50	2.83	2.21	1.61	1.38	0.79	0.67	0.45
w_B	0.1731	0.1731	0.1731	0.1731	0.1731	0.1731	0.1731	0.1731	0.1731
T/K	305.13	307.02	308.99	311.12	313.02	314.13	316.57	317.06	318.07
P/MPa	4.19	3.44	2.74	2.10	1.53	1.33	0.72	0.60	0.37
w_B	0.1731	0.1731	0.1870	0.1870	0.1870	0.1870	0.1870	0.1870	0.1870
T/K	318.62	319.03	306.14	308.19	310.02	312.12	314.16	316.11	318.15
P/MPa	0.33	0.25	4.91	4.09	3.41	2.77	2.20	1.71	1.28
w_B	0.1870	0.1870	0.1870	0.1870	0.1870	0.1870	0.1870	0.1870	0.1870
T/K	320.59	322.07	323.99	393.13	391.16	389.13	386.15	383.14	380.09
P/MPa	0.79	0.52	0.28	5.35	4.76	4.21	3.46	2.77	2.10
w_B	0.1870	0.2201	0.2201	0.2201	0.2201	0.2201	0.2201	0.2201	0.2201
T/K	377.14	303.08	305.14	307.11	309.12	311.16	313.18	315.18	316.18
P/MPa	1.43	4.91	4.08	3.30	2.59	1.94	1.36	0.88	0.65
w_B	0.2201	0.2201	0.2201	0.2201	0.2201	0.2201	0.2201	0.2201	0.2201
T/K	317.31	318.23	393.15	391.12	389.19	386.15	383.13	380.11	377.13
P/MPa	0.45	0.23	4.98	4.41	3.91	3.22	2.56	1.93	1.31
w_B	0.2201	0.2201	0.2201	0.2201	0.2201	0.2201	0.2201	0.2201	0.2201
T/K	317.31	318.23	393.15	391.12	389.19	386.15	383.13	380.11	377.13
P/MPa	0.36	0.16	4.79	4.29	3.75	3.13	2.44	1.76	1.15

Polymer (B): **polystyrene** **1987RAN**
Characterization: M_w/g.mol^{-1} = 200000,
Pressure Chemical Company, Pittsburgh, PA
Solvent (A): **bis(2-ethylhexyl) phthalate** $C_{24}H_{38}O_4$ **117-81-7**

Type of data: cloud points (UCST-behavior)

φ_B	0.0177	0.0263	0.0348	0.0432	0.0512	0.0828
T/K	274.65	276.65	277.95	278.55	278.55	277.65

Polymer (B): **polystyrene** **1987RAN**
Characterization: M_w/g.mol^{-1} = 280000,
Pressure Chemical Company, Pittsburgh, PA
Solvent (A): **bis(2-ethylhexyl) phthalate** $C_{24}H_{38}O_4$ **117-81-7**

Type of data: cloud points (UCST-behavior)

φ_B	0.0177	0.0263	0.0348	0.0432	0.0512	0.0828
T/K	277.75	279.35	279.75	280.55	279.95	279.75

Polymer (B): **polystyrene** **1987RAN**
Characterization: M_w/g.mol^{-1} = 470000,
Pressure Chemical Company, Pittsburgh, PA
Solvent (A): **bis(2-ethylhexyl) phthalate** $C_{24}H_{38}O_4$ **117-81-7**

Type of data: cloud points (UCST-behavior)

φ_B	0.0177	0.0263	0.0348	0.0432	0.0512	0.0828
T/K	281.15	281.95	282.15	282.05	281.45	280.95

Polymer (B): **polystyrene** **1987RAN**
Characterization: M_w/g.mol^{-1} = 900000,
Pressure Chemical Company, Pittsburgh, PA
Solvent (A): **bis(2-ethylhexyl) phthalate** $C_{24}H_{38}O_4$ **117-81-7**

Type of data: cloud points (UCST-behavior)

φ_B	0.0177	0.0263	0.0348	0.0432	0.0512	0.0828
T/K	283.05	283.15	283.05	282.85	282.75	281.65

Polymer (B): **polystyrene** **1987RAN**
Characterization: M_w/g.mol^{-1} = 1800000,
Pressure Chemical Company, Pittsburgh, PA
Solvent (A): **bis(2-ethylhexyl) phthalate** $C_{24}H_{38}O_4$ **117-81-7**

Type of data: cloud points (UCST-behavior)

φ_B	0.0177	0.0263	0.0348	0.0432	0.0512	0.0828
T/K	285.15	284.85	284.55	283.95	283.65	281.65

Polymer (B): **polystyrene** **1996GIR**
Characterization: M_n/g.mol^{-1} = 24000
Solvent (A): **bisphenol-A diglycidyl ether** $C_{21}H_{24}O_4$ **1675-54-3**

Type of data: cloud points (UCST-behavior)

φ_B	0.0554	0.1101	0.1643	0.2178
T/K	309.1	313.0	315.5	315.0

Polymer (B): **polystyrene** **2004ZUC**
Characterization: M_n/g.mol^{-1} = 28400, Polymer Source, Inc., Quebec, Canada
Solvent (A): **bisphenol-A diglycidyl ether** $C_{21}H_{24}O_4$ **1675-54-3**

Type of data: cloud points (UCST-behavior)

φ_B	0.0277	0.0554	0.1101	0.1643
T/K	301.5	310	317	320

Polymer (B): **polystyrene** **1996GIR**
Characterization: M_n/g.mol^{-1} = 76800
Solvent (A): **bisphenol-A diglycidyl ether** $C_{21}H_{24}O_4$ **1675-54-3**

Type of data: cloud points (UCST-behavior)

φ_B	0.0554	0.1101	0.1643	0.2178	0.3232	0.3750
T/K	334.37	336.14	336.00	334.37	330.15	326.35

Polymer (B): **polystyrene** **2004ZUC**
Characterization: M_n/g.mol^{-1} = 83000, M_w/g.mol^{-1} = 86700,
Polymer Source, Inc., Quebec, Canada
Solvent (A): **bisphenol-A diglycidyl ether** $C_{21}H_{24}O_4$ **1675-54-3**

Type of data: cloud points (UCST-behavior)

φ_B	0.0277	0.0554	0.1101	0.1643	0.2708
T/K	331.74	335.45	338.55	339.05	337.03

Polymer (B): **polystyrene** **2004ZUC**
Characterization: M_n/g.mol^{-1} = 217000, M_w/g.mol^{-1} = 228000,
Polymer Source, Inc., Quebec, Canada
Solvent (A): **bisphenol-A diglycidyl ether** $C_{21}H_{24}O_4$ **1675-54-3**

Type of data: cloud points (UCST-behavior)

φ_B	0.0277	0.0554	0.1101	0.1643
T/K	345.33	349.4	350.8	350.38

Polymer (B): **polystyrene** **1996GIR**
Characterization: M_n/g.mol^{-1} = 282100
Solvent (A): **bisphenol-A diglycidyl ether** $C_{21}H_{24}O_4$ **1675-54-3**

continued

continued

Type of data: cloud points (UCST-behavior)

φ_B	0.0554	0.1101	0.1643	0.2178
T/K	348.2	348.77	346.56	345.7

Polymer (B): **polystyrene** **1963ORO**
Characterization: M_n/g.mol^{-1} = 404000, M_w/g.mol^{-1} = 406000
Solvent (A): **1-bromobutane** C_4H_9Br **109-65-9**

Type of data: cloud points (UCST-behavior)

w_B 0.01 T/K 248.15

Polymer (B): **polystyrene** **1963ORO**
Characterization: M_n/g.mol^{-1} = 404000, M_w/g.mol^{-1} = 406000
Solvent (A): **1-bromodecane** $C_{10}H_{21}Br$ **112-29-8**

Type of data: cloud points (UCST-behavior)

w_B 0.01 T/K 268.15

Polymer (B): **polystyrene** **1963ORO**
Characterization: M_n/g.mol^{-1} = 404000, M_w/g.mol^{-1} = 406000
Solvent (A): **1-bromooctane** $C_8H_{17}Br$ **111-83-1**

Type of data: cloud points (UCST-behavior)

w_B 0.01 T/K 265.15

Polymer (B): **polystyrene** **1963ORO**
Characterization: M_n/g.mol^{-1} = 404000, M_w/g.mol^{-1} = 406000
Solvent (A): **2-bromooctane** $C_8H_{17}Br$ **557-35-7**

Type of data: cloud points (UCST-behavior)

w_B 0.01 T/K 298.15

Polymer (B): **polystyrene** **1950JE1**
Characterization: M_η/g.mol^{-1} = 62600, fractionated in the laboratory
Solvent (A): **butanedioic acid dimethyl ester** $C_6H_{10}O_4$ **106-65-0**

Type of data: cloud points, precipitation threshold (UCST-behavior)

z_B 0.118 T/K 335.15

Polymer (B): **polystyrene** **1991OPS**
Characterization: M_n/g.mol^{-1} = 3700, M_w/g.mol^{-1} = 4000, M_z/g.mol^{-1} = 4600, Pressure Chemical Company, Pittsburgh, PA
Solvent (A): **1-butanol** $C_4H_{10}O$ **71-36-3**

continued

continued

Type of data: critical point (UCST-behavior)

$w_{B, crit}$	0.358	T_{crit}/K	383.45

Polymer (B): **polystyrene** **1991OPS**
Characterization: M_w/g.mol^{-1} = 4800,
Pressure Chemical Company, Pittsburgh, PA
Solvent (A): **1-butanol** $C_4H_{10}O$ **71-36-3**

Type of data: critical point (UCST-behavior)

$w_{B, crit}$	0.367	T_{crit}/K	381.45

Polymer (B): **polystyrene** **1991OPS**
Characterization: M_w/g.mol^{-1} = 9000,
Pressure Chemical Company, Pittsburgh, PA
Solvent (A): **1-butanol** $C_4H_{10}O$ **71-36-3**

Type of data: critical point (UCST-behavior)

$w_{B, crit}$	0.284	T_{crit}/K	407.45

Polymer (B): **polystyrene** **1950JE1**
Characterization: M_η/g.mol^{-1} = 62600, fractionated in the laboratory
Solvent (A): **1-butanol** $C_4H_{10}O$ **71-36-3**

Type of data: cloud points, precipitation threshold (UCST-behavior)

z_B	0.030	T/K	454.15

Polymer (B): **polystyrene** **1995PFO**
Characterization: M_n/g.mol^{-1} = 219800, M_w/g.mol^{-1} = 233000,
Polysciences, Inc., Warrington, PA
Solvent (A): **butyl acetate** $C_6H_{12}O_2$ **123-86-4**

Type of data: cloud points (LCST-behavior)

w_B	0.0800	0.1050	0.1298
T/K	497.85	494.95	494.35

Polymer (B): **polystyrene** **1995PFO**
Characterization: M_n/g.mol^{-1} = 545500, M_w/g.mol^{-1} = 600000,
Polysciences, Inc., Warrington, PA
Solvent (A): **butyl acetate** $C_6H_{12}O_2$ **123-86-4**

Type of data: cloud points (LCST-behavior)

w_B	0.0101	0.0248	0.0403	0.0557	0.0703	0.0841	0.0977	0.1163	0.1298
T/K	494.55	492.25	490.75	490.35	489.95	490.35	489.25	489.25	489.75

Polymer (B): **polystyrene** **1991BAE**
Characterization: M_n/g.mol^{-1} = 94300, M_w/g.mol^{-1} = 100000,
Pressure Chemical Company, Pittsburgh, PA
Solvent (A): ***tert*-butyl acetate** $C_6H_{12}O_2$ **540-88-5**

Type of data: cloud points (UCST-behavior)

w_B	0.005	0.010	0.030	0.050	0.071	0.101	0.152	0.202	0.300
T/K	240.65	244.95	250.55	252.65	253.55	254.65	255.65	253.05	247.05

Type of data: cloud points (LCST-behavior)

w_B	0.005	0.010	0.030	0.050	0.071	0.101	0.152	0.202	0.300
T/K	438.65	433.15	425.65	422.95	423.15	422.05	422.45	423.25	435.15

Polymer (B): **polystyrene** **1991BAE**
Characterization: M_n/g.mol^{-1} = 220000, M_w/g.mol^{-1} = 233000,
Pressure Chemical Company, Pittsburgh, PA
Solvent (A): ***tert*-butyl acetate** $C_6H_{12}O_2$ **540-88-5**

Type of data: cloud points (UCST-behavior)

w_B	0.005	0.010	0.031	0.051	0.071	0.102	0.151
T/K	260.35	264.55	269.05	269.45	270.25	270.55	267.25

Type of data: cloud points (LCST-behavior)

w_B	0.005	0.010	0.031	0.051	0.071	0.102	0.151	0.191
T/K	413.05	409.25	403.55	403.35	403.15	403.35	406.45	409.25

Polymer (B): **polystyrene** **1991BAE**
Characterization: M_n/g.mol^{-1} = 545500, M_w/g.mol^{-1} = 600000,
Pressure Chemical Company, Pittsburgh, PA
Solvent (A): ***tert*-butyl acetate** $C_6H_{12}O_2$ **540-88-5**

Type of data: cloud points (UCST-behavior)

w_B	0.005	0.010	0.021	0.030	0.051	0.067	0.101
T/K	270.45	273.55	277.05	278.65	277.35	276.65	274.85

Type of data: cloud points (LCST-behavior)

w_B	0.005	0.010	0.021	0.030	0.051	0.067	0.101
T/K	398.05	394.55	391.95	391.55	391.05	391.75	394.15

Polymer (B): **polystyrene** **1950JE1**
Characterization: M_η/g.mol^{-1} = 62600, fractionated in the laboratory
Solvent (A): **butyl stearate** $C_{22}H_{44}O_2$ **123-95-5**

Type of data: cloud points, precipitation threshold (UCST-behavior)

z_B 0.130 T/K 387.15

Polymer (B): **polystyrene** **1963ORO**
Characterization: M_n/g.mol^{-1} = 404000, M_w/g.mol^{-1} = 406000
Solvent (A): **4-*tert*-butyl-ethylbenzene** $C_{12}H_{18}$ **7364-19-4**

Type of data: cloud points (UCST-behavior)

w_B	0.01	T/K	259.15

Polymer (B): **polystyrene** **1963ORO**
Characterization: M_n/g.mol^{-1} = 404000, M_w/g.mol^{-1} = 406000
Solvent (A): **1-chlorodecane** $C_{10}H_{21}Cl$ **1002-69-3**

Type of data: cloud points (UCST-behavior)

w_B	0.01	T/K	276.15

Polymer (B): **polystyrene** **1991OPS**
Characterization: M_n/g.mol^{-1} = 18400, M_w/g.mol^{-1} = 19200, M_z/g.mol^{-1} = 24500, Pressure Chemical Company, Pittsburgh, PA
Solvent (A): **1-chlorododecane** $C_{12}H_{25}Cl$ **112-52-7**

Type of data: critical point (UCST-behavior)

$w_{B, crit}$	0.242	T_{crit}/K	274.65

Polymer (B): **polystyrene** **1963ORO**
Characterization: M_n/g.mol^{-1} = 404000, M_w/g.mol^{-1} = 406000
Solvent (A): **1-chlorododecane** $C_{12}H_{25}Cl$ **112-52-7**

Type of data: cloud points (UCST-behavior)

w_B	0.01	T/K	325.15

Polymer (B): **polystyrene** **1991OPS**
Characterization: M_n/g.mol^{-1} = 18400, M_w/g.mol^{-1} = 19200, M_z/g.mol^{-1} = 24500, Pressure Chemical Company, Pittsburgh, PA
Solvent (A): **1-chlorohexadecane** $C_{16}H_{33}Cl$ **4860-03-1**

Type of data: critical point (UCST-behavior)

$w_{B, crit}$	0.254	T_{crit}/K	337.05

Polymer (B): **polystyrene** **1963ORO**
Characterization: M_n/g.mol^{-1} = 404000, M_w/g.mol^{-1} = 406000
Solvent (A): **1-chlorohexadecane** $C_{16}H_{33}Cl$ **4860-03-1**

Type of data: cloud points (UCST-behavior)

w_B	0.01	T/K	433.15

Polymer (B): **polystyrene** **1963ORO**
Characterization: M_n/g.mol^{-1} = 404000, M_w/g.mol^{-1} = 406000
Solvent (A): **1-chlorononane** **$C_9H_{19}Cl$** **2473-01-0**

Type of data: cloud points (UCST-behavior)

w_B	0.01	T/K	254.15

Polymer (B): **polystyrene** **1991OPS**
Characterization: M_n/g.mol^{-1} = 18400, M_w/g.mol^{-1} = 19200, M_z/g.mol^{-1} = 24500, Pressure Chemical Company, Pittsburgh, PA
Solvent (A): **1-chlorooctadecane** **$C_{18}H_{37}Cl$** **3386-33-2**

Type of data: critical point (UCST-behavior)

$w_{B,\,crit}$	0.261	T_{crit}/K	365.55

Polymer (B): **polystyrene** **1963ORO**
Characterization: M_n/g.mol^{-1} = 404000, M_w/g.mol^{-1} = 406000
Solvent (A): **1-chlorooctane** **$C_8H_{17}Cl$** **111-85-3**

Type of data: cloud points (UCST-behavior)

w_B	0.01	T/K	223.15

Polymer (B): **polystyrene** **1991OPS**
Characterization: M_n/g.mol^{-1} = 18400, M_w/g.mol^{-1} = 19200, M_z/g.mol^{-1} = 24500, Pressure Chemical Company, Pittsburgh, PA
Solvent (A): **1-chlorotetradecane** **$C_{14}H_{29}Cl$** **2425-54-9**

Type of data: critical point (UCST-behavior)

$w_{B,\,crit}$	0.248	T_{crit}/K	309.35

Polymer (B): **polystyrene** **1963ORO**
Characterization: M_n/g.mol^{-1} = 404000, M_w/g.mol^{-1} = 406000
Solvent (A): **1-chloroundecane** **$C_{11}H_{23}Cl$** **2473-03-2**

Type of data: cloud points (UCST-behavior)

w_B	0.01	T/K	301.15

Polymer (B): **polystyrene** **1976KL2**
Characterization: M_n/g.mol^{-1} = 8000, M_w/g.mol^{-1} = 41000, M_z/g.mol^{-1} = 122000, BASF AG, Germany
Solvent (A): **cyclohexane** **C_6H_{12}** **110-82-7**

Type of data: coexistence data (UCST-behavior)

T/K	298.85	298.85	298.85	298.85	298.85	298.55	298.55
w_B (total)	0.0101	0.0100	0.0508	0.0748	0.1036	0.1252	0.1252
w_B (sol phase)	0.0069	0.0078	0.0358	0.0552	0.0456	0.0936	0.0886
w_B (gel phase)	0.195	0.209	0.179	0.167	0.147	0.141	0.138

Polymer (B): **polystyrene** **1991BAE**
Characterization: M_n/g.mol^{-1} = 19200, M_w/g.mol^{-1} = 20400,
Pressure Chemical Company, Pittsburgh, PA
Solvent (A): **cyclohexane** C_6H_{12} **110-82-7**

Type of data: cloud points (UCST-behavior)

w_B	0.005	0.010	0.029	0.048	0.069	0.147	0.204	0.293	0.400
T/K	276.25	277.15	277.65	278.05	279.15	279.75	279.75	279.05	277.65

Polymer (B): **polystyrene** **2003SIP**
Characterization: M_n/g.mol^{-1} = 23600, M_w/g.mol^{-1} = 25000,
Pressure Chemical Company, Pittsburgh, PA
Solvent (A): **cyclohexane** C_6H_{12} **110-82-7**

Type of data: cloud points (UCST-behavior)

φ_B	0.034	0.044	0.058	0.069	0.084	0.098	0.110	0.122	0.133
T/K	278.32	278.85	279.60	279.84	280.02	280.15	280.23	280.27	280.30
φ_B	0.144	0.153	0.166	0.181	0.197	0.209	0.223		
T/K	280.31	280.28	280.22	280.12	280.03	279.93	279.74		

Type of data: critical point (UCST-behavior)

$\varphi_{B, crit}$ 0.142 T_{crit}/K 280.29

Polymer (B): **polystyrene** **1972KEN**
Characterization: M_n/g.mol^{-1} = 23500, M_w/g.mol^{-1} = 30200, M_z/g.mol^{-1} = 40100,
Pressure Chemical Company, Pittsburgh, PA
Solvent (A): **cyclohexane** C_6H_{12} **110-82-7**

Type of data: critical point (UCST-behavior)

$w_{B, crit}$ 0.200 T_{crit}/K 282.7

Polymer (B): **polystyrene** **1972KEN**
Characterization: M_n/g.mol^{-1} = 27000, M_w/g.mol^{-1} = 35400, M_z/g.mol^{-1} = 45500,
Pressure Chemical Company, Pittsburgh, PA
Solvent (A): **cyclohexane** C_6H_{12} **110-82-7**

Type of data: critical point (UCST-behavior)

$w_{B, crit}$ 0.179 T_{crit}/K 284.6

Polymer (B): **polystyrene** **1973SA1**
Characterization: M_n/g.mol^{-1} = 34900, M_w/g.mol^{-1} = 37000,
Pressure Chemical Company, Pittsburgh, PA
Solvent (A): **cyclohexane** C_6H_{12} **110-82-7**

Type of data: cloud points (UCST-behavior)

continued

continued

w_B	0.0581	0.0837	0.1078	0.1607	0.2309	0.2379
T/K	284.93	285.52	285.38	285.61	285.31	285.10

Type of data: cloud points (LCST-behavior)

w_B	0.0455	0.0671	0.0899	0.1336	0.1910	0.1955
T/K	511.55	510.99	511.04	510.75	513.05	513.44

Polymer (B): **polystyrene** **1981HAS**
Characterization: M_n/g.mol^{-1} = 45000, M_w/g.mol^{-1} = 45300, M_z/g.mol^{-1} =45600, Toyo Soda Manufacturing Co., Japan
Solvent (A): **cyclohexane** C_6H_{12} **110-82-7**

Type of data: coexistence data (UCST-behavior)

T/K	287.15	287.29
φ_B (sol phase)	0.0682	0.149
φ_B (gel phase)	0.1514	0.149

Type of data: critical point (UCST-behavior)

$\varphi_{B, crit}$ 0.149 T_{crit}/K 287.29

Polymer (B): **polystyrene** **1991KAW**
Characterization: M_w/g.mol^{-1} = 48000, Pressure Chemical Company, Pittsburgh, PA
Solvent (A): **cyclohexane** C_6H_{12} **110-82-7**

Type of data: cloud points (UCST-behavior)

φ_B	0.020	0.030	0.040	0.050	0.060	0.080	0.100
T/K	285.61	286.63	287.17	287.52	287.74	287.94	287.96

Polymer (B): **polystyrene** **1971SCH, 1972KEN**
Characterization: M_n/g.mol^{-1} = 49000, M_w/g.mol^{-1} = 51000, M_z/g.mol^{-1} = 55000, Pressure Chemical Company, Pittsburgh, PA
Solvent (A): **cyclohexane** C_6H_{12} **110-82-7**

Type of data: spinodal points (UCST-behavior)

w_B	0.0258	0.0435	0.0764	0.0903	0.1204	0.1415	0.1721	0.2029
T/K	279.55	284.75	287.35	288.25	288.95	289.05	288.95	288.45

Type of data: critical point (UCST-behavior)

$w_{B, crit}$ 0.146 T_{crit}/K 288.85

Polymer (B): **polystyrene** **1990STR**
Characterization: M_n/g.mol^{-1} = 49000, M_w/g.mol^{-1} = 51000, Pressure Chemical Company, Pittsburgh, PA
Solvent (A): **cyclohexane** C_6H_{12} **110-82-7**

continued

continued

Type of data: spinodal points (UCST-behavior)

w_B	0.0258	0.0435	0.0764	0.0903	0.1204	0.1415	0.1721	0.2029	0.0756
T/K	279.55	284.75	287.35	288.25	288.95	289.05	288.95	288.45	287.35

w_B	0.0894	0.1194	0.1404	0.1707	0.2013
T/K	288.25	288.95	289.05	288.95	288.45

Polymer (B): **polystyrene** **1972KEN**
Characterization: M_n/g.mol^{-1} = 55000, M_w/g.mol^{-1} = 61500, M_z/g.mol^{-1} = 70500,
Pressure Chemical Company, Pittsburgh, PA
Solvent (A): **cyclohexane** C_6H_{12} **110-82-7**

Type of data: critical point (UCST-behavior)

$w_{B,\,crit}$ 0.143 T_{crit}/K 290.45

Polymer (B): **polystyrene** **1984GIL**
Characterization: M_n/g.mol^{-1} = 58000, M_w/g.mol^{-1} = 280000
Solvent (A): **cyclohexane** C_6H_{12} **110-82-7**

Type of data: cloud points (UCST-behavior)

w_B	0.00994	0.01538	0.01930	0.03276	0.03906	0.04876	0.05032	0.05663	0.05777
T/K	300.78	300.81	300.90	300.55	300.43	300.02	299.97	299.59	299.68

w_B	0.06764	0.06780	0.07524	0.07662	0.08636	0.09681	0.10001	0.10767	0.10907
T/K	299.28	299.23	298.94	298.88	298.61	298.27	298.18	298.10	297.98

w_B	0.11000	0.11074	0.11192	0.11264	0.11445	0.11522	0.12600	0.13043	0.13382
T/K	297.80	297.79	297.85	297.74	297.66	297.65	297.39	297.33	297.25

w_B	0.14929	0.15200	0.15488	0.19015	0.20770
T/K	296.79	296.74	296.69	295.67	295.22

Type of data: critical point (UCST-behavior)

$w_{B,\,crit}$ 0.110 T_{crit}/K 297.80

Type of data: coexistence data

w_B (total)	0.05377	0.05377	0.05377	0.05377	0.05377	0.05377	0.05377	0.05377
T/K	296.14	298.61	298.08	297.07	296.11	295.13	294.15	293.14
w_B (sol phase)	0.04240	0.03938	0.03512	0.03958	0.02719	0.02216	0.02115	0.01910
w_B (gel phase)	0.18340			0.23229	0.25138			0.29202

w_B (total)	0.07545	0.07545	0.07545	0.07545	0.07545	0.07545	0.07545	0.07545
T/K	298.75	298.44	298.11	297.66	297.15	296.65	295.15	293.15
w_B (sol phase)	0.07117	0.06146	0.05779	0.05305	0.04695	0.04265	0.03462	0.02615
w_B (gel phase)	0.14005	0.15455	0.17290	0.18647	0.20827	0.21815	0.25138	0.28680

continued

continued

Demixing temperature	Fractionation during demixing	
	Sol phase	Gel phase
T/ K	M_n/ g mol^{-1}	M_n/ g mol^{-1}
298.75	56680	102900
298.44	55900	106000
298.11	51800	112140
297.66	45770	117130
297.15	43470	119020
296.65	31630	122600
296.14		128700
295.15	26280	131900
294.16	26100	137700
293.15	23600	149300

w_B (total)	0.10008	0.10008	0.10008	0.10008	0.10008	0.10008	0.10008	0.10008
T/K	298.00	297.65	297.12	296.67	296.14	295.65	295.14	294.14
w_B (sol phase)	0.09120	0.07013	0.06170	0.05405	0.05081	0.04615	0.04315	0.03789
w_B (gel phase)	0.12868	0.15630	0.18340	0.20389	0.21709	0.22671	0.24182	0.26517

Demixing temperature	Fractionation during demixing	
	Sol phase	Gel phase
T/ K	M_n/ g mol^{-1}	M_n/ g mol^{-1}
297.65	43330	75030
297.12	41120	78530
296.67	39780	82890
296.14	36020	84840
295.65	32940	
295.14	26700	
294.14	23330	99780

w_B (total)	0.12544	0.12544	0.12544	0.12544	0.12544	0.12544	0.12544	0.12544
T/K	297.35	297.15	296.65	296.15	295.65	295.15	294.15	293.15
w_B (sol phase)	0.08955	0.08193	0.06888	0.06170	0.05529	0.04982	0.04369	0.03938
w_B (gel phase)	0.14491	0.15564	0.17960	0.19999	0.21580	0.23229	0.26627	0.28056

continued

continued

Demixing temperature	Fractionation during demixing	
	Sol phase	Gel phase
T/ K	M_n/ g mol^{-1}	M_n/ g mol^{-1}
297.15	44160	70870
296.65	39460	82200
296.15	37930	88050
295.65		88580
295.15	31530	
293.15		105460

w_B (total)	0.15180	0.15180	0.15180	0.15180	0.15180	0.15180	0.15180	0.15180
T/K	296.11	295.62	295.13	294.63	294.15	293.64	293.14	292.14
w_B (sol phase)	0.07308	0.06553	0.05504	0.05156	0.05032	0.04720	0.04415	0.03938
w_B (gel phase)	0.17974	0.20074	0.21815	0.23208	0.24504	0.25684	0.27105	0.29275

Polymer (B): **polystyrene** **1960DE2**
Characterization: M_n/g.mol^{-1} = 69000, M_w/g.mol^{-1} = 80000, Dow Chemical Company, Midland
Solvent (A): **cyclohexane** C_6H_{12} **110-82-7**

Type of data: critical point (UCST-behavior)

$\varphi_{B,\,crit}$ 0.068 T_{crit}/K 292.39

Polymer (B): **polystyrene** **1998TER**
Characterization: M_w/g.mol^{-1} = 85000, 4 arms
Solvent (A): **cyclohexane** C_6H_{12} **110-82-7**

Type of data: critical point (UCST-behavior)

$\varphi_{B,\,crit}$ 0.1040 T_{crit}/K 289.60

Type of data: coexistence data (UCST-behavior)

T/K	289.56	289.55	289.55	289.50	289.37	289.18	288.97	288.65
φ_B (sol phase)	0.0774	0.0752	0.0746	0.0627	0.0540	0.0432	0.0365	0.0283
φ_B (gel phase)	0.1307	0.1340	0.1354	0.1483	0.1584	0.1769	0.1884	0.2063

T/K	287.98	287.70	286.86	286.35
φ_B (sol phase)	0.0204	0.0196	0.0129	0.0107
φ_B (gel phase)	0.2266	0.2360	0.2573	0.2730

Comments: The total feed concentration of the polymer is near the critical concentration.

continued

continued

Type of data: spinodal points (UCST-behavior)

φ_B	0.03577	0.05837	0.07252	0.08656	0.10010	0.11260	0.11330	0.12960	0.13540
T/K	286.12	288.68	289.23	289.56	289.60	289.56	289.52	289.39	289.27

φ_B	0.15640	0.16210
T/K	288.98	288.68

Polymer (B): **polystyrene** **1999CHO1, 1999CHO2**
Characterization: M_n/g.mol^{-1} = 83200, M_w/g.mol^{-1} = 190500,
Aldrich Chem. Co., Inc., Milwaukee, WI
Solvent (A): **cyclohexane** C_6H_{12} **110-82-7**

Type of data: cloud points (UCST-behavior)

φ_B	0.0063	0.0118	0.0194	0.0319	0.0362	0.0472	0.0527	0.0570	0.0590
T/K	298.63	299.44	299.35	299.73	299.62	299.15	299.09	298.86	298.66

φ_B	0.0616	0.0688	0.0719	0.0806	0.0817	0.0912	0.1046
T/K	298.75	298.55	298.53	298.24	298.24	298.02	297.70

Polymer (B): **polystyrene** **1999CHO3**
Characterization: M_n/g.mol^{-1} = 83200, M_w/g.mol^{-1} = 190500,
Aldrich Chem. Co., Inc., Milwaukee, WI
Solvent (A): **cyclohexane** C_6H_{12} **110-82-7**

Type of data: cloud points (UCST-behavior)

φ_B	0.0102	0.0168	0.0325	0.0452	0.0475	0.0624	0.0742	0.0932	0.1167
T/K	298.58	298.74	299.01	299.01	298.92	298.82	298.64	298.14	297.82

Type of data: cloud points (LCST-behavior)

φ_B	0.0102	0.0168	0.0325	0.0452	0.0475	0.0624	0.0742	0.0932	0.1167
T/K	497.45	497.26	496.95	496.89	496.91	497.33	497.44	497.62	498.18

Polymer (B): **polystyrene** **1999OCH**
Characterization: M_n/g.mol^{-1} = 86500, M_w/g.mol^{-1} = 90000,
Pressure Chemical Company, Pittsburgh, PA
Solvent (A): **cyclohexane** C_6H_{12} **110-82-7**

Type of data: cloud points (UCST-behavior)

w_B	0.0025	0.0050	0.0060	0.0076	0.0101	0.0120	0.0169	0.0231	0.0574
T/K	285.10	287.11	287.58	288.51	288.80	289.30	290.11	290.96	292.40

w_B	0.0941	0.1136	0.1292	0.1562	0.1995	0.2395
T/K	292.78	292.88	292.91	292.47	291.45	289.77

Polymer (B): **polystyrene** **1999CHO1, 1999CHO2**
Characterization: M_n/g.mol^{-1} = 87200, M_w/g.mol^{-1} = 171300,
Aldrich Chem. Co., Inc., Milwaukee, WI
Solvent (A): **cyclohexane** **C_6H_{12}** **110-82-7**

Type of data: cloud points (UCST-behavior)

φ_B	0.0063	0.0118	0.0194	0.0319	0.0362	0.0472	0.0527	0.0570	0.0590
T/K	298.63	299.44	299.35	299.73	299.62	299.15	299.09	298.86	298.66

φ_B	0.0616	0.0688	0.0719	0.0806	0.0817	0.0912	0.1046
T/K	298.75	298.55	298.53	298.24	298.24	298.02	297.70

Polymer (B): **polystyrene** **1999CHO3**
Characterization: M_n/g.mol^{-1} = 87200, M_w/g.mol^{-1} = 171300,
Aldrich Chem. Co., Inc., Milwaukee, WI
Solvent (A): **cyclohexane** **C_6H_{12}** **110-82-7**

Type of data: cloud points (UCST-behavior)

φ_B	0.0164	0.0238	0.0325	0.0425	0.0533	0.0615	0.0746	0.0939	0.1078
T/K	298.34	298.79	298.98	298.85	298.75	298.65	298.34	297.94	297.58

Type of data: cloud points (LCST-behavior)

φ_B	0.0164	0.0238	0.0325	0.0425	0.0533	0.0615	0.0746	0.0939	0.1078
T/K	497.93	497.73	496.75	496.76	496.99	497.14	497.64	498.21	499.69

Polymer (B): **polystyrene** **1997KIT**
Characterization: M_n/g.mol^{-1} = 86000, M_w/g.mol^{-1} = 239000,
Styron 666, Asahi Chemical Industry Co., Ltd., Japan
Solvent (A): **cyclohexane** **C_6H_{12}** **110-82-7**

Type of data: cloud points (UCST-behavior)

φ_B	0.0006	0.0012	0.0020	0.0035	0.0054	0.0074	0.0101	0.0152	0.0177
T/K	297.52	298.23	298.78	299.50	299.99	300.16	300.40	300.56	300.53

φ_B	0.0202	0.0296	0.0403	0.0502	0.0550	0.0603	0.0648	0.0701	0.0749
T/K	300.44	300.24	300.00	299.75	299.63	299.52	299.45	299.37	299.25

φ_B	0.0803	0.0906	0.1008	0.1109	0.1211	0.1306	0.1410	0.1509
T/K	299.17	298.98	298.73	298.49	298.24	297.96	297.61	297.22

Type of data: critical point (UCST-behavior)

$\varphi_{B,\,crit}$ 0.0694 T_{crit}/K 299.38

Type of data: coexistence data (UCST-behavior)

φ_B (total)	0.0694	(critical concentration)						
T/K	299.38	299.352	299.347	299.337	299.317	299.293	299.251	299.198
φ_B (sol phase)	0.0694	0.0565	0.0543	0.0542	0.0525	0.0496	0.0472	0.0450
φ_B (gel phase)	0.0694	0.0825	0.0847	0.0857	0.0879	0.0906	0.0947	0.0977

continued

continued

T/K	299.116	298.996	298.837	298.641	298.231	297.726	296.984
φ_B (sol phase)	0.0416	0.0386	0.0352	0.0330	0.0271	0.0228	0.0186
φ_B (gel phase)	0.1023	0.1077	0.1145	0.1218	0.1344	0.1474	0.1643

Polymer (B): **polystyrene** **1972KEN**
Characterization: M_n/g.mol^{-1} = 91000, M_w/g.mol^{-1} = 93000, M_z/g.mol^{-1} = 96000, Pressure Chemical Company, Pittsburgh, PA
Solvent (A): **cyclohexane** C_6H_{12} **110-82-7**

Type of data: critical point (UCST-behavior)

$w_{B, crit}$ 0.117 T_{crit}/K 293.65

Polymer (B): **polystyrene** **1973SA1**
Characterization: M_n/g.mol^{-1} = 91700, M_w/g.mol^{-1} = 97200, Pressure Chemical Company, Pittsburgh, PA
Solvent (A): **cyclohexane** C_6H_{12} **110-82-7**

Type of data: cloud points (UCST-behavior)

w_B	0.0280	0.0527	0.0775	0.1055	0.1258	0.1523
T/K	291.86	293.17	293.51	293.64	293.48	293.36

Type of data: cloud points (LCST-behavior)

w_B	0.0224	0.0428	0.0638	0.0879	0.1048	0.1288
T/K	503.81	502.65	502.21	501.91	502.15	502.55

Polymer (B): **polystyrene** **1991BAE**
Characterization: M_n/g.mol^{-1} = 94300, M_w/g.mol^{-1} = 100000, Pressure Chemical Company, Pittsburgh, PA
Solvent (A): **cyclohexane** C_6H_{12} **110-82-7**

Type of data: cloud points (UCST-behavior)

w_B	0.005	0.010	0.030	0.049	0.063	0.081	0.103	0.144	0.197
T/K	288.45	290.05	292.95	293.45	293.95	293.95	293.75	293.65	293.05

w_B	0.300
T/K	291.45

Polymer (B): **polystyrene** **1993BAE**
Characterization: M_n/g.mol^{-1} = 94000, M_w/g.mol^{-1} = 100000, Pressure Chemical Company, Pittsburgh, PA
Solvent (A): **cyclohexane** C_6H_{12} **110-82-7**

Type of data: cloud points (LCST-behavior)

w_B	0.005	0.007	0.010	0.020	0.030	0.050	0.072	0.102	0.152
T/K	511.95	510.25	508.25	504.75	502.95	503.85	502.25	502.15	501.95

w_B	0.191	0.251
T/K	502.35	503.55

Polymer (B): **polystyrene** **1999BUN**
Characterization: M_n/g.mol^{-1} = 93000, M_w/g.mol^{-1} = 101400,
M_z/g.mol^{-1} = 111900, BASF AG, Germany
Solvent (A): **cyclohexane** C_6H_{12} **110-82-7**

Type of data: cloud points

w_B	0.114	was kept constant						
T/K	496.35	498.35	501.45	505.45	511.05	515.15	522.45	530.95
P/bar	20.6	23.5	27.6	33.9	43.6	51.0	61.0	74.0

Type of data: coexistence data

T/K = 530.95

The total feed concentration of the homogeneous system before demixing is: w_B = 0.1138.

P/ bar	w_B Gel phase	w_B Sol phase	
74.0		0.1138	(cloud point)
63.8	0.2553	0.0184	
55.5	0.2749	0.0073	
41.9	0.3500		
32.0	0.3600		

Polymer (B): **polystyrene** **2002HOR**
Characterization: M_n/g.mol^{-1} = 93000, M_w/g.mol^{-1} = 101400,
M_z/g.mol^{-1} = 111900, BASF AG, Germany
Solvent (A): **cyclohexane** C_6H_{12} **110-82-7**

Type of data: cloud points

w_B	0.107	was kept constant							
T/K	496.09	498.61	501.22	503.56	506.14	508.65	511.10	513.56	516.10
P/bar	24	26	32	34	38	41	44	48	51

Polymer (B): **polystyrene** **1999CHO1, 1999CHO2**
Characterization: M_n/g.mol^{-1} = 98100, M_w/g.mol^{-1} = 209000,
Aldrich Chem. Co., Inc., Milwaukee, WI
Solvent (A): **cyclohexane** C_6H_{12} **110-82-7**

Type of data: cloud points (UCST-behavior)

φ_B	0.0035	0.0082	0.0220	0.0319	0.0368	0.0416	0.0510	0.0550	0.0598
T/K	299.43	299.79	300.49	300.75	300.69	300.30	299.98	300.13	299.63

φ_B	0.0705	0.0891	0.0914	0.1207	0.1287	0.1507
T/K	299.34	299.11	299.07	298.13	297.85	297.49

Polymer (B): **polystyrene** **1999CHO3**
Characterization: M_n/g.mol^{-1} = 98100, M_w/g.mol^{-1} = 209000,
Aldrich Chem. Co., Inc., Milwaukee, WI
Solvent (A): **cyclohexane** **C_6H_{12}** **110-82-7**

Type of data: cloud points (UCST-behavior)

φ_B	0.0092	0.0185	0.0312	0.0357	0.0479	0.0628	0.0734	0.0881	0.1171
T/K	299.21	299.64	300.22	300.25	300.02	299.73	299.52	298.98	297.69

Type of data: cloud points (LCST-behavior)

φ_B	0.0092	0.0185	0.0312	0.0357	0.0479	0.0628	0.0734	0.0881	0.1171
T/K	495.62	494.76	493.25	493.27	493.85	494.87	495.43	496.26	497.45

Polymer (B): **polystyrene** **1981HAS**
Characterization: M_n/g.mol^{-1} = 102000, M_w/g.mol^{-1} = 103000,
M_z/g.mol^{-1} = 104000, Toyo Soda Manufacturing Co., Japan
Solvent (A): **cyclohexane** **C_6H_{12}** **110-82-7**

Type of data: coexistence data (UCST-behavior)

T/K	287.15	290.15	293.79
φ_B (sol phase)	0.0023	0.0068	0.115
φ_B (gel phase)	0.2899	0.2364	0.115

Type of data: critical point (UCST-behavior)

$\varphi_{B, crit}$ 0.115 T_{crit}/K 293.79

Polymer (B): **polystyrene** **1984GIL**
Characterization: M_n/g.mol^{-1} = 100000, M_w/g.mol^{-1} = 3240000
Solvent (A): **cyclohexane** **C_6H_{12}** **110-82-7**

Type of data: cloud points (UCST-behavior)

w_B	0.01230	0.01393	0.01894	0.02092	0.02570	0.02817	0.03738	0.03959	0.05044
T/K	301.00	301.08	301.22	301.44	301.37	301.27	301.25	301.13	301.06
w_B	0.05765	0.06851	0.07520	0.08377	0.08731	0.08908	0.09010	0.09312	0.09429
T/K	300.84	300.61	300.48	300.23	300.17	300.10	300.03	300.04	300.02
w_B	0.09730	0.10456	0.11328	0.12379	0.13574	0.15016	0.16344	0.17088	0.19137
T/K	299.94	299.69	299.48	299.54	299.14	298.93	298.55	298.25	297.92

Type of data: critical point (UCST-behavior)

$w_{B, crit}$ 0.0943 T_{crit}/K 300.00

Type of data: coexistence data (UCST-behavior)

w_B (total)	0.09943	0.09943	0.09943	0.09943	0.09943	0.09943	0.09943	0.09943
T/K	299.85	299.63	298.65	298.15	296.15	295.65	294.15	293.15
w_B (sol phase)	0.07349	0.06066	0.04140	0.03588		0.02115	0.01618	0.01420
w_B (gel phase)	0.11237	0.13594	0.17918	0.19723	0.24798	0.26108	0.28886	

continued

continued

Demixing temperature	Fractionation during demixing	
	Sol phase	Gel phase
T/K	M_n/g mol^{-1}	M_n/g mol^{-1}
299.85	90610	107370
299.63	87530	114000
298.65	73580	121600
298.15	69200	127610
296.15		137030
295.65	55820	139780
294.15	42040	
293.15	33240	

w_B (total)	0.12544	0.12544	0.12544	0.12544	0.12544	0.12544	0.12544	0.12544
T/K	299.15	298.65	298.15	297.68	297.15	296.65	296.15	293.65
w_B (sol phase)	0.05803	0.04840	0.04115	0.03487	0.03210	0.02983	0.02745	0.01618
w_B (gel phase)	0.14431	0.16683	0.18647	0.20410	0.21815	0.22986	0.24504	0.29719
w_B (total)	0.15054	0.15054	0.15054	0.15054	0.15054	0.15054	0.15054	0.15054
T/K	298.33	298.12	297.92	297.63	297.15	296.13	295.12	294.12
w_B (sol phase)	0.06338	0.05504	0.05231	0.04696	0.04240	0.03286	0.02641	0.02267
w_B (gel phase)	0.16400	0.17290	0.17897	0.19020	0.20786	0.23656	0.26402	0.28600

Polymer (B): **polystyrene** **1986KRU**
Characterization: M_n/g.mol^{-1} = 105700, M_w/g.mol^{-1} = 111000
Solvent (A): **cyclohexane** C_6H_{12} **110-82-7**

Type of data: cloud points (UCST-behavior)

w_B	0.041	0.060	0.081	0.104	0.105	0.110	0.119	0.125	0.135
T/K	293.78	294.18	294.37	294.47	294.49	294.48	294.45	294.44	294.42

Polymer (B): **polystyrene** **2003SIP**
Characterization: M_n/g.mol^{-1} = 116000, M_w/g.mol^{-1} = 123000,
Pressure Chemical Company, Pittsburgh, PA
Solvent (A): **cyclohexane** C_6H_{12} **110-82-7**

Type of data: cloud points (UCST-behavior)

φ_B	0.009	0.014	0.021	0.024	0.030	0.037	0.048	0.055	0.064
T/K	291.69	292.80	293.32	293.55	293.78	294.11	294.33	294.46	294.56
φ_B	0.068	0.073	0.080	0.087	0.091	0.094	0.102	0.105	0.108
T/K	294.57	294.58	294.59	294.60	294.61	294.62	294.61	294.60	294.59

continued

continued

φ_B	0.118	0.122	0.129	0.144	0.155	0.170
T/K	294.57	294.54	294.42	294.18	293.97	293.45

Type of data: critical point (UCST-behavior)

$\varphi_{B, crit}$ 0.093 T_{crit}/K 294.61

Polymer (B): **polystyrene** **1960DE2**
Characterization: M_n/g.mol^{-1} = 118000, M_w/g.mol^{-1} = 124000,
Dow Chemical Company, Midland
Solvent (A): **cyclohexane** C_6H_{12} **110-82-7**

Type of data: critical point (UCST-behavior)

$\varphi_{B, crit}$ 0.056 T_{crit}/K 294.97

Polymer (B): **polystyrene** **1986KRU**
Characterization: M_n/g.mol^{-1} = 130400, M_w/g.mol^{-1} = 142100
Solvent (A): **cyclohexane** C_6H_{12} **110-82-7**

Type of data: cloud points (UCST-behavior)

w_B	0.01558	0.02061	0.02986	0.04418	0.04856	0.06062	0.06598	0.08358	0.08820
T/K	292.13	292.72	293.28	293.73	293.75	293.80	293.85	293.69	293.75

w_B	0.09387	0.10578	0.10646	0.10958	0.11303	0.11787	0.15582	0.19295
T/K	293.62	293.58	293.52	293.48	293.42	293.29	292.80	292.43

Type of data: critical point (UCST-behavior)

$w_{B, crit}$ 0.113 T_{crit}/K 293.41

Type of data: coexistence data (UCST-behavior)

w_B (total)	0.11787	0.11787	0.11787	0.11787	0.11787	0.11787	0.11787
T/K	293.25	293.15	293.06	292.95	292.67	292.15	291.65
w_B (sol phase)	0.0812	0.0686	0.0607	0.0564	0.0452	0.0361	0.0366
w_B (gel phase)	0.1488	0.1651	0.1768	0.1786	0.1985	0.2165	0.2309

Polymer (B): **polystyrene** **1998TER**
Characterization: M_w/g.mol^{-1} = 155000, 4 arms
Solvent (A): **cyclohexane** C_6H_{12} **110-82-7**

Type of data: critical point (UCST-behavior)

$\varphi_{B, crit}$ 0.0832 T_{crit}/K 294.13

Type of data: coexistence data (UCST-behavior)

T/K	294.06	294.04	293.85	293.65	293.35	292.95	292.33	291.63
φ_B (sol phase)	0.0603	0.0528	0.0399	0.0310	0.0240	0.0184	0.0134	0.0101
φ_B (gel phase)	0.1077	0.1183	0.1332	0.1494	0.1647	0.1817	0.2024	0.2190

continued

continued

T/K	290.76	289.61
φ_B (sol phase)	0.0072	0.0022
φ_B (gel phase)	0.2398	0.2637

Comments: The total feed concentration of the polymer is near the critical concentration.

Type of data: spinodal points (UCST-behavior)

φ_B	0.02341	0.03475	0.04694	0.05892	0.07185	0.08319	0.09495	0.10750	0.12000
T/K	290.23	292.78	293.56	293.99	294.16	294.07	293.94	293.90	293.90

φ_B	0.13720	0.14940	0.17990	0.18750
T/K	293.26	292.74	290.74	290.36

Polymer (B): **polystyrene** **2006GAR**
Characterization: M_n/g.mol^{-1} = 130000, M_w/g.mol^{-1} = 286000
Solvent (A): **cyclohexane** C_6H_{12} **110-82-7**

Type of data: cloud points (UCST-behavior)

w_B	0.0094	0.0130	0.0130	0.0194	0.0194	0.0227	0.0227	0.0480	0.0480
T/K	297.07	297.95	298.70	298.95	299.02	298.38	299.00	298.35	298.45

Polymer (B): **polystyrene** **1992HE2**
Characterization: M_n/g.mol^{-1} = 128000, M_w/g.mol^{-1} = 320000, BASF AG, Germany
Solvent (A): **cyclohexane** C_6H_{12} **110-82-7**

Type of data: coexistence data (UCST-behavior)

T/K = 299.15

w_B (total)	0.050	0.092	0.120
w_B (sol phase)	0.028	0.043	0.049
w_B (gel phase)	0.179	0.167	0.1156

Polymer (B): **polystyrene** **1960DE2**
Characterization: M_n/g.mol^{-1} = 147000, M_w/g.mol^{-1} = 153000, Dow Chemical Company, Midland
Solvent (A): **cyclohexane** C_6H_{12} **110-82-7**

Type of data: critical point (UCST-behavior)

$\varphi_{B,\ crit}$ 0.050 T_{crit}/K 296.34

Polymer (B): **polystyrene** **1990STR**
Characterization: M_n/g.mol^{-1} = 153700, M_w/g.mol^{-1} = 166000, Pressure Chemical Company, Pittsburgh, PA
Solvent (A): **cyclohexane** C_6H_{12} **110-82-7**

Type of data: spinodal points (UCST-behavior)

continued

continued

w_B	0.0186	0.0407	0.0620	0.0821	0.1012	0.1243	0.1496	0.0617	0.0817
T/K	289.95	294.25	296.05	296.35	296.45	296.25	295.55	296.05	296.35

w_B	0.1008	0.1230	0.1489
T/K	296.45	296.25	295.55

Polymer (B): **polystyrene** **1971SCH, 1972KEN**
Characterization: M_n/g.mol^{-1} = 154000, M_w/g.mol^{-1} = 163000,
Pressure Chemical Company, Pittsburgh, PA
Solvent (A): **cyclohexane** C_6H_{12} **110-82-7**

Type of data: spinodal points (UCST-behavior)

w_B	0.0186	0.0407	0.0620	0.0821	0.1012	0.1243	0.1496
T/K	289.95	294.25	296.05	296.35	296.45	296.25	295.55

Type of data: critical point (UCST-behavior)

$w_{B, crit}$ 0.099 T_{crit}/K 296.6

Polymer (B): **polystyrene** **1987PER**
Characterization: M_w/g.mol^{-1} = 171000
Solvent (A): **cyclohexane** C_6H_{12} **110-82-7**

Type of data: cloud points (UCST-behavior)

c_B/(mg/ml)	1.07	1.07	0.921	0.922	0.368	0.353	0.114	0.114
T/K	291.15	291.05	290.75	290.40	289.95	289.15	287.15	286.95

c_B/(mg/ml)	0.0556	0.0375	0.0197	0.0169
T/K	286.15	285.65	284.40	284.00

Polymer (B): **polystyrene** **1997KIT**
Characterization: M_n/g.mol^{-1} = 185300, M_w/g.mol^{-1} = 189000,
Styron 666, Asahi Chemical Industry Co., Ltd., Japan
Solvent (A): **cyclohexane** C_6H_{12} **110-82-7**

Type of data: critical point (UCST-behavior)

$\varphi_{B, crit}$ 0.0669 T_{crit}/K 296.83

Type of data: coexistence data (UCST-behavior)

φ_B (total) 0.0669 (critical concentration)

T/K	296.83	296.804	296.786	296.760	296.718	296.671	296.590	296.521
φ_B (sol phase)	0.0669	0.0477	0.0456	0.0418	0.0376	0.0353	0.0309	0.0287
φ_B (gel phase)	0.0669	0.0863	0.0886	0.0932	0.0981	0.1020	0.1095	0.1128

T/K	296.276	295.977	295.670	294.845
φ_B (sol phase)	0.0226	0.0176	0.0135	0.0081
φ_B (gel phase)	0.1249	0.1374	0.1477	0.1705

Polymer (B): **polystyrene** **1996RON**
Characterization: M_n/g.mol^{-1} = 183000, M_w/g.mol^{-1} = 190000,
Tosoh Corp., Ltd., Tokyo, Japan
Solvent (A): **cyclohexane** **C_6H_{12}** **110-82-7**

Type of data: cloud points (UCST-behavior)

w_B	0.0138	0.0245	0.0306	0.0322	0.0401	0.0488	0.0609	0.0749	0.0826
T/K	295.52	296.25	296.53	296.61	296.85	296.97	297.10	297.15	297.17
w_B	0.0867	0.1051	0.1096	0.1198	0.1477	0.1537	0.1663	0.1692	0.2318
T/K	297.16	297.10	297.07	297.00	296.75	296.67	296.47	296.45	295.27

Type of data: critical point (UCST-behavior)

$w_{B, crit}$ 0.0826 T_{crit}/K 297.17

Type of data: spinodal points (UCST-behavior)

w_B	0.0343	0.0479	0.0720	0.0800	0.0900	0.0982	0.1032	0.1209	0.1370
T/K	294.05	295.85	297.05	297.15	297.05	296.95	296.75	296.05	295.25

Polymer (B): **polystyrene** **1999OCH**
Characterization: M_n/g.mol^{-1} = 183000, M_w/g.mol^{-1} = 190000,
Tosoh Corp., Ltd., Tokyo, Japan
Solvent (A): **cyclohexane** **C_6H_{12}** **110-82-7**

Type of data: cloud points (UCST-behavior)

w_B	0.0022	0.0051	0.0065	0.0098	0.0132	0.0207	0.0294	0.0661	0.1014
T/K	291.48	293.15	293.45	294.55	294.86	295.60	296.22	297.16	297.35
w_B	0.1245	0.1685	0.2258						
T/K	297.13	294.89	292.73						

Polymer (B): **polystyrene** **1973SA1**
Characterization: M_n/g.mol^{-1} = 189000, M_w/g.mol^{-1} = 200000,
Pressure Chemical Company, Pittsburgh, PA
Solvent (A): **cyclohexane** **C_6H_{12}** **110-82-7**

Type of data: cloud points (UCST-behavior)

w_B	0.0318	0.0574	0.0971	0.1026	0.1338	0.1541
T/K	296.43	296.90	297.07	296.98	296.87	296.48

Type of data: cloud points (LCST-behavior)

w_B	0.0270	0.0479	0.0824	0.0883	0.1116	0.1277
T/K	497.45	497.13	497.25	497.18	497.82	498.52

Polymer (B): **polystyrene** **1975NAK**
Characterization: M_n/g.mol^{-1} = 196000, M_w/g.mol^{-1} = 200000,
Pressure Chemical Company, Pittsburgh, PA

continued

continued

Solvent (A): **cyclohexane** C_6H_{12} **110-82-7**

Type of data: coexistence data (UCST-behavior)

T_{crit}/K = 296.99

φ_B (total)	0.0618	0.0618	0.0618	0.0618	0.0618	0.0618	0.0618	0.0618
$(T_{crit} - T)$/K	1.097	0.642	0.400	0.242	0.143	0.083	0.050	0.034
φ_B (sol phase)	0.0162	0.0222	0.0269	0.0325	0.0372	0.0429	0.0464	0.0489
φ_B (gel phase)	0.1515	0.1344	0.1229	0.1124	0.1051	0.0979	0.0929	0.0897
φ_B (total)	0.0618	0.0618	0.0618	0.0618	0.0618	0.0688	0.0688	0.0688
$(T_{crit} - T)$/K	0.022	0.012	0.008	0.005	0.003	1.985	1.145	0.972
φ_B (sol phase)	0.0515	0.0551	0.0566	0.0590	0.0603	0.0088	0.0137	0.0169
φ_B (gel phase)	0.0863	0.0824	0.0803	0.0791	0.0769	0.1782	0.1532	0.1489
φ_B (total)	0.0688	0.0688	0.0688	0.0688	0.0688	0.0688	0.0688	0.0688
$(T_{crit} - T)$/K	0.794	0.586	0.394	0.284	0.227	0.151	0.123	0.102
φ_B (sol phase)	0.0197	0.0229	0.0267	0.0308	0.0333	0.0374	0.0385	0.0404
φ_B (gel phase)	0.1412	0.1323	0.1237	0.1162	0.1118	0.1062	0.1030	0.1006
φ_B (total)	0.0688	0.0688	0.0688	0.0688	0.0688	0.0688	0.0715	0.0715
$(T_{crit} - T)$/K	0.068	0.044	0.029	0.017	0.010	0.003	1.984	1.435
φ_B (sol phase)	0.0436	0.0466	0.0494	0.0528	0.0556	0.0605	0.0113	0.0137
φ_B (gel phase)	0.0964	0.0919	0.0891	0.0848	0.0827	0.0778	0.1701	0.1598
φ_B (total)	0.0715	0.0715	0.0715	0.0715	0.0715	0.0715	0.0715	0.0715
$(T_{crit} - T)$/K	1.088	0.873	0.667	0.518	0.371	0.269	0.170	0.111
φ_B (sol phase)	0.0167	0.0197	0.0226	0.0250	0.0269	0.0314	0.0355	0.0389
φ_B (gel phase)	0.1504	0.1423	0.1329	0.1282	0.1207	0.1150	0.1073	0.1017
φ_B (total)	0.0715	0.0715	0.0715	0.0715	0.0715	0.0715	0.0715	0.0715
$(T_{crit} - T)$/K	0.087	0.068	0.058	0.047	0.031	0.020	0.012	0.008
φ_B (sol phase)	0.0412	0.0432	0.0444	0.0457	0.0487	0.0515	0.0543	0.0558
φ_B (gel phase)	0.0983	0.0962	0.0942	0.0923	0.0897	0.0863	0.0833	0.0812
φ_B (total)	0.0715	0.0752	0.0752	0.0752	0.0752	0.0752	0.0752	0.0752
$(T_{crit} - T)$/K	0.002	1.964	1.370	1.020	0.758	0.556	0.402	0.259
φ_B (sol phase)	0.0615	0.0077	0.0132	0.0165	0.0201	0.0237	0.0274	0.0318
φ_B (gel phase)	0.0769	0.1786	0.1613	0.1491	0.1391	0.1301	0.1222	0.1145
φ_B (total)	0.0752	0.0752	0.0752	0.0752	0.0752	0.0752	0.0752	
$(T_{crit} - T)$/K	0.144	0.086	0.053	0.033	0.020	0.014	0.011	
φ_B (sol phase)	0.0365	0.0417	0.0451	0.0481	0.0511	0.0532	0.0545	
φ_B (gel phase)	0.1049	0.0987	0.0940	0.0904	0.0874	0.0846	0.0835	

Polymer (B): **polystyrene** **1973KU1, 1973KU2**

Characterization: M_n/g.mol^{-1} = 196000, M_w/g.mol^{-1} = 200000,
Pressure Chemical Company, Pittsburgh, PA

Solvent (A): **cyclohexane** C_6H_{12} **110-82-7**

Type of data: cloud points (UCST-behavior)

continued

continued

φ_B	0.0779	T/K	296.99	(threshold point)
φ_B	0.0799	T/K	296.93	(critical point)

Polymer (B): **polystyrene** **1973KU1**
Characterization: M_n/g.mol^{-1} = 197000, M_w/g.mol^{-1} = 206000, Pressure Chemical Company, Pittsburgh, PA
Solvent (A): **cyclohexane** C_6H_{12} **110-82-7**

Type of data: cloud points (UCST-behavior)

φ_B	0.0680	T/K	297.09	(threshold point)
φ_B	0.0781	T/K	297.07	(critical point)

Polymer (B): **polystyrene** **1972KEN**
Characterization: M_n/g.mol^{-1} = 200000, M_w/g.mol^{-1} = 286000, M_z/g.mol^{-1} = 438000, Pressure Chemical Company, Pittsburgh, PA
Solvent (A): **cyclohexane** C_6H_{12} **110-82-7**

Type of data: critical point (UCST-behavior)

$w_{B,\text{crit}}$ 0.095 T_{crit}/K 298.7

Polymer (B): **polystyrene** **1986KRU**
Characterization: M_n/g.mol^{-1} = 198500, M_w/g.mol^{-1} = 335500
Solvent (A): **cyclohexane** C_6H_{12} **110-82-7**

Type of data: cloud points (UCST-behavior)

w_B	0.0297	0.0386	0.0388	0.0495	0.0580	0.0743	0.0785	0.0810	0.0840
T/K	298.31	298.37	298.30	298.29	298.21	298.08	297.97	298.05	297.99
w_B	0.0870	0.0940	0.0995	0.1089	0.1264	0.1288	0.1450		
T/K	297.97	297.81	297.79	297.65	297.42	297.30	297.10		

Type of data: critical point (UCST-behavior)

$w_{B,\text{crit}}$ 0.087 T_{crit}/K 297.95

Polymer (B): **polystyrene** **1998TER**
Characterization: M_w/g.mol^{-1} = 329000, 4 arms
Solvent (A): **cyclohexane** C_6H_{12} **110-82-7**

Type of data: critical point (UCST-behavior)

$\varphi_{B,\text{crit}}$ 0.0590 T_{crit}/K 297.92

Type of data: coexistence data (UCST-behavior)

T/K	297.85	297.79	297.66	297.45	297.22	296.86	296.36	295.65
φ_B (sol phase)	0.0386	0.0323	0.0260	0.0198	0.0138	0.0089	0.0051	0.0033
φ_B (gel phase)	0.0769	0.0879	0.1001	0.1146	0.1268	0.1389	0.1558	0.1745

continued

continued

T/K	294.77	293.75
φ_B (sol phase)	0.0008	0.0006
φ_B (gel phase)	0.1954	0.2137

Comments: The total feed concentration of the polymer is near the critical concentration.

Type of data: spinodal points (UCST-behavior)

φ_B	0.01223	0.02368	0.03617	0.04669	0.05730	0.06932	0.08251	0.09551
T/K	293.90	296.74	297.49	297.80	298.02	297.75	297.57	297.31

Polymer (B): **polystyrene** **1965REH**
Characterization: M_n/g.mol^{-1} = 210000, M_w/g.mol^{-1} = 346000
Solvent (A): **cyclohexane** C_6H_{12} **110-82-7**

Type of data: cloud points (UCST-behavior)

w_B	0.01	0.02	0.03	0.04	0.05	0.06	0.07	0.08	0.09
T/K	300.65	301.05	301.05	300.95	300.85	300.75	300.65	300.45	300.25

w_B	0.10	0.13	0.15	0.18	0.20
T/K	300.05	299.65	299.35	298.55	298.05

Type of data: coexistence data (UCST-behavior)

T/K	292.15	293.15	294.15	295.15	296.15	297.15	298.15	299.15
w_B (total)	0.02	0.02	0.02	0.02	0.02	0.02	0.02	0.02
w_B (sol phase)	0.0034	0.0036	0.0039	0.0052	0.0058	0.0075	0.0100	0.0126
w_B (gel phase)		0.331		0.289		0.247		0.194

T/K	300.15	292.15	293.15	294.15	295.15	296.15	297.15	298.15
w_B (total)	0.02	0.06	0.06	0.06	0.06	0.06	0.06	0.06
w_B (sol phase)	0.0147	0.0050	0.0058	0.0073	0.0090	0.0115	0.0145	0.0190
w_B (gel phase)	0.168		0.329		0.287		0.243	

T/K	299.15	300.15	300.65	292.15	293.15	294.15	295.15	296.15
w_B (total)	0.06	0.06	0.06	0.10	0.10	0.10	0.10	0.10
w_B (sol phase)	0.0255	0.0340	0.0507	0.0050	0.0072	0.0090	0.0106	0.0132
w_B (gel phase)	0.191	0.160		0.348	0.328	0.305	0.285	0.264

T/K	297.15	298.15	299.15	299.95	300.00	292.15	293.15	294.15
w_B (total)	0.10	0.10	0.10	0.10	0.10	0.15	0.15	0.15
w_B (sol phase)	0.0182	0.0238	0.0345	0.0510	0.0532	0.0070	0.0075	0.0115
w_B (gel phase)	0.244	0.216	0.183	0.142	0.126	0.346	0.326	0.304

T/K	295.15	296.15	297.15	298.15	298.65	292.15	293.15	294.15
w_B (total)	0.15	0.15	0.15	0.15	0.15	0.20	0.20	0.20
w_B (sol phase)	0.0120	0.0160	0.0196	0.0270	0.0320	0.0082	0.0093	0.0102
w_B (gel phase)	0.283	0.260	0.236	0.207		0.343	0.323	0.302

T/K	295.15	296.15	297.15	297.55
w_B (total)	0.20	0.20	0.20	0.20
w_B (sol phase)	0.0130	0.0180	0.0243	0.0280
w_B (gel phase)	0.280	0.256	0.233	0.220

continued

continued

Demixing temperature T/K	w_B total	Fractionation during demixing Sol phase M_n/g mol^{-1}	Gel phase M_n/g mol^{-1}
298.85	0.10	114000	243000
299.35	0.10	121000	221000
299.85	0.10	138000	229000

Polymer (B): **polystyrene** **1968RE1**
Characterization: M_n/g.mol^{-1} = 210000, M_w/g.mol^{-1} = 346000
Solvent (A): **cyclohexane** C_6H_{12} **110-82-7**

Type of data: cloud points (UCST-behavior)

w_B	0.025	T/K	301.05	(threshold point)
w_B	0.092	T/K	300.20	(critical point)

Polymer (B): **polystyrene** **1960DE2**
Characterization: M_n/g.mol^{-1} = 221000, M_w/g.mol^{-1} = 239000,
Dow Chemical Company, Midland
Solvent (A): **cyclohexane** C_6H_{12} **110-82-7**

Type of data: critical point (UCST-behavior)

$\varphi_{B, crit}$ 0.046 T_{crit}/K 298.10

Polymer (B): **polystyrene** **1960DE2**
Characterization: M_n/g.mol^{-1} = 248000, M_w/g.mol^{-1} = 253000,
Dow Chemical Company, Midland
Solvent (A): **cyclohexane** C_6H_{12} **110-82-7**

Type of data: critical point (UCST-behavior)

$\varphi_{B, crit}$ 0.043 T_{crit}/K 298.24

Polymer (B): **polystyrene** **1995IKI**
Characterization: M_n/g.mol^{-1} = 269000, M_w/g.mol^{-1} = 274000,
Polymer Standard Services GmbH, Mainz, Germany
Solvent (A): **cyclohexane** C_6H_{12} **110-82-7**

Type of data: critical point (UCST-behavior)

$\varphi_{B, crit}$ 0.063 T_{crit}/K 298.98

Polymer (B): **polystyrene** **1962DEB**
Characterization: M_n/g.mol^{-1} = 282000, M_w/g.mol^{-1} = 350000,
Dow Chemical Company, Midland
Solvent (A): **cyclohexane** **C_6H_{12}** **110-82-7**

Type of data: critical point (UCST-behavior)

$\varphi_{B, crit}$ 0.0125 T_{crit}/K 302.33

Polymer (B): **polystyrene** **1986KRU**
Characterization: M_n/g.mol^{-1} = 341400, M_w/g.mol^{-1} = 423300
Solvent (A): **cyclohexane** **C_6H_{12}** **110-82-7**

Type of data: cloud points (UCST-behavior)

w_B	0.01188	0.02453	0.03284	0.04382	0.05883	0.05891	0.07814	0.08003	0.08250
T/K	298.43	298.80	298.81	298.84	298.79	298.73	298.58	298.56	298.56
w_B	0.08665	0.08812	0.09654	0.10110	0.11438	0.12081	0.13236	0.14853	
T/K	298.50	298.38	298.35	298.18	297.95	297.80	297.48	296.98	

Type of data: critical point (UCST-behavior)

$w_{B, crit}$ 0.078 T_{crit}/K 298.85

Type of data: coexistence data (UCST-behavior)

w_B (total)	0.0800	0.0800	0.0800	0.0800	0.0800	0.0800	0.0800	0.0800
T/K	298.48	298.25	297.88	297.44	296.94	296.43	295.90	295.37
w_B (sol phase)	0.0506	0.0384	0.0288	0.0227	0.0172	0.0124	0.0098	0.0080
w_B (gel phase)	0.0978	0.1256	0.1463	0.1681	0.1921	0.2093	0.2240	0.2375

Polymer (B): **polystyrene** **1977WO3**
Characterization: M_w/g.mol^{-1} = 390000,
Pressure Chemical Company, Pittsburgh, PA
Solvent (A): **cyclohexane** **C_6H_{12}** **110-82-7**

Type of data: cloud points (UCST-behavior)

w_B	0.02064	0.03074	0.03812	0.04975	0.05983	0.06643	0.07302	0.08077	0.08460
T/K	299.15	299.44	299.73	299.77	299.87	300.02	300.04	300.12	300.16
w_B	0.09046	0.12030	0.12030	0.13618	0.15324				
T/K	300.08	299.92	299.94	299.74	299.78				

Type of data: critical point (UCST-behavior)

$w_{B, crit}$ 0.0846 T_{crit}/K 300.16

Polymer (B): **polystyrene** **1972KEN**
Characterization: M_n/g.mol^{-1} = 375000, M_w/g.mol^{-1} = 394000, M_z/g.mol^{-1} =
423000, Pressure Chemical Company, Pittsburgh, PA
Solvent (A): **cyclohexane** **C_6H_{12}** **110-82-7**

continued

continued

Type of data: critical point (UCST-behavior)

$w_{B, crit}$ 0.070 T_{crit}/K 300.7

Polymer (B): **polystyrene** **2003SIP**
Characterization: M_n/g.mol^{-1} = 377400, M_w/g.mol^{-1} = 400000,
Pressure Chemical Company, Pittsburgh, PA
Solvent (A): **cyclohexane** C_6H_{12} **110-82-7**

Type of data: cloud points (UCST-behavior)

φ_B	0.004	0.006	0.008	0.010	0.012	0.014	0.017	0.019	0.022
T/K	297.26	297.70	297.99	298.25	298.39	298.55	298.76	298.88	299.01
φ_B	0.024	0.027	0.030	0.032	0.035	0.038	0.041	0.044	0.048
T/K	299.09	299.17	299.21	299.23	299.26	299.28	299.30	299.33	299.36
φ_B	0.050	0.053	0.058	0.062	0.066	0.069	0.072	0.076	0.079
T/K	299.38	299.37	299.36	299.34	299.29	299.22	299.17	299.08	299.04
φ_B	0.081	0.084	0.086	0.089	0.092	0.094			
T/K	298.98	298.89	298.85	298.78	298.68	298.62			

Type of data: critical point (UCST-behavior)

$\varphi_{B, crit}$ 0.053 T_{crit}/K 299.37

Polymer (B): **polystyrene** **1973SA1**
Characterization: M_n/g.mol^{-1} = 378000, M_w/g.mol^{-1} = 400000,
Pressure Chemical Company, Pittsburgh, PA
Solvent (A): **cyclohexane** C_6H_{12} **110-82-7**

Type of data: cloud points (UCST-behavior)

w_B	0.0111	0.0226	0.0273	0.0405	0.0552	0.0817
T/K	299.22	299.98	300.00	300.18	300.36	300.24

Type of data: cloud points (LCST-behavior)

w_B	0.0043	0.0197	0.0249	0.0687
T/K	497.29	495.27	495.07	494.67

Polymer (B): **polystyrene** **1990GOE**
Characterization: M_w/g.mol^{-1} = 400000,
Pressure Chemical Company, Pittsburgh, PA
Solvent (A): **cyclohexane** C_6H_{12} **110-82-7**

Type of data: critical point (UCST-behavior)

$\varphi_{B, crit}$ 0.0512 T_{crit}/K 299.59

Polymer (B): **polystyrene** **1973KU1**
Characterization: M_n/g.mol^{-1} = 392000, M_w/g.mol^{-1} = 415000,
Pressure Chemical Company, Pittsburgh, PA
Solvent (A): **cyclohexane** C_6H_{12} **110-82-7**

Type of data: cloud points (UCST-behavior)

φ_B	0.0430	T/K	300.13	(threshold point)
φ_B	0.0523	T/K	300.12	(critical point)

Polymer (B): **polystyrene** **1987PER**
Characterization: M_w/g.mol^{-1} = 422000
Solvent (A): **cyclohexane** C_6H_{12} **110-82-7**

Type of data: cloud points (UCST-behavior)

c_B/(mg/ml)	1.16	0.943	0.522	0.522	0.588	0.149	0.149	0.0384
T/K	297.55	297.15	296.45	296.15	295.90	294.70	294.25	292.55

Polymer (B): **polystyrene** **1968RE1**
Characterization: M_n/g.mol^{-1} = 440000, M_w/g.mol^{-1} = 470000,
M_z/g.mol^{-1} = 620000
Solvent (A): **cyclohexane** C_6H_{12} **110-82-7**

Type of data: cloud points (UCST-behavior)

w_B	0.035	T/K	301.05	(threshold point)
w_B	0.084	T/K	300.75	(critical point)

Polymer (B): **polystyrene** **1971SCH, 1972KEN, 1990STR**
Characterization: M_n/g.mol^{-1} = 435000, M_w/g.mol^{-1} = 520000,
Pressure Chemical Company, Pittsburgh, PA
Solvent (A): **cyclohexane** C_6H_{12} **110-82-7**

Type of data: spinodal points (UCST-behavior)

w_B	0.0175	0.0289	0.0450	0.0540	0.0635	0.0832	0.1046
T/K	297.55	299.65	300.45	300.55	300.45	300.15	299.55

Polymer (B): **polystyrene** **1984TSU**
Characterization: M_n/g.mol^{-1} = 474000, M_w/g.mol^{-1} = 498000,
Toyo Soda Manufacturing Co., Japan
Solvent (A): **cyclohexane** C_6H_{12} **110-82-7**

Type of data: critical point (UCST-behavior)

$\varphi_{B, crit}$	0.047	T_{crit}/K	300.60

Polymer (B): **polystyrene** **1972KEN**
Characterization: M_n/g.mol^{-1} = 490000, M_w/g.mol^{-1} = 527000, M_z/g.mol^{-1} = 593000, Pressure Chemical Company, Pittsburgh, PA
Solvent (A): **cyclohexane** **C_6H_{12}** **110-82-7**

Type of data: critical point (UCST-behavior)

$w_{B,\,crit}$ 0.064 T_{crit}/K 301.2

Polymer (B): **polystyrene** **1960DE2**
Characterization: M_n/g.mol^{-1} = 522000, M_w/g.mol^{-1} = 569000, Dow Chemical Company, Midland
Solvent (A): **cyclohexane** **C_6H_{12}** **110-82-7**

Type of data: critical point (UCST-behavior)

$\varphi_{B,\,crit}$ 0.028 T_{crit}/K 300.94

Polymer (B): **polystyrene** **1981WOL**
Characterization: M_n/g.mol^{-1} = 545500, M_w/g.mol^{-1} = 600000, Pressure Chemical Company, Pittsburgh, PA
Solvent (A): **cyclohexane** **C_6H_{12}** **110-82-7**

Type of data: cloud points (UCST-behavior)

w_B	0.0143	0.0303	0.0422	0.0491	0.0501	0.0602	0.0718	0.0815
T/K	300.89	301.29	301.36	301.36	301.37	301.43	301.39	301.28

Type of data: cloud points (UCST-behavior)

w_B 0.060 was kept constant at the critical concentration

T/K	301.34	301.11	300.95	300.96	301.12	301.43	301.68
P/bar	1	37	88	152	187	260	300

Type of data: critical point (UCST-behavior)

$w_{B,\,crit}$ 0.060 T_{crit}/K 301.43

Polymer (B): **polystyrene** **1991BAE**
Characterization: M_n/g.mol^{-1} = 566000, M_w/g.mol^{-1} = 610000, Pressure Chemical Company, Pittsburgh, PA
Solvent (A): **cyclohexane** **C_6H_{12}** **110-82-7**

Type of data: cloud points (UCST-behavior)

w_B	0.005	0.010	0.020	0.029	0.049	0.069
T/K	300.45	301.15	301.25	301.45	301.15	300.75

Polymer (B): **polystyrene** **1986KRU**
Characterization: M_n/g.mol^{-1} = 601200, M_w/g.mol^{-1} = 769500
Solvent (A): **cyclohexane** **C_6H_{12}** **110-82-7**

continued

continued

Type of data: cloud points (UCST-behavior)

w_B	0.0139	0.0227	0.0302	0.0351	0.0426	0.0443	0.0514	0.0581	0.0754
T/K	300.40	300.57	300.48	300.57	300.43	300.45	300.44	300.44	299.97
w_B	0.0879	0.0943	0.0960	0.1120	0.1240	0.1254	0.1330	0.1406	
T/K	299.89	299.75	299.76	299.44	299.30	299.05	298.92	298.67	

Type of data: critical point (UCST-behavior)

$w_{B,\,crit}$ 0.057 T_{crit}/K 300.39

Type of data: coexistence data (UCST-behavior)

w_B (total)	0.0570	0.0570	0.0570	0.0570	0.0570	0.0570	0.0570	0.0570
T/K	300.19	300.18	300.14	300.09	300.00	299.92	299.88	299.77
w_B (sol phase)	0.0359	0.0346	0.0318	0.0293	0.0277	0.0224	0.0224	0.0216
w_B (gel phase)	0.0834	0.0870	0.0886	0.0950	0.0987	0.1108	0.1113	0.1170
w_B (total)	0.0570	0.0570	0.0570	0.0570	0.0570	0.0570		
T/K	299.57	299.47	299.36	299.07	298.66	298.17		
w_B (sol phase)	0.0164	0.0159	0.0147	0.0119	0.0105	0.0077		
w_B (gel phase)	0.1319	0.1305	0.1368	0.1454	0.1594	0.1722		

Polymer (B): **polystyrene** **1973SA1**
Characterization: M_n/g.mol^{-1} = 610000, M_w/g.mol^{-1} = 670000,
Pressure Chemical Company, Pittsburgh, PA
Solvent (A): **cyclohexane** C_6H_{12} **110-82-7**

Type of data: cloud points (UCST-behavior)

w_B	0.0078	0.0155	0.0294	0.0372	0.0676	0.0738
T/K	300.16	300.55	301.21	301.24	301.02	301.16

Type of data: cloud points (LCST-behavior)

w_B	0.0061	0.0119	0.0244	0.0302	0.0555	0.0620
T/K	491.86	492.14	491.64	491.79	492.21	492.49

Polymer (B): **polystyrene** **1973KU1**
Characterization: M_n/g.mol^{-1} = 667000, M_w/g.mol^{-1} = 680000,
Pressure Chemical Company, Pittsburgh, PA
Solvent (A): **cyclohexane** C_6H_{12} **110-82-7**

Type of data: cloud points (UCST-behavior)

φ_B	0.0410	T/K	301.177	(threshold point)
φ_B	0.0446	T/K	301.175	(critical point)

Polymer (B): **polystyrene** **1960DE2**
Characterization: M_n/g.mol^{-1} = 1000000, M_w/g.mol^{-1} = 1190000,
Dow Chemical Company, Midland

continued

continued

Solvent (A):	**cyclohexane**	**C_6H_{12}**	**110-82-7**

Type of data: critical point (UCST-behavior)

$\varphi_{B, crit}$ 0.020 T_{crit}/K 302.15

Polymer (B):	**polystyrene**		**1991YOK**
Characterization:	M_w/g.mol^{-1} = 1200000 (188000 per arm), star-shaped, 6.3 arms		
Solvent (A):	**cyclohexane**	**C_6H_{12}**	**110-82-7**

Type of data: critical point (UCST-behavior)

$\varphi_{B, crit}$ 0.0393 T_{crit}/K 301.15

Polymer (B):	**polystyrene**		**1987PER**
Characterization:	M_w/g.mol^{-1} = 1260000		
Solvent (A):	**cyclohexane**	**C_6H_{12}**	**110-82-7**

Type of data: cloud points (UCST-behavior)

c_B/(mg/ml)	1.43	0.628	0.602	0.416	0.255	0.159	0.138	0.0366
T/K	301.55	300.85	300.45	300.25	300.15	299.60	299.55	298.45

Polymer (B):	**polystyrene**		**1972KEN**
Characterization:	M_n/g.mol^{-1} = 1250000, M_w/g.mol^{-1} = 1500000, M_z/g.mol^{-1} = 1700000, Pressure Chemical Company, Pittsburgh, PA		
Solvent (A):	**cyclohexane**	**C_6H_{12}**	**110-82-7**

Type of data: critical point (UCST-behavior)

$w_{B, crit}$ 0.042 T_{crit}/K 303.2

Polymer (B):	**polystyrene**		**1998TER**
Characterization:	M_w/g.mol^{-1} = 1430000, 4 arms		
Solvent (A):	**cyclohexane**	**C_6H_{12}**	**110-82-7**

Type of data: critical point (UCST-behavior)

$\varphi_{B, crit}$ 0.0325 T_{crit}/K 302.80

Type of data: coexistence data (UCST-behavior)

T/K	302.73	302.72	302.68	302.67	302.63	302.47	302.40	302.15
φ_B (sol phase)	0.0134	0.0133	0.0121	0.0118	0.0113	0.0080	0.0080	0.0050
φ_B (gel phase)	0.0532	0.0540	0.0585	0.0588	0.0612	0.0697	0.0720	0.0799

T/K	301.93	301.35	301.14	300.45	299.35	298.05	297.10
φ_B (sol phase)	0.0050	0.0025	0.0018	0.0010	0.0001		
φ_B (gel phase)	0.0879	0.1053	0.1090	0.1257	0.1502	0.1753	0.1916

Comments: The total feed concentration of the polymer is near the critical concentration.

continued

continued

Type of data: spinodal points (UCST-behavior)

φ_B	0.00605	0.00990	0.01016	0.01368	0.01884	0.02550	0.03341	0.04081	0.04860
T/K	299.80	301.30	301.43	302.11	302.43	302.66	302.76	302.66	302.57

φ_B	0.05649	0.06412	0.07265	0.08418	0.09214
T/K	302.30	302.07	301.34	300.75	299.72

Polymer (B): **polystyrene** **1978NAK**
Characterization: M_n/g.mol^{-1} = 1515000, M_w/g.mol^{-1} = 1560000,
Pressure Chemical Company, Pittsburgh, PA
Solvent (A): **cyclohexane** C_6H_{12} **110-82-7**

Type of data: coexistence data (UCST-behavior)

T_{crit}/K = 303.65

φ_B (total)	0.0321	0.0321	0.0321	0.0321	0.0321	0.0321	0.0321	0.0321
$(T_{crit} - T)$/K	0.015	0.022	0.029	0.038	0.048	0.060	0.088	0.142
φ_B (sol phase)	0.0229	0.0218	0.0210	0.0203	0.0195	0.0184	0.0167	0.0146
φ_B (gel phase)	0.0412	0.0426	0.0439	0.0452	0.0460	0.0475	0.0499	0.0539

φ_B (total)	0.0321	0.0321	0.0321	0.0321	0.0321	0.0321	0.0321	0.0321
$(T_{crit} - T)$/K	0.209	0.280	0.402	0.550	0.750	0.995	1.382	1.976
φ_B (sol phase)	0.0127	0.0114	0.0092	0.0078	0.0065	0.0052	0.0041	0.0022
φ_B (gel phase)	0.0581	0.0616	0.0673	0.0733	0.0796	0.0881	0.0986	0.1131

φ_B (total)	0.0321	0.0321	0.0321	0.0321	0.0324	0.0324	0.0324	0.0324
$(T_{crit} - T)$/K	2.983	4.476	7.000	1.040	0.014	0.023	0.032	0.042
φ_B (sol phase)	0.0018	0.0012	0.0009	0.0009	0.0234	0.0217	0.0209	0.0199
φ_B (gel phase)	0.1360	0.1650	0.2096	0.2721	0.0413	0.0432	0.0443	0.0454

φ_B (total)	0.0324	0.0324	0.0324	0.0324	0.0324	0.0324	0.0324	0.0324
$(T_{crit} - T)$/K	0.065	0.082	0.100	0.125	0.182	0.256	0.390	0.530
φ_B (sol phase)	0.0181	0.0172	0.0164	0.0155	0.0134	0.0121	0.0094	0.0074
φ_B (gel phase)	0.0481	0.0498	0.0513	0.0532	0.0566	0.0609	0.0670	0.0730

φ_B (total)	0.0324	0.0324	0.0324	0.0324	0.0324	0.0324	0.0324	0.0324
$(T_{crit} - T)$/K	0.724	0.990	1.350	1.883	3.010	4.493	6.760	10.780
φ_B (sol phase)	0.0062	0.0047	0.0032	0.0019	0.0015	0.0011	0.0009	0.0009
φ_B (gel phase)	0.0794	0.0874	0.0972	0.1103	0.1366	0.1652	0.2083	0.2690

Polymer (B): **polystyrene** **2003SIP**
Characterization: M_n/g.mol^{-1} = 1538500, M_w/g.mol^{-1} = 2000000,
Pressure Chemical Company, Pittsburgh, PA
Solvent (A): **cyclohexane** C_6H_{12} **110-82-7**

Type of data: cloud points (UCST-behavior)

φ_B	0.003	0.004	0.006	0.007	0.009	0.011	0.012	0.014	0.015
T/K	301.83	302.33	302.51	302.65	302.71	302.76	302.80	302.83	302.84

continued

continued

φ_B	0.017	0.018	0.020	0.021	0.023	0.024	0.026	0.029	0.031
T/K	302.86	302.87	302.88	302.89	302.90	302.91	302.92	302.92	302.92
φ_B	0.033	0.035	0.037	0.038	0.040	0.043	0.045	0.047	0.050
T/K	302.91	302.90	302.88	302.86	302.82	302.78	302.73	302.64	302.58

Type of data: critical point (UCST-behavior)

$\varphi_{B, crit}$ 0.028 T_{crit}/K 302.92

Polymer (B): **polystyrene** **1991YOK**
Characterization: M_w/g.mol^{-1} = 2050000 (188000 per arm), star-shaped, 11.1 arms
Solvent (A): **cyclohexane** C_6H_{12} **110-82-7**

Type of data: critical point (UCST-behavior)

$\varphi_{B, crit}$ 0.0371 T_{crit}/K 301.55

Polymer (B): **polystyrene** **1973SA1**
Characterization: M_n/g.mol^{-1} = 2455000, M_w/g.mol^{-1} = 2700000,
Pressure Chemical Company, Pittsburgh, PA
Solvent (A): **cyclohexane** C_6H_{12} **110-82-7**

Type of data: cloud points (UCST-behavior)

w_B	0.0164	0.0234	0.0366	0.0522	0.0624	0.1062
T/K	304.03	304.06	304.24	304.06	303.87	302.74

Type of data: cloud points (LCST-behavior)

w_B	0.0158	0.0198	0.0301	0.0518
T/K	489.09	488.76	489.07	491.24

Polymer (B): **polystyrene** **1987PER**
Characterization: M_w/g.mol^{-1} = 3840000
Solvent (A): **cyclohexane** C_6H_{12} **110-82-7**

Type of data: cloud points (UCST-behavior)

c_B/(mg/ml)	1.23	0.643	0.565	0.362	0.127	0.166	0.0715	0.0715
T/K	303.55	303.15	302.80	302.80	302.30	302.10	302.00	301.65

Polymer (B): **polystyrene** **1987PER**
Characterization: M_w/g.mol^{-1} = 6770000
Solvent (A): **cyclohexane** C_6H_{12} **110-82-7**

Type of data: cloud points (UCST-behavior)

c_B/(mg/ml)	0.322	0.304	0.322	0.131	0.0568	0.0364
T/K	303.30	303.20	303.05	302.65	302.40	302.10

Polymer (B): **polystyrene** **2003SIP**
Characterization: M_n/g.mol^{-1} = 11680000, M_w/g.mol^{-1} = 13200000
Solvent (A): **cyclohexane** C_6H_{12} **110-82-7**

Type of data: cloud points (UCST-behavior)

φ_B	0.0010	0.0015	0.0024	0.0037	0.0057	0.0070	0.0084	0.0097	0.0112
T/K	304.18	304.62	304.86	304.98	305.12	305.16	305.25	305.30	305.32

φ_B	0.0130	0.0147	0.0179	0.0195	0.0209	0.0224
T/K	305.35	305.33	305.30	305.26	305.26	305.23

Type of data: critical point (UCST-behavior)

$\varphi_{B, crit}$ 0.014 T_{crit}/K 305.39

Polymer (B): **polystyrene** **1987PER**
Characterization: M_w/g.mol^{-1} = 20600000
Solvent (A): **cyclohexane** C_6H_{12} **110-82-7**

Type of data: cloud points (UCST-behavior)

c_B/(mg/ml)	0.135	0.0463	0.0479	0.0315	0.0187	0.00508	0.00155
T/K	304.50	304.25	303.70	303.50	303.35	303.30	302.40

Polymer (B): **polystyrene** **2006GAR**
Characterization: M_n/g.mol^{-1} = 3940, M_w/g.mol^{-1} = 4215
Solvent (A): **cyclohexanol** $C_6H_{12}O$ **108-93-0**

Type of data: cloud points (UCST-behavior)

w_B	0.0127	0.0197	0.0335	0.0405	0.0522	0.0870	0.1005	0.1252	0.1507
T/K	300.87	304.72	310.42	312.91	315.01	321.04	320.24	319.39	318.54

w_B	0.2026
T/K	317.91

Polymer (B): **polystyrene** **2006GAR**
Characterization: M_n/g.mol^{-1} = 70400, M_w/g.mol^{-1} = 78800
Solvent (A): **cyclohexanol** $C_6H_{12}O$ **108-93-0**

Type of data: cloud points (UCST-behavior)

w_B	0.0070	0.0111	0.0193	0.0195	0.0334	0.0413	0.0469	0.0839	0.0963
T/K	347.14	347.39	348.89	349.30	350.04	351.65	351.80	351.73	351.82

w_B	0.1214
T/K	351.95

Polymer (B): **polystyrene** **1963ORO**
Characterization: M_n/g.mol^{-1} = 404000, M_w/g.mol^{-1} = 406000
Solvent (A): **cyclohexanol** $C_6H_{12}O$ **108-93-0**

continued

continued

Type of data: cloud points (UCST-behavior)

w_B	0.01	T/K	358.15

Polymer (B): **polystyrene** **2006GAR**
Characterization: M_n/g.mol^{-1} = 585500, M_w/g.mol^{-1} = 849000
Solvent (A): **cyclohexanol** $C_6H_{12}O$ **108-93-0**

Type of data: cloud points (UCST-behavior)

w_B	0.0084	0.0190	0.0358	0.0399	0.0550
T/K	353.29	353.50	353.34	353.20	353.27

Polymer (B): **polystyrene** **1990IWA**
Characterization: M_n/g.mol^{-1} = 42600, M_w/g.mol^{-1} = 43000,
Tosoh Corp., Ltd., Tokyo, Japan
Solvent (A): **cyclopentane** C_5H_{10} **287-92-3**

Type of data: cloud points (UCST-behavior)

w_B	0.0426	0.0775	0.1186	0.1596	0.2014	0.2534
T/K	266.35	267.65	267.95	267.95	267.75	267.25

Polymer (B): **polystyrene** **1993IWA**
Characterization: M_n/g.mol^{-1} = 42600, M_w/g.mol^{-1} = 43000,
Tosoh Corp., Ltd., Tokyo, Japan
Solvent (A): **cyclopentane** C_5H_{10} **287-92-3**

Type of data: cloud points (LCST-behavior)

w_B	0.0307	0.0420	0.0821	0.1091	0.1482	0.1708	0.2071
T/K	456.65	455.55	454.35	454.25	454.05	454.15	454.65

Polymer (B): **polystyrene** **1990IWA**
Characterization: M_n/g.mol^{-1} = 97000, M_w/g.mol^{-1} = 98900,
Tosoh Corp., Ltd., Tokyo, Japan
Solvent (A): **cyclopentane** C_5H_{10} **287-92-3**

Type of data: cloud points (UCST-behavior)

w_B	0.01058	0.03922	0.08175	0.12450	0.16690	0.21450
T/K	271.85	274.75	275.45	275.45	275.15	274.35

Polymer (B): **polystyrene** **1993IWA**
Characterization: M_n/g.mol^{-1} = 97000, M_w/g.mol^{-1} = 98900,
Tosoh Corp., Ltd., Tokyo, Japan
Solvent (A): **cyclopentane** C_5H_{10} **287-92-3**

Type of data: cloud points (LCST-behavior)

continued

continued

w_B	0.0090	0.0312	0.0679	0.0972	0.1087	0.1507	0.1937
T/K	451.05	447.05	445.95	445.85	445.65	446.05	447.45

Polymer (B): **polystyrene** **1993IWA**
Characterization: M_n/g.mol^{-1} = 166900, M_w/g.mol^{-1} = 171900, Tosoh Corp., Ltd., Tokyo, Japan
Solvent (A): **cyclopentane** C_5H_{10} **287-92-3**

Type of data: cloud points (LCST-behavior)

w_B	0.0099	0.0154	0.0392	0.0552	0.0695	0.0885	0.1217	0.1518	0.1813
T/K	445.65	443.95	442.35	441.95	441.95	442.25	442.35	442.75	443.65

Polymer (B): **polystyrene** **1990IWA**
Characterization: M_n/g.mol^{-1} = 172000, M_w/g.mol^{-1} = 184000, Tosoh Corp., Ltd., Tokyo, Japan
Solvent (A): **cyclopentane** C_5H_{10} **287-92-3**

Type of data: cloud points (UCST-behavior)

w_B	0.01209	0.04137	0.08126	0.12310	0.16640	0.22050
T/K	275.95	278.55	278.75	278.65	278.05	276.75

Polymer (B): **polystyrene** **1990IWA**
Characterization: M_n/g.mol^{-1} = 347000, M_w/g.mol^{-1} = 354000, Tosoh Corp., Ltd., Tokyo, Japan
Solvent (A): **cyclopentane** C_5H_{10} **287-92-3**

Type of data: cloud points (UCST-behavior)

w_B	0.01041	0.03235	0.06277	0.09443	0.13360	0.18090
T/K	281.15	282.55	283.25	282.85	282.15	280.85

Polymer (B): **polystyrene** **1993IWA**
Characterization: M_n/g.mol^{-1} = 347000, M_w/g.mol^{-1} = 354000, Tosoh Corp., Ltd., Tokyo, Japan
Solvent (A): **cyclopentane** C_5H_{10} **287-92-3**

Type of data: cloud points (LCST-behavior)

w_B	0.0112	0.0200	0.0326	0.0609	0.0918	0.1169	0.1698
T/K	438.95	438.05	437.05	436.95	437.55	437.95	440.25

Polymer (B): **polystyrene** **1990IWA**
Characterization: M_n/g.mol^{-1} = 783000, M_w/g.mol^{-1} = 791000, Tosoh Corp., Ltd., Tokyo, Japan
Solvent (A): **cyclopentane** C_5H_{10} **287-92-3**

Type of data: cloud points (UCST-behavior)

continued

continued

w_B	0.00982	0.03196	0.05951	0.09395	0.12520	0.15660
T/K	285.25	286.05	285.95	285.55	284.85	283.75

Polymer (B): **polystyrene** **1993IWA**
Characterization: M_n/g.mol^{-1} = 783000, M_w/g.mol^{-1} = 791000,
Tosoh Corp., Ltd., Tokyo, Japan
Solvent (A): **cyclopentane** C_5H_{10} **287-92-3**

Type of data: cloud points (LCST-behavior)

w_B	0.0098	0.0182	0.0278	0.0419	0.0581	0.0931	0.1342
T/K	435.15	434.45	434.05	434.15	434.05	434.75	436.85

Polymer (B): **polystyrene** **1990IWA**
Characterization: M_n/g.mol^{-1} = 1238000, M_w/g.mol^{-1} = 1300000,
Tosoh Corp., Ltd., Tokyo, Japan
Solvent (A): **cyclopentane** C_5H_{10} **287-92-3**

Type of data: cloud points (UCST-behavior)

w_B	0.00912	0.02550	0.03693	0.06421	0.08349
T/K	286.75	287.15	287.15	286.95	286.65

Polymer (B): **polystyrene** **1963ORO**
Characterization: M_n/g.mol^{-1} = 404000, M_w/g.mol^{-1} = 406000
Solvent (A): **decahydronaphthalene** $C_{10}H_{18}$ **91-17-8**

Type of data: cloud points (UCST-behavior)

w_B 0.01 T/K 283.15

Polymer (B): **polystyrene** **1995PFO**
Characterization: M_n/g.mol^{-1} = 86500, M_w/g.mol^{-1} = 90000,
Polysciences, Inc., Warrington, PA
Solvent (A): ***trans*-decahydronaphthalene** $C_{10}H_{18}$ **493-02-7**

Type of data: cloud points (UCST-behavior)

w_B	0.0049	0.0099	0.0247	0.0398	0.0540	0.0699	0.0847	0.0996	0.1153
T/K	275.25	276.55	278.75	280.75	281.05	281.25	281.15	281.25	281.35
w_B	0.1306	0.1984							
T/K	281.45	281.45							

Polymer (B): **polystyrene** **1995PFO**
Characterization: M_n/g.mol^{-1} = 219800, M_w/g.mol^{-1} = 233000,
Polysciences, Inc., Warrington, PA
Solvent (A): ***trans*-decahydronaphthalene** $C_{10}H_{18}$ **493-02-7**

continued

continued

Type of data: cloud points (UCST-behavior)

w_B	0.0999	0.1494	0.1947
T/K	287.15	286.75	285.95

Polymer (B): **polystyrene** **2003JI1, 2004JIA**
Characterization: M_n/g.mol^{-1} = 135000, M_w/g.mol^{-1} = 270000,
Pressure Chemical Company, Pittsburgh, PA
Solvent (A): ***trans*-decahydronaphthalene** $C_{10}H_{18}$ **493-02-7**

Type of data: cloud points (UCST-behavior)

φ_B	0.042	0.084	0.128	0.172	0.216	0.042	0.084	0.128	0.172
T/K	279.0	282.0	284.9	284.0	273.0	279.7	283.0	285.5	284.3
P/bar	1	1	1	1	1	100	100	100	100
φ_B	0.216	0.042	0.084	0.128	0.172	0.216	0.042	0.084	0.128
T/K	274.8	280.7	283.7	286.1	284.9	275.9	281.5	284.4	286.6
P/bar	100	200	200	200	200	200	300	300	300
φ_B	0.172	0.216	0.042	0.084	0.128	0.172	0.216	0.042	0.084
T/K	285.5	277.0	282.3	285.1	287.2	286.1	278.1	283.2	285.7
P/bar	300	300	400	400	400	400	400	500	500
φ_B	0.128	0.172	0.216	0.042	0.084	0.128	0.172	0.216	0.042
T/K	287.8	286.7	279.2	284.0	286.5	288.4	287.3	280.3	284.8
P/bar	500	500	500	600	600	600	600	600	700
φ_B	0.084	0.128	0.172	0.216	0.042	0.084	0.128	0.172	0.216
T/K	287.1	289.0	287.9	281.5	285.7	287.8	289.6	288.5	282.6
P/bar	700	700	700	700	800	800	800	800	800

Comments: The volume fractions were determined at 1 bar.

Polymer (B): **polystyrene** **1991OPS**
Characterization: M_w/g.mol^{-1} = 4800,
Pressure Chemical Company, Pittsburgh, PA
Solvent (A): **n-decane** $C_{10}H_{22}$ **124-18-5**

Type of data: critical point (LCST-behavior)

$w_{B,\ crit}$ 0.379 T_{crit}/K 360.95

Polymer (B): **polystyrene** **1991OPS**
Characterization: M_n/g.mol^{-1} = 3700, M_w/g.mol^{-1} = 4000, M_z/g.mol^{-1} = 4600,
Pressure Chemical Company, Pittsburgh, PA
Solvent (A): **1-decanol** $C_{10}H_{22}O$ **112-30-1**

Type of data: critical point (UCST-behavior)

$w_{B,\ crit}$ 0.389 T_{crit}/K 375.15

Polymer (B): **polystyrene** **1991OPS**
Characterization: M_w/g.mol^{-1} = 4800,
Pressure Chemical Company, Pittsburgh, PA
Solvent (A): **1-decanol** **$C_{10}H_{22}O$** **112-30-1**

Type of data: critical point (UCST-behavior)

$w_{B,\ crit}$ 0.400 T_{crit}/K 374.15

Polymer (B): **polystyrene** **1991OPS**
Characterization: M_w/g.mol^{-1} = 9000,
Pressure Chemical Company, Pittsburgh, PA
Solvent (A): **1-decanol** **$C_{10}H_{22}O$** **112-30-1**

Type of data: critical point (UCST-behavior)

$w_{B,\ crit}$ 0.313 T_{crit}/K 398.15

Polymer (B): **polystyrene** **1950JE1**
Characterization: M_η/g.mol^{-1} = 62600, fractionated in the laboratory
Solvent (A): **diethyl malonate** **$C_7H_{12}O_4$** **105-53-3**

Type of data: cloud points, precipitation threshold (UCST-behavior)

z_B 0.122 T/K 291.15

Polymer (B): **polystyrene** **1990STA**
Characterization: M_n/g.mol^{-1} = 100000, M_w/g.mol^{-1} = 102000,
Pressure Chemical Company, Pittsburgh, PA
Solvent (A): **diethyl malonate** **$C_7H_{12}O_4$** **105-53-3**

Type of data: cloud points (UCST-behavior)

w_B 0.09744 T/K 276.675

Polymer (B): **polystyrene** **1987GRU**
Characterization: M_n/g.mol^{-1} = 105900, M_w/g.mol^{-1} = 107000,
Toyo Soda Manufacturing Co., Japan
Solvent (A): **diethyl malonate** **$C_7H_{12}O_4$** **105-53-3**

Type of data: critical point (UCST-behavior)

$w_{B,\ crit}$ 0.09469 T_{crit}/K 280.15

Polymer (B): **polystyrene** **1963ORO**
Characterization: M_n/g.mol^{-1} = 404000, M_w/g.mol^{-1} = 406000
Solvent (A): **diethyl malonate** **$C_7H_{12}O_4$** **105-53-3**

Type of data: cloud points (UCST-behavior)

w_B 0.01 T/K 305.15

Polymer (B): **polystyrene** **1986SA2**
Characterization: M_n/g.mol^{-1} = 15800, M_w/g.mol^{-1} = 16700,
Pressure Chemical Company, Pittsburgh, PA
Solvent (A): **diethyl oxalate** **$C_6H_{10}O_4$** **95-92-1**

Type of data: critical point (UCST-behavior)

$\varphi_{B,\ crit}$ = 0.1578 T_{crit}/K = 261.96 $(dT/dP)_{P=1}$/K.atm^{-1} = +0.00126

Polymer (B): **polystyrene** **1986SA2**
Characterization: M_n/g.mol^{-1} = 47200, M_w/g.mol^{-1} = 50000,
Pressure Chemical Company, Pittsburgh, PA
Solvent (A): **diethyl oxalate** **$C_6H_{10}O_4$** **95-92-1**

Type of data: critical point (UCST-behavior)

$\varphi_{B,\ crit}$ = 0.1302 T_{crit}/K = 280.05 $(dT/dP)_{P=1}$/K.atm^{-1} = −0.0045

Polymer (B): **polystyrene** **1950JE1**
Characterization: M_η/g.mol^{-1} = 62600, fractionated in the laboratory
Solvent (A): **diethyl oxalate** **$C_6H_{10}O_4$** **95-92-1**

Type of data: cloud points, precipitation threshold (UCST-behavior)

z_B	0.092	T/K	259.15

Polymer (B): **polystyrene** **1986SA2**
Characterization: M_n/g.mol^{-1} = 100000, M_w/g.mol^{-1} = 110000,
Pressure Chemical Company, Pittsburgh, PA
Solvent (A): **diethyl oxalate** **$C_6H_{10}O_4$** **95-92-1**

Type of data: critical point (UCST-behavior)

$\varphi_{B,\ crit}$ = 0.0927 T_{crit}/K = 292.59 $(dT/dP)_{P=1}$/K.atm^{-1} = −0.0105

Polymer (B): **polystyrene** **1995HAA**
Characterization: M_n/g.mol^{-1} = 290000, M_w/g.mol^{-1} = 300000,
Pressure Chemical Company, Pittsburgh, PA
Solvent (A): **diethyl oxalate** **$C_6H_{10}O_4$** **95-92-1**

Type of data: critical point (UCST-behavior)

$\varphi_{B,\ crit}$ = 0.075 T_{crit}/K = 308.2

Polymer (B): **polystyrene** **1986SA2**
Characterization: M_n/g.mol^{-1} = 545500, M_w/g.mol^{-1} = 600000,
Pressure Chemical Company, Pittsburgh, PA
Solvent (A): **diethyl oxalate** **$C_6H_{10}O_4$** **95-92-1**

Type of data: critical point (UCST-behavior)

$\varphi_{B,\ crit}$ = 0.0427 T_{crit}/K = 309.96 $(dT/dP)_{P=1}$/K.atm^{-1} = −0.019

Polymer (B): **polystyrene** **1994SAT**
Characterization: M_n/g.mol^{-1} = 545500, M_w/g.mol^{-1} = 600000,
Pressure Chemical Company, Pittsburgh, PA
Solvent (A): **diethyl oxalate** $C_6H_{10}O_4$ **95-92-1**

Type of data: critical point (UCST-behavior)

$w_{B,\,crit}$ = 0.0466 T_{crit}/K = 312.78

Polymer (B): **polystyrene** **1995HAA**
Characterization: M_w/g.mol^{-1} = 2000000,
Pressure Chemical Company, Pittsburgh, PA
Solvent (A): **diethyl oxalate** $C_6H_{10}O_4$ **95-92-1**

Type of data: critical point (UCST-behavior)

$\varphi_{B,\,crit}$ = 0.025 T_{crit}/K = 322.2

Polymer (B): **polystyrene** **2001HE1**
Characterization: M_n/g.mol^{-1} = 111900, M_w/g.mol^{-1} = 235000,
Styron 686E, Dow Benelux NV
Solvent (A): **diisodecyl phthalate** $C_{28}H_{46}O_4$ **26761-40-0**

Type of data: cloud points (UCST-behavior)

w_B	0.05	0.10	0.15	0.20	0.25	0.28	0.35	0.40
T/K	332.5	329.9	329.5	329.5	323.0	321.8	313.5	308.0

Polymer (B): **polystyrene** **2001IM1**
Characterization: M_n/g.mol^{-1} = 1160, M_w/g.mol^{-1} = 1240,
Pressure Chemical Company, Pittsburgh, PA
Solvent (A): **2,2-dimethylbutane** C_6H_{14} **75-83-2**

Type of data: cloud points (UCST-behavior)

w_B	0.110	0.140	0.170	0.200	0.235
T/K	303.6	307.7	309.8	310.4	310.2

Polymer (B): **polystyrene** **2001IM1**
Characterization: M_n/g.mol^{-1} = 1160, M_w/g.mol^{-1} = 1240,
Pressure Chemical Company, Pittsburgh, PA
Solvent (A): **2,3-dimethylbutane** C_6H_{14} **79-29-8**

Type of data: cloud points (UCST-behavior)

w_B	0.080	0.110	0.140	0.170	0.200	0.240	0.313
T/K	282.6	284.4	285.2	285.2	284.7	283.8	281.9

Polymer (B): **polystyrene** **1950JE1**
Characterization: M_η/g.mol^{-1} = 62600, fractionated in the laboratory

continued

continued

Solvent (A): **dimethyl malonate** $C_5H_8O_4$ **108-59-8**

Type of data: cloud points, precipitation threshold (UCST-behavior)

z_B 0.150 T/K 409.15

Polymer (B): **polystyrene** **1950JE1**
Characterization: M_η/g.mol^{-1} = 62600, fractionated in the laboratory
Solvent (A): **dimethyl oxalate** $C_4H_6O_4$ **553-90-2**

Type of data: cloud points, precipitation threshold (UCST-behavior)

z_B 0.100 T/K 453.15

Polymer (B): **polystyrene** **2003SIP**
Characterization: M_n/g.mol^{-1} = 23600, M_w/g.mol^{-1} = 25000,
Pressure Chemical Company, Pittsburgh, PA
Solvent (A): **dodecadeuterocyclohexane** C_6D_{12} **1735-17-7**

Type of data: cloud points (UCST-behavior)

φ_B	0.036	0.044	0.053	0.070	0.088	0.102	0.104	0.121	0.130
T/K	281.76	282.16	282.60	283.07	283.31	283.42	283.42	283.47	283.49

φ_B	0.149	0.160	0.172	0.183	0.192	0.205	0.218	0.229
T/K	283.46	283.41	283.34	283.29	283.22	283.11	282.95	282.85

Type of data: critical point (UCST-behavior)

$\varphi_{B,\,crit}$ 0.138 T_{crit}/K 283.48

Polymer (B): **polystyrene** **2003SIP**
Characterization: M_n/g.mol^{-1} = 116000, M_w/g.mol^{-1} = 123000,
Pressure Chemical Company, Pittsburgh, PA
Solvent (A): **dodecadeuterocyclohexane** C_6D_{12} **1735-17-7**

Type of data: cloud points (UCST-behavior)

φ_B	0.014	0.019	0.026	0.036	0.043	0.052	0.058	0.067	0.077
T/K	295.96	296.53	297.02	297.39	297.63	297.88	298.01	298.06	298.08

φ_B	0.084	0.093	0.103	0.115	0.123	0.140	0.158
T/K	298.09	298.10	298.10	298.04	297.91	297.65	297.34

Type of data: critical point (UCST-behavior)

$\varphi_{B,\,crit}$ 0.094 T_{crit}/K 298.10

Polymer (B): **polystyrene** **2003SIP**
Characterization: M_n/g.mol^{-1} = 377400, M_w/g.mol^{-1} = 400000,
Pressure Chemical Company, Pittsburgh, PA
Solvent (A): **dodecadeuterocyclohexane** C_6D_{12} **1735-17-7**

continued

continued

Type of data: cloud points (UCST-behavior)

φ_B	0.009	0.011	0.013	0.015	0.019	0.022	0.025	0.029	0.033
T/K	302.05	302.35	302.54	302.67	302.86	302.93	303.00	303.09	303.18
φ_B	0.037	0.041	0.045	0.049	0.052	0.057	0.061	0.065	0.070
T/K	303.24	303.27	303.29	303.30	303.31	303.31	303.27	303.23	303.17
φ_B	0.073	0.076	0.079	0.083	0.086	0.089	0.093	0.097	0.101
T/K	303.14	303.10	303.04	302.99	302.93	302.88	302.81	302.71	302.61

Type of data: critical point (UCST-behavior)

$\varphi_{B, crit}$ 0.054 T_{crit}/K 303.31

Polymer (B): **polystyrene** **1990GOE**
Characterization: M_w/g.mol^{-1} = 400000,
Pressure Chemical Company, Pittsburgh, PA
Solvent (A): **dodecadeuterocyclohexane** C_6D_{12} **1735-17-7**

Type of data: critical point (UCST-behavior)

$\varphi_{B, crit}$ 0.0512 T_{crit}/K 303.59

Polymer (B): **polystyrene** **2003SIP**
Characterization: M_n/g.mol^{-1} = 1538500, M_w/g.mol^{-1} = 2000000,
Pressure Chemical Company, Pittsburgh, PA
Solvent (A): **dodecadeuterocyclohexane** C_6D_{12} **1735-17-7**

Type of data: cloud points (UCST-behavior)

φ_B	0.005	0.008	0.012	0.014	0.017	0.019	0.022	0.025	0.028
T/K	306.94	307.03	307.20	307.25	307.31	307.35	307.37	307.38	307.39
φ_B	0.030	0.033	0.036	0.039	0.041	0.043	0.045	0.047	0.049
T/K	307.39	307.38	307.37	307.35	307.33	307.29	307.24	307.17	307.07

Type of data: critical point (UCST-behavior)

$\varphi_{B, crit}$ 0.030 T_{crit}/K 307.38

Polymer (B): **polystyrene** **2003SIP**
Characterization: M_n/g.mol^{-1} = 11680000, M_w/g.mol^{-1} = 13200000,
Polymer Laboratories, Inc., Amherst, MA
Solvent (A): **dodecadeuterocyclohexane** C_6D_{12} **1735-17-7**

Type of data: cloud points (UCST-behavior)

φ_B	0.0015	0.0023	0.0033	0.0054	0.0065	0.0074	0.0083	0.0093	0.0103
T/K	309.19	309.41	309.56	309.76	309.80	309.84	309.86	309.88	309.91
φ_B	0.0113	0.0123	0.0132	0.0140	0.0150	0.0159	0.0168	0.0176	0.0185
T/K	309.92	309.93	309.93	309.92	309.90	309.90	309.88	309.88	309.85

continued

continued

Type of data: critical point (UCST-behavior)

$\varphi_{B, crit}$ 0.013 T_{crit}/K 309.92

Polymer (B): **polystyrene** **1991OPS**
Characterization: M_w/g.mol^{-1} = 4800,
Pressure Chemical Company, Pittsburgh, PA
Solvent (A): **n-dodecane** $C_{12}H_{26}$ **112-40-3**

Type of data: critical point (LCST-behavior)

$w_{B, crit}$ 0.389 T_{crit}/K 368.65

Polymer (B): **polystyrene** **1991OPS**
Characterization: M_n/g.mol^{-1} = 3700, M_w/g.mol^{-1} = 4000, M_z/g.mol^{-1} = 4600,
Pressure Chemical Company, Pittsburgh, PA
Solvent (A): **1-dodecanol** $C_{12}H_{26}O$ **112-53-8**

Type of data: critical point (UCST-behavior)

$w_{B, crit}$ 0.401 T_{crit}/K 379.75

Polymer (B): **polystyrene** **1991OPS**
Characterization: M_w/g.mol^{-1} = 4800,
Pressure Chemical Company, Pittsburgh, PA
Solvent (A): **1-dodecanol** $C_{12}H_{26}O$ **112-53-8**

Type of data: critical point (UCST-behavior)

$w_{B, crit}$ 0.406 T_{crit}/K 377.55

Polymer (B): **polystyrene** **1991OPS**
Characterization: M_w/g.mol^{-1} = 9000,
Pressure Chemical Company, Pittsburgh, PA
Solvent (A): **1-dodecanol** $C_{12}H_{26}O$ **112-53-8**

Type of data: critical point (UCST-behavior)

$w_{B, crit}$ 0.317 T_{crit}/K 401.75

Polymer (B): **polystyrene** **2001HE1, 2001HE2**
Characterization: M_n/g.mol^{-1} = 111900, M_w/g.mol^{-1} = 235000,
Styron 686E, Dow Benelux NV
Solvent (A): **1-dodecanol** $C_{12}H_{26}O$ **112-53-8**

Type of data: cloud points (UCST-behavior)

w_B	0.08	0.13	0.26	0.27	0.38	0.38	0.47	0.48	0.66
T/K	445.65	444.05	440.55	439.15	436.85	435.45	428.08	426.05	393.25

w_B	0.68	0.69	0.73	0.81
T/K	396.85	403.55	383.95	361.28

Polymer (B): **polystyrene** **1950JE1**
Characterization: M_η/g.mol^{-1} = 62600, fractionated in the laboratory
Solvent (A): **ethyl acetate** $C_4H_8O_2$ **141-78-6**

Type of data: cloud points, precipitation threshold (UCST-behavior)

z_B	0.080	T/K	242.65

Polymer (B): **polystyrene** **1991BAE**
Characterization: M_n/g.mol^{-1} = 94300, M_w/g.mol^{-1} = 100000,
Pressure Chemical Company, Pittsburgh, PA
Solvent (A): **ethyl acetate** $C_4H_8O_2$ **141-78-6**

Type of data: cloud points (LCST-behavior)

w_B	0.005	0.007	0.010	0.020	0.049	0.069	0.102	0.149	0.196
T/K	447.65	446.35	444.65	441.15	438.65	437.55	437.75	437.75	439.25

w_B	0.299	0.489
T/K	446.45	472.35

Polymer (B): **polystyrene** **1991BAE**
Characterization: M_n/g.mol^{-1} = 220000, M_w/g.mol^{-1} = 233000,
Pressure Chemical Company, Pittsburgh, PA
Solvent (A): **ethyl acetate** $C_4H_8O_2$ **141-78-6**

Type of data: cloud points (LCST-behavior)

w_B	0.005	0.007	0.010	0.020	0.031	0.051	0.070	0.098	0.149
T/K	432.55	431.65	429.85	427.35	426.65	424.95	425.85	426.45	428.35

w_B	0.200
T/K	432.15

Polymer (B): **polystyrene** **1991BAE**
Characterization: M_n/g.mol^{-1} = 545500, M_w/g.mol^{-1} = 600000,
Pressure Chemical Company, Pittsburgh, PA
Solvent (A): **ethyl acetate** $C_4H_8O_2$ **141-78-6**

Type of data: cloud points (LCST-behavior)

w_B	0.002	0.007	0.010	0.020	0.030	0.039	0.050	0.069
T/K	427.65	423.65	423.35	421.75	420.35	420.65	420.85	421.65

Polymer (B): **polystyrene** **1963DEB**
Characterization: M_n/g.mol^{-1} = 118000, M_w/g.mol^{-1} = 124000,
Dow Chemical Company, Midland
Solvent (A): **ethylcyclohexane** C_8H_{16} **1678-91-7**

Type of data: critical point (UCST-behavior)

$\varphi_{B,\,crit}$	0.066	T_{crit}/K	325.94

Polymer (B): **polystyrene** **1963DEB**
Characterization: M_n/g.mol^{-1} = 147000, M_w/g.mol^{-1} = 153000,
Dow Chemical Company, Midland
Solvent (A): **ethylcyclohexane** C_8H_{16} **1678-91-7**

Type of data: critical point (UCST-behavior)

$\varphi_{B, crit}$ 0.0585 T_{crit}/K 327.81

Polymer (B): **polystyrene** **1963DEB**
Characterization: M_n/g.mol^{-1} = 221000, M_w/g.mol^{-1} = 239000,
Dow Chemical Company, Midland
Solvent (A): **ethylcyclohexane** C_8H_{16} **1678-91-7**

Type of data: critical point (UCST-behavior)

$\varphi_{B, crit}$ 0.0525 T_{crit}/K 330.52

Polymer (B): **polystyrene** **1963DEB**
Characterization: M_n/g.mol^{-1} = 522000, M_w/g.mol^{-1} = 569000,
Dow Chemical Company, Midland
Solvent (A): **ethylcyclohexane** C_8H_{16} **1678-91-7**

Type of data: critical point (UCST-behavior)

$\varphi_{B, crit}$ 0.0360 T_{crit}/K 335.17

Polymer (B): **polystyrene** **1991OPS**
Characterization: M_w/g.mol^{-1} = 4800,
Pressure Chemical Company, Pittsburgh, PA
Solvent (A): **n-heptane** C_7H_{16} **142-82-5**

Type of data: critical point (LCST-behavior)

$w_{B, crit}$ 0.361 T_{crit}/K 365.85

Polymer (B): **polystyrene** **1963ORO**
Characterization: M_n/g.mol^{-1} = 404000, M_w/g.mol^{-1} = 406000
Solvent (A): **n-hexadecane** $C_{16}H_{34}$ **544-76-3**

Type of data: cloud points (UCST-behavior)

w_B 0.01 T/K 463.15

Polymer (B): **polystyrene** **1991OPS**
Characterization: M_n/g.mol^{-1} = 3700, M_w/g.mol^{-1} = 4000, M_z/g.mol^{-1} = 4600,
Pressure Chemical Company, Pittsburgh, PA
Solvent (A): **1-hexadecanol** $C_{16}H_{34}O$ **36653-82-4**

Type of data: critical point (UCST-behavior)

$w_{B, crit}$ 0.405 T_{crit}/K 386.25

Polymer (B): **polystyrene** **1992SZY**
Characterization: M_n/g.mol^{-1} = 6600, M_w/g.mol^{-1} = 7800,
Scientific Polymer Products, Inc., Ontario, NY
Solvent (A): **1,1,1,3,3,3-hexadeutero-2-propanone** **C_3D_6O** **666-52-4**

Type of data: cloud points (UCST-behavior)

w_B	0.072	0.080	0.090	0.100	0.110	0.117	0.130	0.149	0.162
φ_B	0.060	0.067	0.075	0.084	0.092	0.098	0.109	0.125	0.137
T/K	277.63	279.18	281.52	282.88	284.58	285.00	286.07	286.75	286.85

w_B	0.170	0.180	0.200	0.220	0.249	0.293	0.310
φ_B	0.143	0.152	0.170	0.187	0.214	0.253	0.269
T/K	286.88	286.84	286.66	286.40	285.86	284.72	284.55

Type of data: cloud points (UCST-behavior)

w_B	0.072	0.090	0.110	0.130	0.149	0.162	0.293
φ_B	0.060	0.075	0.092	0.109	0.125	0.137	0.253
T/K	277.63	281.52	284.58	286.07	286.75	286.85	284.72
$(dT/dP)_{P=0}$/K.MPa^{-1}	−1.70	−1.40	−1.30	−1.27	−1.25	−1.26	−1.31

Type of data: spinodal points (UCST-behavior)

w_B	0.072	0.090	0.100	0.110	0.117	0.130	0.149	0.162	0.170
φ_B	0.060	0.075	0.084	0.092	0.098	0.109	0.125	0.137	0.143
T/K	276.60	279.73	282.38	283.30	284.25	285.01	285.80	286.50	286.69

w_B	0.180	0.200	0.220	0.249	0.293	0.310
φ_B	0.152	0.170	0.187	0.214	0.253	0.269
T/K	286.65	286.56	286.40	285.64	284.01	284.39

Type of data: spinodal points (UCST-behavior)

w_B	0.072	0.090	0.110	0.162	0.293
φ_B	0.060	0.075	0.092	0.137	0.253
T/K	276.60	279.73	283.30	286.50	284.01
$(dT/dP)_{P=0}$/K.MPa^{-1}	−1.63	−1.35	−1.23	−1.27	−1.26

Polymer (B): **polystyrene** **1992SZY**
Characterization: M_n/g.mol^{-1} = 10750, M_w/g.mol^{-1} = 11500,
Scientific Polymer Products, Inc., Ontario, NY
Solvent (A): **1,1,1,3,3,3-hexadeutero-2-propanone** **C_3D_6O** **666-52-4**

Type of data: cloud points (UCST-behavior)

w_B	0.090	0.100	0.120	0.140	0.160	0.179	0.200	0.220	0.240
φ_B	0.075	0.092	0.099	0.115	0.132	0.149	0.167	0.185	0.203
T/K	280.95	282.15	283.59	284.14	284.16	283.98	283.78	283.23	283.02

continued

continued

w_B	0.271	0.308
φ_B	0.231	0.267
T/K	282.45	280.71

Type of data: cloud points (UCST-behavior)

w_B	0.090	0.140	0.220	0.240
φ_B	0.075	0.115	0.185	0.203
T/K	280.95	284.14	283.23	283.02
$(dT/dP)_{P=0}$/K.MPa^{-1}	−1.25	−1.29	−1.29	−1.32

Type of data: spinodal points (UCST-behavior)

w_B	0.090	0.100	0.120	0.140	0.160	0.179	0.200	0.220	0.240
φ_B	0.075	0.092	0.099	0.115	0.132	0.149	0.167	0.185	0.203
T/K	280.48	281.72	283.40	284.00	283.86	293.68	283.59	283.23	282.93

w_B	0.271	0.308
φ_B	0.231	0.267
T/K	282.39	280.58

Type of data: spinodal points (UCST-behavior)

w_B	0.090	0.140	0.220	0.240
φ_B	0.075	0.115	0.185	0.203
T/K	280.48	284.00	283.23	282.93
$(dT/dP)_{P=0}$/K.MPa^{-1}	−1.28	−1.43	−1.29	−1.37

Polymer (B): **polystyrene** **1995LUS**
Characterization: M_n/g.mol^{-1} = 12750, M_w/g.mol^{-1} = 13500,
Pressure Chemical Company, Pittsburgh, PA
Solvent (A): **1,1,1,3,3,3-hexadeutero-2-propanone** **C_3D_6O** **666-52-4**

Type of data: cloud points

w_B	0.101	0.101	0.101	0.101	0.101	0.101	0.101	0.101	0.114
T/K	328.09	332.09	334.55	336.41	337.92	339.34	380.82	379.15	331.96
P/MPa	2.33	1.65	1.05	0.81	0.56	0.45	1.34	0.92	2.25
w_B	0.114	0.114	0.114	0.114	0.114	0.114	0.114	0.114	0.123
T/K	334.66	337.41	340.13	342.71	378.77	380.57	371.78	374.31	332.14
P/MPa	1.72	1.43	1.04	0.92	1.22	1.39	0.93	1.01	2.67
w_B	0.123	0.123	0.123	0.123	0.123	0.123	0.123	0.123	0.123
T/K	333.44	337.59	341.36	344.53	346.66	350.05	352.68	358.35	364.62
P/MPa	2.35	1.64	1.27	1.11	0.86	0.85	0.90	1.31	1.78
w_B	0.123	0.131	0.131	0.131	0.131	0.131	0.131	0.131	0.131
T/K	368.55	334.51	339.31	343.41	345.83	351.16	356.50	358.36	363.62
P/MPa	2.46	2.20	1.59	1.18	1.13	0.84	0.82	0.91	1.06

continued

continued

w_B	0.131	0.131	0.131	0.147	0.147	0.147	0.147	0.147	0.147
T/K	366.71	369.29	372.79	332.28	336.35	340.25	344.21	345.44	348.79
P/MPa	1.08	1.28	1.68	2.48	1.81	1.25	1.01	0.84	0.72
w_B	0.147	0.147	0.147	0.147	0.147	0.147	0.147	0.163	0.163
T/K	351.74	352.11	354.97	358.01	363.66	367.13	371.58	336.06	338.37
P/MPa	0.51	0.49	0.45	0.45	0.53	0.71	0.92	1.77	1.45
w_B	0.163	0.163	0.163	0.163	0.163	0.163	0.163	0.163	0.194
T/K	341.10	345.26	347.69	352.92	358.34	362.11	366.28	368.08	337.28
P/MPa	1.09	0.74	0.59	0.39	0.37	0.41	0.51	0.59	1.54
w_B	0.194	0.194	0.194	0.194	0.194	0.194	0.194	0.204	0.204
T/K	341.73	340.81	344.61	344.60	342.87	339.09	333.17	330.56	333.13
P/MPa	1.05	1.12	0.75	0.76	0.88	1.31	2.24	2.77	2.24
w_B	0.204	0.204	0.204	0.204	0.204	0.204	0.204	0.204	0.204
T/K	333.13	338.53	341.31	346.48	349.67	355.46	358.33	362.83	365.49
P/MPa	2.23	1.38	1.11	0.66	0.49	0.32	0.33	0.40	0.49
w_B	0.204	0.204	0.204	0.229	0.229	0.229	0.229	0.229	0.229
T/K	369.33	372.62	379.32	329.85	332.52	335.22	340.82	343.52	346.05
P/MPa	0.68	0.88	1.39	2.49	2.01	1.54	0.83	0.58	0.39
w_B	0.229	0.229	0.229	0.229	0.229	0.229			
T/K	348.33	370.51	372.95	376.66	379.86	384.97			
P/MPa	0.26	0.55	0.69	0.97	1.23	1.72			

critical concentration: $w_{B, crit} = 0.200$

Polymer (B): **polystyrene-d8** **1995LUS**
Characterization: M_n/g.mol^{-1} = 10300, M_w/g.mol^{-1} = 10500, completely deuterated, Polymer Laboratories, Amherst, MA
Solvent (A): **1,1,1,3,3,3-hexadeutero-2-propanone** **C_3D_6O** **666-52-4**

Type of data: cloud points

w_B	0.219	0.219	0.219	0.219
T/K	266.69	267.22	268.05	268.78
P/MPa	2.61	2.03	1.22	0.47

Polymer (B): **polystyrene-d8** **1995LUS**
Characterization: M_n/g.mol^{-1} = 25400, M_w/g.mol^{-1} = 26900, completely deuterated, Polymer Laboratories, Amherst, MA
Solvent (A): **1,1,1,3,3,3-hexadeutero-2-propanone** **C_3D_6O** **666-52-4**

Type of data: cloud points

w_B	0.120	0.120	0.120	0.120	0.120	0.120	0.120	0.140	0.140
T/K	330.21	336.97	343.05	350.07	357.01	363.63	370.11	330.34	337.04
P/MPa	4.04	3.22	2.79	2.57	2.61	2.62	3.18	4.45	3.59

continued

continued

w_B	0.140	0.140	0.140	0.140	0.140	0.150	0.150	0.150	0.150
T/K	343.20	350.05	356.97	363.50	369.96	330.76	337.14	343.16	350.01
P/MPa	3.12	2.87	2.66	3.07	3.40	4.52	3.70	3.22	2.96

w_B	0.150	0.150	0.150	0.160	0.160	0.160	0.160	0.160	0.160
T/K	357.04	363.44	370.06	336.24	336.23	342.24	348.99	355.63	363.30
P/MPa	2.96	3.13	3.46	3.86	3.83	3.33	3.04	3.00	3.18

w_B	0.160	0.170	0.170	0.170	0.170	0.170	0.170	0.185	0.185
T/K	369.95	337.45	343.67	350.09	356.66	363.67	370.10	329.96	336.98
P/MPa	3.51	3.67	3.30	3.05	3.05	3.23	3.57	4.04	3.23

w_B	0.185	0.185	0.185	0.185	0.185	0.193	0.193	0.193	0.193
T/K	343.41	350.03	356.99	363.52	369.66	330.21	337.13	343.25	349.95
P/MPa	2.84	2.67	2.75	3.02	3.36	3.95	3.16	2.80	2.62

w_B	0.193	0.193	0.193	0.205	0.205	0.205	0.205	0.205	0.205
T/K	357.07	363.73	370.00	357.26	363.67	370.07	329.97	336.99	343.12
P/MPa	2.71	2.93	3.32	2.65	2.89	3.27	3.86	3.11	2.73

w_B	0.205	0.212	0.212	0.212	0.212	0.212	0.212	0.212	0.217
T/K	349.98	330.01	336.89	342.35	349.87	357.01	363.25	369.96	330.04
P/MPa	2.57	3.85	3.08	2.70	2.50	2.60	2.82	3.22	3.33

w_B	0.217	0.217	0.217	0.217	0.217	0.217	0.225	0.225	0.225
T/K	337.00	343.13	350.01	357.16	363.54	370.02	329.98	336.89	342.98
P/MPa	2.83	2.57	2.47	2.55	2.79	3.18	3.59	2.85	2.51

w_B	0.225	0.225	0.225	0.225	0.239	0.239	0.239	0.239	0.239
T/K	350.08	357.37	363.48	370.03	331.21	341.82	345.75	350.19	357.17
P/MPa	2.38	2.47	2.73	3.13	3.12	2.33	2.22	2.21	2.31

w_B	0.239	0.239	0.259	0.259	0.259	0.259	0.259	0.259	0.259
T/K	363.37	369.78	337.02	340.82	347.67	352.05	358.22	363.07	368.57
P/MPa	2.56	2.97	1.86	1.69	1.54	1.58	1.79	2.05	2.41

critical concentration: $w_{B,\,crit} = 0.193$

Polymer (B): **polystyrene** **2001IM1**
Characterization: M_n/g.mol^{-1} = 1160, M_w/g.mol^{-1} = 1240,
Pressure Chemical Company, Pittsburgh, PA
Solvent (A): **n-hexane** C_6H_{14} **110-54-3**

Type of data: cloud points (UCST-behavior)

w_B	0.063	0.083	0.123	0.162	0.176	0.192	0.219	0.243
T/K	263.2	267.4	271.2	274.6	274.9	274.7	274.4	273.8

Polymer (B): **polystyrene** **1950JE1**
Characterization: M_η/g.mol^{-1} = 62600, fractionated in the laboratory
Solvent (A): **hexanoic acid** $C_6H_{12}O_2$ **142-62-1**

Type of data: cloud points, precipitation threshold (UCST-behavior)

z_B 0.056 T/K 448.15

Polymer (B): **polystyrene** **1991OPS**
Characterization: M_n/g.mol^{-1} = 3700, M_w/g.mol^{-1} = 4000, M_z/g.mol^{-1} = 4600, Pressure Chemical Company, Pittsburgh, PA
Solvent (A): **1-hexanol** **$C_6H_{14}O$** **111-27-3**

Type of data: critical point (UCST-behavior)

$w_{B, crit}$ 0.370 T_{crit}/K 372.15

Polymer (B): **polystyrene** **1991OPS**
Characterization: M_w/g.mol^{-1} = 4800, Pressure Chemical Company, Pittsburgh, PA
Solvent (A): **1-hexanol** **$C_6H_{14}O$** **111-27-3**

Type of data: critical point (UCST-behavior)

$w_{B, crit}$ 0.380 T_{crit}/K 373.45

Polymer (B): **polystyrene** **1991OPS**
Characterization: M_w/g.mol^{-1} = 9000, Pressure Chemical Company, Pittsburgh, PA
Solvent (A): **1-hexanol** **$C_6H_{14}O$** **111-27-3**

Type of data: critical point (UCST-behavior)

$w_{B, crit}$ 0.295 T_{crit}/K 398.05

Polymer (B): **polystyrene** **1950JE1**
Characterization: M_η/g.mol^{-1} = 62600, fractionated in the laboratory
Solvent (A): **1-hexanol** **$C_6H_{14}O$** **111-27-3**

Type of data: cloud points, precipitation threshold (UCST-behavior)

z_B 0.100 T/K 425.15

Polymer (B): **polystyrene** **1950JE1**
Characterization: M_η/g.mol^{-1} = 62600, fractionated in the laboratory
Solvent (A): **3-hexanol** **$C_6H_{14}O$** **623-37-0**

Type of data: cloud points, precipitation threshold (UCST-behavior)

z_B 0.138 T/K 396.65

Polymer (B): **polystyrene** **1995PFO**
Characterization: M_n/g.mol^{-1} = 219800, M_w/g.mol^{-1} = 233000, Polysciences, Inc., Warrington, PA
Solvent (A): **hexyl acetate** **$C_8H_{16}O_2$** **142-92-7**

Type of data: cloud points (LCST-behavior)

w_B	0.0768	0.1048	0.1304
T/K	544.75	542.35	542.65

Polymer (B): **polystyrene** **1995PFO**
Characterization: M_n/g.mol^{-1} = 545500, M_w/g.mol^{-1} = 600000,
Polysciences, Inc., Warrington, PA
Solvent (A): **hexyl acetate** $C_8H_{16}O_2$ **142-92-7**

Type of data: cloud points (LCST-behavior)

w_B	0.0100	0.0301	0.0496	0.0694	0.0898	0.1088	0.1306
T/K	544.05	542.85	540.85	539.45	539.15	539.25	538.85

Polymer (B): **polystyrene** **1972ZE2**
Characterization: M_w/g.mol^{-1} = 51000, M_w/M_n < 1.1
Pressure Chemical Company, Pittsburgh, PA
Solvent (A): **methyl acetate** $C_3H_6O_2$ **79-20-9**

Type of data: cloud points

T/K = 423.35 w_B = 0.023 – 0.024 $(dT/dP)_{P=1}$/K.bar^{-1} = +0.47 (LCST-behavior)

Polymer (B): **polystyrene** **1950JE1**
Characterization: M_η/g.mol^{-1} = 62600, fractionated in the laboratory
Solvent (A): **methyl acetate** $C_3H_6O_2$ **79-20-9**

Type of data: cloud points, precipitation threshold (UCST-behavior)

z_B 0.110 T/K 326.15

Polymer (B): **polystyrene** **1991BAE**
Characterization: M_n/g.mol^{-1} = 94300, M_w/g.mol^{-1} = 100000,
Pressure Chemical Company, Pittsburgh, PA
Solvent (A): **methyl acetate** $C_3H_6O_2$ **79-20-9**

Type of data: cloud points (UCST-behavior)

w_B	0.003	0.008	0.018	0.039	0.068	0.078
T/K	301.25	303.60	305.85	305.45	300.15	299.25

Type of data: cloud points (LCST-behavior)

w_B	0.003	0.008	0.018	0.039	0.068	0.078
T/K	402.50	398.15	397.70	397.95	398.50	398.95

Polymer (B): **polystyrene** **1972ZE2**
Characterization: M_w/g.mol^{-1} = 97200, M_w/M_n < 1.1
Pressure Chemical Company, Pittsburgh, PA
Solvent (A): **methyl acetate** $C_3H_6O_2$ **79-20-9**

Type of data: cloud points

T/K = 275.15 w_B = 0.023 – 0.024 $(dT/dP)_{P=1}$/K.bar^{-1} = –0.018 (UCST-behavior)
T/K = 415.15 w_B = 0.023 – 0.024 $(dT/dP)_{P=1}$/K.bar^{-1} = +0.45 (LCST-behavior)

Polymer (B): **polystyrene** **1972ZE2**
Characterization: M_w/g.mol^{-1} = 160000, M_w/M_n < 1.1
Pressure Chemical Company, Pittsburgh, PA
Solvent (A): **methyl acetate** **$C_3H_6O_2$** **79-20-9**

Type of data: cloud points

T/K = 281.45 w_B = 0.023 – 0.024 $(dT/dP)_{P=1}$/K.bar^{-1} = –0.041 (UCST-behavior)
T/K = 409.95 w_B = 0.023 – 0.024 $(dT/dP)_{P=1}$/K.bar^{-1} = +0.44 (LCST-behavior)

Polymer (B): **polystyrene** **1999OCH**
Characterization: M_n/g.mol^{-1} = 183000, M_w/g.mol^{-1} = 190000,
Tosoh Corp., Ltd., Tokyo, Japan
Solvent (A): **methyl acetate** **$C_3H_6O_2$** **79-20-9**

Type of data: cloud points (UCST-behavior)

w_B	0.0103	0.0204	0.0268	0.0304	0.0368	0.0420	0.0448	0.0528	0.0741
T/K	289.41	295.53	297.54	298.91	300.08	300.40	300.42	300.33	299.33

Polymer (B): **polystyrene** **1972ZE2**
Characterization: M_w/g.mol^{-1} = 498000, M_w/M_n < 1.1
Pressure Chemical Company, Pittsburgh, PA
Solvent (A): **methyl acetate** **$C_3H_6O_2$** **79-20-9**

Type of data: cloud points

T/K = 294.15 w_B = 0.023 – 0.024 $(dT/dP)_{P=1}$/K.bar^{-1} = –0.056 (UCST-behavior)
T/K = 400.55 w_B = 0.023 – 0.024 $(dT/dP)_{P=1}$/K.bar^{-1} = +0.46 (LCST-behavior)

Polymer (B): **polystyrene** **1972ZE2**
Characterization: M_w/g.mol^{-1} = 670000, M_w/M_n < 1.1
Pressure Chemical Company, Pittsburgh, PA
Solvent (A): **methyl acetate** **$C_3H_6O_2$** **79-20-9**

Type of data: cloud points

T/K = 296.45 w_B = 0.023 – 0.024 $(dT/dP)_{P=1}$/K.bar^{-1} = –0.060 (UCST-behavior)
T/K = 397.85 w_B = 0.023 – 0.024 $(dT/dP)_{P=1}$/K.bar^{-1} = +0.47 (LCST-behavior)

Polymer (B): **polystyrene** **1972ZE2**
Characterization: M_w/g.mol^{-1} = 860000, M_w/M_n < 1.1
Pressure Chemical Company, Pittsburgh, PA
Solvent (A): **methyl acetate** **$C_3H_6O_2$** **79-20-9**

Type of data: cloud points

T/K = 299.25 w_B = 0.023 – 0.024 $(dT/dP)_{P=1}$/K.bar^{-1} = –0.073 (UCST-behavior)
T/K = 396.15 w_B = 0.023 – 0.024 $(dT/dP)_{P=1}$/K.bar^{-1} = +0.47 (LCST-behavior)

Polymer (B): **polystyrene** **1972ZE2**
Characterization: M_w/g.mol^{-1} = 1800000, M_w/M_n < 1.1
Pressure Chemical Company, Pittsburgh, PA
Solvent (A): **methyl acetate** $C_3H_6O_2$ **79-20-9**

Type of data: cloud points

T/K = 303.15 w_B = 0.023 – 0.024 $(dT/dP)_{P=1}$/K.bar^{-1} = –0.082 (UCST-behavior)
T/K = 391.75 w_B = 0.023 – 0.024 $(dT/dP)_{P=1}$/K.bar^{-1} = +0.48 (LCST-behavior)

Polymer (B): **polystyrene** **1950JE1**
Characterization: M_η/g.mol^{-1} = 62600, fractionated in the laboratory
Solvent (A): **3-methyl-1-butanol** $C_5H_{12}O$ **123-51-3**

Type of data: cloud points, precipitation threshold (UCST-behavior)

z_B 0.034 T/K 444.15

Polymer (B): **polystyrene** **1991SZY**
Characterization: M_n/g.mol^{-1} = 2230, M_w/g.mol^{-1} = 2510,
Pressure Chemical Company, Pittsburgh, PA
Solvent (A): **methylcyclohexane** C_7H_{14} **108-87-2**

Type of data: cloud points (UCST-behavior)

w_B	0.05	0.10	0.20	0.25	0.30
T/K	250	255	254	252	248

Type of data: cloud points (LCST-behavior)

w_B	0.05	0.10	0.20	0.25	0.30
T/K	525	518	509	507	516

Polymer (B): **polystyrene** **1991SZY**
Characterization: M_n/g.mol^{-1} = 4780, M_w/g.mol^{-1} = 5110,
Pressure Chemical Company, Pittsburgh, PA
Solvent (A): **methylcyclohexane** C_7H_{14} **108-87-2**

Type of data: cloud points (UCST-behavior)

w_B	0.05	0.10	0.20	0.25	0.30
T/K	262	267	269	267	267

Type of data: cloud points (LCST-behavior)

w_B	0.05	0.10	0.20	0.25	0.30
T/K	507	500	499	501	502

Polymer (B): **polystyrene** **1991SZY**
Characterization: M_n/g.mol^{-1} = 6570, M_w/g.mol^{-1} = 7820,
Scientific Polymer Products, Inc., Ontario, NY

continued

continued

Solvent (A): **methylcyclohexane** C_7H_{14} **108-87-2**

Type of data: cloud points (UCST-behavior)

w_B	0.05	0.10	0.20	0.25	0.30
T/K	271	272	270	270	271

Type of data: cloud points (LCST-behavior)

w_B	0.05	0.10	0.20	0.25	0.30
T/K	499	495	493	491	493

Polymer (B): **polystyrene** **1980DO1**
Characterization: M_n/g.mol^{-1} = 9630, M_w/g.mol^{-1} = 10200,
Pressure Chemical Company, Pittsburgh, PA
Solvent (A): **methylcyclohexane** C_7H_{14} **108-87-2**

Type of data: coexistence data (UCST-behavior)

φ_B (total) 0.1990 was kept constant (critical concentration)

T/K	285.71	285.631	285.547	285.394	285.122	284.801	284.023	282.262
φ_B (sol phase)	0.1990	0.1576	0.1442	0.1299	0.1133	0.1009	0.0812	0.0594
φ_B (gel phase)	0.1990	0.2424	0.2558	0.2712	0.2906	0.3068	0.3348	0.3757

Type of data: critical point (UCST-behavior)

$\varphi_{B,\text{crit}}$ 0.1990 T_{crit}/K 285.71

Polymer (B): **polystyrene** **1993HOS**
Characterization: M_n/g.mol^{-1} = 9630, M_w/g.mol^{-1} = 10200,
Pressure Chemical Company, Pittsburgh, PA
Solvent (A): **methylcyclohexane** C_7H_{14} **108-87-2**

Type of data: cloud points (UCST-behavior)

w_B 0.1972 was kept constant

T/K	285.72	285.58	285.29	285.00	284.58	284.38	284.06	283.92	283.80
P/MPa	0.10	1.18	3.43	6.28	11.28	14.51	19.91	24.61	28.83
T/K	283.73	283.70	283.73	283.83	283.93	284.03	284.16	284.30	284.48
P/MPa	34.32	40.31	45.11	50.50	55.31	59.53	63.84	67.86	72.37
T/K	284.65	284.83	285.03						
P/MPa	76.59	80.41	84.93						

Type of data: critical double point

T_{dcrit}/K = 283.71 P_{dcrit}/MPa = 38.3

Polymer (B): **polystyrene** **2002ZHO**
Characterization: M_n/g.mol^{-1} = 9900, M_w/g.mol^{-1} = 10100,
Polymer Laboratories, Inc., Amherst, MA

continued

continued

Solvent (A):	**methylcyclohexane**	**C_7H_{14}**	**108-87-2**

Type of data: critical point (UCST-behavior)

$\varphi_{B, crit}$ 0.204 T_{crit}/K 283.632

Polymer (B):	**polystyrene**	**1991SZY**
Characterization:	M_n/g.mol^{-1} = 10750, M_w/g.mol^{-1} = 11500, Scientific Polymer Products, Inc., Ontario, NY	
Solvent (A):	**methylcyclohexane** **C_7H_{14}**	**108-87-2**

Type of data: cloud points (UCST-behavior)

w_B	0.05	0.10	0.125	0.20	0.25	0.30
T/K	291	294	295	294	294	292

w_B	0.092	0.140
T/K	293	294
$(dT/dP)_{P=0}$/K.MPa^{-1}	−0.22	−0.26

Type of data: cloud points (LCST-behavior)

w_B	0.05	0.10	0.125	0.20	0.25	0.30
T/K	489	484	481	480	480	481

Polymer (B):	**polystyrene**	**1991SHE**
Characterization:	M_n/g.mol^{-1} = 12300, M_w/g.mol^{-1} = 13000, Pressure Chemical Company, Pittsburgh, PA	
Solvent (A):	**methylcyclohexane** **C_7H_{14}**	**108-87-2**

Type of data: critical point (UCST-behavior)

$w_{B, crit}$ = 0.238 T_{crit}/K = 291.340

Polymer (B):	**polystyrene**	**2004WIL**
Characterization:	M_n/g.mol^{-1} = 13210, M_w/g.mol^{-1} = 14000, Pressure Chemical Company, Pittsburgh, PA	
Solvent (A):	**methylcyclohexane** **C_7H_{14}**	**108-87-2**

Type of data: cloud points (UCST-behavior)

w_B	0.0986	0.1425	0.1748	0.1901	0.2027	0.2030	0.2174	0.2341	0.2865
T/K	289.2	292.2	292.9	292.1	291.8	291.9	291.8	291.8	291.6

Polymer (B):	**polystyrene**	**1980DO1**
Characterization:	M_n/g.mol^{-1} = 15200, M_w/g.mol^{-1} = 16100, Pressure Chemical Company, Pittsburgh, PA	
Solvent (A):	**methylcyclohexane** **C_7H_{14}**	**108-87-2**

Type of data: coexistence data (UCST-behavior)

continued

continued

φ_B (total)	0.1719	was kept constant	(critical concentration)					
T/K	295.98	295.967	295.950	295.928	295.906	295.880	295.807	295.700
φ_B (sol phase)	0.1719	0.1465	0.1402	0.1341	0.1300	0.1256	0.1167	0.1079
φ_B (gel phase)	0.1719	0.1973	0.2058	0.2115	0.2152	0.2209	0.2304	0.2420
T/K	295.460	295.246	294.935	294.635	294.161	293.458		
φ_B (sol phase)	0.0942	0.0870	0.0762	0.0716	0.0624	0.0534		
φ_B (gel phase)	0.2577	0.2690	0.2825	0.2940	0.3088	0.3263		

Type of data: critical point (UCST-behavior)

$\varphi_{B,\,crit}$ 0.1719 T_{crit}/K 295.98

Polymer (B): **polystyrene** **1993HOS**
Characterization: M_n/g.mol^{-1} = 15200, M_w/g.mol^{-1} = 16100,
Pressure Chemical Company, Pittsburgh, PA
Solvent (A): **methylcyclohexane** C_7H_{14} **108-87-2**

Type of data: cloud points (UCST-behavior)

w_B	0.0786	0.0786	0.0786	0.0786	0.0786	0.0786	0.0786	0.0786	0.0786
T/K	295.69	295.37	295.05	294.75	294.47	294.17	293.91	293.63	293.24
P/MPa	0.10	1.86	3.82	5.79	7.55	9.81	11.96	14.32	18.93
w_B	0.0786	0.0786	0.0786	0.0786	0.0786	0.0786	0.0786	0.0786	0.0786
T/K	293.03	292.82	292.65	292.51	292.38	292.32	292.30	292.33	292.39
P/MPa	21.97	25.20	28.93	33.05	38.15	43.44	49.13	53.64	58.15
w_B	0.0786	0.0786	0.0786	0.0786	0.0998	0.0998	0.0998	0.0998	0.0998
T/K	292.45	292.55	292.73	292.88	296.05	294.96	294.55	294.28	294.08
P/MPa	63.06	67.67	72.86	77.47	0.10	6.67	9.32	11.57	13.53
w_B	0.0998	0.0998	0.0998	0.0998	0.0998	0.0998	0.0998	0.0998	0.0998
T/K	293.82	293.54	293.30	293.11	292.81	292.67	292.60	292.53	292.55
P/MPa	15.98	18.93	22.56	27.36	32.75	37.27	42.66	48.05	52.86
w_B	0.0998	0.0998	0.0998	0.0998	0.0998	0.0998	0.0998	0.1287	0.1287
T/K	292.63	292.68	292.81	292.95	293.13	293.35	293.51	296.32	295.17
P/MPa	57.86	62.27	67.27	72.77	77.57	83.55	87.28	0.10	6.86
w_B	0.1287	0.1287	0.1287	0.1287	0.1287	0.1287	0.1287	0.1287	0.1287
T/K	294.90	294.58	294.47	294.10	293.89	293.59	293.42	293.20	293.03
P/MPa	8.83	10.98	12.75	15.30	17.85	21.57	24.61	28.24	32.36
w_B	0.1287	0.1287	0.1287	0.1287	0.1287	0.1287	0.1287	0.1287	0.1287
T/K	292.92	292.83	292.78	292.78	292.81	292.89	292.96	293.10	293.24
P/MPa	36.48	41.48	46.78	51.68	56.29	60.90	65.51	70.41	74.73
w_B	0.1287	0.1721	0.1721	0.1721	0.1721	0.1721	0.1721	0.1721	0.1721
T/K	293.34	296.21	295.91	295.44	295.12	294.85	294.47	294.12	293.88
P/MPa	78.26	0.10	1.86	4.22	6.47	8.53	11.57	14.12	16.87

continued

continued

w_B	0.1721	0.1721	0.1721	0.1721	0.1721	0.1721	0.1721	0.1721	0.1721
T/K	293.67	293.43	293.23	293.05	292.89	292.78	292.70	292.68	292.68
P/MPa	19.42	22.36	25.99	29.62	33.44	37.76	42.46	47.27	51.78
w_B	0.1721	0.1721	0.1721	0.1721	0.1721	0.1721	0.2044	0.2044	0.2044
T/K	292.72	292.78	292.85	292.95	293.07	293.21	296.27	295.92	295.48
P/MPa	56.19	60.61	64.82	68.84	72.96	77.28	0.10	1.77	4.31
w_B	0.2044	0.2044	0.2044	0.2044	0.2044	0.2044	0.2044	0.2044	0.2044
T/K	295.25	294.96	294.70	294.40	294.13	293.90	293.68	293.44	293.16
P/MPa	5.88	7.85	9.81	12.26	14.51	16.77	19.61	22.85	27.16
w_B	0.2044	0.2044	0.2044	0.2044	0.2044	0.2044	0.2044	0.2044	0.2044
T/K	293.02	292.85	292.77	292.70	292.70	292.75	292.81	292.92	293.05
P/MPa	31.19	35.79	40.70	46.29	51.78	56.88	61.78	66.69	71.78
w_B	0.2044	0.2044	0.2044	0.2461	0.2461	0.2461	0.2461	0.2461	0.2461
T/K	293.20	293.35	293.51	295.79	294.57	294.23	294.00	293.73	293.47
P/MPa	76.49	81.20	86.00	0.10	7.06	9.12	11.38	13.73	15.98
w_B	0.2461	0.2461	0.2461	0.2461	0.2461	0.2461	0.2461	0.2461	0.2461
T/K	293.22	292.95	292.68	292.52	292.35	292.24	292.16	292.20	292.20
P/MPa	18.83	22.85	27.07	31.97	36.19	41.78	48.44	49.03	53.84
w_B	0.2461	0.2461	0.2461	0.2461	0.2461	0.2461	0.2461	0.2952	0.2952
T/K	292.25	292.35	292.45	292.57	292.74	292.87	293.03	294.79	293.70
P/MPa	58.55	63.74	68.65	73.55	78.16	82.08	85.51	0.10	6.57
w_B	0.2952	0.2952	0.2952	0.2952	0.2952	0.2952	0.2952	0.2952	0.2952
T/K	293.42	293.20	292.93	292.70	292.44	292.23	291.95	291.73	291.55
P/MPa	8.53	10.59	12.75	15.20	18.24	21.28	25.60	30.60	35.50
w_B	0.2952	0.2952	0.2952	0.2952	0.2952	0.2952	0.2952	0.2952	0.2952
T/K	291.45	291.40	291.40	291.45	291.53	291.65	291.78	291.92	292.13
P/MPa	40.40	45.50	50.41	54.92	61.78	66.78	71.69	75.90	81.49

Type of data: critical double point

T_{dcrit}/K = 291.43 P_{dcrit}/MPa = 48.6

Polymer (B): **polystyrene** **1984DOB**
Characterization: M_n/g.mol^{-1} = 16200, M_w/g.mol^{-1} = 17200,
Pressure Chemical Company, Pittsburgh, PA
Solvent (A): **methylcyclohexane** C_7H_{14} **108-87-2**

Type of data: coexistence data (UCST-behavior)

φ_B (total)	0.1689	was kept constant	(critical concentration)					
T/K	296.75	296.710	296.678	296.625	296.523	296.350	296.071	295.586
φ_B (sol phase)	0.1689	0.1352	0.1293	0.1207	0.1112	0.0993	0.0874	0.0741
φ_B (gel phase)	0.1689	0.2032	0.2116	0.2199	0.2308	0.2458	0.2614	0.2826
T/K	294.821	293.491	291.625	289.620	286.650			
φ_B (sol phase)	0.0602	0.0447	0.0319	0.0220	0.0140			
φ_B (gel phase)	0.3069	0.3400	0.3751	0.4063	0.4437			

Polymer (B): **polystyrene** **1980DO1**
Characterization: M_n/g.mol^{-1} = 16300, M_w/g.mol^{-1} = 17300,
Pressure Chemical Company, Pittsburgh, PA
Solvent (A): **methylcyclohexane** C_7H_{14} **108-87-2**

Type of data: coexistence data (UCST-behavior)

φ_B (total) 0.1669 was kept constant (critical concentration)

T/K	296.053	295.987	295.898	295.676	295.266	294.475	292.865	289.630
φ_B (sol phase)	0.1272	0.1182	0.1097	0.0961	0.0809	0.0649	0.0453	0.0264
φ_B (gel phase)	0.2080	0.2178	0.2272	0.2439	0.2638	0.2916	0.3292	0.3831

Type of data: critical point (UCST-behavior)

$\varphi_{B, crit}$ 0.1669 T_{crit}/K 296.13

Polymer (B): **polystyrene** **1994VA1**
Characterization: M_n/g.mol^{-1} = 16500, M_w/g.mol^{-1} = 17500,
Pressure Chemical Company, Pittsburgh, PA
Solvent (A): **methylcyclohexane** C_7H_{14} **108-87-2**

Type of data: cloud points (UCST-behavior)

w_B	0.09208	0.09374	0.12272	0.12389	0.15492	0.18131	0.18487	0.21412	0.23075
T/K	296.628	296.604	297.290	297.283	297.522	297.622	297.644	297.662	297.645
w_B	0.23366	0.24135	0.24524	0.27272	0.27350	0.30123	0.30267		
T/K	297.637	297.468	297.581	297.289	297.271	296.980	296.983		
w_B	0.15	0.15	0.15	0.15	0.15	0.15	0.15	0.15	0.15
T/K	296.751	296.044	295.337	294.739	294.433	294.200	293.892	293.953	294.203
P/bar	48.54	90.14	144.85	210.13	249.33	309.56	385.50	638.82	743.95
w_B	0.15	0.27	0.27	0.27	0.27	0.27	0.27	0.27	0.27
T/K	294.490	296.418	295.687	295.080	294.603	294.257	293.944	293.603	293.653
P/bar	828.10	53.06	98.49	151.24	201.30	249.66	306.13	428.80	645.79
w_B	0.27	0.27	0.30	0.30	0.30	0.30	0.30	0.30	0.30
T/K	293.944	294.323	296.448	295.705	294.949	294.449	293.935	293.690	293.492
P/bar	758.94	857.25	37.87	76.82	134.13	185.03	252.98	311.90	377.40
w_B	0.30	0.30	0.30	0.09374	0.12389	0.15492	0.18487	0.21412	0.23366
T/K	293.444	293.694	294.041	295.053	295.673	295.885	295.925	295.954	295.915
P/bar	620.16	726.60	822.60	100	100	100	100	100	100
w_B	0.24524	0.27350	0.30267	0.09374	0.12389	0.15492	0.18487	0.21412	0.23366
T/K	295.822	295.623	295.406	294.014	294.675	294.853	294.871	294.882	294.857
P/bar	100	100	100	200	200	200	200	200	200
w_B	0.24524	0.15492	0.18487	0.21412	0.23366	0.24524	0.15492	0.18487	0.21412
T/K	294.727	294.853	294.871	294.882	294.857	294.727	294.853	294.871	294.882
P/bar	200	200	200	200	200	200	200	200	200

continued

continued

w_B	0.23366	0.24524	0.18487	0.21412	0.23366	0.24524	0.27350	0.30267	0.09111
T/K	294.857	294.727	293.915	293.928	293.912	293.804	293.631	293.425	293.105
P/bar	200	200	400	400	400	400	400	400	500
w_B	0.12100	0.14958	0.17954	0.20815	0.23883	0.26815	0.29882	0.09374	0.12389
T/K	293.615	293.800	293.844	293.846	293.717	293.556	293.417	293.409	293.890
P/bar	500	500	500	500	500	500	500	700	700
w_B	0.15492	0.18487	0.21412	0.23366	0.24524	0.27350	0.30267	0.09374	0.12389
T/K	294.039	294.045	294.059	294.031	293.991	293.792	293.578	293.726	294.200
P/bar	700	700	700	700	700	700	700	800	800
w_B	0.15492	0.18487	0.21412	0.23366	0.24524	0.27350	0.30267	0.09208	0.12272
T/K	294.353	294.344	294.362	294.330	294.223	294.117	293.944	293.744	294.234
P/bar	800	800	800	800	800	800	800	800	800
w_B	0.18131	0.23075	0.24135	0.27272	0.30123	0.14889	0.20957		
T/K	294.393	294.366	294.249	294.137	293.969	294.389	294.402		
P/bar	800	800	800	800	800	800	800		

Type of data: cloud points (hour glass shaped)

T/K = 293.75

w_B	0.09346	0.12511	0.12575	0.09717	0.30657	0.27626	0.24389	0.27265	0.30324
P/bar	242.30	390.04	628.77	807.35	750.55	691.49	458.09	366.24	307.93

Type of data: cloud points (hour glass shaped)

T/K = 293.75

w_B	0.09047	0.12062	0.12013	0.08932	0.29871	0.26886	0.23907	0.26842	0.29852
P/bar	797.75	617.66	379.31	237.63	298.77	358.60	448.76	678.21	739.72

Type of data: cloud points (hour glass shaped)

T/K = 293.85

w_B	0.09098	0.12146	0.15103	0.18096	0.18206	0.15130	0.12167	0.09115	0.29925
P/bar	829.50	672.32	608.93	530.56	501.39	433.99	346.08	225.87	280.15
w_B	0.26965	0.23964	0.23028	0.21022	0.20997	0.23038	0.23969	0.26923	0.29875
P/bar	333.30	384.09	437.78	497.12	544.41	604.05	664.62	723.37	774.25

Type of data: cloud points (LCP and UCP branch)

T/K = 293.95

w_B	0.09764	0.12742	0.15562	0.18612	0.21626	0.23594	0.24565	0.27551	0.30541
P/bar	867.74	722.59	667.50	646.92	650.97	664.81	712.02	761.05	803.69
w_B	0.09764	0.12742	0.15562	0.18612	0.21626	0.23594	0.24565	0.27551	0.30541
P/bar	215.73	322.72	373.59	391.89	395.86	386.19	346.92	310.12	265.02

continued

continued

Type of data: cloud points (LCP and UCP branch)

T/K = 293.95

w_B	0.09048	0.12000	0.15048	0.18044	0.20992	0.22902	0.23962	0.26936	0.29906
P/bar	851.13	717.32	658.03	639.52	635.14	655.21	703.69	754.97	821.17

w_B	0.09048	0.12000	0.15048	0.18044	0.20992	0.22902	0.23962	0.26936	0.29906
P/bar	207.21	314.20	365.48	383.05	383.37	376.80	339.77	304.79	258.84

Type of data: cloud points (LCP and UCP branch)

T/K = 293.95

w_B	0.09071	0.12052	0.15092	0.18080	0.21015	0.24062	0.27109	0.29993
P/bar	852.80	719.97	658.02	642.46	637.21	704.12	758.15	822.49

w_B	0.09071	0.12052	0.15092	0.18080	0.21015	0.24062	0.27109
P/bar	208.48	316.63	366.78	386.02	384.63	342.01	304.55

Type of data: cloud points (LCP and UCP branch)

T/K = 294.15

w_B	0.09764	0.12742	0.15562	0.18612	0.21626	0.23594	0.24565	0.27551	0.30541
P/bar	907.31	784.78	745.01	734.67	733.53	739.27	775.31	812.97	852.68

w_B	0.09764	0.12742	0.15562	0.18612	0.21626	0.23594	0.24565	0.27551	0.30541
P/bar	189.20	279.93	314.50	324.57	326.57	317.03	293.78	270.23	232.38

Type of data: cloud points (LCP and UCP branch)

T/K = 294.35

w_B	0.09764	0.12742	0.15562	0.18612	0.21626	0.23594	0.24565	0.27551	0.30541
P/bar	955.28	842.03	805.16	802.03	797.84	804.64	831.39	859.88	891.35

w_B	0.09764	0.12742	0.15562	0.18612	0.21626	0.23594	0.24565	0.27551	0.30541
P/bar	161.74	246.23	274.68	280.64	282.68	276.23	257.06	235.71	204.83

Type of data: cloud points (UCP behavior)

T/K = 295.15

w_B	0.09764	0.12742	0.15562	0.18612	0.21626	0.23594	0.24565	0.27551	0.30541
P/bar	92.17	151.23	168.52	172.56	176.52	168.00	160.15	141.73	122.22

Type of data: cloud points (UCP behavior)

T/K = 296.55

w_B	0.09764	0.12742	0.15562	0.18612	0.21626	0.23594	0.24565	0.27551	0.30541
P/bar	11.69	44.21	59.46	62.41	66.45	62.94	57.16	40.75	32.56

Polymer (B): **polystyrene** **1999NAK**
Characterization: M_n/g.mol^{-1} = 17650, M_w/g.mol^{-1} = 18700,
Pressure Chemical Company, Pittsburgh, PA
Solvent (A): **methylcyclohexane** C_7H_{14} **108-87-2**

Type of data: coexistence data (UCST-behavior)

φ_B (total) 0.1549 was kept constant (critical concentration)

T/K	296.557	296.550	296.544	296.532	296.517	296.494	296.448	296.403
φ_B (sol phase)	0.1549	0.1466	0.1410	0.1366	0.1330	0.1269	0.1200	0.1146
φ_B (gel phase)	0.1549	0.1822	0.1899	0.1941	0.1987	0.2040	0.2120	0.2172
T/K	296.362	296.306	296.228	296.067	295.852	295.521	295.195	295.015
φ_B (sol phase)	0.1111	0.1068	0.1014	0.0936	0.0855	0.0762	0.0692	0.0658
φ_B (gel phase)	0.2226	0.2271	0.2338	0.2439	0.2557	0.2691	0.2799	0.2863

Type of data: critical point (UCST-behavior)

$\varphi_{B,\,crit}$ 0.1549 T_{crit}/K 296.557

Polymer (B): **polystyrene** **1980DO1**
Characterization: M_n/g.mol^{-1} = 19100, M_w/g.mol^{-1} = 20200,
Pressure Chemical Company, Pittsburgh, PA
Solvent (A): **methylcyclohexane** C_7H_{14} **108-87-2**

Type of data: coexistence data (UCST-behavior)

φ_B (total) 0.1590 was kept constant (critical concentration)

T/K	298.95	298.884	298.846	298.773	298.687	298.453	297.949	297.304
φ_B (sol phase)	0.1590	0.1234	0.1160	0.1078	0.1003	0.0873	0.0712	0.0593
φ_B (gel phase)	0.1590	0.1977	0.2044	0.2127	0.2226	0.2395	0.2633	0.2836
T/K	296.025	293.750						
φ_B (sol phase)	0.0442	0.0279						
φ_B (gel phase)	0.3150	0.3528						

Type of data: critical point (UCST-behavior)

$\varphi_{B,\,crit}$ 0.1590 T_{crit}/K 298.95

Polymer (B): **polystyrene** **1993WEL**
Characterization: M_n/g.mol^{-1} = 21400, M_w/g.mol^{-1} = 22000
Solvent (A): **methylcyclohexane** C_7H_{14} **108-87-2**

Type of data: spinodal points (UCST-behavior)

w_B	0.055	0.055	0.055	0.055	0.055	0.055	0.055	0.055	0.055
T/K	298.20	297.99	297.39	297.11	296.64	296.24	295.75	295.41	294.93
P/bar	5.42	9.64	43.97	62.05	90.96	120.48	160.84	195.78	253.61
w_B	0.055	0.055	0.055	0.055	0.055	0.094	0.094	0.094	0.094
T/K	294.61	294.34	294.29	294.32	294.51	300.25	299.75	299.45	299.05
P/bar	315.66	402.41	481.32	551.81	682.53	18.07	39.76	56.63	80.12

continued

continued

w_B	0.094	0.094	0.094	0.094	0.094	0.094	0.094	0.094	0.094
T/K	298.68	298.35	298.02	297.53	297.08	296.86	296.68	296.13	296.17
P/bar	100.60	124.70	146.38	152.41	239.15	283.13	304.22	509.04	653.61

w_B	0.200	0.200	0.200	0.200	0.200	0.200	0.200	0.200	0.200
T/K	301.09	300.76	300.47	300.09	300.13	299.42	298.93	298.41	298.39
P/bar	24.10	28.31	42.17	62.05	63.25	100.00	134.94	184.34	189.16

w_B	0.200	0.200	0.200	0.200	0.200
T/K	297.74	297.34	297.09	296.89	297.04
P/bar	256.63	333.73	424.70	533.73	680.12

Type of data: critical point (UCST-behavior)

$w_{B, crit}$ = 0.200 T_{crit}/K = 300.85 (at P/bar = 1)

Polymer (B): **polystyrene** **1991SHE**
Characterization: M_n/g.mol^{-1} = 21700, M_w/g.mol^{-1} = 23000,
Pressure Chemical Company, Pittsburgh, PA
Solvent (A): **methylcyclohexane** C_7H_{14} **108-87-2**

Type of data: critical point (UCST-behavior)

$w_{B, crit}$ = 0.197 T_{crit}/K = 302.268

Polymer (B): **polystyrene** **1991SHE**
Characterization: M_n/g.mol^{-1} = 27400, M_w/g.mol^{-1} = 29000,
Pressure Chemical Company, Pittsburgh, PA
Solvent (A): **methylcyclohexane** C_7H_{14} **108-87-2**

Type of data: critical point (UCST-behavior)

$w_{B, crit}$ = 0.184 T_{crit}/K = 306.097

Polymer (B): **polystyrene** **1998KOA**
Characterization: M_n/g.mol^{-1} = 29100, M_w/g.mol^{-1} = 31600,
Aldrich Chem. Co., Inc., Milwaukee, WI
Solvent (A): **methylcyclohexane** C_7H_{14} **108-87-2**

Type of data: cloud points

w_B	0.0323	0.0323	0.0323	0.0323	0.0323	0.0323	0.0323	0.0494	0.0494
T/K	299.38	298.99	298.58	298.25	297.92	297.60	297.14	301.30	301.01
P/MPa	1.05	3.05	5.05	7.05	9.05	11.05	13.05	1.05	2.05

w_B	0.0494	0.0494	0.0494	0.0494	0.0494	0.0494	0.0494	0.0494	0.0494
T/K	300.78	300.60	300.41	300.22	300.03	299.85	299.68	299.50	299.19
P/MPa	3.05	4.05	5.05	6.05	7.05	8.05	9.05	10.05	12.05

continued

continued

w_B	0.0494	0.0726	0.0726	0.0726	0.0726	0.0726	0.0726	0.0726	0.0726
T/K	298.95	302.53	302.29	302.04	301.82	301.61	301.41	301.25	301.05
P/MPa	14.05	1.05	2.05	3.05	4.05	5.05	6.05	7.05	8.05
w_B	0.0726	0.0726	0.0726	0.0726	0.0882	0.0882	0.0882	0.0882	0.0882
T/K	300.88	300.73	300.42	300.13	303.46	303.39	302.93	302.72	302.47
P/MPa	9.05	10.05	12.05	14.05	1.05	2.05	3.05	4.05	5.05
w_B	0.0882	0.0882	0.0882	0.0882	0.0882	0.0882	0.0882	0.1097	0.1097
T/K	302.37	302.10	301.90	301.79	301.52	301.24	300.90	304.05	303.48
P/MPa	6.05	7.05	8.05	9.05	10.05	12.05	14.05	1.05	3.05
w_B	0.1097	0.1097	0.1097	0.1097	0.1097	0.1281	0.1281	0.1281	0.1281
T/K	303.06	302.69	302.25	301.88	301.52	304.44	303.88	303.45	302.97
P/MPa	5.05	7.05	9.05	11.05	13.05	1.05	3.05	5.05	7.05
w_B	0.1281	0.1281	0.1281	0.1281	0.1519	0.1519	0.1519	0.1519	0.1519
T/K	302.56	302.24	301.93	301.68	305.03	304.44	304.00	303.53	303.11
P/MPa	9.05	11.05	13.05	14.05	1.05	3.05	5.05	7.05	9.05
w_B	0.1519	0.1519	0.1519	0.1821	0.1821	0.1821	0.1821	0.2060	0.2060
T/K	302.62	302.35	302.17	305.45	304.41	304.00	303.53	305.42	304.95
P/MPa	11.05	13.05	14.05	1.05	5.05	7.05	9.05	1.05	3.05
w_B	0.2060	0.2060	0.2060	0.2060	0.2060	0.2251	0.2251	0.2251	0.2251
T/K	304.53	303.89	303.52	303.15	302.80	305.43	304.96	304.35	303.91
P/MPa	5.05	7.05	9.05	11.05	13.05	1.05	3.05	5.05	7.05
w_B	0.2251	0.2251	0.2251	0.2634	0.2634	0.2634	0.2634	0.2634	0.2634
T/K	303.45	303.08	302.75	305.03	304.44	303.90	303.51	303.02	302.70
P/MPa	9.05	11.05	13.05	1.05	3.05	5.05	7.05	9.05	11.05
w_B	0.2634								
T/K	302.32								
P/MPa	13.05								

Polymer (B): **polystyrene** **2002ZHO**
Characterization: M_n/g.mol^{-1} = 29800, M_w/g.mol^{-1} = 30400, Polymer Laboratories, Inc., Amherst, MA
Solvent (A): **methylcyclohexane** C_7H_{14} **108-87-2**

Type of data: critical point (UCST-behavior)

$\varphi_{B,\,crit}$ 0.143 T_{crit}/K 304.334

Polymer (B): **polystyrene** **1980DO1**
Characterization: M_n/g.mol^{-1} = 32900, M_w/g.mol^{-1} = 34900, Pressure Chemical Company, Pittsburgh, PA
Solvent (A): **methylcyclohexane** C_7H_{14} **108-87-2**

Type of data: coexistence data (UCST-behavior)

continued

continued

φ_B (total)	0.1293	was kept constant	(critical concentration)					
T/K	309.00	308.964	308.842	308.588	307.930	306.848	305.694	304.040
φ_B (sol phase)	0.1293	0.1010	0.0838	0.0713	0.0529	0.0379	0.0287	0.0205
φ_B (gel phase)	0.1293	0.1594	0.1772	0.1969	0.2241	0.2542	0.2780	0.3057

Type of data: critical point (UCST-behavior)

$\varphi_{B, crit}$ 0.1293 T_{crit}/K 309.00

Polymer (B): **polystyrene** **1993HOS**
Characterization: M_n/g.mol^{-1} = 33000, M_w/g.mol^{-1} = 34900,
Pressure Chemical Company, Pittsburgh, PA
Solvent (A): **methylcyclohexane** C_7H_{14} **108-87-2**

Type of data: cloud points (UCST-behavior)

w_B	0.1292	was kept constant							
T/K	308.87	308.48	308.10	307.80	307.44	307.08	306.73	306.33	305.86
P/MPa	0.10	1.27	2.65	3.82	5.20	6.67	8.24	9.90	12.36
T/K	305.51	305.23	304.77	304.53	304.22	303.90	303.58	303.25	302.97
P/MPa	14.42	16.38	19.61	21.97	24.42	27.65	31.77	36.68	43.44
T/K	302.76	302.71	302.71	302.81	302.91	303.05	303.25		
P/MPa	51.39	58.94	65.51	72.37	78.36	84.53	90.61		

Type of data: critical double point

T_{dcrit}/K = 302.86 P_{dcrit}/MPa = 61.6

Polymer (B): **polystyrene** **1973SA1**
Characterization: M_n/g.mol^{-1} = 34900, M_w/g.mol^{-1} = 37000,
Pressure Chemical Company, Pittsburgh, PA
Solvent (A): **methylcyclohexane** C_7H_{14} **108-87-2**

Type of data: cloud points (UCST-behavior)

w_B	0.05174	0.09153	0.10677	0.14357	0.18093	0.21368
T/K	306.554	308.107	308.229	308.553	308.553	308.380

Type of data: cloud points (LCST-behavior)

w_B	0.05174	0.09153	0.10677	0.14357	0.18093	0.21368
T/K	520.167	519.003	518.831	519.132	520.263	519.493

Polymer (B): **polystyrene** **1996RON**
Characterization: M_n/g.mol^{-1} = 37500, M_w/g.mol^{-1} = 37900,
Tosoh Corp., Ltd., Tokyo, Japan
Solvent (A): **methylcyclohexane** C_7H_{14} **108-87-2**

Type of data: cloud points (UCST-behavior)

continued

continued

w_B	0.0129	0.0517	0.0733	0.1132	0.1322	0.1532	0.1604	0.2169	0.2363
T/K	300.99	308.31	309.35	310.17	310.39	310.53	310.52	310.05	309.69

w_B	0.2794	0.3151
T/K	308.75	308.69

Type of data: critical point (UCST-behavior)

$w_{B, crit}$ 0.1532 T_{crit}/K 310.53

Type of data: spinodal points (UCST-behavior)

w_B	0.0500	0.0600	0.0670	0.0820	0.0850	0.1180	0.1440	0.1570	0.1950
T/K	301.05	304.05	305.35	307.25	307.45	309.75	310.45	310.55	309.85

w_B	0.2340	0.3310
T/K	308.45	301.35

Polymer (B): **polystyrene** **1980DO1**
Characterization: M_n/g.mol^{-1} = 43800, M_w/g.mol^{-1} = 46400,
Pressure Chemical Company, Pittsburgh, PA
Solvent (A): **methylcyclohexane** C_7H_{14} **108-87-2**

Type of data: coexistence data (UCST-behavior)

φ_B (total) 0.1170 was kept constant (critical concentration)

T/K	312.61	312.576	312.545	312.505	312.444	312.334	312.158	311.820
φ_B (sol phase)	0.1170	0.0925	0.0864	0.0816	0.0769	0.0689	0.0623	0.0520
φ_B (gel phase)	0.1170	0.1429	0.1492	0.1545		0.1707	0.1812	0.1969

T/K	311.045	309.775	307.600
φ_B (sol phase)	0.0390	0.0273	0.0167
φ_B (gel phase)	0.2212	0.2501	0.2863

Type of data: critical point (UCST-behavior)

$\varphi_{B, crit}$ 0.1170 T_{crit}/K 312.61

Polymer (B): **polystyrene** **2002ZHO**
Characterization: M_n/g.mol^{-1} = 47400, M_w/g.mol^{-1} = 48800,
Polymer Laboratories, Inc., Amherst, MA
Solvent (A): **methylcyclohexane** C_7H_{14} **108-87-2**

Type of data: critical point (UCST-behavior)

$\varphi_{B, crit}$ 0.119 T_{crit}/K 311.997

Polymer (B): **polystyrene** **1988SC2**
Characterization: M_n/g.mol^{-1} = 54500, M_w/g.mol^{-1} = 297000, M_z/g.mol^{-1} = 695000
Solvent (A): **methylcyclohexane** C_7H_{14} **108-87-2**

Type of data: cloud points (UCST-behavior)

continued

continued

w_B	0.0088	0.019	0.028	0.039	0.044	0.053	0.059	0.069	0.079
T/K	327.98	331.96	332.22	332.01	331.70	331.53	330.97	330.80	330.22

w_B	0.111	0.121	0.135	0.139	0.148
T/K	328.73	328.60	328.16	327.68	327.34

Type of data: critical point (UCST-behavior)

$w_{B, crit}$ 0.115 T_{crit}/K 328.66

Polymer (B): **polystyrene** **1998KOA**
Characterization: M_n/g.mol^{-1} = 64000, M_w/g.mol^{-1} = 250000,
Novacor Technology and Research Corporation
Solvent (A): **methylcyclohexane** C_7H_{14} **108-87-2**

Type of data: cloud points (UCST-behavior)

w_B	0.0211	0.0211	0.0211	0.0211	0.0211	0.0211	0.0330	0.0330	0.0330
T/K	329.81	329.35	328.50	327.87	327.24	326.63	330.76	330.37	329.87
P/MPa	3.95	4.90	6.55	7.95	9.25	10.85	2.40	3.15	4.00

w_B	0.0330	0.0330	0.0330	0.0330	0.0511	0.0511	0.0511	0.0511	0.0511
T/K	329.01	328.09	327.43	326.35	330.91	330.35	329.46	328.62	327.71
P/MPa	5.65	7.60	9.05	11.80	1.45	2.40	3.95	5.60	7.60

w_B	0.0511	0.0511	0.0511	0.0648	0.0648	0.0648	0.0648	0.0648	0.0648
T/K	326.91	326.23	325.75	330.11	330.09	329.04	328.98	327.83	326.85
P/MPa	9.50	11.30	12.60	2.00	2.00	3.90	4.00	6.30	8.60

w_B	0.0648	0.0648	0.0648	0.0971	0.0971	0.0971	0.0971	0.0971	0.0971
T/K	326.37	325.83	325.26	329.34	328.59	327.81	327.05	326.30	325.47
P/MPa	9.75	11.15	12.85	1.60	3.05	4.55	6.10	7.70	9.80

w_B	0.0971	0.1360	0.1360	0.1360	0.1360	0.1360	0.1360	0.1360	0.1360
T/K	324.86	328.36	328.29	327.68	327.33	326.63	325.86	325.15	324.47
P/MPa	11.55	1.50	1.60	3.00	3.50	4.85	6.55	8.25	10.10

Polymer (B): **polystyrene** **2002ZHO**
Characterization: M_n/g.mol^{-1} = 67300, M_w/g.mol^{-1} = 68600,
Polymer Laboratories, Inc., Amherst, MA
Solvent (A): **methylcyclohexane** C_7H_{14} **108-87-2**

Type of data: critical point (UCST-behavior)

$\varphi_{B, crit}$ 0.104 T_{crit}/K 316.798

Polymer (B): **polystyrene** **2004WIL**
Characterization: M_n/g.mol^{-1} = 86540, M_w/g.mol^{-1} = 90000,
Pressure Chemical Company, Pittsburgh, PA
Solvent (A): **methylcyclohexane** C_7H_{14} **108-87-2**

continued

continued

Type of data: cloud points (UCST-behavior)

w_B	0.0353	0.0960	0.1030	0.1304	0.1311	0.1551	0.2022	0.2447
T/K	316.3	319.5	320.1	319.9	320.1	319.9	320.1	318.7

Polymer (B): **polystyrene** **1971KAG**
Characterization: M_w/g.mol^{-1} = 97200
Solvent (A): **methylcyclohexane** C_7H_{14} **108-87-2**

Type of data: cloud points (UCST-behavior)

w_B 0.00277 T/K 318.25

Polymer (B): **polystyrene** **1973SA1**
Characterization: M_n/g.mol^{-1} = 91700, M_w/g.mol^{-1} = 97200,
Pressure Chemical Company, Pittsburgh, PA
Solvent (A): **methylcyclohexane** C_7H_{14} **108-87-2**

Type of data: cloud points (UCST-behavior)

w_B	0.05296	0.08144	0.10475	0.11738	0.16132	0.16878
T/K	320.908	321.333	321.660	321.499	321.287	321.086

Type of data: cloud points (LCST-behavior)

w_B	0.05296	0.08144	0.10475	0.11738	0.16132	0.16878
T/K	505.798	505.863	505.810	505.989	506.937	507.525

Polymer (B): **polystyrene** **2002ZHO**
Characterization: M_n/g.mol^{-1} = 94800, M_w/g.mol^{-1} = 97600,
Polymer Laboratories, Inc., Amherst, MA
Solvent (A): **methylcyclohexane** C_7H_{14} **108-87-2**

Type of data: critical point (UCST-behavior)

$\varphi_{B,\,crit}$ 0.092 T_{crit}/K 319.654

Polymer (B): **polystyrene** **1980DO1**
Characterization: M_n/g.mol^{-1} = 103000, M_w/g.mol^{-1} = 109000,
Pressure Chemical Company, Pittsburgh, PA
Solvent (A): **methylcyclohexane** C_7H_{14} **108-87-2**

Type of data: coexistence data (UCST-behavior)

φ_B (total) 0.0840 was kept constant (critical concentration)

T/K	322.71	322.695	322.677	322.634	322.548	322.386	321.999	321.412
φ_B (sol phase)	0.0840	0.0694	0.0635	0.0582	0.0518	0.0443	0.0344	0.0263
φ_B (gel phase)	0.0840	0.0992	0.1052	0.1118	0.1203	0.1306	0.1471	0.1639

continued

continued

T/K	320.221	317.540	313.040
φ_B (sol phase)	0.0169	0.0077	0.0022
φ_B (gel phase)	0.1924	0.2352	0.2919

Type of data: critical point (UCST-behavior)

$\varphi_{B,\ crit}$ 0.0840 T_{crit}/K 322.71

Polymer (B): **polystyrene** **1997END**
Characterization: M_n/g.mol^{-1} = 143000, M_w/g.mol^{-1} = 405000, BASF AG, Ludwigshafen, Germany
Solvent (A): **methylcyclohexane** **C_7H_{14}** **108-87-2**

Type of data: cloud points (UCST-behavior, measured by PPICS)

w_B	0.0067	0.0067	0.0067	0.0067	0.0067	0.0067	0.0067	0.0067	0.0067
T/K	331.18	330.54	330.06	329.47	329.02	328.57	328.04	327.23	326.43
P/bar	17.71	19.64	20.60	23.92	30.00	37.97	47.02	60.22	74.35
w_B	0.0067	0.0067	0.0067	0.0067	0.0067	0.0067	0.0067	0.0067	0.0067
T/K	325.95	325.50	325.04	324.51	323.82	323.12	322.63	322.29	321.79
P/bar	84.72	92.67	108.28	122.42	143.15	160.12	174.25	187.45	211.95
w_B	0.0067	0.0067	0.0067	0.0067	0.0067	0.0067	0.0067	0.0067	0.0067
T/K	321.47	321.04	320.61	320.33	319.54	319.01	318.41	318.05	317.24
P/bar	225.15	243.05	270.39	297.72	331.65	362.75	393.85	417.41	462.65
w_B	0.018	0.018	0.018	0.018	0.018	0.018	0.018	0.018	0.018
T/K	331.16	330.65	330.05	329.58	329.25	328.78	328.10	327.56	326.81
P/bar	12.55	13.26	19.64	29.31	36.41	44.64	55.16	66.85	82.16
w_B	0.018	0.018	0.018	0.018	0.018	0.018	0.018	0.018	0.018
T/K	326.02	324.96	324.09	323.36	322.98	322.23	321.65	321.15	320.48
P/bar	101.19	127.38	157.74	181.35	196.63	217.61	247.92	269.92	294.99
w_B	0.018	0.018	0.018	0.018	0.018	0.025	0.025	0.025	0.025
T/K	320.00	319.12	318.54	317.64	317.11	331.38	330.80	330.13	329.50
P/bar	325.99	356.25	386.56	426.54	456.84	11.57	18.18	29.14	40.89
w_B	0.025	0.025	0.025	0.025	0.025	0.025	0.025	0.025	0.025
T/K	329.02	328.49	327.67	326.99	326.37	325.88	325.40	324.94	324.17
P/bar	45.25	52.50	73.34	80.95	99.21	115.74	123.63	145.09	169.36
w_B	0.025	0.025	0.025	0.025	0.025	0.025	0.025	0.025	0.025
T/K	323.41	322.93	322.39	321.68	321.01	320.40	320.13	319.70	319.48
P/bar	182.49	196.35	223.73	253.42	279.26	300.20	312.85	337.30	346.84
w_B	0.025	0.025	0.025	0.025	0.025	0.025	0.0458	0.0458	0.0458
T/K	319.13	318.62	318.16	317.74	317.09	316.18	331.22	330.73	330.33
P/bar	364.87	393.05	414.95	442.06	490.77	592.86	27.39	34.03	38.51

continued

continued

w_B	0.0458	0.0458	0.0458	0.0458	0.0458	0.0458	0.0458	0.0458	0.0458
T/K	329.71	329.35	328.60	328.51	327.92	327.41	326.85	326.17	325.63
P/bar	51.45	61.25	71.41	73.00	87.75	102.49	117.62	137.17	152.61
w_B	0.0458	0.0458	0.0482	0.0482	0.0482	0.0482	0.0482	0.0482	0.0482
T/K	325.23	324.84	332.61	332.11	331.63	331.26	330.98	330.23	329.49
P/bar	160.12	170.49	9.84	14.09	19.33	27.87	30.81	44.51	57.70
w_B	0.0482	0.0482	0.0482	0.0482	0.0482	0.0482	0.0482	0.0482	0.0482
T/K	328.66	328.30	327.77	327.44	326.98	326.50	325.87	324.89	324.44
P/bar	73.68	84.18	91.10	112.05	126.19	139.38	153.52	174.26	194.05
w_B	0.0482	0.0482	0.0482	0.0482	0.0482	0.0482	0.0482	0.0482	0.0482
T/K	324.26	323.40	322.92	322.45	322.18	321.59	320.96	320.52	320.22
P/bar	201.59	218.55	235.52	256.06	273.62	299.65	329.25	354.85	376.00
w_B	0.0482	0.0482	0.0482	0.0482	0.0482	0.0482	0.0482	0.0482	0.0482
T/K	320.15	319.94	319.41	318.92	318.51	317.96	317.53	317.11	316.94
P/bar	373.11	388.53	410.81	435.32	459.50	500.35	558.78	613.44	658.68
w_B	0.0482	0.0482	0.0482	0.0682	0.0682	0.0682	0.0682	0.0682	0.0682
T/K	316.84	316.79	316.68	331.79	331.34	330.90	329.89	329.11	328.22
P/bar	692.61	719.94	754.81	19.69	19.69	29.12	49.85	57.39	78.12
w_B	0.0682	0.0682	0.0682	0.0682	0.0682	0.0682	0.0682	0.0682	0.0682
T/K	327.73	327.41	326.45	325.90	325.30	324.79	324.12	323.57	323.25
P/bar	84.72	91.32	126.19	143.15	163.89	177.08	197.82	209.13	218.55
w_B	0.0682	0.0682	0.0682	0.0682	0.0682	0.0682	0.0682	0.0682	0.0682
T/K	322.50	321.92	321.43	321.06	320.52	319.85	319.35	318.93	318.40
P/bar	243.05	273.21	293.95	318.45	352.38	376.88	409.40	439.80	478.50
w_B	0.0682	0.0682							
T/K	318.12	317.82							
P/bar	517.31	551.88							

Type of data: cloud points (UCST-behavior)

w_B	0.011	0.011	0.011	0.011	0.011	0.011	0.011	0.011	0.011
T/K	330.17	329.93	329.88	329.57	329.38	329.30	329.00	328.84	328.77
P/bar	7.0	12.0	13.2	19.2	23.0	24.2	30.6	33.6	35.3
w_B	0.011	0.011	0.011	0.011	0.011	0.011	0.011	0.011	0.011
T/K	328.49	328.37	328.26	327.96	327.80	327.67	327.39	327.32	327.03
P/bar	40.8	43.5	45.8	51.6	55.3	57.7	64.2	65.6	71.3
w_B	0.011	0.011	0.011	0.011	0.011	0.011	0.011	0.011	0.011
T/K	326.71	326.67	326.62	326.47	326.38	326.17	326.15	326.09	325.97
P/bar	78.8	79.8	81.0	84.0	86.0	90.8	91.3	92.5	95.0
w_B	0.011	0.060	0.060	0.060	0.060	0.060	0.060	0.060	0.060
T/K	325.63	331.66	331.47	331.28	331.09	330.91	330.71	330.50	330.30
P/bar	103.3	23.0	26.0	28.4	31.8	34.4	37.6	41.1	44.6

continued

continued

w_B	0.060	0.060	0.060	0.060	0.060	0.060	0.060	0.060	0.060
T/K	330.09	329.96	329.72	329.51	329.30	329.13	328.94	328.78	328.52
P/bar	48.3	50.5	55.0	58.7	62.7	66.2	70.1	73.3	78.5
w_B	0.060	0.060	0.060	0.060	0.060	0.060	0.060	0.120	0.120
T/K	328.29	328.11	327.90	327.74	327.52	327.31	327.12	329.96	329.76
P/bar	83.7	87.5	92.4	96.0	101.0	106.5	111.0	11.6	14.8
w_B	0.120	0.120	0.120	0.120	0.120	0.120	0.120	0.120	0.120
T/K	329.55	329.35	329.14	328.94	328.75	328.56	328.35	328.16	327.97
P/bar	17.7	21.0	24.3	28.0	31.0	34.3	38.3	42.3	45.8
w_B	0.120	0.120	0.120	0.120	0.120	0.120	0.120	0.120	0.120
T/K	327.75	327.57	327.38	327.19	326.98	326.79	326.57	326.39	326.16
P/bar	51.0	54.7	58.8	63.2	68.0	72.5	78.0	82.3	88.5
w_B	0.120	0.120	0.187	0.187	0.187	0.187	0.187	0.187	0.187
T/K	325.98	325.78	327.54	327.33	327.12	326.92	326.73	326.55	326.35
P/bar	93.6	99.0	10.1	13.2	17.0	19.7	23.3	27.0	31.0
w_B	0.187	0.187	0.187	0.187	0.187	0.187	0.187	0.187	0.187
T/K	326.15	325.97	325.75	325.55	325.34	325.14	324.94	324.78	324.58
P/bar	35.1	39.5	44.5	49.6	54.8	60.2	66.0	71.0	77.0
w_B	0.187	0.187	0.187						
T/K	324.35	324.13	323.91						
P/bar	84.0	91.0	99.0						

Type of data: cloud points (LCST-behavior)

w_B	0.065	0.065	0.065	0.065	0.065	0.065	0.065	0.065	0.065
T/K	499.45	502.12	504.64	507.12	509.61	512.10	514.59	517.04	519.50
P/bar	18.0	21.2	24.5	27.5	30.6	33.6	37.05	39.8	42.9
w_B	0.187	0.187	0.187	0.187	0.187	0.187	0.187		
T/K	503.80	506.47	508.41	511.16	513.88	516.08	518.85		
P/bar	13.2	16.1	18.4	21.7	24.7	27.6	30.8		

Type of data: coexistence data (liquid-liquid-vapor three phase equilibrium)

w_B	0.0113	0.0516	0.0600	0.0655	0.0989	0.1207	0.1520	0.18699
T/K	500.92	495.35	493.67	493.90	492.92	496.35	498.78	502.30
P/bar	12.78	11.47	10.81	10.77	9.82	10.68	10.84	11.29

Polymer (B): **polystyrene** **2002ZHO**

Characterization: M_n/g.mol^{-1} = 165000, M_w/g.mol^{-1} = 168000, Polymer Laboratories, Inc., Amherst, MA

Solvent (A): **methylcyclohexane** C_7H_{14} **108-87-2**

Type of data: critical point (UCST-behavior)

$\varphi_{B,\,crit}$ 0.072 T_{crit}/K 325.117

Polymer (B): **polystyrene** **1980DO1**
Characterization: M_n/g.mol^{-1} = 171000, M_w/g.mol^{-1} = 181000,
Pressure Chemical Company, Pittsburgh, PA
Solvent (A): **methylcyclohexane** C_7H_{14} **108-87-2**

Type of data: coexistence data (UCST-behavior)

φ_B (total)	0.0701	was kept constant	(critical concentration)					
T/K	327.00	326.991	326.962	326.928	326.872	326.760	326.507	326.050
φ_B (sol phase)	0.0701	0.0579	0.0522	0.0000	0.0435	0.0379	0.0309	0.0235
φ_B (gel phase)	0.0701	0.0823	0.0897	0.0941	0.1002	0.1077	0.1200	0.1347

T/K	325.207	323.729
φ_B (sol phase)	0.0160	0.0094
φ_B (gel phase)	0.1553	0.1827

Type of data: critical point (UCST-behavior)

$\varphi_{B,\,crit}$ 0.0701 T_{crit}/K 327.00

Polymer (B): **polystyrene** **1973SA1**
Characterization: M_n/g.mol^{-1} = 189000, M_w/g.mol^{-1} = 200000,
Pressure Chemical Company, Pittsburgh, PA
Solvent (A): **methylcyclohexane** C_7H_{14} **108-87-2**

Type of data: cloud points (UCST-behavior)

w_B	0.03418	0.05521	0.07677	0.08971	0.10148	0.13479
T/K	326.460	327.087	327.245	327.375	327.158	326.983

Type of data: cloud points (LCST-behavior)

w_B	0.03418	0.05521	0.07677	0.08971	0.10148	0.13479
T/K	500.808	499.698	499.880	499.765	500.238	500.890

Polymer (B): **polystyrene** **2002ZHO**
Characterization: M_n/g.mol^{-1} = 200000, M_w/g.mol^{-1} = 204000,
Polymer Laboratories, Inc., Amherst, MA
Solvent (A): **methylcyclohexane** C_7H_{14} **108-87-2**

Type of data: critical point (UCST-behavior)

$\varphi_{B,\,crit}$ 0.070 T_{crit}/K 326.907

Polymer (B): **polystyrene** **2002ZHO**
Characterization: M_n/g.mol^{-1} = 317000, M_w/g.mol^{-1} = 330000,
Polymer Laboratories, Inc., Amherst, MA
Solvent (A): **methylcyclohexane** C_7H_{14} **108-87-2**

Type of data: critical point (UCST-behavior)

$\varphi_{B,\,crit}$ 0.057 T_{crit}/K 329.130

Polymer (B): **polystyrene** **1973SA1**
Characterization: M_n/g.mol^{-1} = 378000, M_w/g.mol^{-1} = 400000,
Pressure Chemical Company, Pittsburgh, PA
Solvent (A): **methylcyclohexane** C_7H_{14} **108-87-2**

Type of data: cloud points (UCST-behavior)

w_B	0.01505	0.02488	0.04274	0.05712	0.07090	0.11625
T/K	331.371	332.096	332.559	332.683	332.577	331.893

Type of data: cloud points (LCST-behavior)

w_B	0.01505	0.02488	0.04274	0.05712	0.07090	0.11625
T/K	495.056	494.999	494.359	494.421	494.482	495.725

Polymer (B): **polystyrene** **1971KAG**
Characterization: M_w/g.mol^{-1} = 411000
Solvent (A): **methylcyclohexane** C_7H_{14} **108-87-2**

Type of data: cloud points (UCST-behavior)

w_B 0.00377 T/K 344.15

Polymer (B): **polystyrene** **1973SA1**
Characterization: M_n/g.mol^{-1} = 610000, M_w/g.mol^{-1} = 670000,
Pressure Chemical Company, Pittsburgh, PA
Solvent (A): **methylcyclohexane** C_7H_{14} **108-87-2**

Type of data: cloud points (UCST-behavior)

w_B	0.01871	0.03657	0.05728	0.06907	0.08368
T/K	333.810	334.389	334.609	334.743	334.166

Type of data: cloud points (LCST-behavior)

w_B	0.01871	0.03657	0.05728	0.06907	0.08368
T/K	492.300	491.952	492.192	492.488	492.961

Polymer (B): **polystyrene** **1980DO1**
Characterization: M_n/g.mol^{-1} = 678000, M_w/g.mol^{-1} = 719000,
Pressure Chemical Company, Pittsburgh, PA
Solvent (A): **methylcyclohexane** C_7H_{14} **108-87-2**

Type of data: coexistence data (UCST-behavior)

φ_B (total) 0.0406 was kept constant (critical concentration)

T/K	334.82	334.803	334.766	334.702	334.633	334.513	334.269	333.831
φ_B (sol phase)	0.0406	0.0305	0.0269	0.0227	0.0207	0.0173	0.0140	0.0095
φ_B (gel phase)	0.0406	0.0516	0.0557	0.0608	0.0644	0.0699	0.0783	0.0900

continued

continued

T/K	332.975	331.586	329.376
φ_B (sol phase)	0.0062	0.0026	0.0006
φ_B (gel phase)	0.1076	0.1314	0.1619

Type of data: critical point (UCST-behavior)

$\varphi_{B,\, crit}$ 0.0406 T_{crit}/K 334.82

Polymer (B): **polystyrene** **1971KAG**
Characterization: M_w/g.mol^{-1} = 860000
Solvent (A): **methylcyclohexane** C_7H_{14} **108-87-2**

Type of data: cloud points (UCST-behavior)

w_B 0.00494 T/K 346.45

Polymer (B): **polystyrene** **1971KAG**
Characterization: M_w/g.mol^{-1} = 1800000
Solvent (A): **methylcyclohexane** C_7H_{14} **108-87-2**

Type of data: cloud points (UCST-behavior)

w_B 0.00694 T/K 348.15

Polymer (B): **polystyrene** **1973SA1**
Characterization: M_n/g.mol^{-1} = 2455000, M_w/g.mol^{-1} = 2700000,
Pressure Chemical Company, Pittsburgh, PA
Solvent (A): **methylcyclohexane** C_7H_{14} **108-87-2**

Type of data: cloud points (UCST-behavior)

w_B	0.01083	0.01920	0.03530	0.04682	0.05740
T/K	339.149	339.645	339.764	340.015	339.453

Type of data: cloud points (LCST-behavior)

w_B	0.01083	0.01920	0.03530	0.04682	0.05740
T/K	488.310	488.605	488.140	488.553	489.025

Polymer (B): **polystyrene** **1995LUS**
Characterization: M_n/g.mol^{-1} = 3910, M_w/g.mol^{-1} = 4140,
Pressure Chemical Company, Pittsburgh, PA
Solvent (A): **methylcyclopentane** C_6H_{12} **96-37-7**

Type of data: cloud points

w_B	0.181	0.181
T/K	259.49	259.62
P/MPa	3.70	1.30

Polymer (B): **polystyrene** **1995LUS**
Characterization: M_n/g.mol^{-1} = 12750, M_w/g.mol^{-1} = 13500,
Pressure Chemical Company, Pittsburgh, PA
Solvent (A): **methylcyclopentane** C_6H_{12} **96-37-7**

Type of data: cloud points

w_B	0.165	0.165	0.165	0.165	0.165
T/K	291.30	291.43	291.53	291.68	291.78
P/MPa	2.00	1.49	1.27	0.79	0.45

Polymer (B): **polystyrene** **1995LUS**
Characterization: M_n/g.mol^{-1} = 21450, M_w/g.mol^{-1} = 22100,
Polymer Laboratories, Inc., Amherst, MA
Solvent (A): **methylcyclopentane** C_6H_{12} **96-37-7**

Type of data: cloud points

w_B	0.153	0.153	0.153	0.153	0.153
T/K	299.99	300.12	300.33	300.57	300.75
P/MPa	2.59	2.24	1.64	1.02	0.51

Polymer (B): **polystyrene** **1995LUS**
Characterization: M_n/g.mol^{-1} = 23600, M_w/g.mol^{-1} = 25000,
Pressure Chemical Company, Pittsburgh, PA
Solvent (A): **methylcyclopentane** C_6H_{12} **96-37-7**

Type of data: cloud points

w_B	0.142	0.142	0.142	0.142	0.148	0.148	0.148	0.148	0.157
T/K	302.58	302.93	303.20	303.38	302.60	302.88	303.18	303.46	302.60
P/MPa	2.20	1.36	0.67	0.23	2.31	1.59	0.84	0.19	2.32
w_B	0.157	0.157	0.157	0.164	0.164	0.164	0.164	0.164	0.165
T/K	302.87	303.24	303.44	302.53	302.76	302.98	303.19	303.43	302.64
P/MPa	1.67	0.77	0.32	2.90	2.25	1.67	1.15	0.55	2.33
w_B	0.165	0.165	0.165	0.177	0.177	0.177	0.177	0.188	0.188
T/K	302.93	303.25	303.47	302.68	302.98	303.28	303.51	302.60	302.90
P/MPa	1.60	0.90	0.38	2.27	1.51	0.83	0.28	2.13	1.45
w_B	0.188	0.188							
T/K	303.16	303.41							
P/MPa	0.84	0.26							

critical concentration: $w_{B,\,crit}$ = 0.170

Polymer (B): **polystyrene** **1995LUS**
Characterization: M_n/g.mol^{-1} = 100300, M_w/g.mol^{-1} = 106300,
Pressure Chemical Company, Pittsburgh, PA

continued

continued

Solvent (A): **methylcyclopentane** C_6H_{12} **96-37-7**

Type of data: cloud points

w_B	0.155	0.155	0.155	0.155	0.155
T/K	325.19	325.42	325.96	326.38	327.01
P/MPa	2.45	2.15	1.63	1.16	0.55

Polymer (B): **polystyrene-d8** **1995LUS**
Characterization: M_n/g.mol^{-1} = 25400, M_w/g.mol^{-1} = 26900, completely deuterated, Polymer Laboratories, Amherst, MA
Solvent (A): **methylcyclopentane** C_6H_{12} **96-37-7**

Type of data: cloud points

w_B	0.179	0.179	0.179	0.179	0.179
T/K	302.48	302.90	303.11	303.20	303.39
P/MPa	2.52	1.51	0.95	0.70	0.23

Polymer (B): **polystyrene** **2001IM1**
Characterization: M_n/g.mol^{-1} = 1160, M_w/g.mol^{-1} = 1240, Pressure Chemical Company, Pittsburgh, PA
Solvent (A): **2-methylpentane** C_6H_{14} **107-83-5**

Type of data: cloud points (UCST-behavior)

w_B	0.080	0.110	0.140	0.170	0.200	0.230
T/K	280.7	284.0	285.0	285.6	285.4	284.6

Polymer (B): **polystyrene** **2001IM1**
Characterization: M_n/g.mol^{-1} = 1160, M_w/g.mol^{-1} = 1240, Pressure Chemical Company, Pittsburgh, PA
Solvent (A): **3-methylpentane** C_6H_{14} **96-14-0**

Type of data: cloud points (UCST-behavior)

w_B	0.080	0.110	0.140	0.170	0.200	0.230	0.277
T/K	274.8	277.2	278.2	278.1	277.6	277.2	276.2

Polymer (B): **polystyrene** **1963ORO**
Characterization: M_n/g.mol^{-1} = 404000, M_w/g.mol^{-1} = 406000
Solvent (A): **4-methyl-2-pentanone** $C_6H_{12}O$ **108-10-1**

Type of data: cloud points (UCST-behavior)

w_B	0.01	T/K	262.15

Polymer (B): **polystyrene** **2000DE1**
Characterization: M_n/g.mol^{-1} = 12740, M_w/g.mol^{-1} = 13500,
Pressure Chemical Company, Pittsburgh, PA
Solvent (A): **nitroethane** $C_2H_5NO_2$ **79-24-3**

Type of data: cloud points

w_B	0.3145	0.3145	0.3145	0.3145	0.3145	0.2867	0.2867	0.2867	0.2867
T/K	277.16	277.27	277.37	277.45	277.55	277.47	277.58	277.67	277.77
P/MPa	4.96	3.71	2.46	1.54	0.36	6.04	4.72	3.71	2.41
w_B	0.2867	0.2867	0.2634	0.2634	0.2634	0.2634	0.2634	0.2634	0.2348
T/K	277.85	277.93	278.08	278.18	278.17	278.29	278.38	278.53	278.59
P/MPa	1.47	0.51	5.56	4.36	4.45	3.16	2.04	0.33	5.33
w_B	0.2348	0.2348	0.2348	0.2348	0.2242	0.2242	0.2242	0.2242	0.2242
T/K	278.69	278.79	278.88	279.01	278.96	279.05	279.16	279.25	279.37
P/MPa	4.14	3.00	1.76	0.27	5.11	4.05	2.66	1.61	0.26
w_B	0.2048	0.2048	0.2048	0.2048	0.2048	0.1799	0.1799	0.1799	0.1799
T/K	279.05	279.16	279.26	279.32	279.42	279.16	279.23	279.36	279.43
P/MPa	4.79	3.49	2.26	1.56	0.40	5.22	4.35	2.83	2.00
w_B	0.1562	0.1562	0.1562	0.1562	0.1340	0.1340	0.1340	0.1340	0.1340
T/K	279.32	279.41	279.49	279.63	279.30	279.40	279.49	279.61	279.75
P/MPa	5.45	4.28	3.31	1.59	5.42	4.28	3.22	1.89	0.15
w_B	0.1179	0.1179	0.1179	0.1801	0.1801	0.1801	0.1801	0.1801	0.1611
T/K	279.19	279.30	279.40	279.19	279.30	279.40	279.45	279.54	279.19
P/MPa	5.72	4.40	3.20	4.59	3.21	2.09	1.52	0.37	5.24
w_B	0.1611	0.1611	0.1611	0.1611	0.1385	0.1385	0.1385	0.1385	0.1385
T/K	279.30	279.39	279.49	279.60	279.49	279.60	279.70	279.80	279.96
P/MPa	3.89	2.72	1.56	0.30	5.76	4.65	3.52	2.26	0.38
w_B	0.1215	0.1215	0.1215	0.0999	0.0999	0.0999	0.0999	0.0999	0.0833
T/K	279.49	279.39	279.50	279.39	279.49	279.60	279.70	279.80	279.30
P/MPa	5.72	5.56	4.39	5.37	4.21	2.94	1.65	0.37	4.99
w_B	0.0833	0.0833	0.0833	0.0833	0.0714	0.0714	0.0714	0.0714	0.0625
T/K	279.40	279.49	279.61	279.66	277.27	277.37	277.46	277.57	276.28
P/MPa	3.84	2.75	1.38	0.67	5.30	4.17	2.96	1.56	4.75
w_B	0.0625	0.0625	0.0625						
T/K	276.38	276.47	276.58						
P/MPa	3.53	2.54	1.07						

Polymer (B): **polystyrene** **2000DE1**
Characterization: M_n/g.mol^{-1} = 29000, M_w/g.mol^{-1} = 30740,
Pressure Chemical Company, Pittsburgh, PA
Solvent (A): **nitroethane** $C_2H_5NO_2$ **79-24-3**

continued

continued

Type of data: cloud points

w_B	0.3649	0.3649	0.3649	0.3649	0.3649	0.1977	0.1977	0.1977	0.1977
T/K	289.05	289.10	289.15	289.26	289.42	296.25	296.44	296.55	296.65
P/MPa	4.49	4.02	3.67	2.84	1.66	5.29	3.70	3.17	2.57
w_B	0.1977	0.1977	0.1977	0.1977	0.1654	0.1654	0.1654	0.1654	0.1654
T/K	296.79	296.88	296.95	296.98	296.65	296.74	296.84	297.04	297.24
P/MPa	1.57	1.03	0.66	0.48	4.90	4.35	3.64	2.39	1.22
w_B	0.1654	0.1503	0.1503	0.1503	0.1503	0.1503	0.1503	0.1503	0.1503
T/K	297.35	296.65	296.74	296.84	297.05	297.24	297.35	297.45	297.49
P/MPa	0.49	5.64	5.05	4.44	3.08	1.90	1.26	0.63	0.34
w_B	0.1503	0.1503	0.1393	0.1393	0.1393	0.1393	0.1393	0.1393	0.1393
T/K	297.54	297.54	296.61	296.66	296.77	296.98	297.05	297.15	297.25
P/MPa	0.13	0.05	4.83	4.49	3.80	2.41	1.94	1.36	0.76
w_B	0.1393	0.1393	0.1388	0.1388	0.1388	0.1388	0.1388	0.1388	0.1388
T/K	297.30	297.35	296.65	296.75	296.95	297.05	297.15	297.24	297.35
P/MPa	0.46	0.12	4.93	4.32	3.09	2.45	1.82	1.22	0.56
w_B	0.1388	0.1388	0.1126	0.1126	0.1126	0.1126	0.1126	0.1126	0.1126
T/K	297.40	297.43	296.48	296.58	296.65	296.86	296.96	297.05	297.10
P/MPa	0.30	0.10	5.25	4.62	4.17	2.72	2.18	1.59	1.28
w_B	0.1126	0.1126	0.1126	0.0977	0.0977	0.0977	0.0977	0.0977	0.0977
T/K	297.17	297.25	297.27	296.33	296.44	296.55	296.74	296.89	297.00
P/MPa	0.79	0.39	0.23	4.79	4.16	3.46	2.21	1.26	0.57
w_B	0.0977	0.0944	0.0944	0.0944	0.0944	0.0944	0.0944	0.0944	0.0944
T/K	297.06	296.24	296.36	296.46	296.67	296.73	296.86	296.92	296.99
P/MPa	0.14	5.17	4.39	3.75	2.43	1.98	1.22	0.84	0.43
w_B	0.0944	0.0861	0.0861	0.0861	0.0861	0.0861	0.0861	0.0861	0.0861
T/K	297.02	296.06	296.17	296.25	296.45	296.64	296.79	296.81	296.87
P/MPa	0.23	5.33	4.62	3.98	2.74	1.56	0.53	0.44	0.04
w_B	0.0727	0.0727	0.0727	0.0727	0.0727	0.0727	0.0622	0.0622	0.0622
T/K	295.57	295.65	295.75	295.95	296.16	296.25	294.89	294.95	295.08
P/MPa	4.85	4.28	3.67	2.33	0.95	0.36	4.89	4.47	3.60
w_B	0.0622	0.0622	0.0622	0.0622					
T/K	295.26	295.43	295.49	295.60					
P/MPa	2.34	1.18	0.73	0.03					

Polymer (B): **polystyrene** **1991KAW**
Characterization: M_w/g.mol^{-1} = 48000,
Pressure Chemical Company, Pittsburgh, PA
Solvent (A): **nitroethane** $C_2H_5NO_2$ **79-24-3**

Type of data: cloud points (UCST-behavior)

φ_B	0.030	0.040	0.050	0.060	0.070	0.100
T/K	299.13	300.58	301.55	302.19	302.81	303.06

Polymer (B): **polystyrene** **2000DE1**
Characterization: M_n/g.mol^{-1} = 86600, M_w/g.mol^{-1} = 90100,
Pressure Chemical Company, Pittsburgh, PA
Solvent (A): **nitroethane** $C_2H_5NO_2$ **79-24-3**

Type of data: cloud points

w_B	0.2265	0.2265	0.2265	0.2265	0.2265	0.2265	0.2265	0.1911	0.1911
T/K	309.43	309.43	309.74	309.73	310.14	310.45	310.68	312.43	312.83
P/MPa	4.96	4.96	3.89	3.90	2.42	1.35	0.44	5.27	3.80
w_B	0.1911	0.1911	0.1911	0.1651	0.1651	0.1651	0.1651	0.1651	0.1454
T/K	313.14	313.44	313.84	314.13	314.45	314.75	315.16	315.39	315.44
P/MPa	2.74	1.63	0.22	5.18	4.04	2.90	1.42	0.56	3.68
w_B	0.1454	0.1454	0.1454	0.1390	0.1390	0.1390	0.1390	0.1390	0.1245
T/K	315.75	316.05	316.39	314.84	315.15	315.44	315.65	316.00	315.15
P/MPa	2.64	1.53	0.31	4.68	3.54	2.47	1.69	0.44	4.66
w_B	0.1245	0.1245	0.1245	0.1245	0.1059	0.1059	0.1059	0.1059	0.1059
T/K	315.44	315.74	316.05	316.30	315.44	315.34	315.74	316.05	316.64
P/MPa	3.67	2.54	1.39	0.54	4.63	5.03	3.57	2.42	0.28
w_B	0.0922	0.0922	0.0922	0.0922	0.0922	0.0633	0.0633	0.0633	0.0633
T/K	315.44	315.75	316.05	316.34	316.74	315.34	315.74	316.05	316.34
P/MPa	4.98	3.86	2.75	1.62	0.14	5.22	3.68	2.56	1.52
w_B	0.0633	0.0519	0.0519	0.0519	0.0519	0.0519	0.0519	0.0440	0.0440
T/K	316.65	315.04	315.03	315.44	315.74	316.05	316.45	314.84	315.15
P/MPa	0.47	5.21	5.23	3.78	2.69	1.55	0.13	4.33	3.08
w_B	0.0440	0.0440	0.0440	0.0337	0.0337	0.0337	0.0337	0.0337	
T/K	315.44	315.55	315.84	313.63	314.04	314.34	314.54	314.94	
P/MPa	1.95	1.47	0.40	4.98	3.51	2.41	1.71	0.19	

Polymer (B): **polystyrene** **2000DE1**
Characterization: M_n/g.mol^{-1} = 125400, M_w/g.mol^{-1} = 129200,
Pressure Chemical Company, Pittsburgh, PA
Solvent (A): **nitroethane** $C_2H_5NO_2$ **79-24-3**

Type of data: cloud points

w_B	0.1322	0.1322	0.1322	0.1322	0.1322	0.1212	0.1212	0.1212	0.1212
T/K	321.54	321.94	322.34	322.74	323.03	321.84	322.23	322.63	322.94
P/MPa	5.44	4.03	2.67	1.34	0.27	4.70	3.45	2.12	1.15
w_B	0.1212	0.1119	0.1119	0.1119	0.1119	0.1119	0.0929	0.0929	0.0929
T/K	323.11	321.94	322.34	322.74	322.84	323.24	321.84	321.93	322.03
P/MPa	0.53	4.61	3.26	1.91	1.59	0.29	4.27	3.92	3.62

continued

continued

w_B	0.0929	0.0929	0.0929	0.0929	0.0929	0.0929	0.0812	0.0812	0.0812
T/K	322.22	322.43	322.63	322.84	323.03	323.20	321.63	321.84	322.03
P/MPa	3.06	2.39	1.76	1.17	0.58	0.11	5.52	4.90	4.30

w_B	0.0812	0.0812	0.0812	0.0812	0.0645	0.0645	0.0645	0.0645	0.0645
T/K	322.44	322.84	323.03	323.44	321.72	322.14	322.53	322.84	323.13
P/MPa	3.10	1.92	1.38	0.29	5.61	4.25	3.02	2.10	1.23

w_B	0.0645	0.0535	0.0535	0.0535	0.0535	0.0535	0.0535	0.0457	0.0457
T/K	323.44	321.43	321.94	322.34	322.54	322.85	323.14	321.04	321.44
P/MPa	0.37	5.73	4.06	2.82	2.21	1.30	0.41	6.13	4.77

w_B	0.0457	0.0457	0.0457	0.0457	0.0457	0.0355	0.0355	0.0355	0.0355
T/K	321.63	322.03	322.43	322.74	322.93	320.74	321.13	321.54	321.94
P/MPa	4.21	2.86	1.65	0.76	0.16	5.84	4.58	3.27	1.98

w_B	0.0355	0.0355	0.0307	0.0307	0.0307	0.0307	0.0307	0.0307	0.0250
T/K	322.13	322.39	319.64	320.04	320.44	320.73	320.94	321.23	318.35
P/MPa	1.35	0.60	5.37	4.14	2.89	1.99	1.24	0.36	4.55

w_B	0.0250	0.0250	0.0250	0.0250
T/K	318.74	319.15	319.44	319.75
P/MPa	3.43	2.23	1.26	0.37

Polymer (B): **polystyrene** **2000DE1**
Characterization: M_n/g.mol^{-1} = 86600, M_w/g.mol^{-1} = 90100,
Pressure Chemical Company, Pittsburgh, PA
Solvent (A): **nitroethane-d5** **$C_2D_5NO_2$** **57817-88-6**

Type of data: cloud points

w_B	0.2291	0.2291	0.2291	0.2291	0.2291	0.2291	0.2291	0.2291	0.1844
T/K	336.41	336.41	336.52	336.61	336.82	337.11	337.43	337.74	338.53
P/MPa	4.32	4.25	3.90	3.60	3.02	2.21	1.37	0.58	4.17

w_B	0.1844	0.1844	0.1844	0.1844	0.1543	0.1543	0.1543	0.1543	0.1543
T/K	338.75	338.94	339.24	339.84	339.02	339.25	339.55	339.85	340.47
P/MPa	3.56	2.96	2.12	0.53	4.17	3.54	2.77	2.03	0.48

w_B	0.1326	0.1326	0.1326	0.1326	0.1326	0.1205	0.1205	0.1205	0.1205
T/K	339.31	339.62	339.95	340.25	340.67	339.35	339.35	339.66	339.50
P/MPa	4.17	3.33	2.45	1.63	0.57	4.57	4.54	3.70	2.91

w_B	0.1205	0.1205	0.0995	0.0995	0.0995	0.0995	0.0995	0.0848	0.0848
T/K	340.34	340.93	339.47	339.86	340.26	340.65	341.11	339.16	339.66
P/MPa	1.92	0.49	4.52	3.50	2.56	1.58	0.49	4.87	3.53

w_B	0.0848	0.0848	0.0848	0.0738	0.0738	0.0738	0.0738	0.0738	0.0669
T/K	340.07	340.44	340.92	339.05	339.54	340.06	340.25	340.85	338.96
P/MPa	2.57	1.62	0.44	4.81	3.51	2.26	1.80	0.37	4.60

w_B	0.0669	0.0669	0.0669	0.0669
T/K	339.48	339.87	340.24	340.63
P/MPa	3.33	2.33	1.34	0.47

Polymer (B): **polystyrene-d8** **2000DE1**
Characterization: M_n/g.mol^{-1} = 26400, M_w/g.mol^{-1} = 27200, completely deuterated, Polymer Laboratories, Amherst, MA
Solvent (A): **nitroethane** $C_2H_5NO_2$ **79-24-3**

Type of data: cloud points

w_B	0.2362	0.2362	0.2362	0.2362	0.2362	0.2362	0.2121	0.2121	0.2121
T/K	268.58	268.58	268.68	268.77	268.88	269.01	268.68	268.68	268.78
P/MPa	5.26	5.19	4.07	3.02	1.78	0.32	5.11	5.11	3.99
w_B	0.2121	0.2121	0.2121	0.1866	0.1866	0.1866	0.1866	0.1866	0.1866
T/K	268.89	268.98	269.11	268.89	268.88	268.98	269.07	269.19	269.28
P/MPa	2.75	1.66	0.34	4.95	4.94	3.84	2.75	1.49	0.48
w_B	0.1644	0.1644	0.1644	0.1644	0.1644	0.1644	0.1453	0.1453	0.1453
T/K	269.07	269.07	269.19	269.28	269.35	269.42	269.07	269.07	269.18
P/MPa	4.44	4.43	3.25	2.14	1.40	0.48	4.52	4.52	3.21
w_B	0.1453	0.1453	0.1453	0.1395	0.1395	0.1395	0.1395	0.1395	0.1395
T/K	269.28	269.34	269.42	269.28	269.28	269.38	269.49	269.54	269.62
P/MPa	2.09	1.41	0.54	4.59	4.55	3.34	1.98	1.33	0.39
w_B	0.1202	0.1202	0.1202	0.1202	0.1202	0.1202	0.1021	0.1021	0.1021
T/K	269.18	269.18	269.29	269.38	269.49	269.57	268.98	268.98	269.07
P/MPa	4.73	4.71	3.52	2.45	1.28	0.39	4.46	4.48	3.32
w_B	0.1021	0.1021	0.1021	0.0874	0.0874	0.0874	0.0874	0.0874	0.0874
T/K	269.17	269.27	269.34	268.54	268.54	268.64	268.74	268.84	268.93
P/MPa	2.33	1.23	0.35	5.06	5.07	3.98	2.77	1.65	0.60
w_B	0.0764	0.0764	0.0764	0.0764	0.0764	0.0764			
T/K	268.13	268.13	268.25	268.34	268.43	268.51			
P/MPa	4.88	4.90	3.58	2.42	1.41	0.49			

Polymer (B): **polystyrene-d8** **2000DE1**
Characterization: M_n/g.mol^{-1} = 83500, M_w/g.mol^{-1} = 85200, completely deuterated, Polymer Laboratories, Amherst, MA
Solvent (A): **nitroethane** $C_2H_5NO_2$ **79-24-3**

Type of data: cloud points

w_B	0.2273	0.2273	0.2273	0.2273	0.2273	0.2273	0.2273	0.2273	0.1992
T/K	285.98	285.98	286.19	286.29	286.38	286.59	286.65	286.85	286.49
P/MPa	5.25	5.27	4.08	3.53	3.02	1.81	1.50	0.46	4.65
w_B	0.1992	0.1992	0.1992	0.1992	0.1992	0.1773	0.1773	0.1773	0.1773
T/K	286.49	286.69	286.90	287.09	287.29	286.99	286.99	287.20	287.40
P/MPa	4.65	3.64	2.49	1.40	0.46	4.74	4.74	3.53	2.43
w_B	0.1773	0.1773	0.1522	0.1522	0.1522	0.1522	0.1522	0.1522	0.1315
T/K	287.61	287.77	287.30	287.30	287.49	287.71	287.92	288.06	287.49
P/MPa	1.35	0.47	4.80	4.82	3.72	2.52	1.33	0.61	4.77

continued

continued

w_B	0.1315	0.1315	0.1315	0.1315	0.1315	0.1158	0.1158	0.1158	0.1158
T/K	287.49	287.70	287.91	288.08	288.27	287.49	287.50	287.70	287.92
P/MPa	4.80	3.56	2.37	1.45	0.41	4.87	4.86	3.61	2.38
w_B	0.1158	0.1158	0.1124	0.1124	0.1124	0.1124	0.1124	0.0995	0.0995
T/K	288.08	288.25	288.06	288.07	288.09	288.12	288.14	287.40	287.40
P/MPa	1.51	0.53	0.67	0.61	0.51	0.36	0.25	4.61	4.62
w_B	0.0995	0.0995	0.0995	0.0995	0.0840	0.0840	0.0840	0.0840	0.0840
T/K	287.61	287.80	287.96	288.11	287.30	287.30	287.49	287.71	287.85
P/MPa	3.42	2.30	1.37	0.56	4.65	4.66	3.49	2.28	1.46
w_B	0.0840	0.0703	0.0703	0.0703	0.0703	0.0703	0.0703	0.0531	0.0531
T/K	288.06	286.99	287.00	287.21	287.40	287.56	287.75	286.28	286.28
P/MPa	0.27	4.82	4.82	3.53	2.42	1.45	0.38	4.92	4.92
w_B	0.0531	0.0531	0.0531	0.0531					
T/K	286.50	286.70	286.85	287.05					
P/MPa	3.64	2.43	1.54	0.34					

Polymer (B): **polystyrene-d8** **2000DE1**
Characterization: M_n/g.mol^{-1} = 83500, M_w/g.mol^{-1} = 85200, completely deuterated, Polymer Laboratories, Amherst, MA
Solvent (A): **nitroethane-d5** **$C_2D_5NO_2$** **57817-88-6**

Type of data: cloud points

w_B	0.1964	0.1964	0.1964	0.1964	0.1964	0.1964	0.1964	0.1964	0.1964
T/K	308.08	308.09	308.18	308.28	308.48	308.76	309.06	309.25	309.65
P/MPa	6.28	6.28	5.85	5.38	4.69	3.58	2.45	1.72	0.24
w_B	0.1741	0.1741	0.1741	0.1741	0.1741	0.1519	0.1519	0.1519	0.1519
T/K	309.65	309.95	310.26	310.51	310.83	310.46	310.78	311.03	311.23
P/MPa	4.85	3.61	2.60	1.67	0.47	4.59	3.33	2.28	1.57
w_B	0.1519	0.1330	0.1330	0.1330	0.1330	0.1330	0.1237	0.1237	0.1237
T/K	311.51	310.68	310.86	311.18	311.37	311.65	310.87	310.87	311.18
P/MPa	0.47	4.53	3.78	2.57	1.70	0.63	4.37	4.36	3.19
w_B	0.1237	0.1237	0.1237	0.1183	0.1183	0.1183	0.1183	0.1183	0.1069
T/K	311.37	311.59	311.81	310.77	310.99	311.27	311.49	311.78	310.90
P/MPa	2.30	1.43	0.61	4.48	3.56	2.43	1.61	0.50	4.72
w_B	0.1069	0.1069	0.1069	0.1069	0.0927	0.0927	0.0927	0.0927	0.0927
T/K	311.18	311.38	311.59	311.92	310.77	311.09	311.27	311.49	311.83
P/MPa	3.53	2.63	1.81	0.55	4.60	3.37	2.59	1.83	0.47
w_B	0.0807	0.0807	0.0807	0.0807	0.0807	0.0715	0.0715	0.0715	0.0715
T/K	310.57	310.77	311.08	311.38	311.73	310.60	310.78	311.10	311.30
P/MPa	5.12	4.34	3.05	1.89	0.50	4.45	3.60	2.35	1.56

Polymer (B): **polystyrene** **1991OPS**
Characterization: M_w/g.mol^{-1} = 4800,
Pressure Chemical Company, Pittsburgh, PA
Solvent (A): **n-octadecane** $C_{18}H_{38}$ **593-45-3**

Type of data: critical point (LCST-behavior)

$w_{B,\,crit}$ 0.417 T_{crit}/K 403.55

Polymer (B): **polystyrene** **1991OPS**
Characterization: M_n/g.mol^{-1} = 3700, M_w/g.mol^{-1} = 4000, M_z/g.mol^{-1} = 4600,
Pressure Chemical Company, Pittsburgh, PA
Solvent (A): **1-octadecanol** $C_{18}H_{38}O$ **112-92-5**

Type of data: critical point (UCST-behavior)

$w_{B,\,crit}$ 0.401 T_{crit}/K 390.55

Polymer (B): **polystyrene** **1950JE1**
Characterization: M_η/g.mol^{-1} = 62600, fractionated in the laboratory
Solvent (A): **1-octadecanol** $C_{18}H_{38}O$ **112-92-5**

Type of data: cloud points, precipitation threshold (UCST-behavior)

z_B 0.120 T/K 448.65

Polymer (B): **polystyrene** **2001IM1**
Characterization: M_n/g.mol^{-1} = 1160, M_w/g.mol^{-1} = 1240,
Pressure Chemical Company, Pittsburgh, PA
Solvent (A): **n-octane** C_8H_{18} **111-65-9**

Type of data: cloud points (UCST-behavior)

w_B	0.100	0.130	0.160	0.190	0.220	0.250	0.285	0.350
T/K	271.5	273.5	275.7	277.2	278.0	278.2	278.2	277.3

Polymer (B): **polystyrene** **1991OPS**
Characterization: M_w/g.mol^{-1} = 4800,
Pressure Chemical Company, Pittsburgh, PA
Solvent (A): **n-octane** C_8H_{18} **111-65-9**

Type of data: critical point (LCST-behavior)

$w_{B,\,crit}$ 0.366 T_{crit}/K 358.05

Polymer (B): **polystyrene** **1991OPS**
Characterization: M_n/g.mol^{-1} = 3700, M_w/g.mol^{-1} = 4000, M_z/g.mol^{-1} = 4600,
Pressure Chemical Company, Pittsburgh, PA
Solvent (A): **1-octanol** $C_8H_{18}O$ **111-87-5**

continued

continued

Type of data: critical point (UCST-behavior)

$w_{B, crit}$ 0.380 T_{crit}/K 372.35

Polymer (B): **polystyrene** **1991OPS**
Characterization: M_w/g.mol^{-1} = 4800,
Pressure Chemical Company, Pittsburgh, PA
Solvent (A): **1-octanol** $C_8H_{18}O$ **111-87-5**

Type of data: critical point (UCST-behavior)

$w_{B, crit}$ 0.392 T_{crit}/K 370.35

Polymer (B): **polystyrene** **1991OPS**
Characterization: M_w/g.mol^{-1} = 9000,
Pressure Chemical Company, Pittsburgh, PA
Solvent (A): **1-octanol** $C_8H_{18}O$ **111-87-5**

Type of data: critical point (UCST-behavior)

$w_{B, crit}$ 0.305 T_{crit}/K 395.05

Polymer (B): **polystyrene** **1950JE1**
Characterization: M_η/g.mol^{-1} = 62600, fractionated in the laboratory
Solvent (A): **1-octene** C_8H_{16} **111-66-0**

Type of data: cloud points, precipitation threshold (UCST-behavior)

z_B 0.145 T/K 355.15

Polymer (B): **polystyrene** **1991OPS**
Characterization: M_w/g.mol^{-1} = 4800,
Pressure Chemical Company, Pittsburgh, PA
Solvent (A): **n-pentadecane** $C_{15}H_{32}$ **629-62-9**

Type of data: critical point (LCST-behavior)

$w_{B, crit}$ 0.402 T_{crit}/K 385.25

Polymer (B): **polystyrene** **2001IM1**
Characterization: M_n/g.mol^{-1} = 1160, M_w/g.mol^{-1} = 1240,
Pressure Chemical Company, Pittsburgh, PA
Solvent (A): **n-pentane** C_5H_{12} **109-66-0**

Type of data: cloud points (UCST-behavior)

w_B	0.100	0.130	0.160	0.190	0.203	0.214	0.229	0.234	0.248
T/K	281.0	283.1	284.9	283.4	285.0	283.8	285.4	295.7	285.5

w_B	0.270	0.298	0.327
T/K	285.6	295.5	284.8

Polymer (B): **polystyrene** **1991OPS**
Characterization: M_n/g.mol^{-1} = 3700, M_w/g.mol^{-1} = 4000, M_z/g.mol^{-1} = 4600, Pressure Chemical Company, Pittsburgh, PA
Solvent (A): **1-pentanol** **$C_5H_{12}O$** **71-41-0**

Type of data: critical point (UCST-behavior)

$w_{B, crit}$ 0.364 T_{crit}/K 375.05

Polymer (B): **polystyrene** **1991OPS**
Characterization: M_w/g.mol^{-1} = 4800, Pressure Chemical Company, Pittsburgh, PA
Solvent (A): **1-pentanol** **$C_5H_{12}O$** **71-41-0**

Type of data: critical point (UCST-behavior)

$w_{B, crit}$ 0.372 T_{crit}/K 373.85

Polymer (B): **polystyrene** **1991OPS**
Characterization: M_w/g.mol^{-1} = 9000, Pressure Chemical Company, Pittsburgh, PA
Solvent (A): **1-pentanol** **$C_5H_{12}O$** **71-41-0**

Type of data: critical point (UCST-behavior)

$w_{B, crit}$ 0.299 T_{crit}/K 397.45

Polymer (B): **polystyrene** **1995PFO**
Characterization: M_n/g.mol^{-1} = 219800, M_w/g.mol^{-1} = 233000, Polysciences, Inc., Warrington, PA
Solvent (A): **pentyl acetate** **$C_7H_{14}O_2$** **628-63-7**

Type of data: cloud points (LCST-behavior)

w_B	0.0791	0.1043	0.1277
T/K	519.65	519.35	519.45

Polymer (B): **polystyrene** **1963ORO**
Characterization: M_n/g.mol^{-1} = 404000, M_w/g.mol^{-1} = 406000
Solvent (A): **pentyl acetate** **$C_7H_{14}O_2$** **628-63-7**

Type of data: cloud points (UCST-behavior)

w_B 0.01 T/K 233.15

Polymer (B): **polystyrene** **1995PFO**
Characterization: M_n/g.mol^{-1} = 545500, M_w/g.mol^{-1} = 600000, Polysciences, Inc., Warrington, PA
Solvent (A): **pentyl acetate** **$C_7H_{14}O_2$** **628-63-7**

continued

continued

Type of data: cloud points (LCST-behavior)

w_B	0.0100	0.0247	0.0400	0.0554	0.0698	0.0859	0.1001	0.1144	0.1306
T/K	517.25	515.85	515.05	513.55	512.65	512.35	512.85	513.05	514.25

Polymer (B): **polystyrene** **1995PFO**
Characterization: M_n/g.mol^{-1} = 1564000, M_w/g.mol^{-1} = 1971000,
Polysciences, Inc., Warrington, PA
Solvent (A): **pentyl acetate** $C_7H_{14}O_2$ **628-63-7**

Type of data: cloud points (LCST-behavior)

w_B	0.0010	0.0050	0.0100	0.0199	0.0400	0.0558	0.0724	0.0979
T/K	512.65	510.25	508.65	507.45	508.35	509.15	509.75	510.05

Polymer (B): **polystyrene** **1982GEE**
Characterization: M_w/g.mol^{-1} = 100000,
Pressure Chemical Company, Pittsburgh, PA
Solvent (A): **1-phenyldecane** $C_{16}H_{26}$ **104-72-3**

Type of data: cloud points (UCST-behavior)

w_B 0.140 T/K 283.60

Polymer (B): **polystyrene** **1982GEE**
Characterization: M_w/g.mol^{-1} = 390000,
Pressure Chemical Company, Pittsburgh, PA
Solvent (A): **1-phenyldecane** $C_{16}H_{26}$ **104-72-3**

Type of data: cloud points (UCST-behavior)

w_B	0.020	0.080
T/K	291.85	293.10

Polymer (B): **polystyrene** **1982GEE**
Characterization: M_w/g.mol^{-1} = 600000,
Pressure Chemical Company, Pittsburgh, PA
Solvent (A): **1-phenyldecane** $C_{16}H_{26}$ **104-72-3**

Type of data: cloud points (UCST-behavior)

w_B 0.060 T/K 295.45

Polymer (B): **polystyrene** **1982GEE**
Characterization: M_w/g.mol^{-1} = 390000,
Pressure Chemical Company, Pittsburgh, PA
Solvent (A): **1-phenyldodecane** $C_{18}H_{30}$ **123-01-3**

Type of data: cloud points (UCST-behavior)

w_B 0.020 T/K 326.65

Polymer (B): **polystyrene** **1963ORO**
Characterization: M_n/g.mol^{-1} = 404000, M_w/g.mol^{-1} = 406000
Solvent (A): **1-phenyldodecane** $C_{18}H_{30}$ **123-01-3**

Type of data: cloud points (UCST-behavior)

w_B	0.01	T/K	323.15

Polymer (B): **polystyrene** **1982GEE**
Characterization: M_w/g.mol^{-1} = 390000,
Pressure Chemical Company, Pittsburgh, PA
Solvent (A): **1-phenyloctane** $C_{14}H_{22}$ **2189-60-8**

Type of data: cloud points (UCST-behavior)

w_B	0.02	T/K	253.15

Polymer (B): **polystyrene** **1991SZY**
Characterization: M_n/g.mol^{-1} = 2230, M_w/g.mol^{-1} = 2510,
Pressure Chemical Company, Pittsburgh, PA
Solvent (A): **2-propanone** C_3H_6O **67-64-1**

Type of data: cloud points (UCST-behavior)

w_B	0.04	0.10	0.11	0.14	0.18	0.225
T/K	228	230	232	231	227	225

Type of data: cloud points (LCST-behavior)

w_B	0.04	0.10	0.11	0.14	0.18	0.225
T/K	485	483	480	482	486	488

Polymer (B): **polystyrene** **1991SZY**
Characterization: M_n/g.mol^{-1} = 4780, M_w/g.mol^{-1} = 5110,
Pressure Chemical Company, Pittsburgh, PA
Solvent (A): **2-propanone** C_3H_6O **67-64-1**

Type of data: cloud points (UCST-behavior)

w_B	0.04	0.11	0.13	0.18	0.225
T/K	245	258	263	261	259

Type of data: cloud points (LCST-behavior)

w_B	0.04	0.11	0.13	0.18	0.225
T/K	461	451	449	448	449

Polymer (B): **polystyrene** **1992SZY**
Characterization: M_n/g.mol^{-1} = 6555, M_w/g.mol^{-1} = 7800,
Scientific Polymer Products, Inc., Ontario, NY
Solvent (A): **2-propanone** C_3H_6O **67-64-1**

continued

continued

Type of data: cloud points

w_B	0.077	0.090	0.112	0.145	0.170	0.200	0.325
φ_B	0.059	0.069	0.087	0.113	0.133	0.158	0.265
T/K	248.40	249.75	252.60	254.20	254.53	254.46	251.65
$(dT/dP)_{P=0}$/K.MPa^{-1}	−1.10	−0.95	−0.76	−0.70	−0.73	−0.78	−0.74

Type of data: spinodal points

w_B	0.077	0.090	0.112	0.145	0.170	0.200	0.325
φ_B	0.059	0.069	0.087	0.113	0.133	0.158	0.265
T/K	247.40	249.50	251.75	253.70	254.29	254.00	251.43
$(dT/dP)_{P=0}$/K.MPa^{-1}	−0.90	−0.76			−0.68		−0.70

Polymer (B): **polystyrene** **1991SZY**
Characterization: M_n/g.mol^{-1} = 6570, M_w/g.mol^{-1} = 7820,
Scientific Polymer Products, Inc., Ontario, NY
Solvent (A): **2-propanone** C_3H_6O **67-64-1**

Type of data: cloud points (UCST-behavior)

w_B	0.04	0.10	0.11	0.13	0.18	0.225
T/K	264	268	269	270	271	267

w_B = 0.20 T/K = 270 $(dT/dP)_{P=0}$/K.MPa^{-1} = −1.10

Type of data: cloud points (LCST-behavior)

w_B	0.04	0.10	0.11	0.13	0.18	0.225
T/K	446	443	440	438	438	438

Polymer (B): **polystyrene** **1995LUS**
Characterization: M_n/g.mol^{-1} = 7340, M_w/g.mol^{-1} = 8000,
Pressure Chemical Company, Pittsburgh, PA
Solvent (A): **2-propanone** C_3H_6O **67-64-1**

Type of data: cloud points

w_B	0.222	0.222	0.222	0.222	0.222
T/K	253.77	253.45	253.19	252.98	252.83
P/MPa	0.07	0.34	0.72	1.02	1.25

critical concentration: $w_{B, crit}$ = 0.220

Polymer (B): **polystyrene-d8** **1995LUS**
Characterization: M_n/g.mol^{-1} = 10300, M_w/g.mol^{-1} = 10500,
completely deuterated, Polymer Laboratories, Amherst, MA
Solvent (A): **2-propanone** C_3H_6O **67-64-1**

Type of data: cloud points (UCST-behavior)

continued

continued

w_B	0.209	0.209	0.209	0.209
T/K	234.95	234.60	235.98	235.25
P/MPa	1.55	2.02	0.15	1.11

Polymer (B): **polystyrene** **1991SZY**
Characterization: M_n/g.mol^{-1} = 10750, M_w/g.mol^{-1} = 11500,
Scientific Polymer Products, Inc., Ontario, NY
Solvent (A): **2-propanone** **C_3H_6O** **67-64-1**

Type of data: cloud points (UCST-behavior)

w_B	0.04	0.10	0.11	0.13	0.18	0.225
T/K	263	270	272	272	273	272

Type of data: cloud points (LCST-behavior)

w_B	0.04	0.10	0.11	0.13	0.18	0.225
T/K	426	419	420	421	421	422

Polymer (B): **polystyrene** **1992SZY**
Characterization: M_n/g.mol^{-1} = 10750, M_w/g.mol^{-1} = 11500,
Scientific Polymer Products, Inc., Ontario, NY
Solvent (A): **2-propanone** **C_3H_6O** **67-64-1**

Type of data: cloud points (UCST-behavior)

w_B	0.090	0.110	0.130	0.146	0.164	0.183	0.202	0.218	0.235
φ_B	0.068	0.084	0.099	0.112	0.126	0.142	0.158	0.171	0.185
T/K	245.15	246.74	247.75	247.92	248.18	248.11	247.87	247.52	247.40

w_B	0.259	0.292	0.334
φ_B	0.206	0.234	0.272
T/K	246.85	246.09	244.09

w_B	0.090	0.146	0.164	0.218	0.292
φ_B	0.068	0.112	0.126	0.171	0.234
T/K	245.15	247.92	248.18	247.52	246.09
$(dT/dP)_{P=0}$/K.MPa^{-1}	−0.74	−0.74	−0.75	−0.75	−0.72

Type of data: spinodal points (UCST-behavior)

w_B	0.090	0.110	0.130	0.146	0.164	0.183	0.202	0.218	0.235
φ_B	0.068	0.084	0.099	0.112	0.126	0.142	0.158	0.171	0.185
T/K	244.87	246.44	247.49	247.77	247.86	247.79	247.67	247.52	247.16

w_B	0.259	0.292	0.334
φ_B	0.206	0.234	0.272
T/K	246.60	246.00	243.90

w_B	0.090	0.146	0.164	0.218	0.292
φ_B	0.068	0.112	0.126	0.171	0.234
T/K	244.87	247.77	247.86	247.52	246.00
$(dT/dP)_{P=0}$/K.MPa^{-1}	−0.78	−0.79	−0.73	−0.75	−0.80

Polymer (B): **polystyrene** **1995LUS**
Characterization: M_n/g.mol^{-1} = 11350, M_w/g.mol^{-1} = 11700,
Polymer Laboratories, Inc., Amherst, MA
Solvent (A): **2-propanone** **C_3H_6O** **67-64-1**

Type of data: cloud points (UCST-behavior)

w_B	0.210	0.210	0.210	0.210
T/K	273.75	274.16	274.68	274.91
P/MPa	1.42	1.07	0.61	0.41

critical concentration: $w_{B,\text{crit}}$ = 0.220

Polymer (B): **polystyrene** **1991SZY**
Characterization: M_n/g.mol^{-1} = 12750, M_w/g.mol^{-1} = 13500,
Pressure Chemical Company, Pittsburgh, PA
Solvent (A): **2-propanone** **C_3H_6O** **67-64-1**

Type of data: cloud points (UCST-behavior)

w_B	0.05	0.10	0.15	0.18	0.25
T/K	280	284	289	287	284

w_B	0.19	0.20	0.23
T/K	287	286	285
$(dT/dP)_{P=0}$/K.MPa^{-1}	−1.70	−1.75	−1.76

Type of data: cloud points (LCST-behavior)

w_B	0.05	0.10	0.15	0.18	0.25
T/K	407	404	396	398	404

w_B = 0.20 T/K = 400 $(dT/dP)_{P=0}$/K.MPa^{-1} = +6.53

Polymer (B): **polystyrene** **1995LUS**
Characterization: M_n/g.mol^{-1} = 12750, M_w/g.mol^{-1} = 13500,
Pressure Chemical Company, Pittsburgh, PA
Solvent (A): **2-propanone** **C_3H_6O** **67-64-1**

Type of data: cloud points

w_B	0.102	0.102	0.102	0.102	0.102	0.102	0.107	0.107	0.107
T/K	281.41	281.42	281.83	282.19	282.42	282.78	283.19	282.37	281.91
P/MPa	1.08	1.11	0.92	0.64	0.65	0.31	0.62	1.13	1.24
w_B	0.107	0.107	0.163	0.163	0.163	0.163	0.163	0.163	0.163
T/K	281.60	281.30	287.13	287.11	286.65	286.44	285.96	285.49	285.27
P/MPa	1.27	1.47	0.08	0.09	0.30	0.42	0.70	0.97	1.12
w_B	0.163	0.197	0.197	0.197	0.197	0.212	0.212	0.212	0.212
T/K	284.71	285.60	285.26	284.89	284.60	286.12	285.42	285.26	284.50
P/MPa	1.45	0.40	0.59	0.84	0.98	0.05	0.44	0.56	0.99

continued

continued

w_B	0.212	0.212	0.229	0.229	0.229	0.229	0.229	0.229	0.229
T/K	284.11	283.63	285.77	285.35	285.17	284.79	284.52	284.14	283.80
P/MPa	1.21	1.55	0.07	0.31	0.41	0.64	0.82	1.04	1.28

w_B	0.264	0.264	0.264	0.264	0.264	0.264	0.264	0.264	0.264
T/K	285.10	285.04	284.82	284.60	284.39	284.07	283.64	283.32	283.23
P/MPa	0.07	0.09	0.20	0.33	0.46	0.66	0.90	1.10	1.16

critical concentration: $w_{B,\text{crit}} = 0.220$

Polymer (B): **polystyrene** **2002RE2**
Characterization: M_n/g.mol^{-1} = 15800, M_w/g.mol^{-1} = 16600,
Scientific Polymer Products, Inc., Ontario, NY
Solvent (A): **2-propanone** C_3H_6O **67-64-1**

Type of data: cloud points (negative pressures were measured using a Berthelot-tube technique)

w_B 0.180 was kept constant

T/K	257.4	258.1	258.2	258.9	262.9	265.0	273.0	280.0	283.0
P/bar	767.0	634.3	600.0	525.0	425.0	370.0	255.0	158.6	125.5

T/K	289.0	293.0	298.0	303.0	307.5	308.0	313.0	323.0	336.5
P/bar	82.5	62.5	32.5	0.4	−15.2	−20.2	−26.1	−32.4	−32.8

T/K	351.5	368.0	380.0	403.5	417.0	430.5
P/bar	−26.1	−10.2	4.4	43.0	61.5	97.0

Polymer (B): **polystyrene** **1995LUS**
Characterization: M_n/g.mol^{-1} = 21450, M_w/g.mol^{-1} = 22100,
Polymer Laboratories, Inc., Amherst, MA
Solvent (A): **2-propanone** C_3H_6O **67-64-1**

Type of data: cloud points

w_B	0.111	0.111	0.111	0.111	0.111	0.111	0.111	0.111	0.111
T/K	321.22	322.81	322.82	326.73	329.25	330.88	365.52	367.63	371.30
P/MPa	1.56	1.18	1.19	0.74	0.47	0.31	0.55	0.75	0.99

w_B	0.111	0.111	0.146	0.146	0.146	0.146	0.146	0.146	0.146
T/K	375.80	378.48	325.87	327.52	330.55	333.20	337.06	340.93	342.45
P/MPa	1.49	1.78	1.55	1.33	0.98	0.75	0.47	0.33	0.32

w_B	0.146	0.146	0.146	0.146	0.146	0.146	0.146	0.146	0.146
T/K	344.05	351.54	353.15	355.25	358.39	362.31	366.80	371.68	374.10
P/MPa	0.26	0.29	0.30	0.45	0.60	0.86	1.18	1.59	1.82

w_B	0.158	0.158	0.158	0.158	0.158	0.158	0.158	0.158	0.158
T/K	325.17	329.19	332.14	340.37	344.55	348.88	354.68	359.23	366.57
P/MPa	1.81	1.29	0.98	0.48	0.39	0.38	0.51	0.71	1.19

continued

continued

w_B	0.158	0.163	0.163	0.163	0.163	0.163	0.163	0.163	0.163
T/K	372.72	327.82	330.90	334.09	337.43	340.81	344.46	346.55	349.25
P/MPa	1.73	1.56	1.20	0.92	0.68	0.55	0.48	0.47	0.47

w_B	0.163	0.163	0.163	0.163	0.163	0.169	0.169	0.169	0.169
T/K	352.96	358.40	362.91	367.21	372.39	327.38	327.25	330.92	334.48
P/MPa	0.54	0.75	1.00	1.30	1.73	1.45	1.43	1.01	0.71

w_B	0.169	0.169	0.169	0.169	0.169	0.169	0.169	0.174	0.174
T/K	337.35	341.57	346.73	351.19	357.06	367.75	375.89	328.21	331.83
P/MPa	0.54	0.39	0.35	0.41	0.59	1.29	2.00	1.38	0.97

w_B	0.174	0.174	0.174	0.174	0.174	0.174	0.174	0.174	0.174
T/K	335.82	339.91	342.81	347.19	351.33	356.23	361.75	367.39	373.22
P/MPa	0.68	0.45	0.37	0.34	0.40	0.58	0.85	1.25	1.75

w_B	0.185	0.185	0.185	0.185	0.185	0.185	0.185	0.185	0.185
T/K	329.45	332.27	335.50	338.67	342.72	345.75	350.24	353.05	358.41
P/MPa	1.23	0.96	0.72	0.55	0.40	0.37	0.38	0.44	0.67

w_B	0.185	0.185	0.185	0.213	0.213	0.213	0.213	0.213	0.213
T/K	363.93	368.03	373.76	326.70	329.17	332.66	335.29	342.10	344.41
P/MPa	0.99	1.29	1.79	1.61	1.31	0.95	0.77	0.46	0.44

w_B	0.213	0.213	0.213	0.213	0.213	0.213	0.249	0.249	0.249
T/K	346.66	352.30	357.91	364.35	369.74	372.95	324.78	328.07	331.15
P/MPa	0.45	0.53	0.73	1.11	1.52	1.80	1.59	1.14	0.79

w_B	0.249	0.249	0.249	0.249	0.249	0.249	0.249	0.273	0.273
T/K	334.45	339.12	354.98	358.42	364.97	371.79	376.92	321.19	324.14
P/MPa	0.51	0.27	0.30	0.46	0.86	1.40	1.89	1.71	1.30

w_B	0.273	0.273	0.273	0.273	0.273	0.273
T/K	327.28	329.31	362.71	365.18	373.24	378.35
P/MPa	0.87	0.63	0.47	0.64	1.29	1.83

critical concentration: $w_{B, crit}$ = 0.22

Polymer (B): **polystyrene** **2002RE2**
Characterization: M_n/g.mol^{-1} = 23800, M_w/g.mol^{-1} = 24700,
Scientific Polymer Products, Inc., Ontario, NY
Solvent (A): **2-propanone** C_3H_6O **67-64-1**

Type of data: cloud points (negative pressures were measured using a Berthelot-tube technique)

w_B 0.018 was kept constant

T/K	251.0	251.5	253.0	253.0	256.3	258.0	263.0	265.0	273.0
P/bar	700.00	590.00	536.50	630.00	455.00	403.00	296.80	260.00	156.00

T/K	277.3	281.8	286.0	293.4	295.0	298.5	299.8	301.0	301.6
P/bar	113.00	78.80	49.30	10.00	0.28	−21.60	−27.70	−26.40	−29.40

continued

continued

T/K	301.9	303.2	304.0	308.0	368.0	375.0	381.0	383.0	390.3
P/bar	−30.30	−35.10	−41.00	−42.00	−26.70	−15.60	−12.40	−6.20	5.70

T/K	401.0	413.0	422.5	433.0
P/bar	20.00	40.00	50.00	73.00

Polymer (B): **polystyrene-d8** **1995LUS**
Characterization: M_n/g.mol^{-1} = 25400, M_w/g.mol^{-1} = 26900,
completely deuterated, Polymer Laboratories, Amherst, MA
Solvent (A): **2-propanone** C_3H_6O **67-64-1**

Type of data: cloud points

w_B	0.132	0.132	0.132	0.132	0.157	0.157	0.157	0.181	0.181
T/K	274.49	275.63	276.62	277.47	275.91	276.65	277.79	276.13	276.79
P/MPa	1.89	1.31	0.73	0.29	1.69	1.30	0.71	1.89	1.53

w_B	0.181	0.214	0.214	0.214	0.214	0.214	0.233	0.233	0.233
T/K	279.62	275.36	276.70	277.34	277.93	279.30	275.02	277.09	278.16
P/MPa	1.14	2.57	1.83	1.49	1.14	0.47	2.48	1.41	0.88

w_B	0.233	0.253	0.253	0.253	0.253	0.270	0.270	0.270	0.270
T/K	279.03	274.58	275.75	277.78	278.61	273.48	274.46	275.94	277.36
P/MPa	0.45	2.23	1.73	0.71	0.35	2.25	1.76	1.03	0.31

w_B	0.300	0.300	0.300	0.300
T/K	270.82	271.26	271.87	272.42
P/MPa	1.11	0.84	0.57	0.28

Polymer (B): **polystyrene** **1994IMR**
Characterization: M_n/g.mol^{-1} = 70300, M_w/g.mol^{-1} = 187000,
Mitsui Toatsu Chemicals, Inc., Japan
Solvent (A): **propionitrile** C_3H_5N **107-12-0**

Type of data: cloud points (negative pressures were measured using a Berthelot-tube technique)

w_B	0.044	0.044	0.044	0.063	0.063	0.063	0.063	0.063	0.104
T/K	316.75	323.15	330.05	335.05	337.35	339.95	341.55	345.55	346.65
P/MPa	3.23	0.02	−1.95	3.77	1.81	0.03	−0.95	−1.15	2.60

w_B	0.104	0.104	0.104	0.104	0.201	0.201	0.201	0.201	0.201
T/K	351.65	359.35	371.15	372.00	347.15	350.65	353.10	356.15	358.55
P/MPa	1.37	0.95	−0.40	−0.50	2.01	1.80	1.39	0.70	0.32

w_B	0.201	0.201	0.201	0.201	0.201	0.201	0.201	0.201	0.201
T/K	362.05	369.80	372.35	377.55	391.55	392.75	396.75	401.10	420.80
P/MPa	0.21	−0.36	−0.42	−0.51	−0.48	−0.31	−0.27	0.01	1.45

Polymer (B): **polystyrene** **1996LUS**
Characterization: M_n/g.mol^{-1} = 7340, M_w/g.mol^{-1} = 8000,
Pressure Chemical Company, Pittsburgh, PA
Solvent (A): **propionitrile** **C_3H_5N** **107-12-0**

Type of data: cloud points

w_B	0.1300	0.1300	0.1300	0.1300	0.1300	0.1300	0.1600	0.1600	0.1600
T/K	267.151	267.623	267.948	268.421	268.936	268.397	268.856	269.421	269.981
P/MPa	3.518	2.869	2.366	1.511	0.917	0.508	3.682	2.903	2.045
w_B	0.1600	0.1600	0.1600	0.1800	0.1800	0.1800	0.1800	0.1800	0.1800
T/K	270.454	270.934	271.222	267.979	268.447	268.999	268.503	269.986	270.522
P/MPa	1.430	0.721	0.369	4.058	3.390	2.688	1.866	1.255	0.427
w_B	0.1900	0.1900	0.1900	0.1900	0.1900	0.1900	0.2100	0.2100	0.2100
T/K	268.939	269.449	269.938	270.553	271.086	271.395	268.077	268.470	268.923
P/MPa	4.139	3.412	2.658	1.979	1.267	0.906	3.025	2.602	4.196
w_B	0.2100	0.2100	0.2100	0.2100	0.2100	0.2200	0.2200	0.2200	0.2200
T/K	269.513	269.977	270.491	270.960	271.482	268.994	268.436	270.010	270.486
P/MPa	3.780	2.820	2.048	1.532	0.840	4.086	3.622	2.461	2.274
w_B	0.2200	0.2200	0.2400	0.2400	0.2400	0.2400	0.2400	0.2400	0.2400
T/K	270.988	271.483	268.877	269.461	270.028	270.518	270.930	271.494	271.950
P/MPa	1.346	0.749	4.896	4.305	3.623	2.952	2.319	1.613	1.180
w_B	0.2400	0.2500	0.2500	0.2500	0.2500	0.2500	0.2600	0.2600	0.2600
T/K	272.245	269.072	269.576	270.128	271.130	271.525	269.531	269.954	270.493
P/MPa	0.688	3.971	3.267	2.763	1.249	0.845	4.404	3.874	3.162
w_B	0.2600	0.2600	0.2600	0.2600	0.2800	0.2800	0.2800	0.2800	0.2800
T/K	271.082	271.468	271.926	272.497	269.840	270.538	270.957	271.445	271.979
P/MPa	2.399	1.901	1.328	0.558	4.802	3.565	2.914	2.111	1.649
w_B	0.2800	0.2800	0.2800	0.3000	0.3000	0.3000	0.3000	0.3000	0.3000
T/K	272.465	272.465	272.784	268.445	269.027	269.466	270.143	270.641	271.018
P/MPa	0.922	1.137	0.742	5.566	4.813	4.417	3.424	2.562	2.075
w_B	0.3000	0.3000	0.3300	0.3300	0.3300	0.3300	0.3300	0.3300	0.3300
T/K	271.438	271.933	267.859	268.475	268.972	269.397	269.948	270.515	271.032
P/MPa	1.653	1.022	4.575	3.748	3.230	2.672	1.964	1.409	0.718
w_B	0.3591	0.3591	0.3591	0.3591	0.3591	0.3591	0.3591		
T/K	268.225	268.594	268.977	269.516	270.019	270.425	270.995		
P/MPa	4.441	3.952	3.393	2.711	1.981	1.421	0.654		

Polymer (B): **polystyrene** **1996LUS**
Characterization: M_n/g.mol^{-1} = 12750, M_w/g.mol^{-1} = 13500,
Pressure Chemical Company, Pittsburgh, PA
Solvent (A): **propionitrile** **C_3H_5N** **107-12-0**

Type of data: cloud points

continued

continued

w_B	0.1600	0.1600	0.1600	0.1600	0.1600	0.1906	0.1906	0.1906	0.1906
T/K	307.292	307.952	308.506	308.947	309.475	308.495	308.969	309.496	309.974
P/MPa	2.554	2.133	1.900	1.487	1.256	1.778	1.459	1.194	0.921

w_B	0.1906	0.2256	0.2256	0.2256	0.2256	0.2256
T/K	310.369	305.873	306.207	307.297	308.274	309.210
P/MPa	0.744	3.351	3.035	2.300	1.674	1.098

Polymer (B): **polystyrene** **1996LUS**
Characterization: M_n/g.mol^{-1} = 15900, M_w/g.mol^{-1} = 16700,
Scientific Polymer Products, Inc., Ontario, NY
Solvent (A): **propionitrile (75% deuterated at CH_2)** **$C_3H_5N/C_3H_3D_2N$**

Type of data: cloud points

w_B	0.1000	0.1000	0.1000	0.1000	0.1000	0.1000	0.1000	0.1000	0.1000
T/K	323.698	323.695	330.240	330.239	337.046	337.039	337.037	445.521	445.513
P/MPa	5.540	5.539	2.922	2.964	0.549	0.445	0.514	1.015	0.882
w_B	0.1000	0.1000	0.1000	0.1000	0.1499	0.1499	0.1499	0.1499	0.1499
T/K	445.513	454.443	463.595	463.592	330.263	330.274	338.051	338.047	343.830
P/MPa	0.820	1.858	2.894	2.834	5.203	5.211	2.282	2.249	0.575
w_B	0.1499	0.1499	0.1499	0.1499	0.1499	0.1499	0.1499	0.2000	0.2000
T/K	343.827	436.837	436.823	445.574	445.555	454.465	454.469	330.240	330.247
P/MPa	0.582	0.716	0.765	1.744	1.735	2.778	2.812	5.238	5.253
w_B	0.2000	0.2000	0.2000	0.2000	0.2000	0.2000	0.2000	0.2000	0.2000
T/K	336.994	336.992	343.826	343.832	436.736	436.730	445.503	445.498	454.431
P/MPa	2.686	2.664	0.659	0.665	0.791	0.803	1.903	1.897	3.008
w_B	0.2402	0.2402	0.2402	0.2402	0.2402	0.2402	0.2402	0.2402	0.2402
T/K	330.365	330.351	330.355	330.343	335.772	335.766	341.227	341.219	342.888
P/MPa	3.865	4.352	4.666	4.714	2.837	2.628	1.061	0.967	0.580
w_B	0.2533	0.2533	0.2533	0.2533	0.2533	0.2533	0.2533	0.2533	0.2533
T/K	330.270	337.948	340.070	340.049	436.726	436.737	445.534	445.527	454.467
P/MPa	4.519	1.847	1.170	1.186	0.712	0.706	1.819	1.820	3.012
w_B	0.2533	0.2610	0.2610	0.2610	0.2610	0.2610	0.2610	0.2610	0.2610
T/K	454.453	462.646	462.497	454.442	454.446	447.535	447.536	441.496	436.768
P/MPa	3.014	3.972	3.975	2.975	2.965	2.122	2.156	1.366	0.843
w_B	0.2610	0.2610	0.2610	0.2610	0.2610	0.2610	0.2610	0.2610	0.2610
T/K	436.755	325.361	325.333	330.352	330.346	335.723	335.723	341.393	341.386
P/MPa	0.815	5.693	5.476	3.577	3.607	2.207	2.140	0.443	0.421

Polymer (B): **polystyrene** **1996LUS**
Characterization: M_n/g.mol^{-1} = 21450, M_w/g.mol^{-1} = 22100,
Polymer Laboratories, Inc., Amherst, MA
Solvent (A): **propionitrile** **C_3H_5N** **107-12-0**

continued

continued

Type of data: cloud points

w_B	0.0600	0.0600	0.0600	0.0600	0.0600	0.0600	0.0600	0.0600	0.0600
T/K	331.770	331.771	337.301	337.300	433.404	433.385	445.537	454.400	454.407
P/MPa	2.413	2.049	0.893	0.534	1.044	0.987	2.223	3.527	3.433

w_B	0.0900	0.0900	0.0900	0.0900	0.0900	0.0900	0.0900	0.0900	0.0900
T/K	336.255	336.243	344.053	344.053	351.039	351.037	421.594	421.573	421.571
P/MPa	4.261	4.383	2.451	2.075	0.676	0.617	0.920	0.955	1.070

w_B	0.0900	0.0900	0.0900	0.0900	0.0900	0.0900	0.0900	0.0900	0.1076
T/K	428.369	428.371	436.952	436.953	445.765	445.767	454.720	454.712	337.250
P/MPa	1.454	1.713	2.403	2.184	3.425	3.430	4.388	4.403	4.400

w_B	0.1076	0.1076	0.1076	0.1076	0.1076	0.1076	0.1076	0.1076	0.1076
T/K	337.250	344.098	344.100	351.095	351.082	351.083	355.350	355.351	420.023
P/MPa	4.353	3.033	2.973	1.946	1.736	1.377	1.009	0.861	1.086

w_B	0.1076	0.1076	0.1076	0.1076	0.1076	0.1076	0.1076	0.1076	0.1076
T/K	428.443	428.439	437.035	437.038	437.036	445.843	445.842	454.781	454.778
P/MPa	1.956	1.930	2.804	2.667	2.922	3.708	3.838	4.900	4.926

w_B	0.1300	0.1300	0.1300	0.1300	0.1300	0.1300	0.1300	0.1600	0.1600
T/K	337.180	337.179	337.173	344.101	344.102	351.080	351.072	337.157	337.142
P/MPa	4.889	4.804	4.633	2.668	2.905	1.375	1.362	5.387	5.140

w_B	0.1600	0.1600	0.1600	0.1600	0.1600	0.1600	0.1600	0.1600	0.1600
T/K	337.138	343.960	343.961	343.965	350.936	350.939	357.997	357.991	360.818
P/MPa	5.109	3.149	3.193	3.084	1.545	1.750	0.568	0.602	0.241

w_B	0.1600	0.1600	0.1600	0.1600	0.1600	0.1600	0.1600	0.1600	0.1600
T/K	360.815	410.364	410.302	419.885	419.884	428.336	428.344	436.922	436.920
P/MPa	0.182	0.616	0.488	1.144	1.179	1.940	1.909	2.878	2.854

w_B	0.1600	0.1600	0.1600	0.1600	0.1899	0.1899	0.1899	0.1899	0.1899
T/K	445.755	445.746	454.689	454.691	336.998	336.994	343.823	343.823	350.807
P/MPa	3.850	3.817	4.729	4.848	4.956	4.915	3.093	3.030	1.606

w_B	0.1899	0.1899	0.1899	0.1899	0.1899	0.1899	0.1899	0.1899	0.1899
T/K	350.806	355.097	355.095	357.836	357.831	360.710	410.135	410.099	419.836
P/MPa	1.601	0.951	0.929	0.576	0.569	0.241	0.352	0.357	1.134

w_B	0.1899	0.1899	0.1899	0.1899	0.1899	0.1899	0.1899	0.1899	0.1899
T/K	419.829	428.285	428.284	436.891	436.895	445.680	445.682	454.629	454.635
P/MPa	1.153	1.968	1.947	2.834	2.818	3.777	3.823	4.881	4.834

w_B	0.2199	0.2199	0.2199	0.2199	0.2199	0.2199	0.2199	0.2199	0.2199
T/K	336.988	336.992	343.807	343.804	350.815	350.801	355.202	355.197	359.272
P/MPa	4.741	5.059	3.274	3.130	1.745	1.659	1.055	1.128	0.476

w_B	0.2199	0.2199	0.2199	0.2199	0.2199	0.2199	0.2199	0.2199	0.2199
T/K	359.268	411.469	411.480	419.692	419.662	428.099	428.101	436.717	436.694
P/MPa	0.502	0.542	0.599	1.424	1.241	1.991	2.026	2.821	2.883

continued

continued

w_B	0.2199	0.2199	0.2199	0.2199	0.2499	0.2499	0.2499	0.2499	0.2499
T/K	445.482	445.484	454.423	454.416	336.980	336.977	343.788	343.813	350.806
P/MPa	3.859	3.922	4.847	4.729	4.546	4.539	2.699	2.680	1.281
w_B	0.2499	0.2499	0.2499	0.2499	0.2499	0.2499	0.2499	0.2499	0.2499
T/K	350.798	355.232	355.226	357.840	357.841	414.134	414.392	419.714	419.704
P/MPa	1.308	0.620	0.584	0.269	0.275	0.498	0.577	1.001	0.995
w_B	0.2499	0.2499	0.2499	0.2499	0.2499	0.2499	0.2499	0.2499	0.2750
T/K	428.141	428.149	436.737	436.713	445.540	445.539	454.457	454.465	337.076
P/MPa	1.857	1.822	2.688	2.699	3.650	3.660	4.732	4.704	4.304
w_B	0.2750	0.2750	0.2750	0.2750	0.2750	0.2750	0.2750	0.2750	0.2750
T/K	337.069	343.873	343.877	345.600	345.581	350.878	350.880	353.764	353.758
P/MPa	4.294	2.477	2.460	2.033	2.042	0.978	0.985	0.489	0.470
w_B	0.2750	0.2750	0.2750	0.2750	0.2750	0.2750	0.2750	0.2750	0.2750
T/K	416.490	416.452	419.623	419.624	428.112	428.111	436.712	436.673	445.469
P/MPa	0.546	0.537	0.935	0.951	1.765	1.758	2.644	2.629	3.582
w_B	0.2750	0.2750	0.2750						
T/K	445.463	454.396	454.380						
P/MPa	3.608	4.670	4.671						

Polymer (B): **polystyrene** **1996LUS**
Characterization: M_n/g.mol^{-1} = 21450, M_w/g.mol^{-1} = 22100,
Polymer Laboratories, Inc., Amherst, MA
Solvent (A): **propionitrile (48% deuterated at CH$_2$)** **$C_3H_5N/C_3H_3D_2N$**

Type of data: cloud points

w_B	0.0800	0.0800	0.0800	0.0800	0.0800	0.0800	0.0800	0.0800	0.0800
T/K	357.740	357.741	364.997	364.993	372.356	372.360	379.893	379.888	387.404
P/MPa	3.687	3.572	2.253	2.670	1.582	1.547	1.306	1.058	0.913
w_B	0.0800	0.0800	0.0800	0.0800	0.0800	0.0800	0.0800	0.0800	0.0800
T/K	387.413	395.342	395.341	403.261	403.256	411.385	411.387	419.576	419.550
P/MPa	0.956	0.836	0.911	0.999	0.952	1.301	1.613	1.865	1.820
w_B	0.0800	0.0800	0.0800	0.0800	0.0800	0.0800	0.1101	0.1101	0.1101
T/K	427.986	427.979	436.500	436.496	445.337	445.331	357.758	357.760	365.093
P/MPa	2.442	2.653	3.267	3.221	4.257	4.172	5.301	5.141	3.882
w_B	0.1101	0.1101	0.1101	0.1101	0.1101	0.1101	0.1101	0.1101	0.1101
T/K	365.090	372.385	372.383	380.021	380.015	387.561	387.560	395.380	395.383
P/MPa	3.782	3.423	3.031	2.269	2.364	2.172	2.038	1.815	1.939
w_B	0.1101	0.1101	0.1101	0.1101	0.1101	0.1101	0.1101	0.1101	0.1101
T/K	403.309	403.290	411.400	411.397	411.393	419.617	419.610	428.072	428.067
P/MPa	1.885	1.872	2.510	2.338	1.591	2.763	2.728	3.441	3.520

continued

continued

w_B	0.1101	0.1101	0.1101	0.1101	0.1200	0.1200	0.1200	0.1200	0.1200
T/K	436.644	436.648	445.454	445.456	357.802	357.800	365.036	365.032	365.033
P/MPa	4.157	4.086	5.234	5.289	5.495	5.274	3.549	4.109	4.973

w_B	0.1200	0.1447	0.1447	0.1447	0.1447	0.1447	0.1447	0.1447	0.1447
T/K	365.042	365.034	365.039	372.386	372.391	379.919	379.920	387.571	387.572
P/MPa	3.935	4.785	4.768	3.573	3.491	2.842	2.670	2.583	2.472

w_B	0.1447	0.1447	0.1447	0.1447	0.1447	0.1447	0.1447	0.1447	0.1447
T/K	395.378	395.380	403.301	403.303	411.419	411.421	419.613	419.608	428.062
P/MPa	2.513	2.350	2.543	2.624	2.724	2.161	3.116	3.078	3.534

w_B	0.1447	0.1447	0.1447	0.1447	0.1447	0.1500	0.1500	0.1500	0.1500
T/K	428.064	436.687	436.684	445.457	445.458	357.796	357.794	365.057	365.038
P/MPa	3.658	4.475	4.358	5.160	5.159	5.693	5.807	4.340	4.329

w_B	0.1500	0.1500	0.1500	0.1500	0.1500	0.1500	0.1500	0.1500	0.1500
T/K	372.399	372.390	379.932	379.930	387.551	387.550	395.365	395.370	403.281
P/MPa	3.342	3.359	2.656	2.674	2.305	2.300	2.155	2.142	2.245

w_B	0.1500	0.1500	0.1500	0.1500	0.1500	0.1500	0.1500	0.1500	0.1500
T/K	403.280	411.410	411.409	411.404	419.630	419.629	419.628	428.095	428.092
P/MPa	2.244	2.580	2.620	2.579	2.986	3.015	2.948	3.544	3.595

w_B	0.1500	0.1500	0.1500	0.1500	0.1500	0.1799	0.1799	0.1799	0.1799
T/K	436.673	436.675	445.478	445.474	445.471	357.783	357.786	365.044	365.038
P/MPa	4.254	4.296	5.091	5.016	5.116	5.770	5.510	4.360	4.312

w_B	0.1799	0.1799	0.1799	0.1799	0.1799	0.1799	0.1799	0.1799	0.1799
T/K	372.394	372.391	379.930	379.930	387.575	387.575	395.401	395.393	403.293
P/MPa	3.393	3.047	2.754	2.796	2.232	2.282	2.201	2.085	2.141

w_B	0.1799	0.1799	0.1799	0.1799	0.1799	0.1799	0.1799	0.1799	0.1799
T/K	403.286	411.442	411.435	419.663	419.656	428.114	428.114	436.551	436.557
P/MPa	2.221	2.485	2.518	2.957	2.944	3.608	3.586	4.288	4.332

w_B	0.1799	0.1799	0.1799	0.2101	0.2101	0.2101	0.2101	0.2101	0.2101
T/K	445.492	445.484	445.491	357.790	357.791	365.030	365.043	372.407	372.406
P/MPa	5.110	5.100	4.893	5.199	5.179	3.898	3.859	2.902	2.888

w_B	0.2101	0.2101	0.2101	0.2101	0.2101	0.2101	0.2101	0.2101	0.2101
T/K	379.933	379.930	387.540	387.541	395.392	395.390	403.305	403.303	411.427
P/MPa	2.254	2.255	1.911	1.888	1.778	1.785	1.901	1.927	2.198

w_B	0.2101	0.2101	0.2101	0.2101	0.2101	0.2101	0.2101	0.2101	0.2101
T/K	411.427	419.658	419.655	428.105	428.104	436.673	436.670	445.498	445.493
P/MPa	2.181	2.655	2.661	3.255	3.302	3.993	3.986	4.817	4.810

w_B	0.2400	0.2400	0.2400	0.2400	0.2400	0.2400	0.2400	0.2400	0.2400
T/K	357.803	357.802	365.054	365.051	372.419	372.416	379.949	379.942	387.470
P/MPa	4.785	4.774	3.474	3.465	2.527	2.534	1.931	1.927	1.596

continued

continued

w_B	0.2400	0.2400	0.2400	0.2400	0.2400	0.2400	0.2400	0.2400	0.2400
T/K	387.467	395.430	395.425	403.326	403.322	411.443	411.441	419.684	419.684
P/MPa	1.594	1.518	1.512	1.648	1.614	1.964	1.965	2.443	2.452

w_B	0.2400	0.2400	0.2400	0.2400	0.2400	0.2400	0.2726	0.2726	0.2726
T/K	428.131	428.133	436.696	436.693	445.529	445.535	350.782	350.783	357.901
P/MPa	3.037	3.072	3.802	3.794	4.751	4.792	4.841	4.879	3.343

w_B	0.2726	0.2726	0.2726	0.2726	0.2726	0.2726	0.2726	0.2726	0.2726
T/K	357.895	365.668	365.646	372.512	372.510	380.038	380.031	387.713	387.690
P/MPa	3.325	2.139	2.089	1.362	1.334	0.790	0.782	0.578	0.573

w_B	0.2726	0.2726	0.2726	0.2726	0.2726	0.2726	0.2726	0.2726	0.2726
T/K	395.511	395.508	403.446	403.441	411.723	411.703	419.644	419.644	428.302
P/MPa	0.628	0.633	0.885	0.886	1.308	1.348	1.848	1.845	2.526

w_B	0.2726	0.2726	0.2726	0.2726	0.2726	0.3073	0.3073
T/K	428.297	436.690	436.684	445.492	445.485	343.805	343.803
P/MPa	2.521	3.343	3.292	4.234	4.123	4.981	4.984

Polymer (B): **polystyrene** **1996LUS**

Characterization: M_n/g.mol^{-1} = 21450, M_w/g.mol^{-1} = 22100, Polymer Laboratories, Inc., Amherst, MA

Solvent (A): **propionitrile (63.55% deuterated at CH_2)** **$C_3H_5N/C_3H_3D_2N$**

Type of data: cloud points

w_B	0.1800	0.1800	0.1800	0.1800	0.1800	0.1800	0.1800	0.1800	0.1800
T/K	372.424	372.414	379.970	379.948	387.582	387.588	395.407	395.408	403.104
P/MPa	5.410	5.410	4.542	4.553	3.998	3.977	3.686	3.706	3.612

w_B	0.1800	0.1800	0.1800	0.1800	0.2098	0.2098	0.2098	0.2098	0.2098
T/K	403.112	411.471	411.469	419.685	372.404	372.415	372.415	372.415	380.378
P/MPa	3.620	3.759	3.766	4.106	4.879	5.401	5.385	4.957	3.992

w_B	0.2098	0.2098	0.2098	0.2098	0.2098	0.2098	0.2098	0.2098	0.2098
T/K	380.344	387.507	387.501	395.424	395.416	403.327	403.325	403.325	411.445
P/MPa	4.004	3.510	3.566	3.340	3.324	3.379	3.625	3.322	3.554

w_B	0.2098	0.2098	0.2098	0.2098	0.2098	0.2098	0.2098	0.2098	0.2098
T/K	411.443	419.645	419.644	428.072	428.080	436.643	436.646	445.447	445.446
P/MPa	3.458	4.211	4.290	4.736	4.737	5.381	5.424	6.063	6.116

w_B	0.2098	0.2401	0.2401	0.2401	0.2401	0.2401	0.2401	0.2401	0.2401
T/K	419.650	365.080	365.078	372.387	372.388	380.020	380.015	387.610	387.607
P/MPa	4.423	5.712	5.711	4.510	4.509	3.644	3.659	3.121	3.115

w_B	0.2401	0.2401	0.2401	0.2401	0.2401	0.2401	0.2401	0.2401	0.2401
T/K	395.885	395.866	403.342	403.343	411.695	411.666	419.709	419.700	428.139
P/MPa	2.872	2.866	2.847	2.856	3.052	3.019	3.505	3.520	4.026

continued

continued

w_B	0.2401	0.2401	0.2401	0.2401	0.2401	0.2401	0.2700	0.2700	0.2700
T/K	428.143	436.720	436.724	445.520	445.520	445.522	365.092	365.087	372.436
P/MPa	4.054	4.717	4.709	5.515	5.536	5.485	5.552	5.597	4.408
w_B	0.2700	0.2700	0.2700	0.2700	0.2700	0.2700	0.2700	0.2700	0.2700
T/K	372.430	379.984	379.988	387.570	387.570	395.407	395.407	403.304	403.303
P/MPa	4.348	3.512	3.514	2.952	2.958	2.683	2.716	2.674	2.684
w_B	0.2700	0.2700	0.2700	0.2700	0.2700	0.2700	0.2700	0.2700	0.2700
T/K	411.453	411.449	419.630	419.630	428.111	428.110	436.743	436.723	445.577
P/MPa	2.944	2.927	3.330	3.310	3.859	3.835	4.525	4.521	5.350
w_B	0.2700	0.2964	0.2964	0.2964	0.2964	0.2964	0.2964	0.2964	0.2964
T/K	445.528	360.685	360.684	365.226	365.183	372.427	372.428	372.431	379.999
P/MPa	5.325	6.047	6.067	5.150	5.140	3.691	3.775	3.817	3.008
w_B	0.2964	0.2964	0.2964	0.2964	0.2964	0.2964	0.2964	0.2964	0.2964
T/K	379.998	387.592	387.596	395.420	395.425	403.292	403.283	411.438	411.424
P/MPa	3.024	2.549	2.550	2.339	2.342	2.398	2.362	2.595	2.602
w_B	0.2964	0.2964	0.2964	0.2964	0.2964	0.2964	0.2964	0.2964	0.2964
T/K	419.485	419.472	428.097	428.098	436.487	436.485	445.468	445.437	454.411
P/MPa	2.981	2.989	3.532	3.559	4.396	4.356	5.259	5.169	6.119

Polymer (B): **polystyrene** **1996LUS**

Characterization: M_n/g.mol^{-1} = 21450, M_w/g.mol^{-1} = 22100, Polymer Laboratories, Inc., Amherst, MA

Solvent (A): **propionitrile (75.0% deuterated at CH_2)** **$C_3H_5N/C_3H_3D_2N$**

Type of data: cloud points

w_B	0.1399	0.1399	0.1399	0.1399	0.1399	0.1399	0.1399	0.1399	0.1399
T/K	373.109	373.104	373.105	380.614	380.623	388.293	388.293	388.293	388.293
P/MPa	5.777	5.735	5.555	4.700	5.179	3.590	4.146	3.389	4.268
w_B	0.1399	0.1399	0.1399	0.1399	0.1399	0.1399	0.1399	0.1399	0.1399
T/K	396.073	396.076	403.969	403.971	412.105	412.145	420.378	420.372	428.821
P/MPa	3.794	3.648	3.725	3.651	3.845	3.741	4.223	3.960	4.385
w_B	0.1399	0.1399	0.1399	0.1399	0.1700	0.1700	0.1700	0.1700	0.1700
T/K	428.847	437.480	437.450	446.484	373.092	373.092	373.095	380.624	380.623
P/MPa	4.616	5.349	5.181	5.891	5.711	5.715	5.698	4.744	4.733
w_B	0.1700	0.1700	0.1700	0.1700	0.1700	0.1700	0.1700	0.1700	0.1700
T/K	388.244	388.245	396.093	396.085	404.061	404.031	413.878	420.348	420.350
P/MPa	4.092	4.100	3.743	3.746	3.629	3.633	3.803	4.084	4.094
w_B	0.1700	0.1700	0.1700	0.1700	0.1700	0.1700	0.2202	0.2202	0.2202
T/K	428.803	428.802	437.412	437.415	446.239	446.209	372.913	372.917	372.915
P/MPa	4.490	4.497	5.099	5.102	5.860	5.855	5.851	5.919	5.835

continued

continued

w_B	0.2202	0.2202	0.2202	0.2202	0.2202	0.2202	0.2202	0.2202	0.2202
T/K	380.578	380.575	388.239	388.241	396.057	396.063	403.989	403.989	403.989
P/MPa	4.838	4.871	4.105	4.367	3.885	3.905	3.762	3.820	3.775

w_B	0.2202	0.2202	0.2202	0.2202	0.2202	0.2202	0.2202	0.2202	0.2202
T/K	403.989	412.142	412.127	420.351	420.337	428.830	428.830	437.450	428.830
P/MPa	3.720	3.826	3.670	4.119	4.095	4.655	4.655	5.143	4.573

w_B	0.2202	0.2299	0.2299	0.2299	0.2299	0.2299	0.2299	0.2299	0.2299
T/K	437.431	372.488	372.492	379.987	387.629	387.629	395.570	395.567	403.590
P/MPa	5.101	6.065	6.082	5.157	4.407	4.262	3.907	3.919	3.799

w_B	0.2299	0.2299	0.2299	0.2299	0.2299	0.2299	0.2299	0.2299	0.2299
T/K	403.606	411.920	411.938	420.160	420.145	428.615	428.604	437.210	437.189
P/MPa	3.729	3.894	3.895	4.171	4.124	4.632	4.699	5.213	5.132

w_B	0.2588	0.2588	0.2588	0.2588	0.2588	0.2588	0.2588	0.2588	0.2588
T/K	380.009	380.009	387.600	387.593	395.371	395.384	403.281	403.320	411.465
P/MPa	4.883	4.868	4.189	4.218	3.789	3.826	3,694	3.742	3.823

w_B	0.2588	0.2588	0.2588	0.2588	0.2588	0.2588	0.2588
T/K	411.469	419.692	419.679	428.165	428.154	436.707	436.691
P/MPa	3.814	4.079	4.082	4.542	4.561	5.134	5.139

Polymer (B): **polystyrene** **1996LUS**

Characterization: M_n/g.mol^{-1} = 23600, M_w/g.mol^{-1} = 25000, Pressure Chemical Company, Pittsburgh, PA

Solvent (A): **propionitrile** **C_3H_5N** **107-12-0**

Type of data: cloud points

w_B	0.1687	0.1687	0.1687	0.1687	0.1687	0.1687	0.1687	0.1687	0.1687
T/K	436.982	436.984	428.529	420.105	420.101	413.475	403.751	403.747	397.076
P/MPa	4.593	4.560	3.946	3.229	3.170	2.763	2.275	2.269	2.077

w_B	0.1687	0.1687	0.1687	0.1687	0.1687	0.1687	0.1687	0.1687	0.1687
T/K	397.085	387.998	387.989	380.095	380.110	380.107	372.753	372.723	366.403
P/MPa	2.044	1.997	2.001	2.174	1.401	2.176	2.601	2.607	3.198

w_B	0.1687	0.1687	0.1687	0.1799	0.1799	0.1799	0.1799	0.1799	0.1799
T/K	366.009	358.121	358.116	365.993	365.996	373.340	373.345	380.978	380.974
P/MPa	3.237	4.333	4.405	4.905	4.924	4.151	3.995	3.521	3.434

w_B	0.1799	0.1799	0.1799	0.1799	0.1799	0.1799	0.1799	0.1799	0.1799
T/K	388.677	388.669	396.524	396.529	396.529	396.529	404.434	404.432	412.646
P/MPa	3.140	2.902	3.080	3.046	2.955	3.013	3.299	3.212	3.456

w_B	0.1799	0.1799	0.1799	0.1799	0.1799	0.1799	0.1799	0.1999	0.1999
T/K	412.646	420.921	420.904	429.373	429.330	437.974	437.977	437.501	437.500
P/MPa	3.388	3.956	3.871	4.423	4.230	5.206	5.240	5.261	5.333

continued

continued

w_B	0.1999	0.1999	0.1999	0.1999	0.1999	0.1999	0.1999	0.1999	0.1999
T/K	437.510	428.865	428.857	420.705	420.674	412.373	412.350	404.353	404.339
P/MPa	5.477	4.726	4.522	4.183	3.909	3.583	3.589	3.244	3.286

w_B	0.1999	0.1999	0.1999	0.1999	0.1999	0.1999	0.1999	0.1999	0.1999
T/K	396.419	396.415	388.628	381.025	381.030	381.030	373.334	373.328	366.011
P/MPa	3.130	3.049	3.073	3.366	4.238	3.367	3.971	3.999	4.599

w_B	0.1999	0.1999	0.2200	0.2200	0.2200	0.2200	0.2200	0.2200	0.2200
T/K	366.011	366.016	365.570	365.567	373.014	373.018	380.604	380.601	388.270
P/MPa	4.717	4.790	4.655	4.630	3.779	3.736	3.156	3.165	2.861

w_B	0.2200	0.2200	0.2200	0.2200	0.2200	0.2200	0.2200	0.2200	0.2200
T/K	388.259	396.102	396.100	403.934	403.933	412.096	412.100	420.560	420.572
P/MPa	2.864	2.800	2.813	2.913	2.932	3.235	3.221	3.710	3.699

w_B	0.2200	0.2200	0.2200	0.2200	0.2308	0.2459	0.2459	0.2459	0.2459
T/K	428.894	428.892	437.472	437.482	437.064	437.175	437.181	428.700	428.636
P/MPa	4.297	4.293	5.005	5.023	4.683	4.651	4.676	3.932	4.050

w_B	0.2459	0.2459	0.2459	0.2459	0.2459	0.2459	0.2459	0.2459	0.2459
T/K	420.286	420.277	420.214	412.202	412.045	404.140	404.135	396.033	396.035
P/MPa	3.452	3.474	3.473	3.035	3.002	2.523	2.446	2.314	2.341

w_B	0.2459	0.2459	0.2459	0.2459	0.2459	0.2459	0.2459	0.2459	0.2459
T/K	388.125	388.129	380.387	380.385	372.911	372.914	365.658	365.666	358.452
P/MPa	2.433	2.422	2.748	2.762	3.258	3.228	4.060	4.232	5.406

w_B	0.2459	0.2459	0.2798	0.2798	0.2798	0.2798	0.2798	0.2798	0.2798
T/K	358.449	358.453	437.980	429.468	429.462	429.464	420.935	420.934	412.739
P/MPa	5.451	5.388	3.211	2.832	2.759	2.532	1.932	1.881	1.004

w_B	0.2798	0.2798	0.2798	0.2971	0.2971	0.2971	0.2971	0.2971	0.2971
T/K	412.710	344.822	344.826	438.267	438.256	429.743	429.731	429.728	421.267
P/MPa	0.969	0.644	0.714	3.767	3.849	3.112	2.980	2.976	2.213

w_B	0.2971	0.2971	0.2971	0.2971	0.2971	0.2971	0.2971	0.2971	0.2971
T/K	421.249	413.028	413.027	413.024	404.747	404.741	396.954	358.883	358.883
P/MPa	2.192	1.618	1.600	1.670	0.963	0.979	0.637	0.757	0.778

w_B	0.2971	0.2971	0.2971	0.2971	0.2971	0.2971	0.2971	0.2971	0.2971
T/K	351.969	351.969	344.974	344.967	406.035	412.749	420.973	429.415	437.999
P/MPa	1.693	1.698	2.988	3.066	0.707	1.360	2.665	2.767	3.618

Polymer (B): **polystyrene-d8** **1996LUS**

Characterization: M_n/g.mol^{-1} = 25400, M_w/g.mol^{-1} = 26700, completely deuterated, Polymer Laboratories, Amherst, MA

Solvent (A): **propionitrile** **C_3H_5N** **107-12-0**

Type of data: cloud points

continued

continued

w_B	0.1100	0.1100	0.1100	0.1100	0.1100	0.1100	0.1100	0.1100	0.1100
T/K	298.411	298.417	300.909	300.912	304.551	304.550	304.548	433.682	433.647
P/MPa	3.855	3.881	2.595	2.622	0.970	0.884	0.945	0.722	0.716

w_B	0.1100	0.1100	0.1100	0.1100	0.1100	0.1100	0.1100	0.1100	0.1100
T/K	436.659	436.648	445.435	445.434	445.430	454.325	454.326	463.495	463.490
P/MPa	1.141	1.135	2.199	2.282	2.249	3.555	3.552	4.830	4.842

w_B	0.1100	0.1100	0.1401	0.1401	0.1401	0.1401	0.1401	0.1401	0.1401
T/K	463.484	463.487	298.413	298.407	301.744	301.746	305.074	307.000	306.987
P/MPa	4.757	4.673	4.434	4.431	2.763	2.740	1.272	0.495	0.501

w_B	0.1401	0.1401	0.1401	0.1401	0.1401	0.1401	0.1401	0.1401	0.1401
T/K	431.523	431.515	436.657	436.664	445.455	445.436	454.385	454.377	463.517
P/MPa	0.613	0.607	1.309	1.286	2.487	2.478	3.733	3.706	4.969

w_B	0.1401	0.1699	0.1699	0.1699	0.1699	0.1699	0.1699	0.1699	0.1699
T/K	463.506	299.564	299.418	301.930	301.903	305.638	305.560	308.075	308.082
P/MPa	4.999	4.609	4.914	3.627	3.589	2.294	1.825	0.860	1.037

w_B	0.1699	0.1699	0.1699	0.1699	0.1699	0.1699	0.1699	0.1699	0.1699
T/K	431.307	431.300	436.635	436.638	445.431	445.424	454.336	455.293	455.293
P/MPa	0.720	0.825	1.418	1.550	2.791	2.760	3.906	3.987	3.981

w_B	0.1699	0.1999	0.1999	0.1999	0.1999	0.1999	0.1999	0.1999	0.1999
T/K	463.488	298.698	298.691	301.242	301.215	304.599	304.609	306.836	306.866
P/MPa	5.198	4.400	4.386	3.097	3.104	1.548	1.572	0.674	0.689

w_B	0.1999	0.1999	0.1999	0.1999	0.1999	0.1999	0.1999	0.1999	0.1999
T/K	433.304	433.303	436.649	436.646	445.412	445.426	454.364	454.359	463.517
P/MPa	0.943	0.991	1.460	1.469	2.663	2.653	3.876	3.881	5.222

w_B	0.1999	0.2301	0.2301	0.2301	0.2301	0.2301	0.2301	0.2301	0.2301
T/K	463.520	294.662	294.663	294.674	298.524	298.527	298.522	301.023	301.014
P/MPa	5.207	4.966	5.030	5.011	2.894	3.026	3.087	1.854	1.889

w_B	0.2301	0.2301	0.2301	0.2301	0.2301	0.2301	0.2301	0.2301	0.2301
T/K	303.396	303.411	303.408	304.645	304.647	431.698	431.701	436.840	436.841
P/MPa	1.239	1.075	1.465	1.047	1.147	0.657	0.672	1.360	1.377

w_B	0.2301	0.2301	0.2301	0.2301	0.2301	0.2301	0.2581	0.2581	0.2581
T/K	445.825	445.819	454.766	454.759	463.931	463.929	294.668	294.670	298.405
P/MPa	2.626	2.600	3.842	3.846	5.346	5.407	5.183	5.193	3.291

w_B	0.2581	0.2581	0.2581	0.2581	0.2581	0.2581	0.2581	0.2581	0.2581
T/K	298.404	300.847	300.853	304.522	304.533	431.546	431.558	435.041	435.039
P/MPa	3.258	2.104	2.169	0.578	0.574	0.559	0.575	1.056	1.060

w_B	0.2581	0.2581	0.2581	0.2581	0.2581	0.2581	0.2581	0.2581
T/K	439.829	439.790	445.466	445.456	454.364	454.367	463.608	463.610
P/MPa	1.704	1.708	2.503	2.497	3.694	3.753	5.042	4.984

Polymer (B): **polystyrene** **1991OPS**
Characterization: M_n/g.mol^{-1} = 3700, M_w/g.mol^{-1} = 4000, M_z/g.mol^{-1} = 4600, Pressure Chemical Company, Pittsburgh, PA
Solvent (A): **1-tetradecanol** $C_{14}H_{30}O$ **112-72-1**

Type of data: critical point (UCST-behavior)

$w_{B,\,crit}$ 0.402 T_{crit}/K 383.25

Polymer (B): **polystyrene** **2001IM1**
Characterization: M_n/g.mol^{-1} = 1160, M_w/g.mol^{-1} = 1240, Pressure Chemical Company, Pittsburgh, PA
Solvent (A): **2,2,4-trimethylpentane** C_8H_{18} **540-84-1**

Type of data: cloud points (UCST-behavior)

w_B	0.080	0.110	0.140	0.170	0.200	0.240	0.270
T/K	256	259	261	261	262	261	261

Polymer (B): **polystyrene** **1950JE1**
Characterization: M_η/g.mol^{-1} = 62600, fractionated in the laboratory
Solvent (A): **vinyl acetate** $C_4H_6O_2$ **108-05-4**

Type of data: cloud points, precipitation threshold (UCST-behavior)

z_B 0.036 T/K 384.15

Polymer (B): **poly(styrene-*co*-acrylonitrile)** **1998SCH, 2000SCH**
Characterization: M_n/g.mol^{-1} = 90000, M_w/g.mol^{-1} = 147000, 25.0 wt% acrylonitrile, Bayer AG, Leverkusen, Germany
Solvent (A): **toluene** C_7H_8 **108-88-3**

Type of data: cloud points (UCST-behavior)

w_B	0.006	0.010	0.020	0.042	0.061	0.080	0.081	0.101	0.141
T/K	324.35	326.85	326.15	321.35	316.85	313.15	313.15	309.35	300.25

Type of data: critical point (UCST-behavior)

$w_{B,\,crit}$ 0.080 T_{crit}/K 313.15

Type of data: coexistence data (tie lines)

The total feed concentration of the copolymer in the homogeneous system is: w_B = 0.080. This is the critical concentration, i.e., sol phase and gel phase are identical phases at the critical point, see the first line in the table below.

continued

continued

Demixing temperature	w_B		Fractionation during demixing			
	Sol phase	Gel phase	Sol phase		Gel phase	
T/K			M_n/ g mol^{-1}	M_w/ g mol^{-1}	M_n/ g mol^{-1}	M_w/ g mol^{-1}
313.15	0.080	0.080	90000	147000	90000	147000
311.65	0.055	0.125	62000	124000	80000	170000
309.35	0.043	0.151	48500	105000	74500	163000
307.05	0.029	0.156	56500	187000	84000	167000
305.15	0.046	0.147	54000	110000	91000	174000
302.95	0.035	0.168	47000	89000	90000	171000
300.95	0.042	0.181	48000	90000	90000	175000
299.15	0.033	0.178	80000	170000	79000	169000

Comments: Apparent M_w and M_n values were determined via polystyrene standards.

Polymer (B): **poly(styrene-*b*-butadiene-*b*-styrene-*b*-butadiene-*b*-styrene)** **2004NIE**
Characterization: M_n/g.mol^{-1} = 109000, M_w/g.mol^{-1} = 120000, both polybutadiene blocks have M_n/g.mol^{-1} = 32000, the middle polystyrene block has M_n/g.mol^{-1} = 9000, both outside polystyrene blocks have M_n/g.mol^{-1} = 13000
Solvent (A): **n-heptane** C_7H_{16} **142-82-5**

Type of data: cloud points (UCST-behavior)

w_B 0.005 T/K 330.15

Polymer (B): **poly(styrene-*co*-methyl methacrylate)** **2006GAR**
Characterization: M_n/g.mol^{-1} = 55200, M_w/g.mol^{-1} = 58500
Solvent (A): **cyclohexanol** $C_6H_{12}O$ **108-93-0**

Type of data: cloud points (UCST-behavior)

w_B	0.0015	0.0062	0.0104	0.0417	0.0602	0.0941	0.1179	0.1589	0.1639
T/K	340.74	346.65	347.14	345.68	345.41	345.61	345.45	345.73	344.80

w_B	0.1830
T/K	345.15

Polymer (B): **poly(styrene-*co*-α-methylstyrene)** **1995PFO**
Characterization: M_n/g.mol^{-1} = 100000, M_w/g.mol^{-1} = 114000, 20.0 mol% styrene

continued

continued

Solvent (A): **butyl acetate** $C_6H_{12}O_2$ **123-86-4**

Type of data: cloud points (UCST-behavior)

w_B	0.0050	0.0100	0.0202	0.0298	0.0403	0.0501	0.0705	0.0893	0.1187
T/K	274.85	283.65	288.35	288.75	288.65	288.85	288.55	286.75	281.95

w_B	0.1405
T/K	279.95

Type of data: cloud points (LCST-behavior)

w_B	0.0050	0.0100	0.0202	0.0298	0.0403	0.0501	0.0705	0.0893	0.1187
T/K	468.45	463.55	458.15	456.85	453.05	454.45	453.95	454.45	455.05

w_B	0.1405
T/K	456.55

Polymer (B): **poly(styrene-*co*-α-methylstyrene)** **1995PFO**
Characterization: M_n/g.mol^{-1} = 100000, M_w/g.mol^{-1} = 114000, 20.0 mol% styrene
Solvent (A): **cyclohexane** C_6H_{12} **110-82-7**

Type of data: cloud points (UCST-behavior)

w_B	0.0050	0.0101	0.0196	0.0297	0.0398	0.0505	0.0532	0.0703	0.1008
T/K	279.35	282.15	284.55	284.55	285.05	285.65	285.85	285.55	285.35

w_B	0.1521
T/K	283.55

Type of data: cloud points (LCST-behavior)

w_B	0.0050	0.0101	0.0196	0.0297	0.0398	0.0505	0.0532	0.0703	0.1008
T/K	494.65	490.55	486.45	484.85	486.55	485.15	487.05	486.35	485.95

w_B	0.1521
T/K	489.25

Polymer (B): **poly(styrene-*co*-α-methylstyrene)** **1995PFO**
Characterization: M_n/g.mol^{-1} = 100000, M_w/g.mol^{-1} = 114000, 20.0 mol% styrene
Solvent (A): **cyclopentane** C_5H_{10} **287-92-3**

Type of data: cloud points (UCST-behavior)

w_B	0.0053	0.0104	0.0241	0.0364	0.0482	0.0526	0.0913	0.1437	0.1876
T/K	281.65	284.65	288.95	290.15	290.95	290.85	290.65	289.65	288.45

Type of data: cloud points (LCST-behavior)

w_B	0.0053	0.0104	0.0241	0.0364	0.0482	0.0526	0.0913	0.1437	0.1876
T/K	431.35	427.55	422.65	422.45	421.55	421.05	421.25	422.25	423.35

Polymer (B): **poly(styrene-*co*-α-methylstyrene)** **1995PFO**
Characterization: M_n/g.mol^{-1} = 100000, M_w/g.mol^{-1} = 114000, 20.0 mol% styrene
Solvent (A): ***trans*-decahydronaphthalene** $C_{10}H_{18}$ **493-02-7**

Type of data: cloud points (UCST-behavior)

w_B	0.0050	0.0099	0.0197	0.0301	0.0402	0.0473	0.0687	0.0845	0.1177
T/K	254.45	259.55	262.05	263.05	263.75	263.75	264.05	264.15	263.95

w_B	0.1385	0.1893
T/K	263.55	263.65

Polymer (B): **poly(styrene-*co*-α-methylstyrene)** **1995PFO**
Characterization: M_n/g.mol^{-1} = 100000, M_w/g.mol^{-1} = 114000, 20.0 mol% styrene
Solvent (A): **hexyl acetate** $C_8H_{16}O_2$ **142-92-7**

Type of data: cloud points (UCST-behavior)

w_B	0.0050	0.0100	0.0202	0.0301	0.0393	0.0496	0.0703	0.0899	0.1192
T/K	277.05	284.25	287.35	288.25	287.35	288.55	287.15	286.75	282.95

w_B	0.1486
T/K	280.95

Type of data: cloud points (LCST-behavior)

w_B	0.0050	0.0100	0.0202	0.0301	0.0393	0.0496	0.0703	0.0899	0.1192
T/K	526.55	524.15	515.85	515.05	514.55	514.45	514.65	514.25	514.45

w_B	0.1486
T/K	514.05

Polymer (B): **poly(styrene-*co*-α-methylstyrene)** **1995PFO**
Characterization: M_n/g.mol^{-1} = 100000, M_w/g.mol^{-1} = 114000, 20.0 mol% styrene
Solvent (A): **pentyl acetate** $C_7H_{14}O_2$ **628-63-7**

Type of data: cloud points (UCST-behavior)

w_B	0.0010	0.0050	0.0107	0.0201	0.0301	0.0401	0.0505	0.0689	0.0840
T/K	274.55	289.75	296.25	298.25	304.25	302.25	300.45	300.65	299.95

w_B	0.1093	0.1527
T/K	297.15	294.05

Type of data: cloud points (LCST-behavior)

w_B	0.0010	0.0050	0.0107	0.0201	0.0301	0.0401	0.0505	0.0689	0.0840
T/K	502.75	493.55	489.55	482.95	481.55	482.45	480.65	481.35	481.55

w_B	0.1093	0.1527
T/K	483.05	485.95

Polymer (B): **poly(vinyl acetate)** **1999BEY**
Characterization: M_n/g.mol^{-1} = 52700, M_w/g.mol^{-1} = 124800, Aldrich Chem. Co., Inc., Milwaukee, WI
Solvent (A): **cyclopentane** C_5H_{10} **287-92-3**

Type of data: cloud points (LCST-behavior)

w_B	0.06	0.06	0.06	0.06	0.06	0.06
T/K	516.55	519.35	524.95	531.45	536.25	541.15
P/bar	316.0	302.0	278.0	254.0	246.0	243.0

Polymer (B): **poly(vinyl acetate)** **1999BEY**
Characterization: M_n/g.mol^{-1} = 52700, M_w/g.mol^{-1} = 124800, Aldrich Chem. Co., Inc., Milwaukee, WI
Solvent (A): **cyclopentene** C_5H_8 **142-29-0**

Type of data: cloud points (LCST-behavior)

w_B	0.06	0.06	0.06	0.06	0.06	0.06
T/K	445.65	451.75	460.45	469.55	478.45	490.85
P/bar	270.6	168.2	115.9	97.2	84.2	98.3

Polymer (B): **poly(vinyl acetate)** **1976ALE**
Characterization: M_w/g.mol^{-1} = 100000
Solvent (A): **methanol** CH_4O **67-56-1**

Type of data: cloud points (UCST-behavior)

w_B	0.0052	0.0130	0.0242	0.0396	0.0860
T/K	265.0	266.7	268.7	269.7	266.8

Polymer (B): **poly(vinyl acetate-*co*-vinyl alcohol)** **1991EAG**
Characterization: synthesized in the laboratory
Solvent (A): **water** H_2O **7732-18-5**

Type of data: cloud points (LCST-behavior)

w_B	0.02	T/K	338.35	for a copolymer of 18.0 mol% vinyl acetate
w_B	0.02	T/K	308.75	for a copolymer of 23.7 mol% vinyl acetate
w_B	0.02	T/K	305.35	for a copolymer of 24.4 mol% vinyl acetate
w_B	0.02	T/K	315.05	for a copolymer of 30.0 mol% vinyl acetate

Polymer (B): **poly(*N*-vinylcaprolactam)** **2002MA1**
Characterization: M_w/g.mol^{-1} = 13000, synthesized in the laboratory
Solvent (A): **deuterium oxide** D_2O **7789-20-0**

Type of data: cloud points (LCST-behavior)

w_B 0.20 T/K 306.75

Polymer (B): **poly(*N*-vinylcaprolactam)** **2003OKH**
Characterization: M_w/g.mol^{-1} = 5600, synthesized in the laboratory
Solvent (A): **water** H_2O **7732-18-5**

Type of data: cloud points (LCST-behavior)

c_B/(g/l) 1.0 T/K 303.15 (in 0.05 M NaCl aqueous solution)

Polymer (B): **poly(*N*-vinylcaprolactam)** **2002MA1**
Characterization: M_w/g.mol^{-1} = 13000, synthesized in the laboratory
Solvent (A): **water** H_2O **7732-18-5**

Type of data: cloud points (LCST-behavior)

w_B	0.01	0.02	0.03	0.05	0.10	0.15	0.20	0.30	0.40
T/K	306.95	306.65	306.35	305.65	305.75	305.95	306.75	307.35	310.05

Polymer (B): **poly(*N*-vinylcaprolactam)** **2002MA1**
Characterization: M_w/g.mol^{-1} = 150000, synthesized in the laboratory
Solvent (A): **water** H_2O **7732-18-5**

Type of data: cloud points (LCST-behavior)

w_B 0.005 T/K 306.45

Polymer (B): **poly(*N*-vinylcaprolactam)** **1994TAG**
Characterization: M_η/g.mol^{-1} = 470000
Solvent (A): **water** H_2O **7732-18-5**

Type of data: cloud points (LCST-behavior)

w_B 0.04 T/K 304.0

Polymer (B): **poly(*N*-vinylcaprolactam)** **2005LAU**
Characterization: M_w/g.mol^{-1} = 330000, synthesized in the laboratory
Solvent (A): **water** H_2O **7732-18-5**

Type of data: cloud points (LCST-behavior)

c_B/(g/l) 1.0 T/K 304.95

Polymer (B): **poly(*N*-vinylcaprolactam)** **2005DUB**
Characterization: synthesized in the laboratory
Solvent (A): **water** H_2O **7732-18-5**

Type of data: cloud points (LCST-behavior)

c_B/(g/l)	1.6	2.2	2.5	3.0	3.7	5.0	7.5
T/K	306.05	305.35	305.45	305.25	305.35	305.95	306.05

Polymer (B): **poly(*N*-vinylcaprolactam)** **2000PE1**
Characterization: synthesized in the laboratory
Solvent (A): **water** **H_2O** **7732-18-5**

Type of data: cloud points (LCST-behavior)

w_B 0.01 T/K 306.15

Polymer (B): **poly(*N*-vinylcaprolactam-*co*-methacrylic acid)** **2003OKH**
Characterization: M_w/g.mol^{-1} = 4100, 9 mol% methacrylic acid
Solvent (A): **water** **H_2O** **7732-18-5**

Type of data: cloud points (LCST-behavior)

c_B/(g/l) 1.0 T/K 303.15 (in 0.05 M NaCl aqueous solution)

Polymer (B): **poly(*N*-vinylcaprolactam-*co*-methacrylic acid)** **2003OKH**
Characterization: 12 mol% methacrylic acid, synthesized in the laboratory
Solvent (A): **water** **H_2O** **7732-18-5**

Type of data: cloud points (LCST-behavior)

c_B/(g/l) 1.0 T/K 310.65 (in 0.05 M NaCl aqueous solution)

Polymer (B): **poly(*N*-vinylcaprolactam-*co*-methacrylic acid)** **2003OKH**
Characterization: 18 mol% methacrylic acid, synthesized in the laboratory
Solvent (A): **water** **H_2O** **7732-18-5**

Type of data: cloud points (LCST-behavior)

c_B/(g/l) 1.0 T/K 313.85 (in 0.05 M NaCl aqueous solution)

Polymer (B): **poly(*N*-vinylcaprolactam-*co*-methacrylic acid)** **2003OKH**
Characterization: M_w/g.mol^{-1} = 4100, 37 mol% methacrylic acid
Solvent (A): **water** **H_2O** **7732-18-5**

Type of data: cloud points (LCST-behavior)

c_B/(g/l) 1.0 T/K 304.15 (in 0.05 M NaCl aqueous solution)

Polymer (B): **poly[*N*-vinylcaprolactam-*g*-poly(ethyleneoxidoxyalkyl methacrylate)]** **2005LAU**
Characterization: M_w/g.mol^{-1} = 71000, 6.3 wt% poly(ethyleneoxidoxyalkyl methacrylate), MAC11EO42, synthesized in the laboratory
Solvent (A): **water** **H_2O** **7732-18-5**

Type of data: cloud points (LCST-behavior)

c_B/(g/l) 1.0 T/K 306.25

Polymer (B): **poly[*N*-vinylcaprolactam-*g*-poly(ethyleneoxidoxyalkyl methacrylate)]** **2005LAU**
Characterization: M_w/g.mol^{-1} = 310000, 13.0 wt% poly(ethyleneoxidoxyalkyl methacrylate), MAC11EO42, synthesized in the laboratory
Solvent (A): **water** H_2O **7732-18-5**

Type of data: cloud points (LCST-behavior)

c_B/(g/l) 1.0 T/K 306.25

Polymer (B): **poly[*N*-vinylcaprolactam-*g*-poly(ethyleneoxidoxyalkyl methacrylate)]** **2005LAU**
Characterization: M_w/g.mol^{-1} = 250000, 15.8 wt% poly(ethyleneoxidoxyalkyl methacrylate), MAC11EO42, synthesized in the laboratory
Solvent (A): **water** H_2O **7732-18-5**

Type of data: cloud points (LCST-behavior)

c_B/(g/l) 1.0 T/K 306.55

Polymer (B): **poly[*N*-vinylcaprolactam-*g*-poly(ethyleneoxidoxyalkyl methacrylate)]** **2005LAU**
Characterization: M_w/g.mol^{-1} = 300000, 18.3 wt% poly(ethyleneoxidoxyalkyl methacrylate), MAC11EO42, synthesized in the laboratory
Solvent (A): **water** H_2O **7732-18-5**

Type of data: cloud points (LCST-behavior)

c_B/(g/l) 1.0 T/K 306.65

Polymer (B): **poly[*N*-vinylcaprolactam-*g*-poly(ethyleneoxidoxyalkyl methacrylate)]** **2005LAU**
Characterization: M_w/g.mol^{-1} = 260000, 34.0 wt% poly(ethyleneoxidoxyalkyl methacrylate), MAC11EO42, synthesized in the laboratory
Solvent (A): **water** H_2O **7732-18-5**

Type of data: cloud points (LCST-behavior)

c_B/(g/l) 1.0 T/K 306.95

Polymer (B): **poly(*N*-vinylcaprolactam-*co*-*N*-vinylamine)** **1994TAG**
Characterization: M_η/g.mol^{-1} = 160000, 3.8 mol% vinyl amine
Solvent (A): **water** H_2O **7732-18-5**

Type of data: cloud points (LCST-behavior)

w_B 0.08 T/K 308.8

Polymer (B): **poly(*N*-vinylcaprolactam-*co*-1-vinylimidazole)** **2000PE1**
Characterization: synthesized in the laboratory
Solvent (A): **water** **H_2O** **7732-18-5**

Type of data: cloud points (LCST-behavior)

w_B	0.01	T/K	310.15	for a copolymer of 9.09 mol% 1-vinylimidazole
w_B	0.01	T/K	308.15	for a copolymer of 16.66 mol% 1-vinylimidazole
w_B	0.01	T/K	310.15	for a copolymer of 33.33 mol% 1-vinylimidazole
w_B	0.01	T/K	312.65	for a copolymer of 50.00 mol% 1-vinylimidazole

Polymer (B): **poly(vinyl chloride)** **1985GE1**
Characterization: M_n/g.mol^{-1} = 64700, M_w/g.mol^{-1} = 75000, fractionated in the laboratory
Solvent (A): **1,2-dimethylbenzene** **C_8H_{10}** **95-47-6**

Type of data: cloud points

w_B	0.08	was kept constant			
T/K	323.45	320.55	316.45	312.45	309.15
P/bar	1	250	500	750	1000

Polymer (B): **poly(vinyl chloride)** **1985GE1**
Characterization: M_n/g.mol^{-1} = 31400, M_w/g.mol^{-1} = 37000, fractionated in the laboratory
Solvent (A): **phenetole** **$C_8H_{10}O$** **103-73-1**

Type of data: cloud points

w_B	0.10	was kept constant			
T/K	295.65	292.85	290.15	288.45	287.05
P/bar	1	250	500	750	1000

Polymer (B): **poly(vinyl chloride)** **1985GE1**
Characterization: M_n/g.mol^{-1} = 64700, M_w/g.mol^{-1} = 75000, fractionated in the laboratory
Solvent (A): **phenetole** **$C_8H_{10}O$** **103-73-1**

Type of data: cloud points

w_B	0.08	was kept constant			
T/K	317.65	313.35	311.05	309.85	309.05
P/bar	1	250	500	750	1000

Polymer (B): **poly(*N*-vinylisobutyramide)** **2002KUN**
Characterization: M_n/g.mol^{-1} = 11000, M_w/g.mol^{-1} = 15400 synthesized and fractionated in the laboratory
Solvent (A): **water** **H_2O** **7732-18-5**

continued

continued

Type of data: cloud points (LCST-behavior, determined by DSC)

w_B	0.00001	0.00005	0.00010	0.00020	0.00040
T/K	318.15	316.25	315.65	315.15	314.85

Polymer (B): **poly(*N*-vinylisobutyramide)** **2000KUN**
Characterization: M_n/g.mol^{-1} = 66000, M_w/g.mol^{-1} = 105600
synthesized and fractionated in the laboratory
Solvent (A): **water** **H_2O** **7732-18-5**

Type of data: cloud points (LCST-behavior)

w_B 0.001 T/K 313.25 (pH=13)

Comments: The pH of the solution was adjusted by adding the appropriate amount of NaOH.

Polymer (B): **poly(*N*-vinylisobutyramide)** **2002KUN**
Characterization: M_n/g.mol^{-1} = 66000, M_w/g.mol^{-1} = 105600
synthesized and fractionated in the laboratory
Solvent (A): **water** **H_2O** **7732-18-5**

Type of data: cloud points (LCST-behavior, determined by DSC)

w_B	0.00001	0.00005	0.00010	0.00020	0.00040
T/K	314.85	314.65	314.25	314.05	313.95

Polymer (B): **poly(*N*-vinylisobutyramide)** **2002KUN**
Characterization: M_n/g.mol^{-1} = 460000, M_w/g.mol^{-1} = 1104000
synthesized and fractionated in the laboratory
Solvent (A): **water** **H_2O** **7732-18-5**

Type of data: cloud points (LCST-behavior, determined by DSC)

w_B	0.00001	0.00005	0.00010	0.00020	0.00040
T/K	313.35	313.05	312.85	312.75	312.55

Polymer (B): **poly(*N*-vinylisobutyramide-*co*-*N*-vinylamine)** **2000KUN**
Characterization: M_n/g.mol^{-1} = 20000, M_w/g.mol^{-1} = 40000
8.0 mol% *N*-vinylamine, synthesized in the laboratory
Solvent (A): **water** **H_2O** **7732-18-5**

Type of data: cloud points (LCST-behavior)

w_B	0.001	0.001	0.001
pH	13	11	10
T/K	320.35	327.75	341.05

Polymer (B): **poly(*N*-vinylisobutyramide-*co*-*N*-vinylamine)** **2000KUN**
Characterization: M_n/g.mol^{-1} = 34000, M_w/g.mol^{-1} = 95200
18.0 mol% *N*-vinylamine, synthesized in the laboratory
Solvent (A): **water** **H_2O** **7732-18-5**

Type of data: cloud points (LCST-behavior)

w_B	0.001	pH	13	T/K	325.85

Polymer (B): **poly(*N*-vinylisobutyramide-*co*-*N*-vinylamine)** **2000KUN**
Characterization: M_n/g.mol^{-1} = 33000, M_w/g.mol^{-1} = 89100
26.0 mol% *N*-vinylamine, synthesized in the laboratory
Solvent (A): **water** **H_2O** **7732-18-5**

Type of data: cloud points (LCST-behavior)

w_B	0.001	pH	13	T/K	334.85

Polymer (B): **poly(*N*-vinylisobutyramide-*co*-*N*-vinylamine)** **2000KUN**
Characterization: M_n/g.mol^{-1} = 31000, M_w/g.mol^{-1} = 77500
36.0 mol% *N*-vinylamine, synthesized in the laboratory
Solvent (A): **water** **H_2O** **7732-18-5**

Type of data: cloud points (LCST-behavior)

w_B	0.001	pH	13	T/K	346.75

Polymer (B): **poly(vinyl methyl ether)** **2005NIE**
Characterization: M_w/g.mol^{-1} = 21000
Solvent (A): **deuterium oxide** **D_2O** **7789-20-0**

Type of data: spinodal points (LCST-behavior)

w_B	0.1	0.2	0.3	0.4	0.5	0.6	0.7	0.8
T/K	306.25	307.35	308.32	304.37	305.15	302.80	298.47	305.20

Polymer (B): **poly(vinyl methyl ether)** **2006NIE**
Characterization: –
Solvent (A): **deuterium oxide** **D_2O** **7789-20-0**

Type of data: spinodal points (UCST-behavior)

w_B	0.6897	0.7620	0.7954	0.8004	0.8492	0.8843	0.9000
T/K	184.95	236.45	248.05	240.45	233.95	242.15	219.95

Type of data: spinodal points (LCST-behavior)

w_B	0.1	0.2	0.3	0.4	0.5	0.6	0.6897	0.7	0.762
T/K	306.85	309.45	309.65	306.95	305.55	303.65	293.95	298.45	301.95
w_B	0.7954	0.8004							
T/K	297.65	302.75							

Polymer (B): **poly(vinyl methyl ether)** **2006LOO**
Characterization: M_n/g.mol^{-1} = 8000, M_w/g.mol^{-1} = 20000
Solvent (A): **water** H_2O **7732-18-5**

Type of data: cloud points (LCST-behavior)

w_B	0.02	0.1	0.2	0.3	0.35	0.4	0.5	0.6	0.7
T/K	310.75	307.35	308.65	309.65	310.45	307.15	302.65	301.55	302.15

Polymer (B): **poly(vinyl methyl ether)** **1990OTA**
Characterization: M_w/g.mol^{-1} = 15000, Toyo Kasei Kogyo Co., Japan
Solvent (A): **water** H_2O **7732-18-5**

Type of data: cloud points (LCST-behavior)

w_B	0.061	0.030
T/K	305.45	305.15

Polymer (B): **poly(vinyl methyl ether)** **2005VER**
Characterization: M_n/g.mol^{-1} = 12000, M_w/g.mol^{-1} = 15000
Solvent (A): **water** H_2O **7732-18-5**

Type of data: cloud points (LCST-behavior)

c_B/(g/l) 1.0 T/K 308.15

Polymer (B): **poly(vinyl methyl ether)** **1990OTA**
Characterization: M_w/g.mol^{-1} = 57000, Toyo Kasei Kogyo Co., Japan
Solvent (A): **water** H_2O **7732-18-5**

Type of data: cloud points (LCST-behavior)

w_B	0.060	0.0375
T/K	305.45	305.45

Polymer (B): **poly(vinyl methyl ether)** **2000PE1**
Characterization: M_w/g.mol^{-1} = 60000, Aldrich Chem. Co., Inc., Milwaukee, WI
Solvent (A): **water** H_2O **7732-18-5**

Type of data: cloud points (LCST-behavior)

w_B 0.01 T/K 305.65

Polymer (B): **poly(vinyl methyl ether)** **1990SCH**
Characterization: M_n/g.mol^{-1} = 83000, M_w/g.mol^{-1} = 155000,
Aldrich Chem. Co., Inc., Milwaukee, WI
Solvent (A): **water** H_2O **7732-18-5**

Type of data: cloud points (LCST-behavior)

c_B/(g/l) 0.4 T/K 306.95

Polymer (B): **poly(vinyl methyl ether-*b*-vinyl isobutyl ether)** **2005VER**
Characterization: M_n/g.mol^{-1} = 9800, M_n/g.mol^{-1}(PVME-block) = 9300, M_n/g.mol^{-1}(PVIBE-block) = 500, M_w/g.mol^{-1} = 11270, 3 mol% vinyl isobutyl ether, PVME160-*b*-PVIBE5
Solvent (A): **water** **H_2O** **7732-18-5**

Type of data: cloud points (LCST-behavior)

c_B/(g/l) 1.0 T/K 305.15

Polymer (B): **poly(vinyl methyl ether-*b*-vinyl isobutyl ether)** **2005VER**
Characterization: M_n/g.mol^{-1} = 5900, M_n/g.mol^{-1}(PVME-block) = 5000, M_n/g.mol^{-1}(PVIBE-block) = 900, M_w/g.mol^{-1} = 6785, 10 mol% vinyl isobutyl ether, PVME85-*b*-PVIBE9
Solvent (A): **water** **H_2O** **7732-18-5**

Type of data: cloud points (LCST-behavior)

c_B/(g/l) 1.0 T/K 312.15

Polymer (B): **poly(vinyl methyl ether-*b*-vinyl isobutyl ether-*b*-vinyl methyl ether)** **2005VER**
Characterization: M_n/g.mol^{-1} = 9500, M_n/g.mol^{-1}(PVME-block) = 4500, M_n/g.mol^{-1}(PVIBE-block) = 1000, M_w/g.mol^{-1} = 11210, 7 mol% vinyl isobutyl ether, PVME65-*b*-PVIBE10-*b*-PVME65
Solvent (A): **water** **H_2O** **7732-18-5**

Type of data: cloud points (LCST-behavior)

c_B/(g/l) 1.0 T/K 314.15

Polymer (B): **poly(*N*-vinyl-*N*-propylacetamide)** **1994TAG**
Characterization: M_η/g.mol^{-1} = 30000
Solvent (A): **water** **H_2O** **7732-18-5**

Type of data: cloud points (LCST-behavior)

w_B 0.02 T/K 313.5 (threshold point)

Polymer (B): **tetra(ethylene glycol)** **2002DER**
Characterization: M/g.mol^{-1} = 194.23, Merck Eurolab AS, Oslo, Norway
Solvent (A): **n-heptane** **C_7H_{16}** **142-82-5**

Type of data: coexistence data (I and II denote the coexisting phases)

T/K	305.65	311.15	316.95	321.95	330.05	338.55	347.95	353.55
w_B (I)	0.003043	0.003557	0.004217	0.004628	0.005599	0.006404	0.007699	0.008702
w_B (II)	0.989732	0.989217	0.988164	0.987877	0.986357	0.984921	0.983136	0.981650

Polymer (B): **tetra(ethylene glycol) dimethyl ether** **1993TRE**
Characterization: M/g.mol^{-1} = 222.28
Solvent (A): **n-decane** $C_{10}H_{22}$ **124-18-5**

Type of data: binodal data (UCST-behavior)

x_B	0.1679	0.1914	0.2083	0.2494	0.2702	0.2762	0.2797	0.2864	0.2931
T/K	285.7	287.6	288.5	290.3	290.8	291.0	291.1	291.2	291.3
x_B	0.3583	0.3844	0.3926	0.4378	0.4400	0.4500	0.5000	0.5346	0.5432
T/K	291.9	291.9	292.0	292.1	292.0	291.9	291.9	291.6	291.5
x_B	0.6065	0.6171	0.6310	0.6363	0.6889	0.7099	0.7262		
T/K	290.5	290.3	289.8	289.7	287.4	286.0	284.9		

Polymer (B): **tetra(ethylene glycol) dimethyl ether** **1993TRE**
Characterization: M/g.mol^{-1} = 222.28
Solvent (A): **n-dodecane** $C_{12}H_{26}$ **112-40-3**

Type of data: binodal data (UCST-behavior)

x_B	0.1362	0.1786	0.2053	0.2225	0.2471	0.2832	0.2941	0.3360	0.3812
T/K	289.0	293.7	296.0	297.5	299.2	300.5	300.9	301.9	302.5
x_B	0.4108	0.4694	0.5072	0.5302	0.6281	0.6511	0.6684	0.6731	0.7050
T/K	302.6	302.6	302.5	302.4	301.7	301.3	300.7	300.6	299.5
x_B	0.7541	0.7645							
T/K	296.7	295.9							

Polymer (B): **tetra(ethylene glycol) dimethyl ether** **1993TRE**
Characterization: M/g.mol^{-1} = 222.28
Solvent (A): **n-hexadecane** $C_{16}H_{34}$ **544-76-3**

Type of data: binodal data (UCST-behavior)

x_B	0.2737	0.3325	0.3430	0.4202	0.4759	0.5140	0.5703	0.6039	0.6360
T/K	318.5	321.0	321.3	323.3	324.1	324.2	324.2	324.2	324.0
x_B	0.6541	0.7170	0.8291	0.8942					
T/K	323.3	321.3	314.4	308.7					

Polymer (B): **tetra(ethylene glycol) dimethyl ether** **2002REA**
Characterization: M/g.mol^{-1} = 222.28
Solvent (A): **n-nonane** C_9H_{20} **111-84-2**

Type of data: coexistence data (I and II denote the coexisting phases)

T/K	278.15	280.15	282.15	284.15	286.15	288.05	289.15	290.15	291.65
x_B (I)	0.877	0.865	0.853	0.840	0.826	0.821	0.789	0.756	0.711
x_B (II)	0.250	0.268	0.285	0.310	0.339	0.354	0.356	0.417	0.463

2.2. Upper critical (UCST) and/or lower critical (LCST) solution temperatures

Polymer (B)	M_n g/mol	M_w g/mol	Solvent (A)	UCST/ K	LCST/ K	Ref.
Cellulose diacetate						
		20000*	benzyl alcohol	362		1977PAN
		60000*	benzyl alcohol	370		1977PAN
		120000*	benzyl alcohol	372		1977PAN
		infinite	benzyl alcohol	376		1977PAN
	59900	75500	2-butanone	279.7	471.5	1982SU2
	87000	106000	2-butanone	284.5	465.0	1982SU2
	140000	185000	2-butanone	290.0	457.9	1982SU2
		infinite	2-butanone	310	433	1982SU2
	30600	37600	2-propanone		457.1	1982SU1
	26900		2-propanone		452.3	1971COW
	59900	75500	2-propanone		448.0	1982SU1
	40700		2-propanone		447.9	1971COW
	48700		2-propanone		444.6	1971COW
	87000	106000	2-propanone		444.2	1982SU1
	140000	185000	2-propanone		440.1	1982SU1
	59300		2-propanone	216.2	438.2	1971COW
	63100		2-propanone		432.6	1971COW
	68700		2-propanone	248.2	414.7	1971COW
	86000		2-propanone	268.2	394.6	1971COW
	86600		2-propanone		389.9	1971COW
		infinite	2-propanone	280		1971COW
		infinite	2-propanone		428	1982SU1
Cellulose nitrate (13.3 wt% N)						
			2-propanone	328	182	1991AKH
Cellulose tricaprylate						
		infinite	*N,N*-dimethylformamide	413		1952MAN
		infinite	3-phenyl-1-propanol	321		1952MAN

Polymer (B)	M_n g/mol	M_w g/mol	Solvent (A)	UCST/ K	LCST/ K	Ref.
Cellulose triacetate						
		20000*	benzyl alcohol	322		1977PAN
		60000*	benzyl alcohol	330		1977PAN
		120000*	benzyl alcohol	335		1977PAN
		infinite	benzyl alcohol	341		1977PAN
	26000		2-propanone	275.0	466.5	1971COW
	33300		2-propanone	274.0	465.0	1971COW
	53000		2-propanone	278.5	462.6	1971COW
	55400		2-propanone	275.0	464.4	1971COW
	62500		2-propanone	288.0	469.0	1971COW
	95200		2-propanone	275.0	461.4	1971COW
	100500		2-propanone	290.0	472.0	1971COW
	158000		2-propanone		458.5	1971COW
	214000		2-propanone	293.0	464.0	1971COW
		infinite	2-propanone	300	454	1971COW
Decamethylcyclopentasiloxane						
	370.77		dodecafluoropentane	329.4		1997MCL
	370.77		tetradecafluorohexane	340.8		1997MCL
Decamethyltetrasiloxane						
	310.69		decafluorobutane	310.2		1997MCL
	310.69		dodecafluoropentane	319.2		1997MCL
	310.69		hexadecafluoroheptane	341.6		1997MCL
	310.69		octadecafluorooctane	351.4		1997MCL
	310.69		tetradecafluorohexane	332.59		1997MCL
Dodecamethylpentasiloxane						
	384.84		decafluorobutane	323.4		1997MCL
	384.84		dodecafluoropentane	331.4		1997MCL
	384.84		hexadecafluoroheptane	354.5		1997MCL
	384.84		tetradecafluorohexane	343.8		1997MCL
Ethyl(hydroxyethyl)cellulose						
		200000	water		338	2000PE1
Gutta Percha						
		103000*	propyl acetate	315.15		1952WAG
		194000*	propyl acetate	318.95		1952WAG
		210000*	propyl acetate	321.15		1952WAG

Polymer (B)	M_n g/mol	M_w g/mol	Solvent (A)	UCST/ K	LCST/ K	Ref.
Hepta(ethylene glycol) monotetradecyl ether						
	690		water		331.75	1984COR
	690		water		332.68	1998KUB
Hexa(ethylene glycol) monodecyl ether						
	424		water		335.25	2005IMA
Hexa(ethylene glycol) monododecyl ether						
	452		water		323.55	1984COR
	452		water		324.45	1991SC3
Hexa(ethylene glycol) monooctyl ether						
	396		water		347.55	1991SC3
Hexamethyldisiloxane						
	162.38		decafluorobutane	274.6		1997MCL
	162.38		dodecafluoropentane	284.7		1997MCL
	162.38		hexadecafluoroheptane	305.6		1997MCL
	162.38		octadecafluorooctane	314.0		1997MCL
	162.38		tetradecafluorohexane	296.95		1997MCL
Hydroxypropylcellulose						
		75000	water		318.45	1971KAG
		100000	water		319.15	1990SCH
		150000	water		324.45	1971KAG
		300000	water		331.25	1971KAG
		300000	water		316.05	1990SCH
		1000000	water		315.05	1990SCH
		infinite	water		343.85	1971KAG
Methylcellulose						
		14000	water		323.15	2000PE1
		70000*	water		324.75	1972KAG
		150000*	water		326.35	1972KAG
		300000*	water		328.75	1972KAG
		infinite	water		332.65	1972KAG

Polymer (B)	M_n g/mol	M_w g/mol	Solvent (A)	UCST/ K	LCST/ K	Ref.
Methyl(hydroxypropyl)cellulose						
(25 mol% methyl, 8 mol% hydroxypropyl substitution)						
		80000*	water		340.15	1974BAB
		160000*	water		337.15	1974BAB
		260000*	water		333.15	1974BAB
(28 mol% methyl, 6.5 mol% hydroxypropyl substitution)						
		40000*	water		337.65	1974BAB
		100000*	water		333.15	1974BAB
		200000*	water		338.65	1974BAB
(29 mol% methyl, 10 mol% hydroxypropyl substitution)						
		45000*	water		345.15	1974BAB
		130000*	water		339.15	1974BAB
		260000*	water		333.65	1974BAB
Natural rubber						
		300000	n-pentane		403	1960FRE
		74500*	2-pentanone	274.45		1952WAG
		119000*	2-pentanone	278.25		1952WAG
		164000*	2-pentanone	279.65		1952WAG
		280000*	2-pentanone	280.75		1952WAG
Octa(ethylene glycol) monododecyl ether						
	538		water		348.65	1984COR
Octamethylcyclotetrasiloxane						
	296.62		dodecafluoropentane	322.3		1997MCL
	296.62		tetradecafluorohexane	333.5		1997MCL
Octamethyltrisiloxane						
	236.53		decafluorobutane	294.9		1997MCL
	236.53		dodecafluoropentane	304.4		1997MCL
	236.53		hexadecafluoroheptane	326.1		1997MCL
	236.53		octadecafluorooctane	335.6		1997MCL
	236.53		tetradecafluorohexane	315.35		1997MCL
Penta(ethylene glycol) monodecyl ether						
	378		water	568	317	1980LAN
	378		water		317.45	2005IMA

Polymer (B)	M_n g/mol	M_w g/mol	**Solvent (A)**	UCST/ K	LCST/ K	Ref.
Penta(ethylene glycol) monododecyl ether						
	406		water		305.06	1985HAM
	406		water		305.15	1991SC3
Penta(ethylene glycol) monooctyl ether						
	350		water		334.85	1991SC3
Phenol-formaldehyde resin (acetylated)						
			2-ethoxyethanol	378.2		1998YA2
Poly(acrylamide-*co*-hydroxypropyl acrylate)						
(50 mol% hydroxypropyl acrylate)			water		348.15	1975TAY
(60 mol% hydroxypropyl acrylate)			water		324.15	1975TAY
(70 mol% hydroxypropyl acrylate)			water		305.15	1975TAY
Poly(acrylic acid)						
		120000	tetrahydrofuran		268.3	1996SAF
Poly(acrylonitrile-*co*-butadiene)						
(18% acrylonitrile)		840000*	ethyl acetate		427	2002VSH
(26% acrylonitrile)		1000000*	ethyl acetate		412	2002VSH
Poly[bis(2,3-dimethoxypropanoxy)phosphazene]						
	1070000	1500000	water		317.15	1996ALL
Poly[bis(2-(2'-methoxyethoxy)ethoxy)phosphazene]						
	667000	1000000	water		338.15	1996ALL
Poly[bis(2,3-bis(2-methoxyethoxy)propanoxy)phosphazene]						
	714000	1000000	water		311.15	1996ALL
Poly[bis(2,3-bis(2-(2'-methoxyethoxy)ethoxy)propanoxy)phosphazene]						
	1420000	1700000	water		322.65	1996ALL
Poly[bis(2,3-bis(2-(2'-(2''-dimethoxyethoxy)ethoxy)ethoxy)propanoxy)phosphazene]						
	857000	1200000	water		334.65	1996ALL

Polymer (B)	M_n g/mol	M_w g/mol	Solvent (A)	UCST/ K	LCST/ K	Ref.
Poly(butadiene-*co*-α-methylstyrene)						
(10% α-methylstyrene)		100000*	ethyl acetate	387	393	2002VSH
Poly(1-butene) (atactic)						
		infinite	anisole	359.4		1961KRI
		infinite	toluene	356.2		1967MOR
Poly(1-butene) (isotactic)						
		infinite	anisole	362.3		1961KRI
		530000*	cyclopentane		498	1981CH2
		530000*	2,2-dimethylbutane		444	1981CH2
		530000*	2,5-dimethylhexane		519	1981CH2
		530000*	3,4-dimethylhexane		559	1981CH2
		530000*	2,3-dimethylpentane		517	1981CH2
		530000*	2,4-dimethylpentane		480	1981CH2
		530000*	3-ethylpentane		523	1981CH2
		530000*	n-heptane		509	1981CH2
		infinite	n-hexane		464	1981CH2
		530000*	2-methylbutane		416	1981CH2
		infinite	n-nonane		564	1981CH2
		530000*	n-octane		540	1981CH2
		infinite	n-pentane		421	1981CH2
		530000*	2,2,3-trimethylbutane		507	1981CH2
Poly(butyl methacrylate)						
	278000	470000	1-butanol	287.15		1986SAN
	278000	470000	n-decane	357.25		1986SAN
	278000	470000	ethanol	315.25		1986SAN
	278000	470000	n-heptane	342.55		1986SAN
	278000	470000	n-octane	345.80		1986SAN
	278000	470000	1-pentanol	286.30		1986SAN
	278000	470000	2-propanol	294.90		1986SAN
		infinite	2-propanol	297.30		1986HER
	278000	470000	2,2,4-trimethylpentane	347.50		1986SAN
Poly(carbon monoxide-*alt*-ethylene)						
		1000000	hexafluoro-2-propanol		453	1991WAK

Polymer (B)	M_n g/mol	M_w g/mol	Solvent (A)	UCST/ K	LCST/ K	Ref.
Poly(2-chlorostyrene)						
		infinite	benzene		298	1970MAT
Poly(4-chlorostyrene)						
		infinite	benzene	274.0		1966KUB
		infinite	2-(butoxyethoxy)ethanol		323.25	1972IZU
		infinite	butyl acetate		502.4	1966KUB
		infinite	*tert*-butyl acetate		338.55	1972IZU
		infinite	chlorobenzene	128.8		1966KUB
		infinite	2-(ethoxyethoxy)ethanol		300.95	1972IZU
		infinite	ethyl acetate		613.2	1966KUB
		infinite	ethylbenzene	283.2		1966KUB
		infinite	ethylbenzene	258.45		1972IZU
		infinite	ethyl chloroacetate	271.35		1972IZU
		infinite	isopropylbenzene	332.15		1972IZU
		infinite	isopropyl chloroacetate	264.95		1972IZU
		infinite	methyl chloroacetate	337.75		1972IZU
		infinite	propyl acetate		908.7	1966KUB
		infinite	2-propyl acetate		348.65	1972IZU
		infinite	tetrachloroethene	317.55		1972IZU
		infinite	tetrachloromethane	323.85		1972IZU
		infinite	toluene	236.8		1966KUB
Poly(*N*-cyclopropylacrylamide)						
			water		330.15	1975TAY
Poly(decyl methacrylate)						
	220000	252000	1-butanol	302.80		1983HER
	390000	468000	1-butanol	304.85		1983HER
	564000	718000	1-butanol	305.95		1983HER
	220000	252000	1-pentanol	276.40		1983HER
	390000	468000	1-pentanol	278.40		1983HER
	564000	718000	1-pentanol	278.85		1983HER
		infinite	1-propanol	283.0		1986HER
	220000	252000	2-propanol	346.85		1986SAN
Poly(diacetone acrylamide-*co*-hydroxyethyl acrylate)						
(70 mol% hydroxyethyl acrylate)			water		284.15	1975TAY
(80 mol% hydroxyethyl acrylate)			water		298.15	1975TAY
(90 mol% hydroxyethyl acrylate)			water		326.15	1975TAY

Polymer (B)	M_n g/mol	M_w g/mol	Solvent (A)	UCST/ K	LCST/ K	Ref.
Poly(*N,N*-diethylacrylamide)						
		<5000	water		305	2004PAN
	9600	13300	water		306.05	2003LES
			water		305.15	1999LIU
		19000	water		305.15	2002MA2
	19200	40300	water		304.15	2003LES
	32500	58500	water		304.05	2003LES
		412000	water		303.65	2001CAI
		412000	water		303.65	2001GAN
	81600	165700	water		303.25	2003LES
	96900	252000	water		302.65	2003LES
	180900	376300	water		302.45	2003LES
	363600	709000	water		301.55	2003LES
	1295000	1547000	water		301.75	2003LES
			water		298.15	1975TAY
Poly(*N,N*-diethylacrylamide-*co*-acrylic acid)						
(5.98 mol% acrylic acid)		319000	water		305.05	2001CAI
(13.22 mol% acrylic acid)		306000	water		304.15	2001CAI
(20.64 mol% acrylic acid)		308000	water		300.15	2001CAI
Poly(*N,N*-diethylacrylamide-*co*-*N*-ethylacrylamide)						
(40 mol% *N*-ethylacrylamide)			water		305.15	1999LIU
Poly(*N,N*-dimethylacrylamide-*co*-2-butoxyethyl acrylate)						
(50 wt% 2-butoxyethyl acrylate)			water		<273.2	1992MUE
Poly(*N,N*-dimethylacrylamide-*co*-butyl acrylate)						
(15 wt% butyl acrylate)			water		346.2	1992MUE
(20 wt% butyl acrylate)			water		323.2	1992MUE
(30 wt% butyl acrylate)			water		294.2	1992MUE
(35 wt% butyl acrylate)			water		281.2	1992MUE
Poly(*N,N*-dimethylacrylamide-*co*-2-ethoxyethyl acrylate)						
(50 wt% 2-ethoxyethyl acrylate)			water		319.2	1992MUE
(75 wt% 2-ethoxyethyl acrylate)			water		285.2	1992MUE

Polymer (B)	M_n g/mol	M_w g/mol	Solvent (A)	UCST/ K	LCST/ K	Ref.
Poly(*N,N*-dimethylacrylamide-*co*-ethyl acrylate)						
(25 wt% ethyl acrylate)			water		347.2	1992MUE
(30 wt% ethyl acrylate)			water		334.2	1992MUE
(50 wt% ethyl acrylate)			water		287.2	1992MUE
(55 wt% ethyl acrylate)			water		<273.2	1992MUE
Poly(*N,N*-dimethylacrylamide-*co*-2-methoxyethyl acrylate)						
(38 mol% 2-methoxyethyl acrylate)			water		353.2	1996ELE
(45 mol% 2-methoxyethyl acrylate)			water		333.2	1996ELE
(50 wt% 2-methoxyethyl acrylate)			water		343.2	1992MUE
(55 mol% 2-methoxyethyl acrylate)			water		315.2	1996ELE
(68 mol% 2-methoxyethyl acrylate)			water		305.2	1996ELE
(70 wt% 2-methoxyethyl acrylate)			water		313.2	1992MUE
(75 wt% 2-methoxyethyl acrylate)			water		309.2	1992MUE
(80 wt% 2-methoxyethyl acrylate)			water		303.2	1992MUE
(82 mol% 2-methoxyethyl acrylate)			water		288.2	1996ELE
(92 mol% 2-methoxyethyl acrylate)			water		283.2	1996ELE
Poly(*N,N*-dimethylacrylamide-*co*-methyl acrylate)						
(30 wt% methyl acrylate)			water		371.2	1992MUE
(40 wt% methyl acrylate)			water		338.2	1992MUE
(50 wt% methyl acrylate)			water		314.2	1992MUE
(55 wt% methyl acrylate)			water		294.2	1992MUE
(60 wt% methyl acrylate)			water		279.2	1992MUE
(70 wt% methyl acrylate)			water		<273.2	1992MUE
Poly(*N,N*-dimethylacrylamide-*co*-propyl acrylate)						
(20 wt% propyl acrylate)			water		353.2	1992MUE
(30 wt% propyl acrylate)			water		337.2	1992MUE
(40 wt% propyl acrylate)			water		294.2	1992MUE
(50 wt% propyl acrylate)			water		281.2	1992MUE
Poly(dimethylsiloxane) (cyclic)						
	9810	10300	2,2-dimethylpropane		433	1987BAR
	14330	14620	2,2-dimethylpropane		430	1987BAR
	18680	19800	2,2-dimethylpropane		428	1987BAR
	9810	10300	tetramethylsilane		448	1987BAR
	9810	10300	tetramethylsilane		448	1994BAR
	14330	14620	tetramethylsilane		445	1987BAR
	14330	14620	tetramethylsilane		445	1994BAR
	18680	19800	tetramethylsilane		443	1987BAR
	18680	19800	tetramethylsilane		443	1994BAR

Polymer (B)	M_n g/mol	M_w g/mol	Solvent (A)	UCST/ K	LCST/ K	Ref.
Poly(dimethylsiloxane)						
		626000*	n-butane		392.95	1972ZE1
		infinite	n-decane		603	1967PAT
	6330	7410	2,2-dimethylpropane		433	1987BAR
	10060	11570	2,2-dimethylpropane		431	1987BAR
	14750	16370	2,2-dimethylpropane		428	1987BAR
	18240	18970	2,2-dimethylpropane		427	1987BAR
	21420	22920	2,2-dimethylpropane		426	1987BAR
	30510	31120	2,2-dimethylpropane		424	1987BAR
		infinite	n-dodecane		643	1967PAT
		570*	ethane		272	1960FRE
		1170*	ethane		280.65	1972ZE1
		3200*	ethane		273.15	1972ZE1
		14200*	ethane		272.15	1972ZE1
		626000*	ethane		259.65	1972ZE1
		100000*	ethoxybenzene	341.99		1982SH1
		infinite	n-heptane		528	1967PAT
		infinite	n-hexadecane		708	1967PAT
		infinite	n-hexane		493	1967PAT
		infinite	n-octane		553	1967PAT
		infinite	n-pentane		453	1967PAT
		203000*	propane		340.15	1972ZE1
		626000*	propane		337.75	1972ZE1
	6330	7410	tetramethylsilane		449	1987BAR
	6330	7410	tetramethylsilane		449	1994BAR
	10060	11570	tetramethylsilane		446	1987BAR
	10060	11570	tetramethylsilane		446	1994BAR
	14750	16370	tetramethylsilane		443	1987BAR
	14750	16370	tetramethylsilane		443	1994BAR
	18240	18970	tetramethylsilane		441	1987BAR
	18240	18970	tetramethylsilane		441	1994BAR
	21420	22920	tetramethylsilane		440	1987BAR
	21420	22920	tetramethylsilane		440	1994BAR
	30510	31120	tetramethylsilane		439	1987BAR
	30510	31120	tetramethylsilane		439	1994BAR
Poly(dimethylsiloxane-*co*-methylphenylsiloxane) (15 wt% methylphenylsiloxane)						
	9100	41200	anisole	291.45		1998SCH
	9100	41200	2-propanone	282.45		1998SCH

Polymer (B)	M_n g/mol	M_w g/mol	Solvent (A)	UCST/ K	LCST/ K	Ref.
Poly(*N*-ethylacrylamide)						
	3300	5400	water		358.65	2003XUE
			water		355.15	1999LIU
	5800	12200	water		355.15	2003XUE
	5900	12400	water		354.35	2003XUE
	6400	18300	water		351.95	2003XUE
			water		347.15	1975TAY
	7400	33600	water		345.95	2003XUE
Poly(*N*-ethylacrylamide-*co*-*N*-isopropylacrylamide)						
(50 mol% *N*-isopropylacrylamide)			water		321.15	1975TAY
(60 mol% *N*-isopropylacrylamide)			water		325.15	1975TAY
(70 mol% *N*-isopropylacrylamide)			water		329.15	1975TAY
(80 mol% *N*-isopropylacrylamide)			water		335.15	1975TAY
(90 mol% *N*-isopropylacrylamide)			water		340.15	1975TAY
Poly(ethyl acrylate)						
		18000*	1-butanol	304.75		1967LLO
		30000*	1-butanol	307.65		1967LLO
		48000*	1-butanol	310.05		1967LLO
		155000*	1-butanol	313.05		1967LLO
		infinite	1-butanol	318.05		1967LLO
		18000*	ethanol	294.25		1967LLO
		30000*	ethanol	298.15		1967LLO
		48000*	ethanol	301.15		1967LLO
		155000*	ethanol	304.75		1967LLO
		infinite	ethanol	310.55		1967LLO
		155000*	methanol	282.75		1967LLO
		245000*	methanol	285.65		1967LLO
		380000*	methanol	287.25		1967LLO
		740000*	methanol	288.55		1967LLO
		infinite	methanol	293.65		1967LLO
		18000*	1-propanol	294.75		1967LLO
		30000*	1-propanol	301.95		1967LLO
		48000*	1-propanol	305.15		1967LLO
		155000*	1-propanol	307.45		1967LLO
		infinite	1-propanol	312.65		1967LLO

Polymer (B)	M_n g/mol	M_w g/mol	Solvent (A)	UCST/ K	LCST/ K	Ref.
Polyethylene (branched)						
	8400	32000	diphenyl ether	384.7		1979KLE
	7000	54000	diphenyl ether	385.4		1979KLE
	8500	70000	diphenyl ether	387.0		1979KLE
	24000	123000	diphenyl ether	396.7		1979KLE
	18000	45000	diphenyl ether	399.0		1979KLE
	11000	160000	diphenyl ether	402.4		1979KLE
	14000	70000	diphenyl ether	403.6		1979KLE
	19000	84000	diphenyl ether	408.1		1979KLE
	25000	385000	diphenyl ether	408.4		1979KLE
	25000	600000	diphenyl ether	408.8		1979KLE
	23000	247000	diphenyl ether	409.0		1979KLE
	17000	274000	diphenyl ether	409.2		1979KLE
	38000	640000	diphenyl ether	409.8		1979KLE
	27000	420000	diphenyl ether	409.9		1979KLE
	29000	165000	diphenyl ether	410.0		1979KLE
	19000	229000	diphenyl ether	410.2		1979KLE
	24000	470000	diphenyl ether	410.6		1979KLE
	31000	800000	diphenyl ether	410.9		1979KLE
	34000	230000	diphenyl ether	411.3		1979KLE
	30000	525000	diphenyl ether	411.4		1979KLE
	29000	219000	diphenyl ether	412.3		1979KLE
	35000	375000	diphenyl ether	412.6		1979KLE
	64000	345000	diphenyl ether	413.1		1979KLE
	65000	425000	diphenyl ether	415.3		1979KLE
Polyethylene (linear)						
		20000*	anisole	368.15		1966NA1
		infinite	anisole	426.65		1966NA1
		20000*	benzyl acetate	459.65		1966NA1
		20000*	benzyl phenyl ether	437.15		1966NA1
		infinite	benzyl phenyl ether	462.65		1966NA1
		20000*	benzyl propionate	436.15		1966NA1
		20000*	biphenyl	377.15		1966NA1
		50900*	biphenyl	383.55		1966NA2
		56800*	biphenyl	384.95		1966NA2
		136000*	biphenyl	390.35		1966NA2
		442100*	biphenyl	394.45		1966NA2
		infinite	biphenyl	400.65		1966NA1
		infinite	biphenyl	400.65		1966NA2

Polymer (B)	M_n g/mol	M_w g/mol	**Solvent (A)**	UCST/ K	LCST/ K	**Ref.**
Polyethylene (linear)						
		13600*	butyl acetate	415	518	1974KUW
		20000*	butyl acetate	431	507	1974KUW
		61100*	butyl acetate	448	497	1974KUW
		64000*	butyl acetate	451	490	1974KUW
		infinite	butyl acetate	483	471	1974KUW
		20000*	4-*tert*-butylphenol	466.15		1966NA1
		134000*	cyclohexane		518	1981CH1
		20000*	cyclohexanone	389.65		1966NA1
		134000*	cyclopentane		472	1981CH1
	36700	49300	n-decane		563.75	1978KO1
	60400	82600	n-decane		560.50	1978KO1
	97700	135900	n-decane		558.40	1978KO1
		20000*	1-decanol	400.15		1966NA1
	7900	92000	1-decanol	409.45		1991OPS
		infinite	1-decanol	426.45		1966NA1
		20000*	dibenzyl ether	448.65		1966NA1
		134000*	3,4-dimethylhexane		515	1981CH1
		134000*	2,2-dimethylpentane		399	1981CH1
		134000*	2,3-dimethylpentane		463	1981CH1
		134000*	2,4-dimethylpentane		395	1981CH1
		infinite	2,4-dimethylpentane		388	1984VAR
	15000	27500	diphenyl ether	405.0		1979KLE
	19000	36000	diphenyl ether	406.6		1979KLE
		14300*	diphenyl ether	398.35		1966NA2
		20000*	diphenyl ether	407.15		1966NA1
	8600	55000	diphenyl ether	410.1		1967KO1
	8600	55000	diphenyl ether	410.1		1979KLE
		37900*	diphenyl ether	410.95		1966NA2
	7900	89000	diphenyl ether	411.2		1979KLE
		50900*	diphenyl ether	414.85		1966NA2
		56800*	diphenyl ether	416.05		1966NA2
	12000	150000	diphenyl ether	416.2		1979KLE
	12000	153000	diphenyl ether	416.4		1967KO1
	8000	177000	diphenyl ether	417.8		1979KLE
		97200*	diphenyl ether	418.95		1966NA2
	38000	550000	diphenyl ether	419.5		1979KLE
	34000	150000	diphenyl ether	420.6		1979KLE
	92000	140000	diphenyl ether	421.1		1967KO1
	92000	140000	diphenyl ether	421.9		1979KLE
		136800*	diphenyl ether	422.75		1966NA2
		204900*	diphenyl ether	425.55		1966NA2

Polymer (B)	M_n g/mol	M_w g/mol	Solvent (A)	UCST/ K	LCST/ K	Ref.
Polyethylene (linear)						
	200000	680000	diphenyl ether	427.8		1979KLE
		infinite	diphenyl ether	437.05		1966NA1
		infinite	diphenyl ether	437.05		1966NA2
		20000*	diphenylmethane	385.15		1966NA1
		21300*	diphenylmethane	384.85		1966NA2
		34900*	diphenylmethane	390.65		1966NA2
		56800*	diphenylmethane	396.55		1966NA2
		76300*	diphenylmethane	397.85		1966NA2
		97200*	diphenylmethane	400.25		1966NA2
		136800*	diphenylmethane	402.25		1966NA2
		infinite	diphenylmethane	415.35		1966NA1
		infinite	diphenylmethane	415.35		1966NA2
	36700	49300	n-dodecane		613.05	1978KO1
	60400	82600	n-dodecane		610.85	1978KO1
	97700	135900	n-dodecane		608.35	1978KO1
		20000*	1-dodecanol	384.65		1966NA1
		21300	1-dodecanol	390.15		1990CHI
	7900	92000	1-dodecanol	394.65		1991OPS
		218000	1-dodecanol	405.15		1990CHI
		infinite	1-dodecanol	410.45		1966NA1
		infinite	1-dodecanol	410.15		1990CHI
		134000*	3-ethylpentane		471	1981CH1
	15000	27500	n-heptane		480.25	1991OPS
	8000	177000	n-heptane		468.15	1991OPS
	36700	49300	n-heptane		464.70	1978KO1
	60400	82600	n-heptane		459.80	1978KO1
	97700	135900	n-heptane		458.10	1978KO1
		134000*	n-heptane		459	1981CH1
	180000		n-heptane		457	1967ORW
		infinite	n-heptane		463	1967PAT
		infinite	n-heptane		447.05	1973HAM
		20000*	1-heptanol	440.15		1966NA1
	15000	27500	n-hexane		442.75	1991OPS
	8000	177000	n-hexane		429.65	1991OPS
	36700	49300	n-hexane		414.65	1978KO1
		134000*	n-hexane		411	1981CH1
	60400	82600	n-hexane		410.55	1978KO1
	97700	135900	n-hexane		408.55	1978KO1
		infinite	n-hexane		406.45	1973HAM
		infinite	n-hexane		406	1967PAT

Polymer (B)	M_n g/mol	M_w g/mol	Solvent (A)	UCST/ K	LCST/ K	Ref.
Polyethylene (linear)						
	180000		n-hexane		400	1967ORW
		1000000*	n-hexane		400	1960FRE
		20000*	1-hexanol	423.15		1966NA1
	7900	92000	1-hexanol	458.15		1991OPS
		20000*	2-methoxynaphthalene	427.65		1966NA1
		20000*	3-methylbutyl acetate	407.15		1966NA1
		134000*	methylcyclohexane		537	1981CH1
		134000*	methylcyclopentane		488	1981CH1
	36700	49300	n-nonane		535.55	1978KO1
	60400	82600	n-nonane		531.90	1978KO1
	97700	135900	n-nonane		529.50	1978KO1
		134000*	n-nonane		531	1981CH1
		20000*	1-nonanol	431.15		1966NA1
		20000*	4-nonylphenol	410.15		1966NA1
		infinite	4-nonylphenol	435.55		1966NA1
	15000	27500	n-octane		519.15	1991OPS
	8000	177000	n-octane		506.05	1991OPS
	36700	49300	n-octane		502.40	1978KO1
	60400	82600	n-octane		499.30	1978KO1
	97700	135900	n-octane		496.90	1978KO1
		134000*	n-octane		496	1981CH1
	180000		n-octane		496	1967ORW
		infinite	n-octane		483.15	1973HAM
		20000*	1-octanol	419.65		1966NA1
	7900	92000	1-octanol	426.65		1991OPS
		infinite	1-octanol	453.25		1966NA1
		20000*	4-octylphenol	424.65		1966NA1
		infinite	4-octylphenol	447.65		1966NA1
		134000*	n-pentane		353	1981CH1
		infinite	n-pentane		353	1973HAM
		20000*	1-pentanol	445.15		1966NA1
		13600*	pentyl acetate	389	553	1974KUW
		20000*	pentyl acetate	396	547	1974KUW
		61100*	pentyl acetate	410	535	1974KUW
		64000*	pentyl acetate	415	533	1974KUW
		175000*	pentyl acetate	421	528	1974KUW
		infinite	pentyl acetate	434	519	1974KUW
		20000*	4-*tert*-pentylphenol	443.65		1966NA1
		infinite	4-*tert*-pentylphenol	463.35		1966NA1

Polymer (B)	M_n g/mol	M_w g/mol	Solvent (A)	UCST/ K	LCST/ K	Ref.
Polyethylene (linear)						
		20000*	phenetole	366.65		1966NA1
		134000*	2,2,4,4-tetramethylpentane		513	1981CH1
	36700	49300	n-tridecane		641.15	1978KO1
	60400	82600	n-tridecane		639.30	1978KO1
	97700	135900	n-tridecane		634.55	1978KO1
		134000*	2,2,3-trimethylbutane		444	1981CH1
		134000*	2,3,4-trimethylhexane		545	1981CH1
		134000*	2,2,4-trimethylpentane		495	1981CH1
	36700	49300	n-undecane		590.05	1978KO1
	60400	82600	n-undecane		586.75	1978KO1
	97700	135900	n-undecane		583.95	1978KO1
Poly(ethylene-*co*-propylene) (33 mol% ethylene)						
		145000*	cyclohexane		534	1981CH1
		145000*	cyclopentane		490	1981CH1
		145000*	2,2-dimethylbutane		428	1981CH1
		145000*	2,3-dimethylbutane		452	1981CH1
		145000*	3,4-dimethylhexane		541	1981CH1
		145000*	2,2-dimethylpentane		472	1981CH1
		145000*	2,3-dimethylpentane		500	1981CH1
		145000*	2,4-dimethylpentane		464	1981CH1
		145000*	3-ethylpentane		511	1981CH1
		145000*	n-heptane		502	1981CH1
		145000*	n-hexane		455	1981CH1
		145000*	2-methylbutane		396	1981CH1
		145000*	methylcyclohexane		558	1981CH1
		145000*	methylcyclopentane		512	1981CH1
		145000*	2-methylhexane		486	1981CH1
		145000*	n-nonane		558	1981CH1
		145000*	n-octane		528	1981CH1
		145000*	n-pentane		409	1981CH1
		145000*	2,2,4,4-tetramethylpentane		539	1981CH1
		145000*	2,2,3-trimethylbutane		500	1981CH1
		145000*	2,2,4-trimethylpentane		503	1981CH1
(43 mol% ethylene)						
	70000	140000	n-hexane		436	1986IRA
	70000	140000	2-methylpentane		474	1986IRA
	70000	140000	n-pentane		441	1986IRA

Polymer (B)	M_n g/mol	M_w g/mol	Solvent (A)	UCST/ K	LCST/ K	Ref.
Poly(ethylene-*co*-propylene)						
(53 mol% ethylene)						
		154000*	2,2-dimethylbutane		407	1981CH1
		154000*	2,3-dimethylbutane		437	1981CH1
		154000*	2,2-dimethylpentane		453	1981CH1
		154000*	2,3-dimethylpentane		488	1981CH1
		154000*	2,4-dimethylpentane		445	1981CH1
		154000*	3-ethylpentane		500	1981CH1
		154000*	n-heptane		493	1981CH1
		154000*	n-hexane		443	1981CH1
		154000*	n-pentane		395	1981CH1
		154000*	2,2,3-trimethylbutane		488	1981CH1
		154000*	2,3,4-trimethylhexane		565	1981CH1
		154000*	2,2,4-trimethylpentane		484	1981CH1
(63 mol% ethylene)						
		236000*	cyclohexane		526	1981CH1
		236000*	cyclopentane		481	1981CH1
		236000*	2,3-dimethylbutane		429	1981CH1
		236000*	3,4-dimethylhexane		530	1981CH1
		236000*	2,2-dimethylpentane		444	1981CH1
		236000*	2,3-dimethylpentane		482	1981CH1
		236000*	2,4-dimethylpentane		434	1981CH1
			2,4-dimethylpentane		428	1984VAR
		236000*	3-ethylpentane		492	1981CH1
		236000*	n-heptane		485	1981CH1
		236000*	n-hexane		436	1981CH1
		236000*	2-methylbutane		348	1981CH1
		236000*	methylcyclopentane		498	1981CH1
		236000*	n-nonane		547	1981CH1
		236000*	n-octane		512	1981CH1
		236000*	n-pentane		387	1981CH1
		236000*	2,2,4,4-tetramethylpentane		528	1981CH1
		236000*	2,2,3-trimethylbutane		479	1981CH1
		236000*	2,2,4-trimethylpentane		479	1981CH1
(75 mol% ethylene)						
		109000*	2,2-dimethylpentane		431	1981CH1
		109000*	2,4-dimethylpentane		425	1981CH1
		109000*	n-heptane		475	1981CH1
		109000*	n-hexane		427	1981CH1
		109000*	n-nonane		542	1981CH1
		109000*	n-octane		509	1981CH1
		109000*	n-pentane		378	1981CH1
		109000*	2,2,4,4-tetramethylpentane		523	1981CH1
		109000*	2,2,4-trimethylpentane		469	1981CH1

Polymer (B)	M_n g/mol	M_w g/mol	Solvent (A)	UCST/ K	LCST/ K	Ref.
Poly(ethylene-*co*-propylene) (81 mol% ethylene)						
		195000*	cyclohexane		522	1981CH1
		195000*	cyclopentane		474	1981CH1
		195000*	2,2-dimethylbutane		381	1981CH1
		195000*	2,3-dimethylbutane		413	1981CH1
		195000*	2,4-dimethylhexane		478	1981CH1
		195000*	2,5-dimethylhexane		466	1981CH1
		195000*	3,4-dimethylhexane		522	1981CH1
		195000*	2,2-dimethylpentane		425	1981CH1
		195000*	2,3-dimethylpentane		471	1981CH1
		195000*	2,4-dimethylpentane		420	1981CH1
		195000*	3-ethylpentane		478	1981CH1
		195000*	n-heptane		468	1981CH1
		195000*	n-hexane		425	1981CH1
		195000*	2-methylbutane		327	1981CH1
		195000*	methylcyclohexane		541	1981CH1
		195000*	methylcyclopentane		493	1981CH1
		195000*	2-methylhexane		453	1981CH1
		195000*	3-methylhexane		459	1981CH1
		195000*	n-nonane		540	1981CH1
		195000*	n-octane		506	1981CH1
		195000*	n-pentane		370	1981CH1
		195000*	2,2,4,4-tetramethylpentane		419	1981CH1
		195000*	2,2,3-trimethylbutane		461	1981CH1
		195000*	2,2,4-trimethylpentane		495	1981CH1
Poly(ethylene-*co*-vinyl acetate)						
(2.3 wt% VA)	52000	465000	diphenyl ether	404.2		1990VAN
(4.0 wt% VA)	47000	280000	diphenyl ether	392.5		1990VAN
(7.1 wt% VA)	34000	460000	diphenyl ether	378.2		1990VAN
(9.5 wt% VA)	53000	350000	diphenyl ether	367.3		1990VAN
(9.7 wt% VA)	55000	490000	diphenyl ether	370.8		1990VAN
(12.1 wt% VA)	66000	300000	diphenyl ether	360.4		1990VAN
(42.6 mol% VA)	14800	41500	methyl acetate	307.0		1986RAE
Poly(ethylene-*co*-vinyl alcohol)						
(87.2 mol% vinyl alcohol)		infinite	water	463.55	285.65	1971SHI
(88.9 mol% vinyl alcohol)		infinite	water	449.15	290.75	1971SHI
(91.0 mol% vinyl alcohol)		infinite	water	428.45	302.95	1971SHI
(94.1 mol% vinyl alcohol)		infinite	water	389.25	324.45	1971SHI

Polymer (B)	M_n g/mol	M_w g/mol	Solvent (A)	UCST/ K	LCST/ K	Ref.
Poly(ethylene glycol)						
		8000*	*tert*-butyl acetate	321.2	464.2	1976SA1
		14400*	*tert*-butyl acetate	340.2	446.2	1976SA1
		21200*	*tert*-butyl acetate	353.2	431.2	1976SA1
		2180*	water	489.7	448.7	1976SA1
		2270*	water	505.2	437.2	1976SA1
		2290*	water	507.7	435.7	1976SA1
	6096	6197			404.77	2000FIS
		14000*	water		399	1970NAK
	10457	11615	water		394.33	2000FIS
		8000*	water		389.4	1976SA1
	12161	15481	water		388.47	2000FIS
	15767	18887	water		385.95	2000FIS
	33700		water		382.00	2000FIS
		14400*	water		380.7	1976SA1
	40800	151000	water		378.25	2000FIS
		21200*	water		376.8	1976SA1
	208200	832700	water		372.86	2000FIS
		719000*	water		372.3	1976SA1
		1020000*	water		371.8	1976SA1
		4000000*	water		369.39	1984FLO
		infinite	water		366.04	2000FIS
Poly[(ethylene glycol) monomethacrylate-*co*-methyl methacrylate]						
(60 mol% methyl methacrylate)			water		328.95	2004ALI
(70 mol% methyl methacrylate)			water		322.95	2004ALI
(76 mol% methyl methacrylate)			water		315.85	2004ALI
Poly{ethylene oxide-*b*-[bis(methoxyethoxyethoxy)-phosphazene]}						
(about 67 mol% ethylene oxide)						
	22000	31500	water		338	2003CHA
Poly(ethylene oxide-*b*-propylene oxide-*b*-ethylene oxide)						
(about 30 mol% EO)		4400	water		286.65	2000LAM

Polymer (B)	M_n g/mol	M_w g/mol	**Solvent (A)**	**UCST/** K	**LCST/** K	**Ref.**
Poly(ethylene oxide-*co*-propylene oxide)						
(20.0 mol% EO)	3400		water		303	2000PE2
(27.0 mol% EO)	3000		water		309	2000PE1
(30.0 mol% EO)	5400		water		313	2000PE2
(33.3 mol% EO)	3000-3500		water		340	1997LIM
(38.5 mol% EO)	5000		water		309	2000PE1
(50.0 mol% EO)	2340	2480	water		338.25	2000LIW
	3640	4040	water		324.85	2000LIW
	3900		water		323	2000PE2
	3000-3500		water		343	1997LIM
(58.8 mol% EO)	3000		water		326.65	2000PE1
(66.7 mol% EO)	3000-3500		water		359	1997LIM
(72.4 mol% EO)		36000	water		333	1991LOU
(79.5 mol% EO)		30800	water		345	1991LOU
		32500	water		345	1991LOU
(86.6 mol% EO)		30100	water		355.5	1991LOU
Polyethylethylene						
	48000	52000	diphenyl ether	411.2		1979KLE
	48000	52000	diphenyl ether	411.2		1980KLE
Poly(*N*-ethylmethacrylamide)						
			water		331.15	1975TAY
Poly(*N*-ethyl-*N*-methylacrylamide)						
		<5000	water		345	2004PAN
Poly(2-ethyl-2-oxazoline)						
		170000*	water		335.15	2003CHR
Poly(2-ethyl-*N*-vinylimidazole)						
			water		311.15	1975TAY
Poly(p-hexylstyrene)						
		infinite	2-butanone	302.6		1987MAG

Polymer (B)	M_n g/mol	M_w g/mol	Solvent (A)	UCST/ K	LCST/ K	Ref.
Poly(2-hydroxyethyl acrylate-*co*-hydroxypropyl acrylate)						
(20 mol% hydroxypropyl acrylate)			water		333.15	1975TAY
(30 mol% hydroxypropyl acrylate)			water		317.15	1975TAY
(40 mol% hydroxypropyl acrylate)			water		307.15	1975TAY
(50 mol% hydroxypropyl acrylate)			water		305.15	1975TAY
(60 mol% hydroxypropyl acrylate)			water		298.15	1975TAY
(70 mol% hydroxypropyl acrylate)			water		296.15	1975TAY
Poly(2-hydroxyethyl methacrylate)						
		77400*	1-butanol	337.25		1969DUS
		133000*	1-butanol	340		1969DUS
		175000*	1-butanol	342		1969DUS
		233600*	1-butanol	343.5		1969DUS
		352200*	1-butanol	345		1969DUS
		557800*	1-butanol	346.5		1969DUS
		infinite	1-butanol	362.2		1969DUS
		233600*	2-butanol	287		1969DUS
		352200*	2-butanol	287		1969DUS
		233600*	2-methyl-1-propanol	342		1969DUS
		352200*	2-methyl-1-propanol	344.5		1969DUS
		77400*	1,2,3-propanetriol	345		1969DUS
		133000*	1,2,3-propanetriol	349		1969DUS
		233600*	1,2,3-propanetriol	355.5		1969DUS
		557800*	1,2,3-propanetriol	357		1969DUS
		77400*	1-propanol	311		1969DUS
		133000*	1-propanol	313		1969DUS
		175000*	1-propanol	315		1969DUS
		233600*	1-propanol	315.5		1969DUS
		352200*	1-propanol	318.5		1969DUS
		557800*	1-propanol	320		1969DUS
		infinite	1-propanol	326.5		1969DUS
Poly(hydroxypropyl acrylate)						
			water		289.15	1975TAY
Poly[*N*-(2-hydroxypropyl)methacrylamide dilactate]						
	6300	10700	water		286.15	2004SOG
Poly[*N*-(2-hydroxypropyl)methacrylamide monolactate]						
	11400	24400	water		338.15	2004SOG

Polymer (B)	M_n g/mol	M_w g/mol	Solvent (A)	UCST/ K	LCST/ K	Ref.
Poly[*N*-(2-hydroxypropyl)methacrylamide monolactate-*co*-*N*-(2-hydroxypropyl) methacrylamide dilactate]						
(25 mol% dilactate)	7500	17600	water		323.65	2004SOG
(49 mol% dilactate)	8100	16900	water		309.65	2004SOG
(74 mol% dilactate)	6800	14000	water		298.65	2004SOG
Polyisobutylene						
		infinite	anisole	377		1951FO1
		72000*	benzene		540.5	1970LID
		2800000*	benzene		535.0	1970LID
		infinite	benzene	297		1951FO1
		703000*	n-butane		264.75	1972ZE1
		1660000*	n-butane		253.85	1972ZE1
		infinite	cycloheptane		572	1969BAR
		1500000*	cyclohexane		412	1960FRE
		infinite	cyclohexane		516	1969BAR
		infinite	cyclooctane		637	1969BAR
		1500000*	cyclopentane		344	1960FRE
		infinite	cyclopentane		461	1969BAR
		infinite	n-decane		540	1967PAT
		infinite	n-decane		535	1969BAR
		1500000*	2,2-dimethylbutane		376	1960FRE
		1500000*	2,3-dimethylbutane		404	1960FRE
		infinite	2,2-dimethylhexane		454	1969BAR
		infinite	2,4-dimethylhexane		458	1969BAR
		infinite	2,5-dimethylhexane		446	1969BAR
		infinite	3,4-dimethylhexane		497	1969BAR
		infinite	2,2-dimethylpentane		404	1969BAR
		infinite	2,3-dimethylpentane		451	1969BAR
		infinite	2,4-dimethylpentane		403	1969BAR
		infinite	3,3-dimethylpentane		451	1969BAR
		infinite	diphenyl ether	306		1951FO1
		infinite	n-dodecane		585	1967PAT
		infinite	n-dodecane		582	1969BAR
		infinite	ethylbenzene	249		1951FO1
		infinite	ethylcyclopentane		524	1969BAR
		infinite	ethyl heptanoate	306		1951FO1
		infinite	ethyl hexanoate	330		1951FO1
		infinite	3-ethylpentane		458	1969BAR
		1500000*	n-heptane		441	1960FRE
		infinite	n-heptane		447	1967PAT
		infinite	n-heptane		442	1969BAR

Polymer (B)	M_n g/mol	M_w g/mol	Solvent (A)	UCST/ K	LCST/ K	Ref.
Polyisobutylene						
		72000*	n-hexane		428.5	1970LID
		1500000*	n-hexane		401	1960FRE
		1660000*	n-hexane		409.65	1972ZE1
		2800000*	n-hexane		407.0	1970LID
		infinite	n-hexane		402	1967PAT
		infinite	n-hexane		402	1969BAR
		6030*	2-methylbutane		357.85	1972ZE1
		703000*	2-methylbutane		330.55	1972ZE1
		1500000*	2-methylbutane		327	1960FRE
		1660000*	2-methylbutane		323.35	1972ZE1
		infinite	2-methylbutane		318	1969BAR
		infinite	methylcyclohexane		526	1969BAR
		infinite	methylcyclopentane		478	1969BAR
		infinite	2-methylheptane		466	1969BAR
		infinite	3-methylheptane		478	1969BAR
		infinite	2-methylhexane		426	1969BAR
		infinite	3-methylhexane		446	1969BAR
		infinite	2-methylpentane		376	1969BAR
		infinite	3-methylpentane		405	1969BAR
		470*	2-methylpropane		387	1960FRE
		72000*	n-octane		506.0	1970LID
		2800000*	n-octane		489.0	1970LID
		1500000*	n-octane		453	1960FRE
		infinite	n-octane		485	1967PAT
		infinite	n-octane		477	1969BAR
		6030*	n-pentane		403.55	1972ZE1
		72000*	n-pentane		373.5	1970LID
		703000*	n-pentane		353.55	1972ZE1
		1660000*	n-pentane		347.35	1972ZE1
		1500000*	n-pentane		348	1960FRE
		2800000*	n-pentane		349.0	1970LID
		infinite	n-pentane		346	1967PAT
		infinite	n-pentane		344	1969BAR
		infinite	phenetole	357		1951FO1
		470*	propane		358	1960FRE
		infinite	propylcyclopentane		547	1969BAR
		infinite	toluene	260		1951FO1
		infinite	2,2,3-trimethylbutane		445	1969BAR
		infinite	2,2,4-trimethylpentane		435	1969BAR

Polymer (B)	M_n g/mol	M_w g/mol	Solvent (A)	UCST/ K	LCST/ K	Ref.
Poly(isobutyl vinyl ether-*co*-2-hydroxyethyl vinyl ether)						
(12 mol% isobutyl vinyl ether)			water		338.15	2004SUG
(20 mol% isobutyl vinyl ether)			water		314.15	2004SUG
(33 mol% isobutyl vinyl ether)			water		289.15	2004SUG
1,4-*cis*-Polyisoprene						
		780000*	2,5-dimethylhexane		474.15	1989BOH
		780000*	3,4-dimethylhexane		520.15	1989BOH
		780000*	2,2-dimethylpentane		445.15	1989BOH
		780000*	2,3-dimethylpentane		484.15	1989BOH
		780000*	2,4-dimethylpentane		442.15	1989BOH
		780000*	n-heptane		488.15	1989BOH
		780000*	n-hexane		434.15	1989BOH
		780000*	3-methylpentane		483.15	1989BOH
		780000*	n-nonane		541.15	1989BOH
		780000*	n-octane		509.15	1989BOH
		780000*	2,2,4,4-tetramethylpentane		518.15	1989BOH
		780000*	2,3,4-trimethylhexane		548.15	1989BOH
		780000*	2,2,4-trimethylpentane		471.15	1989BOH
1,4-*trans*-Polyisoprene						
		180000*	2,5-dimethylhexane		451.15	1989BOH
		180000*	3,4-dimethylhexane		521.15	1989BOH
		180000*	2,2-dimethylpentane		405.15	1989BOH
		180000*	2,3-dimethylpentane		460.15	1989BOH
		180000*	2,4-dimethylpentane		404.15	1989BOH
		180000*	n-heptane		467.15	1989BOH
		180000*	n-hexane		407.15	1989BOH
		180000*	3-methylpentane		473.15	1989BOH
		180000*	n-nonane		540.15	1989BOH
		180000*	n-octane		503.15	1989BOH
		180000*	2,2,4,4-tetramethylpentane		519.15	1989BOH
		180000*	2,3,4-trimethylhexane		548.15	1989BOH
		780000*	2,2,4-trimethylpentane		441.15	1989BOH
Poly(*N*-isopropylacrylamide)						
		300000*	deuterium oxide		306.85	2001KUJ
	2300	2700	water		305.85	2002FRE
		<5000	water		307	2004PAN
	5400	14000	water		307.45	1990SCH
	11000	76000	water		307.25	1990SCH
	49000	160000	water		307.05	1990SCH

Polymer (B)	M_n g/mol	M_w g/mol	**Solvent (A)**	**UCST/** K	**LCST/** K	**Ref.**
Poly(*N*-isopropylacrylamide)						
	70000	144000	water		306.4	2002RE1
	70000	224000	water		305.95	2001DJO
	73000	400000	water		306.35	1990SCH
	160000	440000	water		306.35	1990SCH
	146000	530000	water		305.85	1990SCH
		1550000*	water		305.65	1990WIN
		300000*	water		304.95	2001KUJ
		7800000*	water		304.55	2002BER
Poly(*N*-isopropylacrylamide-*co*-acrylamide)						
(15 mol% acrylamide)		3100000	water		315.15	1994MUM
(30 mol% acrylamide)		4500000	water		326.15	1994MUM
(45 mol% acrylamide)		3900000	water		347.15	1994MUM
Poly(*N*-isopropylacrylamide-*co*-1-deoxy-1-methacrylamido-D-glucitol)						
(12.9 mol% glucitol)	78000	170000	water		311.3	2002RE1
(13.3 mol% glucitol)	28600	56000	water		359.6	2002RE1
(13.7 mol% glucitol)	51600	110000	water		314.9	2002RE1
(14.0 mol% glucitol)	145000	432000	water		307.5	2002RE1
Poly(*N*-isopropylacrylamide-*co*-*N,N*-dimethylacrylamide)						
(10.7 mol% *N,N*-dimethylacrylamide)			water		310.05	2006SHE
(13.0 mol% *N,N*-dimethylacrylamide)			water		309.15	2003BAR
(14.7 mol% *N,N*-dimethylacrylamide)			water		311.25	2006SHE
(17.3 mol% *N,N*-dimethylacrylamide)			water		312.95	2006SHE
(21.2 mol% *N,N*-dimethylacrylamide)			water		314.25	2006SHE
(27.0 mol% *N,N*-dimethylacrylamide)			water		312.15	2003BAR
(30.0 mol% *N,N*-dimethylacrylamide)			water		315.15	2003BAR
(31.4 mol% *N,N*-dimethylacrylamide)			water		319.15	2006SHE
(50.0 mol% *N,N*-dimethylacrylamide)			water		323.15	2003BAR
(60.0 mol% *N,N*-dimethylacrylamide)			water		336.15	2003BAR
(66.0 mol% *N,N*-dimethylacrylamide)			water		345.15	2003BAR
Poly(*N*-isopropylacrylamide-*co*-2-hydroxyethyl methacrylate-*co*-acrylic acid)						
(92.0 mol% NIPAM, 5.0 mol% HEMA, 3.0 mol% acrylic acid)						
		160000	water (pH = 7.4)		308.15	2005LEE
(92.0 mol% NIPAM, 4.3 mol% HEMA, 3.7 mol% acrylic acid)						
		120000	water (pH = 7.4)		309.75	2005LEE
(89.8 mol% NIPAM, 4.2 mol% HEMA, 6.0 mol% acrylic acid)						
		140000	water (pH = 7.4)		316.25	2005LEE

Polymer (B)	M_n g/mol	M_w g/mol	Solvent (A)	UCST/ K	LCST/ K	Ref.
Poly(*N*-isopropylacrylamide-*co*-2-hydroxyethyl methacrylate lactate-*co*-acrylic acid)						
(84.4 mol% NIPAM, 9.4 mol% HEMA lactate, 6.2 mol% acrylic acid)						
		230000	water (pH = 7.4)		308.05	2005LEE
(81.7 mol% NIPAM, 12.3 mol% HEMA lactate, 6.0 mol% acrylic acid)						
		240000	water (pH = 7.4)		301.55	2005LEE
(75.8 mol% NIPAM, 18.5 mol% HEMA lactate, 5.7 mol% acrylic acid)						
		160000	water (pH = 7.4)		296.65	2005LEE
Poly(*N*-isopropylacrylamide-*co*-*N*-isopropylmethacrylamide)						
(10.56 mol% *N*-isopropylmethacrylamide)						
	55300	177000	water		307.15	2001DJO
(20.30 mol% *N*-isopropylmethacrylamide)						
	42800	137000	water		308.45	2001DJO
(30.00 mol% *N*-isopropylmethacrylamide)						
	28800	92000	water		309.75	2001DJO
(39.99 mol% *N*-isopropylmethacrylamide)						
	23100	74000	water		311.05	2001DJO
(50.22 mol% *N*-isopropylmethacrylamide)						
	18800	60000	water		312.95	2001DJO
(59.89 mol% *N*-isopropylmethacrylamide)						
	23100	74000	water		314.65	2001DJO
(69.86 mol% *N*-isopropylmethacrylamide)						
	17500	56000	water		315.55	2001DJO
(79.81 mol% *N*-isopropylmethacrylamide)						
	16600	53000	water		317.35	2001DJO
(89.99 mol% *N*-isopropylmethacrylamide)						
	14700	47000	water		318.75	2001DJO
Poly(*N*-isopropylacrylamide-*b*-propylene glycol-*b*-*N*-isopropylacrylamide)						
	6600	12700	water		305.15	2005CH1
Poly(*N*-isopropylmethacrylamide)						
	6250	20000	water		319.95	2001DJO
Poly{*N*-isopropylmethacrylamide-*b*-poly[(*N*-acetylimino)ethylene]}						
(80 wt% *N*-isopropylacrylamide)						
	5500		water		306.2	2003DAV

Polymer (B)	M_n g/mol	M_w g/mol	Solvent (A)	UCST/ K	LCST/ K	Ref.
Poly{*N*-isopropylmethacrylamide-*g*-poly[(*N*-acetylimino)ethylene]}						
(75 wt% *N*-isopropylacrylamide)						
	6030		water		306.2	2003DAV
(73 wt% *N*-isopropylacrylamide)						
	41400		water		308.7	2003DAV
(65 wt% *N*-isopropylacrylamide)						
	16000		water		308.7	2003DAV
Poly(2-isopropyl-2-oxazoline)						
	1700	1785	water		345.65	2004DIA
	2400	2500	water		335.95	2004DIA
	4300	4430	water		324.45	2004DIA
	5700	5870	water		321.25	2004DIA
Poly(DL-lactide)						
		infinite	dibutyl phthalate	358.8		2004LE1
		infinite	dihexyl phthalate	456.7		2004LE1
		infinite	dipentyl phthalate	415.8		2004LE1
Poly(L-lactide)						
		infinite	dihexyl phthalate	453.4		2004LE1
		infinite	dipentyl phthalate	413.3		2004LE1
Poly(maleic anhydride-*alt*-diethylene glycol)						
	2200	4200	styrene	352.15		1992LEC
	2380	4590	styrene	339.15		1992LEC
Poly[*N*-(3-methoxypropyl)acrylamide]						
			water		283.15	1975TAY
Poly(methyl methacrylate)						
		127000*	acetonitrile	267.15		1962FOX
		147000*	acetonitrile	271.65		1962FOX
		590000*	acetonitrile	294.65		1962FOX
		970000*	acetonitrile	303.15		1962FOX
		infinite	acetonitrile	318.15		1962FOX
		infinite	acetonitrile	314	480	1976CO1
		50000*	1-butanol	353.25		1950JE2
		90000*	1-butanol	353.85		1950JE2

Polymer (B)	M_n g/mol	M_w g/mol	Solvent (A)	UCST/ K	LCST/ K	Ref.
Poly(methyl methacrylate)						
		180000*	1-butanol	355.05		1950JE2
		400000*	1-butanol	356.65		1950JE2
		infinite	2-butanone		482	1976CO1
		infinite	1-chlorobutane	320	463	1976CO1
		970000*	2,2-dimethyl-3-pentanone	301.55		1962FOX
		infinite	2,2-dimethyl-3-pentanone	308.15		1962FOX
		127000*	2,4-dimethyl-3-pentanone	280.15		1962FOX
		308000*	2,4-dimethyl-3-pentanone	299.85		1962FOX
		970000*	2,4-dimethyl-3-pentanone	308.15		1962FOX
		infinite	2,4-dimethyl-3-pentanone	319.15		1962FOX
	200000	264000	2-ethoxyethanol	312.15		1998CHA
		77000	ethyl acetate	290	533	1996VS3
		127000*	2-ethylbutanal	264.65		1962FOX
		590000*	2-ethylbutanal	280.85		1962FOX
		970000*	2-ethylbutanal	284.85		1962FOX
		infinite	2-ethylbutanal	295.15		1962FOX
		infinite	3-heptanone	307.7		1986HER
		15900*	4-heptanone	246.15		1962FOX
		66400*	4-heptanone	276.35		1962FOX
		147000*	4-heptanone	286.15		1962FOX
		308000*	4-heptanone	292.55		1962FOX
		970000*	4-heptanone	299.95		1962FOX
		infinite	4-heptanone	306.95		1962FOX
		infinite	4-heptanone	309	533	1976CO1
		infinite	3-hexanone		522	1976CO1
		infinite	methyl acetate		451	1976CO1
		50000*	1-methyl-4-isopropyl-benzene	400.15		1950JE2
		90000*	1-methyl-4-isopropyl-benzene	402.45		1950JE2
		180000*	1-methyl-4-isopropyl-benzene	406.75		1950JE2
		400000*	1-methyl-4-isopropyl-benzene	413.65		1950JE2
		55400*	2-octanone	281.15		1962FOX
		135000*	2-octanone	299.15		1962FOX
		970000*	2-octanone	317.65		1962FOX
		1400000*	2-octanone	321.15		1962FOX
		infinite	2-octanone	325.15		1962FOX
	19830	21600	3-octanone	285.675		1992XIA
	24230	26900	3-octanone	293.613		1996XIA

Polymer (B)	M_n g/mol	M_w g/mol	Solvent (A)	UCST/ K	LCST/ K	Ref.
Poly(methyl methacrylate)						
	46300	48600	3-octanone	307.870		1996XIA
	57030	58700	3-octanone	308.63		1992XIA
		127000*	3-octanone	315.55		1962FOX
	91350	95000	3-octanone	316.883		1996XIA
	118200	125300	3-octanone	319.045		1992XIA
		147000*	3-octanone	319.65		1962FOX
	137500	143000	3-octanone	323.125		1996XIA
	217200	225900	3-octanone	324.447		1992XIA
		308000*	3-octanone	325.65		1962FOX
	218300	227000	3-octanone	326.325		1996XIA
		860000*	3-octanone	333.15		1962FOX
	572400	595300	3-octanone	329.88		1992XIA
	573000	596000	3-octanone	333.459		1996XIA
		970000*	3-octanone	334.15		1962FOX
		infinite	3-octanone	345.15		1962FOX
		infinite	3-octanone	345.15		1992XIA
		infinite	3-octanone	346.85		1996XIA
		infinite	3-pentanone		506	1976CO1
		50000*	1-propanol	349.95		1950JE2
		90000*	1-propanol	350.65		1950JE2
		180000*	1-propanol	352.05		1950JE2
		400000*	1-propanol	355.25		1950JE2
		infinite	2-propanone		439	1976CO1
	200000	264000	tetra(ethylene glycol)	390.15		1998CHA
		400000*	toluene	225.35		1950JE2
		50000*	trichloromethane	231.15		1950JE2
		180000*	trichloromethane	232.65		1950JE2
		400000*	trichloromethane	233.65		1950JE2
	200000	264000	tri(ethylene glycol)	407.15		1998CHA
Poly(methyl methacrylate) (isotactic)						
		infinite	acetonitrile	301	461	1976CO1
		infinite	2-butanone		464	1976CO1
		infinite	1-chlorobutane	309	454	1976CO1
		infinite	4-heptanone	319	522	1976CO1
		infinite	3-hexanone	279	511	1976CO1
		infinite	methyl acetate		441	1976CO1
		infinite	3-pentanone		497	1976CO1
		infinite	2-propanone		428	1976CO1

Polymer (B)	M_n g/mol	M_w g/mol	Solvent (A)	UCST/ K	LCST/ K	Ref.
Poly(4-methyl-1-pentene) (isotactic)						
		infinite	biphenyl	467.8		1973TAN
		152000*	n-butane		388	1981CH2
		152000*	cyclopentane		505	1981CH2
		152000*	2,2-dimethylbutane		462	1981CH2
		152000*	2,2-dimethylpentane		499	1981CH2
		152000*	2,4-dimethylpentane		499	1981CH2
		infinite	diphenyl ether	483.2		1973TAN
		infinite	diphenylmethane	449.8		1973TAN
		152000*	3-ethylpentane		532	1981CH2
		152000*	n-heptane		522	1981CH2
		152000*	n-hexane		487	1981CH2
		152000*	2-methylbutane		431	1981CH2
		152000*	n-nonane		579	1981CH2
		152000*	n-octane		553	1981CH2
		152000*	n-pentane		441	1981CH2
		152000*	2,2,3-trimethylbutane		521	1981CH2
Poly(α-methylstyrene)						
	58500	61400	butyl acetate	262.05	457.15	1995PFO
	58500	61400	cyclohexane	281.5	490.0	1995PFO
	99100	113000	cyclohexane	293.55		1991LEE
	129000	257000	cyclohexane	295.35		1991LEE
	137500	165000	cyclohexane	295.55		1991LEE
	178700	227000	cyclohexane	296.45		1991LEE
		289000	cyclohexane		473	1995PFO
	219700	290000	cyclohexane	297.75		1991LEE
	273500	424000	cyclohexane	298.25		1991LEE
	349000	604000	cyclohexane	300.55		1991LEE
	26000	31200	cyclopentane	276.7	435.95	1995PFO
		289000	*trans*-decahydro-naphthalene	273		1995PFO
	69500	76500	hexyl acetate	285.05	508.15	1995PFO
	19700	20100	methylcyclohexane	305.70		1999PRU
	32300	32900	methylcyclohexane	318.3		1999PRU
	72000	75600	methylcyclohexane	328.9		1999PRU
	32300	32900	methylcyclohexane	317.72		2001PEN
		infinite	methylcyclohexane	357		1999PRU
	25000	29000	pentyl acetate	251.6	500.6	1995PFO
	58500	61400	pentyl acetate	287.1	484.6	1995PFO

Polymer (B)	M_n g/mol	M_w g/mol	Solvent (A)	UCST/ K	LCST/ K	Ref.
Poly(2-methyl-5-vinylpyridine)						
		56000*	butyl acetate	270.75		1965GEC
		73000*	butyl acetate	276.55		1965GEC
		165000*	butyl acetate	279.55		1965GEC
		420000*	butyl acetate	286.55		1965GEC
		600000*	butyl acetate	287.95		1965GEC
		infinite	butyl acetate	294.95		1965GEC
		30000*	ethyl butyrate	315.65		1965GEC
		145000*	ethyl butyrate	315.85		1965GEC
		197000*	ethyl butyrate	316.45		1965GEC
		222000*	ethyl butyrate	318.15		1965GEC
		263000*	ethyl butyrate	319.05		1965GEC
		infinite	ethyl butyrate	323.15		1965GEC
		56000*	ethyl propionate	287.15		1965GEC
		123000*	ethyl propionate	291.95		1965GEC
		276000*	ethyl propionate	292.55		1965GEC
		335000*	ethyl propionate	293.55		1965GEC
		400000*	ethyl propionate	294.35		1965GEC
		infinite	ethyl propionate	298.55		1965GEC
		87000*	3-methylbutyl acetate	306.55		1965GEC
		152000*	3-methylbutyl acetate	309.15		1965GEC
		193000*	3-methylbutyl acetate	312.55		1965GEC
		275000*	3-methylbutyl acetate	314.75		1965GEC
		320000*	3-methylbutyl acetate	318.55		1965GEC
		infinite	3-methylbutyl acetate	322.15		1965GEC
		56000*	4-methyl-2-pentanone	293.45		1965GEC
		123000*	4-methyl-2-pentanone	294.15		1965GEC
		276000*	4-methyl-2-pentanone	295.85		1965GEC
		335000*	4-methyl-2-pentanone	299.95		1965GEC
		400000*	4-methyl-2-pentanone	302.45		1965GEC
		infinite	4-methyl-2-pentanone	310.55		1965GEC
		56000*	2-methylpropyl acetate	305.25		1965GEC
		87000*	2-methylpropyl acetate	306.65		1965GEC
		170000*	2-methylpropyl acetate	312.35		1965GEC
		320000*	2-methylpropyl acetate	315.05		1965GEC
		infinite	2-methylpropyl acetate	326.15		1965GEC
		30000*	pentyl acetate	311.35		1965GEC
		66000*	pentyl acetate	315.95		1965GEC
		126000*	pentyl acetate	316.35		1965GEC
		165000*	pentyl acetate	316.95		1965GEC
		233000*	pentyl acetate	317.75		1965GEC
		483000*	pentyl acetate	318.65		1965GEC
		infinite	pentyl acetate	321.35		1965GEC

Polymer (B)	M_n g/mol	M_w g/mol	Solvent (A)	UCST/ K	LCST/ K	Ref.
Poly(2-methyl-5-vinylpyridine)						
		56000*	propionitrile	256.25		1965GEC
		100000*	propionitrile	259.95		1965GEC
		193000*	propionitrile	261.65		1965GEC
		284000*	propionitrile	262.35		1965GEC
		370000*	propionitrile	264.35		1965GEC
		440000*	propionitrile	265.35		1965GEC
		infinite	propionitrile	269.55		1965GEC
		56000*	propyl acetate	276.45		1965GEC
		152000*	propyl acetate	282.65		1965GEC
		336000*	propyl acetate	285.45		1965GEC
		370000*	propyl acetate	285.85		1965GEC
		440000*	propyl acetate	286.55		1965GEC
		690000*	propyl acetate	287.85		1965GEC
		infinite	propyl acetate	292.45		1965GEC
		87000*	propyl propionate	303.45		1965GEC
		113000*	propyl propionate	305.35		1965GEC
		181000*	propyl propionate	312.15		1965GEC
		412000*	propyl propionate	317.45		1965GEC
		infinite	propyl propionate	331.15		1965GEC
		100000*	tetrahydronaphthalene	313.75		1965GEC
		123000*	tetrahydronaphthalene	314.75		1965GEC
		165000*	tetrahydronaphthalene	316.25		1965GEC
		233000*	tetrahydronaphthalene	316.95		1965GEC
		275000*	tetrahydronaphthalene	317.25		1965GEC
		335000*	tetrahydronaphthalene	317.65		1965GEC
		infinite	tetrahydronaphthalene	322.65		1965GEC
Poly(1-pentene) (isotactic)						
		4500000*	cyclopentane		502	1981CH2
		4500000*	2,2-dimethylbutane		457	1981CH2
		4500000*	3,4-dimethylhexane		>569	1981CH2
		4500000*	2,2-dimethylpentane		502	1981CH2
		4500000*	2,3-dimethylpentane		529	1981CH2
		4500000*	2,4-dimethylpentane		493	1981CH2
		4500000*	3-ethylpentane		537	1981CH2
		4500000*	n-heptane		522	1981CH2
		4500000*	n-hexane		482	1981CH2
		4500000*	2-methylbutane		422	1981CH2
		4500000*	n-octane		556	1981CH2
		4500000*	n-pentane		433	1981CH2
		4500000*	2,2,4-trimethylpentane		527	1981CH2

Polymer (B)	M_n g/mol	M_w g/mol	Solvent (A)	UCST/ K	LCST/ K	Ref.
Poly(*N*-propylacrylamide)						
			water		298.15	2004MAO
Polypropylene (atactic)						
		infinite	diethyl ether		383	1974CO3
		infinite	diphenyl ether	426.5		1959KIN
		242000*	n-heptane		511	1981CH1
		infinite	n-heptane		483	1974CO3
		infinite	n-hexane		441	1974CO3
		242000*	2-methylbutane		413	1981CH1
		242000*	methylcyclohexane		564	1981CH1
		20000	n-pentane		425	1960FRE
		infinite	n-pentane		397	1974CO3
Polypropylene (isotactic)						
		28000*	benzyl phenyl ether	429.2		1968NA1
		infinite	benzyl phenyl ether	455.0		1968NA1
		28000*	benzyl propionate	405.2		1968NA1
		infinite	benzyl propionate	430.7		1968NA1
		28000*	biphenyl	388.2		1968NA1
		infinite	biphenyl	398.3		1968NA1
		28000*	1-butanol	395.2		1968NA1
		infinite	1-butanol	420.4		1968NA1
		28000*	4-*tert*-butylphenol	413.2		1968NA1
		infinite	4-*tert*-butylphenol	439.2		1968NA1
		242000*	cyclohexane		540	1981CH1
		242000*	cyclopentane		495	1981CH1
		28000*	dibenzyl ether	433.2		1968NA1
		infinite	dibenzyl ether	456.4		1968NA1
		242000*	2,2-dimethylbutane		441	1981CH1
		242000*	2,3-dimethylbutane		465	1981CH1
		242000*	3,4-dimethylhexane		553	1981CH1
		242000*	2,2-dimethylpentane		489	1981CH1
		242000*	2,3-dimethylpentane		513	1981CH1
		242000*	2,4-dimethylpentane		481	1981CH1
		28000*	diphenyl ether	395.2		1968NA1
		infinite	diphenyl ether	416.0		1968NA1
		infinite	diphenyl ether	418.4		1959KIN
		28000*	diphenylmethane	389.7		1968NA1
		242000*	3-ethylpentane		520	1981CH1
		28000*	4-ethylphenol	457.2		1968NA1

Polymer (B)	M_n g/mol	M_w g/mol	Solvent (A)	UCST/ K	LCST/ K	Ref.
Polypropylene (isotactic)						
		242000*	n-heptane		511	1981CH1
		242000*	n-hexane		470	1981CH1
		28000*	4-isooctylphenol	383.2		1968NA1
		242000*	2-methylbutane		413	1981CH1
		28000*	3-methylbutyl benzyl ether	384.2		1968NA1
		242000*	methylcyclohexane		564	1981CH1
		242000*	methylcyclopentane		518	1981CH1
		28000*	4-methylphenol	479.2		1968NA1
		28000*	2-methyl-1-propanol	395.2		1968NA1
		242000*	n-nonane		571	1981CH1
		242000*	n-octane		542	1981CH1
		28000*	4-octylphenol	379.2		1968NA1
		242000*	n-pentane		422	1981CH1
		2000000	n-pentane		409	1960FRE
		infinite	4-*tert*-pentylphenol	414.0		1968NA1
		242000*	2,2,4,4-tetramethylpentane		548	1981CH1
		242000*	2,2,3-trimethylbutane		511	1981CH1
		242000*	2,3,4-trimethylhexane		585	1981CH1
		242000*	2,2,4-trimethylpentane		510	1981CH1
Poly(propylene glycol)						
	500		n-hexane	317.15		1972LIR
	1000		n-hexane	288.15		1977VSH
	1800		n-hexane	257.15		1977VSH
	3100		n-hexane	257.15		1977VSH
	6600		n-hexane	258.15		1977VSH
	500		water		318.2	1975TA4
	575		water		318.2	1974BES
Poly(pyrrolidinoacrylamide)						
		<5000	water		345	2004PAN
Polystyrene						
	34900	37000	benzene		538.7	1973SA2
	91700	97200	benzene		532.5	1973SA2
	189000	200000	benzene		530.5	1973SA2
	378000	400000	benzene		528.3	1973SA2
	610000	670000	benzene		527.0	1973SA2
	2455000	2700000	benzene		525.0	1973SA2
		infinite	benzene		523	1973SA2

Polymer (B)	M_n g/mol	M_w g/mol	Solvent (A)	UCST/ K	LCST/ K	Ref.
Polystyrene						
		200000	bis(2-ethylhexyl) phthalate	279.05		1987RAN
		280000	bis(2-ethylhexyl) phthalate	280.55		1987RAN
		280000	bis(2-ethylhexyl) phthalate	281		1974VER
		280000	bis(2-ethylhexyl) phthalate	281		1984RAN
		335000	bis(2-ethylhexyl) phthalate	281.15		1987RAN
		470000	bis(2-ethylhexyl) phthalate	281.95		1987RAN
		900000	bis(2-ethylhexyl) phthalate	283		1984RAN
		900000	bis(2-ethylhexyl) phthalate	283.05		1987RAN
		1800000	bis(2-ethylhexyl) phthalate	285		1984RAN
		1800000	bis(2-ethylhexyl) phthalate	285.15		1987RAN
		1800000	bis(2-ethylhexyl) phthalate	286		1974VER
		2050000	bis(2-ethylhexyl) phthalate	285		1974VER
	404000	406000	1-bromobutane	248.15		1963ORO
	404000	406000	1-bromodecane	268.15		1963ORO
	404000	406000	1-bromooctane	265.15		1963ORO
	404000	406000	2-bromooctane	298.15		1963ORO
		62600*	butanedioic acid dimethyl ester	335.15		1950JE1
	3700	4000	1-butanol	383.45		1991OPS
		4800	1-butanol	381.45		1991OPS
		9000	1-butanol	407.45		1991OPS
		62600*	1-butanol	454.15		1950JE1
	34900	37000	2-butanone		463.2	1973SA2
	91700	97200	2-butanone		448.8	1973SA2
		110000	2-butanone		447	1974NA1
		97000	2-butanone		445.8	1973BAB
		212000	2-butanone		440	1974NA1
	189000	200000	2-butanone		441.1	1973SA2
		200000	2-butanone		439.2	1973BAB
		342000	2-butanone		436	1974NA1
	378000	400000	2-butanone		434.5	1973SA2
		533000	2-butanone		433	1974NA1
	610000	670000	2-butanone		431.8	1973SA2
		860000	2-butanone		428.0	1973BAB
	2455000	2700000	2-butanone		425.7	1973SA2
		2000000	2-butanone		422.9	1973BAB
		11000000	2-butanone		422	1974NA1
		infinite	2-butanone		418.8	1974NA1
		infinite	2-butanone		418.05	1973BAB
		infinite	2-butanone		417	1975KON
		90000*	butyl acetate		528	1974BAT
		180000*	butyl acetate		520.5	1974BAT

Polymer (B)	M_n g/mol	M_w g/mol	Solvent (A)	UCST/ K	LCST/ K	Ref.
Polystyrene						
		570000*	butyl acetate		514	1974BAT
	219800	233000	butyl acetate		494	1995PFO
	545500	600000	butyl acetate		489	1995PFO
		infinite	2-butyl acetate	210	442	1975KON
	34900	37000	*tert*-butyl acetate	230.6	443.7	1975KON
	94300	100000	*tert*-butyl acetate	253.6	422.0	1991BAE
	104000	110000	*tert*-butyl acetate	250.0	417.9	1975KON
	187000	200000	*tert*-butyl acetate	256.2	410.1	1975KON
	220000	233000	*tert*-butyl acetate	270.5	403.3	1991BAE
	609000	670000	*tert*-butyl acetate	270.8	394.6	1975KON
	583000	670000	*tert*-butyl acetate	270.8	393.7	1976SA2
	545500	600000	*tert*-butyl acetate	278.6	391.0	1991BAE
	1320000	1450000	*tert*-butyl acetate	276.7	387.1	1976SA2
	2455000	2700000	*tert*-butyl acetate	280.7	382.5	1976SA2
	3140000	3450000	*tert*-butyl acetate	281.8	381.2	1976SA2
		3450000	*tert*-butyl acetate	281.8	380.3	1975KON
		infinite	*tert*-butyl acetate	296.1	359.3	1976SA2
		infinite	*tert*-butyl acetate	296	357	1975KON
	404000	406000	4-*tert*-butyl-ethylbenzene	259.15		1963ORO
		62600*	butyl stearate	387.15		1950JE1
	404000	406000	1-chlorodecane	276.15		1963ORO
	18400	19200	1-chlorododecane	274.65		1991OPS
	404000	406000	1-chlorododecane	325.15		1963ORO
	18400	19200	1-chlorohexadecane	337.05		1991OPS
	404000	406000	1-chlorohexadecane	433.15		1963ORO
	404000	406000	1-chlorononane	254.15		1963ORO
	18400	19200	1-chlorooctadecane	365.55		1991OPS
	404000	406000	1-chlorooctane	223.15		1963ORO
	18400	19200	1-chlorotetradecane	309.35		1991OPS
	404000	406000	1-chloroundecane	301.15		1963ORO
	46400	51000	cyclodecane	278.9		1986CO1
	139000	153000	cyclodecane	282.5		1986CO1
	243000	267000	cyclodecane	284.1		1986CO1
	782000	860000	cyclodecane	286.6		1986CO1
		1000000	cyclodecane	287.8		1986CO1
		infinite	cyclodecane	289		1986CO1
	46400	51000	cycloheptane	276.2		1986CO1
	139000	153000	cycloheptane	282.2		1986CO1
	243000	267000	cycloheptane	283.9		1986CO1
	782000	860000	cycloheptane	286.9		1986CO1
		1000000	cycloheptane	288.7		1986CO1
		infinite	cycloheptane	290		1986CO1

Polymer (B)	M_n g/mol	M_w g/mol	Solvent (A)	UCST/ K	LCST/ K	Ref.
Polystyrene						
	19200	20400	cyclohexane	279.6		1991BAE
	23600	25000	cyclohexane	280.29		2003SIP
	23500	30200	cyclohexane	282.7		1972KEN
	27000	35400	cyclohexane	284.6		1972KEN
	27000	35400	cyclohexane	284.60		1980IRV
	34900	37000	cyclohexane	285.41		1975SAE
		37000	cyclohexane	285.49		2000KOI
	34900	37000	cyclohexane	285.6	510.9	1973SA1
	45000	45300	cyclohexane	287.29		1981HAS
	45000	45300	cyclohexane	287.29		1984TSU
		48000	cyclohexane	288		1991KAW
	49000	51000	cyclohexane	288.85		1970KON
	49000	51000	cyclohexane	288.9		1972KEN
	49000	51000	cyclohexane	288.85		1980IRV
	55000	61500	cyclohexane	290.5		1972KEN
	55000	62000	cyclohexane	290.45		1980IRV
		43000*	cyclohexane	291.39		1965ALL
		75000	cyclohexane	291.91		2000KOI
	78100	82000	cyclohexane	292.1		1962HAM
	69000	80000	cyclohexane	292.39		1960DE2
		43000*	cyclohexane	292.6		1952SHU
		43000*	cyclohexane	293		1951FO2
	91000	93000	cyclohexane	293.7		1972KEN
	91000	93000	cyclohexane	293.65		1980IRV
	91700	97200	cyclohexane	293.5	502.1	1973SA1
	102000	103000	cyclohexane	293.75		1981HAS
	94300	100000	cyclohexane	294		1991BAE
	111000	111000	cyclohexane	294		1974DER
	104000	110000	cyclohexane	294.38		1975SAE
	116000	123000	cyclohexane	294.61		2003SIP
		89000*	cyclohexane	294.75		1965ALL
	118000	124000	cyclohexane	294.97		1960DE2
	147000	153000	cyclohexane	296.25		1960DE1
	147000	153000	cyclohexane	296.34		1960DE2
	105000	154000	cyclohexane	296.4		1990RA2
	130400	142100	cyclohexane	296.41		1986KRU
	154000	166000	cyclohexane	296.6		1970KON
	154000	166000	cyclohexane	296.6		1972KEN
	154000	166000	cyclohexane	296.6		1974DER
	154000	166000	cyclohexane	296.60		1980IRV
	188700	200000	cyclohexane	296.81		1973KU2
	185300	189000	cyclohexane	296.83		1997KIT

Polymer (B)	M_n g/mol	M_w g/mol	Solvent (A)	UCST/ K	LCST/ K	Ref.
Polystyrene						
	196000	200000	cyclohexane	296.93		1973KU1
	196000	200000	cyclohexane	296.93		1973KU2
	196000	200000	cyclohexane	296.99		1975NAK
	189000	200000	cyclohexane	297.0	496.9	1973SA1
	193000	200000	cyclohexane	297		1974DER
		87000*	cyclohexane	297		1951FO2
	197000	206000	cyclohexane	297.07		1973KU1
		89000*	cyclohexane	297.1		1952SHU
	183000	190000	cyclohexane	297.17		1996RON
	58500	280000	cyclohexane	297.8		1984GIL
	198500	335500	cyclohexane	297.95		1986KRU
		92000*	cyclohexane	298		1951FO2
	221000	239000	cyclohexane	298.10		1960DE2
	247000	267000	cyclohexane	298.1		1962HAM
	248000	253000	cyclohexane	298.24		1960DE2
	314100	423300	cyclohexane	298.58		1986KRU
	301400	424000	cyclohexane	298.6		1990RA2
	200000	286000	cyclohexane	298.7		1972KEN
	200000	286000	cyclohexane	298.70		1980IRV
	269000	274000	cyclohexane	298.98		1995IKI
		215000	cyclohexane	299		1977RIG
	377400	400000	cyclohexane	299.37		2003SIP
	86000	239000	cyclohexane	299.38		1997KIT
		400000	cyclohexane	299.59		1990GOE
		360000*	cyclohexane	300		1951FO2
	100000	324000	cyclohexane	300.00		1984GIL
	100000	324000	cyclohexane	300.0		1990RA2
	392000	415000	cyclohexane	300.12		1973KU1
	128000	320000	cyclohexane	300.15		1992HE2
		390000	cyclohexane	300.16		1977WO3
	210000	346000	cyclohexane	300.20		1965REH
	210000	346000	cyclohexane	300.20		1968RE1
	378000	400000	cyclohexane	300.3	494.7	1973SA1
	601200	769500	cyclohexane	300.39		1986KRU
	474000	498000	cyclohexane	300.60		1984TSU
	375000	394000	cyclohexane	300.7		1972KEN
	375000	394000	cyclohexane	300.70		1980IRV
	440000	470000	cyclohexane	300.75		1968RE1
	522000	569000	cyclohexane	300.94		1960DE2
		250000*	cyclohexane	300.95		1965ALL
	583000	670000	cyclohexane	300.97		1975SAE
	566000	610000	cyclohexane	301		1991BAE

Polymer (B)	M_n g/mol	M_w g/mol	Solvent (A)	UCST/ K	LCST/ K	Ref.
Polystyrene						
		250000*	cyclohexane	301.1		1952SHU
	490000	527000	cyclohexane	301.15		1970KON
	490000	527000	cyclohexane	301.15		1972KEN
	490000	527000	cyclohexane	301.15		1974DER
	490000	527000	cyclohexane	301.15		1980IRV
	667000	680000	cyclohexane	301.17		1973KU1
	545500	600000	cyclohexane	301.34		1981WOL
	610000	670000	cyclohexane	301.1	491.7	1973SA1
	1000000	1190000	cyclohexane	302.15		1960DE2
	282000	350000	cyclohexane	302.33		1963DEB
	1538500	2000000	cyclohexane	302.92		2003SIP
		540000*	cyclohexane	303		1951FO2
	1250000	1500000	cyclohexane	303.2		1972KEN
	1320000	1450000	cyclohexane	303.27		1975SAE
	1515000	1560000	cyclohexane	303.65		1978NAK
	2455000	2700000	cyclohexane	304.2	488.6	1973SA1
		1300000*	cyclohexane	304.35		1965ALL
		1270000*	cyclohexane	304.6		1952SHU
		1270000*	cyclohexane	305		1951FO2
	11680000	13200000	cyclohexane	305.39		2003SIP
	404000	406000	cyclohexane	306.15		1963ORO
		3300000	cyclohexane	306.15		1995VSH
		4350000	cyclohexane	308.55		1976SLA
		infinite	cyclohexane	306.51		2003SIP
		infinite	cyclohexane	307	486	1973SA1
		infinite	cyclohexane	307		1951FO2
		infinite	cyclohexane	307.65		1973CAN
		84000*	cyclohexanol	350		1953SHU
		236000*	cyclohexanol	353.5		1953SHU
		881000*	cyclohexanol	356		1953SHU
		5500000*	cyclohexanol	357.5		1953SHU
	404000	406000	cyclohexanol	358.15		1963ORO
		infinite	cyclohexanol	358.9		1953SHU
	46400	51000	cyclooctane	275.2		1986CO1
	139000	153000	cyclooctane	279.5		1986CO1
	243000	267000	cyclocctane	281.2		1986CO1
	782000	860000	cyclocctane	284.4		1986CO1
		1000000	cyclooctane	285.5		1986CO1
		infinite	cyclooctane	286		1986CO1
	34900	37000	cyclopentane	267.0	455.3	1973SA2
	42600	43000	cyclopentane	267.95		1990IWA
	42600	43000	cyclopentane		454.05	1993IWA
		89000*	cyclopentane		451.15	1965ALL

Polymer (B)	M_n g/mol	M_w g/mol	Solvent (A)	UCST/ K	LCST/ K	Ref.
Polystyrene						
	91700	97200	cyclopentane	275.2	445.5	1973SA2
	97000	98900	cyclopentane	275.45		1990IWA
	97000	98900	cyclopentane		445.25	1993IWA
		89000*	cyclopentane		445.15	1965ALL
	166900	171900	cyclopentane		441.95	1993IWA
	172000	184000	cyclopentane	278.75		1990IWA
	189000	200000	cyclopentane	280.9	440.0	1973SA2
		250000*	cyclopentane		437.15	1965ALL
	347000	354000	cyclopentane	283.25		1990IWA
	347000	354000	cyclopentane		436.9	1993IWA
	378000	400000	cyclopentane	284.7	435.4	1973SA2
	610000	670000	cyclopentane	285.9	433.8	1973SA2
	783000	791000	cyclopentane	286.05		1990IWA
	545500	600000	cyclopentane	286.3		1981WOL
	783000	791000	cyclopentane		434.0	1993IWA
	1238000	1300000	cyclopentane	287.15		1990IWA
	2455000	2700000	cyclopentane	289.9	429.5	1973SA2
		1300000*	cyclopentane		423.15	1965ALL
		infinite	cyclopentane	293	427	1973SA2
	404000	406000	decahydronaphthalene	283.15		1963ORO
		infinite	decahydronaphthalene	287.95		1973CAN
	34900	37000	*trans*-decahydronaphthalene	274.65		1976NAK
	86500	90000	*trans*-decahydronaphthalene	281.5		1995PFO
	91500	97000	*trans*-decahydronaphthalene	281.95		1976NAK
		100000	*trans*-decahydronaphthalene	282		1985KRA
	104000	110000	*trans*-decahydronaphthalene	283		1977WO2
	189000	200000	*trans*-decahydronaphthalene	285.15		1976NAK
	144000	164000	*trans*-decahydronaphthalene	287		1993ARN
	219800	233000	*trans*-decahydronaphthalene	287.5		1995PFO
	79000	305000	*trans*-decahydronaphthalene	288		1993ARN
	377000	400000	*trans*-decahydronaphthalene	288.05		1976NAK
	355000	390000	*trans*-decahydronaphthalene	289		1977WO2
	609000	970000	*trans*-decahydronaphthalene	289.15		1976NAK
		1700000	*trans*-decahydronaphthalene	290.4		1985KRA
	1540000	2000000	*trans*-decahydronaphthalene	291		1977WO2
	2455000	2700000	*trans*-decahydronaphthalene	291.45		1976NAK
		infinite	*trans*-decahydronaphthalene	293		1977WO2
		infinite	*trans*-decahydronaphthalene	294		1976NAK
		infinite	*trans*-decahydronaphthalene	294	>360	1975KON
		4800	n-decane	360.95		1991OPS
	3700	4000	1-decanol	375.15		1991OPS
		4800	1-decanol	374.15		1991OPS
		9000	1-decanol	398.15		1991OPS

Polymer (B)	M_n g/mol	M_w g/mol	Solvent (A)	UCST/ K	LCST/ K	Ref.
Polystyrene						
		180000*	decyl acetate		657	1974BAT
		570000*	decyl acetate		650	1974BAT
	4530	4800	diethyl ether		407.3	1972SIO
	9440	10000	diethyl ether		353.0	1972SIO
	19200	20400	diethyl ether	228.4	314.5	1976SA2
	19200	20400	diethyl ether	230	315	1974CO1
	18700	19800	diethyl ether	235.6	314.5	1972SIO
	34900	37000	diethyl malonate	262.7		1975KON
	100000	102000	diethyl malonate	276.3		1988TVE
	100000	102000	diethyl malonate	277.1		1990STA
	91700	97200	diethyl malonate	278.3		1975KON
	105900	107000	diethyl malonate	280.15		1987GRU
	196000	200000	diethyl malonate	283.9		1979HAM
	187000	200000	diethyl malonate	285.8	589.6	1975KON
	274600	288800	diethyl malonate	291.05		1994SON
		62600*	diethyl malonate	291.15		1950JE1
	377000	400000	diethyl malonate	293.3	586.2	1975KON
	583000	670000	diethyl malonate	296.3	584.1	1975KON
	1524000	1921000	diethyl malonate	301.75		1994SON
	2455000	2700000	diethyl malonate	301.8	580.9	1975KON
	404000	406000	diethyl malonate	305.15		1963ORO
		infinite	diethyl malonate	308.75		1973CAN
		infinite	diethyl malonate	309		1975KON
		62600*	diethyl oxalate	259.15		1950JE1
	15800	16700	diethyl oxalate	261.96		1986SA2
	47200	50000	diethyl oxalate	280.05		1986SA2
	100000	110000	diethyl oxalate	292.59		1986SA2
	290000	300000	diethyl oxalate	308.2		1995HAA
	545500	600000	diethyl oxalate	309.96		1986SA2
	545500	600000	diethyl oxalate	312.78		1994SAT
	2000000	2000000	diethyl oxalate	322.2		1995HAA
	48100	51000	dimethoxymethane		413.7	1972SIO
	91700	97200	dimethoxymethane		406.1	1972SIO
	151000	160000	dimethoxymethane		401.2	1972SIO
	388000	411000	dimethoxymethane		395.0	1972SIO
	811000	860000	dimethoxymethane		391.6	1972SIO
		110000	1,4-dimethylcyclohexane	374	494	1984CO3
		240000	1,4-dimethylcyclohexane	387	482	1984CO3
		860000	1,4-dimethylcyclohexane	402	466	1984CO3
		2000000	1,4-dimethylcyclohexane	417	452	1984CO3
		62600*	dimethyl malonate	409.15		1950JE1
		62600*	dimethyl oxalate	453.15		1950JE1

Polymer (B)	M_n g/mol	M_w g/mol	Solvent (A)	UCST/ K	LCST/ K	Ref.
Polystyrene						
	23600	25000	dodecadeuterocyclohexane	283.48		2003SIP
		131000	dodecadeuterocyclohexane	297.5		1975STR
	116000	123000	dodecadeuterocyclohexane	298.10		2003SIP
		310000	dodecadeuterocyclohexane	302.2		1975STR
	377400	400000	dodecadeuterocyclohexane	303.31		2003SIP
		400000	dodecadeuterocyclohexane	303.59		1990GOE
		625000	dodecadeuterocyclohexane	304.8		1975STR
	1538500	2000000	dodecadeuterocyclohexane	307.38		2003SIP
	11680000	13200000	dodecadeuterocyclohexane	309.92		2003SIP
		infinite	dodecadeuterocyclohexane	312.5		1975STR
		25000	dodecadeuteromethyl-cyclopentane	310.07		1995LUS
		106300	dodecadeuteromethyl-cyclopentane	339.41		1995LUS
		4800	n-dodecane	368.65		1991OPS
	3700	4000	1-dodecanol	379.75		1991OPS
		4800	1-dodecanol	377.55		1991OPS
		9000	1-dodecanol	401.75		1991OPS
		infinite	dodecyl acetate	285.2		2000IMR
		90000*	ethyl acetate		461	1974BAT
		180000*	ethyl acetate		452.6	1974BAT
	34900	37000	ethyl acetate	204.1	451.0	1974SAE
		570000*	ethyl acetate		448	1974BAT
	94300	100000	ethyl acetate		437.5	1991BAE
	104000	110000	ethyl acetate	213.9	435.4	1974SAE
	189000	200000	ethyl acetate	216.5	430.6	1974SAE
	220000	233000	ethyl acetate		424.9	1991BAE
	583000	670000	ethyl acetate	222.9	421.4	1974SAE
	545500	600000	ethyl acetate		420.3	1991BAE
	2455000	2700000	ethyl acetate	226.5	415.7	1974SAE
		infinite	ethyl acetate	229	412	1974SAE
		infinite	ethyl acetate		399	1975KON
		62600*	ethyl acetate	242.65		1950JE1
	34900	37000	ethyl butyrate		541.9	1975KON
	104000	110000	ethyl butyrate		490.8	1975KON
	187000	200000	ethyl butyrate		486.5	1975KON
	583000	670000	ethyl butyrate		479.9	1975KON
	2455000	2700000	ethyl butyrate	180.3		1975KON
		3450000	ethyl butyrate	180.6	474.8	1975KON
		infinite	ethyl butyrate		471	1975KON
		43600*	ethylcyclohexane	320		1951FO2
	118000	124000	ethylcyclohexane	325.94		1963DEB

Polymer (B)	M_n g/mol	M_w g/mol	Solvent (A)	UCST/ K	LCST/ K	Ref.
Polystyrene						
		92000*	ethylcyclohexane	326		1951FO2
	147000	153000	ethylcyclohexane	327.81		1963DEB
	221000	239000	ethylcyclohexane	330.52		1963DEB
	522000	569000	ethylcyclohexane	335.17		1963DEB
		360000*	ethylcyclohexane	336		1951FO2
		1270000*	ethylcyclohexane	339		1951FO2
		infinite	ethylcyclohexane	342.95		1963DEB
		infinite	ethylcyclohexane	343		1951FO2
	2000	2200	ethyl formate	230		1975KON
	3640	4000	ethyl formate	244	480	1975KON
	9440	10000	ethyl formate	272	451	1975KON
	11300	11600	ethyl formate	273.7	435.5	1997IMR
	19250	20400	ethyl formate	294	428	1975KON
	34900	37000	ethyl formate	316	410	1974SAE
	21400	22000	ethyl formate	296.1	420.4	1997IMR
	1920	2030	n-heptane	311	515	1983CO1
	4530	4800	n-heptane	359	477	1983CO1
		4800	n-heptane	365.85		1991OPS
	404000	406000	n-hexadecane	463.15		1963ORO
	3700	4000	1-hexadecanol	386.25		1991OPS
	4780	5110	1,1,1,3,3,3-hexadeutero-2-propanone	259	444	1991SZY
	5500	5770	1,1,1,3,3,3-hexadeutero-2-propanone	270	436	1991SZY
	6600	7800	1,1,1,3,3,3-hexadeutero-2-propanone	286	428	1991SZY
		8000	1,1,1,3,3,3-hexadeutero-2-propanone	286.77		1995LUS
	6600	7800	1,1,1,3,3,3-hexadeutero-2-propanone	286.9		1992SZY
	10750	11500	1,1,1,3,3,3-hexadeutero-2-propanone	290	408	1991SZY
		13500	1,1,1,3,3,3-hexadeutero-2-propanone	350.32	367.34	1995LUS
	850	900	n-hexane	296	490	1983CO1
	1920	2030	n-hexane	318	470	1983CO1
		62600*	hexanoic acid	448.15		1950JE1
	3700	4000	1-hexanol	372.15		1991OPS
		4800	1-hexanol	373.45		1991OPS
		9000	1-hexanol	398.05		1991OPS
		62600*	1-hexanol	425.15		1950JE1
		62600*	3-hexanol	396.65		1950JE1

Polymer (B)	M_n g/mol	M_w g/mol	Solvent (A)	UCST/ K	LCST/ K	Ref.
Polystyrene						
		90000*	hexyl acetate		578	1974BAT
		570000*	hexyl acetate		560	1974BAT
	2000	2200	methyl acetate	213.9		1975KON
	3640	4000	methyl acetate	218.8	484.5	1975KON
	9440	10000	methyl acetate	242.4	459.2	1975KON
	19250	20400	methyl acetate	254.6	445.4	1975KON
		90000*	methyl acetate		441	1974BAT
	34900	37000	methyl acetate	266.6	434.0	1974SAE
		180000*	methyl acetate		432	1974BAT
		48000	methyl acetate		428	1965MYR
		59000	methyl acetate		427	1965MYR
		51000	methyl acetate		423.35	1972ZE2
		64900	methyl acetate		423	1965MYR
		570000*	methyl acetate		421	1974BAT
	91300	105000	methyl acetate		420	1970DEL
		97200	methyl acetate	275.15	415.15	1972ZE2
	104000	110000	methyl acetate	284.2	415.7	1974SAE
	226000	260000	methyl acetate		415	1970DEL
		160000	methyl acetate	281.45	409.95	1972ZE2
	189000	200000	methyl acetate	289.7	409.9	1974SAE
		195000	methyl acetate		409	1965MYR
		270000	methyl acetate		405	1965MYR
	461000	530000	methyl acetate	290	403	1970DEL
		390000	methyl acetate		401	1965MYR
		498000	methyl acetate	294.15	400.55	1972ZE2
		670000	methyl acetate	296.45	397.85	1972ZE2
		770000	methyl acetate	300	400	1993WAK
		770000	methyl acetate		397.6	1991BAE
	583000	670000	methyl acetate	301.5	398.4	1974SAE
		860000	methyl acetate	299.25	396.15	1972ZE2
	91300	1050000	methyl acetate	299	395	1970DEL
		1800000	methyl acetate	303.15	391.75	1972ZE2
	2260000	2600000	methyl acetate	303	380	1970DEL
	2455000	2700000	methyl acetate	311.0	389.2	1974SAE
		infinite	methyl acetate	316	387	1974SAE
		infinite	methyl acetate	322	370	1970DEL
		infinite	methyl acetate	324	370	1975KON
		62600*	methyl acetate	326.15		1950JE1
		62600*	3-methyl-1-butanol	444.15		1950JE1

Polymer (B)	M_n g/mol	M_w g/mol	Solvent (A)	UCST/ K	LCST/ K	Ref.
Polystyrene						
	34900	37000	3-methylbutyl acetate	199.4	529.2	1974SAE
	104000	110000	3-methylbutyl acetate	210.1	510.1	1974SAE
	189000	200000	3-methylbutyl acetate	212.3	505.1	1974SAE
	583000	670000	3-methylbutyl acetate	218.2	499.2	1974SAE
	2455000	2700000	3-methylbutyl acetate	220.8	497.0	1974SAE
		infinite	3-methylbutyl acetate	224	493	1974SAE
	668	761	methylcyclohexane	189		1996IM1
	1160	1241	methylcyclohexane	214		1996IM1
	1586	1681	methylcyclohexane	230		1996IM1
	2294	2500	methylcyclohexane	246		1996IM1
	3775	4000	methylcyclohexane	261		1996IM1
	5505	5780	methylcyclohexane	271		1996IM1
	8500	9000	methylcyclohexane	279		1994VA1
	8490	9000	methylcyclohexane	281.20		1982SH2
	9900	10100	methylcyclohexane	283.632		2002ZHO
	9630	10200	methylcyclohexane	285.71		1980DO1
	12300	13000	methylcyclohexane	291.34		1991SHE
	15200	16100	methylcyclohexane	295.98		1980DO1
		17500	methylcyclohexane	296.05		1992HE1
	16300	17300	methylcyclohexane	296.13		1980DO1
	16500	17500	methylcyclohexane	296.32		1982SH2
	17650	18700	methylcyclohexane	296.557		1999NAK
	16200	17200	methylcyclohexane	296.75		1984DOB
	16500	17500	methylcyclohexane	297		1994VA1
	19100	20200	methylcyclohexane	298.95		1980DO1
		20400	methylcyclohexane	299	531	1974CO2
	21400	22000	methylcyclohexane	300.85		1993WEL
	21700	23000	methylcyclohexane	302.27		1991SHE
	29130	30000	methylcyclohexane	302	497	1996IM1
	29800	30400	methylcyclohexane	304.334		2002ZHO
	25900	28500	methylcyclohexane	305		1994VA1
	27400	29000	methylcyclohexane	306.10		1991SHE
		37000	methylcyclohexane	307	523	1974CO2
	32900	34900	methylcyclohexane	309.00		1980DO1
	34900	37000	methylcyclohexane	309.65		1976NOS
	34900	37000	methylcyclohexane	309.7	518.8	1973SA1
	37500	37900	methylcyclohexane	310.53		1996RON
		42500	methylcyclohexane	311.55		1992HE1
	47400	48800	methylcyclohexane	311.997		2002ZHO
	43800	46400	methylcyclohexane	312.61		1980DO1
	47200	50000	methylcyclohexane	313		1994VA1

Polymer (B)	M_n g/mol	M_w g/mol	Solvent (A)	UCST/ K	LCST/ K	Ref.
Polystyrene						
	67250	68600	methylcyclohexane	316.798		2002ZHO
		97200	methylcyclohexane	318.25		1971KAG
	94760	97600	methylcyclohexane	319.654		2002ZHO
		86300	methylcyclohexane	320.05		1992HE1
	87400	90000	methylcyclohexane		490	1996IM1
	91700	97200	methylcyclohexane	321.8	505.9	1973SA1
	103000	109000	methylcyclohexane	322.71		1980DO1
	104000	110000	methylcyclohexane	323.24		1982SH2
	164700	168000	methylcyclohexane	325.117		2002ZHO
		156000	methylcyclohexane	326	505	1974CO2
		175000	methylcyclohexane	326.65		1992HE1
	200000	204000	methylcyclohexane	326.907		2002ZHO
	171000	181000	methylcyclohexane	327.00		1980DO1
	189000	200000	methylcyclohexane	327.4	499.9	1973SA1
	54500	297000	methylcyclohexane	328.66		1988SC2
	317300	330000	methylcyclohexane	329.130		2002ZHO
	220000	233000	methylcyclohexane	329.92		1982SH2
	220000	233000	methylcyclohexane	330		1991CHU
	378000	400000	methylcyclohexane	332.7	494.6	1973SA1
		670000	methylcyclohexane	336	494	1974CO2
	610000	670000	methylcyclohexane	334.5	492.3	1973SA1
	678000	719000	methylcyclohexane	334.8		1991SHE
	678000	719000	methylcyclohexane	334.82		1980DO1
	820000	900000	methylcyclohexane	336		1991CHU
	1200000	1260000	methylcyclohexane	336.97		1982SH2
	1564300	1971000	methylcyclohexane	336	476	1996IM1
	1660000	1860000	methylcyclohexane	339		1991CHU
	2455000	2700000	methylcyclohexane	339.6	488.4	1973SA1
	16700000	20000000	methylcyclohexane	343	470	1996IM1
	404000	406000	methylcyclohexane	344.15		1963ORO
		411000	methylcyclohexane	344.15		1971KAG
		860000	methylcyclohexane	347.45		1971KAG
		1800000	methylcyclohexane	348.15		1971KAG
		infinite	methylcyclohexane	345	486	1974CO2
		infinite	methylcyclohexane	345	464.5	1996IM1
		infinite	methylcyclohexane	350.65		1971KAG
	2230	2510	methylcyclopentane	255	507	1991SZY
		4100	methylcyclopentane	259.69		1995LUS
	5500	5770	methylcyclopentane	269	500	1991SZY
	6570	7820	methylcyclopentane	270	491	1991SZY
		13500	methylcyclopentane	291.93		1995LUS

Polymer (B)	M_n g/mol	M_w g/mol	Solvent (A)	UCST/ K	LCST/ K	Ref.
Polystyrene						
	10750	11500	methylcyclopentane	295	480	1991SZY
		22100	methylcyclopentane	300.94		1995LUS
		25000	methylcyclopentane	303.65		1995LUS
		106300	methylcyclopentane	327.52		1995LUS
		110000	methylcyclopentane	322	445	1984CO3
		240000	methylcyclopentane	329	438	1984CO3
		860000	methylcyclopentane	338	428	1984CO3
		2000000	methylcyclopentane	342	422	1984CO3
		infinite	methylcyclopentane	348	417	1984CO3
	404000	406000	4-methyl-2-pentanone	262.15		1963ORO
	34900	37000	2-methylpropyl acetate	197.9	487.4	1974SAE
	104000	110000	2-methylpropyl acetate	210.4	468.5	1974SAE
	189000	200000	2-methylpropyl acetate	212.9	463.3	1974SAE
	583000	670000	2-methylpropyl acetate	220.9	453.9	1974SAE
	2455000	2700000	2-methylpropyl acetate	223.1	449.0	1974SAE
		infinite	2-methylpropyl acetate	227	445	1974SAE
		48000	nitroethane	303.1		1991KAW
	404000	406000	nitroethane	341.15		1963ORO
		4800	n-octadecane	403.55		1991OPS
	3700	4000	1-octadecanol	390.55		1991OPS
		62600*	1-octadecanol	448.65		1950JE1
	1920	2030	n-octane	309		1983CO1
	4530	4800	n-octane	353	527	1983CO1
		4800	n-octane	358.05		1991OPS
	3700	4000	1-octanol	372.35		1991OPS
		4800	1-octanol	370.35		1991OPS
		9000	1-octanol	395.05		1991OPS
		62600*	1-octene	355.15		1950JE1
		4800	n-pentadecane	385.25		1991OPS
		1100	n-pentane	292		1988KI2
	3700	4000	1-pentanol	375.05		1991OPS
		4800	1-pentanol	373.85		1991OPS
		9000	1-pentanol	397.45		1991OPS
	404000	406000	pentyl acetate	233.15		1963ORO
		90000*	pentyl acetate		552	1974BAT
		180000*	pentyl acetate		544.2	1974BAT
		570000*	pentyl acetate		538	1974BAT
	219800	233000	pentyl acetate		519	1995PFO
	545500	600000	pentyl acetate		512	1995PFO
	1564000	1971000	pentyl acetate		507	1995PFO

Polymer (B)	M_n g/mol	M_w g/mol	Solvent (A)	UCST/ K	LCST/ K	Ref.
Polystyrene						
		100000	1-phenyldecane	283.60		1982GEE
		390000	1-phenyldecane	293.10		1982GEE
		600000	1-phenyldecane	295.45		1982GEE
		600000	1-phenyldecane	295.5		1981WOL
	404000	406000	1-phenyldodecane	323.15		1963ORO
	2230	2510	2-propanone	232	480	1991SZY
	4530	4800	2-propanone	222	465	1972SIO
	5500	5770	2-propanone	251	452	1991SZY
	7620	8000	2-propanone	253.68		1995LUS
	6600	7800	2-propanone	254.6		1992SZY
	4780	5110	2-propanone	263	448	1991SZY
	6570	7820	2-propanone	271	438	1991SZY
	9720	10300	2-propanone	271	414	1972SIO
	10750	11500	2-propanone	273	419	1991SZY
	12750	13500	2-propanone	285.34		1995LUS
	12750	13500	2-propanone	289	396	1991SZY
	21500	22100	2-propanone	341.2	351.7	1995LUS
	7620	8000	propionitrile	273		1996LUS
	12750	13500	propionitrile	312		1996LUS
	21400	22000	propionitrile	359	401	1996LUS
	21500	22100	propionitrile	364	400	1996LUS
	34900	37000	propyl acetate		484.1	1974SAE
		90000*	propyl acetate		500	1974BAT
		180000*	propyl acetate		489.5	1974BAT
		570000*	propyl acetate		479.5	1974BAT
	104000	110000	propyl acetate	183.7	469.0	1974SAE
	189000	200000	propyl acetate	185.5	464.8	1974SAE
	583000	670000	propyl acetate	189.6	458.2	1974SAE
	2455000	2700000	propyl acetate	191.0	454.2	1974SAE
		infinite	propyl acetate	193	451	1974SAE
	9440	10000	2-propyl acetate		468.5	1974SAE
	34900	37000	2-propyl acetate	206.6	436.7	1974SAE
	104000	110000	2-propyl acetate	220.9	414.2	1974SAE
	189000	200000	2-propyl acetate	225.6	407.7	1974SAE
	583000	670000	2-propyl acetate	235.5	395.2	1974SAE
	2455000	2700000	2-propyl acetate	240.3	385.9	1974SAE
	1028000	1100000	2-propyl acetate	248	380	1980RIC
		infinite	2-propyl acetate	246	380	1974SAE
		infinite	2-propyl acetate	250	365	1975KON
	3700	4000	1-tetradecanol	383.25		1991OPS

Polymer (B)	M_n g/mol	M_w g/mol	Solvent (A)	UCST/ K	LCST/ K	Ref.
Polystyrene						
	34900	37000	toluene		567.2	1973SA1
	91700	97200	toluene		559.9	1973SA1
	189000	200000	toluene		557.2	1973SA1
	378000	400000	toluene		554.9	1973SA1
	610000	670000	toluene		553.1	1973SA1
	2455000	2700000	toluene		552.0	1973SA1
		infinite	toluene		550	1973SA1
		62600*	vinyl acetate	384.15		1950JE1
Polystyrene (three-arm star)						
		93000	cyclohexane	292.2	503.0	1979COW
		131000	cyclohexane	294.0	499.7	1979COW
		230000	cyclohexane	297.1	496.8	1979COW
		265000	cyclohexane	297.4	495.7	1979COW
		410000	cyclohexane	300.9	493.6	1979COW
Polystyrene (four-arm star)						
		85000	cyclohexane	289.60		1998TER
		155000	cyclohexane	294.13		1998TER
		329000	cyclohexane	297.92		1998TER
		1430000	cyclohexane	302.80		1998TER
Polystyrene (star-shaped)						
		125000	cyclohexane	288.1	504.8	1979COW
		206000	cyclohexane	294.1	498.9	1979COW
		605000	cyclohexane	299.3	493.8	1979COW
		790000	cyclohexane	299.9	492.2	1979COW
		960000	cyclohexane	300.4	491.9	1979COW
		1200000	cyclohexane	301.15		1991YOK
		2050000	cyclohexane	301.55		1991YOK
Poly(styrene-*co*-acrylonitrile)						
(21.1 wt% acrylonitrile)		infinite	toluene	325.4		1972TE2
(23.2 wt% acrylonitrile)		infinite	toluene	355.1		1972TE2
(25.0 wt% acrylonitrile)		147000	toluene	313.15		2000SCH
(51.0 mol% acrylonitrile)		347000	ethyl acetate		344.15	1982MAN
(51.0 mol% acrylonitrile)		457000	ethyl acetate		337.35	1982MAN
(51.0 mol% acrylonitrile)		794000	ethyl acetate		328.65	1982MAN
(51.0 mol% acrylonitrile)		912000	ethyl acetate		326.65	1982MAN
(51.0 mol% acrylonitrile)		1365000	ethyl acetate		323.95	1982MAN
(51.0 mol% acrylonitrile)		2240000	ethyl acetate		320.15	1982MAN

Polymer (B)	M_n g/mol	M_w g/mol	Solvent (A)	UCST/ K	LCST/ K	Ref.
Poly(styrene-*g*-cellulose diacetate) (77.4 wt% grafted polystyrene)						
		480000	*N,N*-dimethylformamide	257	402	1982GOL
		750000	*N,N*-dimethylformamide	262	399	1982GOL
		1100000	*N,N*-dimethylformamide	266	389	1982GOL
		1200000	*N,N*-dimethylformamide	271	366	1982GOL
		480000	tetrahydrofuran		367	1982GOL
		750000	tetrahydrofuran		363	1982GOL
		1100000	tetrahydrofuran		361	1982GOL
		1200000	tetrahydrofuran		355	1982GOL
Poly(styrene-*co*-methyl methacrylate)						
(52.0 mol% styrene)		infinite	cyclohexanol	334.65		1970KOT
Poly(styrene-*co*-α-methylstyrene)						
(20 mol% styrene)	100000	114000	butyl acetate	288.85	453.05	1995PFO
(20 mol% styrene)	100000	114000	cyclohexane	285.85	484.85	1995PFO
(20 mol% styrene)	100000	114000	cyclopentane	290.95	421.05	1995PFO
(20 mol% styrene)	100000	114000	*trans*-decahydronaphthalene	264.15		1995PFO
(20 mol% styrene)	100000	114000	hexyl acetate	288.55	514.15	1995PFO
(20 mol% styrene)	100000	114000	pentyl acetate	303.15	480.65	1995PFO
Poly(tetrafluoroethylene-*alt*-trifluoronitrosomethane)						
		infinite	1,1,2-trichloro-1,2,2-trifluoroethane	301.6		1961MOR
Poly(trimethylene oxide)						
		infinite	cyclohexane	300		1976CHI
Poly(*N*-vinylacetamide-*co*-vinyl acetate)						
(58 mol% vinyl acetate)	30000	57000	water		340.15	2003YA1
(63 mol% vinyl acetate)	27000	48600	water		323.15	2003YA1
(69 mol% vinyl acetate)	26000	49400	water		307.15	2003YA1
(71 mol% vinyl acetate)	22000	39600	water		295.15	2003YA1
(78 mol% vinyl acetate)	26000	46800	water		282.15	2003YA1
Poly(vinyl alcohol)						
		40000	water		514	1971TAG
		46000	water		515	1969AND
		84000	water		504	1969AND

Polymer (B)	M_n g/mol	M_w g/mol	**Solvent (A)**	UCST/ K	LCST/ K	Ref.
Poly(vinyl alcohol-*co*-vinyl butyrate)						
(7.5 mol% butyralized PVA)		infinite	water	408.0	298.25	1984SHI
(9.9 mol% butyralized PVA)		infinite	water		296.45	1984SHI
(12.7 mol% butyralized PVA)		infinite	water		287.55	1984SHI
Poly(*N*-vinylamine-*co*-*N*-vinylcaprolactam)						
(3.8 mol% vinylamine)		160000*	water		308.8	1994TAG
Poly(*N*-vinylcaprolactam)						
		5600	water		303.15	2003OKH
		150000	water		306.45	2002MA1
			water		306.15	2000PE1
			water		305.25	2005DUB
		330000	water		304.95	2005LAU
		high	water		304.55	1999LAU
		470000*	water		304.0	1994TAG
Poly(*N*-vinylcaprolactam-*co*-methacrylic acid)						
(9 mol% methacrylic acid)			water		303.15	2003OKH
(12 mol% methacrylic acid)			water		310.65	2005DUB
(18 mol% methacrylic acid)			water		313.85	2005DUB
(37 mol% methacrylic acid)			water		304.15	2003OKH
Poly(vinyl chloride)						
		55000	dibutyl phthalate	353		1983TAG
		55000	tricresyl phosphate	383		1983TAG
		85000*	dimethyl phthalate	355		2002SAF
Poly(*N*-vinylformamide-*co*-vinyl acetate)						
(60 mol% vinyl acetate)	24000	45600	water		310.15	2003YA1
(66 mol% vinyl acetate)	25000	47500	water		291.15	2003YA1
(73 mol% vinyl acetate)	23000	50600	water		277.15	2003YA1
Poly(*N*-vinylisobutyramide)						
	66000	105600	water		313.25	2000KUN

Polymer (B)	M_n g/mol	M_w g/mol	Solvent (A)	UCST/ K	LCST/ K	Ref.
Poly(vinyl methyl ether)						
	46500	98600	deuterium oxide		307.2	1994OKA
		infinite	deuterium oxide		304.0	1994OKA
	12000	15000	water		308.15	2005VER
		60000	water		305.65	2000PE1
	83000	155000	water		306.95	1990SCH
Poly(*N*-vinyl-*N*-propylacetamide)						
		30000*	water		313.5	1994TAG
Tetra(ethylene glycol) diethyl ether						
	250		water		363	1972NAK
Tetra(ethylene glycol) monodecyl ether						
	334		water	573	294	1980LAN
	334		water		293.65	1991SC3
Tetra(ethylene glycol) monododecyl ether						
	362		water		279.75	1991SC3
Tetra(ethylene glycol) monohexyl ether						
	278		water		339.25	1991SC3
Tetra(ethylene glycol) monooctyl ether						
	306		water		313.45	1984COR
	306		water		313.95	1991SC3
	306		water		314.11	1999SHI
Tri(ethylene glycol) monohexyl ether						
	234		water		317.85	1984COR
	234		water		319.15	1991SC3
Tri(ethylene glycol) monooctyl ether						
	262		water		289.15	1991SC3

Comments: The asterisk * denotes a viscosity average M_η in the M_w-column.
LCST and/or UCST values in dependence on pressure are summarized in 2005WOH.

2.3. Table of binary systems where data were published only in graphical form as phase diagrams or related figures

Polymer (B)	Solvent (A)	Ref.
Cellulose diacetate		
	benzyl alcohol	1977PAN
	benzyl alcohol	1987PAN
	2-butanone	1982SU2
	dimethyl phthalate	1985ZAR
	2-propanone	1971COW
	2-propanone	1982SU1
	tetrahydrofuran	1978GOL
Cellulose nitrate		
	dibutyl phthalate	1985RAB
	2-propanone	1991AKH
Cellulose triacetate		
	benzyl alcohol	1977PAN
	benzyl alcohol	1987PAN
	dichloromethane	1978GOL
	ethyl acetate	2005BLA
	nitromethane	1987PAN
	2-propanone	1971COW
	2-propanone	2003VAR
Cellulose tricaprylate		
	N,N-dimethylformamide	1952MAN
	3-phenyl-1-propanol	1952MAN
Cellulose tricarbanilate		
	cyclohexanol	1975TA2
	5-nonanone	1975TA2

Polymer (B)	Solvent (A)	Ref.
Ethyl(hydroxyethyl)cellulose		
	formamide	1991SA1
	water	1985FUR
	water	1991SA1
	water	2000BAD
	water	2000JOA
	water	2003LUT
	water	2004OLS
Gelatine		
	water	1987PAN
	water	2005GUP
Hepta(ethylene glycol) monododecyl ether		
	water	1981HEU
	water	2002ZHE
Hepta(ethylene glycol) monotetradecyl ether		
	water	1984COR
	water	1998KUB
Hexa(ethylene glycol) monodecyl ether		
	water	2005IMA
Hexa(ethylene glycol) monododecyl ether		
	water	1991SC3
	water	2002ZHE
Hexa(ethylene glycol) monooctyl ether		
	water	1991SC3
Hydroxypropylcellulose		
	N,N-dimethylacetamide	1983CON
	water	1971KAG
	water	1983CON
	water	1985FUR
	water	1989FOR
	water	1991ROB
	water	1995GUI
	water	2001KUN

Polymer (B)	Solvent (A)	Ref.
Methylcellulose		
	water	1997CHE
	water	2001TAK
	water	2002LIL
	water	2005SP1
	water	2006SCH
Methyl(hydroxypropyl)cellulose		
	water	1974BAB
	water	1999KIT
	water	2003SCH
Natural rubber		
	dibutyl phthalate	1986IKA
	dibutyl sebacate	1986IKA
	5,8,11,13,16,19-hexaoxatricosane	1986IKA
	2-pentanone	1952WAG
Nylon-6		
	benzyl alcohol	1988PUS
	ε-caprolactam	1986MAL
	sulfuric acid	1985KON
Nylon-12		
	N,N-dimethylacetamide	1986MAL
	dimethylsulfoxide	1986MAL
	dodecalactam	1986MAL
Octa(ethylene glycol) monohexadecyl ether		
	water	2005HAM
Octa(ethylene glycol) monooctadecyl ether		
	water	2005HAM
Octa(ethylene glycol) monotetradecyl ether		
	water	2005HAM
Penta(ethylene glycol) monodecyl ether		
	water	2005IMA

Polymer (B)	Solvent (A)	Ref.
Penta(ethylene glycol) monododecyl ether		
	water	1985HAM
	water	1991SC3
Penta(ethylene glycol) monooctyl ether		
	water	1991SC3
Phenol-formaldehyde resin (acetylated)		
	2-ethoxyethanol	1998YA2
Poly(acrylamide-*co-N,N*-dimethylaminoethyl methacrylate)		
	water	1997CHO
Poly(acrylamide-*co*-hydroxypropyl acrylate)		
	water	1975TAY
Poly(acrylic acid)		
	1,4-dioxane	1990COW
	tetrahydrofuran	1996SAF
Poly(acrylic acid-*co-N*-isopropylacrylamide)		
	water	1993OTA
	water	1997YOO
	water	2000BO1
	water	2000YAM
	water	2004MAE
	water	2006WEN
Poly(acrylic acid-*co*-methyl acrylate)		
	water	1993TA1
Poly(acrylic acid-*co*-2-methyl-5-vinylpyridine)		
	water	1985VED
Poly(acrylic acid-*co*-nonyl acrylate)		
	bisphenol-A-diglycidyl ether	1999MIK
	bisphenol-A-diglycidyl ether	2004MIK
	diethylene glycol bis(methacyl-oxyethylene carbonate)	2004MIK

Polymer (B)	Solvent (A)	Ref.
Poly(acrylonitrile)		
	N,N-dimethylformamide	1978UGL
	dimethylsulfoxide	2005CH2
Poly(acrylonitrile-*co*-butadiene)		
	1,4-butanediyl methacrylate	1985ZHI
	dibutyl phthalate	1984TAG
	dibutyl phthalate	1986IKA
	dibutyl sebacate	1984TAG
	dibutyl sebacate	1986IKA
	4,4'-dicyanate-1,1'-diphenylethane	1995BOR
	ethyl acetate	2002VSH
	ethyl acetate	2004VSH
	5,8,11,13,16,19-hexaoxatricosane	1984TAG
	5,8,11,13,16,19-hexaoxatricosane	1986IKA
Poly(*N*-acryloylpyrrolidine-*co*-vinylferrocene)		
	water	1994KUR
Poly(allylamine)-*g*-poly(*N*-isopropylacrylamide)		
	water	2005GAO
Polyarylate		
	1,2-dichloroethane	1975TA1
	trichloromethane	1975TA1
Poly(m-benzamide)		
	water	2006SUG
Poly(γ-benzyl-L-glutamate)		
	dichloroacetic acid	1985KON
	1,2-dichloroethane	1990JAC
	N,N-dimethylformamide	1985KON
Poly[bis(2,3-dimethoxypropanoxy)phosphazene]		
	water	1996ALL
Poly[bis(2-(2'-methoxyethoxy)ethoxy)phosphazene]		
	water	1996ALL

Polymer (B)	Solvent (A)	Ref.
Poly{[bis(ethyl glycinat-*N*-yl)phosphazene]-*b*-ethylene oxide}		
	water	2002CH2
Poly[bis(2,3-bis(2-methoxyethoxy)propanoxy) phosphazene]		
	water	1996ALL
Poly[bis(2,3-bis(2-(2'-methoxyethoxy)ethoxy) propanoxy)phosphazene]		
	water	1996ALL
Poly[bis(2,3-bis(2-(2'-(2"-dimethoxyethoxy) ethoxy)ethoxy)propanoxy)phosphazene]		
	water	1996ALL
Polybutadiene		
	1-butanol	2000PEK
	dipentyl phthalate	1975CHA
	n-hexane	1974DEL
	n-hexane	1983GUT
	1-hexanol	2000PEK
	2-methylhexane	1974DEL
	1-pentanol	2000PEK
	2,2,3-trimethylbutane	1974DEL
	2,2,4-trimethylpentane	1974DEL
Polybutadiene (hydrogenated)		
	diphenyl ether	1995ZRY
	diphenylmethane	1995ZRY
Poly(butadiene-*co*-α-methylstyrene)		
	butyl acetate	2004VSH
	ethyl acetate	2002VSH
	ethyl acetate	2004VSH
Poly(1-butene)		
	anisole	1961KRI
	n-hexane	1981CH2
	toluene	1967MOR

Polymer (B)	**Solvent (A)**	**Ref.**
Poly(*N-tert*-butylacrylamide-*co*-*N,N*-dimethylacrylamide)		
	water	1999LIU
Poly(*N-tert*-butylacrylamide-*co*-*N*-ethylacrylamide)		
	water	1999LIU
Poly(butylene oxide-*b*-ethylene oxide)		
	water	2001HAM
	water	2002CH1
	water	2002SON
	water	2004KEL
	water	2005CHA
Poly(butylene oxide-*b*-ethylene oxide-*b*-butylene oxide)		
	water	2004KEL
Poly(butyl methacrylate)		
	1-butanol	1986SAN
	2-butanone	1990KYO
	n-decane	1986SAN
	dioctyl phthalate	2006LIC
	ethanol	1986SAN
	n-heptane	1986SAN
	n-hexadecane	2006LIC
	n-octane	1986SAN
	1-pentanol	1986SAN
	2-propanol	1986HER
	2-propanol	1986SAN
	2-propanol	1987JEL
	2,2,4-trimethylpentane	1986SAN
Poly(p-*tert*-butylstyrene-*b*-dimethylsiloxane-*b*-p-*tert*-butylstyrene)		
	2-butanone	1986KUE
	1-nitropropane	1986KUE

Polymer (B)	Solvent (A)	Ref.
Poly(ε-caprolactone)		
	toluene	2002LEE
Poly(ε-caprolactone-*b*-ethylene glycol-*b*-ε-caprolactone)		
	water	2005BAE
	water	2006LUC
Polycarbonate bisphenol-A		
	bis(2-ethylhexyl) adipate	1981SUV
	bis(2-ethylhexyl) azelate	1981SUV
	bis(2-ethylhexyl) 1,10-decane-dicarboxylate	1981SUV
	bis(2-ethylhexyl) sebacate	1981SUV
	bis(2-ethylhexyl) succinate	1981SUV
	dibutyl phthalate	1987IKA
	pentachlorobiphenyl	1987IKA
Poly(2-chlorostyrene)		
	benzene	1970MAT
Poly(4-chlorostyrene)		
	2-(2-butoxyethoxy)ethanol	1972IZU
	tert-butyl acetate	1972IZU
	2-(2-ethoxyethoxy)ethanol	1972IZU
	ethylbenzene	1972IZU
	ethyl chloroacetate	1972IZU
	isopropylbenzene	1972IZU
	isopropylbenzene	1977TA2
	isopropyl chloroacetate	1972IZU
	methyl acetate	1972IZU
	methyl chloroacetate	1972IZU
	2-methylpropyl acetate	1980VSH
	2-propyl acetate	1972IZU
	tetrachloroethene	1972IZU
	tetrachloromethane	1966KUB
	tetrachloromethane	1972IZU
	toluene	1977TA2

Polymer (B)	**Solvent (A)**	**Ref.**
Poly(*N*-cyclopropylacrylamide)		
	water	2001MA1
Poly(decyl methacrylate)		
	1-butanol	1983HER
	1-butanol	1986SAN
	cyclopentane	1983MA2
	n-heptane	1983MA2
	n-hexane	1983MA2
	n-pentane	1983MA2
	1-pentanol	1983HER
	1-pentanol	1986HER
	2-propanol	1986SAN
	toluene	1983MA2
	2,2,4-trimethylpentane	1983MA1
	2,2,4-trimethylpentane	1983MA2
	2,2,4-trimethylpentane	1997WOL
Poly(diacetone acrylamide-*co*-acrylamide)		
	water	1975TAY
Poly(diacetone acrylamide-*co*-hydroxyethyl acrylate)		
	water	1975TAY
Poly(*N,N*-diethylacrylamide)		
	water	1990MAR
	water	2002MA2
	water	2003MA2
Poly(*N,N*-diethylacrylamide-*co*-acrylamide)		
	water	1999LIU
Poly(*N,N*-diethylacrylamide-*co*-acrylic acid)		
	water	2001CAI
	water	2004MAE
Poly(*N,N*-diethylacrylamide-*co*-*N,N*-dimethyl-acrylamide)		
	water	1999LIU

Polymer (B)	Solvent (A)	Ref.
Poly(*N,N*-diethylacrylamide-*co*-*N*-ethylacrylamide)		
	water	1999LIU
Poly(*N,N*-diethylacrylamide-*co*-methacrylic acid)		
	water	2003LIU
Poly(*N,N*-dimethylacrylamide-*co*-glycidyl methacrylate)		
	water	2003YIN
Poly(*N,N*-dimethylacrylamide-*co*-2-methoxyethyl acrylate)		
	water	1996ELE
	water	1997ELE
Poly(*N,N*-dimethylacrylamide-*co*-*N*-phenylacrylamide)		
	water	1996MIY
Poly(*N,N*-dimethylaminoethyl methacrylate)		
	water	2006VER
Poly(*N,N*-dimethylaminoethyl methacrylate-*co*-*N*-vinylcaprolactam)		
	water	2006VER
Poly(2,6-dimethyl-1,4-phenylene oxide)		
	ε-caprolactam	1971SMO
	cyclohexanol	1995BE2
	cyclohexanol	1996BE2
	decahydronaphthalene	1978JAN
	decahydronaphthalene	1995BE2
	toluene	1972EM2
	toluene	1973EM1
	toluene	1973EM2

Polymer (B)	**Solvent (A)**	**Ref.**
Poly(dimethylsiloxane)		
	n-butane	1972ZE1
	2-butanone	1996VS1
	2-butanone	1996VS2
	4-cyano-4'-pentyl-biphenyl	2001GOG
	ethane	1972ZE1
	ethoxybenzene	1982SH1
	propane	1972ZE1
Poly(dimethylsiloxane-*b*-ethylene oxide-*b*-dimethylsiloxane)		
	water	1990MAA
Poly(dimethylsiloxane-*co*-methylphenylsiloxane)		
	anisole	2000SCH
	2-propanone	2000SCH
Poly(dimethylsiloxane-*b*-1,1,3,3-tetramethyl-disiloxanylethylene)		
	ethoxybenzene	2000AZU
Poly(1,3-dioxolane)		
	water	1992BEN
Poly(divinyl ether-*alt*-maleic anhydride)		
	water	2004VOL
	water	2005IZU
Polyetherimide		
	bisphenol-A diglycidyl ether	2004GIA
	1-methyl-2-pyrrolidinone	1992VIA
Polyethersulfone		
	dichloromethane	1980GHA
	trichloromethane	1979BLA
Poly[2-(2-ethoxy)ethoxyethyl vinyl ether]		
	water	2005MA2

Polymer (B)	Solvent (A)	Ref.
Poly[2-(2-ethoxy)ethoxyethyl vinyl ether]-*b*-(2-methoxyethyl vinyl ether)		
	deuterium oxide	2006OSA
Poly[*N*-(3-ethoxypropyl)acrylamide]		
	water	2005UGU
Poly(*N*-ethylacrylamide)		
	water	1975TAY
	water	2003XUE
Poly(*N*-ethylacrylamide-*co*-*N*-isopropyl-acrylamide)		
	water	1975TAY
Poly(ethyl acrylate)		
	1-butanol	1967LLO
	ethanol	1967LLO
	methanol	1967LLO
	1-propanol	1967LLO
Polyethylene		
	anisole	1966NA1
	benzyl acetate	1966NA1
	benzyl phenyl ether	1966NA1
	benzyl propionate	1966NA1
	biphenyl	1966NA1
	biphenyl	1966NA2
	biphenyl	1987SCH
	bis(2-ethylhexyl) sebacate	1996GOR
	n-butane	1963EHR
	n-butane	1975HOR
	n-butane	1994KI2
	butyl acetate	1974KUW
	4-*tert*-butylphenol	1966NA1
	cyclohexanone	1966NA1
	cyclopentane	1999BEY
	cyclopentane	2000BEY
	cyclopentene	1999BEY
	cyclopentene	2000BEY
	decanedioic acid dialkylester	1996GOR
	1-decanol	1966NA1

Polymer (B)	Solvent (A)	Ref.
Polyethylene		
	dibenzyl ether	1966NA1
	1,2-dichlorobenzene	2005MA1
	dichlorodifluoroethene	1975HOR
	dichlorodifluoromethane	1975HOR
	diethyl ether	1974NA2
	diisodecyl phthalate	2004MAT
	1,4-dimethylbenzene	1946RIC
	1,4-dimethylbenzene	1986VS2
	2,2-dimethylpentane	1984VAR
	2,4-dimethylpentane	1988BAR
	diphenyl ether	1966NA1
	diphenyl ether	1966NA2
	diphenyl ether	1967KO1
	diphenyl ether	1967KO2
	diphenyl ether	1967KO3
	diphenyl ether	1968KON
	diphenyl ether	1987SCH
	diphenyl ether	1993AER
	diphenylmethane	1966NA1
	diphenylmethane	1966NA2
	1-dodecanol	1966NA1
	1-dodecanol	1990CHI
	fluorotrichloromethane	1975HOR
	n-heptane	1967ORW
	n-heptane	1973HAM
	n-heptane	1975HOR
	n-heptane	1980KL1
	n-heptane	1996LOO
	1-heptanol	1966NA1
	n-hexane	1967ORW
	n-hexane	1969GOR
	n-hexane	1973HAM
	n-hexane	1975HOR
	n-hexane	1980KL1
	n-hexane	1989HAE
	n-hexane	1990KEN
	n-hexane	1996LOO
	1-hexanol	1966NA1
	2-methoxynaphthalene	1966NA1
	3-methylbutyl acetate	1966NA1
	2-methylpropane	1975HOR
	nitrobenzene	1946RIC

Polymer (B)	Solvent (A)	Ref.
Polyethylene		
	1-nonanol	1966NA1
	4-nonylphenol	1966NA1
	octamethylcyclotetrasiloxane	2002CH3
	n-octane	1967ORW
	n-octane	1973HAM
	n-octane	1975HOR
	n-octane	1980KL1
	n-octane	1996LOO
	1-octanol	1966NA1
	1-octanol	1983MUR
	4-octylphenol	1966NA1
	n-pentane	1963EHR
	n-pentane	1973HAM
	n-pentane	1975HOR
	n-pentane	1992KIR
	n-pentane	1993KIR
	n-pentane	1994KI2
	n-pentane	1998ZHU
	n-pentane	2001LI2
	n-pentane	2003ZH2
	1-pentanol	1966NA1
	pentyl acetate	1946RIC
	4-*tert*-pentylphenol	1966NA1
	perfluoro-2-butene	1975HOR
	phenetole	1966NA1
	styrene	2004LIS
	1,1,2-trichlorotrifluoroethane	1975HOR
Poly(ethylene-*co*-acrylic acid)		
	dioctyl phthalate	2005ZHO
	dioctyl phthalate	2006ZHO
Poly(ethylene-*co*-acrylic acid)-*g*-poly(ethylene glycol) monomethyl ether		
	dioctyl phthalate	2005ZHO
	dioctyl phthalate	2006ZHO
Poly(ethylene-*co*-1-butene)		
	n-heptane	1996LOO

Polymer (B)	**Solvent (A)**	**Ref.**
Poly(ethylene-*co*-1-hexene)		
	n-heptane	1996LOO
Poly(ethylene-*co*-methyl acrylate)		
	n-hexane	1994LOS
Poly(ethylene-*co*-4-methyl-1-pentene)		
	n-heptane	1996LOO
Poly(ethylene-*co*-1-octene)		
	cyclohexane	1996LOO
	n-heptane	1996LOO
	n-hexane	1996LOO
	2-methylpentane	1996LOO
Poly(ethylene-*co*-propylene)		
	2,4-dimethylpentane	1984VAR
	n-heptane	1996LOO
	n-hexane	1986IRA
	2-methylbutane	1986IRA
	methylcyclopentane	1986IRA
	3-methylpentane	1986IRA
	n-pentane	1986IRA
Poly(ethylene-*co*-vinyl acetate)		
	ε-caprolactam	1971SMO
	cyclopentane	1999BEY
	cyclopentane	2000BEY
	cyclopentene	1999BEY
	cyclopentene	2000BEY
	diphenyl ether	1990VAN
	2-heptanone	1986HAE
	tetraethoxysilane	2006CHA
Poly(ethylene-*co*-vinyl alcohol)		
	1,2,3-propanetriol	2003SH1
	1,2,3-propanetriol	2003SH2
	1,2,3-propanetriol	2003SH3
	2-propanol	1990CHE
	water	1971SHI

Polymer (B)	Solvent (A)	Ref.
Poly(ethylene glycol)		
	tert-butyl acetate	1976SA1
	water	1957MAL
	water	1970NAK
	water	1976SA1
	water	1977SAE
	water	1980MED
	water	1987KLE
	water	1994BOR
	water	1996FIS
	water	2004DOR
Poly(ethylene glycol-*b*-ε-caprolactone)		
	water	2006LUC
Poly(ethylene glycol-*b*-*N*-isopropylacrylamide)		
	water	2005MOT
	water	2005ZH3
Poly(ethylene glycol-*co*-*N*-isopropylacrylamide)		
	water	1997TOP
Poly(ethylene glycol) diacetyl ether		
	water	1980MED
Poly(ethylene glycol) didodecyl ether		
	water	1996FRA
Poly(ethylene glycol) dimethyl ether		
	4-ethoxybenzylidene-4'-butyl-aniline	1978KRO
	water	1980MED
	water	2004DOR
Poly(ethylene glycol) dioctadecyl ether		
	water	1996FRA
Poly(ethylene glycol) monomethyl ether		
	4-ethoxybenzylidene-4'-butyl-aniline	1978KRO
	water	2004DOR

Polymer (B)	Solvent (A)	Ref.
Poly(ethylene glycol) monododecyl ether		
	water	1980KUR
	water	1981HEU
	water	1988FUJ
	water	2005YAM
Poly(ethylene glycol) mono(p-nonylphenyl) ether		
	water	2003BAL
Poly(ethylene glycol) mono(p-octylphenyl) ether		
	water	1969GOL
Poly(ethylene glycol) trisiloxanemethyl ether		
	water	1993HEM
Poly(ethylene glycol-*b*-4-vinylpyridine-*b*-*N*-isopropylacrylamide)		
	water	2005ZH4
Poly(ethylene oxide)		
	deuterium oxide	2002HAM
	deuteroethanol	2006HOD
	ethanol	2006HOD
	4-ethoxybenzylidene-4'-*N*-butylaniline	1978KRO
	ethylbenzene	1979TAG
	toluene	1979TAG
	water	1959BAI
	water	1974TAG
	water	1978VSH
	water	1992COO
	water	1994BOR
	water	2002HAM
	water	2006OKA
Poly(ethylene oxide-*co*-alkylglycidyl ether)		
	water	2001LI1
Poly(ethylene oxide) biscarboxymethyl		
	water	2002RAC

Polymer (B)	Solvent (A)	Ref.
Poly(ethylene oxide-*b*-1,2-butylene oxide)		
	water	2001HAM
	water	2002CH1
	water	2002SON
	water	2004KEL
	water	2005CHA
Poly(ethylene oxide-*co*-1,2-butylene oxide)		
	water	2000SAH
Poly(ethylene oxide-*b*-1,2-butylene oxide-*b*-ethylene oxide)		
	water	2002SON
Poly[ethylene oxide-*co*-(dialkoxymethyl)-propylglycidyl ether]		
	water	2001LI1
Poly(ethylene oxide-*b*-dimethylsiloxane-*b*-ethylene oxide)		
	water	1990MAA
	water	1992YAN
Poly(ethylene oxide-*b*-L-lactide-*b*-ethylene oxide)		
	water	1999JEO
Poly[ethylene oxide-*b*-(DL-lactide-*co*-glycolide)-*b*-ethylene oxide]		
	water	2004PA1
Poly(ethylene oxide-*b*-propylene oxide)		
	formamide	1990SAM
	water	1990SAM
	water	1990SIM

Polymer (B)	Solvent (A)	Ref.
Poly(ethylene oxide-*co*-propylene oxide)		
	acetic acid	1993JOH
	butanoic acid	1993JOH
	propionic acid	1993JOH
	water	1981MED
	water	1987ANA
	water	1990SAM
	water	1991LOU
	water	1993JOH
	water	1994ZH1
	water	1999JOH
	water	1999PER
	water	2000PE2
	water	2003CAM
Poly(ethylene oxide-*b*-propylene oxide-*b*-ethylene oxide)		
	formamide	1990SAM
	water	1990SAM
	water	1990SIM
	water	1992MAL
	water	1995ZHA
	water	2000LAM
	water	2001DES
	water	2002DES
	water	2004VAR
Poly(ethylene oxide-*b*-styrene oxide)		
	water	2003YAN
Poly(2-ethylhexyl acrylate)		
	4-cyano-4'-pentyl-biphenyl	2003SLI
Poly(ethyl methacrylate)		
	1-butanol	1996COO
	1-decanol	1998BE1
Poly(2-ethyl-2-oxazoline)		
	water	1988LIN
	water	2003CHR
Poly(2-ethyl-2-oxazoline-*b*-ε-caprolactone)		
	water	2000KIM

Polymer (B)	Solvent (A)	Ref.
Poly(glycidol-*b*-propylene oxide-*b*-glycidol)	water	2006HAL
Poly(p-hexylstyrene)	2-butanone	1987MAG
Poly(2-hydroxyethyl acrylate-*co*-hydroxypropyl acrylate)	water	1975TAY
Poly(2-hydroxyethyl acrylate-*co*-vinyl butyl ether)	water	2006MUN
Poly(2-hydroxyethyl methacrylate-*g*-ethylene glycol)	water	2005ZH1
Poly(2-hydroxyethyl methacrylate)-b-(*N*-isopropyl-acrylamide)	water	2006CAO
Poly(*N*-(1-hydroxymethyl)propylmethacrylamide)	water	2005SET
Poly(hydroxypropyl acrylate)	water	1975TAY
Polyisobutylene	benzene	1970LID
	n-butane	1972ZE1
	diisobutyl ketone	1952SHU
	n-hexane	1970LID
	n-hexane	1972ZE1
	2-methylbutane	1965ALL
	2-methylbutane	1972ZE1
	n-octane	1970LID
	n-pentane	1962BAK
	n-pentane	1970LID
	n-pentane	1972ZE1
	n-pentane	1997WOL
	n-pentane	2002JOU
	propane	1972ZE1
	toluene	1981COH

Polymer (B)	**Solvent (A)**	**Ref.**
Poly(isobutyl vinyl ether-*co*-2-hydroxyethyl vinyl ether)		
	water	2004SUG
Polyisoprene		
	1,4-butanediyl methacrylate	1985ZHI
	1,4-dioxane	1985TAK
Poly(*N*-isopropylacrylamide)		
	deuterium oxide	2004MAO
	deuterium oxide	2004SHI
	water	1989FUJ
	water	1989IN1
	water	1990INO
	water	1990MAR
	water	1993OTA
	water	1997BOU
	water	1997KUN
	water	1997SU1
	water	1998ZHE
	water	1999KUN
	water	2000AFR
	water	2001MA1
	water	2001YA2
	water	2003BAD
	water	2003MAO
	water	2003MA2
	water	2003MIL
	water	2003STI
	water	2004DU2
	water	2004MAO
	water	2004SAN
	water	2004SHI
	water	2005DUR
	water	2005STA
	water	2005TIE
	water	2006DUA
	water	2006KUJ
	water	2006OKA
	water	2006PLU
	water	2006XIA

Polymer (B)	Solvent (A)	Ref.
Poly(*N*-isopropylacrylamide-*co*-acrylamide)	water	1987PRI
Poly(*N*-isopropylacrylamide-*co*-6-acrylaminohexanoic acid)	water	2000KUC
Poly(*N*-isopropylacrylamide-*co*-3-acrylaminopropanoic acid)	water	2000KUC
Poly(*N*-isopropylacrylamide-*co*-11-acrylaminoundecanoic acid)	water	2000KUC
Poly(*N*-isopropylacrylamide-*co*-acrylic acid)	water	1993OTA
	water	1997YOO
	water	2000BO1
	water	2000YAM
	water	2004MAE
	water	2006WEN
Poly(*N*-isopropylacrylamide-*co*-acrylic acid-*co*-ethyl methacrylate)	water	2005TIE
Poly(*N*-isopropylacrylamide-*co*-acryloyloxypropylphosphinic acid)	water	2004NON
Poly(*N*-isopropylacrylamide-*co*-*N*-butylacrylamide)	water	1987PRI
Poly(*N*-isopropylacrylamide-*co*-*N*-*tert*-butylacrylamide)	water	1987PRI

Polymer (B)	Solvent (A)	Ref.
Poly(*N*-isopropylacrylamide-*co*-butyl acrylate)	water	2003MA1
Poly(*N*-isopropylacrylamide-*co*-butyl methacrylate)	water	1993FEI
Poly(*N*-isopropylacrylamide-*co*-1-deoxy-1-methacryl-amide-D-glucitol)	water	2001GOM
Poly(*N*-isopropylacrylamide-*co*-*N,N*-dimethyl-aminopropylmethacrylamide)	water	2000BO2
Poly(*N*-isopropylacrylamide-*co*-*N*-ethylacrylamide)	water	1987PRI
Poly(*N*-isopropylacrylamide-*co*-ethyl acrylate)	water	2003MA1
Poly(*N*-isopropylacrylamide-*b*-ethylene oxide)	water	2004NED
Poly(*N*-isopropylacrylamide-*g*-ethylene oxide)	water	2005BIS
Poly[*N*-isopropylacrylamide-*co*-(2-hydroxyisopropyl)acrylamide]	water	2006MAE
Poly(*N*-isopropylacrylamide-*co*-*N*-hydroxymethylacrylamide)	water	2006DIN
Poly(*N*-isopropylacrylamide-*co*-3H-imidazole-4-carbodithioic acid 4-vinylbenzyl ester)	water	2005CAR

Polymer (B)	**Solvent (A)**	**Ref.**
Poly(*N*-isopropylacrylamide-*co*-*N*-isopropyl-methacrylamide)	water	2005STA
Poly(*N*-isopropylacrylamide-*co*-maleic acid)	water	2004WEI
Poly[*N*-isopropylacrylamide-*co*-(p-methacryl-amido)-acetophenone thiosemicarbazone]	water	2005LIC
Poly(*N*-isopropylacrylamide-*co*-methacrylic acid)	water	2006YIN
Poly(*N*-isopropylacrylamide-*co*-*N*-methacryloyl-L-leucine)	water	2000BIG
Poly(*N*-isopropylacrylamide-*co*-8-methacryloyl-oxyoctanoic acid methyl ester)	water	2004SAL
Poly(*N*-isopropylacrylamide-*co*-5-methacryloyl-oxypentanoic acid methyl ester)	water	2004SAL
Poly(*N*-isopropylacrylamide-*co*-11-methacryloyl-oxyundecanoic acid methyl ester)	water	2004SAL
Poly[*N*-isopropylacrylamide-*co*-methoxy-poly(ethylene glycol) monomethacrylate]	water	2006KI1
	water	2006KI2
Poly(*N*-isopropylacrylamide-*co*-*N*-methylacryl-amide)	water	1987PRI

Polymer (B)	Solvent (A)	Ref.
Poly(*N*-isopropylacrylamide-*co*-methyl acrylate)	water	2003MA1
Poly(*N*-isopropylacrylamide-*co*-octadecyl acrylate)	water	2000SHI
Poly(*N*-isopropylacrylamide-*co*-4-pentenoic acid)	water	1999KUN
Poly(*N*-isopropylacrylamide-*co*-*N*-propyl-acrylamide)	water	2004MAO
Poly(*N*-isopropylacrylamide-*co*-propylacrylic acid)	water	2006YIN
Poly(*N*-isopropylacrylamide-*b*-propylene oxide-*b*-*N*-isopropylacrylamide)	water	2004HAS
Poly(*N*-isopropylacrylamide-*co*-vinyl acetate)	water	2003MA1
Poly(*N*-isopropylacrylamide-*co*-1-vinylimidazole)	water	2005BIS
Poly(*N*-isopropylacrylamide-*co*-vinyl laurate)	water	2005CAO
Poly(*N*-isopropylacrylamide-*co*-p-vinyl-phenylboronic acid)	water	2005CIM
Poly(*N*-isopropylacrylamide-*co*-1-vinyl-2-pyrrolidinone)	water	2006GEE

Polymer (B)	**Solvent (A)**	**Ref.**
Poly(*N*-isopropylmethacrylamide)		
	water	1989FUJ
	water	2001MA2
	water	2005SP2
	water	2005STA
	water	2006KIR
Poly(2-isopropyl-2-oxazoline) (end functionalized)		
	water	2004PA2
Poly(DL-lactide)		
	dibutyl phthalate	2003LEE
	dibutyl phthalate	2004LE1
	dihexyl phthalate	2003LE1
	dihexyl phthalate	2004LE1
	dipentyl phthalate	2003LE1
	dipentyl phthalate	2004LE1
Poly(L-lactide)		
	dihexyl phthalate	2003LEE
	dihexyl phthalate	2004LE1
	dipentyl phthalate	2003LEE
	dipentyl phthalate	2004LE1
Poly(L-lysine isophthalamide)		
	water	2005YUE
Poly(maleic anhydride-*alt*-*tert*-butylstyrene)-*g*-poly(ethylene glycol) monomethyl ether		
	water	2002YIN
Poly(maleic anhydride-*alt*-diethylene glycol)		
	styrene	1992LEC
Poly(maleic anhydride-*alt*-neopentyl glycol)		
	styrene	1992LEC

Polymer (B)	**Solvent (A)**	**Ref.**
Poly(maleic anhydride-*alt*-styrene)-*g*-poly(ethylene glycol) monomethyl ether	water	2002YIN
Poly(methacrylic acid)	water	1974TAN
Poly[methacrylic acid-*co*-butyl methacrylate-*co*-poly(ethylene glycol) monomethyl ether methacrylate]	water	2005JON
Poly(methacrylic acid-*co*-glycidyl methacrylate-*co*-poly(ethylene glycol) monomethyl ether methacrylate)	water	2005JON
Poly(methacrylic acid-*co*-lauryl methacrylate-*co*-poly(ethylene glycol) monomethyl ether methacrylate)	water	2005JON
Poly[methoxydi(ethylene glycol) acrylate]	water	2006HU1
Poly[methoxypoly(ethylene glycol)-*b*-ε-caprolactone]	water	2006KI2
Poly[methoxytri(ethylene glycol) acrylate-*b*-4-vinylbenzyl methoxytris(oxyethylene) ether]	water	2006HU2
Poly(methyl acrylate-*co*-methyl methacrylate-*co*-acrylic acid)	octanoic acid	1995ADA

Polymer (B)	Solvent (A)	Ref.
Poly(methyl methacrylate)		
	acetontrile	1962FOX
	acetonitrile	1976CO1
	acetonitrile	1981HOR
	acetonitrile	1996VS3
	bis(2-chloropropyl) 2-ethylhexyl phosphate	1986VS1
	bis(1,3-dichloroisopropyl) isodecyl phosphate	1986VS1
	bromoethyl bromopropyl phosphate	1986VS1
	1-butanol	1978VSH
	1-butanol	1996COO
	chloroethyl phosphate	1986VS1
	1-butanol	1950JE2
	1-butanol	1991VAN
	1-butanol	1994ARN
	1-butanol	1994VA2
	2-butanone	1976CO1
	1-chlorobutane	1976CO1
	1-chlorobutane	1981HOR
	1-chlorobutane	1988LEE
	cyclohexanol	1991VAN
	cyclohexanol	2002MA4
	4-cyano-4'-heptylbiphenyl	1992AHN
	1-decanol	1998BE1
	2,2-dimethyl-3-pentanone	1962FOX
	2,4-dimethyl-3-pentanone	1962FOX
	1,2-ethanediol	1998CHA
	ethyl acetate	1981VSH
	ethyl acetate	1996VS3
	ethyl acetate	2006RUS
	2-ethylbutanal	1962FOX
	3-heptanone	1986HER
	4-heptanone	1962FOX
	4-heptanone	1976CO1
	3-hexanone	1976CO1
	isopropylbenzene	1979BUR
	methanol	2000MOE
	methyl acetate	1976CO1
	1-methyl-4-isopropylbenzene	1950JE2
	2-methyl-1-propanol	1990TS2
	2-octanone	1962FOX

Polymer (B)	Solvent (A)	Ref.
Poly(methyl methacrylate)		
	3-octanone	1962FOX
	4'-octyl-4-biphenylcarbonitrile	2006CRA
	organic phosphates	1982TA1
	3-pentanone	1976CO1
	1-propanol	1950JE2
	2-propanone	1976CO1
	sulfolane	1985CAN
	sulfolane	1986CAN
	sulfolane	1990TS1
	tetra(ethylene glycol)	1998CHA
	toluene	1950JE2
	trichloromethane	1950JE2
	tri(ethylene glycol)	1998CHA
	tris-(2-chloroethyl) phosphate	1982TA2
Poly(methyl methacrylate) (isotactic)		
	1-butanol	1990BRO
	2-butanone	1990BRO
	1,2-dimethylbenzene	1990BRO
Poly(methyl methacrylate-*co*-methacrylic acid)		
	dibutyl phthalate	1991TA1
	didodecyl phthalate	1991TA1
	dioctyl phthalate	1991TA1
	tricresyl phosphate	1991TA1
Poly(4-methyl-1-pentene)		
	biphenyl	1973TAN
	diisopropylbenzene	1987WIL
	diphenyl ether	1973TAN
	diphenylmethane	1973TAN
Poly(methylphenylsiloxane)		
	4-cyano-4'-heptyl-biphenyl	2003GOG
	4-cyano-4'-octyl-biphenyl	2003GOG
	4-cyano-4'-pentyl-biphenyl	2003GOG

Polymer (B)	Solvent (A)	Ref.
Poly(α-methylstyrene)		
	1-chlorobutane	1975COW
	cyclohexane	1975COW
	cyclohexane	1991LEE
	cyclopentane	1986SA1
	methylcyclohexane	1975COW
	methylcyclohexane	1983SAE
	methylcyclohexane	1993ZHE
	propylene oxide	1975COW
Poly(2-methyl-5-vinylpyridine)		
	butyl acetate	1965GEC
	4-methyl-2-pentanone	1965GEC
Poly(2-methyl-5-vinyltetrazole)		
	formamide	1997KIZ
Poly(nonyl acrylate-*co*-2-methyl-5-vinyltetrazole)		
	nonyl acrylate	2002MIK
Poly(3,6,9,12,15,18,21,24-octaoxatriacontanyl acrylate)		
	water	1987JAH
Polyol (hyperbranched)		
	water	1979SIM
	water	2001JAN
Poly[4-(olgiooxyethylene)styrene]		
	water	2006HU1
Polypeptide		
	water	2000TAM
	water	2003YA2
	water	2004MEY
Poly[perfluoroalkylacrylate-*co*-poly(ethylene oxide) methacrylate]		
	water	2006SHA

Polymer (B)	**Solvent (A)**	**Ref.**
Poly(2-phenoxyethyl acrylate)		
	4-cyano-4'-pentylbiphenyl	2001BOU
Poly(*N*-propylacrylamide)		
	water	1990INO
	water	2001MA1
	water	2004MAO
	water	2005VA1
Polypropylene		
	benzyl acetate	1968NA1
	4-benzylphenol	1968NA1
	benzyl phenyl ether	1968NA1
	benzyl propionate	1968NA1
	biphenyl	1968NA1
	N,N-bis(2-hydroxyethyl)-tallowamine	1991LLO
	N,N-bis(2-hydroxyethyl)-tallowamine	1992KIM
	1-butanol	1968NA1
	butyl acetate	1993VS2
	4-*tert*-butylphenol	1968NA1
	chlorobenzene	1976TA2
	decahydronaphthalene	1963RED
	n-decane	1963RED
	1-decene	1963RED
	dibenzyl ether	1968NA1
	dibutyl phthalate	1998LEE
	dibutyl phthalate	2006YA1
	1,2-dichlorobenzene	1963RED
	diethyl ether	1974CO3
	diheptyl phthalate	1992LEE
	diheptyl phthalate	1998LEE
	dihexyl phthalate	1992LEE
	dihexyl phthalate	1998LEE
	1,4-dimethylbenzene	1963RED
	dioctyl phthalate	1992LEE
	dioctyl phthalate	2006YA1
	diphenyl ether	1963RED
	diphenyl ether	1968NA1
	diphenyl ether	1994LAX
	diphenyl ether	1994MCG

Polymer (B)	Solvent (A)	Ref.
Polypropylene		
	diphenyl ether	1995MCG
	diphenylmethane	1968NA1
	4-ethylphenol	1968NA1
	n-heptane	1974CO3
	n-hexane	1974CO3
	4-isooctylphenol	1968NA1
	3-methylbutyl acetate	1963RED
	3-methylbutyl acetate	1993VS2
	3-methylbutyl benzyl ether	1968NA1
	4-methylphenol	1968NA1
	2-methyl-1-propanol	1968NA1
	methyl salicylate	2000MA1
	2-naphthol	1968NA1
	4-octylphenol	1968NA1
	n-pentane	1974CO3
	n-pentane	1998KI2
	phenetol	1963RED
	phenol	1968NA1
	phenyl acetate	1963RED
	2-phenylphenol	1968NA1
	4-phenylphenol	1968NA1
	1,2,4-trichlorobenzene	1963RED
Poly(propylene glycol)		
	n-decane	1986FIR
	n-dodecane	1986FIR
	n-hexane	1972LIR
	n-hexane	1972TA1
	n-hexane	1976VSH
	n-hexane	1977VSH
	n-hexane	1986FIR
	n-hexane	1986VS3
	n-octane	1986VS3
	n-octane	1986FIR
	water	1957MAL
	water	1969KUZ
	water	1972TA1
	water	1974BES
	water	1974TAN
	water	1975TA3
	water	1975TA4
	water	1977VSH

Polymer (B)	Solvent (A)	Ref.
Poly(propylene glycol)		
	water	1979SIM
	water	1980BIL
	water	1980VSH
	water	1986FIR
	water	1986VS3
	water	1988ZGA
	water	1996GAL
	water	2001KIN
Poly(propylene oxide)		
	n-pentane	1965ALL
	propane	1972ZE2
Poly(propylene oxide-*b*-ethylene oxide-*b*-propylene oxide)		
	water	2004DER
Poly(*N*-propylmethacrylamide)		
	water	2001MA2
Polystyrene		
	acetaldehyde	1997REB
	benzene	1970AND
	benzene	1973SA2
	benzene	1981VSH
	bis(2-ethylhexyl) phthalate	1974VER
	bis(2-ethylhexyl) phthalate	1980WOL
	bis(2-ethylhexyl) phthalate	1986VS2
	bis(2-ethylhexyl) phthalate	1994VSH
	bisphenol-A diglycidyl ether	2004ZUC
	bisphenol-A diglycidyl ether	2005RIC
	bisphenol-A diglycidyl ether	2006RIC
	butanedioic acid dimethyl ester	1950JE1
	1-butanol	1950JE1
	1-butanol	1983HER
	2-butanone	1973BAB
	2-butanone	1973SA2
	2-butanone	1974NA1
	2-butanone	1990KYO

Polymer (B)	Solvent (A)	Ref.
Polystyrene		
	tert-butyl acetate	1976SA2
	tert-butyl acetate	1984SCH
	tert-butyl acetate	1991VSH
	tert-butyl acetate	1997VSH
	butyl stearate	1950JE1
	carbon disulfide	1983TAN
	carbon disulfide	1986FRA
	1-chlorobutane	1976CO2
	1-chlorododecane	1991OPS
	1-chlorohexadecane	1991OPS
	1-chlorooctadecane	1991OPS
	1-chlorotetradecane	1991OPS
	4-cyano-4'-heptylbiphenyl	1992AHN
	4-cyano-4'-heptylbiphenyl	1998RIC
	4-cyano-4'-octylbiphenyl	2006DEM
	cyclodecane	1986CO1
	cycloheptane	1986CO1
	cyclohexane	1951FO2
	cyclohexane	1952SHU
	cyclohexane	1960DE2
	cyclohexane	1962DEB
	cyclohexane	1962HAM
	cyclohexane	1967SCH
	cyclohexane	1967REH
	cyclohexane	1968RE1
	cyclohexane	1968RE2
	cyclohexane	1968TAG
	cyclohexane	1970KON
	cyclohexane	1971BOR
	cyclohexane	1971KUW
	cyclohexane	1972BOR
	cyclohexane	1972SCH
	cyclohexane	1972YAK
	cyclohexane	1973CAN
	cyclohexane	1973KRA
	cyclohexane	1973KU1
	cyclohexane	1973KU2
	cyclohexane	1973SA1
	cyclohexane	1974DER
	cyclohexane	1975KOJ
	cyclohexane	1975STR
	cyclohexane	1975SAE

Polymer (B)	Solvent (A)	Ref.
Polystyrene		
	cyclohexane	1976CO2
	cyclohexane	1976TA1
	cyclohexane	1976VSH
	cyclohexane	1977GOR
	cyclohexane	1977RIG
	cyclohexane	1977TA1
	cyclohexane	1977WO3
	cyclohexane	1978VSH
	cyclohexane	1979COW
	cyclohexane	1980HOS
	cyclohexane	1980IRV
	cyclohexane	1980NAK
	cyclohexane	1980SWI
	cyclohexane	1981VSH
	cyclohexane	1981WOL
	cyclohexane	1982KAJ
	cyclohexane	1984EIN
	cyclohexane	1984KAM
	cyclohexane	1984TAK
	cyclohexane	1984TSU
	cyclohexane	1985EIN
	cyclohexane	1986VS3
	cyclohexane	1987IC1
	cyclohexane	1989HAE
	cyclohexane	1990HEI
	cyclohexane	1990RA2
	cyclohexane	1991SAF
	cyclohexane	1991VSH
	cyclohexane	1991YOK
	cyclohexane	1992HE2
	cyclohexane	1993VS1
	cyclohexane	1993VS2
	cyclohexane	1994SON
	cyclohexane	1995IKI
	cyclohexane	1995VSH
	cyclohexane	1997WOL
	cyclohexane	1999HOO
	cyclohexane	2000KOI
	cyclohexanol	1950JE1
	cyclohexanol	1953SHU
	cyclohexanol	1988HIK
	cyclohexanol	1995SON

Polymer (B)	Solvent (A)	Ref.
Polystyrene		
	cyclohexanol	1999RUD
	cyclooctane	1986CO1
	cyclopentane	1965ALL
	cyclopentane	1973SA2
	cyclopentane	1978ISH
	cyclopentane	1981WOL
	cyclopentane	1986SA1
	cyclopentane	1990IWA
	cyclopentane	1993IWA
	cyclopentane	1997WOL
	decahydronaphthalene	1973CAN
	decahydronaphthalene	1975AND
	decahydronaphthalene	1976ALE
	decahydronaphthalene	1976AND
	decahydronaphthalene	1977AND
	decahydronaphthalene	1977KOB
	decahydronaphthalene	1977TA1
	decahydronaphthalene	1987ARN
	decahydronaphthalene	1991SAF
	decahydronaphthalene	1981VSH
	decahydronaphthalene	1991VSH
	trans-decahydronaphthalene	1975AND
	trans-decahydronaphthalene	1976NAK
	trans-decahydronaphthalene	1977WO1
	trans-decahydronaphthalene	1977WO2
	trans-decahydronaphthalene	1988KR1
	trans-decahydronaphthalene	1988KR2
	trans-decahydronaphthalene	1984WOL
	trans-decahydronaphthalene	1985WOL
	trans-decahydronaphthalene	1993ARN
	trans-decahydronaphthalene	2003JI1
	trans-decahydronaphthalene	2004JIA
	n-decane	2001IM2
	n-decane	2002IMR
	diethyl ether	1972SIO
	diethyl ether	1976SA2
	diethyl ether	1997WOL
	diethyl malonate	1950JE1
	diethyl malonate	1973CAN
	diethyl malonate	1979HAM
	diethyl malonate	1981WO

Polymer (B)	Solvent (A)	Ref.
Polystyrene		
	diethyl malonate	1988TVE
	diethyl malonate	1994SON
	diethyl oxalate	1950JE1
	diethyl oxalate	1986SA2
	diethyl oxalate	1994SAT
	diisodecyl phthalate	1982NOJ
	dimethoxymethane	1972SIO
	1,4-dimethylcyclohexane	1984CO3
	1,4-dimethylcyclohexane	1999IM1
	1,4-dimethylcyclohexane	2002IMR
	dimethyl malonate	1950JE1
	dimethyl oxalate	1950JE1
	dioctyl phthalate	1984RAN
	dioctyl phthalate	2006LIC
	dodecadeuterocyclohexane	1975STR
	dodecadeuterocyclohexane	1987IC2
	dodecadeuterocyclohexane	1987MEL
	dodecadeuterocyclohexane	1989MEL
	n-dodecane	2001IM2
	n-dodecane	2002IMR
	1-dodecanol	2001HE2
	dodecyl acetate	2000IMR
	4-ethoxybenzylidene-4'-*N*-butylaniline	1978KRO
	ethyl acetate	1950JE1
	ethyl acetate	1976CO2
	ethyl acetate	1981VSH
	ethyl acetate	2006RUS
	ethylbenzene	1970AND
	ethylbenzene	1977TA1
	ethylbenzene	1981VSH
	ethylcyclohexane	1963DEB
	ethylcyclohexane	1977KRA
	ethyl formate	1975KON
	ethyl formate	1997IMR
	n-heptane	1983CO1
	n-hexadecane	1991OPS
	n-hexadecane	2006LIC
	1,1,1,3,3,3-hexadeutero-2-propanone	1991SZY
	n-hexane	1983CO1

Polymer (B)	Solvent (A)	Ref.
Polystyrene		
	n-hexane	1997XIO
	hexanoic acid	1950JE1
	1-hexanol	1950JE1
	3-hexanol	1950JE1
	methyl acetate	1950JE1
	methyl acetate	1965MYR
	methyl acetate	1970DEL
	methyl acetate	1972ZE2
	methyl acetate	1982ZRI
	methyl acetate	1983KUB
	methyl acetate	1993WAK
	methyl acetate	1997IMR
	3-methyl-1-butanol	1950JE1
	methylcyclohexane	1971KAG
	methylcyclohexane	1973SA1
	methylcyclohexane	1976NOS
	methylcyclohexane	1980DO2
	methylcyclohexane	1980NAK
	methylcyclohexane	1982SH2
	methylcyclohexane	1983SAE
	methylcyclohexane	1988KI1
	methylcyclohexane	1991CHU
	methylcyclohexane	1991KIE
	methylcyclohexane	1992HE1
	methylcyclohexane	1993DOB
	methylcyclohexane	1993HOS
	methylcyclohexane	1993WEL
	methylcyclohexane	1994VA1
	methylcyclohexane	1996DOB
	methylcyclohexane	1996IM1
	methylcyclohexane	1997END
	methylcyclohexane	1998SZY
	methylcyclohexane	1999HOO
	methylcyclohexane	2000XIO
	methylcyclohexane	2002SHR
	methylcyclopentane	1984CO3
	methylcyclopentane	1991SZY
	4,4'-methylene-bis(2,6-diethylaniline)	2006RIC
	nitroethane	2000DE2
	nitromethane	1987CO2
	1-nitropropane	1983TAN

Polymer (B)	Solvent (A)	Ref.
Polystyrene		
	n-octadecane	1991OPS
	1-octadecanol	1948JEN
	1-octadecanol	1950JE1
	1-octadecanol	1991OPS
	n-octane	1983CO1
	n-octane	2001IM2
	n-octane	2002IMR
	1-octene	1950JE1
	n-pentane	1972HOR
	n-pentane	1988KI1
	n-pentane	1988KI2
	n-pentane	1988KI3
	n-pentane	1991KIE
	n-pentane	1994KI1
	n-pentane	2001LI2
	1-pentanol	1983HER
	1-phenyldecane	1981WOL
	1-phenyldecane	1982GEE
	1-phenyldecane	1997WOL
	2-propanone	1972SIO
	2-propanone	1972ZE2
	2-propanone	1991SZY
	2-propanone	1993REB
	2-propanone	1999IM1
	2-propanone	2006FAN
	propionitrile	1998IMR
	propyl acetate	1974SAE
	2-propyl acetate	1974SAE
	n-tetradecane	2001IM2
	n-tetradecane	2002IMR
	toluene	1973SA1
	toluene	1993VS1
	vinyl acetate	1950JE1
Poly(styrene-*co*-acrylonitrile)		
	2-butanone	1968TER
	cyclohexane	1968TER
	ethyl acetate	1982MAN
	toluene	1972TE1
	toluene	1972TE2
	toluene	2000SCH
	toluene	2002WOL
	toluene	2003LOS

Polymer (B)	**Solvent (A)**	**Ref.**
Poly(styrene-*b*-butyl methacrylate)		
	dioctyl phthalate	2006LIC
	n-hexadecane	2006LIC
	2-propanol	1994SIQ
Poly(styrene-*co*-butyl methacrylate)		
	2-butanone	1990KYO
Poly(styrene-*g*-cellulose diacetate)		
	2-propanone	1982GOL
	N,N-dimethylformamide	1982GOL
	tetrahydrofuran	1982GOL
Poly(styrene-*b*-isoprene)		
	dibutyl phthalate	2002LOD
	diethyl phthalate	2002LOD
	dimethyl phthalate	2002LOD
	propane	2006WIN
Poly(styrene-*alt*-maleic anhydride)		
	water	2006QIU
Polysulfone		
	dimethylsulfoxide	1991MIK
	dimethylsulfoxide	1994MIK
Polytetrafluoroethylene		
	perfluorocarbon ether	1995TUM
	perfluorodecalin	1995TUM
	tetradecafluorohexane	1995TUM
	1,1,2-trichloro-1,2,2-trifluoroethane	1995TUM
Poly(trimethylene oxide)		
	cyclohexane	1976CHI
Poly(urethane)		
	N,N-dimethylformamide	1972YAK

Polymer (B)	Solvent (A)	Ref.
Poly(urethane amine)		
	water	2005IHA
Poly(*N*-vinylacetamide-*co*-acrylic acid)		
	water	2006MOR
Poly(*N*-vinylacetamide-*co*-methyl acrylate)		
	water	2004MOR
Poly(*N*-vinylacetylamide-*co*-vinyl acetate)		
	water	2003SET
Poly(*N*-vinylacetamide-*co*-*N*-vinylisobutyramide)		
	water	1997SU2
Poly(vinyl acetate)		
	cyclopentane	1999BEY
	cyclopentane	2000BEY
	cyclopentene	1999BEY
	cyclopentene	2000BEY
	ethanol	1977KOB
	3-heptanone	1980ISH
	methanol	1968TAG
	methanol	1972TA2
	methanol	1975TA2
	methanol	1977ALE
	2-propanol	1982ZRI
	tetrachchloromethane	1991SAF
Poly(vinyl acetate-*co*-vinyl alcohol)		
	water	1974AND
	water	1978NIK
	water	1997PAE
Poly(vinyl acetate-*co*-*N*-vinylcaprolactam)		
	water	1993PAS
Poly(vinyl acetate-*co*-1-vinyl-2-pyrrolidinone)		
	water	1974TAN

Polymer (B)	Solvent (A)	Ref.
Poly(vinyl alcohol)		
	1,2-ethanediol	1963REH
	1,2-ethanediol	1991STO
	1,5-pentanediol	1991STO
	water	1963REH
	water	1969AND
	water	1971TAG
	water	1974AND
	water	1978NIK
	water	1985FUR
	water	1997PAE
Poly(vinyl alcohol) acetal		
	water	1975TAY
Poly(vinyl alcohol-*co*-vinyl butyrate)		
	water	1974TAN
	water	1984SHI
Poly(*N*-vinylamine-*co*-*N*-vinylcaprolactam)		
	water	1991TA2
	water	1994TAG
Poly(1-vinyl-5-aminotetrazole)		
	formamide	1997KIZ
Poly(*N*-vinylcaprolactam)		
	water	1990TAG
	water	1991TA2
	water	1993TA2
	water	1994TAG
	water	1999LAU
	water	2000ME2
	water	2002MA1
	water	2004DU1
	water	2006VER
Poly(*N*-vinylcaprolactam-*co*-acrylic acid)		
	water	2003SHT

Polymer (B)	Solvent (A)	Ref.
Poly(*N*-vinylcaprolactam-*co*-methacrylic acid)		
	water	2002MAK
Poly(*N*-vinylcaprolactam)-*g*-poly(ethylene oxide)		
	water	2004DU1
	water	2005KJO
Poly(*N*-vinylcaprolactam-*co*-*N*-vinyl-*N*-methylacetamide)		
	water	1993PAS
Poly(vinyl chloride)		
	γ-butyrolactone	1989BOO
	cyclohexanol	1996SOE
	dibutyl phthalate	1983TAG
	dimethyl phthalate	2002SAF
	diphenyl 4-*tert*-butylphenyl phosphate	1980SUV
	nitrobenzene	1994SOE
	pentachlorodiphenyl	1980SUV
	propylene carbonate	1996SOE
	tri(4-*tert*-butylphenyl) phosphate	1980SUV
	tricresyl phosphate	1983TAG
	triphenyl phosphate	1980SUV
Poly(*N*-vinyl-3,6-dibromo carbazole)		
	bromobenzene	1984BAR
	2-chlorophenol	1984BAR
	4-chloro-3-methylphenol	1984BAR
	1,2-dichlorobenzene	1984BAR
Poly(vinyl ethyl ether)		
	water	1966NAG
Poly(*N*-vinylformamide-*co*-vinyl acetate)		
	water	2003SET
Poly(vinylidene fluoride)		
	tetra(ethylene glycol) dimethyl ether	2000HON

Polymer (B)	Solvent (A)	Ref.
Poly(*N*-vinylisobutyramide-*co*-*N*-vinylamine)		
	water	2000KUN
Poly(*N*-vinylisobutyramide)		
	water	1997KUN
	water	1997SU1
	water	1999KUN
Poly(*N*-vinylisobutyramide-*co*-*N*-vinylvaleramide)		
	water	1997SU2
Poly(vinyl methyl ether)		
	deuterium oxide	2005VA2
	water	1988TAN
	water	1997SC1
	water	1998MOE
	water	2000ME1
	water	2001MA3
	water	2001YA1
	water	2003SWI
	water	2005VA1
	water	2006VAN
Poly(4-vinylphenol)		
	N,N-dimethyloctadecylamine	1999AKI
Poly(*N*-vinyl-*N*-propylacetamide)		
	water	1991TA2
	water	1994TAG
Poly(1-vinylpyrazole)		
	toluene	1991SAF
Poly(1-vinyl-1,2,3-triazole)		
	water	2003TSY
Poly(1-vinyl-1,2,4-triazole)		
	water	2003TSY
Poly(vinyltrimethylsilane)		
	dichloromethane	1978IVA

3. LIQUID-LIQUID EQUILIBRIUM DATA OF TERNARY POLYMER SOLUTIOS

3.1. Cloud-point and/or coexistence data of one polymer and two solvents

Polymer (B): **cellulose** **2004ECK**
Characterization: M_n/g.mol^{-1} = 60600, M_w = 203000, M_z = 499000, Solucell 500, Lenzing, Austria
Solvent (A): ***N,N*-dimethylacetamide** C_4H_9NO **127-19-5**
Solvent (C): **2-propanone** C_3H_6O **67-64-1**

Comments: LiCl is added to *N,N*-dimethylacetamide for solving cellulose.

Type of data: cloud points

T/K = 298.15

w_A	0.4059	0.4449	0.4728	0.5007	0.5307	0.5755
w_B	0.0003	0.0002	0.0004	0.0005	0.0006	0.0009
w_C	0.5938	0.5549	0.5268	0.4987	0.4687	0.4236

Type of data: coexistence data (tie lines)

T/K = 298.15

Total system			Sol phase			Gel phase		
w_A	w_B	w_C	w_A	w_B	w_C	w_A	w_B	w_C
0.4167	0.0007	0.5826	0.4170	0.00005	0.5830	0.4177	0.0172	0.5651
0.4507	0.0006	0.5487	0.4509	0.0002	0.5489	0.4526	0.0096	0.5378
0.5107	0.0006	0.4887	0.4889	0.00017	0.5109	0.4901	0.0098	0.5001
0.5297	0.0006	0.4697	0.5298	0.0004	0.4698	0.5329	0.0040	0.4631

Polymer (B): **dextran** **1993NEU**
Characterization: M_n/g.mol^{-1} = 705, M_w/g.mol^{-1} = 988, Pfeifer & Langen, Dormagen, Germany
Solvent (A): **water** H_2O **7732-18-5**
Solvent (C): **ethanol** C_2H_6O **64-17-5**

Type of data: coexistence data (tie lines)

continued

continued

T/K = 293.15

Total system			Top phase			Bottom phase		
w_A	w_B	w_C	w_A	w_B	w_C	w_A	w_B	w_C
0.358	0.144	0.498	0.362	0.119	0.519	0.313	0.440	0.247
0.310	0.134	0.556	0.322	0.056	0.622	0.246	0.571	0.183
0.240	0.103	0.657	0.247	0.019	0.734	0.212	0.636	0.152
0.209	0.090	0.701	0.215	0.011	0.774	0.175	0.712	0.113
0.178	0.075	0.747	0.182	0.006	0.812	0.158	0.746	0.096
0.141	0.059	0.800	0.143	0.001	0.856	0.108	0.788	0.104

Polymer (B): **dextran** **1993NEU**
Characterization: M_n/g.mol^{-1} = 6357, M_w/g.mol^{-1} = 11189,
Pfeifer & Langen, Dormagen, Germany
Solvent (A): **water** **H_2O** **7732-18-5**
Solvent (C): **ethanol** **C_2H_6O** **64-17-5**

Type of data: coexistence data (tie lines)

T/K = 293.15

Total system			Top phase			Bottom phase		
w_A	w_B	w_C	w_A	w_B	w_C	w_A	w_B	w_C
0.540	0.060	0.400	0.561	0.027	0.412	0.407	0.394	0.199
0.497	0.055	0.448	0.517	0.008	0.475	0.351	0.482	0.167
0.452	0.050	0.498	0.468	0.002	0.530	0.292	0.544	0.164
0.405	0.045	0.550	0.418	0.001	0.581	0.260	0.603	0.137
0.361	0.040	0.599	0.371	0.000	0.629	0.235	0.646	0.119
0.313	0.035	0.652	0.320	0.000	0.680	0.218	0.695	0.087

Polymer (B): **dextran** **1995NEU**
Characterization: M_n/g.mol^{-1} = 8800, M_w/g.mol^{-1} = 12300,
Pfeifer & Langen, Dormagen, Germany
Solvent (A): **water** **H_2O** **7732-18-5**
Solvent (C): **ethanol** **C_2H_6O** **64-17-5**

Type of data: coexistence data (tie lines)

T/K = 293.15

continued

continued

Total system			Top phase			Bottom phase		
w_A	w_B	w_C	w_A	w_B	w_C	w_A	w_B	w_C
0.569	0.063	0.368	0.569	0.060	0.371	0.509	0.225	0.266
0.567	0.063	0.370	0.566	0.056	0.378	0.493	0.250	0.257
0.563	0.062	0.375	0.563	0.052	0.385	0.474	0.269	0.257
0.558	0.062	0.380	0.559	0.047	0.394	0.451	0.317	0.232
0.554	0.061	0.385	0.557	0.041	0.402	0.430	0.339	0.231
0.549	0.061	0.390	0.554	0.038	0.408	0.430	0.341	0.229
0.540	0.060	0.400	0.544	0.031	0.425	0.406	0.382	0.212

Fractionation of dextran (B) during demixing (the lines correspond with those above):

Top phase		Bottom phase	
M_n/(g/mol)	M_w/(g/mol)	M_n/(g/mol)	M_w/(g/mol)
7500	11100	12115	18900
7357	10300	11830	18100
7299	10000	11844	16700
6838	9300	11500	16100
6929	8800	11460	15700
6176	8400	10929	15300
5789	7700	10519	14200

Polymer (B): **dextran** **1995NEU**
Characterization: M_n/g.mol^{-1} = 20700, M_w/g.mol^{-1} = 58000,
Pfeifer & Langen, Dormagen, Germany
Solvent (A): **water** H_2O **7732-18-5**
Solvent (C): **ethanol** C_2H_6O **64-17-5**

Type of data: coexistence data (tie lines)

T/K = 293.15

Total system			Top phase			Bottom phase		
w_A	w_B	w_C	w_A	w_B	w_C	w_A	w_B	w_C
0.636	0.034	0.330	0.639	0.031	0.330	0.599	0.137	0.264
0.632	0.033	0.335	0.634	0.027	0.339	0.567	0.183	0.250
0.627	0.033	0.340	0.628	0.024	0.348	0.576	0.174	0.250

continued

continued

Total system			Top phase			Bottom phase		
w_A	w_B	w_C	w_A	w_B	w_C	w_A	w_B	w_C
0.616	0.033	0.351	0.627	0.022	0.351	0.543	0.210	0.247
0.608	0.032	0.360	0.618	0.015	0.367	0.514	0.252	0.234
0.598	0.032	0.370	0.603	0.014	0.383	0.485	0.299	0.216
0.589	0.031	0.380	0.601	0.011	0.388	0.468	0.321	0.211
0.579	0.031	0.390	0.589	0.009	0.402	0.442	0.344	0.214
0.570	0.030	0.400	0.582	0.008	0.410	0.439	0.341	0.220
0.616	0.069	0.315	0.624	0.054	0.322	0.543	0.205	0.252
0.612	0.068	0.320	0.625	0.046	0.329	0.530	0.223	0.247
0.607	0.068	0.325	0.627	0.039	0.334	0.514	0.249	0.237
0.603	0.067	0.330	0.622	0.032	0.346	0.490	0.282	0.228
0.598	0.066	0.336	0.616	0.029	0.355	0.491	0.279	0.230
0.594	0.066	0.340	0.615	0.025	0.360	0.462	0.323	0.215
0.585	0.065	0.350	0.606	0.020	0.374	0.440	0.347	0.213
0.576	0.064	0.360	0.596	0.017	0.387	0.437	0.358	0.205
0.567	0.063	0.370	0.588	0.014	0.398	0.415	0.395	0.190
0.558	0.062	0.380	0.577	0.013	0.410	0.419	0.394	0.187
0.549	0.061	0.390	0.569	0.010	0.421	0.388	0.431	0.181
0.540	0.060	0.400	0.562	0.008	0.430	0.381	0.448	0.171
0.594	0.105	0.301	0.610	0.077	0.313	0.539	0.209	0.252
0.591	0.104	0.305	0.616	0.054	0.330	0.518	0.240	0.242
0.586	0.104	0.310	0.615	0.045	0.340	0.493	0.272	0.235
0.582	0.103	0.315	0.614	0.038	0.348	0.473	0.302	0.225
0.578	0.102	0.320	0.611	0.034	0.355	0.466	0.311	0.223
0.574	0.101	0.325	0.607	0.027	0.366	0.460	0.321	0.219
0.569	0.101	0.330	0.603	0.027	0.370	0.453	0.325	0.222
0.561	0.099	0.340	0.594	0.033	0.373	0.450	0.332	0.218
0.552	0.098	0.350	0.585	0.017	0.398	0.424	0.379	0.197
0.544	0.096	0.360	0.575	0.014	0.411	0.405	0.393	0.202
0.535	0.095	0.370	0.570	0.011	0.419	0.381	0.435	0.184
0.527	0.093	0.380	0.561	0.011	0.428	0.366	0.455	0.179
0.518	0.092	0.390	0.554	0.008	0.438	0.362	0.465	0.173
0.510	0.090	0.400	0.549	0.007	0.444	0.357	0.468	0.175

Fractionation of dextran (B) during demixing (the lines correspond with those above):

Top phase		Bottom phase	
M_n/(g/mol)	M_w/(g/mol)	M_n/(g/mol)	M_w/(g/mol)
17529	44700	45168	134600

continued

continued

Top phase M_n/(g/mol)	Top phase M_w/(g/mol)	Bottom phase M_n/(g/mol)	Bottom phase M_w/(g/mol)
16061	37100	44897	109100
15138	33000	43197	105400
14106	29200	43172	98000
11036	21300	43147	85000
10615	19000	41443	83300
9817	16100	38557	77500
8616	13700	37565	72500
8224	12500	35369	71800
16480	41200	33345	94700
15310	34600	36016	92200
13665	30200	30111	81300
12587	25300	34093	80800
11658	22500	34052	79000
10538	19600	33901	75600
10000	17000	33378	74100
8820	14200	32706	71300
8650	14100	30264	68700
7483	11300	30717	68500
7055	10300	29955	66500
6389	9200	28565	65700
16208	38900	28768	79400
12952	27200	30314	77300
11724	23800	30413	73600
10674	20600	29259	71100
10160	19000	30172	70300
9195	16000	29356	68400
9257	16200	28708	68900
9721	17400	28898	68200
7500	11700	28584	66600
6933	10400	27563	65600
6370	9300	27257	64600
6181	8900	26276	62800
6000	8400	25228	60800
5971	8300	25184	61700

Polymer (B): **dextran** **1993NEU**
Characterization: M_n/g.mol^{-1} = 27081, M_w/g.mol^{-1} = 40350,
Pfeifer & Langen, Dormagen, Germany
Solvent (A): **water** H_2O **7732-18-5**
Solvent (C): **ethanol** C_2H_6O **64-17-5**

continued

continued

Type of data: coexistence data (tie lines)

T/K = 293.15

Total system			Top phase			Bottom phase		
w_A	w_B	w_C	w_A	w_B	w_C	w_A	w_B	w_C
0.592	0.066	0.342	0.619	0.018	0.363	0.471	0.316	0.213
0.567	0.063	0.370	0.593	0.007	0.400	0.413	0.388	0.199
0.540	0.060	0.400	0.559	0.002	0.439	0.377	0.454	0.169
0.496	0.055	0.449	0.516	0.001	0.483	0.351	0.496	0.153
0.450	0.050	0.500	0.465	0.000	0.535	0.311	0.551	0.138
0.426	0.051	0.523	0.439	0.000	0.561	0.279	0.593	0.128
0.359	0.040	0.601	0.372	0.000	0.628	0.245	0.644	0.111

Polymer (B): **dextran** **2002LAA**
Characterization: M_w/g.mol^{-1} = 60000, Fluka AG, Buchs, Switzerland
Solvent (A): **water** H_2O **7732-18-5**
Solvent (C): **ethanol** C_2H_6O **64-17-5**

Type of data: cloud points

T/K = 283.15

w_A	0.477	0.496	0.527	0.540	0.563	0.578	0.600	0.614	0.624
w_B	0.330	0.304	0.252	0.237	0.205	0.180	0.145	0.121	0.104
w_C	0.193	0.200	0.221	0.223	0.232	0.242	0.255	0.264	0.271
w_A	0.632	0.641	0.647	0.652	0.657	0.661			
w_B	0.091	0.077	0.067	0.059	0.051	0.045			
w_C	0.276	0.282	0.286	0.289	0.292	0.294			

T/K = 298.15

w_A	0.523	0.593	0.603	0.611	0.617	0.620	0.623	0.625	0.627
w_B	0.244	0.115	0.090	0.075	0.064	0.055	0.049	0.044	0.040
w_C	0.233	0.293	0.306	0.315	0.320	0.324	0.328	0.331	0.333
w_A	0.628	0.629	0.445	0.479	0.504	0.537	0.558	0.572	0.581
w_B	0.036	0.034	0.369	0.312	0.270	0.212	0.175	0.148	0.129
w_C	0.335	0.337	0.185	0.209	0.226	0.251	0.267	0.280	0.290
w_A	0.589	0.600	0.607	0.613	0.617	0.622	0.625		
w_B	0.114	0.092	0.077	0.067	0.059	0.050	0.043		
w_C	0.297	0.308	0.316	0.321	0.325	0.328	0.332		

T/K = 343.15

continued

continued

w_A	0.449	0.494	0.536	0.546	0.553	0.559	0.565	0.569	0.576
w_B	0.386	0.300	0.205	0.176	0.154	0.137	0.124	0.113	0.096
w_C	0.165	0.206	0.260	0.277	0.292	0.304	0.311	0.317	0.328

w_A	0.580	0.584	0.587	0.593	0.595
w_B	0.083	0.074	0.066	0.055	0.050
w_C	0.337	0.342	0.347	0.353	0.355

Polymer (B): **dextran** **1993NEU**
Characterization: M_n/g.mol^{-1} = 108045, M_w/g.mol^{-1} = 403008, Pfeifer & Langen, Dormagen, Germany
Solvent (A): **water** H_2O **7732-18-5**
Solvent (C): **ethanol** C_2H_6O **64-17-5**

Type of data: coexistence data (tie lines)

T/K = 293.15

Total system			Top phase			Bottom phase		
w_A	w_B	w_C	w_A	w_B	w_C	w_A	w_B	w_C
0.616	0.069	0.315	0.652	0.009	0.339	0.512	0.257	0.231
0.589	0.065	0.346	0.621	0.003	0.376	0.457	0.334	0.209
0.538	0.060	0.402	0.560	0.001	0.439	0.376	0.478	0.146
0.495	0.055	0.450	0.514	0.000	0.486	0.328	0.539	0.133
0.449	0.050	0.501	0.461	0.000	0.539	0.308	0.575	0.117

Polymer (B): **poly(*N*,*N*-diethylacrylamide)** **2002MA2**
Characterization: M_w/g.mol^{-1} = 19000, synthesized in the laboratory
Solvent (A): **water** H_2O **7732-18-5**
Solvent (C): **methanol** CH_4O **67-56-1**

Type of data: cloud points (LCST-behavior)

w_B	0.005	0.005	0.005	0.005	0.005	0.005
φ_C	0.00	0.10	0.20	0.30	0.40	0.50
T/K	305.25	306.25	306.95	309.95	320.05	338.35

Polymer (B): **poly(dimethylsiloxane)** **2002SCH**
Characterization: M_n/g.mol^{-1} = 50000, M_w/g.mol^{-1} = 74000, ρ_B (298.15 K) = 0.97 g/cm^3, Wacker GmbH, Germany
Solvent (A): **ethanol** C_2H_6O **64-17-5**
Solvent (C): **toluene** C_7H_8 **108-88-3**

Type of data: cloud points

continued

continued

T/K = 303.15

w_A	0.4777	0.4487	0.4427	0.4311	0.4213	0.4163	0.4094	0.3969	0.3899
w_B	0.0069	0.0299	0.0393	0.0528	0.0648	0.0797	0.0886	0.1044	0.1180
w_C	0.5154	0.5213	0.5180	0.5161	0.5139	0.5041	0.5020	0.4988	0.4921
w_A	0.3888	0.3742	0.3603	0.1119					
w_B	0.1293	0.1457	0.1744	0.8881					
w_C	0.4819	0.4800	0.4652	0.0000					

T/K = 313.15

w_A	0.5359	0.4953	0.4878	0.4782	0.4682	0.4545	0.4556	0.4467	0.4444
w_B	0.0057	0.0264	0.0383	0.0518	0.0617	0.0720	0.0830	0.0994	0.1076
w_C	0.4584	0.4783	0.4739	0.4700	0.4701	0.4735	0.4614	0.4539	0.4480
w_A	0.4306	0.4221	0.4185	0.4125	0.3982	0.1350			
w_B	0.1273	0.1376	0.1483	0.1570	0.1814	0.8650			
w_C	0.4422	0.4403	0.4332	0.4305	0.4204	0.0000			

T/K = 323.15

w_A	0.5825	0.5655	0.5515	0.5326	0.5351	0.5226	0.5141	0.5103	0.5047
w_B	0.0062	0.0139	0.0218	0.0341	0.0422	0.0550	0.0648	0.0704	0.0812
w_C	0.4114	0.4206	0.4266	0.4333	0.4227	0.4224	0.4211	0.4193	0.4141
w_A	0.4973	0.4887	0.4783	0.4669	0.4470	0.4329	0.1623		
w_B	0.0913	0.1066	0.1208	0.1446	0.1703	0.1937	0.8377		
w_C	0.4114	0.4047	0.4009	0.3885	0.3827	0.3734	0.0000		

T/K = 333.15

w_A	0.6165	0.6381	0.6044	0.5969	0.5896	0.5791	0.5683	0.5634	0.5581
w_B	0.0133	0.0054	0.0206	0.0276	0.0372	0.0474	0.0580	0.0673	0.0760
w_C	0.3701	0.3564	0.3750	0.3755	0.3732	0.3735	0.3737	0.3693	0.3659
w_A	0.5511	0.5456	0.5329	0.5264	0.5177	0.5100	0.5020	0.4966	0.1868
w_B	0.0866	0.0958	0.1060	0.1218	0.1313	0.1443	0.1557	0.1684	0.8132
w_C	0.3623	0.3586	0.3611	0.3518	0.3510	0.3457	0.3423	0.3351	0.0000

Polymer (B): **poly(dimethylsiloxane)** **2001WUE, 2002WUE**
Characterization: M_n/g.mol^{-1} = 50000, M_w/g.mol^{-1} = 75000,
Wacker GmbH, Germany
Solvent (A): **toluene** **C_7H_8** **108-88-3**
Solvent (C): **ethanol** **C_2H_6O** **64-17-5**

Type of data: cloud points

T/K = 303.15

φ_A	0.4927	0.5010	0.4986	0.4980	0.4970	0.4885	0.4873	0.4855	0.4801
φ_B	0.0058	0.0255	0.0336	0.0452	0.0556	0.0686	0.0763	0.0902	0.1022
φ_C	0.5015	0.4735	0.4678	0.4568	0.4474	0.4429	0.4364	0.4243	0.4177

continued

continued

φ_A	0.4708	0.4704	0.4580
φ_B	0.1122	0.1268	0.1525
φ_C	0.4170	0.4028	0.3895

critical concentrations: $\varphi_{A,\,crit}$ = 0.454, $\varphi_{B,\,crit}$ = 0.156, $\varphi_{C,\,crit}$ = 0.390

T/K = 313.15

φ_A	0.4357	0.4728	0.4574	0.4541	0.4514	0.4524	0.4568	0.4456	0.4396
φ_B	0.0048	0.0142	0.0224	0.0325	0.0441	0.0526	0.0616	0.0710	0.0853
φ_C	0.5595	0.5130	0.5202	0.5134	0.5045	0.4950	0.4816	0.4834	0.4751
φ_A	0.4343	0.4301	0.4291	0.4229	0.4209	0.4128	0.3626	0.4181	0.4196
φ_B	0.0925	0.1097	0.1189	0.1283	0.1361	0.1578	0.2130	0.1647	0.1551
φ_C	0.4732	0.4602	0.4520	0.4488	0.4430	0.4294	0.4244	0.4172	0.4253

critical concentrations: $\varphi_{A,\,crit}$ = 0.417, $\varphi_{B,\,crit}$ = 0.160, $\varphi_{C,\,crit}$ = 0.423

T/K = 323.15

φ_A	0.3892	0.3989	0.4055	0.4132	0.4033	0.4041	0.4035	0.4022	0.3979
φ_B	0.0052	0.0117	0.0184	0.0288	0.0357	0.0466	0.0550	0.0598	0.0690
φ_C	0.6056	0.5894	0.5761	0.5580	0.5610	0.5493	0.5415	0.5380	0.5331
φ_A	0.3961	0.3906	0.3879	0.3703	0.3773	0.3735	0.3659		
φ_B	0.0777	0.0910	0.1034	0.1065	0.1243	0.1470	0.1680		
φ_C	0.5262	0.5184	0.5087	0.5232	0.4984	0.4795	0.4661		

critical concentrations: $\varphi_{A,\,crit}$ = 0.363, $\varphi_{B,\,crit}$ = 0.167, $\varphi_{C,\,crit}$ = 0.470

T/K = 333.15

φ_A	0.3493	0.3354	0.3546	0.3555	0.3540	0.3550	0.3560	0.3523	0.3497
φ_B	0.0111	0.0045	0.0172	0.0231	0.0312	0.0398	0.0488	0.0567	0.0641
φ_C	0.6396	0.6601	0.6282	0.6214	0.6148	0.6052	0.5952	0.5910	0.5862
φ_A	0.3468	0.3438	0.3471	0.3389	0.3388	0.3344	0.3318	0.3254	
φ_B	0.0732	0.0811	0.0900	0.1036	0.1119	0.1233	0.1332	0.1444	
φ_C	0.5800	0.5751	0.5629	0.5575	0.5493	0.5423	0.5350	0.5302	

critical concentrations: $\varphi_{A,\,crit}$ = 0.316, $\varphi_{B,\,crit}$ = 0.173, $\varphi_{C,\,crit}$ = 0.511

Polymer (B): **polyester (hyperbranched, aliphatic)** **2003SE2**
Characterization: M_n/g.mol^{-1} = 5650, M_w/g.mol^{-1} = 10500,
hydroxyl functionalities are esterified with saturated fatty acids, Boltorn H3200, Perstorp Specialty Chemicals AB, Perstorp, Sweden
Solvent (A): **water** H_2O **7732-18-5**
Solvent (C): **tetrahydrofuran** C_4H_8O **109-99-9**

Type of data: cloud points

T/K = 321.15

continued

continued

w_A	0.0638	0.0737	0.0789	0.0796	0.0831	0.0868	0.1046
w_B	0.1883	0.1461	0.1179	0.1004	0.0879	0.0746	0.0445
w_C	0.7479	0.7802	0.8032	0.8200	0.8290	0.8386	0.8509

T/K = 334.15

w_A	0.0569	0.0637	0.0736	0.0805	0.0856	0.0887	0.0892	0.0992	0.1031
w_B	0.2401	0.2144	0.1646	0.1414	0.1214	0.1121	0.1018	0.0732	0.0664
w_C	0.7030	0.7219	0.7618	0.7781	0.7930	0.7992	0.8090	0.8276	0.8305

Type of data: coexistence data (tie lines)

T/K = 321.15

Phase I			Phase II		
w_A	w_B	w_C	w_A	w_B	w_C
0.0121	0.7589	0.2290	0.8553	0.0014	0.1434
0.0136	0.6757	0.3107	0.8041	0.0011	0.1948
0.0180	0.5947	0.3873	0.7317	0.0031	0.2652
0.0216	0.5173	0.4611	0.6763	0.0019	0.3218
0.0232	0.4722	0.5046	0.6325	0.0014	0.3660
0.0255	0.4584	0.5162	0.5876	0.0019	0.4105
0.0274	0.4327	0.5399	0.4916	0.0049	0.5035
0.0259	0.4212	0.5529	0.4195	0.0002	0.5803
0.0272	0.4132	0.5596	0.3519	0.0073	0.6408
0.0304	0.3920	0.5776	0.2758	0.0002	0.7240
0.0361	0.3468	0.6171	0.2283	0.0065	0.7652
0.0370	0.3199	0.6431	0.1953	0.0108	0.7939
0.0372	0.3031	0.6597	0.1321	0.0196	0.8483

T/K = 334.15

Phase I			Phase II		
w_A	w_B	w_C	w_A	w_B	w_C
0.0052	0.8940	0.1008	0.9226	0.0010	0.0763
0.0131	0.7204	0.2665	0.8669	0.0012	0.1319
0.0131	0.6850	0.3019	0.8441	0.0013	0.1546
0.0159	0.6100	0.3740	0.8093	0.0005	0:1901
0.0247	0.5013	0.4740	0.7129	0.0024	0.2847
0.0320	0.4382	0.5298	0.5968	0.0010	0.4022
0.0306	0.3801	0.5893	0.4323	0.0040	0.5637
0.0338	0.3425	0.6237	0.2820	0.0119	0.7061
0.0335	0.3195	0.6470	0.1398	0.0514	0.8088

Polymer (B): **polyetherimide** **1989ROE**
Characterization: M_w/g.mol^{-1} = 32800, ρ = 1.27 g/cm^3,
Ultem 1000, General Electric, Bergen, The Netherlands
Solvent (A): **water** H_2O **7732-18-5**
Solvent (C): **1-methyl-2-pyrrolidinone** C_5H_9NO **872-50-4**

Type of data: cloud points

T/K = 294

w_A	0.055	0.050	0.050	0.049	0.048	0.049	0.046	0.044	0.037
w_B	0.004	0.016	0.037	0.057	0.076	0.079	0.094	0.117	0.246
w_C	0.941	0.931	0.913	0.894	0.876	0.872	0.860	0.839	0.717

Polymer (B): **polyethersulfone** **1999BAR**
Characterization: M_w/g.mol^{-1} = 49000, ρ_B (298 K) = 1.37 g/cm^3, Tg/K = 498,
Ultrason E 6020 P, BASF AG, Ludwigshafen, Germany
Solvent (A): ***N,N*-dimethylformamide** C_3H_7NO **68-12-2**
Solvent (C): **ethanol** C_2H_6O **64-17-5**

Type of data: cloud points

T/K = 293.15

w_A	0.7730	0.7657
w_B	0.0323	0.0405
w_C	0.1947	0.1938

T/K = 303.15

w_A	0.7730	0.7657
w_B	0.0323	0.0405
w_C	0.1947	0.1938

T/K = 313.15

w_A	0.7730	0.7657
w_B	0.0323	0.0405
w_C	0.1947	0.1938

T/K = 323.15

w_A	0.7747	0.7657
w_B	0.0324	0.0405
w_C	0.1929	0.1938

Polymer (B): **polyethersulfone** **1999BAR**
Characterization: M_w/g.mol^{-1} = 49000, ρ_B (298 K) = 1.37 g/cm^3, Tg/K = 498,
Ultrason E 6020 P, BASF AG, Ludwigshafen, Germany
Solvent (A): ***N,N*-dimethylformamide** C_3H_7NO **68-12-2**
Solvent (C): **2-propanol** C_3H_8O **67-63-0**

continued

continued

Type of data: cloud points

T/K = 293.15

w_A	0.7779	0.7747	0.7751	0.7736	0.7657	0.7586	0.7422	0.7280	0.7008
w_B	0.0052	0.0098	0.0119	0.0196	0.0321	0.0402	0.0683	0.0827	0.1230
w_C	0.2169	0.2155	0.2130	0.2068	0.2022	0.2012	0.1895	0.1893	0.1762

T/K = 303.15

w_A	0.7779	0.7771	0.7780	0.7759	0.7678	0.7607	0.7427	0.7303	0.7008
w_B	0.0052	0.0099	0.0120	0.0196	0.0322	0.0403	0.0683	0.0829	0.1230
w_C	0.2169	0.2130	0.2100	0.2045	0.2000	0.1990	0.1890	0.1868	0.1762

T/K = 313.15

w_A	0.7814	0.7798	0.7802	0.7775	0.7693	0.7623	0.7435	0.7309	0.7008
w_B	0.0052	0.0099	0.0120	0.0197	0.0323	0.0404	0.0684	0.0830	0.1230
w_C	0.2134	0.2103	0.2078	0.2028	0.1984	0.1973	0.1881	0.1861	0.1762

T/K = 323.15

w_A	0.7829	0.7823	0.7819	0.7790	0.7713	0.7640	0.7450	0.7318	0.7025
w_B	0.0052	0.0099	0.0120	0.0197	0.0324	0.0405	0.0685	0.0831	0.1233
w_C	0.2119	0.2078	0.2061	0.2013	0.1963	0.1955	0.1865	0.1851	0.1742

Polymer (B): **polyethersulfone** **1999BAR**
Characterization: M_w/g.mol^{-1} = 49000, ρ_B (298 K) = 1.37 g/cm^3, Tg/K = 498, Ultrason E 6020 P, BASF AG, Ludwigshafen, Germany
Solvent (A): ***N,N*-dimethylformamide** C_3H_7NO **68-12-2**
Solvent (C): **2-propanone** C_3H_6O **67-64-1**

Type of data: cloud points

T/K = 293.15

w_A	0.5589	0.5589	0.5586	0.5544	0.5496	0.5439	0.5345	0.5299	0.5144
w_B	0.0102	0.0177	0.0200	0.0292	0.0380	0.0478	0.0618	0.0685	0.0912
w_C	0.4309	0.4234	0.4214	0.4164	0.4124	0.4083	0.4037	0.4016	0.3944

T/K = 303.15

w_A	0.5667	0.5670	0.5664	0.5615	0.5572	0.5504	0.5412	0.5367	0.5201
w_B	0.0104	0.0179	0.0203	0.0296	0.0385	0.0484	0.0625	0.0693	0.0922
w_C	0.4229	0.4151	0.4133	0.4089	0.4043	0.4012	0.3963	0.3940	0.3877

critical concentrations: $w_{A,\,crit}$ = 0.5264, $w_{B,\,crit}$ = 0.0840, $w_{C,\,crit}$ = 0.3896

T/K = 313.15

w_A	0.5759	0.5748	0.5749	0.5699	0.5650	0.5589	0.5490	0.5444	0.5273
w_B	0.0105	0.0182	0.0206	0.0301	0.0391	0.0491	0.0634	0.0703	0.0935
w_C	0.4136	0.4070	0.4045	0.4000	0.3959	0.3920	0.3876	0.3853	0.3792

continued

continued

T/K = 323.15

w_A	0.5690	0.5853	0.5844	0.5787	0.5747	0.5673	0.5576	0.5525	0.5353
w_B	0.0023	0.0107	0.0209	0.0305	0.0397	0.0499	0.0644	0.0714	0.0949
w_C	0.4287	0.4040	0.3947	0.3908	0.3856	0.3828	0.3780	0.3761	0.3698

Polymer (B): **polyethersulfone** **1999BAR, 2000BA1**
Characterization: M_w/g.mol^{-1} = 49000, ρ_B (298 K) = 1.37 g/cm^3, Tg/K = 498, Ultrason E 6020 P, BASF AG, Ludwigshafen, Germany
Solvent (A): ***N,N*-dimethylformamide** **C_3H_7NO** **68-12-2**
Solvent (C): **water** **H_2O** **7732-18-5**

Type of data: cloud points

T/K = 293.15

w_A	0.9249	0.9249	0.9174	0.9133	0.9004	0.8273
w_B	0.0055	0.0099	0.0187	0.0244	0.0386	0.1185
w_C	0.0696	0.0652	0.0639	0.0623	0.0610	0.0542

T/K = 303.15

w_A	0.9225	0.9232	0.9155	0.9103	0.8982	0.8641	0.8264
w_B	0.0055	0.0099	0.0187	0.0243	0.0380	0.0763	0.1174
w_C	0.0720	0.0669	0.0658	0.0654	0.0638	0.0596	0.0562

T/K = 313.15

w_A	0.9205	0.9213	0.9132	0.9068	0.8961	0.8627	0.8253
w_B	0.0055	0.0099	0.0186	0.0242	0.0374	0.0753	0.1156
w_C	0.0740	0.0688	0.0682	0.0690	0.0665	0.0620	0.0591

T/K = 323.15

w_A	0.9159	0.9200	0.9108	0.9056	0.8938	0.8610	0.8244
w_B	0.0054	0.0099	0.0186	0.0241	0.0369	0.0741	0.1142
w_C	0.0787	0.0701	0.0706	0.0703	0.0693	0.0649	0.0614

Type of data: coexistence data (tie lines)

T/K = 303.15

Total system			Sol phase			Gel phase		
w_A	w_B	w_C	w_A	w_B	w_C	w_A	w_B	w_C
0.899	0.038	0.063	0.897	0.035	0.068	0.800	0.140	0.060
0.892	0.039	0.069	0.914	0.017	0.069	0.725	0.222	0.053
0.888	0.039	0.073	0.907	0.014	0.079	0.648	0.298	0.054
0.878	0.046	0.076	0.905	0.011	0.084	0.647	0.297	0.056

Plait-point composition: w_C = 0.059 + w_B = 0.080 + w_A = 0.861

Polymer (B): **polyethersulfone** **1988ZEM**
Characterization: M_n/g.mol^{-1} = 24700, M_w/g.mol^{-1} = 46400, M_z/g.mol^{-1} = 73300, ρ_B (298 K) = 1.370 g/cm^3, Victrex 4100, I.C.I. US, Inc.
Solvent (A): **water** H_2O **7732-18-5**
Solvent (C): **1-methyl-2-pyrrolidinone** C_5H_9NO **872-50-4**

Type of data: cloud points

T/K = 298.15

w_A	0.133	0.129	0.116	0.114	0.105	0.098	0.088	0.076
w_B	0.009	0.026	0.049	0.089	0.134	0.180	0.228	0.370
w_C	0.858	0.845	0.835	0.797	0.761	0.722	0.684	0.554

Type of data: coexistence data (tie lines)

T/K = 298.15

Sol phase			Gel phase		
w_A	w_B	w_C	w_A	w_B	w_C
0.1596	0.0139	0.8300	0.06	0.38	0.56
0.1875	0.0025	0.8100	0.11	0.50	0.39

Polymer (B): **polyetherurethane** **1987GER**
Characterization: M_η/g.mol^{-1} = 95000, made from [poly(ethylene glycol adipate), M = 2000] : 4,4'-diphenylmethanediisocyanate : 1,4-butanediol = 1:3:2
Solvent (A): **water** H_2O **7732-18-5**
Solvent (C): ***N,N*-dimethylformamide** C_3H_7NO **68-12-2**

Type of data: cloud points

T/K = 295

φ_A	0.0567	0.0526	0.0445	0.0404	0.0363	0.0322	0.0282	0.0241	0.0200
φ_B	0.0239	0.0284	0.0404	0.0486	0.0592	0.0738	0.0957	0.1346	0.2250
φ_A	0.0646	0.0699	0.0831	0.0917	0.1028	0.1182	0.1417	0.1844	0.2827
φ_B	0.0174	0.0142	0.0088	0.0066	0.0048	0.0032	0.0020	0.0013	0.0019

Type of data: spinodal points

T/K = 295

φ_A	0.0584	0.0564	0.0531	0.0520	0.0512	0.0511	0.0523	0.0559	0.0657
φ_B	0.0226	0.0253	0.0333	0.0392	0.0470	0.0583	0.0758	0.1081	0.1862
φ_A	0.0631	0.0666	0.0766	0.0837	0.0934	0.1073	0.1291	0.1689	0.2564
φ_B	0.0187	0.0169	0.0142	0.0132	0.0125	0.0121	0.0124	0.0142	0.0243

Polymer (B): **polyethylene** **2001TOR**
Characterization: M_n/g.mol^{-1} = 43000, M_w/g.mol^{-1} = 105000, M_z/g.mol^{-1} = 190000
HDPE, DSM, Geleen, The Netherlands
Solvent (A): **n-hexane** C_6H_{14} **110-54-3**
Solvent (C): **1-octene** C_8H_{16} **111-66-0**

Type of data: cloud points

w_A	0.7385	0.7385	0.7385	0.7385	0.7385	0.7385	0.7385	0.7385	0.7385
w_B	0.1794	0.1794	0.1794	0.1794	0.1794	0.1794	0.1794	0.1794	0.1794
w_C	0.0821	0.0821	0.0821	0.0821	0.0821	0.0821	0.0821	0.0821	0.0821
T/K	441.7	451.9	462.3	472.6	462.4	472.6	483.0	493.1	503.3
P/MPa	0.92	1.10	1.33	1.55	2.47	3.44	5.22	6.45	7.61
w_A	0.7385	0.7385	0.5932	0.5932	0.5932	0.5932	0.5932	0.5932	0.5932
w_B	0.1794	0.1794	0.1441	0.1441	0.1441	0.1441	0.1441	0.1441	0.1441
w_C	0.0821	0.0821	0.2627	0.2627	0.2627	0.2627	0.2627	0.2627	0.2627
T/K	513.5	523.2	452.6	462.4	472.6	483.0	472.6	483.	493.4
P/MPa	8.71	9.71	0.93	1.10	1.32	1.58	2.07	3.44	4.72
w_A	0.5932	0.5932	0.5932	0.3210	0.3210	0.3210	0.3210	0.3210	0.3210
w_B	0.1441	0.1441	0.1441	0.1971	0.1971	0.1971	0.1971	0.1971	0.1971
w_C	0.2627	0.2627	0.2627	0.4819	0.4819	0.4819	0.4819	0.4819	0.4819
T/K	503.5	513.3	523.9	472.5	482.7	493.1	503.1	503.1	513.4
P/MPa	5.99	7.09	8.22	1.00	1.22	1.46	1.69	2.52	3.67
w_A	0.3210	0.3210	0.2156	0.2156	0.2156	0.2156	0.2156	0.2156	0.2156
w_B	0.1971	0.1971	0.1324	0.1324	0.1324	0.1324	0.1324	0.1324	0.1324
w_C	0.4819	0.4819	0.6520	0.6520	0.6520	0.6520	0.6520	0.6520	0.6520
T/K	523.3	534.0	482.6	493.1	503.2	513.4	503.2	513.2	523.1
P/MPa	4.72	5.83	1.02	1.25	1.40	1.66	1.73	2.85	3.84
w_A	0.2156	0.2156							
w_B	0.1324	0.1324							
w_C	0.6520	0.6520							
T/K	534.4	545.0							
P/MPa	4.99	5.98							

Polymer (B): **poly(ethylene glycol)** **2000SPI**
Characterization: M_n/g.mol^{-1} = 200, Sigma Chemical Co., Inc., St. Louis, MO
Solvent (A): **dichloromethane** CH_2Cl_2 **75-09-2**
Solvent (C): **n-heptane** C_7H_{16} **142-82-5**

Type of data: cloud points

T/K = 223.15

w_A	0.645	0.668	0.674	0.629	0.609
w_B	0.063	0.112	0.163	0.289	0.325
w_C	0.292	0.220	0.163	0.082	0.066

continued

continued

T/K = 258.15

w_A	0.592	0.583	0.551	0.521	0.430	0.520
w_B	0.122	0.266	0.363	0.398	0.012	0.084
w_C	0.286	0.151	0.086	0.081	0.558	0.396

T/K = 298.15

w_A	0.332	0.475	0.564	0.555	0.457	0.439	0.569	0.563	0.526
w_B	0.015	0.049	0.194	0.289	0.469	0.034	0.172	0.152	0.128
w_C	0.653	0.476	0.242	0.156	0.074	0.527	0.259	0.284	0.346

w_A	0.515	0.511
w_B	0.099	0.083
w_C	0.385	0.406

Polymer (B): **poly(ethylene glycol)** **2000SPI**
Characterization: M_n/g.mol^{-1} = 3350, Fluka AG, Buchs, Switzerland
Solvent (A): **dichloromethane** CH_2Cl_2 **75-09-2**
Solvent (C): **n-heptane** C_7H_{16} **142-82-5**

Type of data: cloud points

T/K = 223.15

w_A	0.644	0.664	0.686	0.654	0.623	0.682	0.639	0.640	0.665
w_B	0.006	0.027	0.117	0.233	0.270	0.184	0.270	0.268	0.065
w_C	0.350	0.310	0.197	0.113	0.108	0.134	0.092	0.092	0.269

T/K = 258.15

w_A	0.596	0.584	0.591	0.587	0.612	0.591	0.565	0.557	0.573
w_B	0.089	0.071	0.236	0.115	0.166	0.094	0.052	0.040	0.064
w_C	0.315	0.345	0.173	0.298	0.222	0.315	0.383	0.403	0.364

w_A	0.548	0.520	0.508	0.570
w_B	0.033	0.015	0.012	0.282
w_C	0.419	0.465	0.480	0.148

T/K = 298.15

w_A	0.543	0.499	0.533	0.507	0.506	0.489	0.516	0.476	0.531
w_B	0.154	0.080	0.113	0.081	0.043	0.343	0.327	0.021	0.254
w_C	0.303	0.421	0.354	0.412	0.451	0.168	0.157	0.504	0.216

w_A	0.540	0.541	0.561	0.568	0.557	0.563	0.553	0.473
w_B	0.285	0.200	0.223	0.174	0.168	0.145	0.153	0.014
w_C	0.175	0.258	0.216	0.258	0.275	0.292	0.294	0.513

T/K = 308.15

w_A	0.467	0.449	0.514	0.497	0.552	0.539	0.519	0.512	0.499
w_B	0.029	0.015	0.033	0.024	0.134	0.105	0.093	0.071	0.294
w_C	0.504	0.536	0.453	0.480	0.313	0.356	0.388	0.416	0.207

continued

continued

w_A	0.523	0.493	0.524
w_B	0.205	0.328	0.232
w_C	0.271	0.179	0.244

Polymer (B): **poly(ethylene glycol)** **2000SPI**
Characterization: M_n/g.mol^{-1} = 10000, Sigma Chemical Co., Inc., St. Louis, MO
Solvent (A): **dichloromethane** CH_2Cl_2 **75-09-2**
Solvent (C): **n-heptane** C_7H_{16} **142-82-5**

Type of data: cloud points

T/K = 223.15

w_A	0.683	0.708	0.684	0.678	0.623
w_B	0.074	0.133	0.166	0.125	0.042
w_C	0.243	0.159	0.149	0.197	0.335

T/K = 258.15

w_A	0.593	0.582	0.543	0.600	0.552	0.593	0.611	0.620	0.613
w_B	0.037	0.267	0.345	0.244	0.028	0.058	0.093	0.123	0.153
w_C	0.369	0.151	0.112	0.156	0.420	0.349	0.296	0.257	0.233

T/K = 298.15

w_A	0.502	0.570	0.529	0.534	0.548	0.548	0.547	0.559	0.517
w_B	0.019	0.104	0.060	0.051	0.196	0.140	0.109	0.091	0.315
w_C	0.479	0.326	0.411	0.415	0.256	0.312	0.344	0.350	0.168
w_A	0.547	0.559	0.569	0.595	0.576	0.557	0.550	0.488	0.482
w_B	0.209	0.163	0.279	0.155	0.144	0.256	0.223	0.335	0.369
w_C	0.244	0.279	0.152	0.249	0.279	0.187	0.227	0.177	0.150
w_A	0.476	0.430	0.376	0.555	0.380	0.157			
w_B	0.003	0.0002	0.0004	0.151	0.364	0.455			
w_C	0.521	0.570	0.624	0.294	0.256	0.387			

Polymer (B): **poly(ethylene glycol)** **1987LOP**
Characterization: M_w/g.mol^{-1} = 12000
Solvent (A): **1,2-ethanediol** $C_2H_6O_2$ **107-21-1**
Solvent (C): **water** H_2O **7732-18-5**

Type of data: cloud points

T/K	391	418	420	430
w_A	0.000	0.010	0.035	0.096
w_B	0.010	0.010	0.010	0.010
w_C	0.990	0.980	0.955	0.894

Polymer (B): **poly(ethylene glycol)** **2000SPI**
Characterization: M_n/g.mol^{-1} = 200, Sigma Chemical Co., Inc., St. Louis, MO
Solvent (A): **methanol** CH_4O **67-56-1**
Solvent (C): **n-heptane** C_7H_{16} **142-82-5**

Type of data: cloud points

T/K = 283.15

w_A	0.713	0.778	0.753	0.779
w_B	0.149	0.089	0.125	0.044
w_C	0.138	0.133	0.122	0.176

T/K = 298.15

w_A	0.724	0.609	0.693
w_B	0.052	0.240	0.127
w_C	0.224	0.151	0.179

T/K = 323.15

w_A	0.641	0.671	0.662	0.657
w_B	0.134	0.077	0.111	0.038
w_C	0.226	0.253	0.227	0.306

Polymer (B): **poly(ethylene glycol)** **2000SPI**
Characterization: M_n/g.mol^{-1} = 3350, Fluka AG, Buchs, Switzerland
Solvent (A): **methanol** CH_4O **67-56-1**
Solvent (C): **n-heptane** C_7H_{16} **142-82-5**

Type of data: cloud points

T/K = 283.15

w_A	0.780	0.721	0.695	0.800	0.763
w_B	0.051	0.091	0.155	0.010	0.072
w_C	0.169	0.188	0.150	0.190	0.166

T/K = 298.15

w_A	0.725	0.704	0.720	0.742	0.676	0.692	0.704	0.722	0.729
w_B	0.055	0.041	0.035	0.025	0.163	0.141	0.124	0.100	0.071
w_C	0.220	0.255	0.245	0.232	0.161	0.167	0.173	0.179	0.201

T/K = 323.15

w_A	0.630	0.608	0.604	0.583	0.610	0.604
w_B	0.042	0.077	0.135	0.191	0.007	0.057
w_C	0.328	0.316	0.261	0.226	0.383	0.339

Polymer (B): **poly(ethylene glycol)** **2000SPI**
Characterization: M_n/g.mol^{-1} = 10000, Sigma Chemical Co., Inc., St. Louis, MO
Solvent (A): **methanol** CH_4O **67-56-1**
Solvent (C): **n-heptane** C_7H_{16} **142-82-5**

continued

continued

Type of data: cloud points

T/K = 283.15

w_A	0.783	0.773
w_B	0.013	0.026
w_C	0.204	0.201

T/K = 298.15

w_A	0.734	0.751	0.732	0.702	0.706	0.716	0.723
w_B	0.014	0.007	0.005	0.087	0.059	0.044	0.027
w_C	0.253	0.242	0.264	0.211	0.235	0.240	0.250

T/K = 323.15

w_A	0.600	0.595	0.622	0.631
w_B	0.010	0.019	0.082	0.041
w_C	0.389	0.385	0.296	0.328

Polymer (B): **poly(ethylene glycol)** **1987LOP**
Characterization: M_w/g.mol^{-1} = 12000
Solvent (A): **1,2,3-propanetriol** $C_3H_8O_3$ **56-81-5**
Solvent (C): **water** H_2O **7732-18-5**

Type of data: cloud points

T/K	391	412	417	423
w_A	0.000	0.010	0.050	0.099
w_B	0.010	0.010	0.010	0.010
w_C	0.990	0.980	0.940	0.891

Polymer (B): **poly(ethylene glycol)** **2000SPI**
Characterization: M_n/g.mol^{-1} = 200, Sigma Chemical Co., Inc., St. Louis, MO
Solvent (A): **trichloromethane** $CHCl_3$ **67-66-3**
Solvent (C): **n-heptane** C_7H_{16} **142-82-5**

Type of data: cloud points

T/K = 223.15

w_A	0.356	0.490	0.595	0.612	0.604	0.455	0.522
w_B	0.018	0.066	0.178	0.219	0.282	0.049	0.092
w_C	0.626	0.444	0.227	0.169	0.114	0.497	0.386

T/K = 258.15

w_A	0.401	0.506	0.564	0.565	0.550	0.478	0.519
w_B	0.023	0.075	0.196	0.270	0.343	0.057	0.102
w_C	0.576	0.419	0.240	0.164	0.107	0.465	0.380

T/K = 273.15

continued

continued

w_A	0.440	0.524	0.559	0.547	0.517	0.495	0.527
w_B	0.028	0.088	0.233	0.313	0.387	0.067	0.116
w_C	0.532	0.388	0.208	0.140	0.096	0.438	0.356

T/K = 298.15

w_A	0.541	0.564	0.538	0.554	0.520	0.527	0.519	0.513	0.484
w_B	0.231	0.142	0.311	0.188	0.348	0.243	0.126	0.089	0.057
w_C	0.227	0.294	0.151	0.258	0.132	0.230	0.355	0.398	0.458

Polymer (B): **poly(ethylene glycol)** **2000SPI**
Characterization: M_n/g.mol^{-1} = 3350, Fluka AG, Buchs, Switzerland
Solvent (A): **trichloromethane** **$CHCl_3$** **67-66-3**
Solvent (C): **n-heptane** **C_7H_{16}** **142-82-5**

Type of data: cloud points

T/K = 258.15

w_A	0.395	0.499	0.557	0.579	0.551	0.476	0.515
w_B	0.021	0.076	0.184	0.265	0.310	0.055	0.104
w_C	0.584	0.424	0.259	0.157	0.139	0.468	0.381

T/K = 273.15

w_A	0.402	0.496	0.552	0.547	0.518	0.477	0.507
w_B	0.024	0.088	0.216	0.300	0.350	0.065	0.116
w_C	0.574	0.415	0.232	0.153	0.132	0.458	0.377

T/K = 298.15

w_A	0.497	0.482	0.523	0.508	0.530	0.519	0.480	0.515
w_B	0.083	0.052	0.201	0.119	0.244	0.168	0.342	0.245
w_C	0.419	0.467	0.276	0.374	0.226	0.312	0.178	0.240

Polymer (B): **poly(ethylene glycol)** **2000SPI**
Characterization: M_n/g.mol^{-1} = 10000, Sigma Chemical Co., Inc., St. Louis, MO
Solvent (A): **trichloromethane** **$CHCl_3$** **67-66-3**
Solvent (C): **n-heptane** **C_7H_{16}** **142-82-5**

Type of data: cloud points

T/K = 258.15

w_A	0.490	0.550	0.579	0.579	0.567	0.533	0.560
w_B	0.018	0.088	0.170	0.181	0.204	0.061	0.107
w_C	0.492	0.362	0.251	0.240	0.229	0.407	0.333

T/K = 273.15

w_A	0.480	0.537	0.558	0.552	0.543	0.527	0.548
w_B	0.020	0.097	0.204	0.199	0.247	0.067	0.118
w_C	0.500	0.366	0.238	0.249	0.210	0.406	0.334

continued

continued

T/K = 298.15

w_A	0.473	0.518	0.507	0.532	0.519	0.512	0.529	0.473	0.526
w_B	0.021	0.095	0.064	0.197	0.125	0.271	0.181	0.329	0.229
w_C	0.506	0.386	0.429	0.271	0.356	0.217	0.290	0.198	0.245
w_A	0.482	0.486	0.470	0.437	0.372	0.525	0.490	0.423	
w_B	0.024	0.332	0.015	0.007	0.003	0.176	0.253	0.349	
w_C	0.494	0.182	0.515	0.555	0.625	0.299	0.256	0.228	

Polymer (B): **poly(ethylene oxide-*b*-propylene oxide)** **1991SA2**
Characterization: M_n/g.mol^{-1} = 3438, 24.8 mol% ethylene oxide, Polysciences, Inc., Warrington, PA
Solvent (A): **water** **H_2O** **7732-18-5**
Solvent (C): ***N*-methylacetamide** **C_3H_7NO** **79-16-3**

Type of data: coexistence data (tie lines)

T/K = 298.15

Top phase			Bottom phase		
w_A	w_B	w_C	w_A	w_B	w_C
0.948	0.000	0.052	0.737	0.199	0.064
0.908	0.000	0.092	0.718	0.199	0.083
0.830	0.000	0.170	0.647	0.200	0.153
0.737	0.000	0.273	0.609	0.144	0.247
0.632	0.000	0.368	0.527	0.115	0.318
0.453	0.000	0.547	0.383	0.112	0.505
0.397	0.000	0.603	0.341	0.096	0.563
0.388	0.000	0.612	0.298	0.127	0.575
0.368	0.000	0.632	0.320	0.098	0.582
0.354	0.000	0.646	0.303	0.113	0.584

Comments: The detection limit for the block copolymer in the top phase was 1 wt%.

Polymer (B): **poly(*N*-ethyl-*N*-methylacrylamide)** **2004PAN**
Characterization: M_w/g.mol^{-1} = <5000, M_w/M_n < 1.5, synthesized in the laboratory by different methods
Solvent (A): **water** **H_2O** **7732-18-5**
Solvent (C): **methanol** **CH_4O** **67-56-1**

Type of data: cloud points (LCST-behavior)

w_B = 0.01 in the binary aqueous solution was kept constant.

continued

continued

Comments: The sample was synthesized by chain transfer polymerization.

c_C/(mol/l)	0.0	0.5	1.0	1.5
T/K	345.2	338.2	336.2	334.2

Comments: The sample was synthesized by anionic polymerization (butyl end group).

c_C/(mol/l)	0.0	0.5	1.0	1.5
T/K	348.2	337.2	336.2	335.2

Polymer (B): **poly(*N*-ethyl-*N*-methylacrylamide)** **2004PAN**
Characterization: M_w/g.mol^{-1} = <5000, M_w/M_n < 1.5,
synthesized in the laboratory by different methods
Solvent (A): **water** H_2O **7732-18-5**
Solvent (C): **2-methyl-1-propanol** $C_4H_{10}O$ **78-83-1**

Type of data: cloud points (LCST-behavior)

w_B = 0.01 in the binary aqueous solution was kept constant.

Comments: The sample was synthesized by chain transfer polymerization.

c_C/(mol/l)	0.000	0.125	0.250	0.375	0.500
T/K	345.2	340.2	336.2	332.2	330.2

Comments: The sample was synthesized by anionic polymerization (butyl end group).

c_C/(mol/l)	0.00	0.25	0.50
T/K	348.2	334.7	321.2

Polymer (B): **poly(*N*-ethyl-*N*-methylacrylamide)** **2004PAN**
Characterization: M_w/g.mol^{-1} = <5000, M_w/M_n < 1.5,
synthesized in the laboratory by different methods
Solvent (A): **water** H_2O **7732-18-5**
Solvent (C): **2-propanol** C_3H_8O **67-63-0**

Type of data: cloud points (LCST-behavior)

w_B = 0.01 in the binary aqueous solution was kept constant.

c_C/(mol/l)	0.0	0.5	1.0	1.5	2.0
T/K	345.2	340.2	340.2	336.7	335.7

Polymer (B): **polyglycerol** **2002SEI**
Characterization: M_n/g.mol^{-1} = 6500, M_w/g.mol^{-1} = 13650,
hyperbranched, acetylated, synthesized in the laboratory
Solvent (A): **water** H_2O **7732-18-5**
Solvent (C): **tetrahydrofuran** C_4H_8O **109-99-9**

Type of data: cloud points

T/K = 295.15

continued

continued

w_A	0.2954	0.3312	0.2515	0.2371	0.2402	0.3760	0.3822	0.3875	0.3996
w_B	0.1018	0.0669	0.1497	0.1694	0.1668	0.0366	0.0315	0.0266	0.0184
w_C	0.6028	0.6019	0.5988	0.5935	0.5930	0.5874	0.5863	0.5859	0.5820

w_A	0.1705	0.1738	0.4366
w_B	0.2489	0.2457	0.0123
w_C	0.5806	0.5805	0.5511

Type of data: coexistence data (tie lines)

T/K = 295.15

Phase I			Phase II		
w_A	w_B	w_C	w_A	w_B	w_C
0.104	0.375	0.521	0.527	0.015	0.458
0.140	0.299	0.561	0.484	0.016	0.500
0.084	0.455	0.461	0.603	0.008	0.389
0.070	0.520	0.410	0.646	0.008	0.346
0.063	0.587	0.350	0.700	0.007	0.293
0.067	0.724	0.209	0.793	0.009	0.198
0.070	0.784	0.146	0.845	0.008	0.147
0.090	0.910	0.000	0.997	0.003	0.000

Polymer (B): **polyisobutylene** **1987GE2, 1989SCH**
Characterization: M_n/g.mol^{-1} = 40700, M_w/g.mol^{-1} = 95000,
Oppanol B15, BASF AG Ludwigshafen, Germany
Solvent (A): **toluene** C_7H_8 **108-88-3**
Solvent (C): **2-butanone** C_4H_8O **78-93-3**

Type of data: coexistence data (tie lines)

T/K = 298.15

Total system			Sol phase			Gel phase		
w_A	w_B	w_C	w_A	w_B	w_C	w_A	w_B	w_C
0.764	0.030	0.206	0.765	0.0267	0.208	0.694	0.122	0.184
0.756	0.030	0.214	0.761	0.0215	0.218	0.633	0.181	0.186
0.749	0.030	0.221	0.758	0.0180	0.224	0.632	0.197	0.172
0.734	0.030	0.236	0.743	0.0130	0.244	0.568	0.238	0.194
0.719	0.030	0.251	0.733	0.0084	0.258	0.557	0.279	0.164
0.704	0.030	0.266	0.721	0.0077	0.271	0.465	0.306	0.229
0.689	0.030	0.281	0.710	0.0055	0.284	0.484	0.339	0.176
0.659	0.030	0.311	0.681	0.0044	0.315	0.441	0.372	0.187

continued

continued

Fractionation during demixing (the lines correspond with those above, M_{GPC} is the maximum value of the elution curve):

Sol phase M_{GPC}/(g/mol)	Gel phase M_{GPC}/(g/mol)
72700	185600
61700	153500
53700	134100
42900	110100
33000	96100
29200	83900
25100	87100
20800	75500

Polymer (B): **poly(*N*-isopropylacrylamide)** **2002MOR**
Characterization: M_n/g.mol^{-1} = 86000, M_w/g.mol^{-1} = 327700,
Solvent (A): **cyclohexane** **C_6H_{12}** **110-82-7**
Solvent (C): **methanol** **CH_4O** **67-56-1**

Type of data: cloud points

c_B/(g/l)	0.0	0.5	1.0	1.5	2.0	2.5
T/K	318.982	320.023	320.651	320.733	321.516	321.368

Comments: The solvent mixture is used at its binary critical demixing concentration, i.e., w_B/w_C = 0.758/0.242.

Polymer (B): **poly(*N*-isopropylacrylamide)** **1990OTA**
Characterization: M_n/g.mol^{-1} = 343000, M_w/g.mol^{-1} = 2300000, synthesized in the laboratory
Solvent (A): **water** **H_2O** **7732-18-5**
Solvent (C): **1-butanol** **$C_4H_{10}O$** **71-36-3**

Type of data: cloud points (LCST-behavior)

w_B	0.014	0.014
w_C	0.0084	0.036
T/K	304.25	292.35

Polymer (B): **poly(*N*-isopropylacrylamide)** **1990OTA**
Characterization: M_n/g.mol^{-1} = 43750, M_w/g.mol^{-1} = 2100000 synthesized in the laboratory
Solvent (A): **water** **H_2O** **7732-18-5**
Solvent (C): **dimethylsulfoxide** **C_2H_6OS** **67-68-5**

continued

continued

Type of data: cloud points (LCST-behavior)

w_B	0.066	0.061	0.058
w_C	0.110	0.174	0.232
T/K	302.65	300.75	296.95

Polymer (B): **poly(*N*-isopropylacrylamide)** **1990OTA**
Characterization: M_n/g.mol^{-1} = 343000, M_w/g.mol^{-1} = 2300000
synthesized in the laboratory
Solvent (A): **water** **H_2O** **7732-18-5**
Solvent (C): **ethanol** **C_2H_6O** **64-17-5**

Type of data: cloud points (LCST-behavior)

w_B	0.014	0.014	0.014
w_C	0.023	0.043	0.119
T/K	306.15	304.95	298.95

Polymer (B): **poly(*N*-isopropylacrylamide)** **1990OTA**
Characterization: M_n/g.mol^{-1} = 343000, M_w/g.mol^{-1} = 2300000
synthesized in the laboratory
Solvent (A): **water** **H_2O** **7732-18-5**
Solvent (C): **methanol** **CH_4O** **67-56-1**

Type of data: cloud points (LCST-behavior)

w_B	0.014	0.014	0.014
w_C	0.016	0.031	0.064
T/K	306.75	306.15	304.75

Polymer (B): **poly(*N*-isopropylacrylamide)** **1990OTA**
Characterization: M_n/g.mol^{-1} = 343000, M_w/g.mol^{-1} = 2300000
synthesized in the laboratory
Solvent (A): **water** **H_2O** **7732-18-5**
Solvent (C): **1,2,3-propanetriol** **$C_3H_8O_3$** **56-81-5**

Type of data: cloud points (LCST-behavior)

w_B	0.014	0.014	0.014
w_C	0.049	0.100	0.150
T/K	305.55	303.95	301.95

Polymer (B): **poly(*N*-isopropylacrylamide)** **1990OTA**
Characterization: M_n/g.mol^{-1} = 343000, M_w/g.mol^{-1} = 2300000
synthesized in the laboratory
Solvent (A): **water** **H_2O** **7732-18-5**
Solvent (C): **1-propanol** **C_3H_8O** **71-23-8**

Type of data: cloud points (LCST-behavior)

continued

continued

w_B	0.014	0.014
w_C	0.029	0.068
T/K	302.85	296.25

Polymer (B): **poly(methyl methacrylate)** **1981HOR**
Characterization: M_n/g.mol^{-1} = 125000, M_w/g.mol^{-1} = 135000, Polymer Laboratories, Inc., Amherst, MA
Solvent (A): **acetonitrile** C_2H_3N **75-05-8**
Solvent (C): **1-chlorobutane** C_4H_9Cl **109-69-3**

Type of data: cloud points

Comments: The experimentally observed cloud-point temperatures depend linearly on the increase in mole fraction of 1-chlorobutane in the cosolvent mixture.

c_B/(g/cm^3)	0.00442	0.0217	0.0369	0.0589	0.0822	0.0947
(dT/dx_C)/K	−0.099	−0.093	−0.096	−0.096	−0.092	−0.094

Type of data: cloud points

Comments: The experimentally observed cloud-point temperatures depend linearly on the increase in mole fraction of acetonitrile in the cosolvent mixture.

c_B/(g/cm^3)	0.0213	0.0436	0.0714	0.0917	0.1220
(dT/dx_A)/K	−0.033	−0.031	−0.031	−0.031	−0.033

Polymer (B): **poly(methyl methacrylate) star polymer** **2002LOS**
Characterization: M_n/g.mol^{-1} = 75000, M_w/g.mol^{-1} = 84000, apparent M_w and M_n were determined by GPC calibrated by PMMA standards, the average arm number is 8, the precurser arm has a M_w/g.mol^{-1} = 11000, and M_w/M_n = 1.05, synthesized in the laboratory
Solvent (A): **methanol** CH_4O **67-56-1**
Solvent (C): **2-propanone** C_3H_6O **67-64-1**

Type of data: cloud points

T/K = 293.15

w_A	0.4914	0.5023	0.5118	0.5165	0.5191	0.5245	0.5175	0.5184	0.5210
w_B	0.1001	0.0848	0.0702	0.0615	0.0560	0.0521	0.0494	0.0467	0.0436
w_C	0.4084	0.4129	0.4180	0.4220	0.4250	0.4234	0.4331	0.4349	0.4354
w_A	0.5210	0.5087	0.5156	0.5226	0.5284	0.5324	0.5347	0.5357	0.5352
w_B	0.0410	0.0725	0.0621	0.0538	0.0473	0.0427	0.0393	0.0365	0.0328
w_C	0.4380	0.4187	0.4223	0.4236	0.4243	0.4249	0.4260	0.4278	0.4320
w_A	0.5372	0.5308	0.5371	0.5429	0.5462	0.5487	0.5508	0.5529	0.5538
w_B	0.0305	0.0469	0.0400	0.0341	0.0302	0.0271	0.0247	0.0225	0.0209
w_C	0.4323	0.4223	0.4229	0.4230	0.4236	0.4242	0.4245	0.4247	0.4253

continued

continued

w_A	0.4134	0.4400	0.4585	0.4772	0.4885	0.5623	0.5515	0.5538	0.5581
w_B	0.2221	0.1741	0.1438	0.1193	0.1025	0.0207	0.0160	0.0140	0.0126
w_C	0.3645	0.3859	0.3978	0.4036	0.4090	0.4170	0.4326	0.4322	0.4293
w_A	0.5669	0.5679	0.5667	0.5694	0.5699	0.5480			
w_B	0.0110	0.0099	0.0091	0.0085	0.0080	0.4520			
w_C	0.4221	0.4222	0.4242	0.4221	0.4221	0.0000			

T/K = 303.15

w_A	0.5890	0.5887	0.5969	0.5972	0.6019	0.6090	0.6138	0.5617	0.5688
w_B	0.0198	0.0160	0.0129	0.0109	0.0094	0.0083	0.0072	0.0431	0.0353
w_C	0.3912	0.3953	0.3901	0.3919	0.3887	0.3827	0.3790	0.3951	0.3960
w_A	0.5746	0.5797	0.5840	0.5874	0.5913	0.5189	0.5309	0.5408	0.5503
w_B	0.0295	0.0240	0.0198	0.0167	0.0147	0.0981	0.0810	0.0675	0.0564
w_C	0.3960	0.3962	0.3962	0.3959	0.3940	0.3830	0.3880	0.3917	0.3932
w_A	0.5548	0.5607	0.5664	0.5698	0.6720				
w_B	0.0480	0.0403	0.0355	0.0316	0.3280				
w_C	0.3971	0.3991	0.3982	0.3986	0.0000				

Polymer (B): **poly(methyl methacrylate) star polymer** **2002LOS**
Characterization: M_n/g.mol^{-1} = 83600, M_w/g.mol^{-1} = 97000, apparent M_w and M_n were determined by GPC calibrated by PMMA standards, the average arm number is 8, the precurser arm has a M_w/g.mol^{-1} = 11000, and M_w/M_n = 1.05
Solvent (A): **methanol** **CH_4O** **67-56-1**
Solvent (C): **2-propanone** **C_3H_6O** **67-64-1**

Type of data: cloud points

T/K = 293.15

w_A	0.4913	0.4908	0.5143	0.5300	0.5358	0.5399	0.5448
w_B	0.0990	0.0807	0.0587	0.0436	0.0378	0.0341	0.0313
w_C	0.4098	0.4285	0.4271	0.4263	0.4264	0.4260	0.4239

T/K = 303.15

w_A	0.5235	0.5419	0.5509	0.5605	0.5645	0.5700	0.5728	0.5801
w_B	0.0842	0.0676	0.0546	0.0452	0.0376	0.0326	0.0294	0.0200
w_C	0.3923	0.3905	0.3945	0.3943	0.3978	0.3973	0.3979	0.3999

Polymer (B): **poly(methyl methacrylate) star polymer** **2002LOS**
Characterization: M_n/g.mol^{-1} = 87800, M_w/g.mol^{-1} = 101000, apparent M_w and M_n were determined by GPC calibrated by PMMA standards, the average arm number is 8, the precurser arm has a M_w/g.mol^{-1} = 11000, and M_w/M_n = 1.05
Solvent (A): **methanol** **CH_4O** **67-56-1**

continued

continued

Solvent (C): **2-propanone** **C_3H_6O** **67-64-1**

Type of data: cloud points

T/K = 293.15

w_A	0.4533	0.4640	0.4882	0.5084	0.5209	0.5260	0.5314	0.5375
w_B	0.1395	0.1159	0.0880	0.0705	0.0607	0.0513	0.0452	0.0359
w_C	0.4072	0.4201	0.4239	0.4211	0.4183	0.4227	0.4234	0.4266

T/K = 303.15

w_A	0.4815	0.5025	0.5229	0.5320	0.5442	0.5553	0.5568	0.5614	0.5639
w_B	0.1350	0.1010	0.0828	0.0692	0.0610	0.0461	0.0417	0.0373	0.0335
w_C	0.3835	0.3966	0.3943	0.3988	0.3948	0.3986	0.4015	0.4013	0.4026

Polymer (B): **poly(methyl methacrylate) star polymer** **2002LOS**
Characterization: M_n/g.mol^{-1} = 131000, M_w/g.mol^{-1} = 152000, apparent M_w and M_n were determined by GPC calibrated by PMMA standards, the average arm number is 6, the precurser arm has a M_w/g.mol^{-1} = 24700, and M_w/M_n = 1.06, synthesized in the laboratory
Solvent (A): **methanol** **CH_4O** **67-56-1**
Solvent (C): **2-propanone** **C_3H_6O** **67-64-1**

Type of data: cloud points

T/K = 293.15

w_A	0.4778	0.4883	0.4961	0.5016	0.5043	0.5091	0.5134	0.5136	0.5173
w_B	0.0981	0.0791	0.0687	0.0575	0.0497	0.0459	0.0422	0.0391	0.0361
w_C	0.4240	0.4325	0.4352	0.4409	0.4459	0.4450	0.4444	0.4473	0.4466

w_A	0.5189
w_B	0.0335
w_C	0.4476

T/K = 303.15

w_A	0.5117	0.5263	0.5332	0.5392	0.5389	0.5434	0.5445	0.5507	0.5508
w_B	0.0951	0.0729	0.0614	0.0532	0.0480	0.0416	0.0378	0.0339	0.0309
w_C	0.3932	0.4008	0.4054	0.4076	0.4132	0.4150	0.4176	0.4154	0.4183

w_A	0.5517
w_B	0.0264
w_C	0.4219

Polymer (B): **poly(methyl methacrylate) star polymer** **2002LOS**
Characterization: M_n/g.mol^{-1} = 131000, M_w/g.mol^{-1} = 152000, apparent M_w and M_n were determined by GPC calibrated by PMMA standards, the average arm number is 6, the precurser arm has a M_w/g.mol^{-1} = 24700, and M_w/M_n = 1.06

continued

continued

Solvent (A):	**methanol**	**CH_4O**	**67-56-1**
Solvent (C):	**2-propanone**	**C_3H_6O**	**67-64-1**

Type of data: cloud points

T/K = 293.15

w_A	0.4695	0.4943	0.4986	0.5183	0.5207	0.5241	0.5273	0.5324
w_B	0.0900	0.0703	0.0588	0.0475	0.0408	0.0373	0.0322	0.0277
w_C	0.4405	0.4354	0.4426	0.4342	0.4386	0.4386	0.4405	0.4400

T/K = 303.15

w_A	0.5216	0.5395	0.5467	0.5493	0.5525	0.5565	0.5605	0.5623
w_B	0.0738	0.0580	0.0477	0.0415	0.0368	0.0307	0.0274	0.0235
w_C	0.4046	0.4025	0.4057	0.4092	0.4107	0.4128	0.4120	0.4142

Polymer (B): **poly(methyl methacrylate) star polymer** **2002LOS**

Characterization: M_n/g.mol^{-1} = 146000, M_w/g.mol^{-1} = 168000, apparent M_w and M_n were determined by GPC calibrated by PMMA standards, the average arm number is 6, the precurser arm has a M_w/g.mol^{-1} = 24700, and M_w/M_n = 1.06, fractionated in the laboratory from the feed product with M_w/g.mol^{-1} = 152000

Solvent (A):	**methanol**	**CH_4O**	**67-56-1**
Solvent (C):	**2-propanone**	**C_3H_6O**	**67-64-1**

Type of data: cloud points

T/K = 293.15

w_A	0.4893	0.5053	0.5116	0.5150	0.5171	0.5203	0.5244	0.5303	0.5334
w_B	0.0905	0.0685	0.0574	0.0502	0.0460	0.0420	0.0382	0.0352	0.0308
w_C	0.4202	0.4262	0.4309	0.4348	0.4370	0.4376	0.4375	0.4345	0.4358
w_A	0.5353								
w_B	0.0288								
w_C	0.4358								

T/K = 303.15

w_A	0.5286	0.5341	0.5414	0.5419	0.5470	0.5494	0.5591	0.5603	0.5629
w_B	0.0768	0.0644	0.0543	0.0481	0.0424	0.0383	0.0329	0.0303	0.0276
w_C	0.3946	0.4015	0.4044	0.4100	0.4106	0.4123	0.4080	0.4093	0.4094
w_A	0.5655								
w_B	0.0250								
w_C	0.4095								

Polymer (B): **poly(methyl methacrylate) star polymer** **2002LOS**
Characterization: M_n/g.mol^{-1} = 196500, M_w/g.mol^{-1} = 224000, apparent M_w and M_n were determined by GPC calibrated by PMMA standards, the average arm number is 5, the precurser arm has a M_w/g.mol^{-1} = 43500, and M_w/M_n = 1.06, synthesized in the laboratory
Solvent (A): **methanol** CH_4O **67-56-1**
Solvent (C): **2-propanone** C_3H_6O **67-64-1**

Type of data: cloud points

T/K = 293.15

w_A	0.4732	0.4897	0.4975	0.5126	0.5105	0.5171	0.5324	0.5315
w_B	0.1008	0.0762	0.0622	0.0477	0.0391	0.0334	0.0286	0.0252
w_C	0.4260	0.4341	0.4403	0.4398	0.4504	0.4496	0.4390	0.4432

T/K = 303.15

w_A	0.5032	0.5102	0.5158	0.5334	0.5379	0.5552	0.5562	0.5594
w_B	0.0851	0.0666	0.0541	0.0441	0.0386	0.0322	0.0287	0.0225
w_C	0.4117	0.4232	0.4301	0.4225	0.4235	0.4126	0.4151	0.4181

Polymer (B): **polypropylene** **2000OLI**
Characterization: M_n/g.mol^{-1} = 48900, M_w/g.mol^{-1} = 245000, 92% isotactic, Polibrasil Resinas S.A., Brazil
Solvent (A): **n-butane** C_4H_{10} **106-97-8**
Solvent (C): **toluene** C_7H_8 **108-88-3**

Type of data: cloud points

w_A	0.8905	0.8905	0.8905	0.8905	0.8905	0.8508	0.8508	0.8508	0.8508
w_B	0.0996	0.0996	0.0996	0.0996	0.0996	0.0983	0.0983	0.0983	0.0983
w_C	0.0099	0.0099	0.0099	0.0099	0.0099	0.0509	0.0509	0.0509	0.0509
T/K	423.15	413.15	403.15	393.15	383.15	423.15	413.15	403.15	393.15
P/bar	108.4	92.3	75.3	56.4	36.9	99.0	84.1	66.9	49.1

w_A	0.8508	0.8012	0.8012	0.8012	0.8012
w_B	0.0983	0.0992	0.0992	0.0992	0.0992
w_C	0.0509	0.0996	0.0996	0.0996	0.0996
T/K	383.15	423.15	413.15	403.15	393.15
P/bar	30.8	87.2	68.8	51.5	33.2

Polymer (B): **polypropylene** **2000OLI**
Characterization: M_n/g.mol^{-1} = 48900, M_w/g.mol^{-1} = 245000, 92% isotactic, Polibrasil Resinas S.A., Brazil
Solvent (A): **1-butene** C_4H_8 **106-98-9**
Solvent (C): **toluene** C_7H_8 **108-88-3**

Type of data: cloud points

continued

continued

w_A	0.8910	0.8910	0.8910	0.8910	0.8910	0.8910	0.8910	0.8512	0.8512
w_B	0.0986	0.0986	0.0986	0.0986	0.0986	0.0986	0.0986	0.0974	0.0974
w_C	0.0104	0.0104	0.0104	0.0104	0.0104	0.0104	0.0104	0.0514	0.0514
T/K	423.15	413.15	403.15	393.15	383.15	373.15	368.15	423.15	413.15
P/bar	127.5	111.0	95.4	75.2	56.6	35.6	26.3	117.7	97.2

w_A	0.8512	0.8512	0.8512	0.8512	0.8022	0.8022	0.8022	0.8022	0.8022
w_B	0.0974	0.0974	0.0974	0.0974	0.0993	0.0993	0.0993	0.0993	0.0993
w_C	0.0514	0.0514	0.0514	0.0514	0.0985	0.0985	0.0985	0.0985	0.0985
T/K	403.15	393.15	383.15	373.15	423.15	413.15	403.15	393.15	383.15
P/bar	78.8	60.6	41.6	21.0	102.8	82.6	66.4	46.2	26.5

Polymer (B): **poly(pyrrolidinoacrylamide)** **2004PAN**
Characterization: M_w/g.mol^{-1} = <5000, M_w/M_n < 1.5,
synthesized in the laboratory by different methods
Solvent (A): **water** H_2O **7732-18-5**
Solvent (C): **methanol** CH_4O **67-56-1**

Type of data: cloud points (LCST-behavior)

w_B = 0.01 in the binary aqueous solution was kept constant.

c_C/(mol/l)	0.0	0.5	1.0	1.5	2.0
T/K	331.2	337.2	335.7	335.7	340.2

Polymer (B): **poly(pyrrolidinoacrylamide)** **2004PAN**
Characterization: M_w/g.mol^{-1} = <5000, M_w/M_n < 1.5,
synthesized in the laboratory by different methods
Solvent (A): **water** H_2O **7732-18-5**
Solvent (C): **2-methyl-1-propanol** $C_4H_{10}O$ **78-83-1**

Type of data: cloud points (LCST-behavior)

w_B = 0.01 in the binary aqueous solution was kept constant.

Comments: The sample was synthesized by chain transfer polymerization.

c_C/(mol/l)	0.000	0.125	0.250	0.375	0.500
T/K	345.2	340.2	337.2	333.2	331.2

Comments: The sample was synthesized by anionic polymerization (butyl end group).

c_C/(mol/l)	0.000	0.125	0.250	0.375	0.500
T/K	331.2	327.7	326.2	324.2	321.2

Polymer (B): **poly(pyrrolidinoacrylamide)** **2004PAN**
Characterization: M_w/g.mol^{-1} = <5000, M_w/M_n < 1.5,
synthesized in the laboratory by different methods
Solvent (A): **water** H_2O **7732-18-5**
Solvent (C): **2-propanol** C_3H_8O **67-63-0**

continued

continued

Type of data: cloud points (LCST-behavior)

w_B = 0.01 in the binary aqueous solution was kept constant.

c_C/(mol/l)	0.0	0.5	1.0	1.5	2.0
T/K	331.2	331.2	331.2	330.2	327.2

Polymer (B): **polystyrene** **1940SCH**
Characterization: M_n/g.mol^{-1} = 500000, synthesized in the laboratory
Solvent (A): **benzene** C_6H_6 **71-43-2**
Solvent (C): **methanol** CH_4O **67-56-1**

Type of data: coexistence data (tie lines)

T/K = 300

Sol phase			Gel phase		
φ_A	φ_B	φ_C	φ_A	φ_B	φ_C
0.549	0.000	0.459	0.498	0.426	0.076
0.250	0.000	0.750	0.261	0.706	0.033
0.000	0.000	1.000	0.000	0.979	0.021

Polymer (B): **polystyrene** **2004ZUC**
Characterization: M_n/g.mol^{-1} = 83000, M_w/g.mol^{-1} = 86700, Polymer Source, Inc., Quebec, Canada
Solvent (A): **bisphenol-A diglycidyl ether** $C_{21}H_{24}O_4$ **1675-54-3**
Solvent (C): **benzylamine** C_7H_9N **100-46-9**

Type of data: cloud points (UCST-behavior)

Comments: The bisphenol-A diglycidyl ether is a commercial product, Der332, Dow. The mixture A + C is the stoichiometric and unreacted mixture.

φ_B	0.0266	0.0539	0.1058	0.1582
T/K	299.6	303.3	305.5	305.23

Polymer (B): **polystyrene** **2004ZUC**
Characterization: M_n/g.mol^{-1} = 217000, M_w/g.mol^{-1} = 228000, Polymer Source, Inc., Quebec, Canada
Solvent (A): **bisphenol-A diglycidyl ether** $C_{21}H_{24}O_4$ **1675-54-3**
Solvent (C): **benzylamine** C_7H_9N **100-46-9**

Type of data: cloud points (UCST-behavior)

Comments: The bisphenol-A diglycidyl ether is a commercial product, Der332, Dow. The mixture A + C is the stoichiometric and unreacted mixture.

φ_B	0.0266	0.0539	0.1058	0.1582
T/K	306.5	309.6	312.55	310.3

Polymer (B): **polystyrene** **2004RIC**
Characterization: M_n/g.mol^{-1} = 163000, M_w/g.mol^{-1} = 325000,
Laqcrene PS 1540N, Atofina, France
Solvent (A): **bisphenol-A diglycidyl ether** $C_{21}H_{24}O_4$ **1675-54-3**
Solvent (C): **4,4'-methylenebis(2,6-diethylaniline)** $C_{21}H_{30}N_2$ **13680-35-8**

Type of data: cloud points

Comments: The bisphenol-A diglycidyl ether is a commercial product of polydispersity 1.15. The mixture A + C is the stoichiometric and unreacted mixture.

φ_B	0.05731	0.09128	0.11375	0.16933	0.22408	0.27801	0.33114	0.38348	0.43506
T/K	355.15	356.15	359.15	358.15	371.15	378.65	382.15	383.15	383.65

φ_B	0.53600	0.63407
T/K	371.15	365.15

Polymer (B): **polystyrene** **2001WUE**
Characterization: M_n/g.mol^{-1} = 133000, M_w/g.mol^{-1} = 233000,
Vestyron 116-30, Huels AG, Marl, Germany
Solvent (A): **2-butanone** C_4H_8O **78-93-3**
Solvent (C): **methanol** CH_4O **67-56-1**

Type of data: cloud points

T/K = 303.15

φ_A	0.8530	0.8444	0.8339	0.8231	0.8142	0.8005
φ_B	0.0091	0.0196	0.0321	0.0464	0.0585	0.0744
φ_C	0.1379	0.1360	0.1340	0.1305	0.1273	0.1251

critical concentrations: $\varphi_{A,\,crit}$ = 0.7827, $\varphi_{B,\,crit}$ = 0.0881, $\varphi_{C,\,crit}$ = 0.1292

T/K = 313.15

φ_A	0.8518	0.8418	0.8290	0.8232	0.8112	0.8002
φ_B	0.0084	0.0198	0.0342	0.0465	0.0608	0.0718
φ_C	0.1398	0.1384	0.1368	0.1303	0.1280	0.1280

critical concentrations: $\varphi_{A,\,crit}$ = 0.7884, $\varphi_{B,\,crit}$ = 0.0843, $\varphi_{C,\,crit}$ = 0.1273

T/K = 323.15

φ_A	0.8530	0.8444	0.8339	0.8231	0.8142	0.8005
φ_B	0.0091	0.0196	0.0321	0.0464	0.0585	0.0744
φ_C	0.1379	0.1360	0.1340	0.1305	0.1273	0.1251

critical concentrations: $\varphi_{A,\,crit}$ = 0.7848, $\varphi_{B,\,crit}$ = 0.0907, $\varphi_{C,\,crit}$ = 0.1245

Type of data: coexistence data (tie lines)

T/K = 303.15

continued

continued

Total system			Sol phase			Gel phase		
φ_A	φ_B	φ_C	φ_A	φ_B	φ_C	φ_A	φ_B	φ_C
0.7680	0.1090	0.1230	0.8270	0.0290	0.1440	0.7060	0.1950	0.0990
0.7580	0.1150	0.1270	0.8260	0.0210	0.1530	0.6810	0.2190	0.1000
0.7440	0.1230	0.1330	0.8170	0.0160	0.1670	0.6260	0.2880	0.0860
0.7220	0.1380	0.1400	0.8120	0.0070	0.1810	0.5720	0.3370	0.0910
0.7080	0.1430	0.1490	0.8000	0.0070	0.1930	0.5050	0.4290	0.0660
0.6920	0.1540	0.1540	0.7860	0.0070	0.2070	0.4560	0.4820	0.0620

T/K = 313.15

Total system			Sol phase			Gel phase		
φ_A	φ_B	φ_C	φ_A	φ_B	φ_C	φ_A	φ_B	φ_C
0.7630	0.1050	0.1320	0.8340	0.0130	0.1530	0.6160	0.2960	0.0880
0.7400	0.1120	0.1480	0.8150	0.0070	0.1780	0.5740	0.3330	0.0930
0.7270	0.1170	0.1560	0.7957	0.0070	0.1973	0.5390	0.3720	0.0890
0.7100	0.1180	0.1720	0.7890	0.0050	0.2060	0.4370	0.4820	0.0810
0.6880	0.1270	0.1850	0.7710	0.0080	0.2210	0.3910	0.4940	0.1150

T/K = 323.15

Total system			Sol phase			Gel phase		
φ_A	φ_B	φ_C	φ_A	φ_B	φ_C	φ_A	φ_B	φ_C
0.7560	0.1140	0.1300	0.8300	0.0130	0.1570	0.6600	0.2440	0.0960
0.7440	0.1180	0.1380	0.8160	0.0080	0.1760	0.6270	0.2820	0.0910
0.7310	0.1210	0.1480	0.8110	0.0070	0.1820	0.5880	0.3200	0.0920
0.7240	0.1210	0.1550	0.8070	0.0060	0.1870	0.5190	0.3970	0.0840
0.7100	0.1230	0.1670	0.7900	0.0060	0.2040	0.4820	0.4160	0.1020
0.6950	0.1260	0.1790	0.7730	0.0050	0.2220	0.4310	0.4880	0.0810

Polymer (B): **polystyrene** **2001WUE**
Characterization: M_n/g.mol^{-1} = 133000, M_w/g.mol^{-1} = 233000,
Vestyron 116-30, Huels AG, Marl, Germany
Solvent (A): **2-butanone** C_4H_8O **78-93-3**
Solvent (C): **1-pentanol** $C_5H_{12}O$ **71-41-0**

Type of data: cloud points

continued

continued

T/K = 303.15

φ_A	0.6297	0.6297	0.6260	0.6230	0.6190	0.6099	0.6041	0.5939
φ_B	0.0042	0.0150	0.0241	0.0368	0.0482	0.0584	0.0688	0.0882
φ_C	0.3661	0.3553	0.3499	0.3402	0.3328	0.3317	0.3271	0.3179

critical concentrations: $\varphi_{A,\,crit}$ = 0.5906, $\varphi_{B,\,crit}$ = 0.0933, $\varphi_{C,\,crit}$ = 0.3161

T/K = 313.15

φ_A	0.6123	0.6116	0.6073	0.6095	0.6047	0.5987	0.5961	0.5831
φ_B	0.0037	0.0146	0.0260	0.0355	0.0438	0.0537	0.0680	0.0734
φ_C	0.3840	0.3738	0.3667	0.3550	0.3515	0.3476	0.3359	0.3435
φ_A	0.5751	0.5654						
φ_B	0.0871	0.0977						
φ_C	0.3378	0.3369						

critical concentrations: $\varphi_{A,\,crit}$ = 0.5473, $\varphi_{B,\,crit}$ = 0.1241, $\varphi_{C,\,crit}$ = 0.3286

T/K = 323.15

φ_A	0.5656	0.5524	0.5381	0.5760	0.5979	0.5965	0.5933	0.5852
φ_B	0.0739	0.0940	0.1023	0.0617	0.0036	0.0154	0.0227	0.0330
φ_C	0.3605	0.3536	0.3596	0.3623	0.3985	0.3881	0.3840	0.3818
φ_A	0.5823	0.5794						
φ_B	0.0420	0.0521						
φ_C	0.3757	0.3685						

critical concentrations: $\varphi_{A,\,crit}$ = 0.5285, $\varphi_{B,\,crit}$ = 0.1309, $\varphi_{C,\,crit}$ = 0.3406

Type of data: coexistence data (tie lines)

Total system			Sol phase			Gel phase		
φ_A	φ_B	φ_C	φ_A	φ_B	φ_C	φ_A	φ_B	φ_C
T/K = 303.15								
0.5750	0.0910	0.3340	0.6110	0.0220	0.3670	0.4920	0.2440	0.2640
0.5650	0.0900	0.3450	0.6070	0.0140	0.3790	0.4300	0.3340	0.2360
0.5570	0.0950	0.3480	0.6010	0.0060	0.3930	0.3980	0.3820	0.2200
0.5440	0.1010	0.3550	0.5890	0.0070	0.4040	0.4010	0.3720	0.2270
0.5290	0.1060	0.3650	0.5740	0.0070	0.4190	0.3280	0.4440	0.2280
T/K = 313.15								
0.5410	0.1160	0.3430	0.5900	0.0200	0.3900	0.4680	0.2560	0.2760
0.5310	0.1170	0.3520	0.5840	0.0130	0.4030	0.4390	0.2980	0.2630
0.5240	0.1220	0.3540	0.5790	0.0100	0.4110	0.4060	0.3480	0.2460
0.5130	0.1230	0.3640	0.5640	0.0110	0.4250	0.3740	0.3890	0.2370
0.5020	0.1270	0.3710	0.5590	0.0070	0.4340	0.3180	0.4590	0.2230

continued

continued

T/K = 323.15

Total system			Sol phase			Gel phase		
φ_A	φ_B	φ_C	φ_A	φ_B	φ_C	φ_A	φ_B	φ_C
0.4910	0.1590	0.3500	0.5580	0.0190	0.4230	0.4100	0.3300	0.2600
0.4790	0.1600	0.3610	0.5460	0.0110	0.4430	0.3670	0.3990	0.2340
0.4620	0.1680	0.3700	0.5280	0.0100	0.4620	0.3450	0.4180	0.2370
0.4470	0.1750	0.3780	0.5190	0.0100	0.4710	0.2680	0.5360	0.1960

Polymer (B): **polystyrene** **2001WUE**
Characterization: M_n/g.mol^{-1} = 133000, M_w/g.mol^{-1} = 233000,
Vestyron 116-30, Huels AG, Marl, Germany
Solvent (A): **2-butanone** C_4H_8O **78-93-3**
Solvent (C): **1-propanol** C_3H_8O **71-23-8**

Type of data: cloud points

T/K = 303.15

φ_A	0.7140	0.7119	0.6982	0.6935	0.6829	0.6727	0.6675	0.6518
φ_B	0.0037	0.0137	0.0429	0.0504	0.0635	0.0745	0.0861	0.0995
φ_C	0.2823	0.2744	0.2589	0.2561	0.2536	0.2528	0.2464	0.2487

critical concentrations: $\varphi_{A,\,crit}$ = 0.6718, $\varphi_{B,\,crit}$ = 0.0843, $\varphi_{C,\,crit}$ = 0.2439

T/K = 313.15

φ_A	0.7047	0.7027	0.6871	0.6786	0.6754	0.6755
φ_B	0.0050	0.0135	0.0399	0.0514	0.0614	0.0715
φ_C	0.2903	0.2838	0.2730	0.2700	0.2632	0.2530

critical concentrations: $\varphi_{A,\,crit}$ = 0.6693, $\varphi_{B,\,crit}$ = 0.0829, $\varphi_{C,\,crit}$ = 0.2478

T/K = 323.15

φ_A	0.7002	0.6945	0.6846	0.6858	0.6594	0.6555	0.6372	0.6824
φ_B	0.0050	0.0164	0.0351	0.0474	0.0728	0.0826	0.0952	0.0354
φ_C	0.2948	0.2891	0.2803	0.2668	0.2678	0.2619	0.2676	0.2822

φ_A	0.6791	0.6683
φ_B	0.0471	0.0609
φ_C	0.2738	0.2708

critical concentrations: $\varphi_{A,\,crit}$ = 0.6490, $\varphi_{B,\,crit}$ = 0.0910, $\varphi_{C,\,crit}$ = 0.2600

Type of data: coexistence data (tie lines)

T/K = 303.15

continued

continued

Total system			Sol phase			Gel phase		
φ_A	φ_B	φ_C	φ_A	φ_B	φ_C	φ_A	φ_B	φ_C
0.6590	0.0780	0.2630	0.6940	0.0080	0.2980	0.5710	0.2290	0.2000
0.6450	0.0840	0.2710	0.6860	0.0080	0.3060	0.4700	0.3850	0.1450
0.6270	0.0880	0.2850	0.6670	0.0060	0.3270	0.4110	0.4660	0.1230
0.6090	0.0980	0.2930	0.6550	0.0060	0.3390	0.3790	0.5110	0.1100
0.5950	0.1060	0.2990	0.6410	0.0080	0.3510	0.3490	0.5310	0.1200
0.5750	0.1150	0.3100	0.6300	0.0050	0.3650	0.2620	0.6280	0.1100

T/K = 313.15

Total system			Sol phase			Gel phase		
φ_A	φ_B	φ_C	φ_A	φ_B	φ_C	φ_A	φ_B	φ_C
0.6450	0.0870	0.2680	0.6800	0.0220	0.2980	0.5200	0.2890	0.1910
0.6290	0.0950	0.2760	0.6780	0.0130	0.3090	0.4620	0.4670	0.0710
0.6180	0.0990	0.2830	0.6740	0.0120	0.3140	0.4450	0.3720	0.1830
0.6020	0.1080	0.2900	0.6620	0.0080	0.3300	0.3760	0.4470	0.1770
0.5830	0.1170	0.3000	0.6420	0.0070	0.3510	0.3300	0.5190	0.1510

T/K = 323.15

Total system			Sol phase			Gel phase		
φ_A	φ_B	φ_C	φ_A	φ_B	φ_C	φ_A	φ_B	φ_C
0.6220	0.1050	0.2730	0.6690	0.0120	0.3190	0.5330	0.2650	0.2020
0.6080	0.1050	0.2870	0.6580	0.0090	0.3330	0.5050	0.2990	0.1960
0.5940	0.1090	0.2970	0.6460	0.0060	0.3480	0.4720	0.3410	0.1870
0.5840	0.1110	0.3050	0.6390	0.0040	0.3570	0.4420	0.3510	0.2070
0.5700	0.1130	0.3170	0.6250	0.0060	0.3690	0.3700	0.4230	0.2070

Polymer (B): **polystyrene** **1993IWA**
Characterization: M_n/g.mol^{-1} = 166900, M_w/g.mol^{-1} = 171900,
Tosoh Corp., Ltd., Tokyo, Japan
Solvent (A): **cyclopentane** C_5H_{10} **287-92-3**
Solvent (C): **cyclohexane** C_6H_{12} **110-82-7**

Type of data: cloud points (LCST-behavior)

continued

continued

w_A	0.8881	0.8810	0.8800	0.8720	0.8605	0.8586	0.8418	0.8399	0.8280
w_B	0.0094	0.0187	0.0204	0.0278	0.0369	0.0410	0.0546	0.0612	0.0736
w_C	0.1025	0.1003	0.0996	0.1002	0.1026	0.1004	0.1036	0.0989	0.0983
T/K	450.55	448.85	448.75	448.15	447.75	447.35	447.55	447.35	447.55
w_A	0.8075	0.7811	0.7501	0.7891	0.7814	0.7670	0.7470	0.7369	0.7219
w_B	0.0921	0.1199	0.1527	0.0102	0.0175	0.0301	0.0509	0.0645	0.0790
w_C	0.1004	0.0990	0.0973	0.2006	0.2010	0.2029	0.2022	0.1986	0.1992
T/K	447.75	447.95	448.65	456.15	454.95	454.15	453.95	453.25	454.25
w_A	0.7024	0.6850	0.6691	0.6469	0.9302	0.8996	0.8856	0.8559	0.8271
w_B	0.1003	0.1155	0.1311	0.1530	0.0402	0.0402	0.0401	0.0411	0.0408
w_C	0.1973	0.1995	0.1998	0.2001	0.0296	0.0602	0.0742	0.1031	0.1320
T/K	454.75	455.05	455.65	456.75	443.55	445.35	446.25	447.65	450.15
w_A	0.7935	0.7536							
w_B	0.0408	0.0404							
w_C	0.1657	0.2060							
T/K	451.55	454.15							

Polymer (B): **polystyrene** **1993IWA**
Characterization: M_n/g.mol^{-1} = 347000, M_w/g.mol^{-1} = 354000,
Tosoh Corp., Ltd., Tokyo, Japan
Solvent (A): **cyclopentane** C_5H_{10} **287-92-3**
Solvent (C): **cyclohexane** C_6H_{12} **110-82-7**

Type of data: cloud points (LCST-behavior)

w_A	0.8951	0.9049	0.8149	0.8657	0.8385	0.8131	0.7672
w_B	0.0100	0.0374	0.0379	0.0385	0.0572	0.0839	0.1370
w_C	0.0949	0.0577	0.1472	0.0958	0.1043	0.1030	0.0959
T/K	444.25	441.75	446.85	443.65	443.65	444.65	445.45

Polymer (B): **polystyrene** **1993IWA**
Characterization: M_n/g.mol^{-1} = 783000, M_w/g.mol^{-1} = 791000,
Tosoh Corp., Ltd., Tokyo, Japan
Solvent (A): **cyclopentane** C_5H_{10} **287-92-3**
Solvent (C): **cyclohexane** C_6H_{12} **110-82-7**

Type of data: cloud points (LCST-behavior)

w_A	0.8900	0.8803	0.8712	0.8424	0.8130	0.7697	0.9343	0.9020	0.8592
w_B	0.0101	0.0196	0.0293	0.0581	0.0861	0.1296	0.0402	0.0402	0.0410
w_C	0.0999	0.1002	0.0995	0.0995	0.1009	0.1008	0.0255	0.0578	0.0998
T/K	441.35	440.25	439.85	440.85	441.55	443.85	436.05	438.35	441.15
w_A	0.8348	0.8048	0.8006						
w_B	0.0398	0.0404	0.0399						
w_C	0.1254	0.1548	0.1596						
T/K	442.25	444.65	444.95						

Polymer (B): **polystyrene** **1988SC2**
Characterization: M_n/g.mol^{-1} = 54500, M_w/g.mol^{-1} = 297000, M_z/g.mol^{-1} = 695000
Solvent (A): **cyclohexane** C_6H_{12} **110-82-7**
Solvent (C): **methylcyclohexane** C_7H_{14} **108-87-2**

Type of data: cloud points (UCST-behavior)

Comments: The weight ratio of w_C/w_A = 3/1 was kept constant.

w_B	0.0096	0.018	0.029	0.040	0.047	0.056	0.072	0.093	0.104
T/K	320.45	321.00	321.15	320.80	320.76	320.25	319.73	319.40	318.64

w_B	0.114	0.121	0.132	0.136	0.154
T/K	318.18	318.06	317.67	317.39	317.43

Comments: The weight ratio of w_C/w_A = 1/1 was kept constant.

w_B	0.0082	0.024	0.024	0.036	0.049	0.059	0.059	0.084	0.102
T/K	311.50	312.16	312.25	312.36	312.16	311.79	311.75	311.80	310.22

w_B	0.103	0.124	0.136	0.142
T/K	310.31	309.85	309.02	308.92

Comments: The weight ratio of w_C/w_A = 1/3 was kept constant.

w_B	0.0096	0.011	0.017	0.032	0.038	0.049	0.056	0.067	0.080
T/K	304.01	304.55	305.22	305.55	305.31	305.08	304.82	304.47	304.07

w_B	0.092	0.102	0.103	0.124	0.126	0.129	0.144
T/K	304.06	303.40	303.40	303.39	303.22	303.15	303.15

Polymer (B): **polystyrene** **1989MER**
Characterization: M_n/g.mol^{-1} = 54500, M_w/g.mol^{-1} = 297000, M_z/g.mol^{-1} = 695000
Solvent (A): **cyclohexane** C_6H_{12} **110-82-7**
Solvent (C): **methylcyclohexane** C_7H_{14} **108-87-2**

Type of data: coexistence data (tie lines)

Comments: The total feed concentration of the polymer in the homogeneous solution is w_B = 0.0511. The weight ratio of w_C/w_A = 1/3 was found to be constant along the coexistence curve in both phases.

T/K	304.88	304.72	304.45	303.79	303.33	302.93	302.22	299.92
w_B (sol phase)	0.0489	0.0451	0.0408	0.0392	0.0365	0.0346	0.0311	0.0198
w_B (gel phase)				0.1673	0.1720	0.1780	0.1869	0.2404

Comments: The total feed concentration of the polymer in the homogeneous solution is w_B = 0.1505. The weight ratio of w_C/w_A = 1/3 was found to be constant along the coexistence curve in both phases.

T/K	301.40	301.31	301.19	300.93	300.71	300.35	299.92
w_B (sol phase)	0.0890	0.0851	0.0789	0.0700	0.0673	0.0605	0.0558
w_B (gel phase)	0.1592	0.1617	0.1696	0.1834	0.1968	0.2084	0.2229

continued

continued

Comments: The total feed concentration of the polymer in the homogeneous solution is $w_B = 0.1335$. The weight ratio of $w_C/w_A = 1/1$ was found to be constant along the coexistence curve in both phases.

T/K	308.93	308.75	308.20	307.75	307.41	307.10	306.65	306.13
w_B (sol phase)	0.0859	0.0786	0.0681	0.0606	0.0557	0.0539	0.0518	0.0441
w_B (gel phase)	0.1442	0.1552	0.1800	0.1989	0.2036	0.2129	0.2236	0.2363

Polymer (B): **polystyrene** **1991KAW**
Characterization: M_w/g.mol^{-1} = 48000,
Pressure Chemical Company, Pittsburgh, PA
Solvent (A): **cyclohexane** C_6H_{12} **110-82-7**
Solvent (C): **nitroethane** $C_2H_5NO_2$ **79-24-3**

Type of data: cloud points (UCST-behavior)

φ_A	0.396	0.493	0.509	0.546	0.591	0.662	0.707	0.749	0.797
φ_B	0.000	0.000	0.000	0.000	0.000	0.000	0.000	0.000	0.000
φ_C	0.604	0.507	0.491	0.454	0.409	0.338	0.293	0.251	0.203
T/K	290.95	294.64	294.95	295.59	295.92	295.98	295.97	295.72	294.90
φ_A	0.848	0.894	0.295	0.392	0.488	0.504	0.541	0.585	0.655
φ_B	0.000	0.000	0.010	0.010	0.010	0.010	0.010	0.010	0.010
φ_C	0.152	0.106	0.695	0.598	0.502	0.486	0.449	0.405	0.335
T/K	292.09	287.11	281.71	290.40	294.47	294.70	295.35	295.75	295.77
φ_A	0.700	0.742	0.789	0.292	0.388	0.483	0.499	0.535	0.579
φ_B	0.010	0.010	0.010	0.020	0.020	0.020	0.020	0.020	0.020
φ_C	0.290	0.248	0.201	0.688	0.592	0.497	0.481	0.445	0.401
T/K	295.70	295.58	295.12	281.00	289.83	294.19	294.42	295.06	295.55
φ_A	0.649	0.693	0.734	0.781	0.831	0.876	0.918	0.958	0.980
φ_B	0.020	0.020	0.020	0.020	0.020	0.020	0.020	0.020	0.020
φ_C	0.331	0.287	0.246	0.199	0.149	0.104	0.062	0.022	0.000
T/K	295.59	295.48	295.44	295.16	294.61	293.95	292.59	289.01	285.61
φ_A	0.000	0.037	0.106	0.289	0.384	0.478	0.494	0.530	0.573
φ_B	0.030	0.030	0.030	0.030	0.030	0.030	0.030	0.030	0.030
φ_C	0.970	0.933	0.864	0.681	0.586	0.492	0.476	0.440	0.397
T/K	299.13	291.23	279.26	280.27	289.26	293.84	294.12	294.74	295.32
φ_A	0.642	0.686	0.727	0.773	0.823	0.867	0.909	0.949	0.970
φ_B	0.030	0.030	0.030	0.030	0.030	0.030	0.030	0.030	0.030
φ_C	0.328	0.284	0.243	0.197	0.147	0.103	0.061	0.021	0.000
T/K	295.43	295.33	295.33	295.13	294.75	294.24	293.02	289.70	286.63
φ_A	0.000	0.036	0.105	0.286	0.380	0.473	0.489	0.524	0.567
φ_B	0.040	0.040	0.040	0.040	0.040	0.040	0.040	0.040	0.040
φ_C	0.960	0.924	0.855	0.674	0.580	0.487	0.471	0.436	0.393
T/K	300.58	292.67	280.71	279.51	288.69	293.43	293.78	294.39	295.08

continued

continued

φ_A	0.636	0.679	0.719	0.765	0.814	0.858	0.900	0.939	0.960
φ_B	0.040	0.040	0.040	0.040	0.040	0.040	0.040	0.040	0.040
φ_C	0.324	0.281	0.241	0.195	0.146	0.102	0.060	0.021	0.000
T/K	295.28	295.24	295.23	295.09	294.77	294.36	293.24	290.11	287.17
φ_A	0.000	0.036	0.104	0.283	0.376	0.468	0.484	0.519	0.561
φ_B	0.050	0.050	0.050	0.050	0.050	0.050	0.050	0.050	0.050
φ_C	0.950	0.914	0.846	0.667	0.574	0.482	0.466	0.431	0.389
T/K	301.55	293.66	281.61	278.73	288.11	292.93	293.39	294.02	294.80
φ_A	0.629	0.672	0.712	0.757	0.806	0.849	0.890	0.929	0.950
φ_B	0.050	0.050	0.050	0.050	0.050	0.050	0.050	0.050	0.050
φ_C	0.321	0.278	0.238	0.193	0.144	0.101	0.060	0.021	0.000
T/K	295.12	295.15	295.14	295.03	294.77	294.40	293.31	290.30	287.52
φ_A	0.000	0.036	0.102	0.280	0.372	0.463	0.478	0.513	0.556
φ_B	0.060	0.060	0.060	0.060	0.060	0.060	0.060	0.060	0.060
φ_C	0.940	0.904	0.838	0.660	0.568	0.477	0.462	0.427	0.384
T/K	302.19	294.27	282.21	277.81	287.53	292.37	292.96	293.65	294.50
φ_A	0.622	0.665	0.704	0.749	0.797	0.840	0.881	0.919	0.940
φ_B	0.060	0.060	0.060	0.060	0.060	0.060	0.060	0.060	0.060
φ_C	0.318	0.275	0.236	0.191	0.143	0.100	0.059	0.021	0.000
T/K	294.95	295.04	295.05	294.97	294.74	294.39	293.30	290.36	287.74
φ_A	0.000	0.364	0.454	0.468	0.502	0.544	0.609	0.650	0.689
φ_B	0.080	0.080	0.080	0.080	0.080	0.080	0.080	0.080	0.080
φ_C	0.920	0.556	0.466	0.452	0.418	0.376	0.311	0.270	0.231
T/K	302.81	286.20	291.20	291.95	292.84	293.79	294.50	294.71	294.82
φ_A	0.733	0.780	0.822	0.862	0.900	0.920	0.000	0.444	0.458
φ_B	0.080	0.080	0.080	0.080	0.080	0.080	0.100	0.100	0.100
φ_C	0.187	0.140	0.098	0.058	0.020	0.000	0.900	0.456	0.442
T/K	294.82	294.66	294.28	293.15	290.32	287.94	303.06	290.05	290.67
φ_A	0.491	0.532	0.596	0.636	0.674	0.717	0.763	0.805	0.843
φ_B	0.100	0.100	0.100	0.100	0.100	0.100	0.100	0.100	0.100
φ_C	0.409	0.368	0.304	0.264	0.226	0.183	0.137	0.095	0.057
T/K	291.78	292.99	293.96	294.30	294.52	294.64	294.54	294.07	292.92
φ_A	0.880	0.900							
φ_B	0.100	0.100							
φ_C	0.020	0.000							
T/K	290.18	287.96							

Polymer (B): **polystyrene** **1995LUS**

Characterization: M_n/g.mol^{-1} = 23600, M_w/g.mol^{-1} = 25000,
Pressure Chemical Company, Pittsburgh, PA

continued

continued

Solvent (A): **methylcyclopentane** C_6H_{12} **96-37-7**
Solvent (C): **dodecadeuteromethylcyclopentane** C_6D_{12} **144120-51-4**

Type of data: cloud points

$x_C/x_A = 0.943$ was kept constant

w_B	0.140	0.140	0.140	0.140	0.140
T/K	308.58	308.83	309.21	309.51	309.76
P/MPa	2.88	2.38	1.62	1.05	0.59

Polymer (B): **polystyrene** **1995LUS**
Characterization: M_n/g.mol^{-1} = 100300, M_w/g.mol^{-1} = 106300,
Pressure Chemical Company, Pittsburgh, PA
Solvent (A): **methylcyclopentane** C_6H_{12} **96-37-7**
Solvent (C): **dodecadeuteromethylcyclopentane** C_6D_{12} **144120-51-4**

Type of data: cloud points

Comments: The mole fraction ratio of C_6D_{12}/C_6H_{12} = 0.943 is kept constant.

$x_C/x_A = 0.943$ was kept constant

w_B	0.134	0.134	0.134	0.134
T/K	336.23	336.75	337.35	338.45
P/MPa	2.13	1.74	1.36	0.64

Polymer (B): **polystyrene-d8** **1995LUS**
Characterization: M_n/g.mol^{-1} = 25400, M_w/g.mol^{-1} = 26900,
completely deuterated, Polymer Laboratories, Amherst, MA
Solvent (A): **methylcyclopentane** C_6H_{12} **96-37-7**
Solvent (C): **dodecadeuteromethylcyclopentane** C_6D_{12} **144120-51-4**

Type of data: cloud points

$x_C/x_A = 0.943$ was kept constant

w_B	0.137	0.137	0.137	0.137
T/K	307.74	308.04	308.32	308.67
P/MPa	2.25	1.64	1.09	0.42

Polymer (B): **polystyrene** **1995LUS**
Characterization: M_n/g.mol^{-1} = 7300, M_w/g.mol^{-1} = 8000,
Pressure Chemical Company, Pittsburgh, PA
Solvent (A): **2-propanone** C_3H_6O **67-64-1**
Solvent (C): **1,1,1,3,3,3-hexadeutero-2-propanone** C_3D_6O **666-52-4**

Type of data: cloud points

$x_A/x_C = 0.546$ was kept constant

w_B	0.219	0.219	0.219	0.219	0.219	0.219
T/K	271.41	271.42	271.27	271.26	270.95	270.67
P/MPa	0.24	0.25	0.42	0.42	0.75	1.07

continued

continued

critical concentration: $w_{B, crit} = 0.210$

$x_A/x_C = 0.670$ was kept constant

w_B	0.211	0.211	0.211	0.211	0.211	0.211	0.211	0.211	0.211
T/K	275.75	275.78	275.72	275.55	275.35	275.10	274.90	274.76	274.54
P/MPa	0.06	0.06	0.07	0.18	0.38	0.63	0.84	1.02	1.27

critical concentration: $w_{B, crit} = 0.208$

Polymer (B): **polystyrene** **1995LUS**
Characterization: M_n/g.mol^{-1} = 12750, M_w/g.mol^{-1} = 13500,
Pressure Chemical Company, Pittsburgh, PA
Solvent (A): **2-propanone** C_3H_6O **67-64-1**
Solvent (C): **1,1,1,3,3,3-hexadeutero-2-propanone** C_3D_6O **666-52-4**

Type of data: cloud points

$x_C/x_A = 0.483$ was kept constant

w_B	0.205	0.205	0.205	0.205	0.205
T/K	302.77	303.49	304.93	306.86	307.34
P/MPa	1.89	1.63	1.27	0.63	0.37

Type of data: spinodal points

w_B	0.205	0.205	0.205	0.205	0.205
T/K	302.77	303.49	304.93	306.86	307.34
P/MPa	1.85	1.58	1.23	0.59	0.32

critical concentration: $w_{B, crit} = 0.200$

Type of data: cloud points

$x_C/x_A = 0.719$ was kept constant

w_B	0.207	0.207	0.207	0.207	0.207	0.207
T/K	313.76	318.51	319.15	316.83	315.61	314.73
P/MPa	1.99	0.58	0.42	1.00	1.39	1.66

Type of data: spinodal points

w_B	0.207	0.207	0.207	0.207	0.207	0.207
T/K	313.76	318.51	319.15	316.83	315.61	314.73
P/MPa	1.95	0.55	0.38	0.97	1.34	1.62

critical concentration: $w_{B, crit} = 0.200$

Type of data: cloud points

$x_C/x_A = 0.952$ was kept constant

w_B	0.206	0.206	0.206	0.206	0.206	0.206	0.206	0.206	0.206
T/K	330.82	330.82	331.84	332.91	333.91	335.96	338.37	340.36	343.28
P/MPa	2.11	2.09	1.86	1.67	1.53	1.15	0.90	0.61	0.34

continued

continued

w_B	0.206	0.206	0.206	0.206
T/K	373.09	375.93	378.69	384.87
P/MPa	0.59	0.77	0.99	1.57

Type of data: spinodal points

w_B	0.206	0.206	0.206	0.206	0.206	0.206	0.206	0.206	0.206
T/K	330.82	330.82	331.84	332.91	333.91	335.96	338.37	340.36	343.28
P/MPa	1.91	1.91	1.73	1.55	1.35	1.05	0.71	0.51	0.23

w_B	0.206	0.206	0.206	0.206
T/K	373.09	375.93	378.69	384.87
P/MPa	0.45	0.69	0.92	1.49

critical concentration: $w_{B, crit} = 0.200$

Polymer (B): **polysulfone** **1990QIN**
Characterization: Union Carbide Corp., New York, NY
Solvent (A): ***N,N*-dimethylacetamide** C_4H_9NO **127-19-5**
Solvent (C): **1-butanol** $C_4H_{10}O$ **71-36-3**

Type of data: cloud points

T/K = 293.15

φ_A	0.6083	0.6247	0.6369	0.6490	0.6583
φ_B	0.1546	0.1350	0.1204	0.1052	0.0933
φ_C	0.2371	0.2403	0.2427	0.2458	0.2484

Polymer (B): **polysulfone** **1990QIN**
Characterization: Union Carbide Corp., New York, NY
Solvent (A): ***N,N*-dimethylacetamide** C_4H_9NO **127-19-5**
Solvent (C): **ethanol** C_2H_6O **64-17-5**

Type of data: cloud points

T/K = 293.15

φ_A	0.6680	0.6927	0.6994	0.7260	0.7335	0.7395	0.7439	0.7481	0.7514
φ_B	0.1263	0.0982	0.0882	0.0610	0.0529	0.0466	0.0417	0.0377	0.0299
φ_C	0.2057	0.2091	0.2124	0.2130	0.2137	0.2139	0.2144	0.2142	0.2187

φ_A	0.7556	0.7587	0.7602	0.7617	0.7681
φ_B	0.0258	0.0226	0.0202	0.0182	0.0166
φ_C	0.2186	0.2187	0.2196	0.2202	0.2204

Polymer (B): **polysulfone** **1990QIN**
Characterization: Union Carbide Corp., New York, NY
Solvent (A): ***N,N*-dimethylacetamide** C_4H_9NO **127-19-5**
Solvent (C): **1-pentanol** $C_5H_{12}O$ **71-41-0**

continued

continued

Type of data: cloud points

T/K = 293.15

φ_A	0.6286	0.6404	0.6529	0.6558	0.6596	0.6628
φ_B	0.1189	0.1038	0.0875	0.0824	0.0780	0.0740
φ_C	0.2525	0.2558	0.2596	0.2619	0.2625	0.2633

Polymer (B): **polysulfone** **1985WIJ**
Characterization: M_n/g.mol^{-1} = 14000, M_w/g.mol^{-1} = 46000, M_z/g.mol^{-1} = 75000, Union Carbide Corp., New York, NY
Solvent (A): ***N,N*-dimethylacetamide** C_4H_9NO **127-19-5**
Solvent (C): **water** H_2O **7732-18-5**

Type of data: cloud points

T/K = 298.15

φ_A	0.9355	0.9399	0.9429	0.9449	0.9468	0.9486
φ_B	0.000018	0.000060	0.000095	0.00015	0.00024	0.00042
φ_C	0.0645	0.0600	0.0570	0.0550	0.0530	0.0510

Polymer (B): **polysulfone** **1987LIS**
Characterization: Type 3500, Union Carbide Corp., New York, NY
Solvent (A): ***N,N*-dimethylacetamide** C_4H_9NO **127-19-5**
Solvent (C): **water** H_2O **7732-18-5**

Type of data: cloud points

T/K = 293.15

w_A	0.766	0.793	0.824	0.821	0.838	0.841	0.854	0.896	0.926
w_B	0.201	0.170	0.148	0.141	0.126	0.120	0.110	0.0644	0.0334
w_C	0.0332	0.0363	0.0359	0.0374	0.0359	0.0385	0.0362	0.0400	0.0406
w_A	0.928	0.931	0.934	0.935	0.937	0.937	0.938	0.939	0.940
w_B	0.0298	0.0266	0.0240	0.0225	0.0211	0.0197	0.0194	0.0180	0.0168
w_C	0.0420	0.0421	0.0424	0.0426	0.0423	0.0429	0.0430	0.0431	0.0428
w_A	0.943	0.946	0.948	0.950	0.950				
w_B	0.0139	0.0103	0.00750	0.00262	0.00050				
w_C	0.0432	0.0435	0.0440	0.0473	0.0499				

T/K = 303.15

w_A	0.802	0.813	0.832	0.840	0.866	0.871	0.878	0.883	0.904
w_B	0.163	0.153	0.151	0.123	0.0939	0.0879	0.0810	0.0757	0.0547
w_C	0.0363	0.0386	0.0370	0.0377	0.0402	0.0400	0.0412	0.0414	0.0410
w_A	0.908	0.916	0.919	0.947	0.948	0.949			
w_B	0.0511	0.0415	0.0387	0.0065	0.0062	0.0047			
w_C	0.0410	0.0427	0.0424	0.0463	0.0460	0.0462			

continued

continued

T/K = 313.15

w_A	0.859	0.866	0.891	0.896	0.901	0.905	0.922	0.925	0.927
w_B	0.0994	0.0931	0.0653	0.0611	0.0552	0.0516	0.0237	0.0334	0.0274
w_C	0.0412	0.0412	0.0437	0.0423	0.0439	0.0437	0.0435	0.0440	0.0453
w_A	0.933	0.936	0.938	0.939	0.942	0.948	0.948		
w_B	0.0216	0.0189	0.0159	0.0150	0.0105	0.0050	0.0045		
w_C	0.0449	0.0466	0.0462	0.0461	0.0468	0.0472	0.0477		

T/K = 323.15

w_A	0.722	0.767	0.787	0.849	0.850	0.856	0.861	0.904	0.910
w_B	0.261	0.194	0.173	0.116	0.106	0.100	0.0947	0.0488	0.0422
w_C	0.0367	0.0389	0.0401	0.0342	0.0435	0.0431	0.0448	0.0469	0.0489
w_A	0.916	0.919	0.920	0.921	0.921	0.926			
w_B	0.0371	0.0340	0.0331	0.0321	0.0318	0.0263			
w_C	0.0473	0.0466	0.0464	0.0468	0.0472	0.0476			

T/K = 333.15

w_A	0.664	0.687	0.776	0.780	0.784	0.794	0.834	0.865	0.892
w_B	0.300	0.192	0.183	0.178	0.174	0.169	0.122	0.0895	0.0684
w_C	0.0358	0.0411	0.0411	0.0416	0.0423	0.0376	0.0469	0.0455	0.0399
w_A	0.895	0.889	0.893	0.929	0.931	0.932	0.934	0.935	0.937
w_B	0.0618	0.0627	0.0595	0.0225	0.0201	0.0181	0.0165	0.0151	0.0121
w_C	0.0400	0.0483	0.0476	0.0483	0.0499	0.0497	0.0495	0.0502	0.0504

Polymer (B): **polysulfone** **1990QIN**
Characterization: Union Carbide Corp., New York, NY
Solvent (A): ***N,N*-dimethylacetamide** C_4H_9NO **127-19-5**
Solvent (C): **water** H_2O **7732-18-5**

Type of data: cloud points

T/K = 293.15

φ_A	0.8121	0.8299	0.8528	0.8689	0.9121	0.9374	0.9378	0.9415	0.9425
φ_B	0.1536	0.1945	0.1107	0.0937	0.0496	0.0228	0.0203	0.0183	0.0172
φ_C	0.0344	0.0357	0.0365	0.0374	0.0383	0.0399	0.0399	0.0402	0.0404
φ_A	0.9439	0.9444	0.9445	0.9455	0.9467	0.9486	0.9511	0.9528	
φ_B	0.0161	0.0150	0.0148	0.0137	0.0128	0.0106	0.0078	0.0057	
φ_C	0.0401	0.0406	0.0407	0.0408	0.0405	0.0408	0.0411	0.0415	

Polymer (B): **polysulfone** **1999BAR**
Characterization: M_w/g.mol^{-1} = 39000, ρ_B (298 K) = 1.24 g/cm^3, T_g/K = 460, Ultrason S 3010, BASF AG, Ludwigshafen, Germany
Solvent (A): ***N,N*-dimethylformamide** C_3H_7NO **68-12-2**
Solvent (C): **ethanol** C_2H_6O **64-17-5**

continued

continued

Type of data: cloud points

T/K = 293.15

w_A	0.8730
w_B	0.0345
w_C	0.0925

T/K = 303.15

w_A	0.8692	0.8577
w_B	0.0344	0.0456
w_C	0.0964	0.0967

T/K = 313.15

w_A	0.8655	0.8537
w_B	0.0342	0.0454
w_C	0.1003	0.1009

T/K = 323.15

w_A	0.8626	0.8510
w_B	0.0341	0.0453
w_C	0.1033	0.1037

Polymer (B): **polysulfone** **1999BAR**
Characterization: M_w/g.mol^{-1} = 39000, ρ_B (298 K) = 1.24 g/cm^3, T_g/K = 460, Ultrason S 3010, BASF AG, Ludwigshafen, Germany
Solvent (A): ***N,N*-dimethylformamide** C_3H_7NO **68-12-2**
Solvent (C): **2-propanol** C_3H_8O **67-63-0**

Type of data: cloud points

T/K = 293.15

w_A	0.8585	0.8591	0.8580	0.8532	0.8408	0.8321	0.7901	0.7446
w_B	0.0049	0.0100	0.0129	0.0256	0.0382	0.0439	0.0863	0.1331
w_C	0.1366	0.1309	0.1291	0.1212	0.1210	0.1240	0.1236	0.1223

T/K = 303.15

w_A	0.8531	0.8555	0.8533	0.8494	0.8382	0.8297	0.7865	0.7433
w_B	0.0049	0.0100	0.0129	0.0255	0.0381	0.0438	0.0859	0.1329
w_C	0.1420	0.1345	0.1338	0.1251	0.1237	0.1265	0.1276	0.1238

T/K = 313.15

w_A	0.8480	0.8514	0.8498	0.8462	0.8346	0.8270	0.7839	0.7395	0.7397
w_B	0.0049	0.0099	0.0128	0.0254	0.0379	0.0436	0.0856	0.1293	0.1323
w_C	0.1471	0.1387	0.1374	0.1284	0.1275	0.1294	0.1305	0.1312	0.1280

continued

continued

T/K = 323.15

w_A	0.8436	0.8474	0.8465	0.8429	0.8309	0.8241	0.7813	0.7384	0.7368
w_B	0.0048	0.0099	0.0128	0.0253	0.0377	0.0435	0.0853	0.1291	0.1318
w_C	0.1516	0.1427	0.1407	0.1318	0.1314	0.1324	0.1334	0.1325	0.1314

Polymer (B): **polysulfone** **1999BAR**
Characterization: M_w/g.mol^{-1} = 39000, ρ_B (298 K) = 1.24 g/cm^3, T_g/K = 460, Ultrason S 3010, BASF AG, Ludwigshafen, Germany
Solvent (A): ***N,N*-dimethylformamide** C_3H_7NO **68-12-2**
Solvent (C): **2-propanone** C_3H_6O **67-64-1**

Type of data: cloud points

T/K = 293.15

w_A	0.7096	0.7247	0.7401	0.7399	0.7392	0.7364	0.7197	0.6963	0.6821
w_B	0.0025	0.0056	0.0095	0.0134	0.0161	0.0179	0.0375	0.0646	0.0750
w_C	0.2879	0.2697	0.2504	0.2467	0.2447	0.2457	0.2428	0.2391	0.2429

w_A	0.6633	0.6417
w_B	0.0934	0.1130
w_C	0.2433	0.2453

T/K = 303.15

w_A	0.6995	0.7157	0.7329	0.7332	0.7330	0.7308	0.7146	0.6906	0.6771
w_B	0.0025	0.0055	0.0094	0.0133	0.0160	0.0177	0.0372	0.0641	0.0745
w_C	0.2980	0.2788	0.2577	0.2535	0.2510	0.2515	0.2482	0.2453	0.2484

w_A	0.6587	0.6365
w_B	0.0927	0.1122
w_C	0.2486	0.2513

T/K = 313.15

w_A	0.6895	0.7074	0.7251	0.7264	0.7266	0.7252	0.7092	0.6854	0.6718
w_B	0.0025	0.0055	0.0093	0.0132	0.0158	0.0176	0.0369	0.0636	0.0739
w_C	0.3080	0.2871	0.2656	0.2604	0.2576	0.2572	0.2539	0.2510	0.2543

w_A	0.6543	0.6324
w_B	0.0921	0.1114
w_C	0.2536	0.2562

T/K = 323.15

w_A	0.6802	0.7001	0.7189	0.7203	0.7207	0.7205	0.7043	0.6809	0.6681
w_B	0.0024	0.0054	0.0092	0.0131	0.0157	0.0175	0.0367	0.0632	0.0735
w_C	0.3174	0.2945	0.2719	0.2666	0.2636	0.2620	0.2590	0.2559	0.2584

w_A	0.6500	0.6281
w_B	0.0915	0.1107
w_C	0.2585	0.2612

continued

continued

Type of data: coexistence data (tie lines)

T/K = 303.15

Total system			Sol phase			Gel phase		
w_A	w_B	w_C	w_A	w_B	w_C	w_A	w_B	w_C
0.6719	0.0751	0.2530	0.6894	0.0680	0.2426	0.6239	0.1326	0.2435
0.6642	0.0764	0.2594	0.6815	0.0573	0.2612	0.5828	0.1613	0.2559
0.6575	0.0748	0.2677	0.6899	0.0471	0.2630	0.5619	0.1859	0.2522

Plait-point composition: $w_A = 0.6355 + w_B = 0.1130 + w_C = 0.2515$

Polymer (B): **polysulfone** **1999BAR, 2000BA1**
Characterization: M_w/g.mol^{-1} = 39000, ρ_B (298 K) = 1.24 g/cm^3, T_g/K = 460, Ultrason S 3010, BASF AG, Ludwigshafen, Germany
Solvent (A): ***N,N*-dimethylformamide** C_3H_7NO **68-12-2**
Solvent (C): **water** H_2O **7732-18-5**

Type of data: cloud points

T/K = 293.15

w_A	0.9689	0.9538	0.9400	0.9005	0.8593	0.8207
w_B	0.0099	0.0262	0.0409	0.0803	0.1224	0.1614
w_C	0.0212	0.0200	0.0191	0.0192	0.0183	0.0179

T/K = 303.15

w_A	0.9675	0.9529	0.9396	0.9006	0.8601	0.8221
w_B	0.0097	0.0257	0.0402	0.0794	0.1203	0.1588
w_C	0.0228	0.0214	0.0202	0.0200	0.0196	0.0191

T/K = 313.15

w_A	0.9660	0.9520	0.9388	0.9008	0.8610	0.8238
w_B	0.0095	0.0251	0.0393	0.0780	0.1182	0.1558
w_C	0.0245	0.0229	0.0219	0.0212	0.0208	0.0204

T/K = 323.15

w_A	0.9642	0.9507	0.9381	0.9009	0.8620	0.8256
w_B	0.0091	0.0244	0.0384	0.0763	0.1157	0.1524
w_C	0.0267	0.0249	0.0235	0.0228	0.0223	0.0220

Type of data: coexistence data (tie lines)

T/K = 303.15

continued

continued

Total system			Sol phase			Gel phase		
w_A	w_B	w_C	w_A	w_B	w_C	w_A	w_B	w_C
0.903	0.076	0.021	0.927	0.049	0.024	0.798	0.177	0.025
0.894	0.083	0.023	0.944	0.037	0.019	0.743	0.234	0.023
0.891	0.082	0.027	0.942	0.024	0.034	0.687	0.292	0.021

Plait-point composition: $w_A = 0.871 + w_B = 0.110 + w_C = 0.019$

Polymer (B): **poly(vinyl alcohol)** **2002YOU**
Characterization: M_w/g.mol^{-1} = 74800, 99.2 mol% saponification, ρ = 1.26 g/cm^3, T_m/K = 499.9, Chang Chun Co., Ltd., Taiwan
Solvent (A): **water** **H_2O** **7732-18-5**
Solvent (C): **dimethylsulfoxide** **C_2H_6OS** **67-68-5**

Type of data: cloud points (LCST-behavior)

T/K = 298.15

w_A	0.65	0.68	0.72	0.75	0.77	0.79	0.80	0.11	0.10
w_B	0.20	0.18	0.15	0.12	0.10	0.07	0.05	0.18	0.15
w_C	0.15	0.14	0.13	0.13	0.13	0.14	0.15	0.71	0.75

w_A	0.09	0.10	0.10	0.11
w_B	0.12	0.10	0.07	0.05
w_C	0.79	0.80	0.83	0.84

φ_A	0.6878	0.7157	0.7522	0.7785	0.7958	0.8121	0.8196	0.1225	0.1110
φ_B	0.1680	0.1504	0.1244	0.0989	0.0820	0.0571	0.0407	0.1590	0.1321
φ_C	0.1442	0.1339	0.1234	0.1226	0.1221	0.1308	0.1397	0.7185	0.7568

φ_A	0.0996	0.1109	0.1099	0.1204
φ_B	0.1054	0.0870	0.0610	0.0434
φ_C	0.7950	0.8021	0.8291	0.8361

Polymer (B): **poly(vinyl alcohol)** **1993MAT**
Characterization: M_w/g.mol^{-1} = 88000, M_w/g.mol^{-1} = 281600, 98% hydrolysis
Solvent (A): **water** **H_2O** **7732-18-5**
Solvent (C): **dimethylsulfoxide** **C_2H_6OS** **67-68-5**

Type of data: spinodal points

w_B	0.01	0.05	0.10	0.01	0.05	0.10
φ_A/φ_C	50/50	50/50	50/50	30/70	30/70	30/70
T/K	296.55	304.35	309.95	293.85	303.15	307.75

Polymer (B):	**poly(*N*-vinylcaprolactam)**		**2002MA1**
Characterization:	M_w/g.mol^{-1} = 13000, synthesized in the laboratory		
Solvent (A):	**deuterium oxide**	**D_2O**	**7789-20-0**
Solvent (C):	**methanol-d4**	**CD_4O**	**811-98-3**

Type of data: cloud points (LCST-behavior)

w_B	0.20	0.20	0.20	0.20	0.20	0.20	0.20
φ_C	0.0	0.10	0.20	0.30	0.40	0.45	0.50
T/K	305.85	307.05	306.75	306.85	308.45	317.45	317.45

Polymer (B):	**poly(*N*-vinylcaprolactam)**		**2002MA1**
Characterization:	M_w/g.mol^{-1} = 13000, synthesized in the laboratory		
Solvent (A):	**water**	**H_2O**	**7732-18-5**
Solvent (C):	**methanol**	**CH_4O**	**67-56-1**

Type of data: cloud points (LCST-behavior)

w_B	0.005	0.005	0.005	0.005	0.005	0.005	0.005	0.005
φ_C	0.00	0.10	0.20	0.30	0.35	0.40	0.425	0.45
T/K	307.85	308.65	308.75	309.75	310.25	312.85	318.85	322.85

Polymer (B):	**poly(vinyl chloride)**		**1978HAY**
Characterization:	Esso 363, Imperial Oil Enterprises		
Solvent (A):	**tetrahydrofuran**	**C_4H_8O**	**109-99-9**
Solvent (C):	**1,2-ethanediol**	**$C_2H_6O_2$**	**107-21-1**

Type of data: cloud points

T/K = 295

w_A	0.6948	0.7094	0.7078	0.7046
w_B	0.0022	0.0046	0.0092	0.0184
w_C	0.303	0.286	0.283	0.277

Polymer (B):	**poly(vinyl chloride)**		**1978HAY**
Characterization:	Esso 363, Imperial Oil Enterprises		
Solvent (A):	**tetrahydrofuran**	**C_4H_8O**	**109-99-9**
Solvent (C):	**ethanol**	**C_2H_6O**	**64-17-5**

Type of data: cloud points

T/K = 295

w_A	0.4854	0.5217	0.5361	0.5682
w_B	0.0016	0.0033	0.0069	0.0148
w_C	0.513	0.475	0.457	0.417

Polymer (B): **poly(vinyl chloride)** **1978HAY**
Characterization: Esso 363, Imperial Oil Enterprises
Solvent (A): **tetrahydrofuran** C_4H_8O **109-99-9**
Solvent (C): **n-heptane** C_7H_{16} **142-82-5**

Type of data: cloud points

T/K = 295

w_A	0.6219	0.6269	0.6337	0.6404	0.6470
w_B	0.0031	0.0061	0.0093	0.0126	0.0160
w_C	0.375	0.367	0.357	0.347	0.337

Polymer (B): **poly(vinyl chloride)** **1978HAY**
Characterization: Esso 363, Imperial Oil Enterprises
Solvent (A): **tetrahydrofuran** C_4H_8O **109-99-9**
Solvent (C): **methanol** CH_4O **67-56-1**

Type of data: cloud points

T/K = 295

w_A	0.5567	0.5585	0.5857	0.5981
w_B	0.0013	0.0025	0.0053	0.0109
w_C	0.442	0.439	0.409	0.391

Polymer (B): **poly(vinyl chloride)** **1978HAY**
Characterization: PV2, Pressure Chemical Company, Pittsburgh, PA
Solvent (A): **tetrahydrofuran** C_4H_8O **109-99-9**
Solvent (C): **methanol** CH_4O **67-56-1**

Type of data: cloud points

T/K = 295

w_A	0.5338	0.5816	0.6050	0.5995
w_B	0.0012	0.0024	0.0050	0.0105
w_C	0.465	0.416	0.390	0.390

Polymer (B): **poly(vinyl chloride)** **1978HAY**
Characterization: PV4, Pressure Chemical Company, Pittsburgh, PA
Solvent (A): **tetrahydrofuran** C_4H_8O **109-99-9**
Solvent (C): **methanol** CH_4O **67-56-1**

Type of data: cloud points

T/K = 295

w_A	0.5449	0.5787	0.5873	0.6131
w_B	0.0011	0.0023	0.0047	0.0099
w_C	0.454	0.419	0.408	0.377

Polymer (B): **poly(vinyl chloride)** **1985GE2**
Characterization: M_n/g.mol^{-1} = 16700, M_w/g.mol^{-1} = 20000, fractionated in the laboratory
Solvent (A): **tetrahydrofuran** **C_4H_8O** **109-99-9**
Solvent (C): **water** **H_2O** **7732-18-5**

Type of data: cloud points (LCST-behavior)

w_A	0.8145	0.8145	0.8145	0.8094	0.8094	0.8094	0.8065	0.8065	0.8065
w_B	0.0795	0.0795	0.0795	0.0790	0.0790	0.0790	0.0787	0.0787	0.0787
w_C	0.1060	0.1060	0.1060	0.1116	0.1116	0.1116	0.1148	0.1148	0.1148
T/K	283.85	286.85	290.25	295.85	299.05	303.45	303.55	306.75	311.55
P/bar	1	500	1000	1	500	1000	1	500	1000

Polymer (B): **poly(vinyl chloride)** **1985GE2**
Characterization: M_n/g.mol^{-1} = 31400, M_w/g.mol^{-1} = 37000, fractionated in the laboratory
Solvent (A): **tetrahydrofuran** **C_4H_8O** **109-99-9**
Solvent (C): **water** **H_2O** **7732-18-5**

Type of data: cloud points (LCST-behavior)

w_A	0.8194	0.8194	0.8194	0.8151	0.8151	0.8151	0.8097	0.8097	0.8097
w_B	0.0800	0.0800	0.0800	0.0796	0.0796	0.0796	0.0789	0.0789	0.0789
w_C	0.1006	0.1006	0.1006	0.1053	0.1053	0.1053	0.1114	0.1114	0.1114
T/K	282.85	286.95	291.35	293.25	298.25	302.55	307.15	310.25	315.65
P/bar	1	500	1000	1	500	1000	1	500	1000

Polymer (B): **poly(vinyl chloride)** **1985GE2**
Characterization: M_n/g.mol^{-1} = 58800, M_w/g.mol^{-1} = 70000, fractionated in the laboratory
Solvent (A): **tetrahydrofuran** **C_4H_8O** **109-99-9**
Solvent (C): **water** **H_2O** **7732-18-5**

Type of data: cloud points (LCST-behavior)

w_A	0.8546	0.8546	0.8546	0.8498	0.8498	0.8498	0.8547	0.8547	0.8547
w_B	0.0200	0.0200	0.0200	0.0200	0.0200	0.0200	0.0199	0.0199	0.0199
w_C	0.1254	0.1254	0.1254	0.1302	0.1302	0.1302	0.1254	0.1254	0.1254
T/K	279.65	283.95	292.65	293.65	302.05	309.65	309.65	319.15	327.15
P/bar	1	500	1000	1	500	1000	1	500	1000
w_A	0.8439	0.8439	0.8439	0.8389	0.8389	0.8389	0.8334	0.8334	0.8334
w_B	0.0407	0.0407	0.0407	0.0405	0.0405	0.0405	0.0403	0.0403	0.0403
w_C	0.1154	0.1154	0.1154	0.1206	0.1206	0.1206	0.1263	0.1263	0.1263
T/K	284.05	289.25	296.05	299.25	305.25	311.55	310.95	319.75	326.15
P/bar	1	500	1000	1	500	1000	1	500	1000

continued

continued

w_A	0.8338	0.8338	0.8338	0.8292	0.8292	0.8292	0.8245	0.8245	0.8245
w_B	0.0613	0.0613	0.0613	0.0609	0.0609	0.0609	0.0607	0.0607	0.0607
w_C	0.1049	0.1049	0.1049	0.1099	0.1099	0.1099	0.1148	0.1148	0.1148
T/K	285.25	287.85	293.15	296.75	301.65	306.05	308.75	314.35	320.45
P/bar	1	500	1000	1	500	1000	1	500	1000
w_A	0.8244	0.8244	0.8244	0.8200	0.8200	0.8200	0.8150	0.8150	0.8150
w_B	0.0952	0.0952	0.0952	0.0999	0.0999	0.0999	0.1056	0.1056	0.1056
w_C	0.0952	0.0952	0.0952	0.0999	0.0999	0.0999	0.1056	0.1056	0.1056
T/K	275.45	277.55	280.35	287.95	290.15	294.45	299.85	304.65	308.45
P/bar	1	500	1000	1	500	1000	1	500	1000

Polymer (B): **poly(vinyl chloride)** **1978HAY**
Characterization: Esso 363, Imperial Oil Enterprises
Solvent (A): **tetrahydrofuran** C_4H_8O **109-99-9**
Solvent (C): **water** H_2O **7732-18-5**

Type of data: cloud points

T/K = 295

w_A	0.84233	0.8496	0.8523	0.8495	0.8441	0.8528
w_B	0.00067	0.0014	0.0027	0.0055	0.0109	0.0222
w_C	0.157	0.149	0.145	0.145	0.145	0.125

Polymer (B): **poly(vinyl methyl ether)** **1990OTA**
Characterization: M_w/g.mol^{-1} = 15000, Toyo Kasei Kogyo Co., Japan
Solvent (A): **water** H_2O **7732-18-5**
Solvent (C): **dimethylsulfoxide** C_2H_6OS **67-68-5**

Type of data: cloud points (LCST-behavior)

w_B = 0.060 w_C = 0.039 T/K = 305.95

Polymer (B): **poly(vinyl methyl ether)** **1990OTA**
Characterization: M_w/g.mol^{-1} = 57000, Toyo Kasei Kogyo Co., Japan
Solvent (A): **water** H_2O **7732-18-5**
Solvent (C): **ethanol** C_2H_6O **64-17-5**

Type of data: cloud points (LCST-behavior)

w_B = 0.060 w_C = 0.024 T/K = 308.05

Polymer (B): **poly(vinyl methyl ether)** **1990OTA**
Characterization: M_w/g.mol^{-1} = 15000, Toyo Kasei Kogyo Co., Japan
Solvent (A): **water** H_2O **7732-18-5**
Solvent (C): **methanol** CH_4O **67-56-1**

Type of data: cloud points (LCST-behavior)

w_B = 0.062 w_C = 0.0165 T/K = 306.85

Polymer (B):	**poly(vinyl methyl ether)**		**1990OTA**
Characterization:	M_w/g.mol^{-1} = 57000, Toyo Kasei Kogyo Co., Japan		
Solvent (A):	**water**	H_2O	**7732-18-5**
Solvent (C):	**2-propanol**	C_3H_8O	**67-63-0**

Type of data: cloud points (LCST-behavior)

w_B = 0.060 w_C = 0.030 T/K = 307.35

Polymer (B):	**pullulan**		**2004ECK**
Characterization:	M_n/g.mol^{-1} = 150000, M_w = 430000, M_z = 1550000, linear polysaccharid mainly composed of α-(1-6) D-maltotriose units, Polymer Standard Service, Mainz, Germany		
Solvent (A):	**water**	H_2O	**7732-18-5**
Solvent (C):	**2-propanol**	C_3H_8O	**67-63-0**

Type of data: cloud points

T/K = 298.15

w_A	0.6927	0.7117
w_B	0.0766	0.0488
w_C	0.2306	0.2395

Polymer (B):	**pullulan**		**2004ECK**
Characterization:	M_n/g.mol^{-1} = 150000, M_w = 430000, M_z = 1550000, linear polysaccharid mainly composed of α-(1-6) D-maltotriose units, Polymer Standard Service, Mainz, Germany		
Solvent (A):	**water**	H_2O	**7732-18-5**
Solvent (C):	**2-propanone**	C_3H_6O	**67-64-1**

Type of data: cloud points

T/K = 298.15

w_A	0.6194	0.6471	0.6752	0.6849	0.6989	0.7109	0.7157	0.7273	0.7339
w_B	0.2377	0.2149	0.1411	0.1211	0.1002	0.0790	0.0835	0.0605	0.0550
w_C	0.1429	0.1379	0.1836	0.1940	0.2009	0.2101	0.2008	0.2122	0.2112
w_A	0.7447	0.7506	0.7517	0.7564	0.7595	0.7620	0.7629	0.7663	0.7710
w_B	0.0428	0.0404	0.0350	0.0300	0.0262	0.0232	0.0204	0.0155	0.0123
w_C	0.2125	0.2090	0.2134	0.2136	0.2143	0.2148	0.2167	0.2182	0.2167
w_A	0.7717	0.7744	0.7746	0.7752					
w_B	0.0095	0.0078	0.0067	0.0058					
w_C	0.2189	0.2178	0.2187	0.2190					

Type of data: coexistence data (tie lines)

T/K = 298.15

continued

continued

Total system			Sol phase			Gel phase		
w_A	w_B	w_C	w_A	w_B	w_C	w_A	w_B	w_C
0.6182	0.0059	0.3759	0.6209	0.0013	0.3778	0.3641	0.4325	0.2033
0.6871	0.0764	0.2364	0.7321	0.0197	0.2482	0.5403	0.2628	0.1969
0.6944	0.0065	0.2992	0.6974	0.0017	0.3009	0.4484	0.3929	0.1587
0.6987	0.0776	0.2236	0.7405	0.0353	0.2242	0.6405	0.1730	0.1865
0.7188	0.0547	0.2265	0.7357	0.0270	0.2373	0.6342	0.1904	0.1754
0.7285	0.0266	0.2449	0.7365	0.0115	0.2520	0.5560	0.2651	0.1789
0.7633	0.0089	0.2278	0.7640	0.0077	0.2283	0.6786	0.1525	0.1689

critical concentrations: $w_{A,\,crit} = 0.692$, $w_{B,\,crit} = 0.111$, $w_{C,\,crit} = 0.197$

Polymer (B): **pullulan** **2004ECK**
Characterization: M_n/g.mol^{-1} = 150000, M_w = 430000, M_z = 1550000, linear polysaccharid mainly composed of α-(1-6) D-maltotriose units, Polymer Standard Service, Mainz, Germany
Solvent (A): **water** H_2O **7732-18-5**
Solvent (C): **tetrahydrofuran** C_4H_8O **109-99-9**

Type of data: cloud points

T/K = 298.15

w_A	0.6545	0.6644	0.6703
w_B	0.0724	0.0573	0.0483
w_C	0.2731	0.2783	0.2814

3.2. Cloud-point and/or coexistence data of one solvent and two polymers

Polymer (B): **cellulose** **1993MAR**
Characterization: M_w/g.mol^{-1} = 150000, DP = 925
Solvent (A): ***N,N*-dimethylacetamide** **C_4H_9NO** **127-19-5**
Polymer (C): **poly(acrylonitrile)**
Characterization: M_w/g.mol^{-1} = 102000, BASF

Type of data: coexistence data (tie lines)

T/K = 293.15 (the solvent contains 7% LiCl)

Total system			Phase I			Phase II		
φ_A	φ_B	φ_C	φ_A	φ_B	φ_C	φ_A	φ_B	φ_C
0.9546	0.0196	0.0258	0.9556	0.0281	0.0163	0.9431	0.0087	0.0482
0.9428	0.0245	0.0327	0.9515	0.0348	0.0137	0.9345	0.0091	0.0564
0.9155	0.0362	0.0483	0.9291	0.0595	0.0114	0.8922	0.0039	0.1039

Polymer (B): **cellulose acetate** **2003SIL**
Characterization: M_n/g.mol^{-1} = 30000, M_w/g.mol^{-1} = 70500, average degree of substitution = 2.5, ρ = 1.30 g/cm^3 at 298.13 K, Aldrich Chem. Co., Inc., Milwaukee, WI
Solvent (A): **tetrahydrofuran** **C_4H_8O** **109-99-9**
Polymer (C): **polystyrene**
Characterization: M_n/g.mol^{-1} = 250000, M_w/g.mol^{-1} = 682500, ρ = 1.03 g/cm^3 at 298.13 K, Proquigel, Brazil

Type of data: coexistence data (tie lines)

T/K = 298.15

Total system			PS-rich phase			CA-rich phase		
w_A	w_B	w_C	w_A	w_B	w_C	w_A	w_B	w_C
0.953	0.028	0.019	0.951	0.020	0.029	0.950	0.039	0.011
0.946	0.032	0.022	0.945	0.020	0.035	0.945	0.047	0.008
0.940	0.035	0.025	0.934	0.021	0.045	0.935	0.058	0.007

critical concentrations: $w_{A,\,crit}$ = 0.958, $w_{B,\,crit}$ = 0.025, $w_{C,\,crit}$ = 0.017

Polymer (B): **dextran** **1986ALB**
Characterization: M_n/g.mol^{-1} = 180000, M_w/g.mol^{-1} = 460000,
Dextran 500, Pharmacia Fine Chemicals, Uppsala, Sweden
Solvent (A): **water** **H_2O** **7732-18-5**
Polymer (C): **hydroxypropyldextran**
Characterization: M_w/g.mol^{-1} = 500000, D.S. = 0.39 per glucose unit

Type of data: coexistence data (tie lines)

T/K = 296.15

Total system			Top phase			Bottom phase		
w_A	w_B	w_C	w_A	w_B	w_C	w_A	w_B	w_C
0.8900	0.0600	0.0500	0.8938	0.0422	0.0640	0.8872	0.0824	0.0304
0.8800	0.0700	0.0500	0.8855	0.0371	0.0774	0.8770	0.0996	0.0230
0.8600	0.0800	0.0600	0.8678	0.0246	0.1076	0.8558	0.1208	0.0234

Polymer (B): **dextran** **1986ALB**
Characterization: M_n/g.mol^{-1} = 280000, M_w/g.mol^{-1} = 2200000,
Dextran 2200, Pharmacia Fine Chemicals, Uppsala, Sweden
Solvent (A): **water** **H_2O** **7732-18-5**
Polymer (C): **hydroxypropyldextran**
Characterization: M_n/g.mol^{-1} = 75000, D.S. = 1.0 per glucose unit

Type of data: coexistence data (tie lines)

T/K = 277.15

Total system			Top phase			Bottom phase		
w_A	w_B	w_C	w_A	w_B	w_C	w_A	w_B	w_C
0.9600	0.0200	0.0200	0.9616	0.0137	0.0247	0.9592	0.0256	0.0152
0.9500	0.0250	0.0250	0.9545	0.0077	0.0378	0.9485	0.0420	0.0095
0.9400	0.0300	0.0300	0.9425	0.0056	0.0519	0.9367	0.0551	0.0082
0.9200	0.0400	0.0400	0.9248	0.0039	0.0713	0.9153	0.0772	0.0075
0.9000	0.0500	0.0500	0.9075	0.0029	0.0896	0.8940	0.0988	0.0072
0.8800	0.0600	0.0600	0.8909	0.0024	0.1067	0.8711	0.1238	0.0051

Polymer (B): **dextran** **1986ALB**
Characterization: M_n/g.mol^{-1} = 280000, M_w/g.mol^{-1} = 2200000,
Dextran 2200, Pharmacia Fine Chemicals, Uppsala, Sweden

continued

continued

Solvent (A): **water** H_2O **7732-18-5**
Polymer (C): **methylcellulose**
Characterization: M_η/g.mol^{-1} = 30000, Methocel 10, Dow Chemical Company

Type of data: coexistence data (tie lines)

T/K = 293.15

Total system			Top phase			Bottom phase		
w_A	w_B	w_C	w_A	w_B	w_C	w_A	w_B	w_C
0.9688	0.0172	0.0140	0.9722	0.0100	0.0178	0.9673	0.0198	0.0129
0.9640	0.0200	0.0160	0.9707	0.0073	0.0220	0.9595	0.0285	0.0120
0.9600	0.0220	0.0180	0.9684	0.0055	0.0261	0.9533	0.0345	0.0122

Polymer (B): **dextran** **1986ALB**
Characterization: M_n/g.mol^{-1} = 280000, M_w/g.mol^{-1} = 2200000,
Dextran 2200, Pharmacia Fine Chemicals, Uppsala, Sweden
Solvent (A): **water** H_2O **7732-18-5**
Polymer (C): **methylcellulose**
Characterization: M_η/g.mol^{-1} = 80000, Methocel 400, Dow Chemical Company

Type of data: coexistence data (tie lines)

T/K = 293.15

Total system			Top phase			Bottom phase		
w_A	w_B	w_C	w_A	w_B	w_C	w_A	w_B	w_C
0.9840	0.0100	0.0060	0.9875	0.0047	0.0078	0.9791	0.0176	0.0033
0.9830	0.0100	0.0070	0.9867	0.0033	0.0100	0.9770	0.0201	0.0029
0.9800	0.0120	0.0080	0.9849	0.0028	0.0123	0.9732	0.0242	0.0026
0.9690	0.0200	0.0110	0.9774	0.0024	0.0202	0.9606	0.0374	0.0020

Polymer (B): **dextran** **1986ALB**
Characterization: M_n/g.mol^{-1} = 280000, M_w/g.mol^{-1} = 2200000,
Dextran 2200, Pharmacia Fine Chemicals, Uppsala, Sweden
Solvent (A): **water** H_2O **7732-18-5**
Polymer (C): **methylcellulose**
Characterization: M_η/g.mol^{-1} = 140000, Methocel 4000, Dow Chemical Company

Type of data: coexistence data (tie lines)

continued

continued

T/K = 277.15

Total system			Top phase			Bottom phase		
w_A	w_B	w_C	w_A	w_B	w_C	w_A	w_B	w_C
0.9890	0.0080	0.0030	0.9906	0.0056	0.0038	0.9862	0.0121	0.0017
0.9896	0.0068	0.0036	0.9906	0.0051	0.0043	0.9849	0.0133	0.0018
0.9872	0.0080	0.0048	0.9899	0.0038	0.0063	0.9814	0.0172	0.0014
0.9810	0.0120	0.0070	0.9863	0.0025	0.0112	0.9735	0.0251	0.0014
0.9830	0.0050	0.0120	0.9845	0.0018	0.0137	0.9703	0.0288	0.0009

T/K = 293.15

Total system			Top phase			Bottom phase		
w_A	w_B	w_C	w_A	w_B	w_C	w_A	w_B	w_C
0.9891	0.0079	0.0030	0.9907	0.0057	0.0036	0.9874	0.0109	0.0017
0.9896	0.0068	0.0036	0.9910	0.0047	0.0043	0.9865	0.0118	0.0017
0.9854	0.0110	0.0036	0.9896	0.0039	0.0065	0.9827	0.0158	0.0015
0.9788	0.0158	0.0054	0.9857	0.0028	0.0115	0.9746	0.0242	0.0012
0.9720	0.0200	0.0080	0.9799	0.0023	0.0178	0.9663	0.0327	0.0010

Polymer (B): **dextran** **1996MIS**
Characterization: M_w/g.mol^{-1} = 515000,
Pharmacia Fine Chemicals, Uppsala, Sweden
Solvent (A): **water** H_2O **7732-18-5**
Polymer (C): **poly(ethylene glycol) monomethyl ether**
Characterization: M_w/g.mol^{-1} = 5000, Sigma Chemical Co., Inc., St. Louis, MO

Type of data: coexistence data (tie lines)

T/K = 298.15

Total system			Top phase			Bottom phase		
w_A	w_B	w_C	w_A	w_B	w_C	w_A	w_B	w_C
0.8729	0.0875	0.0396	0.9063	0.0000	0.0937	0.8518	0.1297	0.0185
0.8440	0.1072	0.0488	0.8967	0.0000	0.1033	0.8067	0.1787	0.0146
0.8135	0.1280	0.0585	0.8745	0.0000	0.1255	0.7556	0.2353	0.0091
0.7885	0.1421	0.0694	0.8530	0.0000	0.1470	0.7123	0.2817	0.0060

Polymer (B):	**dextran**	**1998KIS**
Characterization:	M_w/g.mol^{-1} = 170000-200000, Nacalai Tesque, Kyoto, Japan	
Solvent (A):	**water** H_2O	**7732-18-5**
Polymer (C):	**poly(*N*-vinylacetamide)**	
Characterization:	M_w/g.mol^{-1} = 400000, synthesized in the laboratory	

Type of data: coexistence data (tie lines)

T/K = 277.15

Total system		Top phase			Bottom phase		
c_B/(mol/l)	c_C/(mol/l)	w_A	w_B	w_C	w_A	w_B	w_C
1.2	3.5	0.935	0.035	0.030	0.936	0.053	0.011
1.7	6.0	0.908	0.017	0.075	0.893	0.102	0.005
2.0	8.1	0.849	0.014	0.137	0.873	0.127	0.000
3.3	10.5	0.852	0.011	0.137	0.815	0.185	0.000

Polymer (B):	**poly(γ-benzyl-L-glutamate)**	**1986SAS**
Characterization:	M_η/g.mol^{-1} = 350000, synthesized in the laboratory	
Solvent (A):	**benzyl alcohol** C_7H_8O	**100-51-6**
Polymer (C):	**polystyrene**	
Characterization:	M_n/g.mol^{-1} = 220000, M_w/g.mol^{-1} = 233000, Pressure Chemical Company, Pittsburgh, PA	

Type of data: coexistence data (tie lines)

T/K = 353.15

Total system			Phase I			Phase II		
φ_A	φ_B	φ_C	φ_A	φ_B	φ_C	φ_A	φ_B	φ_C
0.950	0.025	0.025	0.950	0.030	0.020	0.995	0.001	0.004
0.920	0.040	0.040	0.920	0.060	0.020	0.920	0.016	0.064
0.870	0.075	0.055	0.881	0.116	0.003	0.859	0.003	0.138
0.874	0.085	0.041	0.883	0.114	0.003	0.856	0.007	0.137

Comments: Additional data for the nematic-isotropic equilibrium in this system are given in the original source.

Polymer (B):	**poly(butyl methacrylate)**	**1996KRA**
Characterization:	M_n/g.mol^{-1} = 325000, M_w/g.mol^{-1} = 335000, fractionated in the laboratory	

continued

continued

Solvent (A): **cyclohexanone** $C_6H_{10}O$ **108-94-1**
Polymer (C): **polystyrene**
Characterization: M_n/g.mol^{-1} = 66000, M_w/g.mol^{-1} = 196000,
J 110 Vestyron, Huels AG, Marl, Germany

Type of data: cloud points

w_A	0.8062	0.8074	0.8101	0.8125	0.8133	0.8151	0.8162	0.8185	0.8208
w_B	0.0174	0.0173	0.0171	0.0169	0.0168	0.0166	0.0165	0.0163	0.0161
w_C	0.1764	0.1753	0.1728	0.1706	0.1699	0.1683	0.1673	0.1652	0.1631
T/K	310.22	313.31	315.84	318.10	323.76	328.67	331.26	338.80	346.82
w_A	0.8235	0.8265	0.8343	0.8348	0.8352	0.8355	0.8370	0.8383	0.8402
w_B	0.0159	0.0156	0.0414	0.0413	0.0412	0.0411	0.0408	0.0404	0.0400
w_C	0.1606	0.1579	0.1243	0.1239	0.1236	0.1234	0.1223	0.1213	0.1199
T/K	358.74	372.93	307.51	311.87	319.94	324.84	330.26	335.47	351.49
w_A	0.8418	0.8434	0.8371	0.8388	0.8398	0.8406	0.8422	0.8432	0.8440
w_B	0.0396	0.0392	0.0815	0.0806	0.0801	0.0797	0.0789	0.0784	0.0780
w_C	0.1187	0.1175	0.0815	0.0806	0.0801	0.0797	0.0789	0.0784	0.0780
T/K	358.70	371.61	310.58	320.31	325.40	329.73	333.96	337.04	344.52
w_A	0.8453	0.8464	0.8478	0.8377	0.8381	0.8389	0.8402	0.8407	0.8415
w_B	0.0774	0.0768	0.0761	0.1185	0.1182	0.1176	0.1167	0.1163	0.1157
w_C	0.0774	0.0768	0.0761	0.0438	0.0437	0.0435	0.0431	0.0430	0.0428
T/K	349.84	351.63	366.54	301.17	309.61	309.04	314.94	318.98	322.67
w_A	0.8441	0.8458	0.8488	0.8523	0.8264	0.8278	0.8288	0.8303	0.8321
w_B	0.1138	0.1126	0.1104	0.1078	0.1562	0.1550	0.1541	0.1527	0.1511
w_C	0.0421	0.0416	0.0408	0.0399	0.0174	0.0172	0.0171	0.0170	0.0168
T/K	329.66	336.25	345.50	356.36	294.78	304.53	306.93	316.33	318.04
w_A	0.8343	0.8365	0.8386	0.8395	0.8411	0.8419	0.8431	0.8450	0.8452
w_B	0.1491	0.1472	0.1453	0.1445	0.1430	0.1423	0.1412	0.1395	0.1393
w_C	0.0166	0.0164	0.0161	0.0161	0.0159	0.0158	0.0157	0.0155	0.0155
T/K	318.75	327.75	338.36	339.51	347.88	349.84	354.64	364.67	367.74

Polymer (B): **poly(butyl methacrylate)** **1996KRA**
Characterization: M_n/g.mol^{-1} = 1780000, M_w/g.mol^{-1} = 2050000,
Roehm GmbH, Darmstadt, Germany
Solvent (A): **cyclohexanone** $C_6H_{10}O$ **108-94-1**
Polymer (C): **polystyrene**
Characterization: M_n/g.mol^{-1} = 66000, M_w/g.mol^{-1} = 196000,
J 110 Vestyron, Huels AG, Marl, Germany

Type of data: cloud points

w_A	0.8738	0.8743	0.8750	0.8754	0.8760	0.8765	0.8769	0.8776	0.8781
w_B	0.0063	0.0063	0.0063	0.0062	0.0062	0.0062	0.0062	0.0061	0.0061
w_C	0.1199	0.1194	0.1188	0.1184	0.1178	0.1173	0.1169	0.1163	0.1158
T/K	292.93	301.43	314.12	315.69	320.46	325.92	332.75	340.78	345.61

continued

continued

w_A	0.8788	0.8801	0.8807	0.8809	0.8812	0.8815	0.8818	0.8822	0.8826
w_B	0.0061	0.0108	0.0107	0.0107	0.0107	0.0107	0.0106	0.0106	0.0106
w_C	0.1151	0.1091	0.1086	0.1084	0.1081	0.1078	0.1076	0.1072	0.1068
T/K	366.38	305.49	319.02	320.84	326.37	330.54	334.79	341.42	350.95
w_A	0.8828	0.8832	0.8835	0.8831	0.8836	0.8843	0.8852	0.8860	0.8868
w_B	0.0105	0.0105	0.0105	0.0292	0.0291	0.0289	0.0287	0.0285	0.0283
w_C	0.1067	0.1063	0.1060	0.0877	0.0873	0.0868	0.0861	0.0855	0.0849
T/K	358.97	366.57	372.77	306.49	307.26	319.93	331.43	337.43	346.40
w_A	0.8877	0.8839	0.8847	0.8856	0.8866	0.8878	0.8888	0.8899	0.8907
w_B	0.0281	0.0581	0.0577	0.0572	0.0567	0.0561	0.0556	0.0551	0.0547
w_C	0.0842	0.0581	0.0577	0.0572	0.0567	0.0561	0.0556	0.0551	0.0547
T/K	359.49	292.82	299.89	308.21	318.48	327.68	335.30	346.12	349.65
w_A	0.8916	0.8762	0.8773	0.8794	0.8811	0.8828	0.8843	0.8863	0.8877
w_B	0.0542	0.0929	0.0920	0.0905	0.0892	0.0879	0.0868	0.0853	0.0842
w_C	0.0542	0.0310	0.0307	0.0302	0.0297	0.0293	0.0289	0.0284	0.0281
T/K	359.37	285.43	290.47	304.47	315.95	325.12	334.73	346.47	364.68
w_A	0.8548	0.8553	0.8566	0.8587	0.8626	0.8666	0.8695	0.8707	0.8358
w_B	0.1321	0.1317	0.1305	0.1286	0.1250	0.1214	0.1188	0.1177	0.1560
w_C	0.0131	0.0130	0.0129	0.0127	0.0124	0.0120	0.0117	0.0116	0.0082
T/K	290.30	297.11	299.57	300.71	318.95	337.39	358.03	365.27	292.22
w_A	0.8446	0.8527	0.8631						
w_B	0.1476	0.1399	0.1301						
w_C	0.0078	0.0074	0.0068						
T/K	331.07	342.87	373.15						

Polymer (B): **poly(butyl methacrylate)** **1996KRA**
Characterization: M_n/g.mol^{-1} = 1780000, M_w/g.mol^{-1} = 2050000,
Roehm GmbH, Darmstadt, Germany
Solvent (A): **cyclohexanone** **$C_6H_{10}O$** **108-94-1**
Polymer (C): **polystyrene**
Characterization: M_n/g.mol^{-1} = 204000, M_w/g.mol^{-1} = 207000,
Polymer Standards Service, Mainz, Germany

Type of data: cloud points

w_A	0.8639	0.8697	0.8719	0.8734	0.8766	0.8790	0.8815	0.8913	0.8921
w_B	0.0068	0.0065	0.0064	0.0063	0.0062	0.0061	0.0059	0.0098	0.0097
w_C	0.1293	0.1238	0.1217	0.1203	0.1172	0.1150	0.1126	0.0989	0.0982
T/K	316.43	332.74	333.94	344.77	357.86	367.38	385.85	300.21	311.84
w_A	0.8929	0.8937	0.8946	0.8958	0.8973	0.8958	0.8964	0.8971	0.8977
w_B	0.0096	0.0096	0.0095	0.0094	0.0092	0.0261	0.0259	0.0257	0.0256
w_C	0.0975	0.0967	0.0959	0.0948	0.0935	0.0782	0.0777	0.0772	0.0767
T/K	319.81	323.41	343.67	361.73	381.54	316.89	321.92	329.95	330.64

continued

continued

w_A	0.8982	0.8984	0.8989	0.8994	0.9002	0.9007	0.9013	0.9019	0.8902
w_B	0.0255	0.0254	0.0253	0.0252	0.0250	0.0248	0.0247	0.0245	0.0549
w_C	0.0764	0.0762	0.0758	0.0755	0.0749	0.0745	0.0740	0.0736	0.0549
T/K	334.88	335.70	343.00	351.87	356.65	364.26	357.93	368.71	297.79
w_A	0.8919	0.8926	0.8934	0.8943	0.8953	0.8973	0.8706	0.8711	0.8714
w_B	0.0541	0.0537	0.0533	0.0529	0.0524	0.0514	0.0971	0.0967	0.0965
w_C	0.0541	0.0537	0.0533	0.0529	0.0524	0.0514	0.0324	0.0322	0.0322
T/K	315.10	320.25	323.97	335.68	350.45	375.73	302.84	305.88	305.46
w_A	0.8721	0.8724	0.8728	0.8736	0.8741	0.8749	0.8757	0.8766	0.8775
w_B	0.0959	0.0957	0.0954	0.0948	0.0944	0.0938	0.0932	0.0926	0.0919
w_C	0.0320	0.0319	0.0318	0.0316	0.0315	0.0313	0.0311	0.0309	0.0306
T/K	316.86	312.82	319.49	320.40	325.35	330.32	333.63	340.81	339.91
w_A	0.8791	0.8794	0.8804	0.8816	0.8333	0.8363	0.8376	0.8384	0.8407
w_B	0.0907	0.0905	0.0897	0.0888	0.1517	0.1490	0.1478	0.1471	0.1450
w_C	0.0302	0.0302	0.0299	0.0296	0.0150	0.0147	0.0146	0.0145	0.0143
T/K	353.90	354.69	362.08	388.50	292.17	302.62	305.90	311.37	319.79
w_A	0.8411	0.8422	0.8448	0.8473	0.8501	0.8526	0.8560	0.8605	0.8654
w_B	0.1446	0.1436	0.1412	0.1390	0.1364	0.1341	0.1310	0.1269	0.1225
w_C	0.0143	0.0142	0.0140	0.0137	0.0135	0.0133	0.0130	0.0126	0.0121
T/K	320.45	324.82	337.94	343.98	359.03	364.72	362.60	377.56	380.29

Polymer (B): **poly(4-chlorostyrene)** **1956KER**
Characterization: ρ = 1.22 g/cm^3 at 298.15 K
Solvent (A): **benzene** **C_6H_6** **71-43-2**
Polymer (C): **poly(4-methoxystyrene)**
Characterization: ρ = 1.12 g/cm^3 at 298.15 K

Type of data: cloud points (threshold point)

T/K = 298.15

φ_A	0.972	φ_B	0.014	φ_C	0.014

Polymer (B): **poly(4-chlorostyrene)** **1955KER, 1956KER**
Characterization: ρ = 1.22 g/cm^3 at 298.15 K
Solvent (A): **benzene** **C_6H_6** **71-43-2**
Polymer (C): **poly(4-methylstyrene)**
Characterization: ρ = 1.03 g/cm^3 at 298.15 K

Type of data: cloud points (threshold point)

T/K = 298.15

φ_A	0.980	φ_B	0.010	φ_C	0.010

Type of data: coexistence data (tie lines)

continued

continued

T/K = 298.15

Total system			Top phase	Bottom phase
w_A	w_B	w_C	w_B	w_B
0.80	0.10	0.10	0.007	0.990

Polymer (B): **poly(4-chlorostyrene)** **1955KER**
Characterization: ρ = 1.22 g/cm^3 at 298.15 K
Solvent (A): **benzene** C_6H_6 **71-43-2**
Polymer (C): **poly(vinyl acetate)**
Characterization: M_w/g.mol^{-1} = 150000

Type of data: coexistence data (tie lines)

T/K = 298.15

Total system			Top phase	Bottom phase
w_A	w_B	w_C	w_B	w_B
0.80	0.10	0.10	0.018	0.960

Polymer (B): **poly(*N,N*-diallylammonioethanoic acid-*co*-sulfur dioxide)** **2002ALM**
Characterization: 50.0 mol% sulfur dioxide, synthesized in the laboratory
Solvent (A): **water** H_2O **7732-18-5**
Polymer (B): **poly(ethylene glycol)**
Characterization: M_w/g.mol^{-1} = 35000, Merck-Schuchardt, Germany

Type of data: coexistence data (tie lines)

T/K = 296.15

		Total system			Top phase			Bottom phase		
KCl mol/l	HCl equiv.	w_A	w_B	w_C	w_A	w_B	w_C	w_A	w_B	w_C
1.5	0.70	0.92083	0.05227	0.02690	0.94675	0.00135	0.0519	0.89686	0.0995	0.00364
1.0	0.70	0.92083	0.05227	0.02690	0.94541	0.00409	0.0505	0.89450	0.1040	0.00150
0.5	0.70	0.92095	0.05215	0.02690	0.94635	0.00425	0.0494	0.89242	0.1060	0.00158

continued

continued

KCl mol/l	HCl equiv.	Total system w_A	w_B	w_C	Top phase w_A	w_B	w_C	Bottom phase w_A	w_B	w_C
0.1	0.70	0.92106	0.05204	0.02690	0.93820	0.01570	0.0461	0.90203	0.0915	0.00647
0.1	0.70	0.89064	0.05468	0.05468	0.91835	0.00175	0.0799	0.83228	0.1660	0.00172
0.1	0.70	0.91559	0.02812	0.05629	0.92375	0.00775	0.0685	0.87930	0.1180	0.00270
0.1	0.70	0.90740	0.07676	0.01584	0.93700	0.01380	0.0492	0.89647	0.1000	0.00353
0.5	0.00	0.80580	0.11140	0.08280	0.78800	0.00000	0.2120	0.81688	0.1810	0.00212
0.5	0.00	0.83861	0.05934	0.10205	0.83100	0.00000	0.1690	0.84960	0.1480	0.00240
0.5	0.00	0.83440	0.06220	0.10340	0.84200	0.00000	0.1580	0.83920	0.1590	0.00180

Polymer (B): **poly(dimethylsiloxane)** **1960ALL**
Characterization: M_n/g.mol^{-1} = 130000, Midland Silicones, Ltd., U.K.
Solvent (A): **tetrachloromethane** **CCl_4** **56-23-5**
Polymer (C): **polyisobutylene**
Characterization: M_n/g.mol^{-1} = 150000, fractionated in the laboratory

Type of data: coexistence data (tie lines)

T/K = 291.15

Phase I φ_A	φ_B	φ_C	Phase II φ_A	φ_B	φ_C
0.961	0.014	0.025	0.961	0.036	0.003
0.962	0.010	0.028	0.961	0.038	0.001
0.960	0.008	0.032	0.961	0.038	0.001
0.954	0.009	0.037	0.955	0.042	0.003
0.937	0.010	0.053	0.939	0.060	0.001
0.929	0.008	0.063	0.930	0.069	0.001

Polymer (B): **polyethersulfone** **1998GOM**
Characterization: M_n/g.mol^{-1} = 14600, M_w/g.mol^{-1} = 38000, Victrex 3600P, I.C.I., Wilton, U.K.
Solvent (A): ***N,N*-dimethylformamide** **C_3H_7NO** **68-12-2**
Polymer (C): **poly(methyl methacrylate)**
Characterization: M_n/g.mol^{-1} = 714000, M_w/g.mol^{-1} = 750000, Polymer Laboratories, Shropshire, U.K.

Type of data: coexistence data (tie lines)

continued

continued

T/K = 298.15

Phase I			Phase II		
φ_A	φ_B	φ_C	φ_A	φ_B	φ_C
0.93792	0.05915	0.00293	0.93844	0.01800	0.04356
0.94456	0.01376	0.04168	0.94624	0.01660	0.03716
0.93078	0.06853	0.00069	0.94035	0.01755	0.04210
0.94627	0.05273	0.00100	0.93474	0.01935	0.04591
0.94299	0.05596	0.00105	0.93583	0.01837	0.04580
0.94039	0.05405	0.00556	0.93034	0.01626	0.05340
0.93342	0.06552	0.00106	0.94046	0.01622	0.04332
0.93817	0.06055	0.00128	0.93514	0.01532	0.04954
0.92448	0.07087	0.00465	0.94911	0.01305	0.03784
0.93593	0.06272	0.00135	0.93791	0.01471	0.04738
0.93726	0.06001	0.00273	0.93603	0.01608	0.04789
0.94174	0.05038	0.00788	0.93748	0.01851	0.04401
0.94084	0.04966	0.00950	0.93980	0.01769	0.04251
0.94469	0.04914	0.00617	0.93073	0.02502	0.04425
0.94246	0.04859	0.00895	0.93401	0.01810	0.04789
0.94311	0.04872	0.00817	0.93255	0.02078	0.04667
0.94326	0.05051	0.00623	0.93475	0.02275	0.04250
0.94135	0.05696	0.00169	0.93587	0.01818	0.04595
0.94320	0.05476	0.00204	0.93173	0.01725	0.05102
0.94319	0.05620	0.00061	0.93303	0.01908	0.04789
0.94222	0.05673	0.00105	0.93518	0.02057	0.04425
0.94281	0.05584	0.00135	0.93301	0.01753	0.04966
0.94470	0.05361	0.00169	0.92908	0.02219	0.04873
0.91713	0.02247	0.06040	0.94026	0.05622	0.00352
0.95570	0.03464	0.00966	0.92908	0.02982	0.04110
0.95261	0.01815	0.02924	0.94328	0.04042	0.01630
0.94036	0.02935	0.03029	0.93905	0.04113	0.01982
0.94373	0.04011	0.01616	0.93431	0.03306	0.03263

Polymer (B): **polyethersulfone** **1998GOM**
Characterization: M_n/g.mol^{-1} = 14600, M_w/g.mol^{-1} = 38000, Victrex 3600P, I.C.I., Wilton, U.K.
Solvent (A): ***N,N*-dimethylformamide** **C_3H_7NO** **68-12-2**
Polymer (C): **polystyrene**
Characterization: M_n/g.mol^{-1} = 17250, M_w/g.mol^{-1} = 18100, Tosoh Corp., Ltd., Tokyo, Japan

Type of data: coexistence data (tie lines)

continued

continued

T/K = 298.15

Phase I			Phase II		
φ_A	φ_B	φ_C	φ_A	φ_B	φ_C
0.75617	0.03298	0.21085	0.90271	0.08198	0.01531
0.74700	0.02551	0.22749	0.90026	0.08301	0.01673
0.75691	0.02444	0.21865	0.89866	0.08459	0.01675
0.73748	0.04106	0.22146	0.90879	0.07695	0.01426
0.74813	0.03288	0.21899	0.90351	0.08102	0.01547
0.73872	0.05911	0.20217	0.90085	0.06972	0.02943
0.71543	0.07358	0.21099	0.90281	0.06843	0.02876
0.64713	0.09371	0.25916	0.90488	0.06699	0.02813
0.67684	0.06433	0.25883	0.91775	0.05804	0.02421
0.76330	0.02709	0.20961	0.89677	0.07391	0.02932
0.81927	0.04421	0.13562	0.89755	0.06884	0.03361
0.82037	0.04410	0.13553	0.89693	0.06925	0.03382
0.82338	0.03882	0.13780	0.90343	0.06521	0.03136
0.82206	0.03901	0.13893	0.90409	0.06455	0.03136
0.83132	0.03200	0.13668	0.89423	0.07146	0.03431
0.65276	0.08028	0.26696	0.90794	0.06570	0.02636
0.70374	0.05999	0.23627	0.90557	0.06774	0.02669
0.70109	0.05452	0.24439	0.90590	0.06836	0.02574
0.63935	0.07444	0.28621	0.90909	0.06617	0.02474
0.63954	0.07896	0.28150	0.90953	0.06569	0.02478
0.83920	0.02030	0.14050	0.90009	0.06825	0.03166
0.78794	0.04990	0.16216	0.92301	0.05231	0.02468
0.81588	0.02787	0.15625	0.90872	0.06374	0.02754
0.81594	0.02886	0.15520	0.90885	0.06325	0.02790
0.80797	0.04146	0.15057	0.90243	0.06812	0.02945

Polymer (B): **poly(ethylene-*co*-vinyl acetate)** **2003CH2**
Characterization: M_n/g.mol^{-1} = 19800, 41.0 wt% vinyl acetate, commercial sample
Solvent (A): **styrene** C_8H_8 **100-42-5**
Polymer (C): **polystyrene**
Characterization: M_n/g.mol^{-1} = 90000, commercial sample

Type of data: cloud points

T/K = 358.15

w_A	0.8800	0.9050	0.9150	0.9050	0.8900
w_B	0.0120	0.0285	0.0425	0.0665	0.0990
w_C	0.1080	0.0665	0.0425	0.0285	0.0110

continued

continued

Type of data: coexistence data (tie lines)

T/K = 358.15

Total system			Top phase	Bottom phase
w_A	w_B	w_C	w_A	w_A
0.850	0.075	0.075	0.882	0.796
0.802	0.099	0.099	0.844	0.746
0.733	0.100	0.167	0.780	0.680
0.700	0.150	0.150	0.736	0.653

Polymer (B): **poly(ethylene glycol)** **1998INO**
Characterization: M_n/g.mol^{-1} = 103000, M_w/g.mol^{-1} = 110000
Solvent (A): ***N,N*-dimethylformamide** **C_3H_7NO** **68-12-2**
Polymer (C): **poly(α,L-glutamate)**
Characterization: M_η/g.mol^{-1} = 155000, transesterificated with triethylene glycol monomethyl ether in the laboratory

Type of data: coexistence data isotropic-isotropic and isotropic-nematic equilibrium

T/K = 298.15

Total system			Isotropic phase I			Isotropic phase II			Anisotropic phase		
φ_A	φ_B	φ_C	φ_A	φ_B	φ_C	φ_A	φ_B	φ_C	φ_A	φ_B	φ_C
0.845	0.030	0.125	0.897	0.088	0.015				0.827	0.001	0.172
0.825	0.050	0.125	0.877	0.113	0.010				0.791	0.002	0.207
0.802	0.049	0.149	0.867	0.121	0.012				0.747	0.001	0.252
0.854	0.018	0.128	0.895	0.087	0.018	0.861	0.007	0.132	0.847	0.003	0.150
0.860	0.010	0.130	0.912	0.069	0.019	0.868	0.004	0.128	0.845	0.001	0.154
0.881	0.030	0.089	0.915	0.068	0.017	0.869	0.006	0.125	0.844	0.004	0.152
0.895	0.030	0.075	0.920	0.061	0.019	0.887	0.010	0.103			
0.914	0.030	0.056	0.933	0.046	0.021	0.902	0.013	0.085			
0.932	0.024	0.044	0.946	0.032	0.022	0.927	0.018	0.055			

Polymer (B): **poly(ethylene glycol)** **1995PAN**
Characterization: M_w/g.mol^{-1} = 400, Aldrich Chem. Co., Inc., Milwaukee, WI
Solvent (A): **water** **H_2O** **7732-18-5**
Polymer (C): **deca(ethylene glycol) monocetyl ether**

Type of data: cloud points (LCST-behavior)

continued

continued

w_A	0.98	0.96	0.94	0.92	0.90	0.88	0.84	0.78
w_B	0.00	0.02	0.04	0.06	0.08	0.10	0.14	0.20
w_C	0.02	0.02	0.02	0.02	0.02	0.02	0.02	0.02
T/K	350.15	350.15	351.15	352.15	352.15	353.15	354.15	357.15

Polymer (B): **poly(ethylene glycol)** **1995PAN**
Characterization: M_w/g.mol^{-1} = 400, Aldrich Chem. Co., Inc., Milwaukee, WI
Solvent (A): **water** H_2O **7732-18-5**
Polymer (C): **deca(ethylene glycol) monooleyl ether**
Characterization: –

Type of data: cloud points (LCST-behavior)

w_A	0.98	0.96	0.94	0.92	0.90	0.88	0.84	0.78
w_B	0.00	0.02	0.04	0.06	0.08	0.10	0.14	0.20
w_C	0.02	0.02	0.02	0.02	0.02	0.02	0.02	0.02
T/K	336.15	335.15	335.15	336.15	336.15	336.15	338.15	340.15

Polymer (B): **poly(ethylene glycol)** **1993GA1**
Characterization: M_n/g.mol^{-1} = 585, M_w/g.mol^{-1} = 600,
Huels AG, Marl, Germany
Solvent (A): **water** H_2O **7732-18-5**
Polymer (C): **dextran**
Characterization: M_n/g.mol^{-1} = 109300, M_w/g.mol^{-1} = 493100,
PL 500 VC, Pfeifer & Langen, Dormagen, Germany

Type of data: coexistence data (tie lines)

T/K = 293.15

Total system			Top phase			Bottom phase		
w_A	w_B	w_C	w_A	w_B	w_C	w_A	w_B	w_C
0.7443	0.1484	0.1073	0.7757	0.1909	0.0334	0.7158	0.1332	0.1510
0.7133	0.1684	0.1183	0.7641	0.2255	0.0104	0.6590	0.1106	0.2304
0.7015	0.1737	0.1248	0.7541	0.2384	0.0075	0.6349	0.1029	0.2622
0.6768	0.1879	0.1353	0.7364	0.2606	0.0030	0.5971	0.0862	0.3167
0.6563	0.1966	0.1471	0.7230	0.2764	0.0006	0.5709	0.0806	0.3485
0.6417	0.2029	0.1554	0.7072	0.2921	0.0007	0.5480	0.0775	0.3745

Fractionation of dextran during demixing (the lines correspond with those above):

continued

continued

Top phase		Bottom phase	
M_n/(g/mol)	M_w/(g/mol)	M_n/(g/mol)	M_w/(g/mol)
67000	119748	113744	514158
45665	71326	107073	451683
38330	68395	101464	447542
31698	57808	102588	448402
30385	34539	100114	430362
22352	30570	100118	446267

Polymer (B): **poly(ethylene glycol)** **1993GA1**
Characterization: M_n/g.mol^{-1} = 1464, M_w/g.mol^{-1} = 1485,
Huels AG, Marl, Germany
Solvent (A): **water** H_2O **7732-18-5**
Polymer (C): **dextran**
Characterization: M_n/g.mol^{-1} = 109300, M_w/g.mol^{-1} = 493100,
PL 500 VC, Pfeifer & Langen, Dormagen, Germany

Type of data: coexistence data (tie lines)

T/K = 293.15

Total system			Top phase			Bottom phase		
w_A	w_B	w_C	w_A	w_B	w_C	w_A	w_B	w_C
0.8287	0.0905	0.0808	0.8567	0.1115	0.0318	0.8079	0.0728	0.1193
0.8245	0.0906	0.0849	0.8567	0.1169	0.0264	0.7977	0.0694	0.1329
0.8033	0.0958	0.1009	0.7526	0.1386	0.0088	0.7546	0.0520	0.1934
0.7835	0.0985	0.1180	0.8412	0.1547	0.0041	0.7252	0.0425	0.2323
0.7610	0.0895	0.1495	0.8350	0.1619	0.0031	0.7110	0.0380	0.2510
0.7493	0.1108	0.1399	0.8170	0.1816	0.0014	0.6745	0.0307	0.2948

Fractionation of dextran during demixing (the lines correspond with those above):

Top phase		Bottom phase	
M_n/(g/mol)	M_w/(g/mol)	M_n/(g/mol)	M_w/(g/mol)
77794	138494	129136	486174
70185	115761	118567	462465
50536	70517	119028	441921
37870	49985	101766	393164
34212	42852	106709	399612
27860	34021	105530	400696

Polymer (B): **poly(ethylene glycol)** **1995FUR**
Characterization: M_n/g.mol^{-1} = 2900, M_w/g.mol^{-1} = 3100,
Wako Pure Chemical Ind., Ltd., Osaka, Japan
Solvent (A): **water** **H_2O** **7732-18-5**
Polymer (C): **dextran**
Characterization: M_n/g.mol^{-1} = 18800, M_w/g.mol^{-1} = 38000,
T40, Pharmacia Inc., Tokyo, Japan

Type of data: coexistence data (tie lines)

T/K = 293.15

Total system			Top phase			Bottom phase		
w_A	w_B	w_C	w_A	w_B	w_C	w_A	w_B	w_C
0.8501	0.0898	0.0601	0.8581	0.1003	0.0416	0.8001	0.0396	0.1603
0.8399	0.0950	0.0651	0.8583	0.1142	0.0275	0.7820	0.0298	0.1882
0.8298	0.1002	0.0700	0.8560	0.1245	0.0195	0.7608	0.0241	0.2151
0.8200	0.1050	0.0750	0.8498	0.1349	0.0153	0.7414	0.0220	0.2366

Fractionation during demixing (the lines correspond with those above):

Top phase				Bottom phase			
PEG		Dextran		PEG		Dextran	
M_n/ g mol^{-1}	M_w/ g mol^{-1}	M_n/ g mol^{-1}	M_w/ g mol^{-1}	M_n/ g mol^{-1}	M_w/ g mol^{-1}	M_n/ g mol^{-1}	M_w/ g mol^{-1}
3000	3100	14000	24400	3000	3100	23700	47800
3000	3100	12800	22900	3100	3200	24400	45100
3000	3100	10300	16700	3000	3100	22700	43800
3000	3100	9300	14500	3000	3100	21700	42000

Comments: The molecular weight distributions for the dextran sample and for the dextran fractions in the coexisting phases are given in the appendix of the original source.

Polymer (B): **poly(ethylene glycol)** **1995FUR**
Characterization: M_n/g.mol^{-1} = 2900, M_w/g.mol^{-1} = 3100,
Wako Pure Chemical Ind., Ltd., Osaka, Japan
Solvent (A): **water** **H_2O** **7732-18-5**
Polymer (C): **dextran**
Characterization: M_n/g.mol^{-1} = 37900, M_w/g.mol^{-1} = 66700,
T70, Pharmacia Inc., Tokyo, Japan

Type of data: coexistence data (tie lines)

continued

continued

T/K = 293.15

Total system			Top phase			Bottom phase		
w_A	w_B	w_C	w_A	w_B	w_C	w_A	w_B	w_C
0.8598	0.0853	0.0549	0.8765	0.1016	0.0219	0.8105	0.0340	0.1555
0.8500	0.0900	0.0600	0.8718	0.1140	0.0142	0.7900	0.0278	0.1822
0.8399	0.0950	0.0651	0.8680	0.1224	0.0096	0.7683	0.0228	0.2089
0.8298	0.1002	0.0700	0.8617	0.1315	0.0068	0.7475	0.0194	0.2331

Fractionation during demixing (the lines correspond with those above):

Top phase				Bottom phase			
PEG		Dextran		PEG		Dextran	
M_n/ g mol^{-1}	M_w/ g mol^{-1}	M_n/ g mol^{-1}	M_w/ g mol^{-1}	M_n/ g mol^{-1}	M_w/ g mol^{-1}	M_n/ g mol^{-1}	M_w/ g mol^{-1}
2900	3000	27500	44500	2900	3000	47800	78700
2900	3000	21600	34600	2900	3000	46600	75900
2900	3000	17700	28600	2900	3000	45400	73700
2900	3000	15400	24300	2900	3000	44400	72400

Polymer (B): **poly(ethylene glycol)** **1995FUR**
Characterization: M_n/g.mol^{-1} = 2900, M_w/g.mol^{-1} = 3100,
Wako Pure Chemical Ind., Ltd., Osaka, Japan
Solvent (A): **water** **H_2O** **7732-18-5**
Polymer (C): **dextran**
Characterization: M_n/g.mol^{-1} = 35600, M_w/g.mol^{-1} = 139800,
Wako Pure Chemical Ind., Ltd., Osaka, Japan

Type of data: coexistence data (tie lines)

T/K = 293.15

Total system			Top phase			Bottom phase		
w_A	w_B	w_C	w_A	w_B	w_C	w_A	w_B	w_C
0.8697	0.0800	0.0503	0.8859	0.0932	0.0209	0.8218	0.0341	0.1431
0.8591	0.0857	0.0552	0.8815	0.1046	0.0139	0.7995	0.0281	0.1724
0.8500	0.0900	0.0600	0.8765	0.1129	0.0106	0.7824	0.0238	0.1938
0.8397	0.0951	0.0652	0.8687	0.1230	0.0083	0.7692	0.0203	0.2105

continued

continued

Fractionation during demixing (the lines correspond with those above):

Top phase				Bottom phase			
PEG		Dextran		PEG		Dextran	
M_n/ g mol^{-1}	M_w/ g mol^{-1}	M_n/ g mol^{-1}	M_w/ g mol^{-1}	M_n/ g mol^{-1}	M_w/ g mol^{-1}	M_n/ g mol^{-1}	M_w/ g mol^{-1}
3000	3100	24400	61000	3000	3100	58500	176700
2900	3100	19200	42500	3000	3100	57500	171100
2900	3100	17000	35400	3000	3100	60300	158800
3000	3100	14400	26600	3000	3100	56100	150300

Comments: The molecular weight distributions for the dextran sample and for the dextran fractions in the coexisting phases are given in the appendix of the original source.

Polymer (B): **poly(ethylene glycol)** **1991HAR**
Characterization: M_n/g.mol^{-1} = 3000,
Wako Pure Chemical Ind., Ltd., Osaka, Japan
Solvent (A): **water** **H_2O** **7732-18-5**
Polymer (C): **dextran**
Characterization: M_n/g.mol^{-1} = 24200, T40, Pharmacia Inc., Tokyo, Japan

Type of data: coexistence data (tie lines)

T/K = 293.15

Total system			Top phase			Bottom phase		
w_A	w_B	w_C	w_A	w_B	w_C	w_A	w_B	w_C
0.7395	0.1718	0.0887	0.816	0.004	0.180	0.681	0.312	0.007
0.7586	0.1577	0.0837	0.829	0.006	0.165	0.702	0.286	0.012
0.7794	0.1503	0.0703	0.847	0.012	0.141	0.736	0.246	0.018
0.8134	0.0999	0.0867	0.857	0.017	0.126	0.756	0.222	0.022
0.8148	0.1025	0.0827	0.858	0.018	0.124	0.762	0.214	0.024
0.8226	0.0991	0.0783	0.862	0.024	0.114	0.778	0.191	0.028
0.7881	0.1099	0.1083	0.832	0.006	0.162	0.706	0.283	0.011
0.7901	0.1098	0.1001	0.838	0.008	0.154	0.720	0.266	0.014
0.7930	0.1081	0.0989	0.841	0.009	0.150	0.726	0.260	0.014
0.8003	0.1107	0.0890	0.849	0.012	0.139	0.743	0.240	0.017
0.8072	0.1060	0.0868	0.850	0.015	0.135	0.749	0.231	0.020
0.8139	0.1068	0.0793	0.853	0.019	0.128	0.765	0.211	0.024
0.8224	0.1031	0.0745	0.861	0.025	0.114	0.777	0.194	0.029

Polymer (B): **poly(ethylene glycol)** **1995FUR**
Characterization: M_n/g.mol^{-1} = 2900, M_w/g.mol^{-1} = 3100,
Wako Pure Chemical Ind., Ltd., Osaka, Japan
Solvent (A): **water** **H_2O** **7732-18-5**
Polymer (C): **dextran**
Characterization: M_n/g.mol^{-1} = 191000, M_w/g.mol^{-1} = 476000,
T-500, Pharmacia Inc., Tokyo, Japan

Type of data: coexistence data (tie lines)

T/K = 293.15

Total system			Top phase			Bottom phase		
w_A	w_B	w_C	w_A	w_B	w_C	w_A	w_B	w_C
0.8841	0.0731	0.0428	0.9002	0.0898	0.0100	0.8443	0.0381	0.1176
0.8800	0.0750	0.0450	0.8996	0.0929	0.0075	0.8376	0.0349	0.1275
0.8699	0.0800	0.0501	0.8921	0.1042	0.0037	0.8162	0.0281	0.1557
0.8499	0.0901	0.0600	0.8799	0.1189	0.0012	0.7823	0.0201	0.1976

Fractionation during demixing (the lines correspond with those above):

Top phase				Bottom phase			
PEG		Dextran		PEG		Dextran	
M_n/ g mol^{-1}	M_w/ g mol^{-1}	M_n/ g mol^{-1}	M_w/ g mol^{-1}	M_n/ g mol^{-1}	M_w/ g mol^{-1}	M_n/ g mol^{-1}	M_w/ g mol^{-1}
3000	3100	118800	184100	2800	3000	222300	548100
2900	3100	101700	156900	2900	3100	197800	516200
2900	3000	78500	122400	2800	3000	211100	494500
2900	3000	55200	82700	2800	3000	199500	476700

Comments: The molecular weight distributions for the dextran sample and for the dextran fractions in the coexisting phases are given in the appendix of the original source.

Polymer (B): **poly(ethylene glycol)** **1991CON**
Characterization: M_n/g.mol^{-1} = 3140, M_w/g.mol^{-1} = 3250,
Huels AG, Marl, Germany
Solvent (A): **water** **H_2O** **7732-18-5**
Polymer (C): **dextran**
Characterization: M_n/g.mol^{-1} = 101000, M_w/g.mol^{-1} = 432000,
Pfeifer & Langen, Dormagen, Germany

Type of data: coexistence data (tie lines)

continued

continued

T/K = 273.15

Total system			Top phase			Bottom phase		
w_A	w_B	w_C	w_A	w_B	w_C	w_A	w_B	w_C
0.7932	0.0851	0.1217	0.8572	0.1416	0.0012	0.6995	0.0102	0.2903
0.8150	0.0768	0.1082	0.8742	0.1247	0.0011	0.7315	0.0121	0.2564
0.8339	0.0707	0.0954	0.8873	0.1105	0.0022	0.7609	0.0137	0.2254
0.8543	0.0638	0.0819	0.8992	0.0952	0.0056	0.7983	0.0205	0.1812
0.8697	0.0562	0.0741	0.9064	0.0799	0.0137	0.8319	0.0290	0.1391
0.8755	0.0543	0.0702	0.9064	0.0730	0.0206	0.8471	0.0335	0.1194

Fractionation of dextran during demixing (the lines correspond with those above):

Top phase		Bottom phase	
M_n/ g mol^{-1}	M_w/ g mol^{-1}	M_n/ g mol^{-1}	M_w/ g mol^{-1}
30758	51281	110148	439931
38251	62399	114351	435486
39384	60514	112625	432249
49206	70023	123445	460035
59550	89688	122235	458967
69080	114406	126831	490284

T/K = 293.15

Total system			Top phase			Bottom phase		
w_A	w_B	w_C	w_A	w_B	w_C	w_A	w_B	w_C
0.7726	0.0374	0.1900	0.8695	0.1274	0.0031	0.7567	0.0200	0.2233
0.7945	0.0847	0.1208	0.8514	0.1480	0.0006	0.7280	0.0132	0.2588
0.8161	0.0755	0.1084	0.8698	0.1288	0.0014	0.7629	0.0179	0.2192
0.8376	0.0709	0.0915	0.8810	0.1138	0.0052	0.7927	0.0251	0.1822
0.8546	0.0647	0.0807	0.8896	0.0968	0.0136	0.8210	0.0359	0.1431
0.8642	0.0593	0.0765	0.8861	0.0792	0.0347	0.8517	0.0497	0.0986
0.8557	0.1119	0.0324	0.8712	0.1272	0.0016	0.7615	0.0184	0.2201

continued

continued

Fractionation of dextran during demixing (the lines correspond with those above):

Top phase M_n/ g mol^{-1}	Top phase M_w/ g mol^{-1}	Bottom phase M_n/ g mol^{-1}	Bottom phase M_w/ g mol^{-1}
35864	66822	109393	429826
38811	47384	108530	425680
44898	55977	108171	414886
47373	66825	114007	430083
58989	89369	119205	447505
78495	154137	117580	473681
43584	57271	120991	444618

T/K = 313.15

Total system			Top phase			Bottom phase		
w_A	w_B	w_C	w_A	w_B	w_C	w_A	w_B	w_C
0.7946	0.0847	0.1207	0.8470	0.1522	0.0008	0.7396	0.0160	0.2444
0.8137	0.0754	0.1109	0.8642	0.1331	0.0027	0.7703	0.0203	0.2094
0.8356	0.0707	0.0937	0.8782	0.1153	0.0065	0.7990	0.0293	0.1717
0.8529	0.0644	0.0827	0.8856	0.0963	0.0181	0.8323	0.0403	0.1274
0.8598	0.0613	0.0789	0.8821	0.0829	0.0350	0.8526	0.0526	0.0948
0.8499	0.1167	0.0334	0.8647	0.1331	0.0022	0.7697	0.0199	0.2104
0.7878	0.0435	0.1687	0.8655	0.1318	0.0027	0.7693	0.0219	0.2088

Fractionation of dextran during demixing (the lines correspond with those above):

Top phase M_n/ g mol^{-1}	Top phase M_w/ g mol^{-1}	Bottom phase M_n/ g mol^{-1}	Bottom phase M_w/ g mol^{-1}
37473	50431	111950	433225
40139	62992	114965	432477
49976	71755	115741	438271
62026	110830	120856	457811
76350	154715	117566	459023
43369	62351	122336	448637
39799	53717	111907	419156

Polymer (B): **poly(ethylene glycol)** **1989DIA**
Characterization: M_w/g.mol^{-1} = 3400, Aldrich Chem. Co., Inc., Milwaukee, WI
Solvent (A): **water** **H_2O** **7732-18-5**
Polymer (C): **dextran**
Characterization: M_n/g.mol^{-1} = 24200, M_w/g.mol^{-1} = 38800,
Pharmacia Fine Chemicals, Piscataway, NJ

Type of data: coexistence data (tie lines)

T/K = 277.15

Total system			Top phase			Bottom phase		
w_A	w_B	w_C	w_A	w_B	w_C	w_A	w_B	w_C
0.8470	0.0650	0.0880	0.8748	0.0882	0.0370	0.8089	0.0328	0.1583
0.8400	0.0670	0.0930	0.8752	0.0962	0.0286	0.7965	0.0277	0.1758
0.8310	0.0690	0.1000	0.8716	0.1071	0.0213	0.7804	0.0239	0.1957
0.8230	0.0710	0.1060	0.8705	0.1119	0.0176	0.7670	0.0209	0.2121

Polymer (B): **poly(ethylene glycol)** **1989DIA**
Characterization: M_w/g.mol^{-1} = 3400, Aldrich Chem. Co., Inc., Milwaukee, WI
Solvent (A): **water** **H_2O** **7732-18-5**
Polymer (C): **dextran**
Characterization: M_n/g.mol^{-1} = 38400, M_w/g.mol^{-1} = 72200,
Pharmacia Fine Chemicals, Piscataway, NJ

Type of data: coexistence data (tie lines)

T/K = 277.15

Total system			Top phase			Bottom phase		
w_A	w_B	w_C	w_A	w_B	w_C	w_A	w_B	w_C
0.8640	0.0630	0.0730	0.8944	0.0824	0.0232	0.8232	0.0306	0.1462
0.8575	0.0645	0.0780	0.8922	0.0896	0.0182	0.8129	0.0282	0.1589
0.8505	0.0655	0.0840	0.8898	0.0954	0.0148	0.8005	0.0261	0.1734
0.8430	0.0670	0.0900	0.8875	0.1007	0.0118	0.7865	0.0242	0.1893

Polymer (B): **poly(ethylene glycol)** **1989DIA**
Characterization: M_w/g.mol^{-1} = 3400, Aldrich Chem. Co., Inc., Milwaukee, WI
Solvent (A): **water** **H_2O** **7732-18-5**
Polymer (C): **dextran**

continued

continued

Characterization: M_n/g.mol^{-1} = 234200, M_w/g.mol^{-1} = 507000, Pharmacia Fine Chemicals, Piscataway, NJ

Type of data: coexistence data (tie lines)

T/K = 277.15

Total system			Top phase			Bottom phase		
w_A	w_B	w_C	w_A	w_B	w_C	w_A	w_B	w_C
0.8930	0.0500	0.0570	0.9206	0.0643	0.0151	0.8735	0.0345	0.0920
0.8650	0.0650	0.0700	0.9049	0.0939	0.0012	0.8042	0.0210	0.1748
0.8500	0.0700	0.0800	0.8965	0.1028	0.0007	0.7790	0.0198	0.2012
0.8400	0.0800	0.0800	0.8845	0.1151	0.0004	0.7573	0.0150	0.2277

Polymer (B): **poly(ethylene glycol)** **1991HAR**
Characterization: M_n/g.mol^{-1} = 3690, M_w/g.mol^{-1} = 3860, Carbowax 3350, Union Carbide Corp., New York, NY
Solvent (A): **water** H_2O **7732-18-5**
Polymer (C): **dextran**
Characterization: M_n/g.mol^{-1} = 37000, M_w/g.mol^{-1} = 74500, T-70, Pharmacia Fine Chemicals, Piscataway, NJ

Type of data: coexistence data (tie lines)

T/K = 298.15

Total system			Top phase			Bottom phase		
w_A	w_B	w_C	w_A	w_B	w_C	w_A	w_B	w_C
0.8631	0.0722	0.0647	0.8743	0.0787	0.0470	0.8361	0.0514	0.1125
0.8631	0.0722	0.0647	0.8768	0.0869	0.0363	0.8288	0.0339	0.1373
0.8579	0.0764	0.0657	0.8830	0.1023	0.0147	0.8159	0.0285	0.1556
0.8527	0.0915	0.0558	0.8799	0.1142	0.0059	0.7849	0.0201	0.1950
0.8311	0.0870	0.0819	0.8681	0.1240	0.0079	0.7643	0.0225	0.2132
0.8203	0.0976	0.0821	0.8579	0.1373	0.0048	0.7367	0.0214	0.2419
0.7883	0.1139	0.0978	0.8359	0.1622	0.0019	0.6919	0.0137	0.2944

Polymer (B): **poly(ethylene glycol)** **1992ZAS**
Characterization: M_n/g.mol^{-1} = 6000, Serva, Heidelberg, Germany
Solvent (A): **water** H_2O **7732-18-5**
Polymer (C): **dextran**
Characterization: M_n/g.mol^{-1} = 28700, M_w/g.mol^{-1} = 57200

continued

continued

Type of data: coexistence data (tie lines)

T/K = 298.15

Top phase			Bottom phase		
w_A	w_B	w_C	w_A	w_B	w_C
0.8475	0.1481	0.0044	0.7329	0.0057	0.2614
0.8609	0.1324	0.0069	0.7592	0.0066	0.2342
0.8845	0.1013	0.0142	0.8034	0.0120	0.1846
0.8923	0.0868	0.0209	0.8297	0.0165	0.1538
0.8970	0.0666	0.0364	0.8610	0.0228	0.1162

Dependence on pH, the total systems contain 0.01 mol/L universal buffer.

Top phase				Bottom phase			
w_A	w_B	w_C	pH	w_A	w_B	w_C	pH
0.8440	0.1508	0.0052	4.32	0.7274	0.0066	0.2660	4.26
0.8786	0.1045	0.0169	4.23	0.8025	0.0105	0.1870	4.19
0.8873	0.0855	0.0272	4.17	0.8314	0.0163	0.1523	4.14
0.8440	0.1508	0.0052	7.89	0.7274	0.0066	0.2660	7.73
0.8786	0.1045	0.0169	7.88	0.8025	0.0105	0.1870	7.76
0.8873	0.0855	0.0272	7.87	0.8314	0.0163	0.1523	7.79
0.8440	0.1508	0.0052	8.22	0.7274	0.0066	0.2660	8.08
0.8786	0.1045	0.0169	8.24	0.8025	0.0105	0.1870	8.17
0.8873	0.0855	0.0272	8.26	0.8314	0.0163	0.1523	8.22

Polymer (B): **poly(ethylene glycol)** **1995GRO, 1995TIN**
Characterization: M_n/g.mol^{-1} = 6230, M_w/g.mol^{-1} = 6480,
Hoechst AG, Frankfurt, Germany
Solvent (A): **water** **H_2O** **7732-18-5**
Polymer (C): **dextran**
Characterization: M_n/g.mol^{-1} = 179347, M_w/g.mol^{-1} = 507000,
T-500, Pharmacia Fine Chemicals, Uppsala, Sweden

Type of data: coexistence data (tie lines)

T/K = 277.15

continued

continued

Total system			Top phase			Bottom phase		
w_A	w_B	w_C	w_A	w_B	w_C	w_A	w_B	w_C
0.9055	0.0425	0.0520	0.9315	0.0623	0.0062	0.8660	0.0179	0.1161
0.8815	0.0501	0.0684	0.9184	0.0800	0.0016	0.8263	0.0102	0.1635
0.8548	0.0651	0.0801	0.8968	0.1028	0.0004	0.7841	0.0056	0.2103
0.8324	0.0699	0.0977	0.8824	0.1173	0.0003	0.7595	0.0043	0.2362

T/K = 283.15

Total system			Top phase			Bottom phase		
w_A	w_B	w_C	w_A	w_B	w_C	w_A	w_B	w_C
0.9072	0.0426	0.0502	0.9302	0.0583	0.0115	0.8772	0.0219	0.1009
0.8901	0.0511	0.0588	0.9216	0.0756	0.0028	0.8378	0.0132	0.1490
0.8695	0.0599	0.0706	0.9087	0.0902	0.0011	0.8042	0.0084	0.1874
0.8395	0.0704	0.0901	0.8856	0.1139	0.0005	0.7653	0.0048	0.2299

T/K = 293.15

Total system			Top phase			Bottom phase		
w_A	w_B	w_C	w_A	w_B	w_C	w_A	w_B	w_C
0.9076	0.0448	0.0476	0.9251	0.0626	0.0123	0.8845	0.0241	0.0914
0.8828	0.0489	0.0683	0.9191	0.0782	0.0027	0.8436	0.0134	0.1430
0.8824	0.0550	0.0626	0.9169	0.0808	0.0023	0.8318	0.0134	0.1548
0.8628	0.0633	0.0739	0.9007	0.0988	0.0005	0.8049	0.0080	0.1871
0.8541	0.0657	0.0802	0.8941	0.1053	0.0006	0.7913	0.0072	0.2015
0.8320	0.0699	0.0981	0.8799	0.1196	0.0005	0.7691	0.0046	0.2263
0.7696	0.1000	0.1304	0.8267	0.1730	0.0003	0.6808	0.0020	0.3172

T/K = 313.15

Total system			Top phase			Bottom phase		
w_A	w_B	w_C	w_A	w_B	w_C	w_A	w_B	w_C
0.8816	0.0507	0.0677	0.9128	0.0842	0.0030	0.8424	0.0151	0.1425
0.8552	0.0648	0.0800	0.8913	0.1082	0.0005	0.8051	0.0093	0.1856
0.8321	0.0699	0.0980	0.8750	0.1247	0.0003	0.7784	0.0050	0.2166
0.8157	0.0756	0.1087	0.8621	0.1377	0.0002	0.7606	0.0024	0.2370

Polymer (B): **poly(ethylene glycol)** **1993MIS**
Characterization: M_w/g.mol^{-1} = 7500
Solvent (A): **water** **H_2O** **7732-18-5**
Polymer (C): **dextran**
Characterization: M_n/g.mol^{-1} = 195300, M_w/g.mol^{-1} = 515000

Type of data: cloud points

T/K = 298.15

w_A	0.888	0.919	0.929	0.941	0.938	0.901	0.854	0.817	0.752
w_B	0.112	0.081	0.070	0.056	0.038	0.020	0.011	0.006	0.002
w_C	0.000	0.000	0.001	0.003	0.024	0.079	0.135	0.177	0.246

Type of data: coexistence data (tie lines)

T/K = 298.15

Total system			Top phase			Bottom phase		
w_A	w_B	w_C	w_A	w_B	w_C	w_A	w_B	w_C
0.8080	0.0622	0.1298	0.875	0.125	0.000	0.731	0.002	0.267
0.8312	0.0529	0.1159	0.887	0.113	0.000	0.768	0.003	0.229
0.8586	0.0438	0.0976	0.907	0.092	0.001	0.808	0.006	0.186
0.8729	0.0397	0.0874	0.917	0.081	0.002	0.832	0.008	0.160
0.8907	0.0325	0.0768	0.928	0.066	0.006	0.870	0.014	0.116

Polymer (B): **poly(ethylene glycol)** **1991HAR**
Characterization: M_n/g.mol^{-1} = 7620, M_w/g.mol^{-1} = 8000,
Sigma Chemical Co., Inc., St. Louis, MO
Solvent (A): **water** **H_2O** **7732-18-5**
Polymer (C): **dextran**
Characterization: M_n/g.mol^{-1} = 182000, M_w/g.mol^{-1} = 500000,
Sigma Chemical Co., Inc., St. Louis, MO

Type of data: coexistence data (tie lines)

T/K = 298.15

Total system			Top phase			Bottom phase		
w_A	w_B	w_C	w_A	w_B	w_C	w_A	w_B	w_C
0.8800	0.0500	0.0700	0.9140	0.0860	0.0000	0.8332	0.0038	0.1630
0.8800	0.0500	0.0700	0.9140	0.0860	0.0000	0.7945	0.0038	0.2017
0.8580	0.0580	0.0840	0.8940	0.1050	0.0010	0.8116	0.0014	0.1870
0.8430	0.0730	0.0840	0.8825	0.1175	0.0000	0.7735	0.0005	0.2260

continued

continued

Total system			Top phase			Bottom phase		
w_A	w_B	w_C	w_A	w_B	w_C	w_A	w_B	w_C
0.8940	0.0440	0.0620	0.9286	0.0704	0.0010	0.8550	0.0049	0.1401
0.8790	0.0590	0.0620	0.9120	0.0880	0.0000	0.8174	0.0014	0.1812
0.9000	0.0500	0.0500	0.9219	0.0781	0.0000	0.8453	0.0057	0.1490
0.8850	0.0650	0.0500	0.9131	0.0869	0.0000	0.8148	0.0019	0.1833
0.9000	0.0600	0.0400	0.9286	0.0704	0.0010	0.8360	0.0040	0.1600
0.8850	0.0750	0.0400	0.9023	0.0970	0.0007	0.8047	0.0018	0.1935

Polymer (B): **poly(ethylene glycol)** **1989DIA**
Characterization: M_w/g.mol^{-1} = 8000, Aldrich Chem. Co., Inc., Milwaukee, WI
Solvent (A): **water** **H_2O** **7732-18-5**
Polymer (C): **dextran**
Characterization: M_n/g.mol^{-1} = 24200, M_w/g.mol^{-1} = 38800,
Pharmacia Fine Chemicals, Piscataway, NJ

Type of data: coexistence data (tie lines)

T/K = 277.15

Total system			Top phase			Bottom phase		
w_A	w_B	w_C	w_A	w_B	w_C	w_A	w_B	w_C
0.8860	0.0390	0.0750	0.8958	0.0469	0.0573	0.8716	0.0280	0.1004
0.8760	0.0420	0.0820	0.9039	0.0672	0.0289	0.8515	0.0159	0.1326
0.8670	0.0460	0.0870	0.9012	0.0787	0.0201	0.8325	0.0117	0.1558
0.8590	0.0490	0.0920	0.8983	0.0855	0.0162	0.8186	0.0072	0.1742

Polymer (B): **poly(ethylene glycol)** **1989DIA**
Characterization: M_w/g.mol^{-1} = 8000, Aldrich Chem. Co., Inc., Milwaukee, WI
Solvent (A): **water** **H_2O** **7732-18-5**
Polymer (C): **dextran**
Characterization: M_n/g.mol^{-1} = 38400, M_w/g.mol^{-1} = 72200
Pharmacia Fine Chemicals, Piscataway, NJ

Type of data: coexistence data (tie lines)

T/K = 277.15

continued

continued

Total system			Top phase			Bottom phase		
w_A	w_B	w_C	w_A	w_B	w_C	w_A	w_B	w_C
0.8905	0.0370	0.0725	0.9170	0.0595	0.0235	0.8653	0.0156	0.1191
0.8700	0.0500	0.0800	0.9083	0.0829	0.0088	0.8207	0.0076	0.1717
0.8590	0.0470	0.0940	0.9059	0.0864	0.0077	0.8123	0.0072	0.1805
0.8440	0.0510	0.1050	0.8974	0.0971	0.0055	0.7917	0.0058	0.2025

Polymer (B): **poly(ethylene glycol)** **2004HOP**
Characterization: M_w/g.mol^{-1} = 8000, Sigma Chemical Co., Inc., St. Louis, MO
Solvent (A): **water** **H_2O** **7732-18-5**
Polymer (C): **dextran**
Characterization: M_n/g.mol^{-1} = 53000, M_w/g.mol^{-1} = 148000, Sigma Chemical Co., Inc., St. Louis, MO

Comments: phosphate buffered solutions, pH = 7.65.

Type of data: cloud points

T/K = 295

w_A	0.9340	0.9303	0.9259	0.9206	0.9128	0.9049	0.8930	0.8816	0.8536
w_B	0.0591	0.0555	0.0518	0.0485	0.0441	0.0389	0.0322	0.0250	0.0152
w_C	0.0068	0.0142	0.0223	0.0309	0.0431	0.0562	0.0748	0.0933	0.1312

Type of data: coexistence data (tie lines)

T/K = 295

Phase I			Phase II		
w_A	w_B	w_C	w_A	w_B	w_C
0.8015	0.0108	0.1876	0.8958	0.0940	0.0102
0.8144	0.0129	0.1727	0.9019	0.0853	0.0128
0.8255	0.0141	0.1605	0.9056	0.0784	0.0160
0.8486	0.0152	0.1361	0.9093	0.0693	0.0214
0.8646	0.0202	0.1151	0.9116	0.0605	0.0280
0.8864	0.0318	0.0817	0.9129	0.0423	0.0448

Polymer (B): **poly(ethylene glycol)** **1989DIA**
Characterization: M_w/g.mol^{-1} = 8000, Aldrich Chem. Co., Inc., Milwaukee, WI
Solvent (A): **water** **H_2O** **7732-18-5**
Polymer (C): **dextran**
Characterization: M_n/g.mol^{-1} = 234200, M_w/g.mol^{-1} = 507000

continued

continued

Type of data: coexistence data (tie lines)

T/K = 277.15

Total system			Top phase			Bottom phase		
w_A	w_B	w_C	w_A	w_B	w_C	w_A	w_B	w_C
0.9189	0.0327	0.0484	0.9423	0.0491	0.0086	0.8954	0.0163	0.0883
0.8964	0.0450	0.0586	0.9286	0.0703	0.0011	0.8516	0.0073	0.1411
0.8674	0.0576	0.0750	0.9070	0.0927	0.0003	0.8084	0.0043	0.1873
0.8500	0.0700	0.0800	0.8916	0.1083	0.0001	0.7809	0.0030	0.2161

Polymer (B): **poly(ethylene glycol)** **1988SZL**
Characterization: M_n/g.mol^{-1} = 8000, Sigma Chemical Co., Inc., St. Louis, MO
Solvent (A): **water** H_2O **7732-18-5**
Polymer (C): **dextran**
Characterization: M_w/g.mol^{-1} = 480000, Sigma Chemical Co., Inc., St. Louis, MO

Type of data: coexistence data (tie lines)

T/K = 298.15

Total system			Top phase			Bottom phase		
w_A	w_B	w_C	w_A	w_B	w_C	w_A	w_B	w_C
0.8966	0.0457	0.0577	0.8670	0.1280	0.0050	0.8690	0.0110	0.1200
0.8244	0.0699	0.1057	0.8800	0.1180	0.0020	0.7580	0.0020	0.2400
0.7872	0.0850	0.1278	0.8500	0.1470	0.0030	0.6970	0.0030	0.3000
0.8180	0.0931	0.0889	0.8696	0.1300	0.0004	0.7390	0.0110	0.2500

Polymer (B): **poly(ethylene glycol)** **1995FUR**
Characterization: M_n/g.mol^{-1} = 8500, M_w/g.mol^{-1} = 8800,
Wako Pure Chemical Ind., Ltd., Osaka, Japan
Solvent (A): **water** H_2O **7732-18-5**
Polymer (C): **dextran**
Characterization: M_n/g.mol^{-1} = 170300, M_w/g.mol^{-1} = 503000,
T-500, Pharmacia Inc., Tokyo, Japan

Type of data: coexistence data (tie lines)

T/K = 293.15

continued

continued

Total system			Top phase			Bottom phase		
w_A	w_B	w_C	w_A	w_B	w_C	w_A	w_B	w_C
0.9200	0.0550	0.0250	0.9314	0.0641	0.0045	0.8723	0.0128	0.1149
0.9100	0.0600	0.0300	0.9249	0.0723	0.0028	0.8563	0.0096	0.1341
0.9000	0.0650	0.0350	0.9181	0.0802	0.0017	0.8410	0.0073	0.1517
0.8900	0.0700	0.0400	0.9107	0.0883	0.0010	0.8234	0.0054	0.1712

Fractionation during demixing (the lines correspond with those above):

Top phase				Bottom phase			
PEG		Dextran		PEG		Dextran	
M_n/ g mol^{-1}	M_w/ g mol^{-1}	M_n/ g mol^{-1}	M_w/ g mol^{-1}	M_n/ g mol^{-1}	M_w/ g mol^{-1}	M_n/ g mol^{-1}	M_w/ g mol^{-1}
8300	8700	76200	176800	8100	8500	202500	520000
8400	8800	68500	171400	8400	8700	188100	492900
8600	8900	59700	147900	8300	8600	182200	487400
8500	8900	47900	78300	8000	8400	177800	467400

Polymer (B): **poly(ethylene glycol)** **1991HAR**
Characterization: M_n/g.mol^{-1} = 8920, M_w/g.mol^{-1} = 11800,
Carbowax 8000, Union Carbide Corp., New York, NY
Solvent (A): **water** **H_2O** **7732-18-5**
Polymer (C): **dextran**
Characterization: M_n/g.mol^{-1} = 167000, M_w/g.mol^{-1} = 509000,
T-500, Pharmacia Fine Chemicals, Piscataway, NJ

Type of data: coexistence data (tie lines)

T/K = 298.15

Total system			Top phase			Bottom phase		
w_A	w_B	w_C	w_A	w_B	w_C	w_A	w_B	w_C
0.9166	0.0344	0.0490	0.9339	0.0498	0.0163	0.9082	0.0233	0.0685
0.9071	0.0430	0.0499	0.9298	0.0668	0.0034	0.8655	0.0170	0.1175
0.8958	0.0449	0.0593	0.9259	0.0723	0.0018	0.8591	0.0142	0.1367
0.8930	0.0570	0.0500	0.9200	0.0790	0.0010	0.8390	0.0110	0.1500
0.8791	0.0507	0.0702	0.9169	0.0824	0.0007	0.8370	0.0100	0.1630
0.8675	0.0586	0.0739	0.9068	0.0929	0.0003	0.8288	0.0061	0.1757
0.8452	0.0716	0.0832	0.8885	0.1114	0.0001	0.7637	0.0058	0.2305

Polymer (B): **poly(ethylene glycol)** **1989SJO**
Characterization: M_w/g.mol^{-1} = 17000-20000, Serva, Heidelberg, Germany
Solvent (A): **water** H_2O **7732-18-5**
Polymer (C): **dextran**
Characterization: M_w/g.mol^{-1} = 40000,
Pharmacia Fine Chemicals, Uppsala, Sweden

Type of data: coexistence data (tie lines)

T/K = 293.15

Top phase			Bottom phase		
w_A	w_B	w_C	w_A	w_B	w_C
0.9159	0.0545	0.0296	0.8794	0.0082	0.1124
0.9107	0.0735	0.0158	0.8503	0.0051	0.1446
0.8926	0.0995	0.0079	0.8160	0.0000	0.1840
0.8184	0.1806	0.0010	0.6916	0.0000	0.3084

T/K = 323.15

w_A	w_B	w_C	w_A	w_B	w_C
0.9135	0.0668	0.0197	0.8828	0.0110	0.1062
0.8959	0.0956	0.0085	0.8495	0.0050	0.1455
0.8753	0.1200	0.0047	0.8216	0.0019	0.1765
0.8553	0.1421	0.0026	0.7936	0.0007	0.2057
0.8221	0.1766	0.0013	0.7526	0.0011	0.2463
0.7949	0.2044	0.0007	0.7128	0.0010	0.2862

T/K = 353.15

w_A	w_B	w_C	w_A	w_B	w_C
0.9048	0.0875	0.0077	0.8926	0.0070	0.1004
0.8855	0.1103	0.0042	0.8671	0.0029	0.1300
			0.8386	0.0031	0.1583
0.8340	0.1646	0.0014	0.8166	0.0004	0.1830
0.7993	0.1998	0.0009	0.7770	0.0004	0.2236
0.7595	0.2400	0.0004	0.7374	0.0004	0.2622

T/K = 373.15

w_A	w_B	w_C	w_A	w_B	w_C
0.8925	0.1033	0.0042	0.9128	0.0086	0.0786
0.8743	0.1231	0.0026	0.8893	0.0051	0.1056
0.8498	0.1486	0.0016	0.8720	0.0030	0.1250
0.8127	0.1845	0.0028	0.8443	0.0027	0.1530
			0.7925	0.0017	0.2058

Polymer (B): **poly(ethylene glycol)** **1996FUR**
Characterization: M_n/g.mol^{-1} = 17200, M_w/g.mol^{-1} = 19900,
Wako Pure Chemical Ind., Ltd., Osaka, Japan
Solvent (A): **water** **H_2O** **7732-18-5**
Polymer (C): **dextran**
Characterization: M_n/g.mol^{-1} = 170500, M_w/g.mol^{-1} = 503000,
T-500, Pharmacia Inc., Tokyo, Japan

Type of data: coexistence data (tie lines)

T/K = 293.15

Total system			Top phase			Bottom phase		
w_A	w_B	w_C	w_A	w_B	w_C	w_A	w_B	w_C
0.9500	0.0350	0.0150	0.9553	0.0367	0.0080	0.9231	0.0088	0.0681
0.9400	0.0400	0.0200	0.9491	0.0467	0.0042	0.9043	0.0053	0.0904
0.9300	0.0500	0.0200	0.9389	0.0592	0.0018	0.8809	0.0049	0.1142
0.9200	0.0550	0.0250	0.9334	0.0656	0.0010	0.8637	0.0041	0.1322

Fractionation during demixing (the lines correspond with those above):

Top phase				Bottom phase			
PEG		Dextran		PEG		Dextran	
M_n/ g mol^{-1}	M_w/ g mol^{-1}	M_n/ g mol^{-1}	M_w/ g mol^{-1}	M_n/ g mol^{-1}	M_w/ g mol^{-1}	M_n/ g mol^{-1}	M_w/ g mol^{-1}
14800	18200	135800	238400	17400	18800	255400	725800
16000	18900	101800	154200	15000	16500	198200	558000
16100	19200	81200	112700			177300	503500
16100	19200	69500	89400			172200	479400

Comments: The molecular weight distributions for the dextran sample and for the dextran fractions in the coexisting phases are given in the appendix of the original source.

Polymer (B): **poly(ethylene glycol)** **1989DIA**
Characterization: M_w/g.mol^{-1} = 20000, Carbowax 20M,
Union Carbide Corp., New York, NY
Solvent (A): **water** **H_2O** **7732-18-5**
Polymer (C): **dextran**
Characterization: M_n/g.mol^{-1} = 24200, M_w/g.mol^{-1} = 38800
Pharmacia Fine Chemicals, Piscataway, NJ,

Type of data: coexistence data (tie lines)

continued

continued

T/K = 277.15

Total system			Top phase			Bottom phase		
w_A	w_B	w_C	w_A	w_B	w_C	w_A	w_B	w_C
0.9160	0.0420	0.0420	0.9215	0.0472	0.0313	0.9134	0.0302	0.0564
0.9050	0.0450	0.0500	0.9156	0.0652	0.0192	0.8967	0.0234	0.0799
0.8850	0.0500	0.0650	0.9065	0.0793	0.0142	0.8602	0.0141	0.1257
0.8650	0.0550	0.0800	0.8973	0.0937	0.0090	0.8296	0.0082	0.1622

Polymer (B): **poly(ethylene glycol)** **1989DIA**
Characterization: M_w/g.mol^{-1} = 20000, Carbowax 20M,
Union Carbide Corp., New York, NY
Solvent (A): **water** **H_2O** **7732-18-5**
Polymer (C): **dextran**
Characterization: M_n/g.mol^{-1} = 38400, M_w/g.mol^{-1} = 72200,
Pharmacia Fine Chemicals, Piscataway, NJ

Type of data: coexistence data (tie lines)

T/K = 277.15

Total system			Top phase			Bottom phase		
w_A	w_B	w_C	w_A	w_B	w_C	w_A	w_B	w_C
0.9175	0.0415	0.0410	0.9279	0.0592	0.0129	0.9064	0.0233	0.0703
0.9020	0.0465	0.0515	0.9204	0.0701	0.0095	0.8796	0.0152	0.1052
0.8860	0.0510	0.0630	0.9125	0.0804	0.0071	0.8525	0.0107	0.1368
0.8650	0.0580	0.0770	0.8987	0.0965	0.0048	0.8225	0.0057	0.1718

Polymer (B): **poly(ethylene glycol)** **1989DIA**
Characterization: M_w/g.mol^{-1} = 20000, Carbowax 20M,
Union Carbide Corp., New York, NY
Solvent (A): **water** **H_2O** **7732-18-5**
Polymer (C): **dextran**
Characterization: M_n/g.mol^{-1} = 234200, M_w/g.mol^{-1} = 507000

Type of data: coexistence data (tie lines)

T/K = 277.15

continued

continued

Total system			Top phase			Bottom phase		
w_A	w_B	w_C	w_A	w_B	w_C	w_A	w_B	w_C
0.9490	0.0310	0.0200	0.9557	0.0365	0.0078	0.9425	0.0187	0.0388
0.9347	0.0260	0.0393	0.9486	0.0486	0.0028	0.9272	0.0150	0.0578
0.9081	0.0335	0.0584	0.9360	0.0629	0.0011	0.8943	0.0104	0.0953
0.8891	0.0395	0.0714	0.9259	0.0737	0.0004	0.8563	0.0064	0.1373

Polymer (B): **poly(ethylene glycol)** **1995GRO, 1995TIN**
Characterization: M_n/g.mol^{-1} = 39005, Hoechst AG, Frankfurt, Germany
Solvent (A): **water** **H_2O** **7732-18-5**
Polymer (C): **dextran**
Characterization: M_n/g.mol^{-1} = 179347, M_w/g.mol^{-1} = 507000

Type of data: coexistence data (tie lines)

T/K = 277.15

Total system			Top phase			Bottom phase		
w_A	w_B	w_C	w_A	w_B	w_C	w_A	w_B	w_C
0.9400	0.0249	0.0351	0.9563	0.0393	0.0044	0.9090	0.0034	0.0876
0.9190	0.0300	0.0510	0.9456	0.0527	0.0017	0.8823	0.0017	0.1160
0.8990	0.0400	0.0610	0.9307	0.0684	0.0009	0.8511	0.0012	0.1477
0.8800	0.0498	0.0702	0.9162	0.0829	0.0009	0.8210	0.0007	0.1783

T/K = 293.15

Total system			Top phase			Bottom phase		
w_A	w_B	w_C	w_A	w_B	w_C	w_A	w_B	w_C
0.9403	0.0248	0.0349	0.9562	0.0382	0.0056	0.9175	0.0050	0.0775
0.9403	0.0264	0.0333	0.9555	0.0391	0.0054	0.9142	0.0040	0.0818
0.9202	0.0298	0.0500	0.9455	0.0524	0.0021	0.8905	0.0021	0.1074
0.9002	0.0402	0.0596	0.9300	0.0691	0.0009	0.8600	0.0011	0.1389
0.8984	0.0408	0.0608	0.9279	0.0714	0.0007	0.8582	0.0011	0.1407
0.8789	0.0507	0.0704	0.9124	0.0870	0.0006	0.8323	0.0009	0.1668
0.8793	0.0507	0.0700	0.9120	0.0875	0.0005	0.8322	0.0007	0.1671
0.8174	0.0801	0.1025	0.8673	0.1321	0.0006	0.7476	0.0008	0.2516

continued

continued

T/K = 313.15

Total system			Top phase			Bottom phase		
w_A	w_B	w_C	w_A	w_B	w_C	w_A	w_B	w_C
0.9203	0.0299	0.0498	0.9375	0.0599	0.0026	0.8965	0.0026	0.1009
0.9000	0.0400	0.0600	0.9249	0.0745	0.0006	0.8768	0.0013	0.1219
0.8798	0.0502	0.0700	0.9078	0.0920	0.0002	0.8461	0.0009	0.1530
0.8601	0.0600	0.0799	0.8911	0.1086	0.0003	0.8249	0.0008	0.1743

Additional data for the system poly(ethylene glycol) (B) + water (A) + dextran (C) can be found in Refs. **1985WAL**, **1986ALB**, **1990FOR**, **1991FO1**, **1991FO2**, **1995ZAS**, and **2005WOH**.

Polymer (B): **poly(ethylene glycol)** **1996LIM**
Characterization: M_n/g.mol^{-1} = 2000, Shanghai Chem. Reagent Factory, PR China
Solvent (A): **water** **H_2O** **7732-18-5**
Polymer (C): **hydroxypropylstarch**
Characterization: M_n/g.mol^{-1} = 10000, Reppe Glykos AB, Vaxjo, Sweden

Type of data: coexistence data (tie lines)

T/K = 298.15

Total system			Bottom phase			Top phase		
w_A	w_B	w_C	w_A	w_B	w_C	w_A	w_B	w_C
0.7238	0.1097	0.1665	0.6285	0.0443	0.3272	0.7871	0.1669	0.0460
0.7336	0.0999	0.1665	0.6486	0.0465	0.3049	0.7912	0.1540	0.0548
0.7433	0.0898	0.1669	0.6760	0.0510	0.2730	0.7936	0.1345	0.0719
0.7510	0.0840	0.1650	0.7080	0.0570	0.2350	0.8000	0.1200	0.0800
0.7650	0.0840	0.1510	0.7240	0.0610	0.2150	0.8020	0.1040	0.0940

Plait-point composition: w_A = 0.769 + w_B = 0.076 + w_C = 0.155

Polymer (B): **poly(ethylene glycol)** **1996LIM**
Characterization: M_n/g.mol^{-1} = 2000, Shanghai Chem. Reagent Factory, PR China
Solvent (A): **water** **H_2O** **7732-18-5**
Polymer (C): **hydroxypropylstarch**
Characterization: M_n/g.mol^{-1} = 20000, Reppe Glykos AB, Vaxjo, Sweden

Type of data: coexistence data (tie lines)

continued

continued

T/K = 298.15

Total system			Bottom phase			Top phase		
w_A	w_B	w_C	w_A	w_B	w_C	w_A	w_B	w_C
0.7300	0.0899	0.1801	0.6568	0.0331	0.3101	0.8000	0.1470	0.0530
0.7400	0.0800	0.1800	0.6829	0.0359	0.2812	0.8090	0.1253	0.0657
0.7480	0.0720	0.1800	0.7100	0.0420	0.2480	0.8056	0.1144	0.0800
0.7524	0.0676	0.1800	0.7351	0.0531	0.2118	0.7991	0.0937	0.1072
0.7788	0.0682	0.1530	0.7500	0.0480	0.2020	0.8092	0.0900	0.1008

Plait-point composition: $w_A = 0.786 + w_B = 0.066 + w_C = 0.148$

Polymer (B): **poly(ethylene glycol)** **1996LIM**
Characterization: M_n/g.mol^{-1} = 4000, Shanghai Chem. Reagent Factory, PR China
Solvent (A): **water** **H_2O** **7732-18-5**
Polymer (C): **hydroxypropylstarch**
Characterization: M_n/g.mol^{-1} = 10000, Reppe Glykos AB, Vaxjo, Sweden

Type of data: coexistence data (tie lines)

T/K = 298.15

Total system			Bottom phase			Top phase		
w_A	w_B	w_C	w_A	w_B	w_C	w_A	w_B	w_C
0.7211	0.1100	0.1689	0.6155	0.0295	0.3550	0.7963	0.1645	0.0392
0.7331	0.1000	0.1669	0.6443	0.0317	0.3240	0.8036	0.1509	0.0455
0.7432	0.0899	0.1669	0.6616	0.0337	0.3047	0.7936	0.1368	0.0527
0.7532	0.0800	0.1666	0.6854	0.0348	0.2798	0.8105	0.1223	0.0652
0.7636	0.0699	0.1665	0.7121	0.0376	0.2503	0.8141	0.1044	0.0815

Plait-point composition: $w_A = 0.791 + w_B = 0.060 + w_C = 0.149$

Polymer (B): **poly(ethylene glycol)** **1996LIM**
Characterization: M_n/g.mol^{-1} = 4000, Shanghai Chem. Reagent Factory, PR China
Solvent (A): **water** **H_2O** **7732-18-5**
Polymer (C): **hydroxypropylstarch**
Characterization: M_n/g.mol^{-1} = 20000, Reppe Glykos AB, Vaxjo, Sweden

Type of data: coexistence data (tie lines)

T/K = 298.15

continued

continued

Total system			Bottom phase			Top phase		
w_A	w_B	w_C	w_A	w_B	w_C	w_A	w_B	w_C
0.7301	0.0899	0.1800	0.6389	0.0152	0.3459	0.8107	0.1564	0.0329
0.7401	0.0800	0.1799	0.6580	0.0152	0.3268	0.8201	0.1470	0.0329
0.7499	0.0700	0.1801	0.6746	0.0183	0.3071	0.8234	0.1334	0.0432
0.7602	0.0598	0.1800	0.7028	0.0151	0.2821	0.8359	0.1141	0.0500
0.7702	0.0498	0.1800	0.7269	0.0167	0.2564	0.8417	0.0986	0.0597

Polymer (B): **poly(ethylene glycol)** **1995BE1**
Characterization: M_n/g.mol^{-1} = 4000, Merck KGaA, Darmstadt, Germany
Solvent (A): **water** H_2O **7732-18-5**
Polymer (C): **hydroxypropylstarch**
Characterization: M_w/g.mol^{-1} = 200000, Reppe Glykos AB, Vaxjo, Sweden

Type of data: coexistence data (tie lines)

T/K = 293.15

Total system			Top phase			Bottom phase		
w_A	w_B	w_C	w_A	w_B	w_C	w_A	w_B	w_C
0.800	0.070	0.130	0.830	0.097	0.073	0.737	0.032	0.231
0.795	0.080	0.125	0.828	0.115	0.057	0.722	0.028	0.250
0.793	0.092	0.115	0.827	0.130	0.043	0.702	0.024	0.274
0.780	0.095	0.125	0.823	0.137	0.040	0.690	0.023	0.287

Polymer (B): **poly(ethylene glycol)** **1996LIM**
Characterization: M_n/g.mol^{-1} = 6000, Shanghai Chem. Reagent Factory, PR China
Solvent (A): **water** H_2O **7732-18-5**
Polymer (C): **hydroxypropylstarch**
Characterization: M_n/g.mol^{-1} = 10000, Reppe Glykos AB, Vaxjo, Sweden

Type of data: coexistence data (tie lines)

T/K = 298.15

Total system			Bottom phase			Top phase		
w_A	w_B	w_C	w_A	w_B	w_C	w_A	w_B	w_C
0.7331	0.0999	0.1670	0.6360	0.0290	0.3350	0.8103	0.1563	0.0334
0.7432	0.0900	0.1668	0.6626	0.0285	0.3089	0.8188	0.1424	0.0388
0.7632	0.0701	0.1667	0.7051	0.0319	0.2630	0.8338	0.1150	0.0512
0.7790	0.0590	0.1620	0.7546	0.0404	0.2050	0.8384	0.1000	0.0616

Polymer (B): **poly(ethylene glycol)** **1996LIM**
Characterization: M_n/g.mol^{-1} = 6000, Shanghai Chem. Reagent Factory, PR China
Solvent (A): **water** **H_2O** **7732-18-5**
Polymer (C): **hydroxypropylstarch**
Characterization: M_n/g.mol^{-1} = 20000, Reppe Glykos AB, Vaxjo, Sweden

Type of data: coexistence data (tie lines)

T/K = 298.15

Total system			Bottom phase			Top phase		
w_A	w_B	w_C	w_A	w_B	w_C	w_A	w_B	w_C
0.7295	0.0902	0.1803	0.6349	0.0137	0.3514	0.8112	0.1618	0.0270
0.7399	0.0800	0.1801	0.6536	0.0158	0.3306	0.8231	0.1471	0.0298
0.7499	0.0699	0.1802	0.6740	0.0158	0.3102	0.8315	0.1342	0.0343
0.7595	0.0599	0.1806	0.6959	0.0165	0.2876	0.8437	0.1175	0.0388
0.7713	0.0482	0.1805	0.7260	0.0200	0.2540	0.8553	0.1025	0.0422

Plait-point composition: w_A = 0.841 + w_B = 0.040 + w_C = 0.119

Polymer (B): **poly(ethylene glycol)** **1988SZL**
Characterization: M_n/g.mol^{-1} = 8000, Sigma Chemical Co., Inc., St. Louis, MO
Solvent (A): **water** **H_2O** **7732-18-5**
Polymer (C): **maltodextrin**
Characterization: M_n/g.mol^{-1} = 1800, derived from corn starch,
Grain Processing Corp., Muscatine, IA

Type of data: coexistence data (tie lines)

T/K = 298.15

Total system			Top phase			Bottom phase		
w_A	w_B	w_C	w_A	w_B	w_C	w_A	w_B	w_C
0.7379	0.0375	0.2246	0.7560	0.0650	0.1790	0.7210	0.0290	0.2500
0.7112	0.0493	0.2395	0.7690	0.0970	0.1340	0.6900	0.0000	0.3100
0.7246	0.0375	0.2379	0.7650	0.0750	0.1600	0.7100	0.0000	0.2900
0.6303	0.0840	0.2857	0.7190	0.1810	0.1000	0.5800	0.0000	0.4200

Polymer (B): **poly(ethylene glycol)** **2000SIL**
Characterization: M_n/g.mol^{-1} = 8044, M_w/g.mol^{-1} = 8768,
Sigma Chemical Co., Inc., St. Louis, MO

continued

continued

Solvent (A):	**water**	H_2O	**7732-18-5**
Polymer (C):	**maltodextrin**		
Characterization:	M_n/g.mol^{-1} = 1540, M_w/g.mol^{-1} = 2017, Loremalt 2030, Companhia Lorenz, Blumenau, SC, Brazil		

Type of data: coexistence data (tie lines)

T/K = 298.15

Total system			Top phase			Bottom phase		
w_A	w_B	w_C	w_A	w_B	w_C	w_A	w_B	w_C
0.5397	0.1122	0.3480	0.5575	0.1559	0.2865	0.4875	0.0081	0.5042
0.5021	0.1297	0.3681	0.5423	0.1857	0.2719	0.4440	0.0060	0.5499
0.4615	0.1505	0.3878	0.5020	0.2542	0.2436	0.4025	0.0053	0.5921
0.4280	0.1669	0.4050	0.4764	0.2908	0.2327	0.3523	0.0089	0.6387

Polymer (B):	**poly(ethylene glycol)**		**2000SIL**
Characterization:	M_n/g.mol^{-1} = 10535, M_w/g.mol^{-1} = 11589, Sigma Chemical Co., Inc., St. Louis, MO		
Solvent (A):	**water**	H_2O	**7732-18-5**
Polymer (C):	**maltodextrin**		
Characterization:	M_n/g.mol^{-1} = 1540, M_w/g.mol^{-1} = 2017, Loremalt 2030, Companhia Lorenz, Blumenau, SC, Brazil		

Type of data: coexistence data (tie lines)

T/K = 298.15

Total system			Top phase			Bottom phase		
w_A	w_B	w_C	w_A	w_B	w_C	w_A	w_B	w_C
0.5464	0.1072	0.3463	0.5784	0.1464	0.2750	0.4894	0.0045	0.5059
0.5184	0.1238	0.3577	0.5634	0.1804	0.2561	0.4427	0.0073	0.5499
0.4605	0.1515	0.3879	0.5042	0.2609	0.2347	0.3984	0.0091	0.5923
0.4291	0.1681	0.4026	0.4792	0.2960	0.2247	0.3817	0.0090	0.6091

Polymer (B):	**poly(ethylene glycol)**		**2000SIL**
Characterization:	M_n/g.mol^{-1} = 1425, M_w/g.mol^{-1} = 1468, Sigma Chemical Co., Inc., St. Louis, MO		
Solvent (A):	**water**	H_2O	**7732-18-5**

continued

continued

Polymer (C): **maltodextrin**
Characterization: M_n/g.mol^{-1} = 2020, M_w/g.mol^{-1} = 4000,
Loremalt 2001, Companhia Lorenz, Blumenau, SC, Brazil

Type of data: coexistence data (tie lines)

T/K = 298.15

Total system			Top phase			Bottom phase		
w_A	w_B	w_C	w_A	w_B	w_C	w_A	w_B	w_C
0.5521	0.1096	0.3382	0.5644	0.1268	0.3086	0.5059	0.0592	0.4348
0.5126	0.1254	0.3618	0.5504	0.1776	0.2718	0.4521	0.0431	0.5046
0.4305	0.1375	0.4318	0.4949	0.2459	0.2591	0.3628	0.0229	0.6142

Polymer (B): **poly(ethylene glycol)** **2000SIL**
Characterization: M_n/g.mol^{-1} = 8044, M_w/g.mol^{-1} = 8768,
Sigma Chemical Co., Inc., St. Louis, MO
Solvent (A): **water** H_2O **7732-18-5**
Polymer (C): **maltodextrin**
Characterization: M_n/g.mol^{-1} = 2020, M_w/g.mol^{-1} = 4000,
Loremalt 2001, Companhia Lorenz, Blumenau, SC, Brazil

Type of data: coexistence data (tie lines)

T/K = 298.15

Total system			Top phase			Bottom phase		
w_A	w_B	w_C	w_A	w_B	w_C	w_A	w_B	w_C
0.5399	0.1102	0.3498	0.6260	0.1884	0.1855	0.4645	0.0000	0.5354
0.5004	0.1323	0.3671	0.5913	0.2315	0.1770	0.4187	0.0000	0.5812
0.4600	0.1505	0.3893	0.5437	0.3019	0.1543	0.3984	0.0000	0.6015
0.4306	0.1681	0.4011	0.4849	0.3653	0.1497	0.3596	0.0012	0.6391

Polymer (B): **poly(ethylene glycol)** **2000SIL**
Characterization: M_n/g.mol^{-1} = 10535, M_w/g.mol^{-1} = 11589,
Sigma Chemical Co., Inc., St. Louis, MO
Solvent (A): **water** H_2O **7732-18-5**
Polymer (C): **maltodextrin**
Characterization: M_n/g.mol^{-1} = 2020, M_w/g.mol^{-1} = 4000,
Loremalt 2001, Companhia Lorenz, Blumenau, SC, Brazil

continued

continued

Type of data: coexistence data (tie lines)

T/K = 298.15

Total system			Top phase			Bottom phase		
w_A	w_B	w_C	w_A	w_B	w_C	w_A	w_B	w_C
0.5536	0.1061	0.3402	0.6308	0.1875	0.1815	0.4686	0.0000	0.5313
0.5169	0.1252	0.3577	0.5966	0.2302	0.1731	0.4239	0.0000	0.5760
0.4760	0.1455	0.3784	0.5468	0.3014	0.1517	0.4184	0.0004	0.5811
0.4370	0.1643	0.3986	0.5272	0.3270	0.1456	0.3608	0.0050	0.6340

Polymer (B): **poly(ethylene glycol)** **2006SAR**
Characterization: M_n/g.mol^{-1} = 6000, Merck KGaA, Darmstadt, Germany
Solvent (A): **water** **H_2O** **7732-18-5**
Polymer (C): **poly(acrylic acid)**
Characterization: M_n/g.mol^{-1} = 2100, Aldrich Chem. Co., Inc., Milwaukee, WI

Type of data: cloud points (LCST-behavior)

T/K = 293.15

w_A	0.6318	0.6516	0.6618	0.6727	0.6809	0.6880	0.6920	0.6947	0.7032
w_B	0.3183	0.2820	0.2644	0.2435	0.2300	0.2193	0.2097	0.2018	0.1834
w_C	0.0499	0.0664	0.0738	0.0838	0.0891	0.0927	0.0983	0.1035	0.1134
w_A	0.7166	0.7226	0.7289	0.7295	0.7304	0.7329	0.7313	0.7336	0.7350
w_B	0.1501	0.1296	0.1180	0.1052	0.0960	0.0843	0.0804	0.0732	0.0677
w_C	0.1333	0.1478	0.1531	0.1653	0.1736	0.1828	0.1883	0.1932	0.1973

T/K = 303.15

w_A	0.6404	0.6608	0.6734	0.7005	0.7063	0.7116	0.7205	0.7266	0.7323
w_B	0.3162	0.2840	0.2631	0.2110	0.1970	0.1835	0.1580	0.1457	0.1337
w_C	0.0434	0.0552	0.0635	0.0885	0.0967	0.1049	0.1215	0.1277	0.1340
w_A	0.7357	0.7401	0.7438	0.7451	0.7445	0.7444	0.7441	0.7440	0.7464
w_B	0.1246	0.1143	0.1009	0.0958	0.0921	0.0902	0.0866	0.0819	0.0802
w_C	0.1397	0.1456	0.1553	0.1591	0.1634	0.1654	0.1693	0.1741	0.1734

T/K = 313.15

w_A	0.6430	0.6705	0.6913	0.7077	0.7204	0.7366	0.7389	0.7420	0.7489
w_B	0.3195	0.2755	0.2453	0.2161	0.1911	0.1651	0.1585	0.1503	0.1324
w_C	0.0375	0.0540	0.0634	0.0762	0.0885	0.0983	0.1026	0.1077	0.1187
w_A	0.7563	0.7582	0.7558	0.7582	0.7557	0.7527			
w_B	0.1031	0.0939	0.0904	0.0833	0.0818	0.0739			
w_C	0.1406	0.1479	0.1538	0.1585	0.1625	0.1734			

Polymer (B): **poly(ethylene glycol)** **2006YA2**
Characterization: M_n/g.mol^{-1} = 4000, M_w/g.mol^{-1} = 4200
Solvent (A): **water** **H_2O** **7732-18-5**
Polymer (C): **poly(ethyleneimine)**
Characterization: M_n/g.mol^{-1} = 10000, M_w/g.mol^{-1} = 25000

Type of data: cloud points (LCST-behavior)

T/K = 298.15

pH = 5.3

w_A	0.9191	0.9296	0.9535	0.9570	0.9565	0.9535	0.9498	0.9378	0.9273
w_B	0.0789	0.0677	0.0400	0.0295	0.0223	0.0180	0.0152	0.0106	0.0068
w_C	0.0020	0.0027	0.0065	0.0135	0.0212	0.0285	0.0350	0.0516	0.0659
w_A	0.9070	0.8981							
w_B	0.0037	0.0036							
w_C	0.0893	0.0983							

pH = 7.5

w_A	0.8968	0.9011	0.9035	0.9082	0.9178	0.9216	0.9251	0.9248	0.9209
w_B	0.0118	0.0121	0.0137	0.0197	0.0297	0.0334	0.0406	0.0541	0.0653
w_C	0.0914	0.0868	0.0828	0.0721	0.0525	0.0450	0.0343	0.0211	0.0138
w_A	0.9163	0.9143							
w_B	0.0738	0.0771							
w_C	0.0099	0.0086							

pH = 9.2

w_A	0.8665	0.8719	0.8810	0.8834	0.8875	0.8910	0.8911	0.8911	0.8903
w_B	0.1204	0.1141	0.0994	0.0917	0.0851	0.0658	0.0620	0.0619	0.0518
w_C	0.0131	0.0140	0.0196	0.0249	0.0274	0.0432	0.0469	0.0470	0.0579
w_A	0.8872	0.8847							
w_B	0.0445	0.0253							
w_C	0.0683	0.0900							

Type of data: coexistence data (tie lines)

T/K = 298.15

Total system			Top phase			Bottom phase		
w_A	w_B	w_C	w_A	w_B	w_C	w_A	w_B	w_C
pH = 5.3								
0.9437	0.0300	0.0263	0.9533	0.0355	0.0112	0.8893	0.0036	0.1071
0.9314	0.0345	0.0341	0.9485	0.0449	0.0066	0.8666	0.0031	0.1303
0.9384	0.0204	0.0412	0.9562	0.0311	0.0127	0.9111	0.0038	0.0851
0.9162	0.0416	0.0422	0.9413	0.0535	0.0052	0.8319	0.0017	0.1664
0.9015	0.0498	0.0487	0.9325	0.0642	0.0033	0.7988	0.0019	0.1993

continued

continued

Total system			Top phase			Bottom phase		
w_A	w_B	w_C	w_A	w_B	w_C	w_A	w_B	w_C
pH = 7.5								
0.9203	0.0400	0.0397	0.9261	0.0449	0.0290	0.8849	0.0093	0.1058
0.9098	0.0458	0.0444	0.9230	0.0600	0.0170	0.8714	0.0063	0.1223
0.9009	0.0546	0.0445	0.9156	0.0755	0.0089	0.8646	0.0058	0.1296
0.8836	0.0619	0.0545	0.8981	0.0934	0.0085	0.8615	0.0060	0.1325
0.8731	0.0679	0.0590	0.8879	0.1048	0.0073	0.8501	0.0061	0.1438
pH = 9.2								
0.8853	0.0576	0.0571	0.8827	0.0887	0.0286	0.8871	0.0280	0.0849
0.8781	0.0621	0.0598	0.8750	0.1081	0.0169	0.8811	0.0220	0.0969
0.8621	0.0699	0.0680	0.8570	0.1317	0.0113	0.8695	0.0125	0.1180
0.8564	0.0749	0.0687	0.8494	0.1408	0.0098	0.8619	0.0113	0.1268
0.8427	0.0799	0.0774	0.8336	0.1569	0.0095	0.8507	0.0085	0.1408

Polymer (B): **poly(ethylene glycol)** **2006CHE**
Characterization: M_n/g.mol^{-1} = 6000
Solvent (A): **water** H_2O **7732-18-5**
Polymer (C): **xanthan**
Characterization: Monsanto

Type of data: coexistence data (tie lines)

T/K = 298.15

Total system			Top phase			Bottom phase		
w_A	w_B	w_C	w_A	w_B	w_C	w_A	w_B	w_C
0.9780	0.0200	0.0020	0.9595	0.0400	0.0005	0.9806	0.0170	0.0024
0.9775	0.0200	0.0025	0.9525	0.0450	0.0025	0.9814	0.0160	0.0026
0.9770	0.0200	0.0030	0.9462	0.0480	0.0058	0.9812	0.0162	0.0028
0.9765	0.0200	0.0035	0.9440	0.0490	0.0070	0.9791	0.0180	0.0029
0.9760	0.0200	0.0040	0.9363	0.0500	0.0137	0.9805	0.0167	0.0028

Polymer (B): **poly(ethylene glycol) dimethyl ether** **1998INO**
Characterization: M_w/g.mol^{-1} = 10000
Solvent (A): ***N,N*-dimethylformamide** C_3H_7NO **68-12-2**
Polymer (C): **poly(α,L-glutamate)**
Characterization: M_η/g.mol^{-1} = 155000, transesterificated with triethylene glycol monomethyl ether in the laboratory

continued

continued

Type of data: coexistence data (isotropic-nematic equilibrium)

T/K = 298.15

Total system			Isotropic phase			Anisotropic phase		
φ_A	φ_B	φ_C	φ_A	φ_B	φ_C	φ_A	φ_B	φ_C
0.842	0.000	0.158	0.861	0.000	0.139	0.840	0.000	0.160
0.846	0.013	0.141	0.862	0.020	0.113	0.840	0.009	0.151
0.848	0.049	0.103	0.841	0.057	0.072	0.804	0.009	0.187
0.829	0.050	0.121	0.867	0.083	0.050	0.795	0.012	0.193
0.835	0.058	0.107	0.868	0.088	0.044	0.792	0.012	0.196
0.811	0.060	0.129	0.861	0.110	0.029	0.769	0.010	0.221

Polymer (B): **poly(ethylene oxide)** **1993MAL**
Characterization: M_n/g.mol^{-1} = 600, Fluka AG, Buchs, Switzerland
Solvent (A): **water** H_2O **7732-18-5**
Polymer (C): **poly(propylene oxide)**
Characterization: M_n/g.mol^{-1} = 400, Fluka AG, Buchs, Switzerland

Type of data: cloud points (LCST-behavior)

T/K = 298.15

w_A	0.200	0.150	0.148	0.149	0.161	0.170	0.184	0.195	0.201
w_B	0.080	0.130	0.208	0.253	0.293	0.330	0.368	0.400	0.440
w_C	0.720	0.720	0.644	0.598	0.546	0.500	0.448	0.405	0.359
w_A	0.213	0.224	0.230	0.250	0.305	0.362	0.434	0.470	0.489
w_B	0.471	0.506	0.540	0.560	0.546	0.500	0.428	0.370	0.333
w_C	0.316	0.270	0.230	0.190	0.149	0.138	0.138	0.160	0.178
w_A	0.500	0.500	0.487	0.470	0.454	0.431	0.405	0.382	0.305
w_B	0.299	0.276	0.256	0.240	0.218	0.201	0.175	0.155	0.103
w_C	0.201	0.224	0.257	0.290	0.328	0.368	0.420	0.463	0.592

T/K = 326.15

w_A	0.254	0.214	0.203	0.205	0.208	0.222	0.236	0.247	0.260
w_B	0.038	0.083	0.123	0.162	0.242	0.277	0.310	0.339	0.370
w_C	0.708	0.703	0.674	0.633	0.550	0.501	0.454	0.414	0.370
w_A	0.270	0.282	0.290	0.304	0.374	0.456	0.560	0.606	0.649
w_B	0.402	0.431	0.460	0.483	0.471	0.419	0.310	0.256	0.210
w_C	0.328	0.287	0.250	0.213	0.155	0.125	0.130	0.138	0.141
w_A	0.678	0.712	0.736	0.753	0.771	0.785	0.800	0.816	0.820
w_B	0.178	0.144	0.118	0.098	0.080	0.066	0.050	0.029	0.013
w_C	0.144	0.144	0.146	0.149	0.149	0.149	0.150	0.155	0.167

Polymer (B): **poly(ethylene oxide)** **1989BAS**
Characterization: M_n/g.mol^{-1} = 3520, M_w/g.mol^{-1} = 3720,
Polysciences, Inc., Warrington, PA
Solvent (A): **water** **H_2O** **7732-18-5**
Polymer (C): **poly(vinyl methyl ether)**
Characterization: M_n/g.mol^{-1} = 51000, M_w/g.mol^{-1} = 110000,
Scientific Polymer Products, Inc., Ontario, NY

Type of data: coexistence data (tie lines)

T/K = 293.15

Total system			Phase I			Phase II		
w_A	w_B	w_C	w_A	w_B	w_C	w_A	w_B	w_C
0.850	0.060	0.090	0.886	0.078	0.036	0.798	0.034	0.168
0.840	0.070	0.090	0.878	0.094	0.028	0.776	0.030	0.194
0.830	0.080	0.090	0.872	0.106	0.022	0.744	0.028	0.228

Polymer (B): **poly(ethylene oxide)** **1989BAS**
Characterization: M_n/g.mol^{-1} = 10100, M_w/g.mol^{-1} = 10800,
Polysciences, Inc., Warrington, PA
Solvent (A): **water** **H_2O** **7732-18-5**
Polymer (C): **poly(vinyl methyl ether)**
Characterization: M_n/g.mol^{-1} = 51000, M_w/g.mol^{-1} = 110000,
Scientific Polymer Products, Inc., Ontario, NY

Type of data: coexistence data (tie lines)

T/K = 293.15

Total system			Phase I			Phase II		
w_A	w_B	w_C	w_A	w_B	w_C	w_A	w_B	w_C
0.900	0.040	0.060	0.920	0.052	0.028	0.856	0.012	0.132
0.870	0.030	0.100	0.916	0.062	0.022	0.836	0.010	0.154
0.850	0.030	0.120	0.910	0.073	0.017	0.810	0.006	0.184
0.860	0.060	0.080	0.900	0.087	0.013	0.772	0.002	0.226

Polymer (B): **poly(ethylene oxide)** **1989BAS**
Characterization: M_n/g.mol^{-1} = 29400, M_w/g.mol^{-1} = 34300,
Fluka AG, Buchs, Switzerland

continued

continued

Solvent (A): **water** H_2O **7732-18-5**
Polymer (C): **poly(vinyl methyl ether)**
Characterization: M_n/g.mol^{-1} = 51000, M_w/g.mol^{-1} = 110000,
Scientific Polymer Products, Inc., Ontario, NY

Type of data: coexistence data (tie lines)

T/K = 293.15

Total system			Phase I			Phase II		
w_A	w_B	w_C	w_A	w_B	w_C	w_A	w_B	w_C
0.910	0.020	0.070	0.939	0.049	0.012	0.890	0.002	0.108
0.890	0.020	0.090	0.930	0.060	0.010	0.869	0.001	0.130
0.870	0.020	0.110	0.921	0.072	0.007	0.850	0.001	0.149

Polymer (B): **poly(ethylene oxide-*b*-dimethylsiloxane)** **2002MAD**
Characterization: M_n/g.mol^{-1} = 460, M_w/g.mol^{-1} = 600, 75.0 wt% ethylene oxide,
Fluka AG, Buchs, Switzerland
Solvent (A): **1,2,3,4-tetrahydronaphthalene** $C_{10}H_{12}$ **119-64-2**
Polymer (C): **poly(ethylene oxide)**
Characterization: M_n/g.mol^{-1} = 21000, M_w/g.mol^{-1} = 27000,
Fluka AG, Buchs, Switzerland

Type of data: cloud points

w_A	0.000	0.000	0.000	0.000	0.000	0.000	0.000	0.000	0.000
w_B	0.999	0.990	0.985	0.980	0.950	0.900	0.800	0.700	0.600
w_C	0.001	0.010	0.015	0.020	0.050	0.100	0.200	0.300	0.400
T/K	321.15	330.15	340.15	370.15	388.15	395.65	402.15	405.15	400.15

w_A	0.000	0.000	0.000	0.2506	0.1010	0.0000
w_B	0.500	0.400	0.250	0.5992	0.7490	0.8500
w_C	0.500	0.600	0.750	0.1501	0.1499	0.1500
T/K	390.65	376.15	338.15	335.15	366.15	371.15

Polymer (B): **poly(ethylene oxide-*b*-dimethylsiloxane)** **2003JI2**
Characterization: M_w/g.mol^{-1} = 1800, 77 mol% ethylene oxide, EO27-b-DMS8
Solvent (A): **toluene** C_7H_8 **108-88-3**
Polymer (C): **poly(ethylene oxide)**
Characterization: M_w/g.mol^{-1} = 35000, Fluka AG, Buchs, Switzerland

Type of data: cloud points

continued

continued

w_A	0.000	0.000	0.000	0.000	0.000	0.000	0.000	0.351	0.586
w_B	0.948	0.900	0.799	0.698	0.596	0.496	0.397	0.645	0.375
w_C	0.052	0.100	0.201	0.302	0.404	0.504	0.603	0.004	0.039
T/K	393.0	396.0	403.0	405.0	400.0	391.0	378.0	308.2	308.2

w_A	0.597	0.621	0.607	0.680	0.694	0.619	0.657	0.643	0.641
w_B	0.325	0.292	0.267	0.179	0.161	0.224	0.168	0.164	0.122
w_C	0.078	0.087	0.126	0.141	0.146	0.156	0.175	0.193	0.237
T/K	308.2	308.2	308.2	308.2	308.2	308.2	308.2	308.2	308.2

w_A	0.659	0.624	0.669	0.626	0.569	0.297	0.311	0.366	0.377
w_B	0.095	0.124	0.074	0.108	0.082	0.682	0.648	0.557	0.498
w_C	0.246	0.252	0.258	0.266	0.348	0.021	0.041	0.077	0.125
T/K	308.2	308.2	308.2	308.2	308.2	318.2	318.2	318.2	318.2

w_A	0.387	0.384	0.381	0.342	0.386	0.251	0.245	0.243	0.212
w_B	0.466	0.431	0.370	0.347	0.435	0.556	0.490	0.411	0.381
w_C	0.147	0.185	0.249	0.311	0.179	0.192	0.275	0.346	0.407
T/K	318.2	318.2	318.2	318.2	318.2	328.2	328.2	328.2	328.2

w_A	0.184	0.199	0.233	0.262	0.241	0.245	0.175	0.197	0.209
w_B	0.390	0.774	0.715	0.629	0.603	0.514	0.734	0.659	0.628
w_C	0.426	0.027	0.052	0.108	0.156	0.241	0.091	0.144	0.163
T/K	328.2	328.2	328.2	328.2	328.2	328.2	333.2	333.2	333.2

w_A	0.196	0.156
w_B	0.597	0.538
w_C	0.207	0.306
T/K	333.2	333.2

Type of data: spinodal points

w_A	0.695	0.695	0.695	0.695	0.695	0.438	0.438	0.438	0.438
w_B	0.169	0.169	0.169	0.169	0.169	0.386	0.386	0.386	0.386
w_C	0.172	0.172	0.172	0.172	0.172	0.176	0.176	0.176	0.176
T/K	309.2	308.2	307.2	306.2	305.2	318.2	317.2	316.2	315.2
P/bar	588	441	342	112	23	617	483	364	159

w_A	0.438	0.255	0.255	0.255	0.255	0.255	0.209	0.209	0.209
w_B	0.386	0.558	0.558	0.558	0.558	0.558	0.626	0.626	0.626
w_C	0.176	0.187	0.187	0.187	0.187	0.187	0.165	0.165	0.165
T/K	314.2	328.2	327.2	326.2	325.2	324.2	333.2	332.2	331.2
P/bar	21	160	223	265	321	383	85	114	163

w_A	0.209	0.209	0.209	0.209	0.209
w_B	0.626	0.626	0.626	0.626	0.626
w_C	0.165	0.165	0.165	0.165	0.165
T/K	330.2	329.2	328.2	327.2	326.2
P/bar	212	258	322	369	408

Polymer (B): **poly(ethylene oxide-*co*-propylene oxide)** **1995BE1**
Characterization: M_n/g.mol^{-1} = 3200, 30 mol% ethylene oxide, Shearwater Polymers, Huntsville, AL
Solvent (A): **water** **H_2O** **7732-18-5**
Polymer (C): **hydroxypropylstarch**
Characterization: M_w/g.mol^{-1} = 200000, Reppe Glykos AB, Vaxjo, Sweden

Type of data: coexistence data (tie lines)

T/K = 293.15

Total system			Top phase			Bottom phase		
w_A	w_B	w_C	w_A	w_B	w_C	w_A	w_B	w_C
0.850	0.090	0.060	0.865	0.101	0.034	0.775	0.030	0.195
0.830	0.090	0.080	0.859	0.115	0.026	0.760	0.028	0.212
0.790	0.100	0.110	0.832	0.148	0.020	0.722	0.019	0.259
0.780	0.100	0.120	0.829	0.153	0.018	0.714	0.020	0.266

Polymer (B): **poly(ethylene oxide-*co*-propylene oxide)** **1994MOD**
Characterization: M_n/g.mol^{-1} = 4000, 50 mol% ethylene oxide, Ucon 50-HB-5100, Union Carbide Corp., New York, NY
Solvent (A): **water** **H_2O** **7732-18-5**
Polymer (C): **hydroxypropylstarch** **1995BE1**
Characterization: M_w/g.mol^{-1} = 200000, Reppe Glykos AB, Vaxjo, Sweden

Type of data: coexistence data (tie lines)

T/K = 293.15

Total system			Top phase			Bottom phase		
w_A	w_B	w_C	w_A	w_B	w_C	w_A	w_B	w_C
0.856	0.075	0.069	0.873	0.088	0.039	0.797	0.015	0.188
0.845	0.080	0.075	0.866	0.101	0.033	0.786	0.017	0.197
0.820	0.080	0.100	0.857	0.115	0.028	0.750	0.006	0.244
0.808	0.100	0.092	0.840	0.139	0.021	0.734	0.008	0.258
0.759	0.125	0.116	0.803	0.185	0.012	0.676	0.005	0.319

Polymer (B): **poly(ethylene oxide-*co*-propylene oxide)** **1998LIM**
Characterization: M_n/g.mol^{-1} = 3000, M_w/g.mol^{-1} = 3300, 33.3 mol% ethylene oxide, Zhejiang Univ. Chem. Factory, PR China

continued

continued

Solvent (A): **water** H_2O **7732-18-5**
Polymer (C): **hydroxypropylstarch**
Characterization: M_n/g.mol^{-1} = 10000, Reppe Glykos AB, Vaxjo, Sweden

Type of data: coexistence data (tie lines)

T/K = 298.15

Total system			Bottom phase			Top phase		
w_A	w_B	w_C	w_A	w_B	w_C	w_A	w_B	w_C
0.8500	0.1000	0.0500	0.7729	0.0250	0.2021	0.8571	0.1080	0.0349
0.8399	0.0999	0.0602	0.7618	0.0222	0.2160	0.8533	0.1139	0.0328
0.8300	0.1000	0.0700	0.7494	0.0204	0.2302	0.8498	0.1200	0.0302
0.8208	0.0995	0.0797	0.7405	0.0190	0.2405	0.8460	0.1260	0.0280
0.8099	0.0999	0.0902	0.7276	0.0172	0.2552	0.8420	0.1321	0.0259
0.8002	0.0999	0.0999	0.7169	0.0159	0.2672	0.8370	0.1380	0.0250
0.7901	0.1000	0.1099	0.7077	0.0152	0.2771	0.8320	0.1441	0.0239
0.7801	0.1000	0.1199	0.6950	0.0140	0.2909	0.8280	0.1489	0.0231

Plait-point composition: w_A = 0.827 + w_B = 0.063 + w_C = 0.110

Polymer (B): **poly(ethylene oxide-*co*-propylene oxide)** **1998LIM**
Characterization: M_n/g.mol^{-1} = 3000, M_w/g.mol^{-1} = 3300, 33.3 mol% ethylene oxide, Zhejiang Univ. Chem. Factory, PR China
Solvent (A): **water** H_2O **7732-18-5**
Polymer (C): **hydroxypropylstarch**
Characterization: M_n/g.mol^{-1} = 20000, Reppe Glykos AB, Vaxjo, Sweden

Type of data: coexistence data (tie lines)

T/K = 298.15

Total system			Bottom phase			Top phase		
w_A	w_B	w_C	w_A	w_B	w_C	w_A	w_B	w_C
0.8503	0.0999	0.0498	0.7872	0.0220	0.1908	0.8597	0.1118	0.0285
0.8401	0.1000	0.0599	0.7699	0.0171	0.2130	0.8561	0.1200	0.0239
0.8299	0.1001	0.0700	0.7567	0.0122	0.2311	0.8519	0.1271	0.0210
0.8201	0.1000	0.0799	0.7480	0.0100	0.2420	0.8469	0.1340	0.0190
0.8099	0.1000	0.0901	0.7336	0.0062	0.2602	0.8431	0.1390	0.0179
0.7998	0.1000	0.1002	0.7235	0.0050	0.2715	0.8361	0.1481	0.0158
0.7902	0.1000	0.1098	0.7145	0.0050	0.2805	0.8323	0.1529	0.0148
0.7796	0.1002	0.1202	0.7068	0.0050	0.2882	0.8260	0.1600	0.0140

Plait-point composition: w_A = 0.832 + w_B = 0.058 + w_C = 0.110

Polymer (B): **poly(ethylene oxide-*co*-propylene oxide)** **1998LIM**
Characterization: M_n/g.mol^{-1} = 3500, M_w/g.mol^{-1} = 3900, 50.0 mol% ethylene oxide, Zhejiang Univ. Chem. Factory, PR China
Solvent (A): **water** **H_2O** **7732-18-5**
Polymer (C): **hydroxypropylstarch**
Characterization: M_n/g.mol^{-1} = 10000, Reppe Glykos AB, Vaxjo, Sweden

Type of data: coexistence data (tie lines)

T/K = 298.15

Total system			Bottom phase			Top phase		
w_A	w_B	w_C	w_A	w_B	w_C	w_A	w_B	w_C
0.8498	0.1000	0.0502	0.7895	0.0303	0.1802	0.8549	0.1070	0.0381
0.8399	0.0999	0.0602	0.7767	0.0282	0.1951	0.8517	0.1141	0.0342
0.8296	0.1000	0.0704	0.7679	0.0251	0.2070	0.8476	0.1202	0.0322
0.8199	0.1001	0.0800	0.7516	0.0222	0.2262	0.8419	0.1281	0.0300
0.8099	0.1002	0.0899	0.7447	0.0212	0.2341	0.8378	0.1342	0.0280
0.8000	0.1000	0.1000	0.7326	0.0201	0.2473	0.8335	0.1403	0.0262
0.7901	0.1000	0.1099	0.7207	0.0201	0.2592	0.8299	0.1451	0.0250
0.7801	0.1000	0.1199	0.7090	0.0200	0.2710	0.8248	0.1511	0.0241

Plait-point composition: w_A = 0.835 + w_B = 0.067 + w_C = 0.098

Polymer (B): **poly(ethylene oxide-*co*-propylene oxide)** **1998LIM**
Characterization: M_n/g.mol^{-1} = 3000, M_w/g.mol^{-1} = 3300, 33.3 mol% ethylene oxide, Zhejiang Univ. Chem. Factory, PR China
Solvent (A): **water** **H_2O** **7732-18-5**
Polymer (C): **hydroxypropylstarch**
Characterization: M_n/g.mol^{-1} = 20000, Reppe Glykos AB, Vaxjo, Sweden

Type of data: coexistence data (tie lines)

T/K = 298.15

Total system			Bottom phase			Top phase		
w_A	w_B	w_C	w_A	w_B	w_C	w_A	w_B	w_C
0.8403	0.0999	0.0598	0.7753	0.0241	0.2006	0.8627	0.1022	0.0351
0.8499	0.1001	0.0500	0.7620	0.0198	0.2182	0.8605	0.1090	0.0305
0.8403	0.0999	0.0704	0.7448	0.0142	0.2410	0.8570	0.1160	0.0270
0.8302	0.0999	0.0598	0.7290	0.0110	0.2600	0.8534	0.1230	0.0236
0.8201	0.1001	0.0699	0.7203	0.0092	0.2705	0.8495	0.1275	0.0230
0.8099	0.1001	0.0798	0.7089	0.0063	0.2848	0.8448	0.1322	0.0230
0.7996	0.1001	0.0899	0.6973	0.0052	0.2975	0.8410	0.1370	0.0220
0.7900	0.1001	0.1099	0.6892	0.0050	0.3058	0.8357	0.1423	0.0220
0.7802	0.0999	0.1199	0.6750	0.0050	0.3200	0.8320	0.1462	0.0218

Polymer (B): **poly(ethylene oxide-*co*-propylene oxide)** **2004PER**
Characterization: M_n/g.mol^{-1} = 3900, 50.0 mol% ethylene oxide, Ucon 50-HB-5100, Union Carbide Corp.
Solvent (A): **water** **H_2O** **7732-18-5**
Polymer (C): **hydroxypropylstarch**
Characterization: M_n/g.mol^{-1} = 100000, Reppe Glykos AB, Vaxjo, Sweden

Type of data: coexistence data (tie lines)

T/K = 295.15

Total system			Top phase			Bottom phase		
w_A	w_B	w_C	w_A	w_B	w_C	w_A	w_B	w_C
0.8319	0.0473	0.1208	0.8517	0.0599	0.0884	0.8091	0.0236	0.1673
0.7902	0.0804	0.1294	0.8414	0.1236	0.0350	0.7222	0.0175	0.2603
0.8032	0.0627	0.1341	0.8467	0.0946	0.0587	0.7504	0.0193	0.2303
0.7899	0.0709	0.1392	0.8431	0.1142	0.0427	0.7313	0.0180	0.2507
0.8048	0.0450	0.1502	0.8502	0.0822	0.0676	0.7801	0.0207	0.1992

Plait-point composition: w_A = 0.8401 + w_B = 0.0369 + w_C = 0.1230

Polymer (B): **poly(ethylene oxide-*co*-propylene oxide)** **2005BOL**
Characterization: M_n/g.mol^{-1} = 1059, M_η/g.mol^{-1} = 1228, 50 mol% ethylene oxide, Dow Chemical Co., San Lorenzo, Argentina
Solvent (A): **water** **H_2O** **7732-18-5**
Polymer (C): **maltodextrin**
Characterization: M_n/g.mol^{-1} = 838, M_η/g.mol^{-1} = 922, Polimerosa, Kasdorf SA, Buenos Aires, Argentina

Type of data: coexistence data (tie lines)

T/K = 297.15

Total system			Top phase			Bottom phase		
w_A	w_B	w_C	w_A	w_B	w_C	w_A	w_B	w_C
0.6610	0.0950	0.2440	0.7360	0.1640	0.1000	0.5919	0.0306	0.3775
0.6730	0.0930	0.2340	0.7480	0.1550	0.0970	0.6070	0.0312	0.3618
0.6860	0.0890	0.2250	0.7470	0.1376	0.1154	0.6330	0.0366	0.3304
0.7020	0.0860	0.2120	0.7552	0.1400	0.1048	0.6604	0.0460	0.2936
0.7240	0.0790	0.1970	0.7660	0.1145	0.1195	0.6835	0.0386	0.2779
0.7370	0.0750	0.1880	0.7696	0.0967	0.1337	0.6974	0.0404	0.2622

Polymer (B): **poly(ethylene oxide-*co*-propylene oxide)** **2004PER**
Characterization: M_n/g.mol^{-1} = 3900, 50.0 mol% ethylene oxide, Ucon 50-HB-5100, Union Carbide Corp.
Solvent (A): **water** **H_2O** **7732-18-5**
Polymer (C): **poly(vinyl acetate-*co*-vinyl alcohol)**
Characterization: M_n/g.mol^{-1} = 10000, 12.0 mol% vinyl acetate, PVA 10000, 88% hydrolyzed, Scientific Polymer Products, Inc., Ontario, NY

Type of data: coexistence data (tie lines)

T/K = 295.15

Total system			Top phase			Bottom phase		
w_A	w_B	w_C	w_A	w_B	w_C	w_A	w_B	w_C
0.7958	0.1047	0.0995	0.8077	0.1273	0.0650	0.7827	0.0214	0.1959
0.7612	0.1323	0.1065	0.7790	0.1931	0.0279	0.7375	0.0174	0.2451
0.7799	0.1104	0.1097	0.7970	0.1617	0.0413	0.7642	0.0180	0.2178
0.7493	0.1327	0.1180	0.7710	0.2067	0.0223	0.7129	0.0168	0.2703

Plait-point composition: w_A = 0.8100 + w_B = 0.0600 + w_C = 0.1300

Polymer (B): **poly(ethylene oxide-*b*-propylene oxide-*b*-ethylene oxide)** **2004TAD, 2005TAD**
Characterization: M_n/g.mol^{-1} = 4750, 37.0 mol% ethylene oxide, (EO)17-(PO)58-(EO)17, P103, Aldrich Chem. Co., Inc., Milwaukee, WI
Solvent (A): **water** **H_2O** **7732-18-5**
Polymer (C): **dextran**
Characterization: M_n/g.mol^{-1} = 8200, M_w/g.mol^{-1} = 11600, Dextran 19, Sigma Chemical Co., Inc., St. Louis, MO

Type of data: coexistence data (tie lines)

T/K = 298.15

Total system			Top phase			Bottom phase		
w_A	w_B	w_C	w_A	w_B	w_C	w_A	w_B	w_C
0.8183	0.1015	0.0802	0.8450	0.0460	0.1090	0.7829	0.1634	0.0537
0.8210	0.0904	0.0886	0.8503	0.0276	0.1221	0.7657	0.1916	0.0427
0.8073	0.1129	0.0798	0.8518	0.0230	0.1252	0.7569	0.2036	0.0395

Polymer (B): **poly(ethylene oxide-*b*-propylene oxide-*b*-ethylene oxide)** **2004TAD, 2005TAD**
Characterization: M_n/g.mol^{-1} = 4750, 37.0 mol% ethylene oxide, (EO)17-(PO)58-(EO)17, P103, Aldrich Chem. Co., Inc., Milwaukee, WI
Solvent (A): **water** **H_2O** **7732-18-5**
Polymer (C): **dextran**
Characterization: M_n/g.mol^{-1} = 236000, M_w/g.mol^{-1} = 410000, Dextran 400, Sigma Chemical Co., Inc., St. Louis, MO

Type of data: cloud points

T/K = 298.15

w_A	0.9001	0.9103
w_B	0.0705	0.0343
w_C	0.0294	0.0554

Type of data: coexistence data (tie lines)

T/K = 298.15

Total system			Top phase			Bottom phase		
w_A	w_B	w_C	w_A	w_B	w_C	w_A	w_B	w_C
0.9081	0.0524	0.0395	0.9219	0.0203	0.0578	0.8687	0.1207	0.0106
0.8893	0.0602	0.0505	0.9079	0.0230	0.0691	0.8488	0.1401	0.0111
0.8570	0.0729	0.0701	0.8735	0.0236	0.1029	0.8152	0.1805	0.0043

Polymer (B): **poly(ethylene oxide-*b*-propylene oxide-*b*-ethylene oxide)** **2004TAD, 2005TAD**
Characterization: M_n/g.mol^{-1} = 6500, 54.4 mol% ethylene oxide, (EO)37-(PO)62-(EO)37, F105, ICI Surfactants, Cleveland, UK
Solvent (A): **water** **H_2O** **7732-18-5**
Polymer (C): **dextran**
Characterization: M_n/g.mol^{-1} = 8200, M_w/g.mol^{-1} = 11600, Dextran 19, Sigma Chemical Co., Inc., St. Louis, MO

Type of data: coexistence data (tie lines)

T/K = 298.15

Total system			Top phase			Bottom phase		
w_A	w_B	w_C	w_A	w_B	w_C	w_A	w_B	w_C
0.7710	0.1169	0.1121	0.7821	0.0022	0.2157	0.7601	0.2239	0.0160
0.8200	0.0800	0.1000	0.8470	0.0083	0.1447	0.7894	0.1835	0.0271
0.7998	0.0901	0.1101	0.8238	0.0025	0.1737	0.7706	0.2085	0.0209

Polymer (B): **poly(ethylene oxide-*b*-propylene oxide-*b*-ethylene oxide)** **2004TAD, 2005TAD**
Characterization: M_n/g.mol^{-1} = 6500, 54.4 mol% ethylene oxide, (EO)37-(PO)62-(EO)37, F105, ICI Surfactants, Cleveland, UK
Solvent (A): **water** **H_2O** **7732-18-5**
Polymer (C): **dextran**
Characterization: M_n/g.mol^{-1} = 236000, M_w/g.mol^{-1} = 410000, Dextran 400, Sigma Chemical Co., Inc., St. Louis, MO

Type of data: cloud points

T/K = 298.15

w_A	0.8991	0.9095
w_B	0.0720	0.0350
w_C	0.0289	0.0555

Type of data: coexistence data (tie lines)

T/K = 298.15

Total system			Top phase			Bottom phase		
w_A	w_B	w_C	w_A	w_B	w_C	w_A	w_B	w_C
0.8957	0.0442	0.0601	0.9025	0.0142	0.0833	0.8685	0.1236	0.0079
0.8712	0.0601	0.0687	0.8833	0.0124	0.1043	0.8464	0.1485	0.0051
0.8994	0.0511	0.0495	0.9115	0.0173	0.0712	0.8845	0.1027	0.0128
0.8250	0.0875	0.0875	0.8364	0.0076	0.1560	0.8203	0.1767	0.0030

Polymer (B): **poly(ethylene oxide-*b*-propylene oxide-*b*-ethylene oxide)** **1999SVE**
Characterization: M_n/g.mol^{-1} = 6500, 56.9 wt% ethylene oxide, (EO)37-(PO)56-(EO)37, Pluronic P105, BASF, Parsippany, NJ
Solvent (A): **water** **H_2O** **7732-18-5**
Polymer (C): **dextran**
Characterization: M_w/g.mol^{-1} = 500000, T500, Amersham Pharmacia Biotech, Uppsala, Sweden

Type of data: coexistence data (tie lines)

Total system			Top phase			Bottom phase		
w_A	w_B	w_C	w_A	w_B	w_C	w_A	w_B	w_C
T/K = 278.15								
0.880	0.064	0.056	0.905	0.090	0.005	0.820	0.010	0.170
T/K = 303.15								
0.880	0.064	0.056	0.835	0.160	0.005	0.895	0.030	0.075

Polymer (B): **poly(ethylene oxide-*b*-propylene oxide-*b*-ethylene oxide)** **1999SVE**
Characterization: M_n/g.mol^{-1} = 8400, 84 mol% ethylene oxide, (EO)76-(PO)29-(EO)76, Pluronic F68, BASF, Parsippany, NJ
Solvent (A): **water** **H_2O** **7732-18-5**
Polymer (C): **dextran**
Characterization: M_w/g.mol^{-1} = 500000, T500, Amersham Pharmacia Biotech, Uppsala, Sweden

Type of data: coexistence data (tie lines)

Total system			Top phase			Bottom phase		
w_A	w_B	w_C	w_A	w_B	w_C	w_A	w_B	w_C
T/K = 278.15								
0.880	0.050	0.070	0.915	0.080	0.005	0.825	0.005	0.170
T/K = 303.15								
0.880	0.050	0.070	0.910	0.085	0.005	0.840	0.005	0.155

Polymer (B): **poly(ethylene oxide-*b*-propylene oxide-*b*-ethylene oxide)** **2004TAD, 2005TAD**
Characterization: M_n/g.mol^{-1} = 8530, 83.5 mol% ethylene oxide, (EO)76-(PO)30-(EO)76, F68, Aldrich Chem. Co., Inc., Milwaukee, WI
Solvent (A): **water** **H_2O** **7732-18-5**
Polymer (C): **dextran**
Characterization: M_n/g.mol^{-1} = 8200, M_w/g.mol^{-1} = 11600, Dextran 19, Sigma Chemical Co., Inc., St. Louis, MO

Type of data: cloud points

T/K = 298.15

w_A	0.8750	0.8649	0.8629	0.8525	0.8401	0.8174	0.8021
w_B	0.1006	0.0826	0.0637	0.0537	0.0405	0.0235	0.0155
w_C	0.0244	0.0525	0.0734	0.0938	0.1194	0.1591	0.1824

Type of data: coexistence data (tie lines)

T/K = 298.15

Total system			Top phase			Bottom phase		
w_A	w_B	w_C	w_A	w_B	w_C	w_A	w_B	w_C
0.8405	0.0700	0.0895	0.8039	0.0236	0.1725	0.8591	0.1035	0.0374
0.8196	0.0799	0.1005	0.7692	0.0101	0.2207	0.8512	0.1307	0.0181
0.8031	0.0899	0.1070	0.7556	0.0070	0.2374	0.8343	0.1481	0.0176

Polymer (B): **poly(ethylene oxide-*b*-propylene oxide-*b*-ethylene oxide)** **2004TAD, 2005TAD**
Characterization: M_n/g.mol^{-1} = 8530, 83.5 mol% ethylene oxide, (EO)76-(PO)30-(EO)76, F68, Aldrich Chem. Co., Inc., Milwaukee, WI
Solvent (A): **water** **H_2O** **7732-18-5**
Polymer (C): **dextran**
Characterization: M_n/g.mol^{-1} = 236000, M_w/g.mol^{-1} = 410000, Dextran 400, Sigma Chemical Co., Inc., St. Louis, MO

Type of data: cloud points

T/K = 298.15

w_A	0.9201	0.9099	0.8892
w_B	0.0404	0.0351	0.0229
w_C	0.0395	0.0550	0.0879

Type of data: coexistence data (tie lines)

T/K = 298.15

Total system			Top phase			Bottom phase		
w_A	w_B	w_C	w_A	w_B	w_C	w_A	w_B	w_C
0.8913	0.0396	0.0691	0.8713	0.0146	0.1141	0.9219	0.0657	0.0124
0.8699	0.0602	0.0699	0.8198	0.0002	0.1800	0.9008	0.0949	0.0043
0.8414	0.0892	0.0694	0.7608	0.0087	0.2305	0.8750	0.1222	0.0028

Polymer (B): **poly(ethylene oxide-*b*-propylene oxide-*b*-ethylene oxide)** **2004TAD, 2005TAD**
Characterization: M_n/g.mol^{-1} = 14000, 83.6 mol% ethylene oxide, (EO)127-(PO)50-(EO)127, P108, ICI Surfactants, Cleveland, UK
Solvent (A): **water** **H_2O** **7732-18-5**
Polymer (C): **dextran**
Characterization: M_n/g.mol^{-1} = 8200, M_w/g.mol^{-1} = 11600, Dextran 19, Sigma Chemical Co., Inc., St. Louis, MO

Type of data: coexistence data (tie lines)

T/K = 298.15

Total system			Top phase			Bottom phase		
w_A	w_B	w_C	w_A	w_B	w_C	w_A	w_B	w_C
0.8499	0.0600	0.0901	0.8433	0.0432	0.1135	0.8595	0.1038	0.0367
0.8389	0.0700	0.0911	0.8283	0.0323	0.1394	0.8463	0.1288	0.0249
0.8210	0.0800	0.0990	0.8017	0.0235	0.1748	0.8330	0.1485	0.0185

Polymer (B): **poly(ethylene oxide-*b*-propylene oxide-*b*-ethylene oxide)** **2004TAD, 2005TAD**
Characterization: M_n/g.mol^{-1} = 14000, 83.6 mol% ethylene oxide, (EO)127-(PO)50-(EO)127, P108, ICI Surfactants, Cleveland, UK
Solvent (A): **water** **H_2O** **7732-18-5**
Polymer (C): **dextran**
Characterization: M_n/g.mol^{-1} = 236000, M_w/g.mol^{-1} = 410000, Dextran 400, Sigma Chemical Co., Inc., St. Louis, MO

Type of data: cloud points

T/K = 298.15

w_A	0.9256	0.9152
w_B	0.0368	0.0157
w_C	0.0376	0.0691

Type of data: coexistence data (tie lines)

T/K = 298.15

Total system			Top phase			Bottom phase		
w_A	w_B	w_C	w_A	w_B	w_C	w_A	w_B	w_C
0.9191	0.0313	0.0496	0.9059	0.0102	0.0839	0.9324	0.0512	0.0164
0.8912	0.0494	0.0594	0.8700	0.0083	0.1217	0.9118	0.0832	0.0050
0.8715	0.0692	0.0593	0.8357	0.0037	0.1606	0.8951	0.1019	0.0030

Polymer (B): **poly(ethylene oxide-*b*-propylene oxide-*b*-ethylene oxide)** **2004TAD, 2005TAD**
Characterization: M_n/g.mol^{-1} = 4800, 83.8 mol% ethylene oxide, (EO)44-(PO)17-(EO)44, F38, ICI Surfactants, Cleveland, UK
Solvent (A): **water** **H_2O** **7732-18-5**
Polymer (C): **dextran**
Characterization: M_n/g.mol^{-1} = 8200, M_w/g.mol^{-1} = 11600, Dextran 19, Sigma Chemical Co., Inc., St. Louis, MO

Type of data: cloud points

T/K = 298.15

w_A	0.8377	0.8508	0.8628	0.8578	0.8631	0.8620	0.8602	0.8506	0.8459
w_B	0.1534	0.1399	0.1187	0.1199	0.1109	0.1101	0.1021	0.0932	0.0863
w_C	0.0089	0.0093	0.0185	0.0223	0.0260	0.0279	0.0377	0.0562	0.0678

w_A	0.8434	0.8365
w_B	0.0817	0.0838
w_C	0.0749	0.0797

continued

continued

Type of data: coexistence data (tie lines)

T/K = 298.15

Total system			Top phase			Bottom phase		
w_A	w_B	w_C	w_A	w_B	w_C	w_A	w_B	w_C
0.8239	0.0853	0.0908	0.7786	0.0000	0.2214	0.8441	0.1087	0.0472
0.7903	0.0998	0.1099	0.7343	0.0000	0.2657	0.8235	0.1548	0.0217
0.7497	0.1303	0.1200	0.6673	0.0000	0.3327	0.7922	0.1950	0.0128
0.7596	0.1202	0.1202	0.6759	0.0000	0.3241	0.8006	0.1849	0.0145

Polymer (B): **poly(ethylene oxide-*b*-propylene oxide-*b*-ethylene oxide)** **2004TAD, 2005TAD**

Characterization: M_n/g.mol^{-1} = 4800, 83.8 mol% ethylene oxide, (EO)44-(PO)17-(EO)44, F38, ICI Surfactants, Cleveland, UK

Solvent (A): **water** **H_2O** **7732-18-5**

Polymer (C): **dextran**

Characterization: M_n/g.mol^{-1} = 236000, M_w/g.mol^{-1} = 410000, Dextran 400, Sigma Chemical Co., Inc., St. Louis, MO

Type of data: cloud points

T/K = 298.15

w_A	0.8973	0.8944	0.8830
w_B	0.0656	0.0456	0.0380
w_C	0.0371	0.0600	0.0790

Type of data: coexistence data (tie lines)

T/K = 298.15

Total system			Top phase			Bottom phase		
w_A	w_B	w_C	w_A	w_B	w_C	w_A	w_B	w_C
0.8805	0.0499	0.0696	0.8527	0.0264	0.1209	0.9092	0.0757	0.0151
0.8600	0.0700	0.0700	0.8016	0.0149	0.1835	0.8997	0.0950	0.0053
0.8407	0.0896	0.0697	0.7693	0.0028	0.2279	0.8712	0.1247	0.0041

Polymer (B): **poly(ethylene terephthalate-*co*-p-hydroxybenzoic acid)** **1988SC1**

Characterization: M_n/g.mol^{-1} = 6000, 27.0 mol% p-hydroxybenzoic acid

continued

continued

Solvent (A): **trichloromethane** $CHCl_3$ **67-66-3**
Polymer (C): **polycarbonate-bisphenol-A**
Characterization: M_n/g.mol^{-1} = 19300, M_w/g.mol^{-1} = 32800

Type of data: cloud points

T/K = 295.15

φ_A	0.8588	0.8554	0.8777	0.8931	0.9095	0.9096	0.9208	0.9218	0.9234
φ_B	0.1412	0.1367	0.1021	0.0782	0.0537	0.0444	0.0371	0.0308	0.0282
φ_C	0.0000	0.0079	0.0202	0.0287	0.0368	0.0459	0.0421	0.0474	0.0484

φ_A	0.9172	0.9161	0.9037	0.8628
φ_B	0.0265	0.0227	0.0186	0.0129
φ_C	0.0563	0.0612	0.0777	0.1243

Polymer (B): **poly(ethylene terephthalate-*co*-p-hydroxy-benzoic acid)** **1988SC1**
Characterization: M_n/g.mol^{-1} = 6000, 35.0 mol% p-hydroxybenzoic acid
Solvent (A): **trichloromethane** $CHCl_3$ **67-66-3**
Polymer (C): **polycarbonate-bisphenol-A**
Characterization: M_n/g.mol^{-1} = 19300, M_w/g.mol^{-1} = 32800

Type of data: cloud points

T/K = 295.15

φ_A	0.9021	0.9005	0.9021	0.9052	0.9386	0.9493	0.9583	0.9673	0.9692
φ_B	0.0979	0.0973	0.0932	0.0844	0.0442	0.0333	0.0225	0.0126	0.0083
φ_C	0.0000	0.0025	0.0047	0.0104	0.0172	0.0174	0.0192	0.0201	0.0225

φ_A	0.9473	0.9292	0.9234
φ_B	0.0091	0.0094	0.0623
φ_C	0.0435	0.0614	0.0143

Polymer (B): **poly(hexyl isocyanate)** **1989SAT**
Characterization: M_w/g.mol^{-1} = 20900, synthesized in the laboratory
Solvent (A): **toluene** C_7H_8 **108-88-3**
Polymer (C): **poly(hexyl isocyanate)**
Characterization: M_w/g.mol^{-1} = 244000, synthesized in the laboratory

Type of data: coexistence data (isotropic-nematic equilibrium)

T/K = 298.15

continued

continued

Total system			Isotropic phase			Anisotropic phase		
w_A	φ_A	$w_B/(w_B+w_C)$	w_A	φ_A	$w_B/(w_B+w_C)$	w_A	φ_A	$w_B/(w_B+w_C)$
0.7598	0.544	0.239	0.766	0.794	0.348	0.739	0.769	0.174
0.7404	0.530	0.417	0.754	0.783	0.666	0.704	0.737	0.190
0.7321	0.443	0.558	0.728	0.759	0.825	0.700	0.733	0.246
0.6898	0.532	0.896	0.700	0.733	1.000	0.667	0.702	0.801
0.6829	0.541	0.771	0.708	0.740	1.000	0.660	0.695	0.598
0.6819	0.463	0.951	0.693	0.726	1.000	0.661	0.696	0.891

Polymer (B): **poly(hexyl isocyanate)** **1989SAT**
Characterization: M_w/g.mol^{-1} = 68000, synthesized in the laboratory
Solvent (A): **toluene** C_7H_8 **108-88-3**
Polymer (C): **poly(hexyl isocyanate)**
Characterization: M_w/g.mol^{-1} = 244000, synthesized in the laboratory

Type of data: coexistence data (isotropic-nematic equilibrium)

T/K = 298.15

Total system			Isotropic phase			Anisotropic phase		
w_A	φ_A	$w_B/(w_B+w_C)$	w_A	φ_A	$w_B/(w_B+w_C)$	w_A	φ_A	$w_B/(w_B+w_C)$
0.7816	0.144	0.267	0.789	0.815	0.389	0.779	0.806	0.248
0.7767	0.390	0.361	0.783	0.809	0.424	0.762	0.790	0.352
0.7743	0.349	0.501	0.777	0.804	0.657	0.771	0.798	0.435
0.7682	0.442	0.604	0.760	0.788	0.804	0.753	0.782	0.476
0.7554	0.561	0.752	0.768	0.796	0.971	0.749	0.778	0.696
0.7573	0.739	0.897	0.764	0.792	0.928	0.750	0.779	0.881
0.7604	0.255	0.950	0.778	0.805	1.000	0.742	0.772	0.919

Polymer (B): **poly(methyl methacrylate)** **1955KER**
Characterization: Lucite 41, Du Pont
Solvent (A): **2-propanone** C_3H_6O **67-64-1**
Polymer (C): **poly(vinyl acetate)**
Characterization: M_w/g.mol^{-1} = 50000, Gelva 45

Type of data: coexistence data (tie lines)

T/K = 298.15

continued

continued

Total system			Top phase	Bottom phase
w_A	w_B	w_C	w_C	w_C
0.80	0.10	0.10	0.560	0.125
0.80	0.10	0.10	0.600	0.106

Polymer (B): **poly(methyl methacrylate)** **1955KER**
Characterization: Lucite 41, Du Pont
Solvent (A): **2-propanone** C_3H_6O **67-64-1**
Polymer (C): **poly(vinyl acetate)**
Characterization: M_w/g.mol^{-1} = 150000, Gelva 15

Type of data: coexistence data (tie lines)

T/K = 298.15

Total system			Top phase	Bottom phase
w_A	w_B	w_C	w_C	w_C
0.80	0.10	0.10	0.710	0.063

Polymer (B): **poly(3-methylstyrene)** **1956KER**
Characterization: ρ = 1.03 g/cm^3 at 298.15 K
Solvent (A): **trichloromethane** $CHCl_3$ **67-66-3**
Polymer (C): **poly(4-methylstyrene)**
Characterization: ρ = 1.02 g/cm^3 at 298.15 K

Type of data: cloud points (threshold point)

T/K = 298.15

φ_A 0.787 φ_B 0.1065 φ_C 0.1065

Polymer (B): **poly(propylene glycol)** **2001SIL**
Characterization: M_n/g.mol^{-1} = 400, Aldrich Chem. Co., Inc., Milwaukee, WI
Solvent (A): **water** H_2O **7732-18-5**
Polymer (C): **maltodextrin**
Characterization: M_n/g.mol^{-1} = 1540, M_w/g.mol^{-1} = 2017,
Loremalt 2030, Companhia Lorenz, Blumenau, SC, Brazil

Type of data: coexistence data (tie lines)

T/K = 298.15

continued

continued

Total system			Top phase			Bottom phase		
w_A	w_B	w_C	w_A	w_B	w_C	w_A	w_B	w_C
0.4402	0.3959	0.1639	0.3546	0.6282	0.0172	0.5095	0.1640	0.3265
0.4680	0.3070	0.2250	0.3558	0.6254	0.0188	0.5073	0.1554	0.3373
0.4350	0.2639	0.3011	0.3169	0.6711	0.0120	0.4536	0.1183	0.4278
0.3580	0.3361	0.3059	0.2372	0.7594	0.0034	0.4010	0.0720	0.5270

Polymer (B): **poly(propylene glycol)** **2001SIL**
Characterization: M_n/g.mol^{-1} = 400, Aldrich Chem. Co., Inc., Milwaukee, WI
Solvent (A): **water** **H_2O** **7732-18-5**
Polymer (C): **maltodextrin**
Characterization: M_n/g.mol^{-1} = 2020, M_w/g.mol^{-1} = 4000,
Loremalt 2001, Companhia Lorenz, Blumenau, SC, Brazil

Type of data: coexistence data (tie lines)

T/K = 298.15

Total system			Top phase			Bottom phase		
w_A	w_B	w_C	w_A	w_B	w_C	w_A	w_B	w_C
0.5121	0.2210	0.2669	0.5138	0.3732	0.1130	0.5269	0.1388	0.3343
0.4693	0.2462	0.2845	0.4321	0.5204	0.0475	0.4652	0.1107	0.4241
0.4161	0.2845	0.2994	0.3408	0.6406	0.0186	0.4370	0.0913	0.4717
0.3640	0.3245	0.3115	0.2712	0.7193	0.0095	0.4096	0.0711	0.5193

Polymer (B): **poly(propylene glycol)** **2001SIL**
Characterization: M_n/g.mol^{-1} = 3500, Aldrich Chem. Co., Inc., Milwaukee
Solvent (A): **water** **H_2O** **7732-18-5**
Polymer (C): **maltodextrin**
Characterization: M_n/g.mol^{-1} = 1540, M_w/g.mol^{-1} = 2017

Type of data: coexistence data (tie lines)

T/K = 298.15

Total system			Top phase			Bottom phase		
w_A	w_B	w_C	w_A	w_B	w_C	w_A	w_B	w_C
0.4350	0.4007	0.1643	0.0341	0.9659	0.0000	0.7057	0.0174	0.2769
0.4681	0.3068	0.2251	0.0288	0.9712	0.0000	0.6535	0.0205	0.3260
0.4338	0.2666	0.2996	0.0373	0.9676	0.0000	0.5655	0.0247	0.4098
0.3852	0.2857	0.3291	0.0344	0.9784	0.0000	0.5086	0.0280	0.4676

Polymer (B): **poly(propylene glycol)** **2001SIL**
Characterization: M_n/g.mol^{-1} = 3500, Aldrich Chem. Co., Inc., Milwaukee, WI
Solvent (A): **water** **H_2O** **7732-18-5**
Polymer (C): **maltodextrin**
Characterization: M_n/g.mol^{-1} = 2020, M_w/g.mol^{-1} = 4000,
Loremalt 2001, Companhia Lorenz, Blumenau, SC, Brazil

Type of data: coexistence data (tie lines)

T/K = 298.15

Total system			Top phase			Bottom phase		
w_A	w_B	w_C	w_A	w_B	w_C	w_A	w_B	w_C
0.5125	0.2203	0.2672	0.0835	0.9159	0.0006	0.6404	0.0162	0.3434
0.4632	0.2497	0.2871	0.0647	0.9351	0.0002	0.5988	0.0164	0.3848
0.4154	0.2856	0.2990	0.0323	0.9676	0.0001	0.5582	0.0166	0.4252
0.3636	0.3257	0.3107	0.0215	0.9784	0.0001	0.5142	0.0182	0.4676

Polymer (B): **polystyrene** **1956KER**
Characterization: ρ = 1.05 g/cm^3 at 298.15 K
Solvent (A): **anisole** **C_7H_8O** **100-66-3**
Polymer (C): **poly(methyl methacrylate)**
Characterization: ρ = 1.18 g/cm^3 at 298.15 K

Type of data: cloud points (threshold point)

T/K = 298.15

φ_A 0.887 φ_B 0.0565 φ_C 0.0565

Polymer (B): **polystyrene** **1963PAX**
Characterization: M_n/g.mol^{-1} = 2720, synthesized in the laboratory
Solvent (A): **benzene** **C_6H_6** **71-43-2**
Polymer (C): **polybutadiene**
Characterization: M_n/g.mol^{-1} = 1100, synthesized in the laboratory

Type of data: cloud points

T/K = 296.15

w_A	0.670	0.642	0.554	0.516
w_B	0.065	0.142	0.266	0.413
w_C	0.265	0.216	0.180	0.071

Polymer (B): **polystyrene** **1956KER**
Characterization: ρ = 1.05 g/cm^3 at 298.15 K
Solvent (A): **benzene** C_6H_6 **71-43-2**
Polymer (C): **poly(4-chlorostyrene)**
Characterization: ρ = 1.22 g/cm^3 at 298.15 K

Type of data: cloud points (threshold point)

T/K = 298.15

φ_A 0.976 φ_B 0.022 φ_C 0.022

Polymer (B): **polystyrene** **1956KER**
Characterization: ρ = 1.05 g/cm^3 at 298.15 K
Solvent (A): **benzene** C_6H_6 **71-43-2**
Polymer (C): **poly(4-methoxystyrene)**
Characterization: ρ = 1.12 g/cm^3 at 298.15 K

Type of data: cloud points (threshold point)

T/K = 298.15

φ_A 0.876 φ_B 0.062 φ_C 0.062

Polymer (B): **polystyrene** **1988OKA**
Characterization: M_w/g.mol^{-1} = 420000, Toyo Soda Manufacturing Co., Japan
Solvent (A): **benzene** C_6H_6 **71-43-2**
Polymer (C): **poly(methyl methacrylate)**
Characterization: M_n/g.mol^{-1} = 1027000, M_w/g.mol^{-1} = 1090000,
synthesized and fractionated in the laboratory

Type of data: cloud points (critical point)

T/K = 303.15

$c_{B+C, crit}$/(g/cm^3) 0.061 $w_{B, crit}/w_{C, crit}$ = 0.47/0.53

Polymer (B): **polystyrene** **1988OKA**
Characterization: M_w/g.mol^{-1} = 1260000, Toyo Soda Manufacturing Co., Japan
Solvent (A): **benzene** C_6H_6 **71-43-2**
Polymer (C): **poly(methyl methacrylate)**
Characterization: M_n/g.mol^{-1} = 1027000, M_w/g.mol^{-1} = 1090000,
synthesized and fractionated in the laboratory

Type of data: cloud points (critical point)

T/K = 303.15

$c_{B+C, crit}$/(g/cm^3) 0.047 $w_{B, crit}/w_{C, crit}$ = 0.38/0.62

Polymer (B): **polystyrene** **1988OKA**
Characterization: M_w/g.mol^{-1} = 3840000, Toyo Soda Manufacturing Co., Japan
Solvent (A): **benzene** C_6H_6 **71-43-2**
Polymer (C): **poly(methyl methacrylate)**
Characterization: M_n/g.mol^{-1} = 1027000, M_w/g.mol^{-1} = 1090000, synthesized and fractionated in the laboratory

Type of data: cloud points (critical point)

T/K = 303.15

$c_{B+C,\,crit}$/(g/cm^3) 0.038 $w_{B,\,crit}/w_{C,\,crit}$ = 0.26/0.74

Polymer (B): **polystyrene** **1988OKA**
Characterization: M_w/g.mol^{-1} = 5480000, Toyo Soda Manufacturing Co., Japan
Solvent (A): **benzene** C_6H_6 **71-43-2**
Polymer (C): **poly(methyl methacrylate)**
Characterization: M_n/g.mol^{-1} = 1027000, M_w/g.mol^{-1} = 1090000, synthesized and fractionated in the laboratory

Type of data: cloud points (critical point)

T/K = 303.15

$c_{B+C,\,crit}$/(g/cm^3) 0.037 $w_{B,\,crit}/w_{C,\,crit}$ = 0.25/0.75

Polymer (B): **polystyrene** **1956KER**
Characterization: ρ = 1.05 g/cm^3 at 298.15 K
Solvent (A): **benzene** C_6H_6 **71-43-2**
Polymer (C): **poly(methyl methacrylate)**
Characterization: ρ = 1.18 g/cm^3 at 298.15 K

Type of data: cloud points (threshold point)

T/K = 298.15

φ_A 0.889 φ_B 0.0555 φ_C 0.0555

Polymer (B): **polystyrene** **1956KER**
Characterization: ρ = 1.05 g/cm^3 at 298.15 K
Solvent (A): **benzonitrile** C_7H_5N **100-47-0**
Polymer (C): **poly(methyl methacrylate)**
Characterization: ρ = 1.18 g/cm^3 at 298.15 K

Type of data: cloud points (threshold point)

T/K = 298.15

φ_A 0.889 φ_B 0.0555 φ_C 0.0555

Polymer (B): **polystyrene** **1989TAK**
Characterization: M_n/g.mol^{-1} = 204000, M_w/g.mol^{-1} = 214000
Solvent (A): **bis(2-ethylhexyl) phthalate** $C_{24}H_{38}O_4$ **117-81-7**
Polymer (C): **polybutadiene**
Characterization: M_n/g.mol^{-1} = 165000, M_w/g.mol^{-1} = 313000,

Type of data: cloud points (threshold point)

T/K = 327.15

w_A 0.9700 w_B 0.0150 w_C 0.0150

T/K = 347.15

w_A 0.9670 w_B 0.0165 w_C 0.0165

Polymer (B): **polystyrene** **1996KUM**
Characterization: M_n/g.mol^{-1} = 204000, M_w/g.mol^{-1} = 214000
Solvent (A): **bis(2-ethylhexyl) phthalate** $C_{24}H_{38}O_4$ **117-81-7**
Polymer (C): **polybutadiene**
Characterization: M_n/g.mol^{-1} = 165000, M_w/g.mol^{-1} = 313000,

Type of data: cloud points (threshold point)

T/K = 349.15

w_A 0.9670 w_B 0.0165 w_C 0.0165

Polymer (B): **polystyrene** **1956KER**
Characterization: ρ = 1.05 g/cm^3 at 298.15 K
Solvent (A): **bromobenzene** C_6H_5Br **108-86-1**
Polymer (C): **poly(methyl methacrylate)**
Characterization: ρ = 1.18 g/cm^3 at 298.15 K

Type of data: cloud points (threshold point)

T/K = 298.15

φ_A 0.885 φ_B 0.0575 φ_C 0.0575

Polymer (B): **polystyrene** **1956KER**
Characterization: ρ = 1.05 g/cm^3 at 298.15 K
Solvent (A): **chlorobenzene** C_6H_5Cl **108-90-7**
Polymer (C): **poly(methyl methacrylate)**
Characterization: ρ = 1.18 g/cm^3 at 298.15 K

Type of data: cloud points (threshold point)

T/K = 298.15

φ_A 0.889 φ_B 0.0555 φ_C 0.0555

Polymer (B): **polystyrene** **1993GE1**
Characterization: M_n/g.mol^{-1} = 13000, M_w/g.mol^{-1} = 13640,
Pressure Chemical Company, Pittsburgh, PA
Solvent (A): **cyclohexane** C_6H_{12} **110-82-7**
Polymer (C): **polybutadiene**
Characterization: M_n/g.mol^{-1} = 22600, M_w/g.mol^{-1} = 24000,
52% 1,4-*trans*, 40% 1,4-*cis*, 8% 1,2-vinyl, 0.45% antioxidant,
Pressure Chemical Company, Pittsburgh, PA

Type of data: coexistence data (tie lines)

T/K = 348.15

Phase I			Phase II		
w_A	w_B	w_C	w_A	w_B	w_C
0.7435	0.0418	0.2147	0.5906	0.4079	0.0015
0.7642	0.0535	0.1823	0.6167	0.3811	0.0021
0.7738	0.0749	0.1513	0.6392	0.3553	0.0055
0.7806	0.1061	0.1133	0.6894	0.2978	0.0128

Polymer (B): **polystyrene** **1999BUN**
Characterization: M_n/g.mol^{-1} = 93000, M_w/g.mol^{-1} = 101400, M_z/g.mol^{-1} = 111900
BASF AG, Germany
Solvent (A): **cyclohexane** C_6H_{12} **110-82-7**
Polymer (C): **polyethylene**
Characterization: M_n/g.mol^{-1} = 13000, M_w/g.mol^{-1} = 89000, M_z/g.mol^{-1} = 600000,
LDPE, DSM Stamylan, DSM, Geleen, The Netherlands

Type of data: cloud points

T/K = 343.75

w_A	0.83095	0.88938	0.91458	0.92780	0.94625	0.95880	0.95611	0.95360
w_B	0.16810	0.11020	0.08440	0.07000	0.04800	0.02220	0.00869	0.00381
w_C	0.00095	0.00042	0.00102	0.00220	0.00575	0.01900	0.03520	0.04260

w_A	0.95020	0.93743	0.94016
w_B	0.00230	0.00077	0.00044
w_C	0.04750	0.06180	0.05940

Polymer (B): **polystyrene** **2000BEH**
Characterization: M_n/g.mol^{-1} = 93000, M_w/g.mol^{-1} = 101400, M_z/g.mol^{-1} = 111900
BASF AG, Germany
Solvent (A): **cyclohexane** C_6H_{12} **110-82-7**

continued

continued

Polymer (C): **polyethylene**
Characterization: M_n/g.mol^{-1} = 13000, M_w/g.mol^{-1} = 89000, M_z/g.mol^{-1} = 600000, LDPE, DSM Stamylan, DSM, Geleen, The Netherlands

Type of data: cloud points (UCST-behavior)

w_A	0.9131	0.9131	0.9131	0.9131	0.9131	0.9156	0.9156	0.9156	0.9156
w_B	0.0797	0.0797	0.0797	0.0797	0.0797	0.0718	0.0718	0.0718	0.0718
w_C	0.0072	0.0072	0.0072	0.0072	0.0072	0.0126	0.0126	0.0126	0.0126
T/K	401.29	403.75	403.96	405.45	404.75	435.05	433.85	433.31	434.92
P/bar	154.1	98.3	64.3	40.5	15.4	241.7	155.5	100.1	52.2

w_A	0.9156
w_B	0.0718
w_C	0.0126
T/K	436.37
P/bar	21.4

Type of data: cloud points (LCST-behavior)

w_A	0.9131	0.9131	0.9131	0.9131	0.9131	0.9131	0.9131	0.9156	0.9156
w_B	0.0797	0.0797	0.0797	0.0797	0.0797	0.0797	0.0797	0.0718	0.0718
w_C	0.0072	0.0072	0.0072	0.0072	0.0072	0.0072	0.0072	0.0126	0.0126
T/K	495.51	501.21	506.00	510.65	516.02	520.97	527.38	465.74	470.30
P/bar	29.8	37.3	41.5	47.2	53.7	59.4	67.1	11.1	13.7

w_A	0.9156	0.9156	0.9156	0.9156	0.9156	
w_B	0.0718	0.0718	0.0718	0.0718	0.0718	
w_C	0.0126	0.0126	0.0126	0.0126	0.0126	
T/K	475.97	480.61	486.19	491.25	496.44	506.12
P/bar	15.5	19.3	24.2	30.1	35.5	46.0

Type of data: cloud points (1:1 polymer blend)

w_A 0.9302 w_B 0.0349 w_C 0.0349 were kept constant

T/K	445.15	449.85	454.98	460.09	465.17	469.07	470.30	471.86	473.65
P/bar	228.1	136.6	91.1	71.2	61.0	60.6	54.1	56.8	56.3

T/K	476.70	485.80	496.17	506.27	516.76	527.38
P/bar	55.6	56.8	61.9	69.5	78.7	88.8

Polymer (B): **polystyrene** **1982GAL**
Characterization: M_n/g.mol^{-1} = 35000, M_w/g.mol^{-1} = 36000, M_z/g.mol^{-1} = 38000, Pressure Chemical Company, Pittsburgh, PA
Solvent (A): **cyclohexane** **C_6H_{12}** **110-82-7**
Polymer (C): **polystyrene**
Characterization: M_n/g.mol^{-1} = 178000, M_w/g.mol^{-1} = 184000, M_z/g.mol^{-1} = 195000, Pressure Chemical Company, Pittsburgh, PA

continued

continued

Comments: The composition of the binary PS mixture is kept constant at w_B/w_C = 0.7293/0.2707. The corresponding molar mass averages are given by M_n = 45000 g/mol, M_w = 76000 g/mol, and M_z = 141000 g/mol.

Type of data: cloud points

$\varphi_B + \varphi_C$	0.0496	0.0521	0.0837	0.0884	0.1052	0.1177	0.1280	0.1402	0.1425
T/K	293.34	293.26	292.67	292.51	291.98	291.58	291.28	290.96	291.03

$\varphi_B + \varphi_C$	0.1465	0.1647	0.1801
T/K	290.88	290.44	289.81

Type of data: spinodal points

$\varphi_B + \varphi_C$	0.0837	0.0884	0.1177	0.1280	0.1402	0.1425	0.1465	0.1647	0.1801
T/K	291.95	292.02	291.56	291.02	290.80	290.76	290.78	290.34	289.72

Polymer (B): **polystyrene** **1987EIN**
Characterization: M_n/g.mol^{-1} = 43200, M_w/g.mol^{-1} = 43600, Toyo Soda Manufacturing Co., Japan
Solvent (A): **cyclohexane** C_6H_{12} **110-82-7**
Polymer (C): **polystyrene**
Characterization: M_n/g.mol^{-1} = 1200000, M_w/g.mol^{-1} = 1260000, Toyo Soda Manufacturing Co., Japan

Type of data: coexistence data (tie lines)

T/K = 286.95

Total system		Sol phase		Gel phase	
φ_B	φ_C	φ_B	φ_C	φ_B	φ_C
0.0736	0.0071	0.06846	0.0000		
0.1800	0.0102			0.1923	0.0114
0.1825	0.0159	0.0517	0.0000	0.2067	0.0183
0.1277	0.0056	0.0800	0.0000	0.1879	0.0139
0.1256	0.0158	0.0671	0.0000	0.1978	0.0331
0.0914	0.0120	0.0728	0.0000	0.1813	0.0717
0.0845	0.0157	0.0608	0.0000	0.1791	0.0781
0.1279	0.0039	0.0687	0.0000	0.1965	0.0077

T/K = 287.05

Total system		Sol phase		Gel phase	
φ_B	φ_C	φ_B	φ_C	φ_B	φ_C
0.1535	0.0134	0.0777	0.0000	0.1964	0.0197

continued

continued

Type of data: three phase equilibrium data

T/K = 286.95

Total system		Phase I		Phase II		Phase III	
φ_B	φ_C	φ_B	φ_C	φ_B	φ_C	φ_B	φ_C
0.1665	0.0081			0.1618	0.0025	0.1848	0.0184
0.1572	0.0052	0.1291	0.0015	0.1678	0.0039	0.1665	0.0115
0.1648	0.0074	0.1162	0.0008	0.1724	0.0039	0.1844	0.0131

T/K = 287.05

Total system		Phase I		Phase II		Phase III	
φ_B	φ_C	φ_B	φ_C	φ_B	φ_C	φ_B	φ_C
0.1647	0.0074			0.1575	0.0022	0.1891	0.0178

Polymer (B): **polystyrene** **1984TSU**
Characterization: M_n/g.mol^{-1} = 44900, M_w/g.mol^{-1} = 45300, Toyo Soda Manufacturing Co., Japan
Solvent (A): **cyclohexane** **C_6H_{12}** **110-82-7**
Polymer (C): **polystyrene**
Characterization: M_n/g.mol^{-1} = 474000, M_w/g.mol^{-1} = 498000, Toyo Soda Manufacturing Co., Japan

Type of data: coexistence data (tie lines)

T/K = 287.15

Total system		Sol phase		Gel phase	
$w_B/(w_B+w_C)$	$\varphi_B+\varphi_C$	φ_B	φ_C	φ_B	φ_C
0.250	0.02043	0.004127	–	0.02500	0.3084
0.250	0.06985	0.009181	–	0.05735	0.2808
0.250	0.1543	0.01035	–	0.07791	0.2529
0.500	0.1261	0.02341	–	0.1312	0.1714
0.500	0.1886	0.02455	–	0.1458	0.1610
0.800	0.04945	0.03103	–	0.1639	0.1217
0.800	0.1281	0.04701	–	0.1909	0.06006
0.900	0.09531	0.05251	–	0.1858	0.03788
0.900	0.1438	0.06570	–	0.1865	0.02456
0.900	0.1793	0.06133	–	0.1947	0.02294
0.950	0.06547	0.05151	–	0.1894	0.04067
0.950	0.1196	0.06449	–	0.1956	0.01773

continued

continued

Total system		Sol phase		Gel phase	
$w_B/(w_B+w_C)$	$\varphi_B+\varphi_C$	φ_B	φ_C	φ_B	φ_C
0.990	0.1379	0.0720	–	0.1863	0.002148
0.990	0.1451	0.0738	–	0.1778	0.001850
T/K = 291.15					
0.250	0.04994	0.01019	–	0.02420	0.2464
0.500	0.04023	0.01745	–	0.05030	0.2090
0.500	0.1482	0.04387	0.000425	0.09793	0.1294
0.800	0.07603	0.05380	0.000445	0.1199	0.09244
0.800	0.09980	0.06945	0.001259	0.1356	0.08025
0.800	0.1446	0.1095	0.01208	0.1261	0.03607
T/K = 295.15					
0.250	0.03497	0.007735	0.000516	0.01695	0.1834
0.250	0.09011	0.01612	0.001696	0.03166	0.1633
0.500	0.06042	0.02658	0.001894	0.04486	0.1407
0.500	0.08129	0.03400	0.003412	0.05575	0.1206
0.500	0.1006	0.04097	0.005413	0.06217	0.1038
0.500	0.1222	0.05034	0.008989	0.06663	0.08167
0.670	0.09009	0.05724	0.02170	0.06747	0.06585
T/K = 298.15					
0.237	0.04458	0.09918	0.006524	0.01359	0.1104
0.237	0.09677	0.02003	0.01485	0.02529	0.08739
0.237	0.07146	0.01440	0.009165	0.02003	0.1008
0.485	0.04944	0.02300	0.02026	0.03154	0.06676

Type of data: critical points (sol phase is equal to gel phase)

T_{crit}/K	$w_B/(w_B+w_C)$	$\varphi_B+\varphi_C$
293.79	0.000	0.047
292.15	0.436	0.074
291.15	0.500	0.082
290.15	0.640	0.100
289.15	0.860	0.131
288.86	0.950	0.152
288.54	0.990	0.126
287.29	1.000	0.108

Polymer (B): **polystyrene** **1981HAS**
Characterization: M_n/g.mol^{-1} = 45000, M_w/g.mol^{-1} = 45300, M_z/g.mol^{-1} = 45600, Toyo Soda Manufacturing Co., Japan
Solvent (A): **cyclohexane** C_6H_{12} **110-82-7**
Polymer (C): **polystyrene**
Characterization: M_n/g.mol^{-1} = 102000, M_w/g.mol^{-1} = 103000, M_z/g.mol^{-1} = 104000, Toyo Soda Manufacturing Co., Japan

Type of data: coexistence data (tie lines)

T/K = 287.15

Total system		Sol phase		Gel phase	
$w_B/(w_B+w_C)$	w_B+w_C	φ_B	φ_C	φ_B	φ_C
0.651	0.0484	0.0157	0.00185	0.1234	0.1539
0.651	0.1265	0.0244	0.00181	0.1452	0.1041
0.501	0.0491	0.01129	0.00185	0.0892	0.1852
0.501	0.1008	0.0160	0.00187	0.1087	0.1524
0.501	0.1773	0.0182	0.00203	0.1209	0.1364
0.850	0.0686	0.0291	0.00144	0.1738	0.0683
0.850	0.1292	0.0372	0.00131	0.1787	0.0390
0.850	0.1674	0.0354	0.00132	0.1715	0.0417
0.000		0.0000	0.00230	0.0000	0.2899
1.000		0.0682	0.00000	0.1514	0.0000

T/K = 290.15

Total system		Sol phase		Gel phase	
$w_B/(w_B+w_C)$	w_B+w_C	φ_B	φ_C	φ_B	φ_C
0.501	0.1236	0.0294	0.01268	0.0790	0.1090
0.501	0.1615	0.0344	0.01504	0.0811	0.0946
0.651	0.0836	0.0351	0.01418	0.0839	0.0886
0.651	0.1131	0.0470	0.01993	0.0836	0.0666
0.651	0.1414	0.0579	0.0255	0.0791	0.0489
0.651	0.0580	0.0257	0.01083	0.0744	0.1086
0.651	0.1328	0.0535	0.0226	0.0803	0.0528
0.651	0.1335	0.0568	0.0264	0.0748	0.0464
0.000		0.0000	0.0068	0.0000	0.2364

continued

continued

Type of data: critical points (sol phase is equal to gel phase)

T_{crit}/K	$w_B/(w_B+w_C)$	w_B+w_C	$\varphi_B+\varphi_C$
293.79	0.000	0.115	0.086
292.15	0.350	0.128	0.096
291.15	0.532	0.135	0.101
290.15	0.685	0.140	0.105
289.15	0.780	0.146	0.110
288.86	0.867	0.151	0.114
288.54	0.885	0.150	0.113
287.29	1.000	0.149	0.112

Polymer (B): **polystyrene** **1973KU1**
Characterization: M_n/g.mol^{-1} = 196000, M_w/g.mol^{-1} = 200000, Pressure Chemical Company, Pittsburgh, PA
Solvent (A): **cyclohexane** C_6H_{12} **110-82-7**
Polymer (C): **polystyrene**
Characterization: M_n/g.mol^{-1} = 667000, M_w/g.mol^{-1} = 680000, Pressure Chemical Company, Pittsburgh, PA

Type of data: cloud points (UCST-behavior)

φ_{B+C}	0.0310	T/K	300.05	(threshold point)
φ_{B+C}	0.0667	T/K	299.60	(critical point)

Comments: The weight ratio of the polystyrenes was kept constant at w_B/w_C=0.52/0.48.

Polymer (B): **polystyrene** **1997SC2**
Characterization: M_n/g.mol^{-1} = 92300, M_w/g.mol^{-1} = 96000
Solvent (A): **cyclohexane** C_6H_{12} **110-82-7**
Polymer (C): **poly(styrene-*b*-dimethylsiloxane)**
Characterization: M_w/g.mol^{-1} = 55000, 50.0 wt% styrene, synthesized in the laboratory

Type of data: cloud points

$w_B + w_C$	0.011	0.040	0.070	0.100	0.126	0.150	0.179
T/K	290.32	293.57	295.07	296.99	301.85	310.82	340.68

Comments: The weight ratio of polystyrene/S-*b*-DMS/50w was kept constant at 7.5/1.

Polymer (B): **polystyrene** **2002FRI**
Characterization: M_n/g.mol^{-1} = 1800, M_w/g.mol^{-1} = 1930, synthesized in the laboratory

continued

continued

Solvent (A): **1,2-dichlorobenzene** $C_6H_4Cl_2$ **95-50-1**
Polymer (C): **polybutadiene (deuterated)**
Characterization: M_n/g.mol^{-1} = 1900, M_w/g.mol^{-1} = 2100, 46% 1,4- (unspecified) and 54% 1,2-content, synthesized in the laboratory

Type of data: binodal points

Comments: The concentration of the polymer mixture is kept constant close to the critical concentration of the polymer blend, i.e., φ_C/φ_B = 0.477/0.523.

φ_A	0.000	0.010	0.022	0.050	0.095	0.198
T/K	354.18	350.83	348.12	343.24	332.66	291.29

φ_A	0.00	0.00	0.00	0.00	0.00	0.05	0.05	0.05	0.05
P/MPa	0.1	50	100	150	200	0.1	50	100	150
T/K	358.65	361.97	365.05	368.21	372.12	344.94	348.59	351.75	354.90

φ_A	0.05	0.20	0.20	0.20	0.20	0.20	0.20
P/MPa	200	0.1	50	100	150	180	200
T/K	358.05	292.47	294.49	298.10	299.45	302.84	301.60

Type of data: spinodal points

φ_A	0.000	0.010	0.022	0.050	0.095	0.198
T/K	353.045	349.561	346.973	342.556	332.305	290.670

φ_A	0.00	0.00	0.00	0.00	0.00	0.05	0.05	0.05	0.05
P/MPa	0.1	50	100	150	200	0.1	50	100	150
T/K	355.750	359.483	362.259	365.682	369.246	343.625	346.447	349.770	353.040

φ_A	0.05	0.20	0.20	0.20	0.20	0.20	0.20
P/MPa	200	0.1	50	100	150	180	200
T/K	356.253	291.126	293.477	295.811	298.476	301.210	301.226

Polymer (B): **polystyrene** **2002FRI**
Characterization: M_n/g.mol^{-1} = 1700, M_w/g.mol^{-1} = 1820, synthesized in the laboratory
Solvent (A): **1,2-dichlorobenzene** $C_6H_4Cl_2$ **95-50-1**
Polymer (C): **polybutadiene (deuterated)**
Characterization: M_n/g.mol^{-1} = 2100, M_w/g.mol^{-1} = 2300, 40% 1,4-*cis*-, 53% 1,4-*trans*-, and 7% 1,2-content

Type of data: binodal points

Comments: The concentration of the polymer mixture is kept constant close to the critical concentration of the polymer blend, i.e., φ_C/φ_B = 0.43/0.57.

φ_A	0.00	0.00	0.00	0.00
P/MPa	0.1	50	100	150
T/K	338.31	342.15	343.10	349.72

continued

continued

Type of data: spinodal points

φ_A	0.00	0.00	0.00	0.00	0.05	0.05	0.05	0.05	0.05
P/MPa	0.1	50	100	150	0.1	50	100	150	200
T/K	336.534	340.520	344.352	349.116	331.779	335.483	339.267	342.280	345.696

φ_A	0.20	0.20	0.20	0.20	0.20
P/MPa	0.1	50	100	150	200
T/K	295.388	298.722	301.750	304.939	308.190

Polymer (B): **polystyrene** **2002FRI**
Characterization: M_n/g.mol^{-1} = 1800, M_w/g.mol^{-1} = 1930, synthesized in the laboratory
Solvent (A): **1,2-dichlorobenzene** $C_6H_4Cl_2$ **95-50-1**
Polymer (C): **polybutadiene (deuterated)**
Characterization: M_n/g.mol^{-1} = 2100, M_w/g.mol^{-1} = 2300, 40% 1,4-*cis*-, 53% 1,4-*trans*-, and 7% 1,2-content

Comments: The concentration of the polymer mixture is kept constant at φ_C/φ_B = 0.43/0.57.

Type of data: binodal points

φ_A	0.000	0.026	0.052	0.109	0.162
T/K	345.01	338.43	331.90	316.66	306.27

Type of data: spinodal points

φ_A	0.000	0.026	0.052	0.109	0.162
T/K	344.976	338.414	331.399	316.643	305.732

Polymer (B): **polystyrene** **2001POS**
Characterization: M_n/g.mol^{-1} = 100000, M_w/g.mol^{-1} = 290000, ρ = 1.047 g/cm^3 at 303.15 K, Degussa-Huels, Germany
Solvent (A): ***N,N*-dimethylacetamide** C_4H_9NO **127-19-5**
Polymer (C): **poly(acrylonitrile)**
Characterization: M_w/g.mol^{-1} = 240000, ρ = 1.175 g/cm^3 at 303.15 K,

Type of data: cloud points

T/K = 298.15

w_A	0.9784	0.9737	0.9786	0.9754	0.9735	0.98037	0.98079	0.9649	0.9786
w_B	0.01086	0.01826	0.00857	0.0149	0.0212	0.0020	0.0030	0.0316	0.00011
w_C	0.01074	0.00804	0.01283	0.0097	0.0053	0.0175	0.0153	0.0035	0.02029

Polymer (B): **polystyrene** **1956KER**
Characterization: ρ = 1.05 g/cm^3 at 298.15 K
Solvent (A): ***N,N*-dimethylaniline** $C_8H_{11}N$ **121-69-7**
Polymer (C): **poly(methyl methacrylate)**
Characterization: ρ = 1.18 g/cm^3 at 298.15 K

continued

continued

Type of data: cloud points (threshold point)

T/K = 298.15

φ_A	0.870	φ_B	0.065	φ_C	0.065

Polymer (B): **polystyrene** **1956KER**
Characterization: ρ = 1.05 g/cm^3 at 298.15 K
Solvent (A): **1,2-dimethylbenzene** C_8H_{10} **95-47-6**
Polymer (C): **poly(methyl methacrylate)**
Characterization: ρ = 1.18 g/cm^3 at 298.15 K

Type of data: cloud points (threshold point)

T/K = 298.15

φ_A	0.957	φ_B	0.0215	φ_C	0.0215

Polymer (B): **polystyrene** **1956KER**
Characterization: ρ = 1.05 g/cm^3 at 298.15 K
Solvent (A): **1,3-dimethylbenzene** C_8H_{10} **108-38-3**
Polymer (C): **poly(methyl methacrylate)**
Characterization: ρ = 1.18 g/cm^3 at 298.15 K

Type of data: cloud points (threshold point)

T/K = 298.15

φ_A	0.971	φ_B	0.0145	φ_C	0.0145

Polymer (B): **polystyrene** **1956KER**
Characterization: ρ = 1.05 g/cm^3 at 298.15 K
Solvent (A): **1,4-dimethylbenzene** C_8H_{10} **106-42-3**
Polymer (C): **poly(methyl methacrylate)**
Characterization: ρ = 1.18 g/cm^3 at 298.15 K

Type of data: cloud points (threshold point)

T/K = 298.15

φ_A	0.975	φ_B	0.0125	φ_C	0.0125

Polymer (B): **polystyrene** **1995GOM**
Characterization: M_n/g.mol^{-1} = 5850, M_w/g.mol^{-1} = 5970, ρ = 1.05 g/cm^3 at 298.15 K, Tosoh Corp., Ltd., Tokyo, Japan
Solvent (A): ***N,N*-dimethylformamide** C_3H_7NO **68-12-2**
Polymer (C): **poly(vinylidene fluoride)**
Characterization: M_n/g.mol^{-1} = 250000, M_w/g.mol^{-1} = 674000, ρ = 1.78 g/cm^3 at 298.15 K, Kynar 721, Penwalt Corp., Oxford, UK

continued

continued

Type of data: coexistence data (tie lines)

T/K = 298.15

Phase I			Phase II		
φ_A	φ_B	φ_C	φ_A	φ_B	φ_C
0.84266	0.15110	0.00624	0.89500	0.06424	0.04076
0.84225	0.15257	0.00518	0.89489	0.06423	0.04088
0.84940	0.14324	0.00736	0.89513	0.06331	0.04156
0.86152	0.12560	0.01288	0.89317	0.06629	0.04044
0.81650	0.17879	0.00471	0.89195	0.04696	0.06109
0.81827	0.17748	0.00425	0.89343	0.04365	0.06282
0.82008	0.17801	0.00191	0.89030	0.04869	0.06101
0.82406	0.16850	0.00744	0.89357	0.04307	0.06336
0.80438	0.19161	0.00401	0.84633	0.06093	0.09274
0.80498	0.19173	0.00329	0.84608	0.06134	0.09258
0.79858	0.19873	0.00269	0.85020	0.05568	0.09412
0.82359	0.17444	0.00197	0.83892	0.06679	0.09429

Polymer (B): **polystyrene** **1996CAM**
Characterization: M_n/g.mol^{-1} = 14700, M_w/g.mol^{-1} = 15000, ρ = 1.05 g/cm^3 at 298.15 K, Tosoh Corp., Ltd., Tokyo, Japan
Solvent (A): ***N,N*-dimethylformamide** **C_3H_7NO** **68-12-2**
Polymer (C): **poly(vinylidene fluoride)**
Characterization: M_n/g.mol^{-1} = 250000, M_w/g.mol^{-1} = 674000, ρ = 1.78 g/cm^3 at 298.15 K, Kynar 721, Penwalt Corp., Oxford, UK

Type of data: coexistence data (tie lines)

T/K = 298.15

Phase I			Phase II		
φ_A	φ_B	φ_C	φ_A	φ_B	φ_C
0.96332	0.02891	0.00778	0.69515	0.01508	0.28977
0.94337	0.03765	0.01898	0.95540	0.00700	0.03760
0.92965	0.04030	0.03004	0.93571	0.02167	0.04262
0.93565	0.04744	0.01691	0.94002	0.02807	0.03192
0.91585	0.06741	0.01674	0.95605	0.00947	0.03448
0.87457	0.07014	0.05529	0.96239	0.00999	0.02762
0.90925	0.08517	0.00557	0.94568	0.01810	0.03622

continued

continued

Phase I			Phase II		
φ_A	φ_B	φ_C	φ_A	φ_B	φ_C
0.89814	0.09603	0.00583	0.95276	0.02557	0.02167
0.89872	0.09907	0.00221	0.94953	0.02445	0.02602
0.89217	0.09938	0.00845	0.95447	0.02015	0.02538
0.89829	0.10021	0.00150	0.95617	0.01463	0.02920
0.89608	0.10027	0.00365	0.95040	0.02143	0.02817
0.87921	0.11063	0.01016	0.96059	0.01802	0.02139
0.84942	0.13268	0.01790	0.94338	0.00504	0.05158
0.86398	0.13535	0.00067	0.94872	0.01778	0.03351
0.83239	0.13864	0.02897	0.96701	0.00782	0.02517
0.83755	0.14241	0.02004	0.95514	0.00830	0.03656
0.78016	0.14427	0.07557	0.95463	0.00516	0.04021
0.79642	0.14575	0.05783	0.95393	0.00636	0.03971
0.84175	0.15151	0.00674	0.91757	0.01105	0.07137
0.82702	0.15846	0.01451	0.94729	0.01090	0.04182
0.70220	0.16673	0.13107	0.96072	0.01085	0.02843
0.79998	0.17259	0.02743	0.95811	0.01344	0.02846
0.80592	0.19333	0.00074	0.88998	0.00342	0.10678
0.74462	0.21893	0.03646	0.94805	0.00603	0.04688
0.64428	0.23153	0.12419	0.96014	0.00508	0.03478

Polymer (B): **polystyrene** **2001GAR**
Characterization: M_n/g.mol^{-1} = 16800, M_w/g.mol^{-1} = 17200,
Polymer Standard Services GmbH, Mainz, Germany
Solvent (A): **1,4-dioxane** $C_4H_8O_2$ **123-91-1**
Polymer (C): **polybutadiene**
Characterization: M_n/g.mol^{-1} = 83200, M_w/g.mol^{-1} = 86500,
Pressure Chemical Company, Pittsburgh, PA

Type of data: coexistence data (tie lines)

T/K = 298.15

Phase I			Phase II		
φ_A	φ_B	φ_C	φ_A	φ_B	φ_C
0.986	0.0014	0.0123	0.962	0.0347	0.00304
0.975	0.0108	0.0147	0.969	0.0298	0.00087
0.973	0.0122	0.0146	0.970	0.0292	0.00073

continued

continued

Phase I			Phase II		
φ_A	φ_B	φ_C	φ_A	φ_B	φ_C
0.969	0.0169	0.0144	0.974	0.0258	0.00063
0.972	0.0151	0.0130	0.971	0.0282	0.00083
0.986	0.00098	0.0126	0.965	0.0340	0.00063
0.957	0.0060	0.0369	0.935	0.0643	0.00110
0.951	0.0128	0.0359	0.936	0.0632	0.00063
0.937	0.0305	0.0322	0.941	0.0578	0.00074
0.944	0.0248	0.0308	0.938	0.0610	0.00067
0.953	0.0180	0.0293	0.934	0.0653	0.00062
0.953	0.0195	0.0273	0.933	0.0660	0.00076
0.961	0.0086	0.0309	0.900	0.0998	0.00074
0.960	0.0110	0.0291	0.898	0.1010	0.00095
0.953	0.0187	0.0280	0.900	0.0990	0.00073
0.956	0.0175	0.0261	0.896	0.1031	0.00101
0.943	0.0316	0.0250	0.904	0.0956	0.00073
0.925	0.0516	0.0231	0.918	0.0807	0.00103

Polymer (B): **polystyrene** **1956KER**
Characterization: $\rho = 1.05$ g/cm^3 at 298.15 K
Solvent (A): **1,4-dioxane** **$C_4H_8O_2$** **123-91-1**
Polymer (C): **poly(methyl methacrylate)**
Characterization: $\rho = 1.18$ g/cm^3 at 298.15 K

Type of data: cloud points (threshold point)

T/K = 298.15

φ_A	0.906	φ_B	0.047	φ_C	0.047

Polymer (B): **polystyrene** **1956KER**
Characterization: $\rho = 1.05$ g/cm^3 at 298.15 K
Solvent (A): **ethyl acetate** **$C_4H_8O_2$** **141-78-6**
Polymer (C): **poly(methyl methacrylate)**
Characterization: $\rho = 1.18$ g/cm^3 at 298.15 K

Type of data: cloud points (threshold point)

T/K = 298.15

φ_A	0.944	φ_B	0.023	φ_C	0.023

Polymer (B): **polystyrene** **1956KER**
Characterization: $\rho = 1.05$ g/cm^3 at 298.15 K

continued

continued

Solvent (A): **ethylbenzene** C_8H_{10} **100-41-4**
Polymer (C): **poly(methyl methacrylate)**
Characterization: ρ = 1.18 g/cm^3 at 298.15 K

Type of data: cloud points (threshold point)

T/K = 298.15

φ_A	0.940	φ_B	0.030	φ_C	0.030

Polymer (B): **polystyrene** **1994MIY**
Characterization: M_n/g.mol^{-1} = 348000, M_w/g.mol^{-1} = 355000, Tosoh Corp., Ltd., Tokyo, Japan
Solvent (A): **hexadeuterobenzene** C_6D_6 **1076-43-3**
Polymer (C): **poly(methyl methacrylate)**
Characterization: M_n/g.mol^{-1} = 297000, M_w/g.mol^{-1} = 327000, Pressure Chemical Company, Pittsburgh, PA

Type of data: cloud points (LCST-behavior)

φ_A	0.9184	0.9181	0.9171	0.9163
φ_B	0.0203	0.0340	0.0436	0.0531
φ_C	0.0613	0.0479	0.0393	0.0306
T/K	315.25	304.55	297.75	300.50

Type of data: spinodal points (LCST-behavior)

φ_A	0.9184	0.9181	0.9171	0.9163
φ_B	0.0203	0.0340	0.0436	0.0531
φ_C	0.0613	0.0479	0.0393	0.0306
T/K	335.85	305.85	299.15	305.95

Polymer (B): **polystyrene** **1984DOB**
Characterization: M_n/g.mol^{-1} = 16200, M_w/g.mol^{-1} = 17200, Pressure Chemical Company, Pittsburgh, PA
Solvent (A): **methylcyclohexane** C_7H_{14} **108-87-2**
Polymer (C): **polystyrene**
Characterization: M_n/g.mol^{-1} = 678000, M_w/g.mol^{-1} = 719000, Pressure Chemical Company, Pittsburgh, PA

Comments: The volume fraction of polystyrene (C) in the polystyrene mixture is 0.1704. The molar mass averages of B+C are M_w/g.mol^{-1} = 137000, M_z/g.mol^{-1} = 645000.

Type of data: coexistence data (UCST-behavior)

φ_{B+C} (total) 0.1815 was kept constant (critical concentration)

T/K	311.726	311.689	311.587	311.429	311.160	310.718	309.869	308.398
φ_{B+C} (sol phase)	0.1673	0.1654	0.1624	0.1585	0.1545	0.1509	0.1463	0.1431
φ_{B+C} (gel phase)	0.1984	0.1999	0.2062	0.2120	0.2196	0.2292	0.2453	0.2664

continued

continued

T/K	306.990	305.820	304.470	303.600	302.050	300.970	299.690	298.750
φ_{B+C} (sol phase)	0.1405	0.1385	0.1377	0.1342	0.1325	0.1269	0.1226	0.1156
φ_{B+C} (gel phase)	0.2796	0.2958	0.3055	0.3142	0.3242	0.3348	0.3423	0.3460
T/K	298.070	297.750	297.690	296.710	295.380	294.090	293.250	292.210
φ_{B+C} (sol phase)	0.1131	0.1048	0.1025	0.0795	0.0550	0.0420	0.0349	0.0301
φ_{B+C} (gel phase)	0.3525	0.3546	0.3536	0.3509	0.3622	0.3793	0.3798	0.4007
T/K	290.550	289.060	287.610					
φ_{B+C} (sol phase)	0.0230	0.0195	0.0146					
φ_{B+C} (gel phase)	0.4149	0.4197	0.4386					

Type of data: critical point (UCST-behavior)

$\varphi_{B+C, crit}$ 0.1815 T_{crit}/K 311.82

Polymer (B): **polystyrene** **1986DOB**
Characterization: M_n/g.mol^{-1} = 16200, M_w/g.mol^{-1} = 17200,
Pressure Chemical Company, Pittsburgh, PA
Solvent (A): **methylcyclohexane** **C_7H_{14}** **108-87-2**
Polymer (C): **polystyrene**
Characterization: M_n/g.mol^{-1} = 678000, M_w/g.mol^{-1} = 719000,
Pressure Chemical Company, Pittsburgh, PA

Type of data: critical point (UCST-behavior)

$\varphi_{B, crit}$ 0.1822 $\varphi_{C, crit}$ 0.0037 T_{crit}/K 305.61

Type of data: three-phase equilibrium data

Comments: The total feed concentrations of the two polystyrenes in the solution are φ_B = 0.1822 and φ_C = 0.0037.

T/K	Upper phase $\varphi_B + \varphi_C$	Middle phase $\varphi_B + \varphi_C$	Lower phase $\varphi_B + \varphi_C$
305.183		0.1841	0.2827
304.484		0.1816	0.2902
303.453		0.1809	0.3024
302.448		0.1805	0.3129
301.457		0.1797	0.3208
301.053		0.1790	0.3237
300.466		0.1788	0.3277
299.962		0.1785	0.3301
299.466		0.1783	0.3319
298.955		0.1775	0.3327
298.451		0.1759	0.3318

continued

continued

T/K	Upper phase $\varphi_B + \varphi_C$	Middle phase $\varphi_B + \varphi_C$	Lower phase $\varphi_B + \varphi_C$
297.958		0.1741	0.3261
297.877		0.1737	0.3216
297.852	0.1378	0.1903	0.3201
297.835	0.1343	0.1969	0.3202
297.795	0.1292	0.2061	0.3187
297.707	0.1182	0.2224	0.3149
297.657	0.1136	0.2298	0.3125
297.605	0.1107	0.2361	0.3101
297.557	0.1077	0.2420	0.3067
297.508	0.1041	0.2522	0.3017
297.487	0.1023	0.2594	0.2958
297.407	0.0971	0.2634	
297.355	0.0958	0.2659	
297.318	0.0928	0.2676	
297.281	0.0912	0.2689	
296.958	0.0817	0.2786	
296.462	0.0668	0.2956	
295.461	0.0525	0.3197	
294.430	0.0424	0.3406	
293.430	0.0348	0.3584	
292.430	0.0298	0.3741	
291.480	0.0252	0.3876	
290.480	0.0208	0.4023	
289.480	0.0168	0.4153	

Type of data: critical point (UCST-behavior)

$\varphi_{B,\,crit}$ 0.2075 $\varphi_{C,\,crit}$ 0.0109 T_{crit}/K 303.85

Type of data: three-phase equilibrium data

Comments: The total feed concentrations of the two polystyrenes in the solution are $\varphi_B = 0.2075$ and $\varphi_C = 0.0109$.

T/K	Upper phase $\varphi_B + \varphi_C$	Middle phase $\varphi_B + \varphi_C$	Lower phase $\varphi_B + \varphi_C$
303.784		0.2172	0.2646
303.324		0.2121	0.2795
302.765		0.2088	0.2889

continued

continued

T/K	Upper phase $\varphi_B + \varphi_C$	Middle phase $\varphi_B + \varphi_C$	Lower phase $\varphi_B + \varphi_C$
301.947		0.2039	0.3016
301.159		0.2010	0.3107
300.167		0.1989	0.3191
299.142		0.1956	0.3251
298.647		0.1945	0.3264
298.134		0.1905	0.3268
297.905		0.1892	0.3251
297.806		0.1875	0.3231
297.756		0.1850	0.3230
297.703		0.1861	0.3216
297.653	0.1364	0.1901	0.3203
297.613	0.1322	0.1986	0.3189
297.558	0.1250	0.2091	0.3182
297.456	0.1164	0.2258	0.3138
297.356	0.1110	0.2384	0.3096
297.303	0.1059	0.2500	0.3025
297.278	0.1032	0.2612	0.2944
297.264	0.1017		0.2867
297.204	0.0939		0.2897
297.154	0.0979		0.2902
297.004	0.0871		0.2947
296.853	0.0825		0.2983
296.456	0.0711		0.3080
295.754	0.0587		0.3232
295.153	0.0514		0.3345
294.153	0.0401		0.3520
293.130	0.0337		0.3680
292.160	0.0285		0.3816
291.180	0.0240		0.3952
290.180	0.0196		0.4077
289.180	0.0167		0.4201

Polymer (B): **polystyrene** **1999NAK**
Characterization: M_n/g.mol^{-1} = 17650, M_w/g.mol^{-1} = 18700, Pressure Chemical Company, Pittsburgh, PA
Solvent (A): **methylcyclohexane** C_7H_{14} **108-87-2**
Polymer (C): **polystyrene**
Characterization: M_n/g.mol^{-1} = 389000, M_w/g.mol^{-1} = 412000, Pressure Chemical Company, Pittsburgh, PA

continued

continued

Type of data: coexistence data (UCST-behavior)

Comments: The volume fraction of polystyrene (C) in the polystyrene mixture is 0.0070.

φ_{B+C} (total)	0.2005	was kept constant	(critical concentration)				
T/K	296.510	296.470	296.447	296.368	296.250	296.050	295.850
φ_{B+C} (sol phase)	0.1516	0.1413	0.1366	0.1259	0.1131	0.1001	0.0911
φ_{B+C} (gel phase)	0.2035	0.2110	0.2129	0.2217	0.2309	0.2437	0.2537
T/K	295.550	295.250					
φ_{B+C} (sol phase)	0.0811	0.0738					
φ_{B+C} (gel phase)	0.2661	0.2778					

Type of data: critical point (UCST-behavior)

$\varphi_{B+C, crit}$ 0.2005 T_{crit}/K 296.550

Type of data: coexistence data (UCST-behavior)

Comments: The volume fraction of polystyrene (C) in the polystyrene mixture is 0.0069.

φ_{B+C} (total)	0.1936	was kept constant	(critical concentration)					
T/K	296.630	296.627	296.609	296.597	296.582	296.543	296.497	296.451
φ_{B+C} (sol phase)	0.1704	0.1659	0.1579	0.1544	0.1471	0.1402	0.1315	0.1255
φ_{B+C} (gel phase)	0.1945	0.1988	0.2025	0.2046	0.2090	0.2136	0.2184	0.2223
T/K	296.403	296.300	296.180	295.973	295.850	295.650	295.330	295.020
φ_{B+C} (sol phase)	0.1201	0.1112	0.1031	0.0925	0.0879	0.0816	0.0735	0.0670
φ_{B+C} (gel phase)	0.2266	0.2345	0.2414	0.2524	0.2576	0.2663	0.2770	0.2869

Type of data: critical point (UCST-behavior)

$\varphi_{B+C, crit}$ 0.1936 T_{crit}/K 296.636

Type of data: coexistence data (UCST-behavior)

Comments: The volume fraction of polystyrene (C) in the polystyrene mixture is 0.0069.

φ_{B+C} (total)	0.1890	was kept constant	(critical concentration)					
T/K	296.681	296.671	296.661	296.649	296.633	296.618	296.591	296.535
φ_{B+C} (sol phase)	0.1759	0.1691	0.1652	0.1603	0.1546	0.1516	0.1434	0.1340
φ_{B+C} (gel phase)	0.1953	0.1980	0.2011	0.2029	0.2052	0.2075	0.2112	0.2174
T/K	296.491	296.468	296.324	296.103	296.082	295.847	295.550	295.198
φ_{B+C} (sol phase)	0.1290	0.1264	0.1131	0.0990	0.0974	0.0884	0.0795	0.0706
φ_{B+C} (gel phase)	0.2215	0.2235	0.2339	0.2468	0.2481	0.2587	0.2702	0.2817

Type of data: critical point (UCST-behavior)

$\varphi_{B+C, crit}$ 0.1890 T_{crit}/K 296.687

continued

continued

Type of data: coexistence data (UCST-behavior)

Comments: The volume fraction of polystyrene (C) in the polystyrene mixture is 0.0070.

φ_{B+C} (total) 0.1835 was kept constant (critical concentration)

T/K	296.698	296.647	296.603	296.550	296.450	296.350	296.250	296.050
φ_{B+C} (sol phase)	0.1801	0.1701	0.1585	0.1452	0.1300	0.1198	0.1116	0.0984
φ_{B+C} (gel phase)	0.2179	0.2091	0.2101	0.2146	0.2228	0.2305	0.2367	0.2480

T/K	295.750	295.454	295.050
φ_{B+C} (sol phase)	0.0863	0.0777	0.0691
φ_{B+C} (gel phase)	0.2613	0.2724	0.2857

Type of data: critical point (UCST-behavior)

$\varphi_{B+C,\ crit}$ 0.1835 T_{crit}/K 296.630

Type of data: coexistence data (UCST-behavior)

Comments: The volume fraction of polystyrene (C) in the polystyrene mixture is 0.0084.

φ_{B+C} (total) 0.1841 was kept constant

T/K	296.948	296.820	296.772	296.740	296.710	296.649	296.600	296.550
φ_{B+C} (sol phase)	0.1825	0.1816	0.1805	0.1780	0.1725	0.1599	0.1485	0.1400
φ_{B+C} (gel phase)	0.2806	0.2612	0.2451	0.2398	0.2229	0.2180	0.2189	0.2213

T/K	296.448	296.350	296.150	295.950	295.650	295.154
φ_{B+C} (sol phase)	0.1266	0.1184	0.1036	0.0932	0.0833	0.0710
φ_{B+C} (gel phase)	0.2283	0.2352	0.2463	0.2556	0.2681	0.2843

Polymer (B): **polystyrene** **2004WIL**
Characterization: M_n/g.mol^{-1} = 86540, M_w/g.mol^{-1} = 90000,
Pressure Chemical Company, Pittsburgh, PA
Solvent (A): **methylcyclohexane** C_7H_{14} **108-87-2**
Polymer (C): **polystyrene**
Characterization: M_n/g.mol^{-1} = 2075, M_w/g.mol^{-1} = 2200,
Pressure Chemical Company, Pittsburgh, PA

Comments: The mass fraction ratio of polymer(B)/polymer(C) = 0.400/0.600 is kept constant.
The molar mass averages of B+C are M_n/g.mol^{-1} = 3676, M_w/g.mol^{-1} = 38600.

Type of data: cloud points (UCST-behavior)

w_{B+C}	0.0088	0.0165	0.0289	0.0418	0.0509	0.0556	0.0710	0.1067	0.1955
T/K	310.6	312.3	313.2	313.3	313.1	313.0	312.5	310.0	302.9

w_{B+C}	0.2855	0.3607
T/K	295.9	289.0

Polymer (B): **polystyrene** **2004WIL**
Characterization: M_n/g.mol^{-1} = 86540, M_w/g.mol^{-1} = 90000,
Pressure Chemical Company, Pittsburgh, PA
Solvent (A): **methylcyclohexane** C_7H_{14} **108-87-2**
Polymer (C): **polystyrene**
Characterization: M_n/g.mol^{-1} = 5450, M_w/g.mol^{-1} = 5780,
Pressure Chemical Company, Pittsburgh, PA

Comments: The mass fraction ratio of polymer(B)/polymer(C) = 0.390/0.610 is kept constant. The molar mass averages of B+C are M_n/g.mol^{-1} = 9190, M_w/g.mol^{-1} = 38600.

Type of data: cloud points (UCST-behavior)

w_{B+C}	0.0098	0.0205	0.0465	0.0580	0.0757	0.1049	0.1166	0.2184	0.2831
T/K	311.3	313.3	313.8	313.4	312.6	310.8	310.1	304.1	299.4

w_{B+C}	0.3325	0.3844
T/K	296.2	291.7

Polymer (B): **polystyrene** **2004WIL**
Characterization: M_n/g.mol^{-1} = 86540, M_w/g.mol^{-1} = 90000,
Pressure Chemical Company, Pittsburgh, PA
Solvent (A): **methylcyclohexane** C_7H_{14} **108-87-2**
Polymer (C): **polystyrene**
Characterization: M_n/g.mol^{-1} = 12260, M_w/g.mol^{-1} = 13000,
Pressure Chemical Company, Pittsburgh, PA

Comments: The mass fraction ratio of polymer(B)/polymer(C) = 0.330/0.670 is kept constant. The molar mass averages of B+C are M_n/g.mol^{-1} = 18380, M_w/g.mol^{-1} = 38600.

Type of data: cloud points (UCST-behavior)

w_{B+C}	0.0222	0.0376	0.0638	0.1037	0.1587	0.2052	0.2813	0.3442
T/K	311.5	313.5	311.8	311.8	309.0	306.9	303.2	300.1

Polymer (B): **polystyrene** **1956KER**
Characterization: ρ = 1.05 g/cm^3 at 298.15 K
Solvent (A): **nitrobenzene** $C_6H_5NO_2$ **98-95-3**
Polymer (C): **poly(methyl methacrylate)**
Characterization: ρ = 1.18 g/cm^3 at 298.15 K

Type of data: cloud points (threshold point)

T/K = 298.15

φ_A	0.895	φ_B	0.0525	φ_C	0.0525

Polymer (B): **polystyrene** **1996LUS**
Characterization: M_n/g.mol^{-1} = 23600, M_w/g.mol^{-1} = 25000,
Pressure Chemical Company, Pittsburgh, PA
Solvent (A): **propionitrile** **C_3H_5N** **107-12-0**
Polymer (C): **polystyrene-d8**
Characterization: M_n/g.mol^{-1} = 25400, M_w/g.mol^{-1} = 26700,
completely deuterated, Polymer Laboratories, Amherst, MA

Comments: The mole fraction ratio of polymer(C)/polymer(B) = 0.50 is kept constant.

Type of data: cloud points

w_{B+C}	0.1300	0.1300	0.1300	0.1300	0.1300	0.1300	0.1300	0.1300	0.1300
T/K	330.309	330.309	330.305	330.305	337.064	337.052	343.871	343.869	351.509
P/MPa	5.017	5.009	4.991	4.996	3.032	3.036	1.604	1.583	0.396

w_{B+C}	0.1300	0.1300	0.1300	0.1300	0.1300	0.1300	0.1300	0.1300	0.1300
T/K	351.391	411.479	411.466	411.477	419.828	419.828	419.828	419.828	419.814
P/MPa	0.402	0.563	0.620	0.699	1.439	1.452	1.476	1.445	1.432

w_{B+C}	0.1300	0.1300	0.1300	0.1300	0.1300	0.1300	0.1300	0.1300	0.1300
T/K	419.814	428.203	428.197	428.197	436.813	436.813	436.817	436.817	436.817
P/MPa	1.436	2.269	2.160	2.165	3.198	3.201	3.195	3.224	3.202

w_{B+C}	0.1300	0.1300	0.1300	0.1300	0.1300	0.1300	0.1300	0.1300	0.1300
T/K	436.817	436.817	445.655	445.650	445.650	445.650	454.681	454.681	454.681
P/MPa	3.191	3.198	4.182	4.202	4.204	4.202	5.246	5.252	5.261

w_{B+C}	0.1300	0.1300	0.1300	0.1599	0.1599	0.1599	0.1599	0.1599	0.1599
T/K	454.658	454.658	454.658	330.291	330.290	337.001	336.998	343.823	343.821
P/MPa	5.273	5.271	5.270	4.938	4.934	3.105	3.018	1.638	1.547

w_{B+C}	0.1599	0.1599	0.1599	0.1599	0.1599	0.1599	0.1599	0.1599	0.1599
T/K	350.797	350.796	411.521	411.522	419.731	419.733	428.235	428.234	436.879
P/MPa	0.509	0.483	0.684	0.689	1.479	1.520	2.356	2.305	3.254

w_{B+C}	0.1599	0.1599	0.1599	0.1599	0.1599	0.1599	0.1599	0.1599	0.1900
T/K	436.875	436.875	436.875	445.699	445.699	445.698	454.664	454.674	330.206
P/MPa	3.138	3.158	3.274	4.353	4.267	4.253	5.276	5.376	5.257

w_{B+C}	0.1900	0.1900	0.1900	0.1900	0.1900	0.1900	0.1900	0.1900	0.1900
T/K	330.206	337.173	337.133	343.716	343.716	351.032	351.030	411.488	411.485
P/MPa	5.453	3.364	3.257	1.992	1.923	0.864	0.764	0.904	1.073

w_{B+C}	0.1900	0.1900	0.1900	0.1900	0.1900	0.1900	0.1900	0.1900	0.1900
T/K	420.694	420.679	428.290	428.292	436.890	436.888	445.686	445.684	454.553
P/MPa	1.878	1.809	2.509	2.585	3.459	3.464	4.439	4.503	5.515

w_{B+C}	0.1900	0.2198	0.2198	0.2198	0.2198	0.2198	0.2198	0.2198	0.2198
T/K	454.575	333.385	333.385	333.415	333.397	340.168	340.181	347.027	345.281
P/MPa	5.559	4.889	4.836	4.657	4.684	3.069	2.770	1.419	1.366

continued

continued

w_{B+C}	0.2198	0.2198	0.2198	0.2198	0.2198	0.2198	0.2198	0.2198	0.2198
T/K	350.852	350.851	411.434	411.433	411.432	419.661	419.655	428.097	428.086
P/MPa	0.375	0.302	0.691	0.879	0.659	1.359	1.458	2.227	2.069

w_{B+C}	0.2198	0.2198	0.2198	0.2198	0.2198	0.2198	0.2499	0.2499	0.2499
T/K	436.660	436.655	445.476	445.468	454.374	454.380	334.038	333.672	333.672
P/MPa	3.172	3.277	4.378	4.166	5.327	5.342	4.333	4.341	4.337

w_{B+C}	0.2499	0.2499	0.2499	0.2499	0.2499	0.2499	0.2499	0.2499	0.2499
T/K	340.277	340.222	346.716	346.689	352.653	352.728	416.714	423.091	423.027
P/MPa	2.553	2.529	1.185	1.177	0.197	0.177	0.685	1.251	1.251

w_{B+C}	0.2499	0.2499	0.2499	0.2499	0.2499	0.2499	0.2499	0.2499
T/K	431.439	431.438	440.118	440.116	449.061	449.009	457.927	458.073
P/MPa	2.096	2.108	2.996	3.006	4.015	4.080	5.175	5.118

Polymer (B): **polystyrene** **1963PAX**
Characterization: M_n/g.mol^{-1} = 2720, synthesized in the laboratory
Solvent (A): **tetrachloromethane** **CCl_4** **56-23-5**
Polymer (C): **polybutadiene**
Characterization: M_n/g.mol^{-1} = 1670, synthesized in the laboratory

Type of data: cloud points

T/K = 296.15

w_A	0.882	0.862	0.819	0.741
w_B	0.023	0.055	0.108	0.231
w_C	0.095	0.083	0.073	0.028

Polymer (B): **polystyrene** **1960ALL**
Characterization: M_n/g.mol^{-1} = 700000, I.C.I. Ltd., U.K.
Solvent (A): **tetrachloromethane** **CCl_4** **56-23-5**
Polymer (C): **polybutadiene**
Characterization: M_n/g.mol^{-1} = 300000, Canadian Polymer Corporation

Type of data: coexistence data (tie lines)

T/K = 291.15

Phase I			Phase II		
φ_A	φ_B	φ_C	φ_A	φ_B	φ_C
0.910	0.080	0.010	0.943	0.007	0.050
0.922	0.067	0.011	0.941	0.010	0.049
0.933	0.054	0.013	0.944	0.013	0.043
0.948	0.035	0.017	0.949	0.017	0.034

Polymer (B): **polystyrene** **1960ALL**
Characterization: M_n/g.mol^{-1} = 1000000, fractionated in the laboratory
Solvent (A): **tetrachloromethane** **CCl_4** **56-23-5**
Polymer (C): **polyisobutylene**
Characterization: M_n/g.mol^{-1} = 1500000, fractionated in the laboratory

Type of data: coexistence data (tie lines)

T/K = 291.15

Phase I			Phase II		
φ_A	φ_B	φ_C	φ_A	φ_B	φ_C
0.9818	0.0127	0.0055	0.9825	0.0012	0.0163
0.9796	0.0170	0.0036	0.9801	0.0011	0.0188
0.9780	0.0194	0.0026	0.9787	0.0009	0.0204
0.9692	0.0304	0.0004	0.9719	0.0010	0.0271

Polymer (B): **polystyrene** **1983NAR**
Characterization: M_n/g.mol^{-1} = 33000, M_w/g.mol^{-1} = 36000,
Pressure Chemical Company, Pittsburgh, PA
Solvent (A): **tetrahydrofuran** **C_4H_8O** **109-99-9**
Polymer (C): **polybutadiene**
Characterization: M_n/g.mol^{-1} = 135000, M_w/g.mol^{-1} = 170000,
47.1% 1,4-*cis*, 44.5% 1,4-*trans*, 8.4% 1,2-*vinyl*

Type of data: cloud points (plait-point composition)

T/K = 296.15

w_A	0.910	w_B	0.0525	w_C	0.0375
φ_A	0.921	φ_B	0.042	φ_C	0.037

Polymer (B): **polystyrene** **1989NAR**
Characterization: M_n/g.mol^{-1} = 33000, M_w/g.mol^{-1} = 36000,
Pressure Chemical Company, Pittsburgh, PA
Solvent (A): **tetrahydrofuran** **C_4H_8O** **109-99-9**
Polymer (C): **polybutadiene**
Characterization: M_n/g.mol^{-1} = 135000, M_w/g.mol^{-1} = 170000,
47.1% 1,4-*cis*, 44.5% 1,4-*trans*, 8.4% 1,2-*vinyl*

Comments: The polymer ratio in the homogeneous feed phase is 1:1 by weight fraction.

Type of data: coexistence data (tie lines)

T/K = 296.15

continued

continued

Total system		Top phase			Bottom phase		
w_A	$w_B + w_C$	φ_A	φ_B	φ_C	φ_A	φ_B	φ_C
0.7916	0.2084	0.817	0.014	0.170	0.809	0.185	0.006
0.7897	0.2103	0.810	0.013	0.177	0.812	0.184	0.004
0.7994	0.2006	0.816	0.021	0.163	0.825	0.174	0.001
0.8659	0.1341	0.877	0.021	0.102	0.886	0.112	0.003
0.8818	0.1182	0.887	0.024	0.089	0.905	0.092	0.003

Polymer (B): **polystyrene** **1999GOM**
Characterization: M_n/g.mol^{-1} = 36800, M_w/g.mol^{-1} = 37900, Tosoh Corp., Ltd., Tokyo, Japan
Solvent (A): **tetrahydrofuran** **C_4H_8O** **109-99-9**
Polymer (C): **polybutadiene**
Characterization: M_n/g.mol^{-1} = 85600, M_w/g.mol^{-1} = 89960, Pressure Chemical Company, Pittsburgh, PA

Type of data: coexistence data (tie lines)

T/K = 298.15

Phase I			Phase II		
φ_A	φ_B	φ_C	φ_A	φ_B	φ_C
0.884	0.0387	0.0778	0.880	0.0745	0.0451
0.893	0.0360	0.0706	0.882	0.0728	0.0451
0.895	0.0356	0.0691	0.886	0.0699	0.0443
0.897	0.0352	0.0676	0.888	0.0682	0.0442
0.899	0.0350	0.0658	0.889	0.0666	0.0441
0.903	0.0341	0.0628	0.891	0.0649	0.0440
0.876	0.0371	0.0872	0.868	0.0872	0.0449
0.878	0.0376	0.0845	0.872	0.0841	0.0444
0.882	0.0367	0.0816	0.873	0.0822	0.0444
0.884	0.0364	0.0800	0.875	0.0807	0.0441
0.886	0.0364	0.0781	0.877	0.0786	0.0445
0.888	0.0363	0.0762	0.881	0.0757	0.0438
0.817	0.0415	0.142	0.795	0.158	0.0463
0.824	0.0405	0.135	0.804	0.149	0.0464
0.819	0.0436	0.138	0.799	0.154	0.0466
0.829	0.0420	0.129	0.812	0.141	0.0478
0.831	0.0423	0.127	0.817	0.135	0.0474
0.835	0.0421	0.123	0.821	0.132	0.0474

Polymer (B): **polystyrene** **1989NAR**
Characterization: M_n/g.mol^{-1} = 104000, M_w/g.mol^{-1} = 110000,
Pressure Chemical Company, Pittsburgh, PA
Solvent (A): **tetrahydrofuran** C_4H_8O **109-99-9**
Polymer (C): **polybutadiene**
Characterization: M_n/g.mol^{-1} = 135000, M_w/g.mol^{-1} = 170000,
47.1% 1,4-*cis*, 44.5% 1,4-*trans*, 8.4% 1,2-*vinyl*

Type of data: cloud points (plait-point composition)

T/K = 296.15

w_A	0.946	w_B	0.028	w_C	0.026
φ_A	0.952	φ_B	0.022	φ_C	0.026

Polymer (B): **polystyrene** **1999GOM**
Characterization: M_n/g.mol^{-1} = 106000, M_w/g.mol^{-1} = 110000,
Tosoh Corp., Ltd., Tokyo, Japan
Solvent (A): **tetrahydrofuran** C_4H_8O **109-99-9**
Polymer (C): **polybutadiene**
Characterization: M_n/g.mol^{-1} = 85600, M_w/g.mol^{-1} = 89960,
Pressure Chemical Company, Pittsburgh, PA

Type of data: coexistence data (tie lines)

T/K = 298.15

Phase I			Phase II		
φ_A	φ_B	φ_C	φ_A	φ_B	φ_C
0.924	0.0273	0.0488	0.906	0.0560	0.0382
0.926	0.0268	0.0474	0.909	0.0533	0.0372
0.930	0.0257	0.0448	0.913	0.0503	0.0367
0.931	0.0253	0.0433	0.915	0.0488	0.0364
0.933	0.0249	0.0417	0.917	0.0474	0.0360
0.935	0.0245	0.0402	0.920	0.0449	0.0349
0.882	0.0355	0.0829	0.846	0.1040	0.0502
0.883	0.0352	0.0814	0.848	0.1020	0.0504
0.886	0.0352	0.0793	0.849	0.0999	0.0507
0.889	0.0346	0.0761	0.864	0.0887	0.0472
0.891	0.0342	0.0745	0.866	0.0868	0.0473
0.895	0.0335	0.0714	0.870	0.0838	0.0467
0.814	0.0396	0.146	0.815	0.138	0.0471
0.816	0.0400	0.144	0.819	0.133	0.0482
0.821	0.0408	0.139	0.823	0.129	0.0482
0.826	0.0403	0.133	0.823	0.128	0.0489
0.828	0.0406	0.131	0.826	0.125	0.0491
0.831	0.0410	0.129	0.828	0.123	0.0495

Polymer (B): **polystyrene** **1989NAR**
Characterization: M_n/g.mol^{-1} = 107800, M_w/g.mol^{-1} = 123600,
Pressure Chemical Company, Pittsburgh, PA
Solvent (A): **tetrahydrofuran** C_4H_8O **109-99-9**
Polymer (C): **polybutadiene**
Characterization: M_n/g.mol^{-1} = 135000, M_w/g.mol^{-1} = 170000,
47.1% 1,4-*cis*, 44.5% 1,4-*trans*, 8.4% 1,2-*vinyl*

Comments: The polymer ratio in the homogeneous feed phase is 1:1 by weight fraction.

Type of data: coexistence data (tie lines)

T/K = 296.15

Total system		Top phase			Bottom phase		
w_A	$w_B + w_C$	φ_A	φ_B	φ_C	φ_A	φ_B	φ_C
0.7949	0.2051	0.809	0.023	0.169	0.823	0.169	0.009
0.8041	0.1959	0.820	0.022	0.159	0.830	0.164	0.007
0.8597	0.1403	0.881	0.011	0.108	0.870	0.126	0.004
0.9112	0.0888	0.924	0.011	0.065	0.920	0.074	0.007
0.9341	0.0659	0.940	0.015	0.045	0.944	0.043	0.014

Polymer (B): **polystyrene** **1993GE1**
Characterization: M_n/g.mol^{-1} = 274600, M_w/g.mol^{-1} = 283300,
Pressure Chemical Company, Pittsburgh, PA
Solvent (A): **tetrahydrofuran** C_4H_8O **109-99-9**
Polymer (C): **poly(butyl methacrylate)**
Characterization: M_n/g.mol^{-1} = 54350, M_w/g.mol^{-1} = 64100,
DuPont Co., Philadelphia, PA

Type of data: coexistence data (tie lines)

T/K = 303.15

Phase I			Phase II		
w_A	w_B	w_C	w_A	w_B	w_C
0.6290	0.0016	0.3694	0.6330	0.3492	0.0178
0.7153	0.0039	0.2808	0.7111	0.2582	0.0307
0.7512	0.0098	0.2390	0.7315	0.2172	0.0512
0.7368	0.0124	0.2507	0.7351	0.2118	0.0531

Polymer (B): **polystyrene** **1993GE1**
Characterization: M_n/g.mol^{-1} = 13000, M_w/g.mol^{-1} = 13640,
Pressure Chemical Company, Pittsburgh, PA
Solvent (A): **tetrahydrofuran** C_4H_8O **109-99-9**
Polymer (C): **poly(methyl methacrylate)**
Characterization: M_n/g.mol^{-1} = 32840, M_w/g.mol^{-1} = 37100,
DuPont Co., Philadelphia, PA

Type of data: coexistence data (tie lines)

T/K = 303.15

Phase I			Phase II		
w_A	w_B	w_C	w_A	w_B	w_C
0.6194	0.3782	0.0024	0.5177	0.0363	0.4460
0.6700	0.3118	0.0182	0.5999	0.0674	0.3327
0.6800	0.2955	0.0245	0.6170	0.0928	0.2902
0.6884	0.2627	0.0490	0.6375	0.1120	0.2506

Polymer (B): **polystyrene** **1993GE1**
Characterization: M_n/g.mol^{-1} = 13000, M_w/g.mol^{-1} = 13640,
Pressure Chemical Company, Pittsburgh, PA
Solvent (A): **tetrahydrofuran** C_4H_8O **109-99-9**
Polymer (C): **poly(methyl methacrylate)**
Characterization: M_n/g.mol^{-1} = 296300, M_w/g.mol^{-1} = 326000,
DuPont Co., Philadelphia, PA

Type of data: coexistence data (tie lines)

T/K = 303.15

Phase I			Phase II		
w_A	w_B	w_C	w_A	w_B	w_C
0.6654	0.3335	0.0011	0.6312	0.0366	0.3322
0.7117	0.2881	0.0002	0.6492	0.0521	0.2987
0.7561	0.2421	0.0019	0.6974	0.0887	0.2139
0.7673	0.2256	0.0071	0.7479	0.1120	0.1401

Polymer (B): **polystyrene** **1993GE1**
Characterization: M_n/g.mol^{-1} = 274600, M_w/g.mol^{-1} = 283300, Pressure Chemical Company, Pittsburgh, PA
Solvent (A): **tetrahydrofuran** **C_4H_8O** **109-99-9**
Polymer (C): **poly(methyl methacrylate)**
Characterization: M_n/g.mol^{-1} = 32840, M_w/g.mol^{-1} = 37100, DuPont Co., Philadelphia, PA

Type of data: coexistence data (tie lines)

T/K = 303.15

Phase I			Phase II		
w_A	w_B	w_C	w_A	w_B	w_C
0.7618	0.2082	0.0300	0.7171	0.0026	0.2803
0.7760	0.1873	0.0367	0.7343	0.0019	0.2638
0.7850	0.1678	0.0472	0.7489	0.0081	0.2430
0.7985	0.1329	0.0686	0.7755	0.0071	0.2174

Polymer (B): **polystyrene** **1993GE1**
Characterization: M_n/g.mol^{-1} = 274600, M_w/g.mol^{-1} = 283300, Pressure Chemical Company, Pittsburgh, PA
Solvent (A): **tetrahydrofuran** **C_4H_8O** **109-99-9**
Polymer (C): **poly(methyl methacrylate)**
Characterization: M_n/g.mol^{-1} = 296300, M_w/g.mol^{-1} = 326000, DuPont Co., Philadelphia, PA

Type of data: coexistence data (tie lines)

T/K = 303.15

Phase I			Phase II		
w_A	w_B	w_C	w_A	w_B	w_C
0.8082	0.1901	0.0017	0.7892	0.0009	0.2098
0.8450	0.1476	0.0074	0.8279	0.0014	0.1707
0.8897	0.1011	0.0092	0.8787	0.0054	0.1159
0.9007	0.0899	0.0094	0.8847	0.0099	0.1054
0.9105	0.0716	0.0179	0.9013	0.0183	0.0804

Polymer (B): **polystyrene** **1956KER**
Characterization: $\rho = 1.05$ g/cm^3 at 298.15 K
Solvent (A): **tetrahydrofuran** C_4H_8O **109-99-9**
Polymer (C): **poly(methyl methacrylate)**
Characterization: $\rho = 1.18$ g/cm^3 at 298.15 K

Type of data: cloud points (threshold point)

T/K = 298.15

φ_A 0.948 φ_B 0.026 φ_C 0.026

Polymer (B): **polystyrene** **1993GE2**
Characterization: M_n/g.mol^{-1} = 13000, M_w/g.mol^{-1} = 13640,
Pressure Chemical Company, Pittsburgh, PA
Solvent (A): **tetrahydrofuran** C_4H_8O **109-99-9**
Polymer (C): **polytetrahydrofuran**
Characterization: M_n/g.mol^{-1} = 300000, M_w/g.mol^{-1} = 339000,
Pressure Chemical Company, Pittsburgh, PA

Type of data: coexistence data (tie lines)

T/K = 303.15

Phase I			Phase II		
w_A	w_B	w_C	w_A	w_B	w_C
0.7107	0.1714	0.1179	0.6845	0.3084	0.0070
0.7266	0.1881	0.0854	0.6986	0.2856	0.0158
0.7293	0.2041	0.0665	0.7058	0.2723	0.0219

Polymer (B): **polystyrene** **1993GE2**
Characterization: M_n/g.mol^{-1} = 274600, M_w/g.mol^{-1} = 283300,
Pressure Chemical Company, Pittsburgh, PA
Solvent (A): **tetrahydrofuran** C_4H_8O **109-99-9**
Polymer (C): **polytetrahydrofuran**
Characterization: M_n/g.mol^{-1} = 37600, M_w/g.mol^{-1} = 40200,
Pressure Chemical Company, Pittsburgh, PA

Type of data: coexistence data (tie lines)

T/K = 303.15

continued

continued

Phase I			Phase II		
w_A	w_B	w_C	w_A	w_B	w_C
0.8413	0.0002	0.1585	0.7580	0.2276	0.0144
0.8625	0.0021	0.1353	0.7994	0.1791	0.0215
0.8788	0.0040	0.1172	0.8192	0.1548	0.0260
0.8914	0.0348	0.0738	0.8748	0.0722	0.0530

Polymer (B): **polystyrene** **1993GE2**
Characterization: M_n/g.mol^{-1} = 274600, M_w/g.mol^{-1} = 283300, Pressure Chemical Company, Pittsburgh, PA
Solvent (A): **tetrahydrofuran** **C_4H_8O** **109-99-9**
Polymer (C): **polytetrahydrofuran**
Characterization: M_n/g.mol^{-1} = 300000, M_w/g.mol^{-1} = 339000, Pressure Chemical Company, Pittsburgh, PA

Type of data: coexistence data (tie lines)

T/K = 303.15

Phase I			Phase II		
w_A	w_B	w_C	w_A	w_B	w_C
0.9167	0.0021	0.0813	0.8708	0.1262	0.0031
0.9259	0.0035	0.0706	0.8853	0.1123	0.0023
0.9407	0.0077	0.0515	0.9089	0.0849	0.0062
0.9449	0.0197	0.0354	0.9273	0.0630	0.0097

Polymer (B): **polystyrene** **1955KER**
Characterization: ρ = 1.05 g/cm^3 at 298.15 K
Solvent (A): **tetrahydrofuran** **C_4H_8O** **109-99-9**
Polymer (C): **poly(vinyl chloride)**
Characterization: Ultron 300

Type of data: coexistence data (tie lines)

T/K = 298.15

Total system			Top phase	Bottom phase
w_A	w_B	w_C	w_C	w_C
0.85	0.075	0.075	0.230	0.860

Polymer (B): **polystyrene** **1963PAX**
Characterization: M_n/g.mol^{-1} = 2720, synthesized in the laboratory
Solvent (A): **toluene** C_7H_8 **108-88-3**
Polymer (C): **polybutadiene**
Characterization: M_n/g.mol^{-1} = 1100, synthesized in the laboratory

Type of data: cloud points

T/K = 296.15

w_A	0.688	0.680	0.654	0.605	0.572	0.490	0.473
w_B	0.031	0.063	0.137	0.196	0.255	0.408	0.473
w_C	0.281	0.257	0.209	0.199	0.173	0.102	0.054

Polymer (B): **polystyrene** **1963PAX**
Characterization: M_n/g.mol^{-1} = 2720, synthesized in the laboratory
Solvent (A): **toluene** C_7H_8 **108-88-3**
Polymer (C): **polybutadiene**
Characterization: M_n/g.mol^{-1} = 1670, synthesized in the laboratory

Type of data: cloud points

T/K = 296.15

w_A	0.739	0.745	0.744	0.705	0.670	0.610	0.589
w_B	0.026	0.050	0.101	0.147	0.196	0.312	0.369
w_C	0.235	0.205	0.155	0.148	0.134	0.078	0.042

Polymer (B): **polystyrene** **1993GE1**
Characterization: M_n/g.mol^{-1} = 13000, M_w/g.mol^{-1} = 13640,
Pressure Chemical Company, Pittsburgh, PA
Solvent (A): **toluene** C_7H_8 **108-88-3**
Polymer (C): **polybutadiene**
Characterization: M_n/g.mol^{-1} = 22600, M_w/g.mol^{-1} = 24000,
52% 1,4-*trans*, 40% 1,4-*cis*, 8% 1,2-vinyl, 0.45% antioxidant,
Pressure Chemical Company, Pittsburgh, PA

Type of data: coexistence data (tie lines)

T/K = 303.15

Phase I			Phase II		
w_A	w_B	w_C	w_A	w_B	w_C
0.7465	0.0501	0.2034	0.6908	0.3007	0.0085
0.7656	0.0730	0.1614	0.7189	0.2632	0.0179
0.7666	0.0555	0.1778	0.7160	0.2694	0.0146
0.7781	0.0775	0.1444	0.7472	0.2348	0.0180

Polymer (B):	**polystyrene**		**1983NAR**
Characterization:	M_n/g.mol^{-1} = 33000, M_w/g.mol^{-1} = 36000, Pressure Chemical Company, Pittsburgh, PA		
Solvent (A):	**toluene**	**C_7H_8**	**108-88-3**
Polymer (C):	**polybutadiene**		
Characterization:	M_n/g.mol^{-1} = 135000, M_w/g.mol^{-1} = 170000, 47.1% 1,4-*cis*, 44.5% 1,4-*trans*, 8.4% 1,2-*vinyl*		

Type of data: cloud points (plait-point composition)

T/K = 296.15

w_A	0.9050	w_B	0.0575	w_C	0.0375
φ_A	0.920	φ_B	0.045	φ_C	0.035

Polymer (B):	**polystyrene**		**1984NA2, 1986NAR**
Characterization:	M_n/g.mol^{-1} = 63600, M_w/g.mol^{-1} = 195000, Pressure Chemical Company, Pittsburgh, PA		
Solvent (A):	**toluene**	**C_7H_8**	**108-88-3**
Polymer (C):	**polybutadiene**		
Characterization:	M_n/g.mol^{-1} = 115000, M_w/g.mol^{-1} = 130000, 38.0% 1,4-*cis*, 53.0% 1,4-*trans*, 9.0% 1,2-*vinyl*		

Comments: The polymer ratio in the homogeneous feed phase is 1:1 by weight fraction.

Type of data: coexistence data (tie lines)

T/K = 296.15

Total system		Top phase			Bottom phase		
w_A	$w_B + w_C$	φ_A	φ_B	φ_C	φ_A	φ_B	φ_C
0.8021	0.1979	0.831	0.024	0.146	0.818	0.157	0.026
0.8027	0.1973	0.824	0.032	0.145	0.827	0.154	0.019
0.8195	0.1805	0.841	0.031	0.127	0.838	0.130	0.032
0.8491	0.1509	0.871	0.018	0.112	0.862	0.116	0.022
0.8833	0.1167	0.901	0.012	0.085	0.891	0.086	0.023

Fractionation of polystyrene (B) during demixing (the lines correspond with those above):

Top phase		Bottom phase	
M_n/(g/mol)	M_w/(g/mol)	M_n/(g/mol)	M_w/(g/mol)
64120	185070	70280	211950
67570	193270	72590	215010
67350	191730	71040	212760
56470	167890	78960	213310
45650	135020	73490	223180

Polymer (B): **polystyrene** **1984NA2, 1986NAR**
Characterization: M_n/g.mol^{-1} = 63600, M_w/g.mol^{-1} = 195000,
Pressure Chemical Company, Pittsburgh, PA
Solvent (A): **toluene** **C_7H_8** **108-88-3**
Polymer (C): **polybutadiene**
Characterization: M_n/g.mol^{-1} = 45000, M_w/g.mol^{-1} = 270000,
high-*cis*, Taktene 1202, Polysar Ltd.

Comments: The polymer ratio in the homogeneous feed phase is 1:1 by weight fraction.

Type of data: coexistence data (tie lines)

T/K = 296.15

Total system		Top phase			Bottom phase		
w_A	$w_B + w_C$	φ_A	φ_B	φ_C	φ_A	φ_B	φ_C
0.8136	0.1864	0.828	0.027	0.145	0.842	0.126	0.032
0.8155	0.1845	0.831	0.030	0.139	0.843	0.137	0.021
0.8451	0.1549	0.860	0.025	0.115	0.868	0.118	0.015
0.8754	0.1246	0.907	0.018	0.076	0.873	0.092	0.035
0.9075	0.0925	0.924	0.015	0.061	0.915	0.069	0.017

Fractionation of polystyrene (B) during demixing (the lines correspond with those above):

Top phase		Bottom phase	
M_n/(g/mol)	M_w/(g/mol)	M_n/(g/mol)	M_w/(g/mol)
64440	201760	81380	210550
65080	205790	73700	213110
60000	188010	71810	213530
47650	164790	76800	222970
46280	142560	75170	222520

Polymer (B): **polystyrene** **1984NA2, 1986NAR**
Characterization: M_n/g.mol^{-1} = 70700, M_w/g.mol^{-1} = 87300,
Pressure Chemical Company, Pittsburgh, PA
Solvent (A): **toluene** **C_7H_8** **108-88-3**
Polymer (C): **polybutadiene**
Characterization: M_n/g.mol^{-1} = 115000, M_w/g.mol^{-1} = 130000,
38.0% 1,4-*cis*, 53.0% 1,4-*trans*, 9.0% 1,2-*vinyl*

Comments: The polymer ratio in the homogeneous feed phase is 1:1 by weight fraction.

Type of data: coexistence data (tie lines)

continued

continued

T/K = 296.15

Total system		Top phase			Bottom phase		
w_A	$w_B + w_C$	φ_A	φ_B	φ_C	φ_A	φ_B	φ_C
0.7982	0.2018	0.836	0.016	0.148	0.806	0.177	0.016
0.8197	0.1803	0.848	0.014	0.138	0.832	0.149	0.019
0.8285	0.1715	0.853	0.017	0.129	0.844	0.144	0.012
0.8591	0.1409	0.873	0.016	0.111	0.878	0.105	0.016
0.8896	0.1104	0.905	0.012	0.084	0.901	0.085	0.014

Fractionation of polystyrene (B) during demixing (the lines correspond with those above):

Top phase		Bottom phase	
M_n/(g/mol)	M_w/(g/mol)	M_n/(g/mol)	M_w/(g/mol)
65020	79020	72040	85740
58170	75020	72120	96910
71860	91070	71190	97920
69700	90210	74740	94970
71510	89020	85320	103370

Polymer (B): **polystyrene** **1984NA2, 1986NAR**
Characterization: M_n/g.mol^{-1} = 70700, M_w/g.mol^{-1} = 87300
Solvent (A): **toluene** **C_7H_8** **108-88-3**
Polymer (C): **polybutadiene**
Characterization: M_n/g.mol^{-1} = 45000, M_w/g.mol^{-1} = 270000,
high-*cis*, Taktene 1202, Polysar Ltd.

Comments: The polymer ratio in the homogeneous feed phase is 1:1 by weight fraction.

Type of data: coexistence data (tie lines)

T/K = 296.15

Total system		Top phase			Bottom phase		
w_A	$w_B + w_C$	φ_A	φ_B	φ_C	φ_A	φ_B	φ_C
0.7994	0.2006	0.845	0.022	0.133	0.799	0.165	0.037
0.8211	0.1789	0.855	0.018	0.127	0.827	0.139	0.034
0.8410	0.1590	0.859	0.020	0.122	0.862	0.131	0.007
0.8743	0.1257	0.882	0.015	0.103	0.896	0.093	0.011
0.9049	0.0951	0.911	0.015	0.073	0.921	0.063	0.016

continued

continued

Fractionation of polystyrene (B) during demixing (the lines correspond with those above):

Top phase		Bottom phase	
M_n/(g/mol)	M_w/(g/mol)	M_n/(g/mol)	M_w/(g/mol)
70740	83320	70380	96980
74080	85920	79750	94630
73960	84530	76570	94480
72560	86330	70190	90800
71300	85620	74160	94940

Polymer (B): **polystyrene** **1983NAR, 1984NA1**
Characterization: M_n/g.mol^{-1} = 94500, M_w/g.mol^{-1} = 100000,
Pressure Chemical Company, Pittsburgh, PA
Solvent (A): **toluene** **C_7H_8** **108-88-3**
Polymer (C): **polybutadiene**
Characterization: M_n/g.mol^{-1} = 135000, M_w/g.mol^{-1} = 170000,
47.1% 1,4-*cis*, 44.5% 1,4-*trans*, 8.4% 1,2-*vinyl*

Type of data: cloud points (plait-point composition)

T/K = 296.15

w_A	0.9250	w_B	0.0375	w_C	0.0375
φ_A	0.934	φ_B	0.030	φ_C	0.036

Polymer (B): **polystyrene** **1976ESK**
Characterization: M_w/g.mol^{-1} = 194000,
Pressure Chemical Company, Pittsburgh, PA
Solvent (A): **toluene** **C_7H_8** **108-88-3**
Polymer (C): **polyisobutylene**
Characterization: M_w/g.mol^{-1} = 156000

Type of data: spinodal points

T/K = 294.15

w_A	0.9579	0.9613	0.9645	0.9634	0.9661	0.9668	0.9680
w_B	0.0068	0.0085	0.0112	0.0123	0.0123	0.0141	0.0188
w_C	0.0353	0.0302	0.0243	0.0243	0.0216	0.0191	0.0132

Type of data: critical point

$w_{A, crit}$ = 0.9634 $w_{B, crit}$ = 0.0123 $w_{C, crit}$ = 0.0243

Polymer (B): **polystyrene** **1976ESK**
Characterization: M_w/g.mol^{-1} = 526000,
Pressure Chemical Company, Pittsburgh, PA
Solvent (A): **toluene** **C_7H_8** **108-88-3**
Polymer (C): **polyisobutylene**
Characterization: M_w/g.mol^{-1} = 670000

Type of data: spinodal points

T/K = 294.15

w_A	0.9764	0.9822	0.9838	0.9837
w_B	0.0036	0.0048	0.0062	0.0063
w_C	0.0200	0.0130	0.0100	0.0100

Type of data: critical point

$w_{A, crit}$ = 0.9838 $w_{B, crit}$ = 0.0062 $w_{C, crit}$ = 0.0100

Polymer (B): **polystyrene** **1976ESK**
Characterization: M_w/g.mol^{-1} = 2400000,
Pressure Chemical Company, Pittsburgh, PA
Solvent (A): **toluene** **C_7H_8** **108-88-3**
Polymer (C): **polyisobutylene**
Characterization: M_w/g.mol^{-1} = 2440000

Type of data: spinodal points

T/K = 294.15

w_A	0.9919	0.9924	0.9927	0.9932	0.9929	0.9930	0.9924
w_B	0.0013	0.0016	0.0022	0.0024	0.0027	0.0028	0.0046
w_C	0.0068	0.0060	0.0051	0.0044	0.0041	0.0042	0.0030

Type of data: critical point

$w_{A, crit}$ = 0.9932 $w_{B, crit}$ = 0.0024 $w_{C, crit}$ = 0.0044

Polymer (B): **polystyrene** **1993GE1**
Characterization: M_n/g.mol^{-1} = 13000, M_w/g.mol^{-1} = 13640,
Pressure Chemical Company, Pittsburgh, PA
Solvent (A): **toluene** **C_7H_8** **108-88-3**
Polymer (C): **polyisoprene**
Characterization: M_n/g.mol^{-1} = 10800, M_w/g.mol^{-1} = 12000,
18% 1,4-*trans*, 76% 1,4-*cis*, 6% 3,4-content, 0.5% antioxidant,
Goodyear Tire and Rubber Co., Akron, OH

Type of data: coexistence data (tie lines)

T/K = 303.15

continued

continued

Phase I			Phase II		
w_A	w_B	w_C	w_A	w_B	w_C
0.6429	0.0001	0.3570	0.5761	0.4103	0.0136
0.6450	0.0001	0.3549	0.5889	0.3849	0.0262
0.7098	0.0010	0.2892	0.6475	0.3390	0.0135
0.7326	0.0021	0.2653	0.6712	0.3124	0.0164
0.7796	0.0145	0.2059	0.7318	0.2454	0.0228
0.7805	0.0119	0.2076	0.7330	0.2421	0.0250

Polymer (B): **polystyrene** **1991TSE**
Characterization: M_n/g.mol^{-1} = 17500, M_w/g.mol^{-1} = 18200, Pressure Chemical Company, Pittsburgh, PA
Solvent (A): **toluene** **C_7H_8** **108-88-3**
Polymer (C): **polyisoprene**
Characterization: M_n/g.mol^{-1} = 31300 (peak molecular weight), M_w/M_n = 1.05, Polymer Laboratories, Inc., Stow, OH

Type of data: coexistence data (tie lines)

Comments: The polymer ratio in the homogeneous feed phase is 1:1 by weight fraction.

T/K = 288.15

Total system		Top phase			Bottom phase		
w_A	$w_B + w_C$	w_A	w_B	w_C	w_A	w_B	w_C
0.7000	0.3000	0.7214	0.0190	0.2596	0.6642	0.3292	0.0066
0.7250	0.2750	0.7445	0.0263	0.2292	0.6862	0.3033	0.0105
0.7500	0.2500	0.7595	0.0392	0.2013	0.7107	0.2680	0.0213

Plait-point composition: w_A = 0.8129 + w_B = 0.1034 + w_C = 0.0837

T/K = 303.15

Total system		Top phase			Bottom phase		
w_A	$w_B + w_C$	w_A	w_B	w_C	w_A	w_B	w_C
0.7000	0.3000	0.7233	0.0198	0.2569	0.6568	0.3317	0.0115
0.7250	0.2750	0.7418	0.0322	0.2260	0.6951	0.2902	0.0147
0.7500	0.2500	0.7585	0.0449	0.1966	0.7189	0.2580	0.0231

Plait-point composition: w_A = 0.8017 + w_B = 0.1105 + w_C = 0.0878

continued

continued

T/K = 318.15

Total system		Top phase			Bottom phase		
w_A	$w_B + w_C$	w_A	w_B	w_C	w_A	w_B	w_C
0.700	0.300	0.7183	0.0234	0.2584	0.6548	0.3316	0.0136
0.725	0.275	0.7412	0.0393	0.2195	0.6904	0.2922	0.0174
0.750	0.250	0.7652	0.0466	0.1882	0.7092	0.2614	0.0294

Plait-point composition: w_A = 0.7630 + w_B = 0.1458 + w_C = 0.0912

Polymer (B): **polystyrene** **1991TSE**
Characterization: M_n/g.mol^{-1} = 17500, M_w/g.mol^{-1} = 18200,
Pressure Chemical Company, Pittsburgh, PA
Solvent (A): **toluene** C_7H_8 **108-88-3**
Polymer (C): **polyisoprene**
Characterization: M_n/g.mol^{-1} = 107900 (peak molecular weight), M_w/M_n = 1.06,
Polymer Laboratories, Inc., Stow, OH

Type of data: coexistence data (tie lines)

Comments: The polymer ratio in the homogeneous feed phase is 1:1 by weight fraction.

T/K = 288.15

Total system		Top phase			Bottom phase		
w_A	$w_B + w_C$	w_A	w_B	w_C	w_A	w_B	w_C
0.7500	0.2500	0.7684	0.0239	0.2077	0.7082	0.2918	0.0000
0.7750	0.2250	0.7907	0.0271	0.1822	0.7383	0.2617	0.0000
0.8000	0.2000	0.8146	0.0344	0.1510	0.7669	0.2322	0.0009

Plait-point composition: w_A = 0.8646 + w_B = 0.0925 + w_C = 0.0429

T/K = 303.15

Total system		Top phase			Bottom phase		
w_A	$w_B + w_C$	w_A	w_B	w_C	w_A	w_B	w_C
0.7500	0.2500	0.7624	0.0351	0.2025	0.7179	0.2821	0.0000
0.7750	0.2250	0.7920	0.0282	0.1798	0.7372	0.2609	0.0019
0.8000	0.2000	0.8115	0.0374	0.1511	0.7690	0.2220	0.0090

Plait-point composition: w_A = 0.8565 + w_B = 0.0914 + w_C = 0.0521

Polymer (B): **polystyrene** **1991TSE**
Characterization: M_n/g.mol^{-1} = 51130, M_w/g.mol^{-1} = 54200,
Pressure Chemical Company, Pittsburgh, PA
Solvent (A): **toluene** **C_7H_8** **108-88-3**
Polymer (C): **polyisoprene**
Characterization: M_n/g.mol^{-1} = 10800 (peak molecular weight), M_w/M_n = 1.07,
Polymer Laboratories, Inc., Stow, OH

Type of data: coexistence data (tie lines)

Comments: The polymer ratio in the homogeneous feed phase is 1:1 by weight fraction.

T/K = 288.15

Total system		Top phase			Bottom phase		
w_A	$w_B + w_C$	w_A	w_B	w_C	w_A	w_B	w_C
0.6000	0.4000	0.6183	0.0000	0.3817	0.5714	0.4077	0.0209
0.7000	0.3000	0.7284	0.0078	0.2638	0.6564	0.3079	0.0357
0.7500	0.2500	0.7716	0.0148	0.2136	0.7115	0.2326	0.0559

Plait-point composition: w_A = 0.8009 + w_B = 0.0837 + w_C = 0.1154

T/K = 303.15

Total system		Top phase			Bottom phase		
w_A	$w_B + w_C$	w_A	w_B	w_C	w_A	w_B	w_C
0.6000	0.4000	0.6195	0.0000	0.3805	0.5677	0.4040	0.0283
0.7000	0.3000	0.7224	0.0091	0.2685	0.6578	0.2949	0.0473
0.7500	0.2500	0.7741	0.0178	0.2081	0.7185	0.2167	0.0648

Plait-point composition: w_A = 0.7907 + w_B = 0.0877 + w_C = 0.1216

Polymer (B): **polystyrene** **1991TSE**
Characterization: M_n/g.mol^{-1} = 51130, M_w/g.mol^{-1} = 54200,
Pressure Chemical Company, Pittsburgh, PA
Solvent (A): **toluene** **C_7H_8** **108-88-3**
Polymer (C): **polyisoprene**
Characterization: M_n/g.mol^{-1} = 31300 (peak molecular weight), M_w/M_n = 1.05,
Polymer Laboratories, Inc., Stow, OH

Type of data: coexistence data (tie lines)

Comments: The polymer ratio in the homogeneous feed phase is 1:1 by weight fraction.

continued

continued

T/K = 288.15

Total system		Top phase			Bottom phase		
w_A	$w_B + w_C$	w_A	w_B	w_C	w_A	w_B	w_C
0.7500	0.2500	0.7785	0.0032	0.2183	0.7151	0.2805	0.0044
0.8000	0.2000	0.8184	0.0079	0.1737	0.7814	0.2186	0.0000
0.8250	0.1750	0.8407	0.0163	0.1430	0.8048	0.1756	0.0196

Plait-point composition: w_A = 0.8703 + w_B = 0.0593 + w_C = 0.0704

T/K = 303.15

Total system		Top phase			Bottom phase		
w_A	$w_B + w_C$	w_A	w_B	w_C	w_A	w_B	w_C
0.7500	0.2500	0.7763	0.0081	0.2156	0.7133	0.2812	0.0055
0.8000	0.2000	0.8180	0.0145	0.1675	0.7715	0.2190	0.0095
0.8250	0.1750	0.8382	0.0233	0.1385	0.8057	0.1699	0.0244

Plait-point composition: w_A = 0.8657 + w_B = 0.0652 + w_C = 0.0691

T/K = 318.15

Total system		Top phase			Bottom phase		
w_A	$w_B + w_C$	w_A	w_B	w_C	w_A	w_B	w_C
0.750	0.250	0.7733	0.0131	0.2136	0.7172	0.2738	0.0090
0.800	0.200	0.8175	0.0189	0.1656	0.7708	0.2161	0.0131
0.825	0.175	0.8374	0.0293	0.1334	0.8076	0.1649	0.0276

Plait-point composition: w_A = 0.8405 + w_B = 0.0861 + w_C = 0.0734

Polymer (B): **polystyrene** **1991TSE**
Characterization: M_n/g.mol^{-1} = 51130, M_w/g.mol^{-1} = 54200,
Pressure Chemical Company, Pittsburgh, PA
Solvent (A): **toluene** **C_7H_8** **108-88-3**
Polymer (C): **polyisoprene**
Characterization: M_n/g.mol^{-1} = 107900 (peak molecular weight), M_w/M_n = 1.06,
Polymer Laboratories, Inc., Stow, OH

Type of data: coexistence data (tie lines)

continued

continued

Comments: The polymer ratio in the homogeneous feed phase is 1:1 by weight fraction.

T/K = 288.15

Total system		Top phase			Bottom phase		
w_A	$w_B + w_C$	w_A	w_B	w_C	w_A	w_B	w_C
0.7500	0.2500	0.7642	0.0010	0.2348	0.7187	0.2813	0.0000
0.8000	0.2000	0.8153	0.0036	0.1811	0.7772	0.2228	0.0000
0.8500	0.1500	0.8628	0.0071	0.1301	0.8367	0.1597	0.0036

Plait-point composition: w_A = 0.9071 + w_B = 0.0507 + w_C = 0.0422

T/K = 318.15

Total system		Top phase			Bottom phase		
w_A	$w_B + w_C$	w_A	w_B	w_C	w_A	w_B	w_C
0.750	0.250	0.7612	0.0031	0.2357	0.7227	0.2773	0.0000
0.800	0.200	0.8123	0.0076	0.1802	0.7838	0.2122	0.0040
0.850	0.150	0.8567	0.0189	0.1264	0.8332	0.1581	0.0087

Plait-point composition: w_A = 0.8850 + w_B = 0.0673 + w_C = 0.0477

Polymer (B): **polystyrene** **1991TSE**
Characterization: M_n/g.mol^{-1} = 87770, M_w/g.mol^{-1} = 93040, Pressure Chemical Company, Pittsburgh, PA
Solvent (A): **toluene** C_7H_8 **108-88-3**
Polymer (C): **polyisoprene**
Characterization: M_n/g.mol^{-1} = 10800 (peak molecular weight), M_w/M_n = 1.07, Polymer Laboratories, Inc., Stow, OH

Type of data: coexistence data (tie lines)

Comments: The polymer ratio in the homogeneous feed phase is 1:1 by weight fraction.

T/K = 288.15

Total system		Top phase			Bottom phase		
w_A	$w_B + w_C$	w_A	w_B	w_C	w_A	w_B	w_C
0.7000	0.3000	0.7195	0.0000	0.2805	0.6624	0.3254	0.0122
0.7500	0.2500	0.7726	0.0044	0.2230	0.6948	0.2818	0.0234
0.7750	0.2250	0.7956	0.0126	0.1918	0.7455	0.2169	0.0376

continued

continued

Plait-point composition: $w_A = 0.8248 + w_B = 0.0787 + w_C = 0.0965$

T/K = 303.15

Total system		Top phase			Bottom phase		
w_A	$w_B + w_C$	w_A	w_B	w_C	w_A	w_B	w_C
0.7000	0.3000	0.7242	0.0070	0.2688	0.6473	0.3344	0.0183
0.7500	0.2500	0.7743	0.0085	0.2172	0.7121	0.2553	0.0326
0.7750	0.2250	0.7897	0.0162	0.1941	0.7416	0.2056	0.0528

Plait-point composition: $w_A = 0.8152 + w_B = 0.0765 + w_C = 0.1083$

Polymer (B): **polystyrene** **1991TSE**
Characterization: M_n/g.mol^{-1} = 87770, M_w/g.mol^{-1} = 93040,
Pressure Chemical Company, Pittsburgh, PA
Solvent (A): **toluene** **C_7H_8** **108-88-3**
Polymer (C): **polyisoprene**
Characterization: M_n/g.mol^{-1} = 31300 (peak molecular weight), M_w/M_n = 1.05,
Polymer Laboratories, Inc., Stow, OH

Type of data: coexistence data (tie lines)

Comments: The polymer ratio in the homogeneous feed phase is 1:1 by weight fraction.

T/K = 288.15

Total system		Top phase			Bottom phase		
w_A	$w_B + w_C$	w_A	w_B	w_C	w_A	w_B	w_C
0.7500	0.2500	0.7746	0.0000	0.2254	0.7100	0.2900	0.0000
0.8000	0.2000	0.8215	0.0011	0.1774	0.7631	0.2317	0.0052
0.8500	0.1500	0.8690	0.0048	0.1262	0.8266	0.1594	0.0140

Plait-point composition: $w_A = 0.8951 + w_B = 0.0517 + w_C = 0.0532$

T/K = 303.15

Total system		Top phase			Bottom phase		
w_A	$w_B + w_C$	w_A	w_B	w_C	w_A	w_B	w_C
0.7500	0.2500	0.7745	0.0000	0.2255	0.7156	0.2844	0.0000
0.8000	0.2000	0.8170	0.0017	0.1813	0.7617	0.2296	0.0087
0.8500	0.1500	0.8654	0.0064	0.1282	0.8237	0.1604	0.0159

continued

continued

Plait-point composition: $w_A = 0.8882 + w_B = 0.0526 + w_C = 0.0592$

T/K = 318.15

Total system		Top phase			Bottom phase		
w_A	$w_B + w_C$	w_A	w_B	w_C	w_A	w_B	w_C
0.750	0.250	0.7708	0.0008	0.2284	0.7123	0.2859	0.0021
0.775	0.225	0.7965	0.0007	0.2028	0.7410	0.2543	0.0047
0.800	0.200	0.8188	0.0027	0.2551	0.7665	0.2220	0.0115

Plait-point composition: $w_A = 0.8640 + w_B = 0.0683 + w_C = 0.0677$

Polymer (B): **polystyrene** **1993GE1**
Characterization: M_n/g.mol^{-1} = 91200, M_w/g.mol^{-1} = 94850,
Pressure Chemical Company, Pittsburgh, PA
Solvent (A): **toluene** **C_7H_8** **108-88-3**
Polymer (C): **polyisoprene**
Characterization: M_n/g.mol^{-1} = 10800, M_w/g.mol^{-1} = 12000,
18% 1,4-*trans*, 76% 1,4-*cis*, 6% 3,4-content, 0.5% antioxidant,
Goodyear Tire and Rubber Co., Akron, OH

Type of data: coexistence data (tie lines)

T/K = 303.15

Phase I			Phase II		
w_A	w_B	w_C	w_A	w_B	w_C
0.7859	0.0012	0.2129	0.7220	0.2472	0.0308
0.8074	0.0055	0.1871	0.7556	0.2129	0.0315

Polymer (B): **polystyrene** **1991TSE**
Characterization: M_n/g.mol^{-1} = 220500, M_w/g.mol^{-1} = 233750,
Pressure Chemical Company, Pittsburgh, PA
Solvent (A): **toluene** **C_7H_8** **108-88-3**
Polymer (C): **polyisoprene**
Characterization: M_n/g.mol^{-1} = 10800 (peak molecular weight), M_w/M_n = 1.07,
Polymer Laboratories, Inc., Stow, OH

Type of data: coexistence data (tie lines)

Comments: The polymer ratio in the homogeneous feed phase is 1:1 by weight fraction.

continued

continued

T/K = 288.15

Total system		Top phase			Bottom phase		
w_A	$w_B + w_C$	w_A	w_B	w_C	w_A	w_B	w_C
0.7500	0.2500	0.7673	0.0000	0.2327	0.7192	0.2553	0.0255
0.7750	0.2250	0.7970	0.0021	0.2009	0.7421	0.2256	0.0323
0.8000	0.2000	0.8115	0.0074	0.1811	0.7699	0.1864	0.0437

Plait-point composition: $w_A = 0.8475 + w_B = 0.0591 + w_C = 0.0934$

Polymer (B): **polystyrene** **1991TSE**
Characterization: M_n/g.mol^{-1} = 220500, M_w/g.mol^{-1} = 233750,
Pressure Chemical Company, Pittsburgh, PA
Solvent (A): **toluene** **C_7H_8** **108-88-3**
Polymer (C): **polyisoprene**
Characterization: M_n/g.mol^{-1} = 31300 (peak molecular weight), M_w/M_n = 1.05,
Polymer Laboratories, Inc., Stow, OH

Type of data: coexistence data (tie lines)

Comments: The polymer ratio in the homogeneous feed phase is 1:1 by weight fraction.

T/K = 288.15

Total system		Top phase			Bottom phase		
w_A	$w_B + w_C$	w_A	w_B	w_C	w_A	w_B	w_C
0.7500	0.2500	0.7712	0.0000	0.2288	0.7124	0.2876	0.0000
0.8500	0.1500	0.8656	0.0009	0.1335	0.8186	0.1769	0.0045
0.8750	0.1250	0.8916	0.0018	0.1066	0.8579	0.1334	0.0087

Plait-point composition: $w_A = 0.9163 + w_B = 0.0399 + w_C = 0.0438$

T/K = 303.15

Total system		Top phase			Bottom phase		
w_A	$w_B + w_C$	w_A	w_B	w_C	w_A	w_B	w_C
0.7500	0.2500	0.7723	0.0000	0.2277	0.7122	0.2878	0.0000
0.8500	0.1500	0.8652	0.0013	0.1335	0.8137	0.1824	0.0039
0.8750	0.1250	0.8907	0.0030	0.1063	0.8549	0.1347	0.0104

continued

continued

Plait-point composition: $w_A = 0.9147 + w_B = 0.0422 + w_C = 0.0431$

T/K = 318.15

Total system		Top phase			Bottom phase		
w_A	$w_B + w_C$	w_A	w_B	w_C	w_A	w_B	w_C
0.750	0.250	0.7711	0.0000	0.2289	0.7114	0.2886	0.0000
0.850	0.150	0.8625	0.0020	0.1355	0.8283	0.1633	0.0104
0.875	0.125	0.8885	0.0071	0.1044	0.8624	0.1174	0.0202

Plait-point composition: $w_A = 0.8930 + w_B = 0.0502 + w_C = 0.0568$

Polymer (B): **polystyrene** **1991TSE**
Characterization: M_n/g.mol^{-1} = 220500, M_w/g.mol^{-1} = 233750, Pressure Chemical Company, Pittsburgh, PA
Solvent (A): **toluene** **C_7H_8** **108-88-3**
Polymer (C): **polyisoprene**
Characterization: M_n/g.mol^{-1} = 107900 (peak molecular weight), M_w/M_n = 1.06, Polymer Laboratories, Inc., Stow, OH

Type of data: coexistence data (tie lines)

Comments: The polymer ratio in the homogeneous feed phase is 1:1 by weight fraction.

T/K = 288.15

Total system		Top phase			Bottom phase		
w_A	$w_B + w_C$	w_A	w_B	w_C	w_A	w_B	w_C
0.7500	0.2500	0.7747	0.0000	0.2253	0.7203	0.2797	0.0000
0.8500	0.1500	0.8669	0.0000	0.1331	0.8218	0.1782	0.0000
0.9000	0.1000	0.9089	0.0014	0.0897	0.8870	0.1098	0.0032

Plait-point composition: $w_A = 0.9521 + w_B = 0.02179 + w_C = 0.0262$

T/K = 318.15

Total system		Top phase			Bottom phase		
w_A	$w_B + w_C$	w_A	w_B	w_C	w_A	w_B	w_C
0.750	0.250	0.7684	0.0000	0.2316	0.7165	0.2835	0.0000
0.850	0.150	0.8647	0.0000	0.1353	0.8242	0.1751	0.0007
0.900	0.100	0.9061	0.0032	0.0908	0.8920	0.1051	0.0029

continued

continued

Plait-point composition: $w_A = 0.9328 + w_B = 0.0340 + w_C = 0.0332$

Polymer (B): **polystyrene** **1956KER**
Characterization: $\rho = 1.05$ g/cm^3 at 298.15 K
Solvent (A): **toluene** C_7H_8 **108-88-3**
Polymer (C): **poly(methyl methacrylate)**
Characterization: $\rho = 1.18$ g/cm^3 at 298.15 K

Type of data: cloud points (threshold point)

T/K = 298.15

φ_A 0.923 φ_B 0.0385 φ_C 0.0385

Polymer (B): **polystyrene** **1985ROB, 1987TS2**
Characterization: M_n/g.mol^{-1} = 18080, M_w/g.mol^{-1} = 19890, Pressure Chemical Company, Pittsburgh, PA
Solvent (A): **trichloromethane** $CHCl_3$ **67-66-3**
Polymer (C): **polybutadiene**
Characterization: M_n/g.mol^{-1} = 33250, M_w/g.mol^{-1} = 36400, 37.5% 1,4-*cis*, 51.1% 1,4-*trans*, 11.4% 1,2-*vinyl*

Type of data: coexistence data (tie lines)

Comments: The polymer ratio in the homogeneous feed phase is 1:1 by weight fraction.

T/K = 298.15

Total system		Top phase			Bottom phase		
w_A	$w_B + w_C$	φ_A	φ_B	φ_C	φ_A	φ_B	φ_C
0.7869	0.2131	0.7405	0.0211	0.2384	0.6827	0.3133	0.0040
0.8029	0.1971	0.7562	0.0301	0.2137	0.7072	0.2878	0.0050
0.8404	0.1596	0.8046	0.0533	0.1421	0.7567	0.2317	0.0116

Type of data: cloud points (plait-point composition)

T/K = 298.15

w_A 0.8600 w_B 0.1098 w_C 0.0302
φ_A 0.8084 φ_B 0.1449 φ_C 0.0467

Polymer (B): **polystyrene** **1985ROB, 1987TS2**
Characterization: M_n/g.mol^{-1} = 18080, M_w/g.mol^{-1} = 19890, Pressure Chemical Company, Pittsburgh, PA

continued

continued

Solvent (A): **trichloromethane** $CHCl_3$ **67-66-3**
Polymer (C): **polybutadiene**
Characterization: M_n/g.mol^{-1} = 104040, M_w/g.mol^{-1} = 114360,
35.9% 1,4-*cis*, 51.5% 1,4-*trans*, 12.6% 1,2-*vinyl*

Type of data: coexistence data (tie lines)

Comments: The polymer ratio in the homogeneous feed phase is 1:1 by weight fraction.

T/K = 298.15

Total system		Top phase			Bottom phase		
w_A	$w_B + w_C$	φ_A	φ_B	φ_C	φ_A	φ_B	φ_C
0.7963	0.2037	0.7368	0.0203	0.2429	0.7080	0.2897	0.0023
0.8186	0.1814	0.7688	0.0310	0.2002	0.7334	0.2594	0.0072
0.8572	0.1428	0.8174	0.0505	0.1321	0.7869	0.2058	0.0073

Type of data: cloud points (plait-point composition)

T/K = 298.15

w_A	0.8787	w_B	0.0898	w_C	0.0315
φ_A	0.8317	φ_B	0.1193	φ_C	0.0490

Polymer (B): **polystyrene** **1985ROB, 1987TS2**
Characterization: M_n/g.mol^{-1} = 18080, M_w/g.mol^{-1} = 19890,
Pressure Chemical Company, Pittsburgh, PA
Solvent (A): **trichloromethane** $CHCl_3$ **67-66-3**
Polymer (C): **polybutadiene**
Characterization: M_n/g.mol^{-1} = 223080, M_w/g.mol^{-1} = 275090,
40.8% 1,4-*cis*, 47.4% 1,4-*trans*, 11.8% 1,2-*vinyl*

Type of data: coexistence data (tie lines)

Comments: The polymer ratio in the homogeneous feed phase is 1:1 by weight fraction.

T/K = 298.15

Total system		Top phase			Bottom phase		
w_A	$w_B + w_C$	φ_A	φ_B	φ_C	φ_A	φ_B	φ_C
0.8418	0.1582	0.8218	0.0159	0.1627	0.7412	0.2587	0.0001
0.8637	0.1363	0.8474	0.0379	0.1147	0.7767	0.2232	0.0001
0.8762	0.1238	0.8543	0.0691	0.0766	0.8057	0.1942	0.0001

continued

continued

Type of data: cloud points (plait-point composition)

T/K = 298.15

w_A	0.8860	w_B	0.09905	w_C	0.01495
φ_A	0.8441	φ_B	0.1325	φ_C	0.0234

Polymer (B): **polystyrene** **1985ROB, 1987TS2**
Characterization: M_n/g.mol^{-1} = 94380, M_w/g.mol^{-1} = 103470,
Pressure Chemical Company, Pittsburgh, PA
Solvent (A): **trichloromethane** **$CHCl_3$** **67-66-3**
Polymer (C): **polybutadiene**
Characterization: M_n/g.mol^{-1} = 33250, M_w/g.mol^{-1} = 36400,
37.5% 1,4-*cis*, 51.1% 1,4-*trans*, 11.4% 1,2-*vinyl*

Type of data: coexistence data (tie lines)

Comments: The polymer ratio in the homogeneous feed phase is 1:1 by weight fraction.

T/K = 298.15

Total system		Top phase			Bottom phase		
w_A	$w_B + w_C$	φ_A	φ_B	φ_C	φ_A	φ_B	φ_C
0.8795	0.1205	0.858	0.004	0.138	0.801	0.191	0.0080
0.8822	0.1178	0.861	0.004	0.135	0.806	0.186	0.0080
0.9176	0.0824	0.895	0.007	0.098	0.867	0.125	0.0080

Type of data: cloud points (plait-point composition)

T/K = 298.15

w_A	0.9375	w_B	0.0300	w_C	0.0325
φ_A	0.9075	φ_B	0.0408	φ_C	0.0517

Polymer (B): **polystyrene** **1985ROB, 1987TS2**
Characterization: M_n/g.mol^{-1} = 94380, M_w/g.mol^{-1} = 103470,
Pressure Chemical Company, Pittsburgh, PA
Solvent (A): **trichloromethane** **$CHCl_3$** **67-66-3**
Polymer (C): **polybutadiene**
Characterization: M_n/g.mol^{-1} = 104040, M_w/g.mol^{-1} = 114360,
35.9% 1,4-*cis*, 51.5% 1,4-*trans*, 12.6% 1,2-*vinyl*

Type of data: coexistence data (tie lines)

Comments: The polymer ratio in the homogeneous feed phase is 1:1 by weight fraction.

T/K = 298.15

continued

continued

Total system		Top phase			Bottom phase		
w_A	$w_B + w_C$	φ_A	φ_B	φ_C	φ_A	φ_B	φ_C
0.8781	0.1219	0.845	0.005	0.150	0.809	0.189	0.002
0.8884	0.1116	0.859	0.011	0.130	0.824	0.171	0.005
0.9013	0.0987	0.878	0.004	0.118	0.839	0.160	0.001

Type of data: cloud points (plait-point composition)

T/K = 298.15

w_A	0.9375	w_B	0.0400	w_C	0.0225
φ_A	0.9096	φ_B	0.0545	φ_C	0.0359

Polymer (B): **polystyrene** **1985ROB, 1987TS2**
Characterization: M_n/g.mol^{-1} = 94380, M_w/g.mol^{-1} = 103470,
Pressure Chemical Company, Pittsburgh, PA
Solvent (A): **trichloromethane** **$CHCl_3$** **67-66-3**
Polymer (C): **polybutadiene**
Characterization: M_n/g.mol^{-1} = 223080, M_w/g.mol^{-1} = 275090,
40.8% 1,4-*cis*, 47.4% 1,4-*trans*, 11.8% 1,2-*vinyl*

Type of data: coexistence data (tie lines)

Comments: The polymer ratio in the homogeneous feed phase is 1:1 by weight fraction.

T/K = 298.15

Total system		Top phase			Bottom phase		
w_A	$w_B + w_C$	φ_A	φ_B	φ_C	φ_A	φ_B	φ_C
0.8670	0.1330	0.855	0.001	0.144	0.774	0.226	0.0001
0.8805	0.1195	0.876	0.001	0.123	0.789	0.211	0.0001
0.9002	0.0998	0.890	0.003	0.107	0.827	0.165	0.008

Type of data: cloud points (plait-point composition)

T/K = 298.15

w_A	0.9382	w_B	0.0377	w_C	0.0241
φ_A	0.9103	φ_B	0.0514	φ_C	0.0383

Polymer (B): **polystyrene** **1985ROB, 1987TS2**
Characterization: M_n/g.mol^{-1} = 260900, M_w/g.mol^{-1} = 295770,
Pressure Chemical Company, Pittsburgh, PA
Solvent (A): **trichloromethane** **$CHCl_3$** **67-66-3**
Polymer (C): **polybutadiene**
Characterization: M_n/g.mol^{-1} = 33250, M_w/g.mol^{-1} = 36400,
37.5% 1,4-*cis*, 51.1% 1,4-*trans*, 11.4% 1,2-*vinyl*

Type of data: coexistence data (tie lines)

Comments: The polymer ratio in the homogeneous feed phase is 1:1 by weight fraction.

T/K = 298.15

Total system		Top phase			Bottom phase		
w_A	$w_B + w_C$	φ_A	φ_B	φ_C	φ_A	φ_B	φ_C
0.8723	0.1277	0.856	0.001	0.144	0.784	0.203	0.013
0.8904	0.1096	0.873	0.001	0.127	0.816	0.183	0.001
0.9154	0.0846	0.901	0.001	0.098	0.855	0.130	0.015

Type of data: cloud points (plait-point composition)

T/K = 298.15

w_A	0.9385	w_B	0.0320	w_C	0.0295
φ_A	0.9095	φ_B	0.0435	φ_C	0.0470

Polymer (B): **polystyrene** **1985ROB, 1987TS2**
Characterization: M_n/g.mol^{-1} = 260900, M_w/g.mol^{-1} = 295770
Solvent (A): **trichloromethane** **$CHCl_3$** **67-66-3**
Polymer (C): **polybutadiene**
Characterization: M_n/g.mol^{-1} = 104040, M_w/g.mol^{-1} = 114360,
35.9% 1,4-*cis*, 51.5% 1,4-*trans*, 12.6% 1,2-*vinyl*

Type of data: coexistence data (tie lines)

Comments: The polymer ratio in the homogeneous feed phase is 1:1 by weight fraction.

T/K = 298.15

Total system		Top phase			Bottom phase		
w_A	$w_B + w_C$	φ_A	φ_B	φ_C	φ_A	φ_B	φ_C
0.8823	0.1177	0.852	0.002	0.146	0.814	0.186	0.0001
0.8987	0.1013	0.872	0.003	0.125	0.839	0.156	0.0050
0.9099	0.0901	0.886	0.004	0.110	0.856	0.143	0.0010

continued

continued

Type of data: cloud points (plait-point composition)

T/K = 298.15

w_A	0.9405	w_B	0.0372	w_C	0.0223
φ_A	0.9137	φ_B	0.0507	φ_C	0.0356

Polymer (B): **polystyrene** **1985ROB, 1987TS2**
Characterization: M_n/g.mol^{-1} = 260900, M_w/g.mol^{-1} = 295770, Pressure Chemical Company, Pittsburgh, PA
Solvent (A): **trichloromethane** **$CHCl_3$** **67-66-3**
Polymer (C): **polybutadiene**
Characterization: M_n/g.mol^{-1} = 223080, M_w/g.mol^{-1} = 275090, 40.8% 1,4-*cis*, 47.4% 1,4-*trans*, 11.8% 1,2-*vinyl*

Type of data: coexistence data (tie lines)

Comments: The polymer ratio in the homogeneous feed phase is 1:1 by weight fraction.

T/K = 298.15

Total system		Top phase			Bottom phase		
w_A	$w_B + w_C$	φ_A	φ_B	φ_C	φ_A	φ_B	φ_C
0.8834	0.1166	0.847	0.001	0.152	0.820	0.171	0.009
0.9149	0.0851	0.889	0.004	0.107	0.866	0.134	0.0001
0.9430	0.0570	0.931	0.001	0.068	0.903	0.097	0.0001

Type of data: cloud points (plait-point composition)

T/K = 298.15

w_A	0.9607	w_B	0.0270	w_C	0.0123
φ_A	0.9430	φ_B	0.0372	φ_C	0.0198

Polymer (B): **polystyrene** **1956KER**
Characterization: ρ = 1.05 g/cm^3 at 298.15 K
Solvent (A): **trichloromethane** **$CHCl_3$** **67-66-3**
Polymer (C): **poly(methyl methacrylate)**
Characterization: ρ = 1.18 g/cm^3 at 298.15 K

Type of data: cloud points (threshold point)

T/K = 298.15

φ_A	0.906	φ_B	0.047	φ_C	0.047

Polymer (B): **polystyrene** **1956KER**
Characterization: ρ = 1.05 g/cm^3 at 298.15 K
Solvent (A): **trichloromethane** **$CHCl_3$** **67-66-3**
Polymer (C): **poly(2-methylstyrene)**
Characterization: ρ = 1.04 g/cm^3 at 298.15 K

Type of data: cloud points (threshold point)

T/K = 298.15

φ_A	0.855	φ_B	0.0725	φ_C	0.0725

Polymer (B): **polystyrene** **1956KER**
Characterization: ρ = 1.05 g/cm^3 at 298.15 K
Solvent (A): **trichloromethane** **$CHCl_3$** **67-66-3**
Polymer (C): **poly(3-methylstyrene)**
Characterization: ρ = 1.05 g/cm^3 at 298.15 K

Type of data: cloud points (threshold point)

T/K = 298.15

φ_A	0.892	φ_B	0.054	φ_C	0.054

Polymer (B): **polystyrene** **1956KER**
Characterization: ρ = 1.05 g/cm^3 at 298.15 K
Solvent (A): **trichloromethane** **$CHCl_3$** **67-66-3**
Polymer (C): **poly(4-methylstyrene)**
Characterization: ρ = 1.02 g/cm^3 at 298.15 K

Type of data: cloud points (threshold point)

T/K = 298.15

φ_A	0.925	φ_B	0.0375	φ_C	0.0375

Polymer (B): **poly(styrene-*co*-acrylonitrile)** **2001POS**
Characterization: M_n/g.mol^{-1} = 70000, M_w/g.mol^{-1} = 320000,
35.0 mol% acrylonitrile, ρ = 1.0347 g/cm^3 at 303.15 K,
BASF Ludwigshafen, Germany
Solvent (A): ***N,N*-dimethylacetamide** **C_4H_9NO** **127-19-5**
Polymer (C): **poly(acrylonitrile)**
Characterization: M_w/g.mol^{-1} = 240000, ρ = 1.175 g/cm^3 at 303.15 K,
Acrids Kehlheim GmbH, Germany

Type of data: cloud points

T/K = 298.15

w_A	0.9587	0.9629	0.9686	0.9633	0.9555	0.9674	0.95096
w_B	0.0288	0.0149	0.0094	0.0220	0.0356	0.0076	0.0441
w_C	0.0125	0.0222	0.0220	0.0147	0.0089	0.0250	0.00494

Polymer (B): **poly(styrene-*co*-acrylonitrile)** **2001POS**
Characterization: M_n/g.mol^{-1} = 70000, M_w/g.mol^{-1} = 320000, 35.0 mol% acrylonitrile, ρ = 1.0347 g/cm^3 at 303.15 K, BASF Ludwigshafen, Germany
Solvent (A): ***N,N*-dimethylacetamide** **C_4H_9NO** **127-19-5**
Polymer (C): **polystyrene**
Characterization: M_n/g.mol^{-1} = 90000, M_w/g.mol^{-1} = 100000, ρ = 1.047 g/cm^3 at 303.15 K, Degussa-Huels, Germany

Type of data: cloud points

T/K = 298.15

w_A	0.9168	0.9155	0.9155	0.91802	0.91651	0.9145	0.9130	0.9106	0.90097
w_B	0.0414	0.0508	0.0590	0.07377	0.06678	0.0370	0.0271	0.01944	0.01493
w_C	0.0418	0.0337	0.0255	0.00821	0.01671	0.0490	0.0599	0.06996	0.0841

Polymer (B): **poly(styrene-*co*-acrylonitrile)** **2002LOS**
Characterization: M_n/g.mol^{-1} = 88600, M_w/g.mol^{-1} = 195000, 48.0 mol% acrylonitrile, BASF Ludwigshafen, Germany
Solvent (A): ***N,N*-dimethylacetamide** **C_4H_9NO** **127-19-5**
Polymer (C): **polystyrene**
Characterization: M_n/g.mol^{-1} = 189500, M_w/g.mol^{-1} = 195000, Polymer Standards, Mainz, Germany

Type of data: cloud points

T/K = 293.15

w_A	0.8011	0.9503	0.9541	0.9510	0.8534	0.9323	0.9467
w_B	0.0014	0.0037	0.0196	0.0306	0.1427	0.0596	0.0381
w_C	0.1974	0.0460	0.0263	0.0184	0.0040	0.0081	0.0152

Type of data: critical point

$w_{A, crit}$ = 0.953 $w_{B, crit}$ = 0.024 $w_{C, crit}$ = 0.023

Polymer (B): **poly(vinyl acetate)** **1955KER**
Solvent (A): **ethyl acetate** **$C_4H_8O_2$** **141-78-6**
Polymer (C): **poly(methyl vinyl ketone)**

Type of data: coexistence data (tie lines)

T/K = 298.15

Total system			Top phase	Bottom phase
w_A	w_B	w_C	w_B	w_B
0.80	0.10	0.10	0.680	0.045

Polymer (B): **poly(vinyl acetate)** **1955KER**
Solvent (A): **2-propanone** C_3H_6O **67-64-1**
Polymer (C): **poly(methyl vinyl ketone)**

Type of data: coexistence data (tie lines)

T/K = 298.15

Total system			Top phase	Bottom phase
w_A	w_B	w_C	w_B	w_B
0.80	0.10	0.10	0.145	0.780

Polymer (B): **poly(vinyl acetate-*co*-vinyl alcohol)** **2004PER**
Characterization: M_n/g.mol^{-1} = 10000, 12.0 mol% vinyl acetate, PVA 10000, 88% hydrolyzed, Scientific Polymer Products, Inc., Ontario, NY
Solvent (A): **water** H_2O **7732-18-5**
Polymer (C): **poly(ethylene glycol)**
Characterization: M_n/g.mol^{-1} = 8000

Type of data: coexistence data (tie lines)

T/K = 298.15

Total system			Top phase			Bottom phase		
w_A	w_B	w_C	w_A	w_B	w_C	w_A	w_B	w_C
0.8640	0.0430	0.0930	0.8780	0.0620	0.0600	0.8540	0.0290	0.1170
0.8530	0.0480	0.0990	0.8810	0.0810	0.0380	0.8300	0.0200	0.1500
0.8420	0.0530	0.1050	0.8770	0.0930	0.0300	0.8100	0.0150	0.1750
0.8210	0.0600	0.1190	0.8650	0.1110	0.0240	0.7790	0.0130	0.2080

Plait-point composition: w_A = 0.8790 + w_B = 0.0800 + w_C = 0.0410

3.3. Table of ternary systems where data were published only in graphical form as phase diagrams or related figures

Polymer (B)	Second and third component	Ref.
Agarose		
	dextran and water	1993MED
	poly(ethylene glycol) and water	1993MED
Amylose		
	amylopectin and water	1987KAL
	dextran and water	1986KAL
	water and xanthan	2004MAN
Amylopectin		
	amylose and water	1987KAL
	gelatine and water	1993DUR
Benzoyl dextran		
	dextran and water	1995LUM
	poly(ethylene glycol) and water	1994LUM
	poly(ethylene oxide-*co*-propylene oxide) and water	1996LUM
	valeryl dextran and water	1994LUM
Cellulose		
	aramide and *N,N*-dimethylacetamide	1992KAM
	dextran and water	2002EDG
	N,N-dimethylacetamide and 2-propanone	2005ECK
Cellulose acetate		
	acetic acid and water	1986AN2
	N,N-dimethylacetamide and poly(ethylene glycol)	1986AN1
	N,N-dimethylformamide and 2-methyl-2,4-pentanediol	2000MA2
	1,4-dioxane and water	1980BRO
	1,4-dioxane and water	1986REU
	1,4-dioxane and water	1989RON
	1,4-dioxane and water	1990BOO

Polymer (B)	Second and third component	Ref.
Cellulose acetate		
	2-ethyl-1,3-hexanediol and 2-methyl-2,4-pentanediol	2003MA3
	methyl acetate and 2-propanol	2002LOS
	polystyrene and tetrahydrofuran	2003SIL
	poly(vinyl acetal) and 2-propanone	1947DOB
	2-propanone and water	1980LEM
	2-propanone and water	1986AN2
	2-propanone and water	1986REU
	2-propanone and water	1986VS2
	2-propanone and water	1987REU
	2-propanone and water	1989VSH
	2-propanone and water	1990BOO
	2-propanone and water	1999BAT
	tetrahydrofuran and water	1986REU
	tetrahydrofuran and water	1990BOO
Cellulose diacetate		
	poly(propylene glycol) adipate and triacetin	1992SUV
	2-propanone and water	1986VS2
	2-propanone and water	1989VSH
Cellulose nitrate		
	2-butanone and polystyrene	1947DOB
	1,2-ethanediol and 2-(2-ethoxyethoxy)ethanol	1988TAG
o-Cresol-formaldehyde resin		
	cyclohexane and 2-propanone	1992YAM
Dextran		
	agarose and water	1993MED
	amylose and water	1986KAL
	benzoyl dextran and water	1995LUM
	bovine serum albumin and water	2006ANT
	cellulose and water	2002EDG
	ethanol and water	2002LAA
	ethyl(hydroxyethyl)cellulose and water	1958ALB
	ficoll and water	1980ZAS
	ficoll and water	1986ZAS
	ficoll and water	1987ZA3
	ficoll and water	1988ZAS
	gelatine and water	1970GRI
	gelatine and water	1972GRI

Polymer (B)	**Second and third component**	**Ref.**
Dextran		
	gelatine and water	1991TOL
	gelatine and water	1997VIN
	gelatine and water	2001AND
	gelatine and water	2001EDE
	gelatine and water	2004ANT
	gelatine and water	2005LUN
	methylcellulose and water	1958ALB
	methylcellulose and water	1980POL
	methyl(hydroxypropyl)cellulose and water	1984HEF
	octa(ethylene oxide) dodecyl ether and water	1996PIC
	penta(ethylene oxide) dodecyl ether and water	1996BE1
	penta(ethylene oxide) dodecyl ether and water	1996PIC
	poly(ethylene glycol) and water	1958ALB
	poly(ethylene glycol) and water	1968EDM
	poly(ethylene glycol) and water	1981SHI
	poly(ethylene glycol) and water	1984BAM
	poly(ethylene glycol) and water	1984HEF
	poly(ethylene glycol) and water	1986GUS
	poly(ethylene glycol) and water	1986ZAS
	poly(ethylene glycol) and water	1987ANA
	poly(ethylene glycol) and water	1987BAS
	poly(ethylene glycol) and water	1987KAN
	poly(ethylene glycol) and water	1987SJO
	poly(ethylene glycol) and water	1987ZA3
	poly(ethylene glycol) and water	1988ABB
	poly(ethylene glycol) and water	1988KAN
	poly(ethylene glycol) and water	1988MAK
	poly(ethylene glycol) and water	1988ZAS
	poly(ethylene glycol) and water	1989HA1
	poly(ethylene glycol) and water	1989HA2
	poly(ethylene glycol) and water	1989ZAS
	poly(ethylene glycol) and water	1990CAB
	poly(ethylene glycol) and water	1990RAG
	poly(ethylene glycol) and water	1993GA2
	poly(ethylene glycol) and water	1993HAR
	poly(ethylene glycol) and water	1993HAY
	poly(ethylene glycol) and water	1994HAR
	poly(ethylene glycol) and water	1995DOE
	poly(ethylene glycol) and water	1996CES
	poly(ethylene glycol) and water	1997SAR
	poly(ethylene glycol) and water	1998NIL
	poly(ethylene glycol) and water	2002ZAV
	poly(ethylene glycol) and water	2004HOP

Polymer (B)	Second and third component	Ref.
Dextran		
	poly(ethylene glycol) and water	2004KAV
	poly(ethylene glycol) and water	2005MOO
	poly(ethylene glycol) and water	2006EDA
	poly(ethylene glycol) monododecyl ether and water	2000SIV
	poly(ethylene oxide) and water	1998NIL
	poly(ethylene oxide) and water	2003EDE
	poly(ethylene oxide) and water	2005OL1
	poly(ethylene oxide-*co*-propylene oxide) and water	1986ALB
	poly(ethylene oxide-*co*-propylene oxide) and water	1994ALR
	poly(ethylene oxide-*co*-propylene oxide) and water	1998PLA
	poly(ethylene oxide-*co*-propylene oxide) and water	1999BER
	poly(ethylene oxide-*co*-propylene oxide) and water	2005OL1
	poly(*N*-isopropylacrylamide) and water	1999KIS
	poly(oxyethylene) sorbitan-monolaurate and water	2000SIV
	poly(oxyethylene) sorbitan-monooleate and water	2000SIV
	poly(vinyl alcohol) and water	1958ALB
	poly(vinyl alcohol) and water	1986ZAS
	poly(vinyl alcohol) and water	1989ZAS
	poly(*N*-vinylcaprolactam-*co*-1-vinylimidazole) and water	1997FRA
	poly(*N*-vinylisobutyramide) and water	1999KIS
	poly(vinyl methyl ether) and water	1984HEF
	poly(1-vinyl-2-pyrrolidinone) and water	1958ALB
	poly(1-vinyl-2-pyrrolidinone) and water	1984HEF
	poly(1-vinyl-2-pyrrolidinone) and water	1986ZAS
	poly(1-vinyl-2-pyrrolidinone) and water	1987ZA3
	poly(1-vinyl-2-pyrrolidinone) and water	1988ZAS
	poly(1-vinyl-2-pyrrolidinone) and water	1989ZAS
	valeryl dextran and water	1995LUM
Ethyl(hydroxyethyl)cellulose		
	1-butanol and water	1990KAR
	1-butanol and water	1996THU
	butanoic acid and water	1997JOH
	dextran and water	1958ALB
	ethanol and water	1990KAR
	ethanol and water	1996THU
	1-heptanol and water	1990KAR
	1-hexanol and water	1990KAR
	1-hexanol and water	1995THU
	1-hexanol and water	1996THU
	methanol and water	1990KAR

Polymer (B)	Second and third component	Ref.
Ethyl(hydroxyethyl)cellulose		
	methanol and water	1996THU
	octa(ethylene glycol) decyl ether and water	2005OL2
	octa(ethylene glycol) dodecyl ether and water	1990KAR
	octa(ethylene glycol) dodecyl ether and water	1992ZHA
	octa(ethylene glycol) dodecyl ether and water	1994ZH1
	penta(ethylene glycol) dodecyl ether and water	1990KAR
	1-pentanol and water	1990KAR
	1-pentanol and water	1996THU
	phenol and water	1997JOH
	1-propanol and water	1990KAR
	1-propanol and water	1996THU
	propanoic acid and water	1997JOH
	tetra(ethylene glycol) dodecyl ether and water	1992ZHA
	tetra(ethylene glycol) dodecyl ether and water	1994ZH1
	tetra(ethylene glycol) dodecyl ether and water	1994ZH2
	Triton X-100 and water	2005OL2
	Triton X-100 and water	2005OL2
Ficoll		
	dextran and water	1980ZAS
	dextran and water	1986ZAS
	dextran and water	1987ZA3
	dextran and water	1988ZAS
Gelatine		
	amylopectin and water	1993DUR
	tert-butanol and water	2005GUP
	dextran and water	1970GRI
	dextran and water	1972GRI
	dextran and water	1991TOL
	dextran and water	1997VIN
	dextran and water	2001AND
	dextran and water	2001EDE
	dextran and water	2004ANT
	dextran and water	2005LUN
	ethanol and water	2005GUP
	maltodextrin and water	2001BUT
	maltodextrin and water	2001LOR
	methanol and water	2005GUP
	methylcellulose and water	1984GRI
	methylcellulose and water	1991TOL
	polysaccharide and water	1997VIN
	1-propanol and water	2005GUP

Polymer (B)	Second and third component	Ref.
Hexa(ethylene glycol) mono dodecyl ether		
	poly(ethylene glycol) and water	2005HO1
	poly(ethylene glycol) and water	2005HO2
Hydroxyethylcellulose		
	poly(ethyleneimine) and water	1993DIS
	poly(ethyleneimine) and water	1994DIS
Hydroxypropylcellulose		
	N,N-dimethylacetamide and poly(vinylidenefluoride)	1992AMB
	poly(acrylic acid) and water	2002LUS
	poly(maleic acid-*co*-acrylic acid) and water	2005BUM
	poly(maleic acid-*co*-styrene) and water	2005BUM
	poly(maleic acid-*co*-vinyl acetate) and water	2005BUM
	1,2,3-propanetriol and water	2001FUJ
	1,2,3-propanetriol and water	2002FUJ
Hydroxypropylstarch		
	poly(ethylene glycol) and water	1986MAT
	poly(ethylene glycol) and water	1986TJE
	poly(ethylene glycol) and water	1989LIN
	poly(ethylene glycol) and water	1996CES
	poly(ethylene glycol) and water	1998ALM
	poly(ethylene oxide-*co*-propylene oxide) and water	1994MOD
	poly(ethylene oxide-*co*-propylene oxide) and water	1997CUN
	poly(ethylene oxide-*co*-propylene oxide) and water	2000PE2
Maltodextrin		
	dextran and water	2001BUT
	gelatine and water	2001LOR
	poly(ethylene glycol) and water	1988SZL
	poly(ethylene glycol) and water	1991RAG
	poly(ethylene glycol) and water	1994ATK
	poly(ethylene glycol) and water	2000SRI
Methylcellulose		
	dextran and water	1958ALB
	dextran and water	1980POL
	gelatine and water	1984GRI
	gelatine and water	1991TOL

Polymer (B)	Second and third component	Ref.
Methyl(hydroxypropyl)cellulose		
	dextran and water	1984HEF
	poly(vinyl alcohol) and water	1993SAK
Nylon-4,6		
	formic acid and water	1996BUL
Nylon-6		
	formic acid and water	1994CHE
	formic acid and water	1996BUL
	poly(p-phenylene-1,3,4-oxadiazol) and sulfuric acid	1992KAM
	poly(p-phenylene terephthalamide) and sulfuric acid	1992KAM
Nylon-6,6		
	formic acid and water	1994CHE
	formic acid and water	1995CH1
	formic acid and water	1995CH2
Nylon-6,10		
	formic acid and water	1994CHE
Octa(ethylene glycol) monododecyl ether		
	poly(ethylene oxide-*co*-propylene oxide) and water	1992ZHA
	poly(ethylene oxide-*co*-propylene oxide) and water	1994ZH1
	ethyl(hydroxyethyl)cellulose and water	1992ZHA
	ethyl(hydroxyethyl)cellulose and water	1994ZH1
Pectin		
	methanol and water	2006THO
Penta(ethylene glycol) monoheptyl ether	n-dodecane and water	1992SAS
Phenol-formaldehyde resin		
	cyclohexane and 2-propanone	1998YA1
	n-hexane and 2-propanone	1998YA1
Phenoxy		
	1,2-dichloroethane and copolyester	1978AHA
	1,1,2,2-tetrachloroethane and copolyester	1978AHA

Polymer (B)	Second and third component	Ref.
Poly(acrylamide)		
	1,4-dioxane and water	2006DAL
	methanol and water	2002NOW
	poly(acrylamide-*co*-*N,N*-dihexylacrylamide) and water	2000JIM
	poly(ethylene glycol) and water	1984HEF
	poly(ethylene glycol) and water	1989PER
	poly(vinyl acetate-*co*-vinyl alcohol) and water	1984HEF
	poly(vinyl alcohol) and water	1982MIN
	poly(vinyl methyl ether) and water	1984HEF
	poly(1-vinyl-2-pyrrolidinone) and water	1989PER
Poly(acrylamide-*co*-*N*-benzyl-acrylamide)		
	poly(ethylene glycol) and water	2003ABU
Poly(acrylamide-*co*-*N,N*-dihexyl-acrylamide)		
	polyacrylamide and water	2000JIM
Poly(acrylamide-*co*-*N*-isopropyl-acrylamide)		
	1,4-dioxane and water	2006DAL
Poly(acrylamide-*co*-4-methoxy styrene)		
	poly(ethylene glycol) and water	2003ABU
Poly(acrylamide-*co*-*N*-phenyl-acrylamide)		
	poly(ethylene glycol) and water	2003ABU
Poly(acrylic acid)		
	hydroxypropylcellulose and water	2002LUS
	2,6-dimethylpyridine and water	2001TOK
	poly(ethylene glycol) and water	2002GUP
Poly(acrylic acid-*co*-2-methyl-5-vinylpyridine)		
	diethyl ether and methanol	1985VED

Polymer (B)	**Second and third component**	**Ref.**
Poly(acrylonitrile)		
	N,N-dimethylacetamide and polystyrene	2001POS
	N,N-dimethylacetamide and polystyrene	2002KUL
	N,N-dimethylacetamide and poly(styrene-*co*-acrylonitrile)	2001POS
	N,N-dimethylacetamide and poly(styrene-*co*-acrylonitrile)	2002KUL
	N,N-dimethylformamide and propylene carbonate	1978UGL
Poly(acrylonitrile-*co*-butadiene)		
	ethyl acetate and poly(butadiene-*co*-α-methylstyrene)	2002VSH
	ethyl acetate and poly(butadiene-*co*-α-methylstyrene)	2004VSH
Polyamic acid		
	1,4-dioxane and water	1985ZHU
	tetrahydrofuran and water	1985ZHU
Polyarylate		
	n-hexane and trichloromethane	1986AN2
	poly(dimethylsiloxane) and trichloromethane	1980ROG
Poly(arylate-*b*-dimethylsiloxane) polyblock copolymer		
	dichloromethane and 1-butanol	1988KAR
	dichloromethane and ethanol	1988KAR
	dichloromethane and methanol	1988KAR
	dichloromethane and 1-propanol	1988KAR
	trichloromethane and 1-butanol	1988KAR
	trichloromethane and ethanol	1988KAR
	trichloromethane and methanol	1988KAR
	trichloromethane and 1-propanol	1988KAR
Poly(arylene sulfonoxide-*b*-butadiene) polyblock copolymer		
	polybutadiene and 1,1,2,2-tetrachloroethane	1988ROG
	polybutadiene and trichloromethane	1988ROG
Polyarylsulfone		
	polyimide and sulfolane	1992JER

Polymer (B)	Second and third component	Ref.
Poly(γ-benzyl-L-glutamate)		
	N,N-dimethylformamide and methanol	1968NA2
	N,N-dimethylformamide and methanol	1971WEE
	dichloroacetic acid and 1,2-dichloroethane	1984SUB
	1,4-dioxane and 2-propanol	1990JAC
	1,4-dioxane and water	1990JAC
Polybutadiene		
	bis(2-ethylhexyl) phthalate and polystyrene	1989TAK
	bis(2-ethylhexyl) phthalate and polystyrene	1996KUM
	cyclohexane and *N,N*-dimethylformamide	1982STA
	cyclohexane and polystyrene	1988EIN
	cyclohexane and polystyrene	1996PET
	1,2-dichlorobenzene and polystyrene	2002FRI
	ethyl acetate and toluene	2004VSH
	poly(arylene sulfonoxide-*b*-butadiene) polyblock copolymer and tetrachloroethane	1988ROG
	poly(arylene sulfonoxide-*b*-butadiene) polyblock copolymer and trichloromethane	1988ROG
	polystyrene and styrene	1978RIG
	polystyrene and tetrahydrofuran	1972WHI
	polystyrene and tetrahydrofuran	1979NAR
	polystyrene and tetrahydrofuran	1980LLO
	polystyrene and tetrahydrofuran	1983NAR
	polystyrene and tetrahydronaphthalene	1973WEL
	polystyrene and tetrahydronaphthalene	1974WEL
	polystyrene and tetrahydronaphthalene	1978LLO
	polystyrene and tetrahydronaphthalene	1981LLO
	polystyrene and toluene	1983NAR
	polystyrene and toluene	1984NA1
	polystyrene and toluene	1984NA2
	polystyrene and toluene	1984HAS
	polystyrene and toluene	1984SAS
	polystyrene and toluene	1986NAR
	polystyrene and toluene	2006RUS
Poly(butadiene-*co*-α-methylstyrene)		
	ethyl acetate and poly(acrylonitrile-*co*-butadiene)	2002VSH
	ethyl acetate and poly(acrylonitrile-*co*-butadiene)	2004VSH
Poly(butyl acrylate)		
	butyl acrylate and poly(vinyl chloride)	1984WAL
	poly(vinyl chloride) and vinyl chloride	1982WAL

Polymer (B)	Second and third component	Ref.
Poly(butyl methacrylate)		
	cyclohexanone and polystyrene	1996KRA
	cyclohexanone and polystyrene	1997KR1
	cyclohexanone and polystyrene	1997KR2
	polystyrene and trichloromethane	1986LIP
Poly(ε-caprolactam)		
	m-cresol and petroleum ether	1978CRA
Polycarbonate bisphenol-A		
	dichloromethane and polycarbonate tetramethylbisphenol-A	1991HEL
	dichloromethane and poly(methyl methacrylate)	1992SAK
	1-methyl-2-pyrrolidinone and water	2000LUC
	polystyrene and trichloromethane	1986LIP
Polycarbonate tetrabromo-bisphenol-A		
	acetophenone and di(ethylene glycol) dibutyl ether	1993BE2
	acetophenone and dioctyl phthalate	1993BE2
	acetophenone and tri(ethylene glycol)	1993BE2
	cyclohexanone and di(ethylene glycol) dibutyl ether	1993BE2
	cyclohexanone and n-dodecane	1993BE2
	cyclohexanone and n-hexadecane	1993BE2
	cyclohexanone and (triethylene glycol)	1993BE2
	di(ethylene glycol) and 1-methyl-2-pyrrolidinone	1993BE2
	di(ethylene glycol) dibutyl ether and 1-methyl-2-pyrrolidinone	1993BE2
	2-ethoxyethanol and 1-methyl-2-pyrrolidinone	1993BE2
	1-methyl-2-pyrrolidinone and tri(ethylene glycol)	1993BE2
Polycarbonate tetramethyl-bisphenol-A		
	dichloromethane and polycarbonate bisphenol-A	1991HEL
Poly(2-chlorostyrene)		
	2-chlorostyrene and polystyrene	2004OKA
Poly(4-chlorostyrene)		
	benzene and polystyrene	1956KER
	tetrachloromethane and toluene	1966KUB

Polymer (B)	Second and third component	Ref.
Poly(4-decylstyrene)		
	2-butanone and 3-heptanone	1987MAG
Poly(*N,N*-diethylacrylamide)		
	tert-butanol and water	2004PAN
	N,N-dimethylformamide and water	2004PAN
	1,4-dioxane and water	2004PAN
	1,4-dioxane and water	2006PAG
	ethanol and water	2004PAN
	methanol and water	2002MA2
	methanol and water	2004PAN
	2-methyl-1-propanol and water	2004PAN
	poly(*N*-isopropylacrylamide) and water	2003MAO
	1-propanol and water	2004PAN
	2-propanol and water	2004PAN
Poly(*N,N*-dimethylacrylamide)		
	1,4-dioxane and water	2004PAG
	2-propanone and water	2004PAG
Poly(*N,N*-dimethylacrylamide-*co*-2-hydroxyethyl methacrylate)		
	water and β-cyclodextrin	1998GOS
Poly(2,6-dimethyl-1,4-phenylene-oxide)		
	ethanol and toluene	1972EM1
Poly(dimethylsiloxane)		
	benzene and poly(methyl methacrylate)	1998MIY
	benzene and polystyrene	1987KA2
	ethanol and toluene	1996MAL
	ethanol and toluene	2002WUE
	ethyl acetate and polystyrene	1975OKA
	methyl acetate and polystyrene	1975OKA
	n-pentane and polyethylene	2002KIR
	phenetole and polyisobutylene	1975OKA
	polyarylate and trichloromethane	1980ROG
	poly(methyl methacrylate) and toluene	1998MIY
	polystyrene and propylbenzene	1982SH3
	polystyrene and styrene	2001GER
	polystyrene and tetrahydrofuran	1987KA1
	polystyrene and toluene	1987KA2

Polymer (B)	Second and third component	Ref.
Poly(divinyl ether-*alt*-maleic anhydride)		
	poly(methacrylic acid) and water	2004VOL
	poly(methacrylic acid) and water	2005IZU
Polyesterurethane		
	dimethylsulfoxide and water	2005HEI
Polyetherimide		
	acetic acid and *N,N*-dimethylacetamide	1999WAN
	acetic acid and *N,N*-dimethylacetamide	2003SHE
	acetic acid and *N,N*-dimethylformamide	1999WAN
	acetic acid and 1-methyl-2-pyrrolidinone	1999WAN
	bisphenol-A diglycidyl ether and polyethersulfone	2004GIA
	bisphenol-A diglycidyl ether and polysulfone	2004GIA
	γ-butyrolactone and tetrahydrofuran	1998MAG
	γ-butyrolactone and tetrahydrofuran	2000MAG
	dichloromethane and 1-methyl-2-pyrrolidinone	2003YOU
	dichloromethane and 1-methyl-2-pyrrolidinone	2006TAO
	di(ethylene glycol) and *N,N*-dimethylacetamide	1999WAN
	di(ethylene glycol) and *N,N*-dimethylformamide	1999WAN
	di(ethylene glycol) and 1-methyl-2-pyrrolidinone	1999WAN
	N,N-dimethylacetamide and 1,2-ethanediol	1999WAN
	N,N-dimethylacetamide and ethanol	1999WAN
	N,N-dimethylacetamide and ethanol	2003SHE
	N,N-dimethylacetamide and methanol	1999WAN
	N,N-dimethylacetamide and 1-propanol	1999WAN
	N,N-dimethylacetamide and 2-propanol	1999WAN
	N,N-dimethylacetamide and propionic acid	1999WAN
	N,N-dimethylacetamide and water	1999WAN
	N,N-dimethylacetamide and water	2003SHE
	N,N-dimethylformamide and water	1999WAN
	N,N-dimethylformamide and 1,2-ethanediol	1999WAN
	N,N-dimethylformamide and ethanol	1999WAN
	N,N-dimethylformamide and methanol	1999WAN
	N,N-dimethylformamide and 1-propanol	1999WAN
	N,N-dimethylformamide and 2-propanol	1999WAN
	N,N-dimethylformamide and propanoic acid	1999WAN
	N,N-dimethylformamide and water	2001ALB
	1,2-ethanediol and 1-methyl-2-pyrrolidinone	1999WAN
	ethanol and 1-methyl-2-pyrrolidinone	1999WAN
	methanol and 1-methyl-2-pyrrolidinone	1999WAN
	1-methyl-2-pyrrolidinone and 1-propanol	1999WAN
	1-methyl-2-pyrrolidinone and 2-propanol	1999WAN

Polymer (B)	Second and third component	Ref.
Polyetherimide		
	1-methyl-2-pyrrolidinone and propanoic acid	1999WAN
	1-methyl-2-pyrrolidinone and water	1991HOU
	1-methyl-2-pyrrolidinone and water	1992VIA
	1-methyl-2-pyrrolidinone and water	1994BO2
	1-methyl-2-pyrrolidinone and water	1999WAN
	1-methyl-2-pyrrolidinone and water	2001ALB
	1-methyl-2-pyrrolidinone and water	2001FER
Polyethersulfone		
	bisphenol-A diglycidyl ether and polyetherimide	2004GIA
	N,N-dimethylacetamide and water	1988SWI
	N,N-dimethylacetamide and water	1991LA1
	N,N-dimethylformamide and ethanol	1988SWI
	N,N-dimethylformamide and methanol	1988SWI
	N,N-dimethylformamide and 1-propanol	1988SWI
	N,N-dimethylformamide and 2-propanone	1988SWI
	N,N-dimethylformamide and tetrachloromethane	1988SWI
	N,N-dimethylformamide and water	1991LA1
	N,N-dimethylformamide and water	2000BA2
	dimethylpropyleneurea and water	1991LA1
	dimethylsulfoxide and water	1988SWI
	dimethylsulfoxide and water	1991LA1
	dimethylsulfoxide and water	1999BAT
	1-methyl-2-pyrrolidinone and water	1988SWI
	1-methyl-2-pyrrolidinone and water	1991LA1
	1-methyl-2-pyrrolidinone and water	1994BO2
	1-methyl-2-pyrrolidinone and water	1999BAI
	polyimide and sulfolane	1991LIA
	1,1,3,3-tetramethylurea and water	1991LA1
Polyetherurethane		
	N,N-dimethylformamide and water	2001LEE
Poly[*N*-(3-ethoxypropyl)-acrylamide]		
	ethanol and water	2005UGU
Poly(ethyl acrylate)		
	toluene and poly(vinyl propionate)	1989BHA

Polymer (B)	Second and third component	Ref.
Poly(ethyl acrylate-*co*-4-vinylpyridine)		
	polystyrene and tetrahydrofuran	1992WAN
Polyethylene		
	2-butoxyethanol and decahydronaphthalene	1989SCH
	2-butoxyethanol and 1,4-dimethylbenzene	1989SCH
	2-butoxyethanol and 1,2,3,4-tetrahydronaphthalene	1989SCH
	decahydronahpthalene and 2-ethyl-1-hexanol	1989SCH
	N,N-dimethylformamide and 1,2-dimethylbenzene	1995AND
	N,N-dimethylformamide and 1,2-dimethylbenzene	1996AND
	2,4-dimethylpentane and polypropylene	1984VAR
	diphenyl ether and polyethylene	1967KO1
	diphenyl ether and polyethylene	1967KO4
	2-ethoxyethanol and 1,2,3,4-tetrahydronaphthalene	1989SCH
	2-ethoxyethanol and 1,4-dimethylbenzene	1989SCH
	2-(ethoxyethoxy)ethanol and 1,2,3,4-tetrahydro-naphthalene	1989SCH
	2-ethyl-1-hexanol and 1,4-dimethylbenzene	1989SCH
	2-ethyl-1-hexanol and 1,2,3,4-tetrahydro-naphthalene	1989SCH
	n-hexane and n-octane	1989HAE
	n-nonane and n-octane	1978KO2
	n-pentane and poly(dimethylsiloxane)	2002KIR
Poly(ethylene-*co*-acrylic acid)-*g*-poly(ethylene glycol) monomethyl ether		
	dioctyl phthalate and poly(ethylene glycol) monomethyl ether	2005ZHO
	dioctyl phthalate and poly(ethylene glycol) monomethyl ether	2006ZHO
Poly(ethylene-*co*-vinyl acetate)		
	2-heptanone and poly(ethylene-*co*-vinyl acetate)	1986HAE
	methyl methacrylate and poly(methyl methacrylate)	2003CH1
	polystyrene and styrene	2003CH1
	polystyrene and styrene	2003CH2

Polymer (B)	Second and third component	Ref.
Poly(ethylene-*co*-vinyl alcohol)		
	dimethylsulfoxide and 2-propanol	2001CHE
	dimethylsulfoxide and water	1997YOU
	dimethylsulfoxide and water	1999YOU
	dimethylsulfoxide and water	2001CHE
	dimethylsulfoxide and water	2002YOU
	1,3-propanediol and 1,2,3-propanetriol	2005SHA
	2-propanol and water	1998YOU
Poly(ethylene glycol)		
	agarose and water	1993MED
	2-butoxyethanol and water	1980DON
	benzoyl dextran and water	1994LUM
	cellulose acetate and *N,N*-dimethylacetamide	1986AN1
	cyclohexane and methanol	1982STA
	dextran and water	1958ALB
	dextran and water	1968EDM
	dextran and water	1981SHI
	dextran and water	1984BAM
	dextran and water	1984HEF
	dextran and water	1986GUS
	dextran and water	1986ZAS
	dextran and water	1987ANA
	dextran and water	1987BAS
	dextran and water	1987KAN
	dextran and water	1987SJO
	dextran and water	1987ZA3
	dextran and water	1988ABB
	dextran and water	1988KAN
	dextran and water	1988MAK
	dextran and water	1988ZAS
	dextran and water	1989HA1
	dextran and water	1989HA2
	dextran and water	1989ZAS
	dextran and water	1990CAB
	dextran and water	1990RAG
	dextran and water	1993GA2
	dextran and water	1993HAR
	dextran and water	1993HAY
	dextran and water	1994HAR
	dextran and water	1995DOE
	dextran and water	1996CES
	dextran and water	1996PIC
	dextran and water	1997SAR

Polymer (B)	Second and third component	Ref.
Poly(ethylene glycol)		
	dextran and water	2002ZAV
	dextran and water	2004HOP
	dextran and water	2004KAV
	dextran and water	2005MOO
	dextran and water	2006EDA
	1,2-ethanediol and water	1987LOP
	1,2-dimethylbenzene and tri(ethylene glycol)	1961OKA
	hexa(ethylene glycol) monododecyl ether and water	2005HO1
	hexa(ethylene glycol) monododecyl ether and water	2005HO2
	hydroxypropyl starch and water	1986MAT
	hydroxypropyl starch and water	1986TJE
	hydroxypropyl starch and water	1989LIN
	hydroxypropyl starch and water	1996CES
	hydroxypropyl starch and water	1998ALM
	maltodextrin and water	1988SZL
	maltodextrin and water	1991RAG
	maltodextrin and water	1994ATK
	maltodextrin and water	2000SRI
	octa(ethylene oxide) dodecyl ether and water	1996PIC
	penta(ethylene oxide) dodecyl ether and water	1996PIC
	poly(acrylamide) and water	1984HEF
	poly(acrylamide) and water	1989PER
	poly(acrylamide-*co*-*N*-benzylacrylamide) and water	2003ABU
	poly(acrylamide-*co*-4-methoxystyrene) and water	2003ABU
	poly(acrylamide-*co*-*N*-phenylacrylamide) and water	2003ABU
	poly(acrylic acid) and water	2002GUP
	poly(diallyaminoethanoate-*co*-dimethylsulfoxide) and water	2004WAZ
	poly(ethylene glycol) monoalkyl ether and water	1995PAN
	poly(ethyleneimine) and water	1993DIS
	poly(ethyleneimine) and water	2002GUP
	poly(ethylene oxide-*co*-propylene oxide) and water	1994ZH1
	poly(*N*-isopropylacrylamide) and water	2004HUE
	polypropylene and tetrahydronaphthalene	1963RED
	poly(propylene glycol) and water	1993MAL
	polystyrene and styrene	2004LIS
	polystyrene and toluene	1995HAR
	poly(vinyl acetate-*co*-vinyl alcohol) and water	1984HEF
	poly(vinyl alcohol) and water	1984INA
	poly(vinyl alcohol) and water	1986INA
	poly(vinyl methyl ether) and water	1984HEF
	poly(1-vinyl-2-pyrrolidinone) and water	1984HEF

Polymer (B)	Second and third component	Ref.
Poly(ethylene glycol)		
	poly(1-vinyl-2-pyrrolidinone) and water	2004INA
	1,2,3-propanetriol and water	1987LOP
	1,2,3-propanetriol and water	1988VSH
	starch and water	1988LAR
	starch-*g*-polyacrylamide and water	2000PIE
	Tween surfactant and water	2004MAH
	valeryl dextran and water	1994LUM
Poly(ethylene glycol) monomethyl ether	dioctyl phthalate and poly(ethylene-*co*-acrylic acid)-*g*-poly(ethylene glycol) monomethyl ether	2005ZHO
	dioctyl phthalate and poly(ethylene-*co*-acrylic acid)-*g*-poly(ethylene glycol) monomethyl ether	2006ZHO
Poly(ethylene glycol) mono(p-nonylphenyl) ether	n-hexane and water	2003BAL
Poly(ethyleneimine)		
	hydroxyethylcellulose and water	1993DIS
	hydroxyethylcellulose and water	1994DIS
	poly(ethylene glycol) and water	1993DIS
	poly(ethylene glycol) and water	2002GUP
Poly(ethylene oxide)		
	benzene and polystyrene	1982LE2
	benzene and poly(styrene-*co*-methycrylic acid)	1982LE2
	2-butoxyethanol and water	1991WOR
	chlorobenzene and n-heptane	2000SPI
	chlorobenzene and water	2002SPI
	dextran and water	1998NIL
	dextran and water	2003EDE
	dextran and water	2005OL1
	dichloromethane and n-heptane	2000SPI
	dichloromethane and water	2002SPI
	2,6-dimethylpyridine and water	2002SHR
	isobutyric acid and water	1995TOK
	isobutyric acid and water	2002SHR
	n-heptane and trichloromethane	2000SPI
	3-methylpentane and nitroethane	2005VEN
	polystyrene and styrene	2004SUT
	poly(ethylene oxide-*b*-dimethylsiloxane) and 1,2,3,4-tetrahydronaphthalene	2002MAD

Polymer (B)	Second and third component	Ref.
Poly(ethylene oxide)		
	poly(ethylene oxide-*b*-dimethylsiloxane) and 1,2,3,4-tetrahydronaphthalene	2003MAD
	polystyrene and tetrahydrofuran	1982LE2
	polystyrene and trichloromethane	1982LE2
	poly(styrene-*co*-methycrylic acid) and tetrahydrofuran	1982LE2
	poly(styrene-*co*-methycrylic acid) and trichloromethane	1982LE2
	poly(1-vinyl-2-pyrrolidinone) and water	1998NIL
	tetra(ethylene glycol) octyl ether and water	1991WOR
	tetrahydrofuran and water	2001SCH
	trichloromethane and water	2002SPI
Poly(ethylene oxide-*b*-dimethyl-siloxane)		
	poly(ethylene oxide) and 1,2,3,4-tetrahydro-naphthalene	2002MAD
	poly(ethylene oxide) and 1,2,3,4-tetrahydro-naphthalene	2003MAD
Poly(ethylene oxide-*b*-isoprene)		
	oligo(ethylene oxide) monododecyl ether and water	2004KUN
Poly(ethylene oxide-*co*-propylene oxide)		
	acetic acid and water	1993JOH
	benzoyl dextran and water	1996LUM
	butanoic acid and water	1993JOH
	dextran and water	1986ALB
	dextran and water	1994ALR
	dextran and water	1998PLA
	dextran and water	1999BER
	dextran and water	2005OL1
	hydroxypropyl starch and water	1994MOD
	hydroxypropyl starch and water	1997CUN
	hydroxypropyl starch and water	2000PE2
	octa(ethylene glycol) dodecyl ether and water	1992ZHA
	octa(ethylene glycol) dodecyl ether and water	1994ZH1
	octa(ethylene glycol) dodecyl ether and water	1994ZH2
	phenol and water	1997JOH

Polymer (B)	Second and third component	Ref.
Poly(ethylene oxide-*co*-propylene oxide)		
	poly(ethylene glycol) and water	1994ZH1
	poly(ethylene oxide-*co*-propylene oxide) and water	1999PER
	poly(*N*-isopropylacrylamide-*co*-1-vinylimidazole) and water	2000PE1
	poly(propylene glycol) and water	1994ZH1
	propanoic acid and water	1993JOH
	tetra(ethylene glycol) dodecyl ether and water	1992ZHA
	tetra(ethylene glycol) dodecyl ether and water	1994ZH1
	tetra(ethylene glycol) dodecyl ether and water	1994ZH2
Poly(ethylene oxide-*b*-propylene oxide)	*N*-methylacetamide and water	1991SA2
Poly(ethylene oxide-*b*-propylene oxide-*b*-ethylene oxide)	1-butanol and water	1997HO2
	1-butanol and water	2001KWO
	dextran and water	1986ALB
	dextran and water	1995SVE
	dextran and water	1999SVE
	1,2-dichloroethane and water	2006LAZ
	ethanol and water	2001KWO
	2-phenylethanol and water	2000FRI
	tributyl phosphate and water	2006CAU
Poly(ethylene oxide-*b*-tetrahydro-furan-*b*-ethylene oxide)		
	1-butanol and water	1997HO2
Poly(ethylene terephthalate-*co*-p-hydroxybenzoic acid)		
	polycarbonate-bisphenol-A and trichloromethane	1988SC1
Poly(ethyl methacrylate)		
	1-butanol and poly(methyl methacrylate)	1996COO
	1-decanol and poly(methyl methacrylate)	1998BE1
Poly(2-ethyl-2-oxazoline)		
	1,4-dioxane and water	1988LIN

Polymer (B)	Second and third component	Ref.
Poly(*N*-ethyl-*N*-methylacrylamide)		
	tert-butanol and water	2004PAN
	methanol and water	2004PAN
	2-methyl-1-propanol and water	2004PAN
	1-propanol and water	2004PAN
Poly(hexafluoroacetone-*co*-vinylidene fluoride)		
	poly(2-ethylhexylacrylate-*co*-acrylic acid-*co*-vinyl acetate) and tetrahydrofuran	1999KAN
Poly(3-hydroxybutanoic acid)		
	N,N-dimethylformamide and trichloromethane	1996MAS
	ethanol and trichloromethane	1996MAS
	tetrahydrofuran and trichloromethane	1996MAS
Poly(3-hydroxybutanoic acid-*co*-3-hydroxypentanoic acid		
	N,N-dimethylformamide and trichloromethane	1996MAS
	ethanol and trichloromethane	1996MAS
	tetrahydrofuran and trichloromethane	1996MAS
Poly(2-hydroxypropyl methacrylate)		
	methanol and water	2004GAR
Poly(hydroxystearic acid-*b*-ethylene oxide-*b*-hydroxystearic acid)	isopropyl myristate and water	2002PLA
Poly(p-hydroxystyrene)		
	cyclohexane and 2-propanone	1992YAM
Polyimide		
	N,N-dimethylacetamide and methanol	2001LEB
	N,N-dimethylacetamide and water	2000CHU
	N,N-dimethylacetamide and water	2002SHI
	N,N-dimethylformamide and ethanol	2001LEB
	N,N-dimethylformamide and methanol	2001LEB
	N,N-dimethylformamide and 2-propanol	2001LEB
	N,N-dimethylformamide and water	2001LEB

Polymer (B)	Second and third component	Ref.
Polyimide		
	N,N-dimethylformamide and water	2002SHI
	N,N-dimethylformamide and water	2003MA4
	dimethylsulfoxide and water	2001KI2
	dimethylsulfoxide and water	2002SHI
	1,4-dioxane and methanol	2001LEB
	ethanol and 1-methyl-2-pyrrolidinone	2000CLA
	methanol and 1-methyl-2-pyrrolidinone	2001LEB
	1-methyl-2-pyrrolidinone and water	2002SHI
	polyarylsulfone and sulfolane	1992JER
	polyethersulfone and sulfolane	1991LIA
	2-propanone and water	2002SHI
	tetrahydrofuran and water	2002SHI
Polyisobutylene		
	benzene and polystyrene	1989TON
	benzene and polystyrene	1990TON
	2-butanone and cyclohexane	1987GE1
	2-butanone and cyclohexane	1989SCH
	2-butanone and n-heptane	1987GE1
	2-butanone and n-heptane	1987GE2
	2-butanone and n-heptane	1989SCH
	2-butanone and toluene	1987GE1
	2-butanone and toluene	1987GE2
	2-butanone and toluene	1989SCH
	2-butanone and dibutyl ether	1987GE1
	2-butanone and dibutyl ether	1989SCH
	cyclohexane and polystyrene	1989TON
	cyclohexane and polystyrene	1990TON
	cyclohexane and 2-propanone	1987GE1
	cyclohexane and 2-propanone	1989SCH
	2,4-dimethylpentane and n-heptane	1984VAR
	diphenyl ether and polypropylene	2002MA5
	n-heptane and 2-propanol	1987GE1
	n-heptane and 2-propanol	1989SCH
	phenetole and poly(dimethylsiloxane)	1975OKA
	polystyrene and tetrahydrofuran	1972WHI
	polystyrene and toluene	1976ESK
	2-propanol and toluene	1987GE1
	2-propanol and toluene	1989SCH

Polymer (B)	Second and third component	Ref.
Polyisoprene		
	benzene and poly(methyl methacrylate)	1959BRI
	benzene and polystyrene	1947DOB
	butyl acetate and poly(methyl methacrylate)	1959BRI
	cyclohexane and poly(α-methylstyrene)	1992HON
	cyclohexane and polystyrene	1987TO1
	cyclohexane and polystyrene	1987TO2
	ethyl acetate and toluene	2004VSH
	polystyrene and tetrahydrofuran	1972WHI
Poly(isoprene-*b*-ethylene oxide)		
	oligo(ethylene oxide) monododecyl ether and water	2004KUN
Poly(*N*-isopropylacrylamide)		
	acetonitrile and water	2004PAN
	1-butanol and water	2004PAN
	tert-butanol and water	2004PAN
	cyclohexane and methanol	2002MOR
	deuterium oxide and water	2004MAO
	dextran and water	1999KIS
	N,N-dimethylformamide and water	2002COS
	N,N-dimethylformamide and water	2004PAN
	dimethylsulfoxide and water	2002COS
	1,4-dioxane and water	1991SC1
	1,4-dioxane and water	2004PAN
	1,4-dioxane and water	2006DAL
	ethanol and water	2002COS
	ethanol and water	2004PAN
	methanol and water	1990WIN
	methanol and water	1991SC1
	methanol and water	2002COS
	methanol and water	2004GAR
	methanol and water	2004PAN
	methanol and water	2005TAO
	2-methyl-1-propanol and water	2004PAN
	poly(*N,N*-diethylacrylamide) and water	2003MAO
	poly(ethylene glycol) and water	2004HUE
	poly(*N*-isopropylmethacrylamide) and water	2005STA
	1,2,3-propanetriol and water	2004PAN
	1-propanol and water	2002COS
	1-propanol and water	2004PAN
	2-propanol and water	2002COS
	2-propanol and water	2004PAN

Polymer (B)	Second and third component	Ref.
Poly(*N*-isopropylacrylamide)		
	2-propanone and water	2002COS
	2-propanone and water	2004PAN
	sulfolane and water	2004PAN
	tetrahydrofuran and water	1991SC1
	tetrahydrofuran and water	2004PAN
Poly(*N*-isopropylacrylamide-*co*-*N,N*-dimethylacrylamide		
	1,4-dioxane and water	2006PAG
Poly(*N*-isopropylacrylamide-*co*-1-vinylimidazole)	poly(ethylene oxide-*co*-propylene oxide) and water	2000PE1
Poly(*N*-isopropylmethacrylamide)		
	deuterium oxide and poly(vinyl methyl ether)	2005SP2
	poly(*N*-isopropylacrylamide) and water	2005STA
Poly(DL-lactide)		
	1,4-dioxane and methanol	1996WI3
	1,4-dioxane and water	1996WI3
	methanol and trichloromethane	1996WI1
	methanol and trichloromethane	1996WI2
	methanol and trichloromethane	1996WI3
	1-methyl-2-pyrrolidinone and water	1996WI3
Poly(L-lactide)		
	1,4-dioxane and methanol	1996WI3
	1,4-dioxane and water	1996WI3
	1,4-dioxane and water	2001HUA
	1,4-dioxane and water	2004TAN
	methanol and trichloromethane	1996WI1
	methanol and trichloromethane	1996WI2
	methanol and trichloromethane	1996WI3
	1-methyl-2-pyrrolidinone and water	1996WI3
Poly(maleic acid-*co*-acrylic acid)		
	hydroxypropylcellulose and water	2005BUM
Poly(maleic acid-*co*-styrene)		
	hydroxypropylcellulose and water	2005BUM

Polymer (B)	Second and third component	Ref.
Poly(maleic acid-*co*-vinyl acetate)		
	hydroxypropylcellulose and water	2005BUM
Poly(methacrylic acid)		
	N,N-dimethylformamide and 1,4-dioxane	1987SIV
	methacrylic acid and water	2002SAX
	poly(divinyl ether-*alt*-maleic anhydride) and water	2004VOL
	poly(divinyl ether-*alt*-maleic anhydride) and water	2005IZU
Poly(methyl acrylate)		
	1,2-dichloroethane and poly(vinyl acetate)	1985NAN
	toluene and poly(vinyl acetate)	1985NAN
	toluene and poly(vinyl acetate)	1989BHA
Poly(methyl methacrylate)		
	acetonitrile and 1-chlorobutane	1981HOR
	benzene and cyclohexane	1978CRA
	benzene and poly(dimethylsiloxane)	1998MIY
	benzene and polyisoprene	1959BRI
	2-butanol and 1-chlorobutane	1975WOL
	1-butanol and poly(ethyl methacrylate)	1996COO
	1-butanol and water	1987CO1
	butyl acetate and n-hexane	1998LAI
	butyl acetate and polyisoprene	1959BRI
	1-chlorobutane and 4-heptanone	1975WOL
	1-chlorobutane and 4-heptanone	1986CO2
	cyclohexane and ethanol	1986CO2
	cyclohexane and methanol	1986CO2
	cyclohexane and toluene	1989BAR
	cyclohexanol and water	2002MA4
	1-decanol and poly(ethyl methacrylate)	1998BE1
	dichloromethane and polycarbonate bisphenol-A	1992SAK
	dimethyl phthalate and poly(styrene-*co*-acrylonitrile)	1977BER
	ethanol and formamide	1981FER
	ethanol and water	1987CO1
	ethyl acetate and polystyrene	2006RUS
	n-hexane and 2-propanone	1998LAI
	n-hexane and trichloromethane	1960ELI
	methanol and water	1987CO1
	methyl methacrylate and poly(ethylene-*co*-vinyl acetate)	2003CH1
	1-methyl-2-pyrrolidinone and 1,2,3-propanetriol	1999GRA

Polymer (B)	Second and third component	Ref.
Poly(methyl methacrylate)		
	1-methyl-2-pyrrolidinone and water	2002LIN
	poly(dimethylsiloxane) and toluene	1998MIY
	polystyrene and styrene	1996CHA
	polystyrene and toluene	1984LAU
	polystyrene and toluene	1985LAU
	polystyrene and toluene	1992VEN
	poly(vinyl chloride) and tetrahydrofuran	1986CH1
	poly(vinyl chloride) and tetrahydrofuran	1986CH2
	1-propanol and water	1987CO1
	2-propanol and water	1987CO1
	2-propanol and water	2000CHE
	2-propanone and water	1998LAI
	tetrahydrofuran and water	2001SCH
	m-xylylene diisocyanate and 4-mercaptomethyl-3,6-dithia-1,8-octanedithiol	2006SOU
Poly(methyl methacrylate-*co*-*N,N*-dimethylacrylamide)		
	m-xylylene diisocyanate and 4-mercaptomethyl-3,6-dithia-1,8-octanedithiol	2006SOU
Poly(methyl methacrylate-*g*-dimethylsiloxane)		
	dimethylsulfoxide and tetrachloroethene	1989STE
	dimethylsulfoxide and tetrachloroethene	2000KAW
Poly(methyl methacrylate-*co*-4-vinylpyridine)		
	poly(styrene-*co*-4-vinylpyridine) and 1,4-dioxane	1977DJA
	poly(styrene-*co*-4-vinylpyridine) and 2-butanone	1977DJA
	poly(styrene-*co*-4-vinylpyridine) and trichloromethane	1977DJA
Poly(α-methylstyrene)		
	cyclohexane and 1,4-*cis*-polyisoprene	1992HON
	cyclopentane and polystyrene	1986SA1
	dibutyl phthalate and polystyrene	1987LIN
	methylcyclohexane and polystyrene	1983SAE
Poly(octyl isocyanate)		
	polystyrene and 1,1,2,2-tetrachloroethane	1979AHA

Polymer (B)	Second and third component	Ref.
Polypropylene		
	2-(2-butoxyethoxy)ethanol and chlorobenzene	1971OGA
	2-(2-butoxyethoxy)ethanol and 1-chloronaphthalene	1971OGA
	2-(2-butoxyethoxy)ethanol and decahydro-naphthalene	1971OGA
	2-(2-butoxyethoxy)ethanol and 1,2-dichlorobenzene	1971OGA
	2-(2-butoxyethoxy)ethanol and n-hexadecane	1971OGA
	2-(2-butoxyethoxy)ethanol and 1,2,3,4-tetrahydro-naphthalene	1971OGA
	decahydronaphthalene and di(ethylene glycol) monobutyl ether acetate	1971OGA
	decahydronaphthalene and di(ethylene glycol) monomethyl ether acetate	1971OGA
	decahydronaphthalene and 2-(2-methoxy-ethoxy)ethanol	1971OGA
	dibutyl phthalate and dioctyl phthalate	2006YA1
	2,4-dimethylpentane and polyethylene	1984VAR
	diphenyl ether and polyisobutylene	2002MA5
	poly(ethylene glycol) and 1,2,3,4-tetrahydro-naphthalene	1963RED
	polystyrene and toluene	1967BER
Poly(propylene glycol)		
	n-alkyl polyglycol ether and water	1986FIR
	butylbenzene and water	1986FIR
	cyclohexane and water	1986FIR
	1,3-diethylurea and water	1967SAI
	1,3-dimethylurea and water	1967SAI
	formamide and water	1967SAI
	n-octane and water	1986FIR
	1-phenylheptane and water	1986FIR
	1-phenylhexane and water	1986FIR
	1-phenyloctane and water	1986FIR
	1-phenylpentane and water	1986FIR
	poly(ethylene glycol) and water	1993MAL
	Triton X-100 and water	1996GAL
Poly(propylene oxide)		
	polystyrene and styrene	1978RIG

Polymer (B)	Second and third component	Ref.
Poly(pyrrolidinoacrylamide)		
	tert-butanol and water	2004PAN
	1,4-dioxane and water	2004PAN
	ethanol and water	2004PAN
	methanol and water	2004PAN
	2-methyl-1-propanol and water	2004PAN
	1-propanol and water	2004PAN
	2-propanol and water	2004PAN
Polystyrene		
	benzene and ethanol	1964TAG
	benzene and methanol	1940SCH
	benzene and methanol	1967SUH
	benzene and methanol	1968SUH
	benzene and methanol	1969KLE
	benzene and methanol	1973CAN
	benzene and methanol	1978CRA
	benzene and poly(butyl methacrylate)	1986LIP
	benzene and poly(4-chlorostyrene)	1956KER
	benzene and poly(dimethylsiloxane)	1987KA2
	benzene and poly(ethylene oxide)	1982LE2
	benzene and polyisobutylene	1989TON
	benzene and polyisobutylene	1990TON
	benzene and polyisoprene	1947DOB
	benzene and poly(methyl methacrylate)	1971HON
	benzene and 2-propanone	1966HEI
	benzene and 2-propanone	1967SUH
	bis(2-ethylhexyl) phthalate and polybutadiene	1989TAK
	bis(2-ethylhexyl) phthalate and polybutadiene	1996KUM
	bisphenol-A diglycidyl ether and 4,4'-methylene-bis(2,6-diethylaniline)	2006RIC
	bromobenzene and n-hexane	1967SUH
	bromobenzene and methanol	1967SUH
	bromobenzene and methanol	1968SUH
	bromobenzene and 2-propanone	1967SUH
	2-bromobutane and n-hexane	1967SUH
	1-bromobutane and methanol	1968SUH
	2-bromobutane and methanol	1967SUH
	2-bromobutane and methanol	1968SUH
	bromoethane and methanol	1968SUH
	1-bromopropane and methanol	1968SUH
	p-bromotoluene and methanol	1968SUH
	2-butanone and cellulose nitrate	1947DOB
	2-butanone and 2-propanone	1970WO1

Polymer (B)	Second and third component	Ref.
Polystyrene		
	2-butanone and 2-propanone	2000STR
	1-butanol and decahydronaphthalene	1991VSH
	tert-butyl acetate and n-hexane	1991VSH
	cellulose acetate and tetrahydrofuran	2003SIL
	chlorobenzene and methanol	1967SUH
	chlorobenzene and methanol	1968SUH
	2-chlorostyrene and poly(2-chlorostyrene)	2004OKA
	cyclohexane and cyclopentane	1993IWA
	cyclohexane and *N,N*-dimethylformamide	1978WOL
	cyclohexane and *N,N*-dimethylformamide	1982STA
	cyclohexane and *N,N*-dimethylformamide	1985STA
	cyclohexane and *N,N*-dimethylformamide	1988EIN
	cyclohexane and *N,N*-dimethylformamide	1992NAK
	cyclohexane and 1-hexanol	1970KLE
	cyclohexane and methanol	1982STA
	cyclohexane and 1-propanol	1991VSH
	cyclohexane and polybutadiene	1988EIN
	cyclohexane and polybutadiene	1996PET
	cyclohexane and polyisobutylene	1989TON
	cyclohexane and polyisobutylene	1990TON
	cyclohexane and polyisoprene	1987TO1
	cyclohexane and polyisoprene	1987TO2
	cyclohexane and polystyrene	1952SHU
	cyclohexane and polystyrene	1970KON
	cyclohexane and polystyrene	1976KL1
	cyclohexane and polystyrene	1985EIN
	cyclohexane and polystyrene	1988SUN
	cyclohexane and polystyrene	2000SUZ
	cyclohexane and poly(styrene-*b*-dimethylsiloxane)	1997SC2
	cyclohexane and styrene	1990RUI
	cyclohexane and water	1991DEE
	cyclohexanone and poly(butyl methacrylate)	1996KRA
	cyclohexanone and poly(butyl methacrylate)	1997KR1
	cyclohexanone and poly(butyl methacrylate)	1997KR2
	cyclopentane and poly(α-methylstyrene)	1986SA1
	n-decane and 2-propanone	1983CO2
	1,3-dibromobutane and n-hexane	1968SUH
	1,3-dibromobutane and methanol	1968SUH
	1,2-dibromoethane and n-hexane	1967SUH
	1,2-dibromoethane and methanol	1967SUH
	1,2-dibromopropane and methanol	1968SUH
	dibutyl phthalate and poly(α-methylstyrene)	1987LIN
	1,2-dichlorobenzene and polybutadiene	2002FRI
	1,2-dichloroethane and methanol	1968SUH

Polymer (B)	Second and third component	Ref.
Polystyrene		
	dichloromethane and methanol	1967SUH
	diethyl ether and methylcyclohexane	1974CO2
	diethyl ether and methylcyclohexane	1984CO2
	diethyl ether and nitromethane	1987CO2
	diethyl ether and orthotrimethyl formiate	1977RIG
	diethyl ether and 2-propanone	1974CO1
	diethyl ether and 2-propanone	1976WOL
	diethyl ether and 2-propanone	1984CO1
	N,N-dimethylacetamide and poly(acrylonitrile)	2001POS
	N,N-dimethylacetamide and poly(acrylonitrile)	2002KUL
	N,N-dimethylacetamide and poly(styrene-*co*-acrylonitrile)	2001POS
	N,N-dimethylacetamide and poly(styrene-*co*-acrylonitrile)	2002KUL
	N,N-dimethylacetamide and poly(styrene-*co*-acrylonitrile)	2003LOS
	1,2-dimethylbenzene and methanol	1968SUH
	1,3-dimethylbenzene and methanol	1968SUH
	1,4-dimethylbenzene and methanol	1968SUH
	1,4-dimethylcyclohexane and 2-propanone	1984CO2
	divinylbenzene and toluene	2003KON
	n-eicosane and 2-propanone	1983CO2
	ethanol and toluene	1962UTR
	ethyl acetate and methanol	1966HEI
	ethyl acetate and poly(dimethylsiloxane)	1975OKA
	ethyl acetate and poly(methyl methacrylate)	2006RUS
	n-heptane and methylcyclohexane	1999IM2
	n-heptane and 2-propanone	1983CO2
	n-hexadecane and 2-propanone	1983CO2
	n-hexane and isopropylbenzene	1967SUH
	n-hexane and 2-propanone	1983CO2
	n-hexane and toluene	1967SUH
	isopropylbenzene and methanol	1967SUH
	isopropylbenzene and methanol	1968SUH
	isopropylbenzene and 2-propanone	1967SUH
	iodomethane and methanol	1968SUH
	methanol and styrene	1968SUH
	methanol and tetrachloromethane	1968SUH
	methanol and toluene	1967SUH
	methanol and toluene	1968SUH
	methanol and trichloromethane	1967SUH
	methanol and trichloromethane	1968SUH

Polymer (B)	Second and third component	Ref.
Polystyrene		
	methyl acetate and poly(dimethylsiloxane)	1975OKA
	methylcyclohexane and poly(α-methylstyrene)	1983SAE
	methylcyclohexane and polystyrene	1980DO3
	methylcyclohexane and polystyrene	1986NAK
	methylcyclohexane and polystyrene	1990SHE
	methylcyclohexane and polystyrene	1993DOB
	methylcyclohexane and 2-propanone	1966HEI
	methylcyclopentane and 2-propanone	1984CO1
	n-octane and 2-propanone	1983CO2
	polybutadiene and styrene	1978RIG
	polybutadiene and tetrahydrofuran	1972WHI
	polybutadiene and tetrahydrofuran	1979NAR
	polybutadiene and tetrahydrofuran	1980LLO
	polybutadiene and tetrahydrofuran	1983NAR
	polybutadiene and 1,2,3,4-tetrahydronaphthalene	1973WEL
	polybutadiene and 1,2,3,4-tetrahydronaphthalene	1974WEL
	polybutadiene and 1,2,3,4-tetrahydronaphthalene	1978LLO
	polybutadiene and 1,2,3,4-tetrahydronaphthalene	1981LLO
	polybutadiene and toluene	1983NAR
	polybutadiene and toluene	1984NA1
	polybutadiene and toluene	1984NA2
	polybutadiene and toluene	1984HAS
	polybutadiene and toluene	1984SAS
	polybutadiene and toluene	1986NAR
	polybutadiene and toluene	2006RUS
	polycarbonate and trichloromethane	1986LIP
	poly(dimethylsiloxane) and propylbenzene	1982SH3
	poly(dimethylsiloxane) and styrene	2001GER
	poly(dimethylsiloxane) and tetrahydrofuran	1987KA1
	poly(dimethylsiloxane) and tetrahydrofuran	1987KA2
	poly(dimethylsiloxane) and toluene	1987KA2
	poly(ethyl acrylate-*co*-4-vinylpyridine) and tetrahydrofuran	1992WAN
	polyethylene and styrene	2004LIS
	poly(ethylene-*co*-propylene-*co*-diene) and styrene	1978RIG
	poly(ethylene-*co*-vinyl acetate) and styrene	2003CH1
	poly(ethylene-*co*-vinyl acetate) and styrene	2003CH2
	poly(ethylene glycol) and toluene	1995HAR
	poly(ethylene oxide) and styrene	2004SUT
	poly(ethylene oxide) and tetrahydrofuran	1982LE2
	poly(ethylene oxide) and trichloromethane	1982LE2
	polyisobutylene and tetrahydrofuran	1972WHI

Polymer (B)	Second and third component	Ref.
Polystyrene		
	polyisobutylene and toluene	1976ESK
	polyisoprene and tetrahydrofuran	1972WHI
	poly(methyl methacrylate) and styrene	1996CHA
	poly(methyl methacrylate) and toluene	1984LAU
	poly(methyl methacrylate) and toluene	1985LAU
	poly(methyl methacrylate) and toluene	1992VEN
	poly(octyl isocyanate) and 1,1,2,2-tetrachloroethane	1979AHA
	polypropylene and toluene	1967BER
	poly(propylene oxide) and styrene	1978RIG
	poly(styrene-*co*-butadiene) and styrene	1975WHI
	poly(styrene-*co*-butadiene) and tetrahydrofuran	1972WHI
	poly(styrene-*co*-butadiene) and toluene	1984HAS
	poly(styrene-*co*-butadiene) and toluene	1984SAS
	poly(vinyl acetal) and trichloromethane	1947DOB
	poly(vinyl methyl ether) and trichloroethene	1977RO2
	poly(vinyl methyl ether) and trichloromethane	1977RO1
	poly(vinyl methyl ether) and trichloromethane	1979LIP
	poly(4-vinylpyridine) and trichloromethane	2006TO1
	poly(1-vinyl-2-pyrrolidinone) and trichloromethane	2006TO1
	2-propanone and toluene	1966HEI
	2-propanone and toluene	1967SUH
	2-propanone and trichloromethane	1970WO2
	2-propanone and n-tridecane	1983CO2
Poly(styrene-*co*-acrylamide)		
	poly(vinyl acetate-*co*-vinyl alcohol) and tetrahydrofuran	1982LE1
Poly(styrene-*co*-acrylic acid)		
	poly(vinyl acetate-*co*-vinyl alcohol) and tetrahydrofuran	1982LE1
Poly(styrene-*co*-acrylonitrile)		
	N,N-dimethylacetamide and poly(acrylonitrile)	2001POS
	N,N-dimethylacetamide and poly(acrylonitrile)	2002KUL
	N,N-dimethylacetamide and polystyrene	2001POS
	N,N-dimethylacetamide and polystyrene	2002KUL
	N,N-dimethylacetamide and polystyrene	2003LOS
	dimethyl phthalate and poly(methyl methacrylate)	1977BER
	methanol and 2-propanone	1974GLO
	poly(styrene-*co*-acrylonitrile) and toluene	1972TE1

Polymer (B)	**Second and third component**	**Ref.**
Poly(styrene-*co*-butadiene)		
	N,N-dimethylacetamide and poly(acrylonitrile)	2001POS
	N,N-dimethylacetamide and poly(acrylonitrile)	2002KUL
	N,N-dimethylacetamide and polystyrene	2001POS
	N,N-dimethylacetamide and polystyrene	2002KUL
	N,N-dimethylacetamide and polystyrene	2003LOS
	polybutadiene and toluene	1989IN2
	polystyrene and styrene	1975WHI
	polystyrene and tetrahydrofuran	1972WHI
	polystyrene and toluene	1984HAS
	polystyrene and toluene	1984SAS
Poly(styrene-*b*-butadiene-*b*-styrene)		
	polybutadiene and 4-methyl-2-pentanone	1997HWA
Poly(styrene-*b*-dimethylsiloxane)		
	cyclohexane and polystyrene	1997SC2
Poly(styrene-*b*-isoprene)		
	N,N-dimethylformamide and methylcyclohexane	1993PO1
Poly(styrene-*co*-methacrylic acid)		
	benzene and poly(ethylene oxide)	1982LE2
	1,4-dioxane and poly(styrene-*co*-4-vinylpyridine)	1983DJA
	poly(ethylene oxide) and tetrahydrofuran	1982LE2
	poly(ethylene oxide) and trichloromethane	1982LE2
	poly(styrene-*co*-4-vinylpyridine) and trichloromethane	1983DJA
	poly(4-vinylpyridine) and trichloromethane	2006TO1
	poly(4-vinylpyridine) and trichloromethane	2006TO2
	poly(1-vinyl-2-pyrrolidinone) and trichloromethane	2006TO1
Poly(styrene-*co*-2-methoxyethyl methacrylate)		
	tetrachloroethene and dimethylsulfoxide	1993PO2
Poly(styrene-*co*-methyl methacrylate)		
	decahydronaphthalene and nitroethane	1994KAW

Polymer (B)	Second and third component	Ref.
Poly(styrene-*co*-vinylphenol)		
	poly(4-vinylpyridine) and trichloromethane	2006TO1
	poly(1-vinyl-2-pyrrolidinone) and trichloromethane	2006TO1
Poly(styrene-*co*-4-vinylpyridine)		
	2-butanone and poly(methyl methacrylate-*co*-4-vinylpyridine)	1977DJA
	1,4-dioxane and poly(styrene-*co*-methacrylic acid)	1983DJA
	1,4-dioxane and poly(methyl methacrylate-*co*-4-vinylpyridine)	1977DJA
	poly(methyl methacrylate-*co*-4-vinylpyridine) and trichloromethane	1977DJA
	poly(styrene-*co*-methacrylic acid) and trichloromethane	1983DJA
	poly(vinyl acetate-*co*-methacrylic acid) and tetrahydrofuran	1982LE1
	poly(vinyl acetate-*co*-vinyl alcohol) and tetrahydrofuran	1982LE1
Poly(styrene-*co*-1-vinyl-2-pyrrolidinone)		
	poly(vinyl acetate-*co*-vinyl alcohol) and tetrahydrofuran	1982LE1
Polysulfone		
	N,N-dimethylacetamide and ethanol	1985WI2
	N,N-dimethylacetamide and water	1985WI2
	N,N-dimethylacetamide and water	1989GAI
	N,N-dimethylacetamide and water	1991LA1
	N,N-dimethylacetamide and water	1993BOO
	N,N-dimethylformamide and ethanol	1988SWI
	N,N-dimethylformamide and methanol	1988SWI
	N,N-dimethylformamide and 1-propanol	1988SWI
	N,N-dimethylformamide and 2-propanone	1988SWI
	N,N-dimethylformamide and tetrachloromethane	1988SWI
	N,N-dimethylformamide and water	1980BRO
	N,N-dimethylformamide and water	1988SWI
	N,N-dimethylformamide and water	1991LA1
	N,N-dimethylformamide and water	2000BA2
	dimethylpropyleneurea and water	1991LA1
	dimethylsulfoxide and water	1991LA1
	ethanol and trichloromethane	1985WI2
	methanol and trichloromethane	1985WI2
	1-methyl-2-pyrrolidinone and water	1988SWI

Polymer (B)	Second and third component	Ref.
Polysulfone		
	1-methyl-2-pyrrolidinone and water	1991LA1
	1-methyl-2-pyrrolidinone and water	1997KIM
	1-methyl-2-pyrrolidinone and water	1999BAI
	1-methyl-2-pyrrolidinone and water	2004LEE
	2-propanol and trichloromethane	1985WI2
	tetrahydrofuran and water	1997KIM
	1,1,3,3-tetramethylurea and water	1991LA1
Polysulfone (carboxylated)		
	N,N-dimethylacetamide and water	1991LA2
	N,N-dimethylformamide and water	1991LA2
	dimethylsulfoxide and water	1991LA2
	1-methyl-2-pyrrolidinone and water	1991LA2
Poly[tetrafluoroethylene-*co*-perfluoro(alkyl vinyl ether)]		
	water and xanthan	2003KOE
Poly(tetrahydrofuran)		
	cyclohexane and methanol	1980DON
Polyurethane		
	N,N-dimethylformamide and ethyl acetate	1997CHA
	N,N-dimethylformamide and water	1977KOE
	N,N-dimethylformamide and water	1999KIM
Poly(*N*-vinylacetamide-*co*-methyl acrylate)		
	1-butanol and water	2004MOR
	ethanol and water	2004MOR
	methanol and water	2004MOR
	1-propanol and water	2004MOR
	2-propanol and water	2004MOR
Poly(vinyl acetate)		
	2-butanone and poly(vinyl acetal)	1947DOB
	1,2-dichloroethane and poly(methyl acrylate)	1985NAN
	2-propanone and poly(vinyl acetal)	1947DOB
	toluene and poly(methyl acrylate)	1985NAN
	toluene and poly(methyl acrylate)	1989BHA
	trichloromethane and poly(vinyl acetal)	1947DOB

Polymer (B)	Second and third component	Ref.
Poly(vinyl acetate-*co*-methacrylic acid)	poly(styrene-*co*-4-vinylpyridine) and tetrahydrofuran	1982LE1
Poly(vinyl acetate-*co*-vinyl alcohol)	1,3-diethylurea and water	1967SAI
	1,3-dimethylurea and water	1967SAI
	formamide and water	1967SAI
	poly(acrylamide) and water	1984HEF
	poly(ethylene glycol) and water	1984HEF
	poly(styrene-*co*-acrylamide) and tetrahydrofuran	1982LE1
	poly(styrene-*co*-acrylic acid) and tetrahydrofuran	1982LE1
	poly(styrene-*co*-4-vinylpyridine) and tetrahydrofuran	1982LE1
	poly(styrene-*co*-1-vinyl-2-pyrrolidinone) and tetrahydrofuran	1982LE1
	poly(vinyl alcohol) and water	1997PAE
Poly(vinyl acetate-*co*-1-vinyl-2-pyrrolidinone)	1,3-dimethylurea and water	1967SAI
Poly(vinyl alcohol)	dextran and water	1958ALB
	dextran and water	1986ZAS
	dextran and water	1989ZAS
	dimethylsulfoxide and water	2002YOU
	dimethylsulfoxide and water	2003TAK
	methyl(hydroxypropyl)cellulose and water	1993SAK
	poly(acrylamide) and water	1982MIN
	poly(ethylene glycol) and water	1984INA
	poly(ethylene glycol) and water	1986INA
	poly(vinyl acetate-*co*-vinyl alcohol) and water	1997PAE
Poly(*N*-vinylcaprolactam)	acetonitrile and water	1996KIR
	2-aminoethanol and water	1996KIR
	tert-butanol and water	1996KIR
	ethanol and water	1996KIR
	methanol and water	1996KIR
	methanol and water	2002MA1
	phenol and water	1996KIR
	1-propanol and water	1996KIR
	2-propanol and water	1996KIR

Polymer (B)	Second and third component	Ref.
Poly(*N*-vinylcaprolactam-*co*-1-vinylimidazole)		
	dextran and water	1997FRA
	starch-*g*-polyacrylamide and water	2000PIE
Poly(vinyl chloride)		
	butyl acrylate and poly(butyl acrylate)	1984WAL
	poly(butyl acrylate) and vinyl chloride	1982WAL
	poly(methyl methacrylate) and tetrahydrofuran	1986CH1
	poly(methyl methacrylate) and tetrahydrofuran	1986CH2
Poly(vinylidene fluoride)		
	N,N-dimethylacetamide and ethanol	2003YEO
	N,N-dimethylacetamide and ethanol	2006ZUO
	N,N-dimethylacetamide and hydroxpropylcellulose	1992AMB
	N,N-dimethylacetamide and methanol	2006ZUO
	N,N-dimethylacetamide and 1,2,3-propanetriol	2003YEO
	N,N-dimethylacetamide and 2-propanol	2006ZUO
	N,N-dimethylacetamide and water	1991BOT
	N,N-dimethylacetamide and water	2003YEO
	N,N-dimethylacetamide and water	2006ZUO
	N,N-dimethylformamide and 1-octanol	1995SOH
	N,N-dimethylformamide and 1-octanol	1999CH1
	N,N-dimethylformamide and water	1991BOT
	N,N-dimethylformamide and water	1999CH1
	N,N-dimethylformamide and water	1999CH2
	N,N-dimethylformamide and water	1999MAT
	N,N-dimethylformamide and water	2003YEO
	dimethylsulfoxide and water	1991BOT
	hexamethylphosphotriamide and water	1991BOT
	1-methyl-2-pyrrolidinone and water	1991BOT
	1-methyl-2-pyrrolidinone and water	2003YEO
	1,1,3,3-tetramethylurea and water	1991BOT
	triethyl phosphate and water	1991BOT
	triethyl phosphate and water	2003YEO
	triethyl phosphate and water	2006LIN
	trimethyl phosphate and water	1991BOT
Poly(*N*-vinylisobutyramide)		
	dextran and water	1999KIS

Polymer (B)	Second and third component	Ref.
Poly(vinyl methyl ether)		
	cyclohexane and methanol	1994TAK
	dextran and water	1984HEF
	deuterium oxide and poly(*N*-isopropyl-methacrylamide)	2005SP2
	methanol and water	1991SC1
	petroleum ether and toluene	1995PET
	poly(ethylene glycol) and water	1984HEF
	polystyrene and trichloroethene	1977RO2
	polystyrene and trichloromethane	1977RO1
	polystyrene and trichloromethane	1979LIP
	poly(1-vinyl-2-pyrrolidinone) and water	1984HEF
	tetrahydrofuran and water	2004BER
Poly(vinyl propionate)		
	toluene and poly(ethyl acrylate)	1989BHA
Poly(1-vinylpyridine)		
	bisphenol-A and poly(1-vinyl-2-pyrrolidinone)	2002LIX1
	cyclohexane and methanol	1979IZU
	poly(1-vinyl-2-pyrrolidinone) and suberic acid	2002LIX2
	poly(1-vinyl-2-pyrrolidinone) and succinic acid	2002LIX2
Poly(2-vinylpyridine)		
	poly(1-vinyl-2-pyrrolidinone) and suberic acid	2002LIX2
	poly(1-vinyl-2-pyrrolidinone) and succinic acid	2002LIX2
Poly(4-vinylpyridine)		
	polystyrene and trichloromethane	2006TO1
	poly(styrene-*co*-methacrylic acid) and trichloromethane	2006TO1
	poly(styrene-*co*-methacrylic acid) and trichloromethane	2006TO2
	poly(styrene-*co*-vinylphenol) and trichloromethane	2006TO1
Poly(1-vinyl-2-pyrrolidinone)		
	bisphenol-A and poly(1-vinylpyridine)	2002LIX1
	dextran and water	1958ALB
	dextran and water	1984HEF
	dextran and water	1986ZAS
	dextran and water	1987ZA3

Polymer (B)	Second and third component	Ref.
Poly(1-vinyl-2-pyrrolidinone)		
	dextran and water	1988ZAS
	dextran and water	1989ZAS
	poly(acrylamide) and water	1989PER
	poly(ethylene glycol) and water	1984HEF
	poly(ethylene glycol) and water	2004INA
	poly(ethylene glycol) monoalkyl ether and water	1996PAN
	poly(ethylene oxide) and water	1998NIL
	polystyrene and trichloromethane	2006TO1
	poly(styrene-*co*-methacrylic acid) and trichloromethane	2006TO1
	poly(styrene-*co*-vinylphenol) and trichloromethane	2006TO1
	poly(vinyl methyl ether) and water	1984HEF
	poly(2-vinylpyidine) and suberic acid	2002LIX2
	poly(2-vinylpyridine) and succinic acid	2002LIX2
Poly(vinyltrimethylsilane)		
	1,2-chlorobenzene and 2-methyl-1-propanol	1980RAB
	dichloromethane and 2-methyl-1-propanol	1980RAB
Starch		
	poly(ethylene glycol) and water	1988LAR
	water and xanthan	2004MAN
Starch-*g*-polyacrylamide		
	poly(ethylene glycol) and water	2000PIE
	poly(*N*-vinylcaprolactam-*co*-1-vinylimidazole) and water	2000PIE
Tetra(ethylene glycol) mono dodecyl ether	ethyl(hydroxyethyl)cellulose and water	1992ZHA
	ethyl(hydroxyethyl)cellulose and water	1994ZH1
	poly(ethylene oxide-*co*-propylene oxide) and water	1992ZHA
	poly(ethylene oxide-*co*-propylene oxide) and water	1994ZH1
Tetra(ethylene glycol) mono octyl ether	poly(ethylene oxide) and water	1991WOR

Polymer (B)	Second and third component	Ref.
Trimethylsilylcellulose		
	dimethylsulfoxide and tetrahydrofuran	1998STO
	dimethylsulfoxide and toluene	1998STO
Valeryl dextran		
	benzoyl dextran and water	1994LUM
	dextran and water	1995LUM
	poly(ethylene glycol) and water	1994LUM
Xanthan		
	amylose and water	2004MAN
	poly[tetrafluoroethylene-*co*-perfluoro(alkyl vinyl ether)] and water	2003KOE
	starch and water	2004MAN

4. LIQUID-LIQUID EQUILIBRIUM DATA OF QUATERNARY POLYMER SOLUTIOS

4.1. Cloud-point and/or coexistence data

Polymer (B): **polyetherimide** **2001ALB**
Characterization: Ultem 1000, General Electric
Solvent (A): **water** H_2O **7732-18-5**
Solvent (C): **1-methyl-2-pyrrolidinone** C_5H_9NO **872-50-4**
Solvent (D): ***N,N*-dimethylformamide** C_3H_7NO **68-12-2**

Type of data: cloud points

T/K = 295-297

w_A	0.03021	0.02600	0.02404	0.02339	0.02209	0.01738	0.01570	0.01440	0.01205
w_B	0.010	0.019	0.039	0.059	0.078	0.010	0.020	0.039	0.059
w_C	0.4799	0.4775	0.4685	0.4588	0.4500	0.2432	0.2411	0.2367	0.2322
w_D	0.4799	0.4775	0.4685	0.4588	0.4500	0.7295	0.7232	0.7100	0.6967
w_A	0.01110	0.01040	0.01295	0.01187	0.01025	0.01099	0.00929	0.00502	0.00420
w_B	0.079	0.099	0.010	0.020	0.040	0.059	0.079	0.020	0.040
w_C	0.2275	0.2227	0.1954	0.1936	0.1900	0.1860	0.1823	0.1706	0.1673
w_D	0.6824	0.6680	0.7816	0.7745	0.7598	0.7440	0.7294	0.8044	0.7885
w_A	0.00296	0.00458	0.00413	0.00320	0.00687	0.00569	0.00426	0.00545	
w_B	0.080	0.020	0.060	0.080	0.010	0.020	0.040	0.060	
w_C	0.1605	0.1463	0.1404	0.1375	0.0983	0.0974	0.0956	0.0935	
w_D	0.7566	0.8291	0.7955	0.7793	0.8848	0.8769	0.8602	0.8411	

Polymer (B): **polyetherimide** **1989ROE**
Characterization: M_w/g.mol^{-1} = 32800, ρ = 1.27 g/cm^3, Ultem 1000
Solvent (A): **water** H_2O **7732-18-5**
Solvent (C): **1-methyl-2-pyrrolidinone** C_5H_9NO **872-50-4**
Polymer (D): **poly(1-vinyl-2-pyrrolidinone)**
Characterization: M_w/g.mol^{-1} = 423000, ρ = 1.22 g/cm^3

Type of data: cloud points

T/K = 294

w_A	0.055	0.053	0.052	0.052	0.048	0.046	0.039	0.035	0.033
w_B	0.006	0.009	0.011	0.013	0.019	0.022	0.029	0.035	0.038
w_C	0.931	0.929	0.926	0.922	0.914	0.910	0.903	0.895	0.891
w_D	0.006	0.009	0.011	0.013	0.019	0.022	0.029	0.035	0.038
w_A	0.029	0.028	0.025	0.021	0.016				
w_B	0.044	0.048	0.056	0.071	0.174				
w_C	0.883	0.876	0.863	0.837	0.636				
w_D	0.044	0.048	0.056	0.071	0.174				

Polymer (B): **polyethylene** **2004AGA**
Characterization: M_w/g.mol^{-1} = 30400, T_m/K = 405, crystallinity = 0.399, ρ = 0.919 g/cm^3
Solvent (A): **1,2-dimethylbenzene** **C_8H_{10}** **95-47-6**
Solvent (C): **1,3-dimethylbenzene** **C_8H_{10}** **108-38-3**
Solvent (D): **1,4-dimethylbenzene** **C_8H_{10}** **106-42-3**

Type of data: cloud points

$x_A/x_C/x_D$ = 0.39/0.15/0.46 was kept constant

w_B	0.0149	0.0540	0.110	0.250	0.360	0.420
T/K	370	375	379	378	379	381

Polymer (B): **polyethylene** **2004AGA**
Characterization: M_w/g.mol^{-1} = 32600, T_m/K = 410, crystallinity = 0.444, ρ = 0.927 g/cm^3
Solvent (A): **1,2-dimethylbenzene** **C_8H_{10}** **95-47-6**
Solvent (C): **1,3-dimethylbenzene** **C_8H_{10}** **108-38-3**
Solvent (D): **1,4-dimethylbenzene** **C_8H_{10}** **106-42-3**

Type of data: cloud points

$x_A/x_C/x_D$ = 0.39/0.15/0.46 was kept constant

w_B	0.0149	0.0540	0.110	0.250	0.360	0.420
T/K	376	377	380	380	382	384

Polymer (B): **poly(ethylene glycol)** **1993GA1**
Characterization: M_n/g.mol^{-1} = 585, M_w/g.mol^{-1} = 600,
Solvent (A): **water** **H_2O** **7732-18-5**
Polymer (C): **dextran**
Characterization: M_n/g.mol^{-1} = 109300, M_w/g.mol^{-1} = 493100
Polymer (D): **poly(ethylene glycol)**
Characterization: M_n/g.mol^{-1} = 2810, M_w/g.mol^{-1} = 3026

Type of data: coexistence data (tie lines)

T/K = 293.15

Total system			Top phase			Bottom phase		
w_B	w_C	w_D	w_B	w_C	w_D	w_B	w_C	w_D
0.0203	0.0631	0.0675	0.0885	0.0101	0.0237	0.0261	0.1606	0.0161
0.0220	0.0738	0.0732	0.1028	0.0058	0.0272	0.0191	0.1977	0.0155
0.0236	0.0893	0.0785	0.1160	0.0018	0.0290	0.0132	0.2343	0.0154
0.0250	0.1038	0.0831	0.1293	0.0005	0.0321	0.0107	0.2637	0.0156
0.0261	0.1175	0.0869	0.1412	0.0010	0.0352	0.0084	0.2881	0.0157
0.0270	0.1324	0.0898	0.1527	0.0008	0.0369	0.0069	0.3102	0.0153

continued

continued

Total system			Top phase			Bottom phase		
w_B	w_C	w_D	w_B	w_C	w_D	w_B	w_C	w_D
0.0370	0.1492	0.0110	0.0173	0.0067	0.0969	0.0114	0.1766	0.0234
0.0306	0.1777	0.0091	0.0160	0.0038	0.1058	0.0101	0.1980	0.0191
0.0265	0.2228	0.0079	0.0170	0.0011	0.1282	0.0078	0.2463	0.0117
0.0252	0.2518	0.0075	0.0173	0.0025	0.1470	0.0079	0.2791	0.0084
0.0416	0.0781	0.0488	0.0551	0.0334	0.0574	0.0460	0.1016	0.0333
0.0438	0.0908	0.0515	0.0635	0.0125	0.0716	0.0373	0.1549	0.0270
0.0516	0.1053	0.0605	0.0785	0.0036	0.0885	0.0296	0.2189	0.0214
0.0360	0.1065	0.0423	0.0747	0.0009	0.1169	0.0241	0.2803	0.0174
0.0818	0.0220	0.0244	0.0271	0.0068	0.0858	0.0189	0.1567	0.0251
0.0940	0.0191	0.0281	0.0301	0.0031	0.0996	0.0184	0.1894	0.0207
0.1133	0.0160	0.0338	0.0355	0.0013	0.1180	0.0192	0.2471	0.0103
0.1184	0.0133	0.0354	0.0368	0.0009	0.1222	0.0192	0.2509	0.0105
0.1291	0.0121	0.0386	0.0398	0.0003	0.1323	0.0200	0.2667	0.0089

Fractionation during demixing (the lines correspond with those above):

Top phase				Bottom phase			
PEG		Dextran		PEG		Dextran	
M_n/ g mol^{-1}	M_w/ g mol^{-1}	M_n/ g mol^{-1}	M_w/ g mol^{-1}	M_n/ g mol^{-1}	M_w/ g mol^{-1}	M_n/ g mol^{-1}	M_w/ g mol^{-1}
1549	2411	59841	115909	1090	1955	123630	511782
1369	2372	50205	134610	937	1775	115613	471896
1469	2414	34081	139665	826	1550	112915	477048
1487	2414	34189	39976	774	1440	111116	480602
1460	2409	37775	68868	667	1309	109415	468984
1502	2423	66268	123180	683	1228	108017	460696
1797	2561	47576	67225	1183	2063	109187	471143
1901	2609	41064	57452	1214	2072	110979	478258
1977	2630	37856	59561	1089	1906	110594	473940
2063	2694	23308	32140	984	1756	110429	478511
906	1653	71440	140824	809	1463	111000	479326
936	1712	49253	75268	733	1267	105901	441200
935	1703	35478	41990	640	993	101988	416771
1068	1886	28927	36634	621	953	99362	428621

continued

continued

Top phase PEG M_n/ g mol^{-1}	Top phase PEG M_w/ g mol^{-1}	Top phase Dextran M_n/ g mol^{-1}	Top phase Dextran M_w/ g mol^{-1}	Bottom phase PEG M_n/ g mol^{-1}	Bottom phase PEG M_w/ g mol^{-1}	Bottom phase Dextran M_n/ g mol^{-1}	Bottom phase Dextran M_w/ g mol^{-1}
1487	2350	62046	91881	1057	1880	156789	619956
1482	2354	47534	67142	991	1782	138075	542165
1503	2379	38062	56423	783	1351	122639	484896
1490	2364	41095	55018	764	1351	122098	503323
1484	2356	35051	39940	794	1273	117152	492948

Polymer (B): **poly(ethylene glycol)** **1991HAR**
Characterization: M_n/g.mol^{-1} = 7620, M_w/g.mol^{-1} = 8000, Sigma Chemical Co., Inc., St. Louis, MO
Solvent (A): **water** **H_2O** **7732-18-5**
Polymer (C): **dextran**
Characterization: M_n/g.mol^{-1} = 182000, M_w/g.mol^{-1} = 500000, Sigma Chemical Co., Inc., St. Louis, MO
Polymer (D): **poly(ethylene glycol)**
Characterization: M_n/g.mol^{-1} = 910, M_w/g.mol^{-1} = 1000, Sigma Chemical Co., Inc., St. Louis, MO

Type of data: coexistence data (tie lines)

T/K = 298.15

Total system				Bottom phase				Top phase			
w_A	w_B	w_C	w_D	w_A	w_B	w_C	w_D	w_A	w_B	w_C	w_D
0.8650	0.0500	0.0700	0.0150	0.8943	0.0860	0.0000	0.0197	0.8065	0.0038	0.1780	0.0117
0.8650	0.0500	0.0700	0.0150	0.8955	0.0860	0.0005	0.0180	0.8418	0.0038	0.1430	0.0114
0.8430	0.0580	0.0840	0.0150	0.8939	0.0987	0.0004	0.0070	0.7834	0.0007	0.2077	0.0082
0.8580	0.0430	0.0840	0.0150	0.9017	0.0807	0.0000	0.0176	0.8088	0.0015	0.1799	0.0098
0.8790	0.0440	0.0620	0.0150	0.9133	0.0696	0.0004	0.0167	0.8242	0.0028	0.1648	0.0082
0.8940	0.0290	0.0620	0.0150	0.9297	0.0520	0.0000	0.0183	0.8763	0.0070	0.1044	0.0123
0.8640	0.0440	0.0620	0.0300	0.9000	0.0650	0.0000	0.0350	0.8109	0.0023	0.1650	0.0218
0.8850	0.0500	0.0500	0.0150	0.9141	0.0700	0.0000	0.0159	0.8249	0.0031	0.1620	0.0100
0.9000	0.0350	0.0500	0.0150	0.9251	0.0560	0.0015	0.0174	0.8622	0.0076	0.1188	0.0114
0.8850	0.0600	0.0400	0.0150	0.9066	0.0780	0.0000	0.0154	0.8252	0.0035	0.1630	0.0083
0.9000	0.0450	0.0400	0.0150	0.9215	0.0627	0.0010	0.0148	0.8471	0.0067	0.1366	0.0096
0.8700	0.0600	0.0400	0.0300	0.8832	0.0782	0.0000	0.0386	0.7968	0.0015	0.1830	0.0187

Polymer (B):	**poly(ethylene glycol)**	**1989KAN**
Characterization:	M_n/g.mol^{-1} = 900, M_w/g.mol^{-1} = 1000 Polysciences, Inc., Warrington, PA	
Solvent (A):	**water** **H_2O**	**7732-18-5**
Polymer (C):	**dextran**	
Characterization:	M_n/g.mol^{-1} = 60700, M_w/g.mol^{-1} = 167000 Sigma Chemical Co., Inc., St. Louis	
Polymer (D):	**poly(ethylene glycol)**	
Characterization:	M_n/g.mol^{-1} = 4760, M_w/g.mol^{-1} = 5000 Polysciences, Inc., Warrington, PA	
Polymer (E):	**poly(ethylene glycol)**	
Characterization:	M_n/g.mol^{-1} = 20750, M_w/g.mol^{-1} = 22000 Polysciences, Inc., Warrington, PA	

Type of data: coexistence data (tie lines)

T/K = 293.15

Comments: The ratio of PEG/dextran in the total system was kept constant at 4.2 wt%/4.2 wt%.

Total system				Bottom phase				Top phase			
w_B	w_C	w_D	w_E	w_B	w_C	w_D	w_E	w_B	w_C	w_D	w_E
0.0126	0.0420	0.0084	0.0210	0.0144	0.0080	0.0132	0.0341	0.0112	0.0800	0.0072	0.0039
0.0000	0.0420	0.0084	0.0336	0.0000	0.0017	0.0140	0.0504	0.0000	0.1080	0.0060	0.0040
0.0063	0.0420	0.0063	0.0294	0.0069	0.00313	0.0097	0.0457	0.0051	0.1020	0.0043	0.0000
0.0000	0.0420	0.0063	0.0357	0.0000	0.0000	0.0100	0.0543	0.0000	0.1183	0.0037	0.0000

Polymer (B):	**poly(ethylene glycol)**	**1993SAR**
Characterization:	M_n/g.mol^{-1} = 3970, M_w/g.mol^{-1} = 4490, M_z/g.mol^{-1} = 4980, Merck KGaA, Darmstadt, Germany	
Solvent (A):	**water** **H_2O**	**7732-18-5**
Polymer (C):	**poly(ethylene glycol)**	
Characterization:	M_n/g.mol^{-1} = 5775, M_w/g.mol^{-1} = 6440, M_z/g.mol^{-1} = 7440, Merck KGaA, Darmstadt, Germany	
Polymer (D):	**poly(ethylene glycol)**	
Characterization:	M_n/g.mol^{-1} = 7385, M_w/g.mol^{-1} = 8640, M_z/g.mol^{-1} = 9760, Aldrich Chem. Co., Inc., Milwaukee, WI	

Type of data: cloud points (closed loop miscibility gap)

$w_B/w_C/w_D$ = 0.389/0.513/0.098 was kept constant

$w_B + w_C + w_D$	0.0105	0.0249	0.0531	0.1077	0.1489	0.2198	0.3012	0.3402
T/K	410.45	407.55	405.35	403.55	402.55	404.85	410.65	416.95
T/K	524.65	543.45	546.65	544.45	544.15	539.45	536.55	534.15

continued

continued

$w_B + w_C + w_D$	0.3914
T/K	408.35
T/K	528.15

Comments: The corresponding molar mass averages are given by M_n = 5000 g/mol, M_w = 6000 g/mol, and M_z = 7080 g/mol.

Polymer (B): **poly(ethylene glycol)** **1993SAR**
Characterization: M_n/g.mol^{-1} = 2750, M_w/g.mol^{-1} = 3110, M_z/g.mol^{-1} = 3545
Aldrich Chem. Co., Inc., Milwaukee, WI
Solvent (A): **water** **H_2O** **7732-18-5**
Polymer (C): **poly(ethylene glycol)**
Characterization: M_n/g.mol^{-1} = 3970, M_w/g.mol^{-1} = 4490, M_z/g.mol^{-1} = 4980
Merck KGaA, Darmstadt, Germany
Polymer (D): **poly(ethylene glycol)**
Characterization: M_n/g.mol^{-1} = 10500, M_w/g.mol^{-1} = 12490, M_z/g.mol^{-1} = 14240
Aldrich Chem. Co., Inc., Milwaukee, WI

Type of data: cloud points (closed loop miscibility gap)

$w_B/w_C/w_D$ = 0.524/0.197/0.279 was kept constant

$w_B + w_C + w_D$	0.0106	0.0246	0.0528	0.0986	0.1440	0.1909	0.2954	0.3405
T/K	402.65	401.05	399.85	401.05	402.05	404.35	409.05	414.55
T/K	536.35	549.45	557.35	552.65	545.55	541.85	531.55	528.85

$w_B + w_C + w_D$	0.3825
T/K	418.55
T/K	523.15

Comments: The corresponding molar mass averages are given by M_n = 3750 g/mol, M_w = 6000 g/mol, and M_z = 9960 g/mol.

Polymer (B): **poly(ethylene glycol)** **1993SAR**
Characterization: M_n/g.mol^{-1} = 2750, M_w/g.mol^{-1} = 3110, M_z/g.mol^{-1} = 3545,
Aldrich Chem. Co., Inc., Milwaukee, WI
Solvent (A): **water** **H_2O** **7732-18-5**
Polymer (C): **poly(ethylene glycol)**
Characterization: M_n/g.mol^{-1} = 3970, M_w/g.mol^{-1} = 4490, M_z/g.mol^{-1} = 4980,
Merck KGaA, Darmstadt, Germany
Polymer (D): **poly(ethylene glycol)**
Characterization: M_n/g.mol^{-1} = 5775, M_w/g.mol^{-1} = 6440, M_z/g.mol^{-1} = 7440,
Merck KGaA, Darmstadt, Germany
Polymer (E): **poly(ethylene glycol)**
Characterization: M_n/g.mol^{-1} = 10500, M_w/g.mol^{-1} = 12490, M_z/g.mol^{-1} = 14240,
Aldrich Chem. Co., Inc., Milwaukee, WI

continued

continued

Type of data: cloud points (closed loop miscibility gap)

$w_B/w_C/w_D/w_E$ = 0.245/0.331/0.264/0.160 was kept constant

$w_B + w_C + w_D + w_E$	0.0101	0.0251	0.0511	0.0972	0.1320	0.2025	0.3011	0.3361
T/K	406.65	401.45	400.95	402.85	405.25	406.15	409.15	411.15
T/K	533.45	549.15	551.85	549.35	544.65	541.15	533.85	532.65

$w_B + w_C + w_D + w_E$	0.3818
T/K	417.05
T/K	527.15

Comments: The corresponding molar mass averages are given by M_n = 4285 g/mol, M_w = 6000 g/mol, and M_z = 8580 g/mol.

Type of data: cloud points (closed loop miscibility gap)

$w_B/w_C/w_D/w_E$ = 0.158/0.584/0.058/0.200 was kept constant

$w_B + w_C + w_D + w_E$	0.0114	0.0245	0.0492	0.0965	0.1487	0.2468	0.2985	0.3485
T/K	407.75	403.15	400.45	401.65	403.55	406.35	408.25	414.75
T/K	531.55	537.45	553.05	550.15	545.15	538.35	534.75	531.05

$w_B + w_C + w_D + w_E$	0.4067
T/K	421.65
T/K	523.55

Comments: The corresponding molar mass averages are given by M_n = 4285 g/mol, M_w = 6000 g/mol, and M_z = 8880 g/mol.

Polymer (B): **polystyrene** **1985TON**
Characterization: M_n/g.mol^{-1} = 10800, M_w/g.mol^{-1} = 11000
Solvent (A): **cyclohexane** C_6H_{12} **110-82-7**
Polymer (C): **polystyrene**
Characterization: M_n/g.mol^{-1} = 44100, M_w/g.mol^{-1} = 44500
Polymer (D): **polystyrene**
Characterization: M_n/g.mol^{-1} = 182000, M_w/g.mol^{-1} = 195000
Polymer (E): **polystyrene**
Characterization: M_n/g.mol^{-1} = 799000, M_w/g.mol^{-1} = 807000

Comments: The polymer mixture is made of 40 wt% (B) + 30 wt% (C) + 20 wt% (D) + 10 wt% (E).

Feed phase	Sol phase				Gel phase			
$(1 - \varphi_A)$	φ_B	φ_C	φ_D	φ_E	φ_B	φ_C	φ_D	φ_E
T/K = 288.15								
0.08647	0.03248	0.02091	0.00322	–	0.04360	0.05558	0.10664	0.05972
0.10068	0.03795	0.02471	0.00581	–	0.05274	0.05921	0.09903	0.06329
0.11490	0.04356	0.02847	0.00772	–	0.05491	0.05565	0.07955	0.05384
0.13100	0.04925	0.03362	0.01319	–	0.06329	0.05891	0.06751	0.05095
0.15331	0.05988	0.04117	0.01968	0.00188	0.06321	0.05216	0.04792	0.03375

continued

continued

Feed phase $(1-\varphi_A)$	Sol phase φ_B	φ_C	φ_D	φ_E	Gel phase φ_B	φ_C	φ_D	φ_E
T/K = 291.15								
0.07774	0.02938	0.02079	0.00800	–	0.03335	0.03859	0.07890	0.06730
0.09449	0.03593	0.02548	0.01154	–	0.04106	0.04159	0.06690	0.06070
0.10636	0.04148	0.02922	0.01390	–	0.04475	0.04282	0.05689	0.05488

Comments: The polymer mixture is composed of 25 wt% (B) + 25 wt% (C) + 25 wt% (D) + 25 wt% (E).

Feed phase $(1-\varphi_A)$	Sol phase φ_B	φ_C	φ_D	φ_E	Gel phase φ_B	φ_C	φ_D	φ_E
T/K = 288.15								
0.07383	0.01733	0.01392	0.00104	–	0.02647	0.04533	0.10941	0.11031
0.09867	0.02298	0.01659	0.00149	–	0.03317	0.05103	0.10360	0.10166
0.13476	0.03171	0.02130	0.00286	–	0.04264	0.05226	0.08645	0.08613
0.17141	0.03830	0.02505	0.00379	–	0.05130	0.05688	0.07606	0.07394
T/K = 291.15								
0.09048	0.02201	0.01851	0.00668	–	0.02548	0.03526	0.07327	0.08931
0.11391	0.02726	0.02245	0.00906	–	0.03152	0.03897	0.06509	0.07960
0.12609	0.02926	0.02476	0.01027	–	0.03372	0.04004	0.06300	0.07267
0.15629	0.03574	0.03005	0.01487	0.00312	0.04098	0.04347	0.05135	0.05732

Comments: The polymer mixture is composed of 10 wt% (B) + 20 wt% (C) + 30 wt% (D) + 40 wt% (E).

Feed phase $(1-\varphi_A)$	Sol phase φ_B	φ_C	φ_D	φ_E	Gel phase φ_B	φ_C	φ_D	φ_E
T/K = 288.15								
0.09583	0.00882	0.01084	0.00050	–	0.01307	0.04046	0.10023	0.12801
0.11109	0.00986	0.01049	0.00044	–	0.01528	0.04656	0.09899	0.12853
0.12852	0.01126	0.01117	0.00050	–	0.01634	0.04894	0.10211	0.13249
T/K = 291.15								
0.06076	0.00616	0.00920	0.00204	–	0.00775	0.02451	0.09161	0.12703
0.08021	0.00778	0.01159	0.00236	–	0.00961	0.02827	0.08621	0.11984
0.11893	0.01115	0.01524	0.00281	–	0.01358	0.03613	0.08328	0.11045
0.14911	0.01355	0.01648	0.00315	–	0.01653	0.04039	0.07702	0.10225

Polymer (B): **polystyrene** **1982GAL**
Characterization: M_n/g.mol^{-1} = 35000, M_w/g.mol^{-1} = 36000, M_z/g.mol^{-1} = 38000, Pressure Chemical Company, Pittsburgh, PA
Solvent (A): **cyclohexane** C_6H_{12} **110-82-7**
Polymer (C): **polystyrene**
Characterization: M_n/g.mol^{-1} = 108000, M_w/g.mol^{-1} = 115000, M_z/g.mol^{-1} = 122000, Pressure Chemical Company, Pittsburgh, PA
Polymer (D): **polystyrene**
Characterization: M_n/g.mol^{-1} = 178000, M_w/g.mol^{-1} = 184000, M_z/g.mol^{-1} = 195000, Pressure Chemical Company, Pittsburgh, PA

Comments: The composition of the ternary PS mixture is kept constant at $w_B/w_C/w_D$ = 0.4984/0.4966/0.0050. The corresponding molar mass averages are given by M_n = 59000 g/mol, M_w = 76000 g/mol, and M_z = 103000 g/mol.

Type of data: cloud points

$\varphi_B + \varphi_C + \varphi_D$	0.0338	0.0446	0.0538	0.0706	0.0750	0.0886	0.0936	0.1016
T/K	291.97	292.35	292.26	292.34	292.27	291.95	292.05	291.86
$\varphi_B + \varphi_C + \varphi_D$	0.1138	0.1230	0.1391	0.1512	0.1577	0.1698	0.1823	0.1973
T/K	291.74	291.61	291.37	291.31	291.24	290.93	290.91	290.54

Type of data: spinodal points

$\varphi_B + \varphi_C + \varphi_D$	0.1016	0.1138	0.1230	0.1391	0.1512	0.1577	0.1698	0.1973
T/K	291.57	291.53	291.53	291.21	291.13	291.11	290.85	290.14

Polymer (B): **polystyrene** **1982GAL**
Characterization: M_n/g.mol^{-1} = 35000, M_w/g.mol^{-1} = 36000, M_z/g.mol^{-1} = 38000, Pressure Chemical Company, Pittsburgh, PA
Solvent (A): **cyclohexane** C_6H_{12} **110-82-7**
Polymer (C): **polystyrene**
Characterization: M_n/g.mol^{-1} = 108000, M_w/g.mol^{-1} = 115000, M_z/g.mol^{-1} = 122000, Pressure Chemical Company, Pittsburgh, PA
Polymer (D): **polystyrene**
Characterization: M_n/g.mol^{-1} = 178000, M_w/g.mol^{-1} = 184000, M_z/g.mol^{-1} = 195000, Pressure Chemical Company, Pittsburgh, PA
Polymer (E): **polystyrene**
Characterization: M_n/g.mol^{-1} = 391000, M_w/g.mol^{-1} = 414000, M_z/g.mol^{-1} = 460000, Pressure Chemical Company, Pittsburgh, PA

Comments: The composition of the ternary PS mixture is kept constant at $w_B/w_C/w_D/w_E$ = 0.6284/0.2693/0.0868/0.0156. The corresponding molar mass averages are given by M_n = 48000 g/mol, M_w = 78000 g/mol, and M_z = 145000 g/mol.

Type of data: cloud points

$\varphi_B + \varphi_C + \varphi_D + \varphi_E$	0.0380	0.0473	0.0538	0.0584	0.0699	0.0744	0.0992	0.1153
T/K	294.15	293.86	293.65	293.80	293.39	293.14	292.29	291.82

continued

continued

$\varphi_B + \varphi_C + \varphi_D + \varphi_E$	0.1337	0.1526	0.1562	0.1725	0.1753	0.1871
T/K	291.31	290.86	290.82	290.44	290.38	289.97

Type of data: spinodal points

$\varphi_B + \varphi_C + \varphi_D + \varphi_E$	0.1153	0.1337	0.1526	0.1562	0.1725	0.1753	0.1871
T/K	291.52	291.21	290.75	290.66	290.33	290.26	289.70

Polymer (B): **polystyrene** **1982GAL**
Characterization: M_n/g.mol^{-1} = 35000, M_w/g.mol^{-1} = 36000, M_z/g.mol^{-1} = 38000, Pressure Chemical Company, Pittsburgh, PA
Solvent (A): **cyclohexane** C_6H_{12} **110-82-7**
Polymer (C): **polystyrene**
Characterization: M_n/g.mol^{-1} = 108000, M_w/g.mol^{-1} = 115000, M_z/g.mol^{-1} = 122000, Pressure Chemical Company, Pittsburgh, PA
Polymer (D): **polystyrene**
Characterization: M_n/g.mol^{-1} = 391000, M_w/g.mol^{-1} = 414000, M_z/g.mol^{-1} = 460000, Pressure Chemical Company, Pittsburgh, PA

Comments: The composition of the ternary PS mixture is kept constant at $w_B/w_C/w_D$ = 0.5719/0.4045/0.0236. The corresponding molar mass averages are given by M_n = 50000 g/mol, M_w = 77000 g/mol, and M_z = 142000 g/mol.

Type of data: cloud points

$\varphi_B + \varphi_C + \varphi_D$	0.0466	0.0536	0.0681	0.0739	0.0941	0.1171	0.1248	0.1361
T/K	294.21	294.10	293.41	293.10	292.41	291.68	291.59	291.26

$\varphi_B + \varphi_C + \varphi_D$	0.1459	0.1528	0.1669	0.1825	0.1905
T/K	291.09	290.97	290.70	290.35	290.00

Type of data: spinodal points

$\varphi_B + \varphi_C + \varphi_D$	0.1171	0.1248	0.1361	0.1459	0.1528	0.1669	0.1825
T/K	291.33	291.36	291.22	290.93	290.77	290.47	290.14

Polymer (B): **polystyrene** **1982GAL**
Characterization: M_n/g.mol^{-1} = 35000, M_w/g.mol^{-1} = 36000, M_z/g.mol^{-1} = 38000
Solvent (A): **cyclohexane** C_6H_{12} **110-82-7**
Polymer (C): **polystyrene**
Characterization: M_n/g.mol^{-1} = 153000, M_w/g.mol^{-1} = 162000, M_z/g.mol^{-1} = 177000, Pressure Chemical Company, Pittsburgh, PA
Polymer (D): **polystyrene**
Characterization: M_n/g.mol^{-1} = 391000, M_w/g.mol^{-1} = 414000, M_z/g.mol^{-1} = 460000, Pressure Chemical Company, Pittsburgh, PA

Comments: The composition of the ternary PS mixture is kept constant at $w_B/w_C/w_D$ = 0.6961/0.2956/0.0083. The corresponding molar mass averages are given by M_n = 46000 g/mol, M_w = 76000 g/mol, and M_z = 144000 g/mol.

continued

continued

Type of data: cloud points

$\varphi_B + \varphi_C + \varphi_D$	0.0357	0.0547	0.0657	0.0861	0.1065	0.1278	0.1448
T/K	292.85	293.01	292.99	292.36	291.78	290.96	290.66

$\varphi_B + \varphi_C + \varphi_D$	0.1551	0.1756	0.1814
T/K	290.34	289.85	289.66

Type of data: spinodal points

$\varphi_B + \varphi_C + \varphi_D$	0.1207	0.1448	0.1551	0.1756	0.1814
T/K	290.90	290.52	290.33	289.76	289.60

Polymer (B): **polystyrene** **1982GAL**
Characterization: M_n/g.mol^{-1} = 35000, M_w/g.mol^{-1} = 36000, M_z/g.mol^{-1} = 38000
Solvent (A): **cyclohexane** C_6H_{12} **110-82-7**
Polymer (C): **polystyrene**
Characterization: M_n/g.mol^{-1} = 153000, M_w/g.mol^{-1} = 162000, M_z/g.mol^{-1} = 177000, Pressure Chemical Company, Pittsburgh, PA
Polymer (D): **polystyrene**
Characterization: M_n/g.mol^{-1} = 394000, M_w/g.mol^{-1} = 508000, M_z/g.mol^{-1} = 593000, Pressure Chemical Company, Pittsburgh, PA

Comments: The composition of the ternary PS mixture is kept constant at $w_B/w_C/w_D$ = 0.6914/0.3046/0.0040. The corresponding molar mass averages are given by M_n = 46000 g/mol, M_w = 76000 g/mol, and M_z = 143000 g/mol.

Type of data: cloud points

$\varphi_B + \varphi_C + \varphi_D$	0.0240	0.0268	0.0283	0.0425	0.0504	0.0637	0.0770	0.0920
T/K	293.08	293.12	293.35	293.48	293.39	293.20	292.77	292.29

$\varphi_B + \varphi_C + \varphi_D$	0.1094	0.1280	0.1386	0.1478	0.1601	0.1858
T/K	291.86	291.18	290.89	290.69	290.56	289.77

Type of data: spinodal points

$\varphi_B + \varphi_C + \varphi_D$	0.1280	0.1386	0.1478	0.1601	0.1858
T/K	290.90	290.80	290.59	290.26	289.63

Polymer (B): **polystyrene** **1995LUS**
Characterization: M_n/g.mol^{-1} = 12750, M_w/g.mol^{-1} = 13500, Pressure Chemical Company, Pittsburgh, PA
Solvent (A): **2-propanone** C_3H_6O **67-64-1**
Solvent (C): **1,1,1,3,3,3-hexadeutero-2-propanone** C_3D_6O **666-52-4**
Polymer (D): **polystyrene**
Characterization: M_n/g.mol^{-1} = 21450, M_w/g.mol^{-1} = 22100, Polymer Laboratories, Inc., Amherst, MA

continued

continued

$x_C/x_A = 0.501$ was kept constant $x_D/x_B = 0.500$ was kept constant

Type of data: cloud points

w_{B+D}	0.210	0.210	0.210	0.210	0.210	0.210	0.210	0.210	0.210
T/K	335.83	340.15	343.69	347.37	351.31	353.49	358.40	362.30	368.51
P/MPa	2.55	2.10	1.81	1.62	1.51	1.47	1.52	1.63	1.90

w_{B+D}	0.210
T/K	378.34
P/MPa	2.58

Type of data: spinodal points

w_{B+D}	0.210	0.210	0.210	0.210	0.210	0.210	0.210	0.210	0.210
T/K	335.83	340.15	343.69	347.37	351.31	353.49	358.40	362.30	368.51
P/MPa	2.46	2.01	1.76	1.56	1.45	1.39	1.49	1.56	1.85

w_{B+D}	0.210
T/K	378.34
P/MPa	2.53

critical concentration: $w_{B+D,\,crit} = 0.210$

$x_C/x_A = 0.501$ was kept constant $x_D/x_B = 0.501$ was kept constant

Type of data: cloud points

w_{B+D}	0.103	0.103	0.103	0.103	0.103	0.103	0.103	0.103	0.103
T/K	325.99	329.97	335.66	341.31	347.11	356.39	364.63	373.45	381.86
P/MPa	2.35	1.76	1.05	0.73	0.54	0.59	0.79	1.35	2.20

w_{B+D}	0.265	0.265	0.265	0.265	0.265	0.265	0.265	0.265
T/K	324.00	328.57	334.53	338.01	366.05	373.78	379.64	384.82
P/MPa	2.39	1.53	0.71	0.38	0.43	1.00	1.52	2.03

Type of data: spinodal points

w_{B+D}	0.103	0.103	0.103	0.103	0.103	0.103	0.103	0.103	0.103
T/K	325.99	329.97	335.66	341.31	347.11	356.39	364.63	373.45	381.86
P/MPa	1.33	1.00	0.46	0.01	0.35	0.27	0.50		1.63

w_{B+D}	0.265	0.265	0.265	0.265	0.265	0.265	0.265	0.265
T/K	324.00	328.57	334.53	338.01	366.05	373.78	379.64	384.82
P/MPa	2.22	1.33	0.62	0.13	0.22	0.84	1.42	1.77

critical concentration: $w_{B+D,\,crit} = 0.210$

$x_C/x_A = 0.501$ was kept constant $x_D/x_B = 0.502$ was kept constant

Type of data: cloud points

w_{B+D}	0.126	0.126	0.126	0.126	0.126	0.126	0.126	0.126	0.126
T/K	331.06	335.10	339.51	343.35	350.12	354.21	356.36	364.60	372.36
P/MPa	2.11	1.65	1.29	1.05	0.67	0.81	0.92	1.15	1.63

continued

continued

w_{B+D}	0.126	0.145	0.145	0.145	0.145	0.145	0.145	0.145	0.145
T/K	380.49	331.95	335.12	337.71	341.63	345.55	351.27	356.14	365.16
P/MPa	2.36	2.51	2.07	1.75	1.43	1.22	1.07	1.13	1.37

w_{B+D}	0.145	0.145	0.167	0.167	0.167	0.167	0.167	0.167	0.167
T/K	372.07	379.54	331.56	335.91	341.02	345.32	350.63	353.45	356.39
P/MPa	1.80	2.40	2.58	1.96	1.48	1.22	1.08	1.07	1.13

w_{B+D}	0.167	0.167	0.167
T/K	365.41	372.96	360.01
P/MPa	1.39	1.66	2.43

Type of data: spinodal points

w_{B+D}	0.126	0.126	0.126	0.126	0.126	0.126	0.126	0.126	0.167
T/K	331.06	335.10	339.51	343.35	350.12	364.60	372.36	380.49	331.56
P/MPa	1.81	1.34	0.67	0.45	0.36	0.52	1.26	1.66	2.21

w_{B+D}	0.167	0.167	0.167	0.167	0.167	0.167	0.167	0.167	0.167
T/K	335.91	341.02	345.32	350.63	353.45	356.39	365.41	372.96	360.01
P/MPa	1.72	1.36	1.13	0.96	0.96	1.00	1.32	1.75	2.32

critical concentration: $w_{B+D, crit} = 0.210$

$x_C/x_A = 0.500$ was kept constant $x_D/x_B = 0.503$ was kept constant

Type of data: cloud points

w_{B+D}	0.245	0.245	0.245	0.245	0.245	0.245	0.245	0.245
T/K	332.25	336.80	343.55	349.28	358.29	363.91	372.41	379.29
P/MPa	2.53	1.93	1.34	1.09	1.06	1.22	1.69	2.25

Type of data: spinodal points

w_{B+D}	0.245	0.245	0.245	0.245	0.245	0.245	0.245	0.245
T/K	332.25	336.80	343.55	349.28	358.29	363.91	372.41	379.29
P/MPa	2.45	1.82	1.27	0.99	0.97	1.07	1.62	2.00

critical concentration: $w_{B+D, crit} = 0.210$

$x_C/x_A = 0.501$ was kept constant $x_D/x_B = 0.505$ was kept constant

Type of data: cloud points

w_{B+D}	0.186	0.186	0.186	0.186	0.186	0.186	0.186	0.186	0.186
T/K	334.35	337.99	341.51	346.27	351.43	354.56	358.37	365.30	372.69
P/MPa	2.27	1.85	1.55	1.27	1.14	1.12	1.16	1.41	1.83

w_{B+D}	0.186	0.236	0.236	0.236	0.236	0.236	0.236	0.236	0.236
T/K	379.72	331.15	335.67	338.49	342.30	347.37	351.79	354.86	358.45
P/MPa	2.40	3.17	2.48	2.16	1.82	1.53	1.40	1.36	1.39

w_{B+D}	0.236	0.236	0.236
T/K	365.87	371.73	376.61
P/MPa	1.63	1.97	2.32

continued

continued

Type of data: spinodal points

w_{B+D}	0.186	0.186	0.186	0.186	0.186	0.186	0.186	0.186	0.186
T/K	334.35	337.99	341.51	346.27	351.43	354.56	358.37	365.30	372.69
P/MPa	2.09	1.72	1.45	1.19	1.05	1.05	1.11	1.35	1.79
w_{B+D}	0.186	0.236	0.236	0.236	0.236	0.236	0.236	0.236	0.236
T/K	379.72	331.15	335.67	338.49	342.30	347.37	351.79	354.86	358.45
P/MPa	2.33	3.02	2.34	2.03	1.71	1.43	1.29	1.27	1.30
w_{B+D}	0.236	0.236	0.236						
T/K	365.87	371.73	376.61						
P/MPa	1.57	1.89	2.27						

critical concentration: $w_{B+D, crit} = 0.210$

Polymer (B): **polysulfone** **2002HAN**
Characterization: M_w/g.mol^{-1} = 30000, Aldrich Chem. Co., Inc., Milwaukee, WI
Solvent (A): **water** H_2O **7732-18-5**
Solvent (C): **1-methyl-2-pyrrolidinone** C_5H_9NO **872-50-4**
Polymer (D): **poly(1-vinyl-2-pyrrolidinone)**
Characterization: M_w/g.mol^{-1} = 55000, Aldrich Chem. Co., Inc., Milwaukee, WI

Type of data: cloud points

T/K = 298.15

w_A	0.000	0.048	0.097
w_B	0.141	0.144	0.145
w_C	0.798	0.771	0.726
w_D	0.061	0.037	0.032

Polymer (B): **poly(vinyl chloride)** **1978HAY**
Characterization: Esso 363, Imperial Oil Enterprises
Solvent (A): **tetrahydrofuran** C_4H_8O **109-99-9**
Solvent (C): **ethanol** C_2H_6O **64-17-5**
Solvent (D): **water** H_2O **7732-18-5**

Type of data: cloud points

T/K = 295

w_A	0.6679	0.6786	0.6882	0.6803
w_B	0.0021	0.0044	0.0088	0.0177
w_{C+D}	0.330	0.317	0.303	0.302

Comments: The nonsolvent mixture is composed of 83 vol% ethanol and 17 vol% water.

Polymer (B):	**poly(vinyl chloride)**		**1978HAY**
Characterization:	Esso 363, Imperial Oil Enterprises		
Solvent (A):	**tetrahydrofuran**	C_4H_8O	**109-99-9**
Solvent (C):	**methanol**	CH_4O	**67-56-1**
Solvent (D):	**water**	H_2O	**7732-18-5**

Type of data: cloud points

T/K = 295

w_A	0.7466	0.7452	0.7493	0.7427
w_B	0.0024	0.0048	0.0097	0.0193
w_{C+D}	0.251	0.250	0.241	0.238

Comments: The nonsolvent mixture is composed of 75 vol% methanol and 25 vol% water.

4.2. Table of quaternary systems where data were published only in graphical form as phase diagrams or related figures

Polymer (B)	Second/third/fourth component	Ref.
Cellulose acetate phthalate		
	ethylcellulose and dichloromethane/methanol	1991SAK
Dextran		
	poly(ethylene glycol)/benzene derivatives and water	1993MIS
	poly(ethylene oxide)/polystyrene and water	2005OL1
	poly(ethylene oxide)/silica and water	2005OL1
	poly(ethylene oxide-*co*-propylene oxide)/polystyrene and water	2005OL1
	poly(ethylene oxide-*co*-propylene oxide)/silica and water	2005OL1
Ethylcellulose		
	cellulose acetate phthalate and dichloromethane/methanol	1991SAK
Poly(acrylic acid)		
	pentanol/toluene and water	2003KOT
Polyarylate		
	poly(dimethylsiloxane) and n-hexane/trichloromethane	1980ROG
Poly(arylene sulfonoxide-*b*-butadiene) polyblock copolymer		
	polybutadiene and n-octane/1,1,2,2-tetrachloroethane	1988ROG
Polybutadiene		
	ethyl acetate/toluene and polyisoprene	2004VSH
	poly(arylene sulfonoxide-*b*-butadiene) polyblock copolymer and n-octane/1,1,2,2-tetrachloroethane	1988ROG

Polymer (B)	**Second/third/fourth component**	**Ref.**
Poly(2,6-dimethyl-1,4-phenyleneoxide)		
	trichloroethene and methanol/1-octanol	1985WI1
Poly(dimethylsiloxane)		
	polyarylate and n-hexane/trichloromethane	1980ROG
Polyetherimide		
	4,4'-diaminodiphenylsulfone/bisphenol-A diglycidyl ether and polysulfone	2004GIA
	acetic acid/*N,N*-dimethylacetamide and water	2003SHE
	N,N-dimethylformamide/1-methyl-2-pyrrolidinone and water	2001ALB
	4,4'-methylenebis(3-chloro 2,6-diethylaniline)/bisphenol-A diglycidyl ether and polysulfone	2004GIA
	1-methyl-2-pyrrolidinone/poly(1-vinyl-2-pyrrolidinone) and water	1994BO2
Polyethersulfone		
	1-methyl-2-pyrrolidinone/polysulfone and water	1999BAI
	1-methyl-2-pyrrolidinone/poly(1-vinyl-2-pyrrolidinone) and water	1994BO1
	1-methyl-2-pyrrolidinone/poly(1-vinyl-2-pyrrolidinone) and water	1994BO2
Poly(ethylene glycol)		
	albumin/dextran and water	2001NER
	dextran/benzene derivatives and water	1993MIS
Poly(ethylene oxide)		
	dextran/polystyrene and water	2005OL1
	dextran/silica and water	2005OL1
Poly(ethylene oxide-*co*-propylene oxide)		
	dextran/polystyrene and water	2005OL1
	dextran/silica and water	2005OL1
Poly(ethylene terephthalate)		
	n-heptane and phenol/1,1,2,2-tetrachloroethane	1962TUR

Polymer (B)	Second/third/fourth component	Ref.
Poly(hydroxystearic acid-*b*-ethylene oxide-*b*-hydroxystearic acid)		
	isopropyl myristate/1,2-hexanediol and water	2002PLA
	isopropyl myristate/1,2-octanediol and water	2002PLA
Polyimide		
	N,N-dimethylformamide/poly(1-vinyl-2-pyrrolidinone) and water	2001KI1
	ethanol and 1-methyl-2-pyrrolidinone/tetrahydrofuran	2000CLA
	1-methyl-2-pyrrolidinone/poly(1-vinyl-2-pyrrolidinone) and water	2001KI1
Polyisoprene		
	ethyl acetate/toluene and polybutadiene	2004VSH
Polystyrene		
	dextran/poly(ethylene oxide) and water	2005OL1
	dextran/poly(ethylene oxide-*co*-propylene oxide) and water	2005OL1
Polysulfone		
	4,4'-diaminodiphenylsulfone/bisphenol-A diglycidyl ether and poly(ether imide)	2004GIA
	4,4'-methylenebis(3-chloro 2,6-diethylaniline)/bisphenol-A diglycidyl ether and poly(ether imide)	2004GIA
	1-methyl-2-pyrrolidinone/polyethersulfone and water	1999BAI
Poly(vinylidene fluoride)		
	N,N-dimethylacetamide/ethanol and water	2003YEO
	N,N-dimethylacetamide/poly(1-vinyl-2-pyrrolidinone) and water	2003YEO
Poly(1-vinyl-2-pyrrolidinone)		
	N,N-dimethylacetamide/poly(vinylidene fluoride) and water	2003YEO
	N,N-dimethylformamide/polyimide and water	2001KI1
	1-methyl-2-pyrrolidinone/polyethersulfone and water	1994BO1
	1-methyl-2-pyrrolidinone/polyethersulfone and water	1994BO2
	1-methyl-2-pyrrolidinone/polyimide and water	2001KI1

5. REFERENCES

1940SCH Schulz, G.V. and Jirgensons, B., Die Abhängigkeit der Löslichkeit vom Molekulargewicht. Über die Löslichkeit makromolekularer Stoffe VIII., *Z. Phys. Chem. B*, 46, 105, 1940.

1946RIC Richards, R.B., The phase equilibria between a crystalline polymer and solvents. I. The effect of polymer chain length on the solubility and swelling of polymers, *Trans. Faraday Soc.*, 42, 10, 1946.

1947DOB Dobry, A. and Boyer-Kawenoki, F., Phase separation in polymer solution, *J. Polym. Sci.*, 2, 90, 1947.

1948JEN Jenckel, E., Dampfdruck und Entmischung an hochmolekularen Lösungen, *Z. Naturforsch.*, 3a, 290, 1948.

1950JE1 Jenckel, E. and Keller, G., Entmischungskurven von Polystyrollösungen, *Z. Naturforsch.*, 5a, 317, 1950.

1950JE2 Jenckel, E. and Gorke, K., Über die Entmischung von Lösungen des Polymethacrylesters, *Z. Naturforsch.*, 5a, 556, 1950.

1951FO1 Fox, T.G. and Flory, P.J., Intrinsic viscosity-temperature relationships for polyisobutylene in various solvents, *J. Amer. Chem. Soc.*, 73, 1909, 1951.

1951FO2 Fox, T.G. and Flory, P.J., Intrinsic viscosity relationships for polystyrene, *J. Amer. Chem. Soc.*, 73, 1915, 1951.

1952MAN Mandelkern, L. and Flory, P.J., Molecular dimensions of cellulose triesters, *J. Amer. Chem. Soc.*, 74, 2517, 1952.

1952SHU Shultz, A.R. and Flory, P.J., Phase equilibria in polymer-solvent systems, *J. Amer. Chem. Soc.*, 74, 4760, 1952.

1952WAG Wagner, H.L. and Flory, P.J., Molecular dimensions of natural rubber and gutta percha, *J. Amer. Chem. Soc.*, 74, 195, 1952.

1953SHU Shultz, A.R. and Flory, P.J., Phase equilibria in polymer-solvent systems. II. Thermodynamic interaction parameters from critical miscibility data, *J. Amer. Chem. Soc.*, 75, 3888, 1953.

1955KER Kern, R.J. and Slocombe, R.J., Phase separation in solutions of vinyl polymers, *J. Polym. Sci.*, 15, 183, 1955.

1956KER Kern, R.J., Component effects in phase separation of polymer-polymer-solvent systems, *J. Polym. Sci.*, 21, 19, 1956.

1957MAL Malcolm, G.N. and Rowlinson, J.S., Thermodynamic properties of aqueous solutions of polyethylene glycol, polypropylene glycol, and dioxane, *Trans. Faraday Soc.*, 53, 921, 1957.

1958ALB Albertson, P.-A., Particle fractionation in liquid two-phase systems, *Biochim. Biophys. Acta*, 27, 378, 1958.

1959BAI Bailey, F.E. and Callard, R.W., Some properties of poly(ethylene oxide) in aqueous solution, *J. Appl. Polym. Sci.*, 1, 56, 1959.

1959BRI Bristow, G.M., Phase separation in rubber-poly(methyl methacrylate)-solvent systems with polystyrene, *J. Appl. Polym. Sci.*, 2, 120, 1959.

1959KIN Kinsinger, J.B. and Wessling, R.A., The Flory (theta) temperature of atactic and isotactic polypropylene, *J. Amer. Chem. Soc.*, 81, 2908, 1959.

1960ALL Allen, G., Gee, G., and Nicholson, J.P., The miscibility of polymers. I. Phase equilibria in systems containing two polymers and a mutual solvent, *Polymer*, 1, 56, 1960.

1960DE1 Debye, P., Coll, H., and Woermann, D., Critical opalescence of polystyrene in cyclohexane, *J. Chem. Phys.*, 32, 939, 1960.

1960DE2 Debye, P., Coll, H., and Woermann, D., Critical opalescence of polystyrene in cyclohexane, *J. Chem. Phys.*, 33, 1746, 1960.

1960ELI Elias, H.-G., Dobler, M., and Wyss, H.-R., Zur Bestimmung der Taktizität von Hochpolymeren in gelöstem Zustand, *J. Polym. Sci.*, 46, 264, 1960.

1960FRE Freeman, P.I. and Rowlinson, J.S., Lower critical points in polymer solutions, *Polymer*, 1, 20, 1960.

1961KRI Krigbaum, W.R., Kurz, J.E., and Smith, P., The conformation of polymer molecules IV. Poly(1-butene), *J. Phys. Chem.*, 65, 1984, 1961.

1961MOR Morneau, G.A., Roth, P.I., and Shultz, A.R., Trifluoronitrosomethane/tetrafluoroethylene elastomers dilute solution properties and molecular weight, *J. Polym. Sci.*, 55, 609, 1961.

1961OKA Okamoto, H. and Sekikawa, K., Phase equilibrium of polydisperse polymer solution and fractionation, *J. Polym. Sci.*, 55, 597, 1961.

1962BAK Baker, C.H., Brown, W.B., Gee, G., Rowlinson, J.S., Stubley, D., and Yeadon, R.E., A study of the thermodynamic properties and phase equilibria of solutions of polyisobutylene in n-pentane, *Polymer*, 3, 215, 1962.

1962DEB Debye, P., Chu, B., and Woermann, D., Critical opalescence of polystyrene in cyclohexane: Range of molecular forces and radius of gyration, *J. Chem. Phys.*, 36, 1803, 1962.

1962FOX Fox, T.G., Properties of dilute polymer solutions III. Intrinsic viscosity/temperature relationships for conventional polymethyl methacrylate, *Polymer*, 3, 111, 1962.

1962HAM Ham, J.S., Bolen, M.C., and Hughes, J.K., The use of high pressure to study polymer-solvent interaction, *J. Polym. Sci.*, 57, 25, 1962.

1962TUR Turska, E. and Utracki, L., Investigations of the phenomena of coacervation. Part II. Coacervation of polyethylene terphthalate, *J. Appl. Polym. Sci.*, 6, 393, 1962.

1962UTR Utracki, L., Investigations of the phenomena of coacervation. Part III. Phase equilibrium in the three-component system toluene-ethanol-polystyrene, *J. Appl. Polym. Sci.*, 6, 399, 1962.

1963DEB Debye, P., Woermann, D., and Chu, B., Critical opalescence of polystyrene in ethylcyclohexane, *J. Polym. Sci.: Part A*, 1, 255, 1963.

1963EHR Ehrlich, P. and Kurpen, J.J., Phase equilibria of polymer-solvent systems at high pressure near their critical loci, polyethylene with n-alkanes, *J. Polym. Sci.: Part A*, 1, 3217, 1963.

1963ORO Orofino, T.A. and Mickey, J.W., Dilute solution properties of linear polystyrene in theta-solvent media, *J. Chem. Phys.*, 38, 2512, 1963.

1963PAX Paxton, T.R., The miscibility of polymers: Interaction parameter for polybutadiene with polystyrene, *J. Appl. Polym. Sci.*, 7, 1499, 1963.

1963RED Redlich, O., Jacobson, A.L., and Fadden, W.H., On the fractionation of polypropylene by coacervation, *J. Polym. Sci. A*, 1, 393, 1963.

1963REH Rehage, G., Quellung, Gelierung und Diffusion bei Kunststoffen, *Kunststoffe*, 53, 605, 1963.

1964TAG Tager, A. A., Dreval, V. E., and Khabarova, K. G., Viscosity of critical mixtures of polymers and low molecular weight liquids (Russ.), *Vysokomol. Soedin.*, 6, 1593, 1964.

1965ALL Allen, G. and Baker, C.H., Lower critical solution phenomena in polymer-solvent systems, *Polymer*, 6, 181, 1965.

1965GEC Gechele, G.B. and Crescentini, L., Phase separation, viscosity, and thermodynamic parameters for poly-2-methyl-5-vinylpyridine diluent system, *J. Polym. Sci.: Part A*, 3, 3599, 1965.

1965MYR Myrat, C.D. and Rowlinson, J.S., The separation and fractionation of polystyrene at a lower critical solution point, *Polymer*, 6, 645, 1965.

1965REH Rehage, G., Möller, D., and Ernst, O., Entmischungserscheinungen in Lösungen von molekularuneinheitlichen Hochpolymeren, *Makromol. Chem.*, 88, 232, 1965.

1966KUB Kubo, K. and Ogino, K., Solution properties of poly(p-chlorostyrene), *Sci. Pap. Coll. Art. Sci. Univ. Tokyo*, 16, 193, 1966.

1966HEI Heil, J.F. and Prausnitz, J.M., Phase equilibria in polymer solutions, *AIChE-J.*, 12, 678, 1966.

1966KUB Kubo, K. and Ogino, K., Solution properties of poly(p-chlorostyrene), *Sci. Pap. Coll. Art. Sci. Univ. Tokyo*, 16, 193, 1966.

1966NAG Nagasawa, M., Asai, Y., and Sugiura, I., Determination of molecular weight distribution of linear polymers by measuring specific heat of solution (Jap.), *Kogyo Kagaku Zasshi*, 69, 1759, 1966.

1966NA1 Nakajima, A., Fujiwara, H., and Hamada, F., Phase relationships and thermodynamic interactions in linear polyethylene-diluent systems, *J. Polym. Sci.: Part A-2*, 4, 507, 1966.

1966NA2 Nakajima, A., Hamada, F., and Hayashi, S., Unperturbed dimensions of polyethylene in theta solvents, *J. Polym. Sci.: Part C*, 15, 285, 1966.

1967BER Berek, D., Lath, D., and Durdovic, V., Phase relations in the ternary system atactic polystyrene-atactic polypropylene-toluene, *J. Polym. Sci.: Part C*, 16, 659, 1967.

1967REH Rehage, G. and Möller, D., Entmischungseffekte in hochmolekularen Lösungen, *J. Polym. Sci.: Part C*, 16, 1787, 1967.

1967KO1 Koningsveld, R., On liquid-liquid phase relationships and fractionation in multicomponent polymer solutions, *Proefschrift Univ. Leiden*, Heerlen, 1967.

1967KO2 Koningsveld, R. and Staverman, A.J., Determination of critical points in multicomponent polymer solutions, *J. Polym. Sci.: Part C*, 16, 1775, 1967.

1967KO3 Koningsveld, R. and Staverman, A.J., Liquid-liquid phase separation in multicomponent polymer solutions IV. Coexistence curves, *Kolloid-Z. Z. Polym.*, 218, 114, 1967.

1967KO4 Koningsveld, R. and Staverman, A.J., Liquid-liquid phase separation in multicomponent polymer solutions. V. Separation into three liquid phases, *Kolloid-Z. Z. Polym.*, 220, 31, 1967.

1967LLO Llopis, J., Albert, A., and Usobinaga P., Studies of poly(ethyl acrylate) in theta solvents, *Eur. Polym. J.*, 3, 259, 1967.

1967MOR Moraglio, G., Gianotti, G., and Danusso, F., Unperturbed molecular dimensions of atactic polybutene-1 at different temperatures, *Eur. Polym. J.*, 3, 251, 1967.

1967ORW Orwoll, R.A. and Flory, P.J., Thermodynamic properties of binary mixtures of n-alkanes, *J. Amer. Chem. Soc.*, 89, 6822, 1967.

1967PAT Patterson, D., Delmas, G., and Somcynsky, T., A comparison of lower critical solution temperatures of some polymer solutions, *Polymer*, 8, 503, 1967.

1967SAI Saito, S. and Otsuka, T., Dissolution of some polymers in aqueous solutions of urea, of its related compounds and tetraalkylammonium salts, *J. Colloid Interface Sci.*, 25, 531, 1967.

1967SCH Scholte, Th.G. and Koningsveld, R., Determination of liquid-liquid phase relations with the ultracentrifuge, *Kolloid-Z. Z. Polym.*, 218, 58, 1967.

1967SUH Suh, K.W. and Clarke, D.H., Cohesive energy densities of polymers from turbidimetric titrations, *J. Polym. Sci.: Part A-1*, 5, 1671, 1967.

1968BON Bondi, A., *Physical Properties of Molecular Crystals, Liquids and Glasses*, J. Wiley & Sons, New York, 1968.

1968EDM Edmond, E. and Ogston, A.G., An approach to the study of phase separation in ternary aqueous systems, *Biochem. J.*, 109, 569, 1968.

1968HES Heskins, M. and Guillet, J.E., Solution properties of poly(*N*-isopropylacrylamide), *J. Macromol. Sci.-Chem. A*, 2, 1441, 1968.

1968KON Koningsveld, R. and Staverman, A.J., Liquid-liquid phase separation in multi-component polymer solutions. I, II, and III., *J. Polym. Sci., Part A-2*, 6, 305, 325, 349, 1968.

1968NA1 Nakajima, A. and Fujiwara, H., Phase relations and thermodynamic interactions in isotactic polypropylene-diluent systems, *J. Polym. Sci.: Part A-2*, 6, 723, 1968.

1968NA2 Nakajima, A., Hayashi, T., and Ohmori, M., Phase equilibria of rodlike molecules in binary solvent systems, *Biopolymers*, 6, 973, 1968.

1968RE1 Rehage, G. and Koningsveld, R., Liquid-liquid phase separation in multicomponent polymer solutions VI. Some errors introduced by the binary approximation, *J. Polym. Sci., Polym. Lett.*, 6, 421, 1968.

1968RE2 Rehage, G. and Wefers, W., A sensitive method for testing monodispersity of high polymers, *J. Polym. Sci.: Part A-2*, 6, 1683, 1968.

1968SUH Suh, K.W. and Liou, D.W., Phase equilibria in polymer liquid-liquid systems, *J. Polym. Sci.: Part A-2*, 6, 813, 1968.

1968TAG Tager, A.A., Anikeeva, A.A., Andreeva, V.M., Gumarova, T.Ya., and Chernoskutov, L.A., Phase equilibrium and light scattering in polymer solutions (Russ.), *Vysokomol. Soedin., Ser. A*, 7, 1661, 1968.

1968TER Teramachi, S. and Nagasawa, M., The fractionation of copolymers by chemical composition, *J. Macromol. Sci.-Chem. A*, 2, 1169, 1968.

1969AND Andreeva, V.M., Tager, A.A., Anikeeva, A.A., and Kuz'mina, T.A., Lower critical temperature of the dissolution of poly(vinyl alcohol) in water (Russ.), *Vysokomol. Soedin., Ser. B*, 11, 555, 1969.

1969BAR Bardin, J.-M. and Patterson, D., Lower critical solution temperatures of polyisobutylene plus isomeric alkanes, *Polymer*, 10, 247, 1969.

1969DUS Dusek, K., Solubility of poly(2-hydroxyethyl methacrylate) in some aliphatic alcohols, *Coll. Czech. Chem. Commun.*, 34, 3309, 1969.

1969GOL Goldfarb, J. and Sepulveda, L., Application of a theory of polymer solutions to the cloud points of nonionic detergents, *J. Colloid Interface Sci.*, 31, 454, 1969.

1969GOR Gordon, M., Chermin, H.A.G., and Koningsveld, R., Liquid-liquid phase seapration in multicomponent polymer solutions. VII. Relations for the spinodal and critical loci, *Macromolecules*, 2, 207, 1969.

1969KLE Klein, J. und Wittenbecher, U., Zum Fällungsverhalten von Mischungen polymer-homologer Substanzen, *Makromol. Chem.*, 122, 1, 1969.

1969KUZ Kuznetsov, V.I., Kogan, V.B., and Vilesova, M.S., Investigation of mutual solubility of polyoxypropylene polyoles and water (Russ.), *Vysokomol. Soedin., Ser. A*, 11, 1330, 1969.

1969WOL Wolf, B.A., Zur Bestimmung der kritischen Konzentration von Polymerlösungen, *Makromol. Chem.*, 128, 284, 1969.

1970AND Andreeva, V.M., Anikeeva, A.A., Vshivkov, S.A., and Tager, A.A., Lower critical temperatures of polystyrene dissolution in benzene and ethylbenzene (Russ.), *Vysokomol. Soedin., Ser. B*, 12, 789, 1970.

1970DEL Delmas, G. and Patterson, D., The molecular weight dependence of lower and upper critical solution temperatures, *J. Polym. Sci.: Part C*, 30, 1, 1970.

1970GRI Grinberg, V.Ya., Shvenke, K.D., and Tolstoguzov, V.E., On phase states in the system gelatin-dextran-water (Russ.), *Izv. Akad. Nauk SSSR, Ser. Khim.*, 1430, 1970.

1970KLE Klein, J. and Schiedermaier, E., Temperature dependent precipitation behavior in polymer-solvent-precipitant systems (Ger.), *Kolloid-Z. Z. Polym.*, 236, 118, 1970.

1970KON Koningsveld, R., Kleintjens, L.A., and Shultz, A.R., Liquid-liquid phase separation in multicomponent polymer solutions. IX. Concentration-dependent pair interaction parameter from critical miscibility data on the system polystyrene-cyclohexane, *J. Polym. Sci.: Part A-2*, 8, 1261, 1970.

1970KOT Kotaka, T., Tanaka, T., Ohnuma, H., Murakami, Y., and Inagaki, H., Dilute solution properties of styrene-methyl methacrylate copolymers with different architecture, *Polym. J.*, 1, 245, 1970.

1970LID Liddell, A.H. and Swinton, F.L., Thermodynamic properties of some polymer solutions at elevated temperatures, *Discuss. Faraday Soc.*, 49, 115, 1970.

1970MAT Matsumura, K., Solution behaviour of poly(o-chlorostyrene), *Polym. J.*, 1, 322, 1970.

1970NAK Nakayama, H., Temperature dependence of the heat of solution of poly(ethylene glycol) and of related compounds, *Bull. Chem. Soc. Japan*, 43, 1683, 1970.

1970WO1 Wolf, B.A., Breitenbach, J.W., and Senftl, H., Upper and lower solubility gaps in the system butanone-acetone-polystyrene, *J. Polym. Sci.: Part C*, 31, 345, 1970.

1970WO2 Wolf, B.A., Breitenbach, J.W., and Senftl, H., Mutual influence of polymers of different molecular weights on solubility. II. Mixed solvent systems with preferential adsorption of precipitant by the polymer (Ger.), *Monatsh. Chem.*, 101, 57, 1970.

1971BOR Borchard, W. and Rehage, G., Critical phenomena in multicomponent polymer solutions, *Adv. Chem. Ser.*, 99, 42, 1971.

1971COW Cowie, J.M.G., Maconnachie, A., and Ranson, R.J., Phase equilibria of cellulose acetate-acetone solutions. The effect of degree of substitution and molecular weight on upper and lower critical solution temperature, *Macromolecules*, 4, 57, 1971.

1971HON Hong, S.-D. and Burns, C.M., Compatibility of polystyrene and poly(methyl methacrylate) in benzene. Effects of molecular weight and temperature, *J. Appl. Polym. Sci.*, 15, 1995, 1971.

1971KAG Kagemoto, A. and Baba, Y., Phase diagrams of polymer solutions (Jap.), *Kobunshi Kagaku*, 28, 784, 1971.

1971KUW Kuwahara, N., Fenby, D.V., Tamsky, M., and Chu, B., Intensity and linewidth studies of the system polystyrene-cyclohexane in the critical region, *J. Chem. Phys.*, 55, 1140, 1971.

1971OGA Ogawa, T. and Hoshino, S., Cloud points in isotactic polypropylene-solvent-nonsolvent systems (Jap.), *Kobunshi Kagaku*, 28, 348, 1971.

1971SCH Scholte, Th.G., Thermodynamic parameters of polymer-solvent systems from light scattering measurements below the theta temperature, *J. Polym. Sci.: Part A-2*, 9, 1553, 1971.

1971SHI Shibatani, K. and Oyanagi, Y., Solution properties of vinyl alcohol-ethylene copolymers in water, *Kobunshi Kagaku*, 28, 361, 1971.

1971SMO Smolders, C.A., Van Aartsen, J.J., and Steenbergen, A., Liquid-liquid phase separation in concentrated solutions of non-crystallizable polymers by spinodal decomposition, *Kolloid-Z. Z. Polym.*, 243. 14, 1971.

1971TAG Tager, A.A., Anikeeva, A.A., Adamova, L.V., Andreeva, V.M., Kuz'mina, T.A., and Tsilipotkina, M.V., Effect of temperature on the solubility of poly(vinyl alcohol) in water (Russ.), *Vysokomol. Soedin., Ser. A*, 13, 659, 1971.

1971WEE Wee, E.L. and Miller, W.G., Liquid crystal-isotropic phase equilibria in the system poly(γ-benzyl-α,L-glutamate)-dimethylformamide, *J. Phys. Chem.*, 76, 1446, 1971.

1971YAM Yamakawa, H., *Modern Theory of Polymer Solutions*, Harper & Row, New York, 1971.

1972BOR Borchard, W., Critical opalescence in solutions of polystyrene and cyclohexane, *Ber. Bunsenges. Phys. Chem.*, 76, 224, 1972.

1972EM1 Emmerik, P.T. van and Smolders, C.A., Phase separation in polymer solutions. I. Liquid-liquid phase separation of poly(2,6-dimethyl-1,4-phenylene oxide) in binary mixtures with toluene and ternary mixtures with toluene and ethyl alcohol, *J. Polym. Sci.: Part C*, 38, 73, 1972.

1972EM2 Emmerik, P.T. van and Smolders, C.A., Phase separation in polymer solutions. II. Determination of thermodynamic parameters of poly(2,6-dimethyl-1,4-phenylene oxide) toluene mixtures, *J. Polym. Sci.: Part C*, 39, 311, 1972.

1972GRI Grinberg, V.Ya. and Tolstoguzov, V.B., Thermodynamic compatibility of gelatine with some D-glucans in aqueous media, *Carbohydr. Res.*, 25, 313, 1972.

1972HOR Horacek, H., Gleichgewichtsdrücke, Löslichkeit und Mischbarkeit des Systems unvernetztes Polystyrol (PS)/Kohlenwasserstoffe (KW) 1. Teil: Der Einfluss des Verzweigungsgrades (n-, iso- und neo-Pentan), *Kolloid-Z. Z. Polym.*, 250, 863, 1972.

1972IZU Izumi, Y. and Miyake, Y., Study of linear poly(p-chlorostyrene)-diluent systems. I. Solubilities, phase relationships, and thermodynamic interactions, *Polym. J.*, 3, 647, 1972.

1972KAG Kagemoto, A., Baba, Y., and Fujishiro, R., Phase equilibrium of the cellulose derivatives determined by differential thermal analysis. 1. Methylcellulose/water, *Makromol. Chem.*, 154, 105, 1972.

1972KEN Kennedy, J.W., Gordon, M., and Koningsveld, R., Generalization of the Flory-Huggins treatment of polymer solutions, *J. Polym. Sci.: Part C*, 39, 43, 1972.

1972KON Koningsveld, R., Polymer solutions and fractionation, in *Polymer Science*, Jenkins, E.D. (ed.), North-Holland, Amsterdam, 1047, 1972.

1972LIR Lirova, B.I., Smolyanskii, A.L., Savchenko, T.A., and Tager, A.A., Spectroscopic study of poly(oxypropylene diol) solutions in n-hexane in the phase separation region (Russ.), *Vysokomol. Soedin., Ser. B*, 14, 265, 1972.

1972NAK Nakayama, H., Thermodynamic properties of an aqueous solution of tetraethylene glycol diethyl ether, *Bull. Chem. Soc. Japan*, 45, 1371, 1972.

1972SCH Scholte, Th.G., Light scattering of concentrated polydisperse polymer solutions, *J. Polym. Sci.: Part C*, 38, 281, 1972.

1972SIO Siow, K.S., Delmas, G., and Patterson, D., Cloud point curves in polymer solutions with adjacent upper and lower critical solution temperatures, *Macromolecules*, 5, 29, 1972.

1972TA1 Tager, A.A., Adamova, L.V., Bessonov, Yu.S., Kuznetsov, V.N., Plyusnina, T.A., Soldatov, V.V., and Tsilipotkina, M.V., Thermodynamic study of oligomeric poly(oxypropylene)diol solutions in water and n-hexane in the precritical region (Russ.), *Vysokomol. Soedin., Ser. A*, 14, 1991, 1972.

1972TA2 Tager, A.A., Anikeeva, A.A., Andreeva, V.M., and Vshivkov, S.A., Poly(vinyl acetate) solutions having upper and lower critical mixing temperatures (Russ.), *Vysokomol. Soedin., Ser. B*, 14, 231, 1972.

1972TE1 Teramachi, S., Tomioka, H., and Mamoru, S., Phase-separation phenomena of copolymer solutions and fractionation of copolymers by chemical composition, *J. Macromol. Sci.-Chem. A*, 6, 97, 1972.

1972TE2 Teramachi, S. and Fujikawa, T., Phase-separation phenomena of random copolymer solution: SAN random copolymer-toluene systems, *J. Macromol. Sci.-Chem. A*, 6, 1393, 1972.

1972WHI White, J.L., Salladay, D.G., Quisenberry, D.O., and Mac Lean, D.L., Gel permeation and thin-layer chromatographic characterization and solution properties of butadiene and styrene homopolymers and copolymers, *J. Appl. Polym. Sci.*, 16, 2811, 1972.

1972YAK Yakhnin, E.D. and Lyubomirova, O.I., Strukturoobrazovanie v napolnennykh rastvorakh amorfnykh polimerov v usloviyakh ikh rassloeniya, *Vysokomol. Soedin, Ser. B*, 14, 254, 1972.

1972ZE1 Zeman, L., Biros, J., Delmas, G., and Patterson, D., Pressure effects in polymer solution phase equilibria. I. The lower critical solution temperature of polyisobutylene and polydimethylsiloxane in lower alkanes, *J. Phys. Chem.*, 76, 1206, 1972.

1972ZE2 Zeman, L. and Patterson, D., Pressure effects in polymer solution phase equilibria. II. Systems showing upper and lower critical solution temperatures, *J. Phys. Chem.*, 76, 1214, 1972.

1973BAB Baba, Y., Fujita, Y., and Kagemoto, A., Thermal properties of atactic polystyrene/ ethyl methyl ketone system determined by a modified differential thermal analysis, *Makromol. Chem.*, 164, 349, 1973.

1973CAN Candau, F., Strazielle, C., and Benoit, H., Influence du degre de ramification sur la temperature 'theta' de quelques polymeres dans differents solvants, *Makromol. Chem.*, 170, 165, 1973.

1973EM1 Emmerik, P.T. van and Smolders, C.A., Phase separation of polymer solutions, *Eur. Polym. J.*, 9, 157, 1973.

1973EM2 Emmerik, P.T. van and Smolders, C.A., Differential scanning calorimetry of poly(2,6-dimethyl-1,4-phenylene oxide)-toluene solutions, *Eur. Polym. J.*, 9, 293, 1973.

1973HAM Hamada, F., Fujisawa, K., and Nakajima, A.: Lower critical solution temperature in linear polyethylene−n-alkane systems, *Polym. J.*, 4, 316, 1973.

1973KRA Krause, S. and Stroud, D.E., Cloud-point curves for polystyrene of low polydispersity in cyclohexane solution, *J. Polym. Sci.: Polym. Phys. Ed.*, 11, 2253, 1973.

1973KU1 Kuwahara, N., Nakata, M., and Kaneko, M., Cloud-point curves of the polystyrene-cyclohexane system near the critical point, *Polymer*, 14, 415, 1973.

1973KU2 Kuwahara, N., Kojima, J., and Kaneko, M., Coexistence curve of the polystyrene-cyclohexane in the critical region, *J. Polym. Sci.: Polym. Phys. Ed.*, 11, 2307, 1973.

1973SA1 Saeki, S., Kuwahara, N., Konno, S., and Kaneko, M., Upper and lower critical solution temperatures in polystyrene solutions, *Macromolecules*, 6, 246, 1973.

1973SA2 Saeki, S., Kuwahara, N., Konno, S., and Kaneko, M., Upper and lower critical solution temperatures in polystyrene solutions II., *Macromolecules*, 6, 589, 1973.

1973TAN Tani, S., Hamada, F., and Nakajima, A., Unperturbed chain dimensions of poly(4-methyl-pentene-1) in theta solvents, *Polym. J.*, 5, 86, 1973.

1973WEL Welygam, D.G. and Burns, C.M., Effect molecular weight distribution on cloud point curves for the system polystyrene/polybutadiene/tetralin, *J. Polym. Sci.: Polym. Lett.*, 11, 339, 1973.

1974AND Andreeva, V.M., Anikeeva, A.A., Tager, A.A., and Kosareva, L.P., Phase equilibrium of aqueous solutions of poly(vinyl alcohol) and products of its acetylation (Russ.), *Vysokomol. Soedin., Ser. B*, 16, 277, 1974.

1974BAB Baba, Y. and Kagemoto, A., Phase diagrams of aqueous solutions of cellulose derivatives (Jap.), *Kobunshi Ronbunshu*, 31, 446, 1974.

1974BAT Bataille, P., Lower critical solution temperatures in alkyl acetates, *J. Chem. Eng. Data*, 19, 224, 1974.

1974BES Bessonov, Yu.S. and Tager, A.A., Thermodynamic investigation of oligomeric polyoxypropylenediol-water system in the demixing region, *Tr. Khim. Khim. Tekhnol.*, 1, 150, 1974.

1974CO1 Cowie, J.M.G. and McEwen, I.J., Upper and lower critical solution temperatures in the cosolvent system acetone (1) + diethyl ether (2) + polystyrene (3), *J. Chem. Soc., Faraday Trans. I*, 70, 171, 1974.

1974CO2 Cowie, J.M.G. and McEwen, I.J., Polymer-cosolvent systems. IV. Upper and lower critical solution temperatures in the system methylcyclohexane-diethyl ether-polystyrene, *Macromolecules*, 7, 291, 1874.

1974CO3 Cowie, J.M.G. and McEwen, I.J., Lower critical solution temperature of polypropylene solutions, *J. Polym. Sci.: Polym. Phys. Ed.*, 12, 441, 1974.

1974DEL Delmas, G. and De Saint-Romain, P., Upper and lower critical solution temperatures in polybutadiene-alkane systems, *Eur. Polym. J.*, 10, 1133, 1974.

1974DER Derham, K.W., Goldsbrough, J., and Gordon, M., Pulse-induced critical scattering (PICS) from polymer solutions, *Pure Appl. Chem.*, 38, 97, 1974.

1974GLO Glöckner, G., Über die Quellbarkeit und Löslichkeit von Styrol/Acrylnitril-Copolymeren, *Faserforsch. Textiltechn.*, 25, 476, 1974.

1974KUW Kuwahara, N., Saeki, S., Chiba, T., and Kaneko, M., Upper and lower critical solution temperatures in polyethylene solutions, *Polymer*, 15, 777, 1974.

1974NA1 Nakajima, A., Hamada, F., Yasue, K., Fujisawa, K., and Shiomi, T., Thermodynamic studies based on corresponding states theory for solutions of polystyrene in ethyl methyl ketone, *Makromol. Chem.*, 175, 197, 1974.

1974NA2 Nakano, S., Dilute solution properties and phase separation of branched low-density polyethylene, *J. Polym. Sci.: Polym. Phys. Ed.*, 12, 1499, 1974.

1974SAE Saeki, S., Konno, S., Kuwahara, N., Nakata, M., and Kaneko, M., Upper and lower critical solution temperatures in polystyrene solutions III. Temperature dependence of the χ_1 parameter, *Macromolecules*, 7, 521, 1974.

1974TAG Tager, A.A., Vshivkov, S.A., Andreeva, V.M., and Sekacheva, T.V., Structural study of solutions of poly(oxyethylene) with lower critical temperatures of mixing (Russ.), *Vysokomol. Soedin., Ser. A*, 16, 9, 1974.

1974TAN Taniguchi, Y., Suzuki, K., and Enomoto, T., The effect of pressure on the cloud point of aqueous polymer solutions, *J. Colloid Interface Sci.*, 46, 511, 1974.

1974VER VerStrate, G. and Philippoff, W., Phase separation in flowing polymer solutions, *J. Polym. Sci.: Polym. Lett. Ed.*, 12, 267, 1974.

1974WEL Welygam, D.G. and Burns, C.M., The effect molecular weight and temperature on phase separation in the ternary system polystyrene/polybutadiene/tetralin, *J. Appl. Polym. Sci.*, 18, 521, 1974.

1975AND Andreeva, V. M., Tager, A. A., and Fominykh, I. S., Study of the interaction of polystyrene with decalin by a light scattering method (Russ.), *Tr. Khim. Khim. Tekhnol.*, (4), 105, 1975.

1975CHA Chalykh, A.E., Use of interference micromethod for the construction of phase fields of phase diagrams in polymer-solvent systems (Russ.), *Vysokomol. Soedin., Ser. A*, 17, 2603, 1975.

1975COW Cowie, J.M.G. and McEwen, I.J., Phase equilibria in quasi-binary poly(α-methylstyrene) solutions, *Polymer*, 16, 244, 1975.

1975HOR Horacek, H., Gleichgewichtsdrücke, Löslichkeit und Mischbarkeit des Systems Polyethylen niedriger Dichte und Kohlenwasserstoffen bzw. halogenierten Kohlenwasserstoffen, *Makromol. Chem., Suppl.*, 1, 415, 1975.

1975KOJ Kojima, J., Kuwahara, N., and Kaneko, M., Light scattering and pseudospinodal curve of the system polystyrene-cyclohexane in the critical region, *J. Chem. Phys.*, 63, 333, 1975.

1975KON Konno, S., Saeki, S., Kuwahara, N., Nakata, M., and Kaneko, M., Upper and lower critical solution temperatures in polystyrene solutions. IV. Role of configurational heat capacity, *Macromolecules*, 8, 799, 1975.

1975NAK Nakata, M., Kuwahara, N., and Kaneko, M., Coexistence curve for polystyrene-cyclohexane near the critical point, *J. Chem. Phys.*, 62, 4278, 1975.

1975OKA Okazawa, T., On the narrow miscibility gap in polymer 1-polymer 2-solvent ternary systems, *Macromolecules*, 8, 371, 1975.

1975SAE Saeki, S., Kuwahara, N., Nakata, M., and Kaneko, M., Pressure dependence of upper critical solutions temperatures in the polystyrene-cyclohexane system, *Polymer*, 16, 445, 1975.

1975STR Strazielle, C. and Benoit, H., Some thermodynamic properties of polymer-solvent systems. Comparison between deuterated and undeuterated systems, *Macromolecules*, 8, 203, 1975.

1975TA1 Tager, A. A., Tsilipotkina, M. V., Nechaeva, O. V., Vasnev, V. A., Salazkin, S. N., Vasil'ev, A. V., and Milykh, L. M., Role of a solvent in the formation of a porous structure of aromatic polymers during synthesis (Russ.), *Vysokomol. Soedin., Ser. A*, 17, 2301, 1975.

1975TA2 Tager, A.A. and Bessonov, Yu.S., Thermodynamic study of solutions of poly(vinyl acetate) and cellulose tricarbanilate in the precipitation region (Russ.), *Vysokomol. Soedin., Ser. A*, 17, 2377, 1975.

1975TA3 Tager, A.A. and Bessonov, Yu.S., Thermochemistry of solutions of polymers and polymer compositions in the phase separation region, *Vysokomol. Soedin., Ser. A*, 17, 2383, 1975.

1975TA4 Tager, A.A., Lirova, B.I., Smolyanskii, A.L., and Plomadil, L.A., Phase equilibriums in systems with a lower critical temperature of mixing studied by an IR spectroscopic method, *Vysokomol. Soedin., Ser. B*, 17, 61, 1975.

1975TAY Taylor, L.D. and Cerankowski, L.D., Preparation of films exhibiting a balanced temperature dependence to permeation by aqueous solutions. A study of lower consolute behavior, *J. Polym. Sci.: Polym. Chem. Ed.*, 13, 2551, 1975.

1975WHI White, J.L. and Patel, R.D., Phase separation conditions in polystyrene-styrene-(butadiene-styrene) copolymer solutions, *J. Appl. Polym. Sci.*, 19, 1775, 1975.

1975WOL Wolf, B.A. and Blaum, G., Measured and calculated solubility of polymers in mixed solvents: Monotony and cosolvency, *J. Polym. Sci.: Polym. Phys. Ed.*, 13, 1115, 1975.

1976ALE Aleksandrova, T.A., Vasserman, A.M., Kovarskii, A.L., and Tager, A.A., Paramagnetic probe study of phase equilibriums in polymer solutions (polystyrene-decalin and polyvinylacetate-methanol) (Russ.), *Vysokomol. Soedin., Ser. B*, 18, 326, 1976.

1976AND Andreeva, V. M., Tager, A. A., Fominykh, P. S., and Zamaraeva, O. L., Light-scattering study of the spinodal phase separation and thermodynamic parameters of reaction of components of moderately concentrated polystyrene solutions (Russ.), *Vysokomol. Soedin., Ser. A*, 18, 286, 1976.

1976CHI Chiu, D.S., Takahashi, Y., and Mark, J.E., Dimensions of poly(trimethylene oxide) chains in a theta-solvent, *Polymer*, 17, 670, 1976.

1976CO1 Cowie, J.M.G. and McEwen, I.J., Influence of microstructure on the upper and lower critical solution temperatures of poly(methyl methacrylate) solutions, *J. Chem. Soc., Faraday Trans. I*, 72, 526, 1976.

1976CO2 Cowie, J.M.G. and McEwen, I.J., Absolute predicition of upper and lower critical solution temperatures in polymer solutions from corresponding states theory: A semiempirical approach, *J. Chem. Soc., Faraday Trans. I*, 72, 1675, 1976.

1976ESK Esker, M.W.J. van den and Vrij, A., Incompatibility of polymer solutions. I. Binodals and spinodals in the system polystyrene + polyisobutylene + toluene, *J. Polym. Sci.: Polym. Phys. Ed.*, 14, 1943, 1976.

1976KL1 Kleintjens, L.A., Schoefferleer, H.M., and Domingo, L., Liquid-liquid phase separation in multicomponent polymer systems XIII. Quasi-binary phase diagrams with three-liquid-phase separation, *Brit. Polym. J.*, 8, 29, 1976.

1976KL2 Kleintjens, L.A., Koningsveld, R., and Stockmayer, W.H., Liquid-liquid phase separation in multicomponent polymer systems XIV. Dilute and concentrated polymer solutions in equilibrium (continued), *Br. Polym. J.*, 8, 144, 1976.

1976NAK Nakata, M., Higashida, S., Kuwahara, N., Saeki, S., and Kaneko, M., Thermodynamic properties of the system polystyrene-*trans*-decalin, *J. Chem. Phys.*, 64, 1022, 1976.

1976NOS Nose, T. and Tan, T.V., Interfacial tension of demixed polystyrene-methylcyclohexane solution, *J. Polym. Sci.: Polym. Lett. Ed.*, 14, 705, 1976.

1976SA1 Saeki, S., Kuwahara, N., Nakata, M., and Kaneko, M., Upper and lower critical solution temperatures in poly(ethylene glycol) solutions, *Polymer*, 17, 685, 1976.

1976SA2 Saeki, S., Kuwahara, N., and Kaneko, M., Pressure dependence of upper and lower critical solution temperatures in polystyrene solutions, *Macromolecules*, 9, 101, 1976.

1976SLA Slagowski, E., Tsai, B., and McIntyre, D., The dimensions of polystyrene near and below the theta temperature, *Macromolecules*, 9, 687, 1976.

1976TA1 Tager, A.A., Andreeva, V.M., Vshivkov, S.A., and Terent'eva, V.P., Determination by the light scattering method of the position of the polystyrene-cyclohexane system spinodal near the lower critical dissolution temperature (Russ.), *Vysokomol. Soedin., Ser. B*, 18, 205, 1976.

1976TA2 Tager, A.A., Vshivkov, S.A., Andreeva, V.M., and Tarasova, R.N., Study of phase equilibrium of solutions of crystallizable polymers. Polypropylene-chlorobenzene system (Russ.), *Vysokomol. Soedin., Ser. B*, 18, 592, 1976.

1976VSH Vshivkov, S.A., Tager, A.A., and Gaifulina, N.B., Study of the phase equilibrium of polymer solutions by the cloud-point method and measurement of the volume of the coexisting phases and their composition (Russ.), *Vysokomol. Soedin., Ser. B*, 18, 25, 1976.

1976WOL Wolf, B.A. and Blaum, G., Pressure influence on true cosolvency, *Makromol. Chem.*, 177, 1073, 1976.

1977ALE Aleksandrova, T. A., Vasserman, A. M., and Tager, A. A., Study of the structure and phase equilibrium of the poly(vinyl acetate)-methanol system by the spin labeling technique (Russ.), *Vysokomol. Soedin., Ser. A*, 19, 137, 1977.

1977AND Andreeva, V. M., Tager, A. A., Tyukova, I. S., and Golenkova, L. F., Study of the spinodal mechanism of the phase separation of polystyrene solutions in decalin (Russ.), *Vysokomol. Soedin., Ser. A*, 19, 2604, 1977.

1977BER Bernstein, R.E., Cruz, C.A., Paul, D.R., and Barlow, J.W., LCST behavior in polymer blends, *Macromolecules*, 10, 681, 1977.

1977DJA Djadoun, S., Goldberg, R.N., and Morawetz, H., Ternary systems containing an acidic copolymer, a basic copolymer, and a solvent 1, *Macromolecules*, 10, 1015, 1977.

1977GOR Gordon, M., Irvine, P., and Kennedy, J.W., Phase diagrams and pulse-induced critical scattering, *J. Polym. Sci.: Polym. Symp.*, 61, 199, 1977.

1977KOB Kobayashi, M., Ikeda, H., and Masuda, Y., Shear viscosities of polymer solutions near the critical point, *Nippon Kagaku Kaishi*, (6), 866, 1977.

1977KOE Koehnen, D.M., Mulder, M.H.V., and Smolders, C.A., Phase separation phenomena during the formation of asymmetric membranes, *J. Appl. Polym. Sci.*, 21, 199, 1977.

1977KRA Kratochvil, J. and Sedlacek, B., A study of the thermodynamic properties of the system polystyrene-ethylcyclohexane by the light scattering method, *Brit. Polym. J.*, 9, 206, 1977.

1977PAN Panina, N.I., Lozgacheva, V.P., and Aver'yanova, V.M., Macromolecular parameters of cellulose acetates near theta-temperature (Russ.), *Vysokomol. Soedin., Ser. B*, 19, 786, 1977.

1977RIG Rigler, J.K., Wolf, B.A., and Breitenbach, J.W., The viscosity of polymer solutions in upper and lower region of critical solution temperature, *Angew. Makromol. Chem.*, 57, 15, 1977.

1977RO1 Robard, A., Patterson, D., and Delmas, G., The '$\Delta\chi$-effect' and polystyrene-poly(vinyl methyl ether) compatibility in solution, *Macromolecules*, 10, 706, 1977.

1977RO2 Robard, A. and Patterson, D., Temperature dependence of polystyrene-poly(vinyl methyl ether) compatibility in trichloroethene, *Macromolecules*, 10, 1021, 1977.

1977SAE Saeki, S., Kuwahara, N., Nakata, M., and Kaneko, M., Phase separation of poly(ethylene glycol)-water-salt systems, *Polymer*, 18, 1027, 1977.

1977TA1 Tager, A.A., Andreeva, V.M., Vshivkov, S.A., and Tjukova, I.S., Phase equilibria of polystyrene solutions, *J. Polym. Sci.: Polym. Symp.*, 61, 283, 1977.

1977TA2 Tager, A.A. and Ikanina, T.V., Thermodynamic study of dilute and concentrated solutions of poly(p-chlorostyrene) (Russ.), *Vysokomol. Soedin., Ser. B*, 19, 192, 1977.

1977VSH Vshivkov, S.A., Tager, A.A., Lantseva, N.V., and Loginova, L., Phase equilibrium and structure of solutions of oligomeric poly(oxypropylene)diols with upper and lower critical solution temperatures (Russ.), *Protsessy Studneobras. Polimern. Sistem.*, (2), 3, 1977.

1977WO1 Wolf, B.A. and Jend, R., Pressure and molecular-weight dependence of polymer solubility in the case of *trans*-decahydronaphthalene-polystyrene, *High Temp.-High Press.*, 9, 561, 1977.

1977WO2 Wolf, B.A. and Jend, R., Über die Möglichkeiten zur Bestimmung von Mischungsenthalpien und -volumina aus der Molekulargewichtsabhängigkeit der kritischen Entmischungstemperaturen und -drücke am Beispiel des Systems *trans*-Decahydronaphthalin/Polystyrol, *Makromol. Chem.*, 178, 1811, 1977.

1977WO3 Wolf, B.A. and Sezen, M.C., Viscometric determination of thermodynamic demixing data for polymer solution (exp. data by B.A. Wolf), *Macromolecules*, 10, 1010, 1977.

1978AHA Aharoni, S.M., Bimodal closed cloud point curves of ternary systems, *Macromolecules*, 11, 277, 1978.

1978CRA Craubner, H., Thermodynamic perturbation theory of phase separation in macromolecular multicomponent systems. 2. Concentration dependence, *Macromolecules*, 11, 1161, 1978.

1978GOL Goloborod'ko, V.I. and Tashmukhamedov, S.A., Fasovoe ravnovesie rastvorov polimerov. Sistemy s verkhnimi i nishnimi kriticheskimi temperaturami smesheniya, *Uzb. Khim. Zh.*, (1), 27, 1978.

1978HAY Hayduk, W. and Bromfiled, H.A., θ-cosolvent solutions for poly(vinyl chloride) by cloud point and osmotic pressure measurements, *J. Appl. Polym. Sci.*, 22, 149, 1978.

1978ISH Ishizawa, M., Kuwahara, N., Nakata, M., Nagayama, W., and Kaneko, M.: Pressure dependence of upper critical solution temperatures in the system polystyrene-cyclopentane, *Macromolecules*, 11, 871, 1978.

1978IVA Ivanova, N.A., Banduryan, S.I., Novikova, T.A., Kostrov, Yu.A., and Iovleva, M.M., Poly(vinyltrimethylsilane)-methylene chloride phase diagram (Russ.), *Vysokomol. Soedin., Ser. B*, 20, 922, 1978.

1978JAN Janeczek, H., Turska, E., Szekely, T., Lengyel, M., and Till, F., D.s.c. studies on the phase separation of solutions of poly(2,6-dimethyl-1,4-phenylene ether) in decalin, *Polymer*, 19, 85, 1978.

1978KO1 Kodama, Y. and Swinton, F.L., Lower critical solution temperature, Part I. Polymethylene in n-alkanes, *Brit. Polym. J.*, 10, 191, 1978.

1978KO2 Kodama, Y. and Swinton, F.L., Lower critical solution temperatures, Part II. Polymethylene in binary n-alkane solvents, *Brit. Polym. J.*, 10, 201, 1978.

1978KRO Kronberg, B., Bassignana, I., and Patterson, D., Phase diagrams of liquid crystal + polymer systems, *J. Phys. Chem.*, 82, 1714, 1978.

1978LLO Lloyd, D.R. and Burns, C.M., Reexamination of cloud point isotherms for the system polystyrene/polybutadiene/tetralin, *J. Appl. Polym. Sci.*, 22, 593, 1978.

1978NAK Nakata, M., Dobashi, T., Kuwahara, N., Kaneko, M., and Chu, B., Coexistence curve and diameter of polystyrene in cyclohexane, *Phys. Rev. A*, 18, 2683, 1978.

1978NIK Nikolaeva, N.D., Stankevich, R.P., and Vilesova, M.S., Nekotorye zakonomernosti zhidkofaznogo razdeleniya rastvorov polivinilovogo spirta, *Termodinam. Kinet. Khim. Protsess., Leningrad*, 1, 87, 1978.

1978RIG Rigler, J.K., Müller, L., and Wolf, B.A., Interaction parameters in some ternary systems styrene/polystyrene/rubber, *Angew. Makromol. Chem.*, 74, 113, 1978.

1978UGL Uglanova, G.G., Gembitskii, L.S., and Verkhotina, L.N., Phase diagram of polyacrylonitrile-dimethylformamide and polyacrylonitrile-dimethylformamide-propylene carbonate systems (Russ.), *Vysokomol. Soedin., Ser. B*, 20, 782, 1978.

1978VSH Vshivkov, S.A., Tager, A.A., and Benkovskii, A.D., Phase equilibrium of polymer solutions in a hydrodynamic field (Russ.), *Vysokomol. Soedin., Ser. B*, 20, 603, 1978.

1978WOL Wolf, B.A. and Willms, M.M., Measured and calculated solubility of polymers in mixed solvents: co-nonsolvency, *Makromol. Chem.*, 179, 2265, 1978.

1979AHA Aharoni, S.M., Rigid backbone polymers. 5. Preliminary ternary phase diagram of poly(isocyanates), *Macromolecules*, 11, 537, 1979.

1979BLA Blackadder, D.A. and Ghavamikia, H., Dissolution of polyethersulphone in chloroform, *Polymer*, 20, 523, 1979.

1979BUR Burshtein, L.L., Malinovskaya, V.P., and Bubnova, L.P., Molecular mobility in polymer solutions near the phase separation of the system (Russ.), *Vysokomol. Soedin., Ser. B*, 21, 663, 1979.

1979COW Cowie, J.M.G., Horta, A., McEwen, I.J., and Prochazka, K., Upper and lower critical solution temperatures for star branched polystyrene in cyclohexane, *Polym. Bull.*, 1, 329, 1979.

1979HAM Hamano, K., Kuwahara, N., and Kaneko, M., Scaled functions of osmotic compressibility and correlation length of polystyrene in diethyl malonate, *Phys. Rev. A*, 20, 1135, 1979.

1979IZU Izumi, Y., Dondos, A., Picot, C., and Benoit, H., Etude des solutions de polymeres dans un melange de solvants au voisinage de leur point critique de demixtion, 1. Systeme methanol/cyclohexane/polyvinyl-2-pyridine, *Makromol. Chem.*, 180, 2483, 1979.

1979KLE Kleintjens, L.A.L., Effects of chain branching and pressure on thermodynamic properties of polymer solutions, *Ph.D. Thesis*, Univ. Essex, 1979.

1979LIP Lipatov, Yu.S. and Nesterov, A.E., Thermodynamics of interactions in polymer mixtures (Russ.), *Kompoz. Polim. Mater.*, 1, 5, 1979.

1979NAR Narasimhan, V., Lloyd, D.R., and Burns, C.M., Phase equilibria of polystyrene/ polybutadiene/tetrahydrofuran using gel permeation chromatography, *J. Appl. Polym. Sci.*, 23, 749, 1979.

1979SIM Simenido, A.V., Medved, Z.N., Denisova, L.L., Tarakanov, O.G., and Starikova, N.A., Some physicochemical properties of the mixtures of polypropylene polyols with water (Russ.), *Vysokomol. Soedin., Ser. A*, 21, 1727, 1979.

1979TAG Tager, A.A., Vshivkov, S.A., and Pridannikova, N.A., Phase equilibrium, structure and thermodynamic stability of the solutions of crystallizable polymers (Russ.), *Vysokomol. Soedin., Ser. A*, 21, 565, 1979.

1980BIL Bilimova, E.S., Gladkovskii, G.A., Golubev, V.M., and Medved, Z.N., Study of phase equilibriums in mixtures of poly(propylene oxide) with water (Russ.), *Vysokomol. Soedin., Ser. A*, 22, 2240, 1980.

1980BRO Broens, L., Altena, F.W., Smolders, C.A., and Koehnen, D.M., Asymmetric membrane structures as a result of phase separation phenomena, *Desalination*, 32, 33, 1980.

1980DO1 Dobashi, T., Nakata, M., and Kaneko, M., Coexistence curve of polystyrene in methylcyclohexane. I. Range of simple scaling and critical exponents, *J. Chem. Phys.*, 72, 6685, 1980.

1980DO2 Dobashi, T., Nakata, M., and Kaneko, M., Coexistence curve of polystyrene in methylcyclohexane. II. Comparison of coexistence curves observed and calculated from classical free energy, *J. Chem. Phys.*, 72, 6692, 1980.

1980DO3 Dobashi, T., Nakata, M., and Kaneko, M., Coexistence curve of the ternary system polystyrene homologues + methylcyclohexane, *Rep. Progr. Polym. Phys. Jpn.*, 23, 9, 1980.

1980DON Dondos, A. and Izumi, Y., Study on polymer solutions in solvent mixtures in the vicinity of the critical point of the solvents, 2., *Makromol. Chem.*, 181, 701, 1980.

1980GHA Ghavamikia, H. and Blackadder, D.A., Dissolution of poly(ethersulphone) in dichloromethane, *Polymer*, 21, 659, 1980.

1980HOS Hosokawa, H., Nakata, M., Dobashi, T., and Kaneko, M., Pressure dependence of the upper critical solution temperature in the system polystyrene + cyclohexane, *Rep. Progr. Polym. Phys. Japan*, 23, 13, 1980.

1980IRV Irvine, P. and Gordon, M., Graph-like state of matter. 14. Statistical thermodynamics of semidilute polymer solution, *Macromolecules*, 13, 761, 1980.

1980ISH Ishihara, N., Ikeda, H., and Masuda, Y., The shear viscosity of the polyvinylacetate-3-heptanone mixture near the critical point, *Nippon Reoroji Gakkaishi*, 8, 39, 1980.

1980KL1 Kleintjens, L.A. and Koningsveld, R., Liquid-liquid phase separation in multicomponent polymer systems XIX. Mean-field lattice-gas treatment of the system n-alkane/linear polyethylene, *Colloid Polym. Sci.*, 258, 711, 1980.

1980KL2 Kleintjens, L.A., Koningsveld, R., and Gordon, M., Liquid-liquid phase separation in multicomponent polymer systems. 18. Effect of short-chain branching, *Macromolecules*, 13, 303, 1980.

1980KUR Kuroiwa, S., Matsuda, H., and Fujimatsu, H., Temperature dependence of solution properties of nonionic surface active agent, *Nippon Kagaku Kaishi*, (3), 362, 1980.

1980LAN Lang, J.C. and Morgan, R.D., Nonionic surfactant mixtures. I. Phase equilibria in $C_{10}E_4$-H_2O and closed-loop coexistence, *J. Chem. Phys.*, 73, 5849, 1980.

1980LEM Lemoyne, C., Friedrich, C., Halary, J.L., Noel, C., and Monnerie, L., Physico-chemical processes occurring during the formation of cellulose diacetate membranes, *J. Appl. Polym. Sci.*, 25, 1883, 1980.

1980LLO Lloyd, D.R., Narasimhan, V., and Burns, C.M., Quantitative analysis of mixed polymer systems by the use of gel permeation chromatography, *J. Liq. Chromatogr.*, 3, 1111, 1980.

1980MED Medved, Z.N. and Bulgakova, M.V., Phase equilibrium in aqueous solutions of poly(ethylene oxide) derivatives possessing eng groups of different hydrophilic character (Russ.), *Zh. Prikl. Khim.*, 53, 1669, 1980.

1980NAK Nakata, M., Dobashi, T., and Kaneko, M., Range of asymptotic critical behavior of polystyrene solutions, *Rep. Progr. Polym. Phys. Japan*, 23, 11, 1980.

1980POL Polyakov, V.I., Grinberg, V.Ya., and Tolstoguzov, V.B., Application of phase-volume-ratio method for determining the phase diagram of water-casein-soybean globulins system, *Polym. Bull.*, 2, 757, 1980.

1980RAB Rabinovich, I.B., Moseeva, E.M., and Maslova, V.A., Diagram of physical states for three-component mixtures of poly(vinyltrimethylsilane) with solvents and a precipitator (Russ.), *Vysokomol. Soedin., Ser. A*, 22, 2206, 1980.

1980RIC Richards, R.W., A corresponding states interpretation of the temperature dependence of the intrinsic viscosity of polystyrene in isopropyl acetate, *Polymer*, 21, 715, 1980.

1980ROG Rogovina, L.Z., Chalykh, A.E., Adamova, L.V., Aliev, A.D., Nekhaenko, E.A., Valetskii, P.M., Slonimskii, G.L., and Tager, A.A., Structure and thermodynamic stability of polyblock copolymers of poly(arylate-dimethylsiloxane) (Russ.), *Vysokomol. Soedin., Ser. A*, 22, 428, 1980.

1980SUV Suvorova, A. I., Andreeva, V. M., Ikanina, T. V., Zyryanova, L. K., Sorkina, I. I., and Tager, A. A., Phase diagrams of poly(vinyl chloride)-plasticizer systems (Russ.), *Vysokomol. Soedin., Ser. B*, 22, 910, 1980.

1980SWI Swislow, G., Sun, S.-T., Nishio, I., and Tanaka, T., Coil-globule phase transition in a single polystyrene chain in cyclohexane, *Phys. Rev. Lett.*, 44, 796, 1980.

1980VSH Vshivkov, S.A. and Tager, A.A., Effect of a hydrodynamic field on the phase equilibrium of polymer solutions (Russ.), *Vysokomol. Soedin., Ser. B*, 22, 110, 1980.

1980VAN Van Konynenburg, P.H. and Scott, R.L., Critical lines and phase equilibria in binary Van-der-Waals mixtures, *Philos. Trans. Roy. Soc.*, 298, 495, 1980.

1980WOL Wolf, B. and Krämer, H., Phase separation of flowing polymer solutions studied by viscosity and by turbidity, *J. Polym. Sci.: Polym. Lett. Ed.*, 18, 789, 1980.

1980ZAS Zaslavsky, B.Yu., Miheeva, L.M., Mestechkina, N.M., Shchyukina, L.G., Chlenov, M.A., Kudryashov, L.I., and Rogozhin, S.V., Use of solute partition for characterization of several aqueous biphasic polymeric systems, *J. Chromatogr.*, 202, 63, 1980.

1981CH1 Charlet, G. and Delmas, G., Thermodynamic properties of polyolefin solutions at high temperature. 1., *Polymer*, 22, 1181, 1981.

1981CH2 Charlet, G., Ducasse, R., and Delmas, G., Thermodynamic properties of polyolefin solutions at high temperature. 2., *Polymer*, 22, 1190, 1981.

1981COH Cohen-Addad, J.P. and Roby, C., NMR study of the demixing process in concentrated polyisobutylene solutions, *J. Polym. Sci., Polym. Phys. Ed.*, 19, 1395, 1981.

1981FER Fernandez-Pierola, I. and Horta, A., Cosolvents of PMMA, *Makromol. Chem.*, 182, 1705, 1981.

1981HAS Hashizume, J., Teramoto, A., and Fujita, H., Phase equilibrium study of the ternary system composed of two monodisperse polystyrenes and cyclohexane, *J. Polym. Sci.: Polym. Phys. Ed.*, 19, 1405, 1981.

1981HEU Heusch, R., Formation of crystalline, liquid-crystalline, and liquid phases during the formation of hydrates of poly(glycol ethers) with different lengths of poly(glycol ether) chains (Ger.), *Makromol. Chem.*, 182, 589, 1981.

1981HOR Horta, A. and Fernandez-Pierola, I., Depression of the critical temperature in PMMA-cosolvent systems, *Polymer*, 22, 783, 1981.

1981LLO Lloyd, D.R., Burns, C.M., and Narasimhan, V., A comparison of experimental techniques for studying polymer-polymer-solvent compatibility, *J. Polym. Sci.: Polym. Lett. Ed.*, 19, 299, 1981.

1981MED Medved, Z.N., Penzel, U., Denosiva, T.A., and Lebedev, V.S., Study of phase equilibrium in mixtures of water and ethylene oxide-propylene oxide cooligomers (Russ.), *Vysokomol. Soedin., Ser. B*, 23, 276, 1981.

1981SHI Shishov, A.K., Krivobokov, V.V., Chubarova, E.V., and Frenkel, S.Ya., Phase separation in aqueous solutions of polyethylene glycol and dextran (Russ.), *Vysokomol. Soedin., Ser. A*, 23, 1197, 1981.

1981SUV Suvorova, A. I., Andreeva, V. M., Ikanina, T. V., and Tager, A. A., Phase diagrams of a plasticized polycarbonate (Russ.), *Vysokomol. Soedin., Ser. B*, 23, 497, 1981.

1981VSH Vshivkov, S.A. and Komolova, N.A., O fazovom ravnovesii polimernykh sistem, *Vysokomol. Soedin., Ser. A*, 23, 2780, 1981.

1981WOL Wolf, B.A. and Geerissen, H., Pressure dependence of the demixing of polymer solutions determined by viscometry, *Colloid Polym. Sci.*, 259, 1214, 1981.

1982GAL Galina, H., Gordon, M., Irvine, P., and Kleintjens, L.A., Pulse-induced critical scattering and the characterization of polymer samples, *Pure Appl. Chem.*, 54, 365, 1982.

1982GEE Geerissen, H. and Wolf, B.A., Phenylalkanes as theta-solvents for polystyrene, *Makromol. Chem., Rapid Commun.*, 3, 17, 1982.

1982GOL Goloborod'ko, V.I., Valatin, S.M., and Tashmukhamedov, I.P., Vliyanie sostava i molekulyarnoi massy privitykh sopolimerov na fasovoe ravnovesie ikh rastvorov, *Uzb. Khim. Zh.*, (3), 33, 1982.

1982KAJ Kajiwara, K., Burchard, W., Kleintjens, L.A., and Koningsveld, R., Pulse induced critical scattering from solutions of randomly crosslinked polystyrene in cyclohexane, *Polym. Bull.*, 7, 191, 1982.

1982LE1 Lecourtier, J., Lafuma, F., and Quivoron, C., Compatibilization of copolymers in solution through interchain hydrogen bonding, *Eur. Polym. J.*, 18, 241, 1982.

1982LE2 Lecourtier, J., Lafuma, F., and Quivoron, C., Study of polymer compatibilization in solution through polymer/polymer interactions: ternary systems poly(ethylene oxide)/ styrene-methacrylic acid copolymers/solvent, *Makromol. Chem.*, 183, 2021, 1982.

1982MAN Mangalam, P. V. and Kalpagam, V., Styrene-acrylonitrile random copolymer in ethyl acetate, *J. Polym. Sci., Polym. Phys. Ed.*, 20, 773, 1982.

1982MIN Minh, L. V., Ohtsuka, T., Soyama, H., and Nose, T., Phase equilibrium of poly(acryl amide)-poly(vinyl alcohol)-water system, *Polym. J.*, 14, 575, 1982.

1982NOJ Nojima, S., Shiroshita, K., and Nose, T., Phase separation process in polymer systems. II. Microscopic studies on a polystyrene and diisodecyl phtalate mixture, *Polym. J.*, 14, 289, 1982.

1982SH1 Shinozaki, K., Abe, M., and Nose, T., Interfacial tension of demixed polymer solutions over a wide range of reduced temperature, *Polymer*, 23, 722, 1982.

1982SH2 Shinozaki, K., Van Tan, T., Saito, Y., and Nose, T., Interfacial tension of demixed polymer solutions near the critical temperature. Polystyrene + methylcyclohexane, *Polymer*, 23, 728, 1982.

1982SH3 Shinozaki, K., Saito, Y., and Nose, T., Interfacial tension between demixed symmetrical polymer solutions: Polystyrene-poly(dimethylsiloxane)-propylbenzene system, *Polymer*, 23, 1937, 1982.

1982STA Staikos, G., Skondras, P., and Dondos, A., Study on polymer solutions in solvent mixtures in the vicinity of the critical point of the solvents, 3. Turbidimetric study of four new ternary systems, *Makromol. Chem.*, 183, 603, 1982.

1982SU1 Suzuki, H., Kamide, K., and Saitoh, M., Lower critical solution temperature study on cellulose diacetate-acetone solutions, *Eur. Polym. J.*, 18, 123, 1982.

1982SU2 Suzuki, H., Muraoka, Y., Saitoh, M., and Kamide, K., Upper and lower critical solution temperatures in 2-butanone solutions of cellulose diacetate, *Brit. Polym. J.*, 14, 23, 1982.

1982TA1 Tager, A.A., Adamova, L.V., Vshivkov, C.A., Ikanina, T.V., Isvozchikova, V.A., and Suvorova, A.I., Phase diagram of the PMMA-phosphate-type plasticizer system and thermodynamic affinity among components (Russ.), *Vysokomol. Soedin., Ser. B*, 24, 731, 1982.

1982TA2 Tager, A.A., Vshivkov, S.A., and Polyak, O.E., Study of the phase equilibrium by refractometry method and the determination of second virial coefficients of the polymer-plasticizer system (Russ.), *Vysokomol. Soedin., Ser. B*, 24, 661, 1982.

1982WAL Walsh, D.J. and Cheng, G.L., The in situ polymerization of vinyl chloride in poly(butyl acrylate), *Polymer*, 23, 1965, 1982.

1982ZRI Zrinyi, M. and Wolfram, E., Experimental study of phase separation phenomena in swollen poly(vinyl acetate) and polystyrene gels near the critical solution temperature, *J. Colloid Interface Sci.*, 90, 34, 1982.

1983CON Conio, G., Bianchi, E., Ciferri, A., Tealdi, A., and Aden, M.A., Mesophase formation and chain rigidity in cellulose and derivatives. 1. (Hydroxypropyl)cellulose in dimethylacetamide, *Macromolecules*, 16, 1264, 1983.

1983CO1 Cowie, J.M.G. and McEwen, I.J., Polymer-cosolvent systems. 5. Upper and lower critical solution temperatures of polystyrene in n-alkanes, *Polymer*, 24, 1445, 1983.

1983CO2 Cowie, J.M.G. and McEwen, I.J., Polymer-cosolvent systems. 6. Phase behavior of polystyrene in binary mixed solvents of acetone with n-alkanes: examples of "classic cosolvency", *Polymer*, 24, 1449, 1983.

1983DJA Djadoun, S., Ternary systems of two interacting copolymers and a solvent, *Polym. Bull.*, 9, 313, 1983.

1983GUT Gutowski, T.G., Suh, N.P., Cangialose, C., and Berube, G.M., A low-energy solvent separation method, *Polym. Eng. Sci.*, 23, 230, 1983.

1983HER Herold, F.K., Schulz, G.V., and Wolf, B.A., Solution properties of poly(decyl methacrylate)s, *Mater. Chem. Phys.*, 8, 243, 1983.

1983KUB Kubota, K., Abbey, K.M., and Chu, B., Static and dynamical properties of a polymer solution with upper and lower critical solution points. NBS 705 polystyrene in methyl acetate, *Macromolecules*, 16, 138, 1983.

1983MA1 Maderek, E., Schulz, G.V, and Wolf, B.A, High temperature demixing of poly(decyl methacrylate) solutions in isooctane and its pressure dependence, *Makromol. Chem.*, 184, 1303, 1983.

1983MA2 Maderek, E., Schulz, G.V., and Wolf, B.A., Lower critical solution temperatures of poly(decyl methacrylate) in hydrocarbons, *Eur. Polym. J.*, 19, 963, 1983.

1983MUR Muraoka, Y., Inagaki, H., and Suzuki, H., Upper and lower critical solution temperatures in 1-octanol solutions of polyethylene, *Br. Polym. J.*, 15, 110, 1983.

1983NAR Narasimhan, V., Huang, R.Y.M., and Burns, C.M., Polymer-polymer interaction parameters of polystyrene and polybutadiene from studies in solutions of toluene or tetrahydrofuran, *J. Polym. Sci.: Polym. Phys. Ed.*, 21, 1993, 1983.

1983SAE Saeki, S., Narita, Y., Tsubokawa, M., and Yamaguchi, T., Phase separation temperatures in the polystyrene-poly(α-methyl styrene)-methylcyclohexane system, *Polymer*, 24, 1631, 1983.

1983TAG Tager, A.A., Bessonov, Yu.S., Ikanina, T.V., Rodionova, T.A., Suvorova, A.I., and El'boim, S.A., Heats of interaction of poly(vinyl chloride) with plasticizers (Russ.), *Vysokomol. Soedin., Ser. A*, 25, 1444, 1983.

1983TAN Tan, H.-M., Moet, A., Hiltner, A., and Baer,E., Thermoreversible gelation of atactic polystyrene solutions, *Macromolecules*, 16, 28, 1983.

1984BAM Bamberger, S., Seaman, G.V.F., Brown, J.A., and Brooks, D.E., The partition of sodium phosphate and sodium chloride in aqueous dextran poly(ethylene glycol) two-phase systems, *J. Colloid Interface Sci.*, 99, 187, 1984.

1984BAR Barrales-Rienda, J.M. and Gomez, P.A.G., Dilute solution properties of poly(*N*-vinyl-3,6-dibromo carbazole) III. Phase equilibrium studies, *Eur. Polym. J.*, 20, 1213, 1984.

1984COR Corti, M., Minero, C., and Degiorgio, V., Cloud point transition in nonionic micellar solutions, *J. Phys. Chem.*, 88, 309, 1984.

1984CO1 Cowie, J.M.G. and McEwen, I.J., Polymer-cosolvent systems. 8. Free volume and contact energies in two cosolvent mixtures for polystyrene and an interpretation in terms of a site model, *Macromolecules*, 17, 755, 1984.

1984CO2 Cowie, J.M.G. and McEwen, I.J., Polymer cosolvent systems. Part 9. The role of contact energies and free volume in two binary solvent mixtures for polystyrene, *J. Chem. Soc., Faraday Trans. I*, 80, 905, 1984.

1984CO3 Cowie, J.M.G. and McEwen, I.J., Upper and lower critical solution temperatures for polystyrene in methylcyclopentane and dimethylcyclohexane, *Polymer*, 25, 1107, 1984.

1984DOB Dobashi, T., Nakata, M., and Kaneko, M., Coexistence curve of polystyrene in methylcyclohexane. III. Asymptotic behavior of ternary system near the plait point, *J. Chem. Phys.*, 80, 948, 1984.

1984FLO Florin, E., Kjellander, R., and Eriksson, J.C., Salt effects on the cloud point of the poly(ethylene oxide) + water system, *J. Chem. Soc., Faraday Trans. I*, 80, 2889, 1984.

1984GIL Gilluck, M., Das Entmischungsgleichgewicht von Cyclohexan + Polystyren-Lösungen, *Dissertation*, TH Merseburg, 1984.

1984GRI Grishchenkova, E.V., Antonov, Yu.A., Braudo, E.E., and Tolstoguzov, V.B., A study of gelatine-methylcellulose compatibility in aqueous media, *Nahrung*, 28, 15, 1984.

1984EIN Einaga, Y., Ohashi, S., Tong, Z., and Fujita, H., Light scattering study on polystyrene in cyclohexane below the theta point, *Macromolecules*, 17, 527, 1984.

1984HAS Hashimoto, T., Sasaki, K., and Kawai, H., Time resolved light scattering studies on the kinetics of phase separation and phase dissolution of polymer blends. 2. Phase separation of ternary mixtures of polymer A, polymer B, and solvent, *Macromolecules*, 17, 2812, 1984.

1984HEF Hefford, R.J., Polymer mixing in aqueous solution, *Polymer*, 25, 979, 1984.

1984INA Inamura, I., Toki, K., Tamae, T., and Araki, T., Effects of molecular weight on the phase equilibrium of a poly(vinyl alcohol)-poly(ethylene glycol)-water system, *Polym. J.*, 16, 657, 1984.

1984KAM Kamide, K., Matsuda, S., Dobashi, T., and Kaneko, M., Cloud point curve and critical point of multicomponent polymer/ solvent system, *Polym. J.*, 16, 839, 1984.

1984LAU Lau, W.W.Y., Burns, C.M., and Huang, R.Y.M., Compatibility of polystyrene and poly(methyl methacrylate) of various molecular weights in toluene, *J. Appl. Polym. Sci.*, 29, 1531, 1984.

1984NA1 Narasimhan, V., Burns, C.M., Huang, R.Y.M., and Lloyd, D.R., Gel permeation chromatography. Use in the determination of polymer-polymer interaction parameters, *Adv. Chem. Ser.*, 206, 3, 1984.

1984NA2 Narasimhan, V., Huang, R.Y.M., and Burns, C.M., Partitioning of polystyrene between incompatible phases and distributional effects in the system polystyrene-polybutadiene-toluene, *J. Polym. Sci.: Polym. Chem. Ed.*, 22, 3895, 1984.

1984RAN Rangel-Nafaile, C., Metzner, A.B., and Wissbrun, K.F., Analysis of stress-induced phase separation in polymer solutions, *Macromolecules*, 17, 1187, 1984.

1984SAN Sander, U., Löslichkeit von Polybutylmethacrylat und Fliessverhalten verdünnter Lösungen, *Diploma Paper*, Johannes Gutenberg University Mainz, 1984.

1984SAS Sasaki, K. and Hashimoto, T., Time resolved light scattering studies on the kinetics of phase separation and phase dissolution of polymer blends. 3. Spinodal decomposition of ternary mixtures of polymer A, polymer B, and solvent, *Macromolecules*, 17, 2818, 1984.

1984SCH Schotsch, K., Wolf, B.A., Jeberien, H.-E., and Klein, J., Concentration dependece of the Flory-Huggins parameter at different thermodynamic conditions, *Makromol. Chem.*, 185, 2169, 1984.

1984SHI Shiomi, T., Imai, K., Watanabe, C., and Miya, M., Thermodynamic and conformational properties of partially butyralized poly(vinyl alcohol) in aqueous solution, *J. Polym. Sci. Polym. Phys. Ed.*, 22, 1305, 1984.

1984SUB Subramanian, R. and Dupre, D.B., Phase transitions in polypeptide liquid crystals: (T, φ) phase diagram including the reentrant isotropic phase, *J. Chem. Phys.*, 81, 4626, 1984.

1984TAG Tager, A. A., Suvorova, A. I., Ikanina, T. V., Litvinova, T. V., Khodosh, T. S., Monstakova, I. M., Maksimova, G. V., and Kutsenko, A. I., Phase diagrams of rubber-plasticizer systems (Russ.), *Vysokomol. Soedin., Ser. B*, 26, 323, 1984.

1984TAK Takahashi, H., Ikeda, H., and Masuda, Y., Phase diagrams of polystyrene-cyclohexane solutions by ultrasonic velocity in the MHz range (Jap.), *Nippon Kagaku Kaishi*, (7), 1177, 1984.

1984TSU Tsuyumoto, M., Einaga, Y., and Fujita, H., Phase equilibrium of the ternary system consisting of two monodisperse polystyrenes and cyclohexane, *Polym. J.*, 16, 229, 1984.

1984VAR Varennes, S., Charlet, G., and Delmas, G., Use of the lower critical solution temperature for the characterization of polymer mixtures and the study of their compatibility. Application to polyethylene, polypropylene, and their copolymers, *Polym. Eng. Sci.*, 24, 98, 1984.

1984WAL Walsh, D.J. and Sham, C.K., In-situ polymerization of n-butyl acrylate in poly(vinyl chloride), *Polymer*, 25, 1023, 1984.

1984WOL Wolf, B.A., Thermodynamic theory of flowing polymer solutions and its application to phase separation, *Macromolecules*, 17, 615, 1984.

1985CAN Caneba, G.T. and Soong, D.S., Polymer membrane formation through the thermal-inversion process. 1. Experimental study of membrane structure formation, *Macromolecules*, 18, 2538, 1985.

1985EIN Einaga, Y., Tong, Z., and Fujita, H., Empirical approach to phase equilibrium behavior of quasi-binary polymer solutions, *Macromolecules*, 18, 2258, 1985.

1985FUR Furusawa, K. and Tagawa, T., Adsorption behavior of water soluble polymers with lower critical solution temperature, *Colloid Polym. Sci.*, 263, 353, 1985.

1985GE1 Geerissen, H., Roos, J., and Wolf, B.A., Continuous fractionation and solution properties of PVC 3. Pressure dependence of the solubility in single solvents, *Makromol. Chem.*, 186, 769, 1985.

1985GE2 Geerissen, H., Roos, J., and Wolf, B.A., Continuous fractionation and solution properties of PVC 4. Pressure dependence of the solubility in a mixed solvent, *Makromol. Chem.*, 186, 777, 1985.

1985HAM Hamano, K., Kuwahara, N., Koyama, T., and Harada, S., Critical behaviors in the two-phase region of a micellar solution, *Phys. Rev. A*, 32, 3168, 1985.

1985KON Konevets, V.I., Andreeva, V.M., Tager, A.A., Ershova, I.A., and Kolesnikova, E.N., Study of the structure of moderately concentrated solutions of some polyamides in the region of compositions preceding the formation of liquid crystals (Russ.), *Vysokomol. Soedin., Ser. A*, 27, 959, 1985.

1985KRA Krämer, H. and Wolf, B.A., On the occurrence of shear-induced dissolution and shear-induced demixing within one and the same polymer/solvent system, *Makromol. Chem., Rapid Commun.*, 6, 21, 1985.

1985LAU Lau, W.W.Y., Burns, C.M., and Huang, R.Y.M., Molecular weights and molecular weight distribution in phase equilibria of the polystyrene-poly(methyl methacrylate)-toluene system, *J. Appl. Polym. Sci.*, 30, 1187, 1985.

1985LUE Luehmann, B., Finkelmann, H., and Rehage, G., Phase behavior and structure of polymer surfactants in aqueous solution. The occurrence of lyotropic nematic phases, *Makromol. Chem.*, 186, 1059, 1985.

1985NAN Nandi, A.K., Mandal, B.M., and Bhattacharyya, S.N., Miscibility of poly(methyl acrylate) and poly(vinyl acetate). Incompatibility in solution and thermodynamic characterization by inverse gas chromatography, *Macromolecules*, 18, 1454, 1985.

1985RAB Rabinovich, I.B., Khlyustova, T.B., and Mochalov, A.N., Calorimetric determination of thermochemical properties and the phase diagram of cellulose nitrate-dibutyl phthalate mixtures (Russ.), *Vysokomol. Soedin., Ser. A*, 27, 525, 1985.

1985ROB Robledo-Muniz, J.G., Tseng, H.S., Lloyd, D.R., and Ward, T.C., Phase behavior studies of the system polystyrene-polybutadiene-chloroform. I. Application of the Flory-Huggins theory, *Polym. Eng. Sci.*, 25, 934, 1985.

1985STA Staikos, G. and Dondos, A., Study on polymer solutions in solvent mixtures in the vicinity of the critical point of the solvents. 4. Saturation phenomena, *Polymer*, 26, 293, 1985.

1985TAK Takano, N., Einaga, Y., and Fujita, H., Phase equilibrium in the binary system polyisoprene + dioxane, *Polym. J.*, 17, 1123, 1985.

1985TON Tong, Z., Einaga, Y., and Fujita, H., Separation factor study of model polystyrene mixtures in cyclohexane, *Macromolecules*, 18, 2264, 1985.

1985VED Vedikhina, L.I., Kurmaeva, A.I., and Barabanov, V.P., Phase separation in solutions of polyampholytes of acrylic acid-2-methyl-5-vinylpyridine copolymers (Russ.), *Vysokomol. Soedin., Ser. A*, 27, 2131, 1985.

1985WAL Walter, H., Brooks, D.E., and Fischer, D., *Partitioning in Aqueous Two-Phase Systems: Theory, Methods, Uses, and Applications to Biotechnology*, Academic Press, New York, 1985.

1985WI1 Wijmans, J.G., Rutten, H.J.J., and Smolders, C.A., Phase separation phenomena in solutions of poly(2,6-dimethyl-1,4-phenyleneoxide) in mixtures of trichloroethylene, 1-octanol, and methanol. Relationship to membrane formation, *J. Polym. Sci.: Polym. Phys. Ed.*, 23, 1941, 1985.

1985WI2 Wijmans, J. G., Kant, J., Mulder, M.H.V., and Smolders, C.A., Phase separation phenomena in solutions of polysulfone in mixtures of a solvent and a nonsolvent. Relationship with membrane formation, *Polymer*, 26, 1539, 1985.

1985ZAR Zarudaeva, S.S., Pet'kov, V.T., Rabinovich, I.B., and Kir'yanov, K.V., Thermodynamics of mixtures of cellulose diacetate with dimethyl phthalate (Russ.), *Vysokomol. Soedin., Ser. B*, 27, 1778, 1985.

1985ZHI Zhil'tsova, L.A., Meshtkovskii, S.M., and Chalykh, A.E., Kinetics of curing rubber-oligoester acrylate blends at thermodynamic equilibrium (Russ.), *Vysokomol. Soedin., Ser. A*, 27, 587, 1985.

1985ZHU Zhubanov, B.A., Solomin, V.A., Lyakh, E.A., and Cherdabaev, A.S., Phase behavior of polyamic acid solutions in cyclic ethers (Russ.), *Vestn. Akad. Nauk Kaz. SSR*, (9) 19, 1985.

1986ALB Albertson, P.-A., *Partition of Cell Particles and Macromolecules*, 3rd ed., J. Wiley & Sons, New York, 1986.

1986AN1 Andreeva, V. M., Budnitskii, G. A., Tager, A. A., Shil'nikova, N. I., Bakunov, V. A., and Maiboroda, L. F., Phase separation and structure of cellulose acetate solutions used for formation of hollow selective fibers (Russ.), *Vysokomol. Soedin., Ser. B*, 28, 606, 1986.

1986AN2 Andreeva, V.M., Tsilipotkina, M.V., Tager, A.A., Safronova, V.A., and Shil'nikova, N.I., Effect of phase separation of cellulose acetate and polyarylate solutions on the porous structure and strength of films cast from them (Russ.), *Vysokomol. Soedin., Ser. A* ,28, 2147, 1986.

1986CAN Caneba, G.T. and Soong, D.S., Determination of binodal compositions of poly(methyl methacrylate)/sulfolane solutions with pulsed-NMR techniques, *Macromolecules*, 19, 369, 1986.

1986CH1 Chalykh, A.E., Sapozhnikova, I.N., Medvedeva, L.I., and Gerasimov, V.K., Mechanism of the formation of disperse phase structure in polymer mixtures during transition from ternary (polymer-polymer-solvent) to binary (polymer-polymer) systems (Russ.), *Vysokomol. Soedin., Ser. A*, 28, 1895, 1986.

1986CH2 Chalykh, A.E., Sapozhnikova, I.N., and Medvedeva, L.I., Mechanism of formation of a phase structure of binary polymeric mixtures from polymer-polymer-solvent ternary systems (Russ.), *Dokl. Akad. Nauk SSSR, Fiz. Khim.*, 288, 939, 1986.

1986CO1 Cowie, J.M.G. and McEwen, I.J., A comparison of the phase behaviour of polystyrene in cycloalkanes and n-alkanes, *Brit. Polym. J.*, 18, 387, 1986.

1986CO2 Cowie, J.M.G., McEwen, I.J., and Garay, M.T., Polymer-cosolvent systems. Synergism and antisynergism of solvent mixtures for poly(methyl methacrylate), *Polym. Commun.*, 27, 122, 1986.

1986DOB Dobashi, T. and Nakata, M., Coexistence curve of polystyrene in methylcyclohexane. IV. Three-phase coexistence curve of ternary systems, *J. Chem. Phys.*, 84, 5775, 1986.

1986FIR Firman, P. and Kahlweit, M., Phase behavior of the ternary system H_2O-oil-polypropylene glycol (PPG), *Colloid Polym. Sci.*, 264, 936, 1986.

1986FRA Francois, J., Gan, J.Y.S., and Guenet, J.M., Sol-gel transition and phase diagram of the system atactic polystyrene-carbon disulfide, *Macromolecules*, 19, 2755, 1986.

1986GUS Gustafsson, A. and Wennerström, H., Aqueous polymer two-phase systems in biotechnology, *Fluid Phase Equil.*, 29, 365, 1986.

1986HER Herold, F.K. and Wolf, B.A., Poly(n-alkylmethacrylate)s: Characterization, good and poor solvents, densities and intrinsic viscosities, *Mater. Chem. Phys.*, 14, 311, 1986.

1986HAE Haegen, R. van der, Liquid-liquid phase separation in mixtures of statistical copolymers, in *Integration of Polymer Science and Technology*, Elsevier, London, New York, 1, 67, 1986.

1986IKA Ikanina, T. V., Suvorova, A. I., and Tager, A.A., Possibility of the antiplasticization of polymers in a rubberlike state (Russ.), *Vysokomol. Soedin., Ser. A*, 28, 817, 1986.

1986INA Inamura, I., Liquid-liquid phase separation and gelation in the poly(vinyl alcohol)-poly(ethylene glycol)-water-system. Dependence on molecular weight of poly(ethylene glycol), *Polym. J.*, 18, 269, 1986.

1986IRA Irani, C.A. and Cozewith, C., Lower critical solution temperature behavior of ethylene-propylene copolymers in multicomponent solvents, *J. Appl. Polym. Sci.*, 31, 1879, 1986.

1986KAL Kalichevsky, M.T., Oorford P.D., and Ring, S.G., The incompatibility of concentrated aqueous solutions of dextran and amylose and its effect on amylose gelation, *Carbohydr.Polym.*, 6, 145, 1986

1986KRU Krüger, B., Untersuchungen zum Entmischungsgleichgewicht von Polymerlösungen, *Dissertation*, TH Merseburg, 1986.

1986KUE Kuecuekyavruz, Z. and Kuecuekyavruz, S., Theta-behaviour of poly(p-*tert*-butylstyrene)-*b*-poly(dimethylsiloxane)-*b*-poly(p-*tert*-butylstyrene), *Makromol. Chem.*, 187, 2469, 1986.

1986LIP Lipatov, Yu.S., Thermodynamic stability of colloid polymer-polymer systems, *Colloid Polym. Sci.*, 264, 377, 1986.

1986MAL Malkin, A.Ya., Kulichikhin, S.G., and Markovich, R.Z., Viscosity and phase stability of solutions of aliphatic polyamides in various solvents (Russ.), *Vysokomol. Soedin., Ser. A*, 28, 1958, 1986.

1986MAT Mattiasson, B. and Ling, T.G.I., Efforts to integrate affinity interactions with conventional separation technologies. Affinity partition using biospecific chromatographic particles in aqueous two-phase systems, *J. Chromatogr.*, 376, 235, 1986.

1986NAK Nakata, M. and Dobashi, T., Coexistence curve of polystyrene in methylcyclohexane. V. Critical behavior of the three-phase coexistence curve, *J. Chem. Phys.*, 84, 5782, 1986.

1986NAR Narasimhan, V., Huang, R.Y.M., and Burns, C.M., Determination of polymer-polymer interaction parameters of incompatible monodisperse and polydisperse polymers in solution, *J. Polym. Sci.: Polym. Symp.*, 74, 265, 1986.

1986PAP Papkov, S.P. and Iovleva, M.M., Experimental analysis of phase equilibrium in the rigid polyamide-sulfuric acid system (Russ.), *Vysokomol. Soedin., Ser. B*, 28, 677, 1986.

1986RAE Rätzsch, M.T., Kehlen, H., Browarzik, D., and Schirutschke, M., Cloud-point curve for the system copoly(ethylene-vinyl acetate) + methyl acetate. Measurement and prediction by continuous thermodynamics, *J. Macromol. Sci.-Chem. A*, 23, 1349, 1986.

1986REU Reuvers, A.J., Altena, F.W., and Smolders, C.A., Demixing and gelation behavior of ternary cellulose acetate solutions, *J. Polym. Sci.: Part B: Polym. Phys.*, 24, 793, 1986.

1986RUS Russo, P.S., Siripanyo, S., Saunders, M.J., and Karasz, F.E., Observation of a porous gel strucutre in poly(p-phenylenebenzobisthiazole)/97% H_2SO_4, *Macromolecules*, 19, 2856, 1986.

1986SA1 Saeki, S., Tsubotani, S., Kominami, H., and Tsubokawa, M., Phase separation temperatures in the ternary system polystyrene/poly(α-methylstyrene)/cyclopentane, *J. Polym. Sci.: Polym. Phys. Ed.*, 24, 325, 1986.

1986SA2 Saeki, S., Kuwahara, N., Hamano, K., Kenmochi, Y., and Yamaguchi, T., Pressure dependence of upper critical solution temperatures in polymer solutions, *Macromolecules*, 19, 2353, 1986.

1986SAN Sander, U. and Wolf, B.A., Solubility of poly(n-alkylmethacrylate)s in hydrocarbons and in alcohols, *Angew. Makromol. Chem.*, 139, 149, 1986.

1986SAS Sasaki, S. and Uzawa, T., Phase behavior of the ternary system solvent, poly(γ-benzyl-L-glutamate) and polystyrene, *Polym. Bull.*, 15, 517, 1986.

1986SCH Schulz, D.N., Peiffer, D.G., Agarwal, P.K., Larabee, J, Kaladas, J.J., Soni, L., Handwerker, B., and Garner, R.T., Phase behavior and solution properties of sulfobetaine polymers, *Polymer*, 27, 1764, 1986.

1986TJE Tjerneld, F., Berner, S., Cajarville, A., and Johansson, G., New aqueous two-phase system based on hydroxypropyl starch useful in enzyme purification, *Enzyme Microb. Technol.*, 8, 417, 1986.

1986VS1 Vshivkov, S.A. and Isakova, I.I., Fasovoe ravnovesie i struktura sistem polimetilmetakrilat-plastifikatory klassa fosfatov, *Vysokomol. Soedin., Ser. A*, 28, 2488, 1986.

1986VS2 Vshivkov, S.A. and Safronov, A.P., Fasovoe ravnovesie rastvorov polimerov v staticheskikh usloviyakh i v rezhime techeniya, *Vysokomol. Soedin., Ser. A*, 28, 2516, 1986.

1986VS3 Vshivkov, S.A., Struktura i termodinamika rastvorov polimerov vblizi verkhnich i nizhnikh kriticheskikh temperatur rastvoreniya, *Vysokomol. Soedin., Ser. A*, 28, 2601, 1986.

1986ZAS Zaslavskii, B.Yu., Bagirov, T.O., Borovskaya, A.A., Gasanova, G.Z., Gulaeva, N.D., Levin, V.Yu., and Rogozhin, S.V., Aqueous biphasic systems formed by nonionic polymers, *Colloid Polym. Sci.*, 264, 1066, 1986.

1987ANA Ananthapadmanabhan, K.P. and Goddard, E.D., The relationship between clouding and aqueous biphase formation in polymer solutions, *Coll. Surfaces*, 25, 393, 1987.

1987ARN Arnauts, J. and Berghmans, H., Amorphous thermoreversible gels of atactic polystyrene, *Polym. Commun.*, 28, 66, 1987.

1987BAR Barbarin-Castillo, J.-M., McLure, I.A., Clarson, S.J., and Semlyen, J.A., Studies of cyclic and linear poly(dimethylsiloxanes): 25. Lower critical threshold temperatures in tetramethyl solvents, *Polym. Commun.*, 28, 212, 1987.

1987BAS Baskir, J.N., Hatton, T.A., and Suter, U.W., Thermodynamics of the separation of biomaterials in two-phase aqueous polymer systems: effect of the phase-forming polymers, *Macromolecules*, 20, 1300, 1987.

1987CO1 Cowie, J.M.G., Mohain, M.A., and McEwen, I.J., Alcohol-water cosolvent systems for poly(methyl methacrylate), *Polymer*, 28, 1569, 1987.

1987CO2 Cowie, J.M.G. and McEwen, I.J., Polymer-cosolvent systems. XI. Polystyrene dissolved in mixtures of diethyl ether and nitromethane, *J. Polym. Sci.: Part B: Polym. Phys.*, 25, 1501, 1987.

1987EIN Einaga, Y., Nakamura, Y., and Fujita, H., Three-phase separation in cyclohexane solutions of binary polystyrene mixtures, *Macromolecules*, 20, 1083, 1987.

1987GE1 Geerissen, H., Schützeichel, P., and Wolf, B.A., Continuous fractionation and solution properties of PIB. I., *J. Appl. Polym. Sci.*, 34, 271, 1987.

1987GE2 Geerissen, H., Schützeichel, P., and Wolf, B.A., Continuous fractionation and solution properties of PIB. II., *J. Appl. Polym. Sci.*, 34, 287, 1987.

1987GER Gerasimov, V.K. and Chalykh, A.E., Diagramma fazovogo sostoyaniya sistemy poliefiruretan-dimetilformamid-voda, *Vysokomol. Soedin., Ser. B*, 29, 234, 1987.

1987GRI Grinberg, V.Ya., Dotdaev, S.Kh., Borisov, Yu.A., and Tolstoguzov, V.B., Possibility of the determination of Flory-Huggins interaction parameters for the polymer-polymer-solvent system from its binodal (Russ.), *Vysokomol. Soedin., Ser. B*, 29, 145, 1987.

1987GRU Gruner, K. and Greer, S.C., Density of polystyrene in diethyl malonate in the one-phase and two-phase regions near the critical solution point, *Macromolecules*, 20, 2238, 1987.

1987IC1 Ichimura, T., Okano, K., Kurita, K., and Wada, E., Small-angle X-ray scattering and coexistence curve of semidilute polymer solution, *Polymer*, 28, 1573, 1987.

1987IC2 Ichimura, T., Okano, K., Kurita, K., and Wada, E., Deuteration effect on the coexistence curve of semidilute polymer solution, *Polym. J.*, 19, 1101, 1987.

1987IKA Ikanina, T.V., Suvorova, A.I., and Tager, A.A., Plasticizer effect on polycarbonate structure (Russ.), *Vysokomol. Soedin., Ser. A*, 29, 1888, 1987.

1987JAH Jahns, E. and Finkelmann, H., Lyotropic liquid crystalline phase behavior of a polymeric amphiphile polymerized via their hydrophilic ends, *Colloid Polym. Sci.*, 265, 304, 1987.

1987JEL Jelich, L.M., Nunes, S.P., Paul, E., and Wolf, B.A., On the cooccurrence of demixing and thermoreversible gelation of polymer solutions. 1. Experimental observations, *Macromolecules*, 20, 1943, 1987.

1987KA1 Kaddour, L.O. and Strazielle, C., Experimental investigations of light scattering by a solution of two polymers, *Polymer*, 28, 459, 1987.

1987KA2 Kaddour, L.O., Anasagasti, M.S., and Strazielle, C., Molecular weight dependence of interaction parameter and demixing concentration in polymer-good solvent systems. Comparison with theory, *Makromol. Chem.*, 188, 2223, 1987.

1987KAL Kalichevsky, M.T. and Ring, S.G., Incompatibility of amylose and amylopectin in aqueous solution, *Carbohydr. Res.*, 162, 323, 1987.

1987KAN Kang, C.H. and Sandler, S.I., Phase behavior of aqueous two-polymer systems, *Fluid Phase Equil.*, 38, 245, 1987.

1987KLE Klenin, V.I., Kolnibolotchuk, N.K., Solonina, N.A., Ivanyuta, Yu.F., and Panina, N.I., Phase analysis of the polyethylene glycol-water system (Russ.), *Vysokomol. Soedin., Ser. A*, 29, 636, 1987.

1987LIN Lin, J.-L. and Roe, R.-J., Relationships among polymer-polymer interaction energy densities and the deuterium isotope effect, *Macromolecules*, 20, 2168, 1987.

1987LIS Li, S., Jiang, C., and Zhang, Y., The investigation of solution thermodynamics for the polysulfone-DMAC-water system, *Desalination*, 62, 79, 1987.

1987LOP Lopyrev, V.A., Shaglaeva, N.S., Tager, A.A., Kogan, B.R., and Gelman, A.S., Phase equilibrium of polyethylene glycol solutions in water-glycerol and water-ethylene glycol binary mixtures (Russ.), *Vysokomol. Soedin., Ser. B*, 29, 503, 1987.

1987MAG Magarik, S.Ya., Filippov, A.P., and D'yakonova, N.V., Temperature dependence of the intrinsic viscosity of polystyrene and poly(alkylstyrene) solutions (Russ.), *Vysokomol. Soedin., Ser. A*, 29, 698, 1987.

1987MEL Mel'nichenko, Yu.B., Shilov, V.V., Lipatov, Yu.S., Bulavin, L.A., and Klepko, V.V., Behavior of a polymer-solvent system near the critical phase separation point (Russ.), *Vysokomol. Soedin., Ser. A*, 29, 1554, 1987.

1987PAN Panina, N.I., Lojev, A.M., Averjanova, V.M., and Zelenev, V.M., Vyzkum fazovych rovnvoah v polymernich systemech, *Plasty Kauch.*, 24, (2), 36, 1987.

1987PER Perzynski, R., Delanti, M., and Adam, M., Experimental study of polymer interaction in a bad solvent, *J. Phys. France*, 48, 115, 1987.

1987PRI Priest, J.H., Murray, S.L., Nelson, R.J., and Hoffman, A.S., Lower critical solution temperatures of aqueous copolymers of *N*-isopropylacrylamide and other *N*-substituted acrylamides, *ACS Symp. Ser.*, 350, 255, 1987.

1987RAN Rangel-Nafaile, C. and Munoz-Lara, J.J., Analysis of the solubility of PS in DOP through various thermodynamic approaches, *Chem. Eng. Commun.*, 53, 177, 1987.

1987REU Reuvers, A.J. and Smolders, C.A., Formation of membranes by means of immersion precipitation. Part II. The mechanism of formation of membranes prepared from the system cellulose acetate-acetone-water, *J. Membrane Sci.*, 34, 67, 1987.

1987SCH Schaaf, P., Lotz, B., and Wittmann, J.C., Liquid-liquid phase separation and crystallization in binary polymer systems, *Polymer*, 28, 193, 1987.

1987SIV Sivadasan, K. and Gundiah, S., Solution behavior of poly(methacrylic acid) in an organic theta solvent, *Polymer*, 28, 1426, 1987.

1987SJO Sjöberg, A., Karlström, G., Wennerström, H., and Tjerneld, F., Compatibility of aqueous poly(ethylene glycol) and dextran at elevated temperatures, *Polym. Commun.*, 28, 263, 1987.

1987TO1 Tong, Z., Einaga, Y., Miyashita, H., and Fujita, H., Phase equilibrium in polymer + polymer + solvent ternary systems. 2. Phase diagram of polystyrene + polyisoprene + cyclohexane, *Macromolecules*, 20, 1888, 1987.

1987TO2 Tong, Z., Einaga, Y., Miyashita, H., and Fujita, H., Phase equilibrium in polymer + polymer + solvent ternary systems III. Polystyrene + polyisoprene + cyclohexane system revisited, *Polym. J.*, 19, 965, 1987.

1987TS1 Tseng, H.-S., Lloyd, D.R., and Ward, T.C., Interaction parameters of polystyrene-polyisoprene-toluene systems at 45°C from gel permeation chromatography, *J. Polym. Sci.: Part B: Polym. Phys.*, 25, 325, 1987.

1987TS2 Tseng, H.-S., Lloyd, D.R., and Ward, T.C., Phase behavior studies of the system polystyrene-polybutadiene-chloroform. II. Modification of the Flory-Huggins theory, *Polym. Eng. Sci.*, 27, 1688, 1987.

1987WIL Williams, J.M. and Moore, J.E., Microcellular foams. Phase behaviour of poly(4-methyl-1-pentene) in diisopropylbenzene, *Polymer*, 28, 1950, 1987.

1987ZA1 Zaslavsky, B.Yu., Miheeva, L.M., Gasanova, G.Z., and Mahmudov, A.U., Influence of inorganic electrolytes on partitioning of non-ionic solutes in an aqueous dextran-poly(ethylene glycol) biphasic system, *J. Chromatogr.*, 392, 95, 1987.

1987ZA2 Zaslavsky, B.Yu., Miheeva, L.M., Gasanova, G.Z., and Mahmudov, A.U., Effect of polymer composition on the relative hydrophobicity of the phases of the biphasic system aqueous dextran-poly(ethylene glycol), *J. Chromatogr.*, 403, 123, 1987.

1987ZA3 Zaslavskii, B.Yu., Mahmudov, A.U., Bagirov, T.O., Borovskaya, A.A., Gasanova, G.Z., Gulaeva, N.D., Levin, V.Yu., Mestechkina, N.M., Miheeva, L.M., and Rogozhin, S.V., Aqueous biphasic systems formed by nonionic polymers. II. Concentration effects of inorganic salts on phase separation, *Colloid Polym. Sci.*, 265, 548, 1987.

1988ABB Abbott, N.L. and Hatton, T.A., Liquid-liquid extraction for protein separations, *Chem. Eng. Progr.*, (8), 31, 1988.

1988BAR Barbalata, A., Bohossian, T., Prochazka, K., and Delmas, G., Characterization of the molecular weight distribution of high-density polyethylene by a new method using the turbidity at a lower critical solution temperature, *Macromolecules*, 21. 3286, 1988.

1988EIN Einaga, Y., A thermodynamic approach to phase equilibrium in ternary polymer solutions, *Bull. Inst. Chem. Res. Kyoto Univ.*, 66, 140, 1988.

1988FUJ Fujimatsu, H., Ogasawara, S., and Kuroiwa, S., Lower critical solution temperature (LCST) and theta temperature of aqueous solutions of nonionic surface active agents of various polyoxyethylene chain lengths, *Colloid Polym. Sci.*, 266, 594, 1988.

1988HIK Hikmet, R.M., Callister, S., and Keller, A., Thermoreversible gelation of atactic polystyrene: Phase transformation and morphology, *Polymer*, 29, 1378, 1988.

1988KAN Kang, C.H. and Sandler, S.I., Effects of polydispersivity on the phase behavior of aqueous two-phase polymer systems, *Macromolecules*, 21, 3088, 1988.

1988KAR Karachevtsev, V.G., Bon, A.I., Dubyaga, V.P., Dubyaga, E.G., Komarova, A.B., and Yakuseva, N.N., Analysis of phase equilibrium curves for silar block copolymer-solvent-precipitant systems (Russ.), *Vysokomol. Soedin., Ser. A*, 30, 737, 1988.

1988KI1 Kiepen, F., Streulichtuntersuchungen an den Systemen Oligostyrol/n-Pentan und Polystyrol/Methylcyclohexan unter hohen Drücken in der Nähe der Mischungslücke, *Dissertation*, Universität Duisburg, 1988.

1988KI2 Kiepen, F. and Borchard, W., Pressure-pulse-induced critical scattering of oligostyrene in n-pentane, *Macromolecules*, 21, 1784, 1988.

1988KI3 Kiepen, F. and Borchard, W., Critical opalescence of polymer solutions at high pressures, *Makromol. Chem.*, 189, 2595, 1988.

1988KIN King, R.S., Blanch, H.W., and Prausnitz, J.M., Molecular thermodynamics of aqueous two-phase systems for bioseparations, *AIChE-J.*, 34, 1585, 1988.

1988KR1 Krämer-Lucas, H., Schenck, H., and Wolf, B.A., Influence of shear on the demixing of polymer solutions. 1. Apparatus and experimental results, *Makromol. Chem.*, 189, 1613, 1988.

1988KR2 Krämer-Lucas, H., Schenck, H., and Wolf, B.A., Influence of shear on the demixing of polymer solutions. 2. Stored energy and theoretical calculations, *Makromol. Chem.*, 189, 1627, 1988.

1988LAR Larsson, M. and Mattiasson, B., Characterization of aqueous two-phase systems based on polydisperse phase forming polymers: enzymatic hydrolysis of starch in a PEG-starch aqueous two-phase system, *Biotechnol. Bioeng.*, 31, 979, 1988.

1988LEE Lee, H.-O., Ban, Y.-B., and Kim, J.-D., Upper and lower consolute solution temperatures of the pseudo-binary mixtures of polydispersed polymethyl methacrylate and n-butylchloride, *Korean J. Chem. Eng.*, 8, 147, 1988.

1988LIN Lin, P., Clash, C., Pearce, E.M., Kwei, T.K., and Aponte, M.A., Solubility and miscibility of poly(ethyl oxazoline), *J. Polym. Sci.: Part B: Polym. Phys.*, 26, 603, 1988.

1988MAK Makhmudov, A.U., Gasanova, G.Z., Zaslavskii, B.Yu., and Mikheeva, L.M., Effect of the polymeric composition on relative water repellency of phases of the dextran-polyethylene glycol aqueous two-phase system (Russ.), in *Fiz. Elementar. Chastits, Atomov i Molekul*, Baku, 90, 1988.

1988MEL Mel'nichenko, Yu.B., Klepko, V.V., and Shilov, V.V., Critical phenomena in a polymer-solvent system. Slow neutron transmission technique, *Polymer*, 29, 1010, 1988.

1988OKA Okada,M., Numasawa,N., and Nose,T., Molecular weight dependence of critical concentration of polymer-polymer-solvent systems, *Polym. Commun.*, 29, 294, 1988.

1988PUS Pustovoit, M.V., Kulichikhin, S.G., Pakhomov, S.I., and Andrianova, G.P., Phase stability of the polyamide 6-benzyl alcohol system (Russ.), *Izv. Vyssh. Uchebn. Zav., Khim. Khim. Teknol.*, 31, (6), 132, 1988.

1988ROG Rogovina, L.Z., Nikiforova, G.G., Martirosov, V.A., Shilov, V.V., Gomza, Yu.P., and Slonimskii, G.L., Parametry termodinamicheskoi nesovmestimosti v sisteme poliarilensul'foksid-polibutadien i ikh svyaz' so svoistvami poliblochnykh sopolimerov na ikh osnove, *Vysokomol. Soedin., Ser. A*, 30, 598, 1988.

1988SC1 Schubert, F., Friedrich, K., Hess, M., and Kosfeld, R., Investigations on phase diagrams of a coil-polymer (PC) and a semiflexible thermotropic mainchain polymer (PET-*co*-PHB) in solution (exp. data by M. Hess), *Mol. Cryst. Liq. Cryst.*, 155, 477, 1988.

1988SC2 Schuster, R., Das Entmischungsgleichgewicht im System Cyclohexan + Methylcyclohexan + Polystyren, *Diploma Paper*, TH Merseburg, 1988.

1988SUN Sundar, G. and Widom, B., Three-phase equilibrium in solutions of polystyrene homologues in cyclohexane, *Fluid Phase Equil.*, 40, 289, 1988.

1988SWI Swinyard, B.T. and Barrie, J.A., Phase separation in non-solvent/*N,N*-dimethylformamide/polyethersulfone and non-solvent/*N,N*-dimethylformamide/polysulfone systems, *Brit. Polym. J.*, 20, 317, 1988.

1988SZL Szlag, D.C. and Giuliano, K.A., A low-cost aqueous two phase system for enzyme extraction, *Biotechnol. Techn.*, 2, 277, 1988.

1988TAG Tager, A. A., Shil'nikova, N. I., Sopin, V. F., and Marchenko, G. N., Phase diagrams of the ternary systems cellulose nitrate-ethylcarbitol-ethylene glycol and cellulose nitrate-formalglycerol-ethylene glycol (Russ.), *Vysokomol. Soedin., Ser. B*, 30, 699, 1988.

1988TAN Tanaka, H. and Nishi, T., Anomalous phase separation behavior in a binary mixture of poly(vinyl methyl ether) and water under deep quench condition, *Jap. J. Appl. Phys.*, 27, L1787, 1988.

1988TVE Tveekrem, J.L., Greer, S.C., and Jacobs, D.T., The dielectric constant near the liquid-liquid critical point for polystyrene in diethyl malonate, *Macromolecules*, 21, 147, 1988.

1988VSH Vshivkov, S.A. and Balashova, M.I., Phase equilibrium of the polyethylene glycol-glycerol-water system in a mechanical field (Russ.), *Vysokomol. Soedin., Ser. B*, 30, 689, 1988.

1988ZAS Zaslavsky, B.Yu., Miheeva, L.M., Aleschko-Ozhevskii, Yu.P., Mahmudov, A.U., Bagirov, T.O., and Garaev, E.S., Distribution of inorganic salts between the coexisting phases of aqueous polymer two-phase systems. Interrelationship between the ionic and polymer composition of the phases, *J. Chromatogr.*, 439, 267, 1988.

1988ZEM Zeman, L. and Tkacik, G., Thermodynamic analysis of a membrane-forming system water/*N*-methyl-2-pyrrolidone/polyethersulfone, *J. Membrane Sci.*, 36, 119, 1988.

1988ZGA Zgadzai, O.E., Maklakov, A.E., Skirda, V.D., and Chalykh, A.E., Self-diffusion and phase separation in aqueous solutions of polyoxypropylenediol (Russ.), *Vysokomol. Soedin., Ser. A*, 30, 104, 1988.

1989BAR Barsukov, L.A., Emel'yanov, D.N., Kamskii, R.A., and Bobykina, N.S., Features of rheological behaviour of polymer solutions in conditions of phase separation (Russ.), *Vysokomol. Soedin., Ser. A*, 31, 1402, 1989.

1989BAS Baskir, J.N., Hatton, T.A., and Suter, U.W., Thermodynamics of the partitioning of biomaterials in two-phase aqueous polymer systems: comparison of lattice model to experimental data, *J. Phys. Chem.*, 93, 2111, 1989.

1989BHA Bhattacharyya, C., Maiti, N., Mandal, B.M., and Bhattacharyya, S.N., Sensitivity of ternary phase diagrams of miscible polymer pairs in common solvents to traces of water: Solvent-poly(ethyl acrylate)-poly(vinyl propionate) and solvent-poly(methyl acrylate)-poly(vinyl acetate) systems, *Macromolecules*, 22, 487, 1989.

1989BOH Bohossian, T., Charlet, G., and Delmas, G., Solution properties and characterization of polyisoprenes at a lower critical solution temperature, *Polymer*, 30, 1695, 1989.

1989BOO Boo, H.-K. and Shaw, M.T., Gelation and interaction in plasticizer/PVC solutions, *J. Vinyl Technol.*, 11, 176, 1989.

1989DIA Diamond, A.D. and Hsu, J.T., Fundamental studies of biomolecule partitioning in aqueous two-phase systems, *Biotechnol. Bioeng.*, 34, 1000, 1989.

1989FOR Fortin, S. and Charlet, G., Phase diagram of aqueous solutions of hydroxypropyl-cellulose, *Macromolecules*, 22, 2286, 1989.

1989FUJ Fujishige, S., Kubota, K., and Ando, I., Phase transition of aqueous solutions of poly(*N*-isopropylacrylamide) and poly(*N*-isopropylmethacrylamide), *J. Phys. Chem.*, 93, 3311, 1989.

1989GAI Gaides, G.E. and McHugh, A.J., Gelation in an amorphous polymer. A discussion of its relation to membrane formation, *Polymer*, 30, 2118, 1989.

1989HAE Haegen, R. van der, Kleintjens, L.A., Opstal, L. van, and Koningsveld, R., Thermodynamics of polymer solutions, *Pure Appl. Chem.*, 61, 159, 1989.

1989HA1 Haynes, C.A., Beynon, R.A., King, R.S., Blanch, H.W., and Prausnitz, J.M., Thermodynamic properties of aqueous polymer solutions. Poly(ethylene glycol)/dextran, *J. Phys. Chem.*, 93, 5612, 1989.

1989HA2 Haynes, C.A., Blanch, H.W., and Praunsitz, J.M., Separation of protein mixtures by extraction: thermodynamic properties of aqueous two-phase polymer systems containing salts and proteins, *Fluid Phase Equil.*, 53, 463, 1989.

1989IN1 Inomata, H., Yagi, Y., Otake, K., Konno, M., and Saito, S., Spinodal decomposition of an aqueous solution of poly(*N*-isopropylacrylamide), *Macromolecules*, 22, 3494, 1989.

1989IN2 Inoue, T. and Ougizawa, T., Characterization of phase behavior in polymer blends by light scattering, *J. Macromol. Sci.-Chem. A*, 26, 147, 1989.

1989KAN Kang, C.-H., Lee, C.-K., and Sandler, S.I., Polydispersity effects on the behavior of aqueous two-phase two-polymer systems, *Ind. Eng. Chem. Res.*, 28, 1537, 1989.

1989LIN Ling, T.G.I., Nilsson, H., and Mattiasson, B., Reppal PES – a starch derivative for aqueous two-phase systems, *Carbohydr. Polym.*, 11, 43, 1989.

1989MER Merten, A.-K., Das Flüssig-Flüssig-Gleichgewicht in quasiternären Systemen Lösungsmittel 1 + Lösungsmittel 2 + Polymer, *Diploma Paper*, TH Merseburg, 1989.

1989NAR Narasimhan, V., Huang, R.Y.M., and Burns, C.M., An approximate method for the determination of polymer-polymer interaction parameters of incompatible polymers in solution, *J. Appl. Polym. Sci.*, 37, 1909, 1989.

1989PER Perrau, M.B., Iliopoulos, I., and Audebert, R., Phase separation of polyelectrolyte/nonionic polymer systems in aqueous solution: effects of salt and charge density, *Polymer*, 30, 2112, 1989.

1989RAE Rätzsch, M.T. and Kehlen, H., Continuous thermodynamics of polymer systems, *Prog. Polym. Sci.*, 14, 1, 1989.

1989ROE Rösink, H.D.W., Microfiltration membrane development and module design, *Proefschrift*, Universiteit Twente, 1989.

1989RON Ronner, J.A., Wassink, S.G., and Smolders, C.A., Investigation of liquid-liquid demixing and aggregate formation in a membrane-forming system by means of pulse-induced critical scattering (PICS), *J. Membrane Sci.*, 42, 27, 1989.

1989SAT Sato, T., Ikeda, N., Itou, T., and Teramoto, A., Phase equilibrium in ternary solutions containing two semiflexible polymers with different lengths, *Polymer*, 30, 311, 1989.

1989SCH Schützeichel, P., Kontinuierliche Polymerfraktionierung und Paarwechselwirkung, *Dissertation*, Johannes Gutenberg Universität Mainz, 1989.

1989SJO Sjöberg, A. and Karlström, G., Temperature dependence of phase equilibria for the system poly(ethylene glycol)/dextran/water. A theoretical and experimental study, *Macromolecules*, 22, 1325, 1989.

1989STE Stejskal, J., Strakova, D., Kratochvil, P., Smith, S.D., and McGrath,J.E., Chemical composition distribution of a graft copolymer prepared from macromonomer. Fractionation in demixing solvents, *Macromolecules*, 22, 861, 1989.

1989TAK Takebe, T., Sawaoka, R., and Hashimoto, T., Shear-induced homogenization of semi-dilute solution of polymer mixture and unmixing after cessation of the shear, *J. Chem. Phys.*, 91, 4369, 1989.

1989TON Tong, Z., Einaga, Y., Kitagawa, T., and Fujita, H., Phase equilibrium in polymer + polymer + solvent ternary systems. 4. Polystyrene + polyisobutylene in cyclohexane and in benzene, *Macromolecules*, 22, 450, 1989.

1989VSH Vshivkov, S.A., Pastukhnova, L.A., and Titov, R.V., Influence of the mechanical field on the phase equilibrium of polyether mixtures and the cellulose diacetate-acetone-water system (Russ.), *Vysokomol. Soedin., Ser. A*, 31, 1404, 1989.

1989ZAS Zaslavsky, B.Yu., Bagirov, T.O., Borovskaya, A.A., Gulaeva, N.D., Miheeva, L.H., Mahmudov, A.U., and Rodnikova, M.N., Structure of water as a key factor of phase separation in aqueous mixtures of two nonionic polymers, *Polymer*, 30, 2104, 1989.

1990BAR Barton, A.F.M., *CRC Handbook of Polymer-Liquid Interaction Parameters and Solubility Parameters*, CRC Press, Boca Raton, 1990.

1990BOO Boomgaard, Th. van den, Boom, R.M., and Smolders, C.A., Diffusion and phase separation in polymer solution during asymmetric membrane formation, *Makromol. Chem., Macromol. Symp.*, 39, 271, 1990.

1990BRO Broecke, Ph. van den and Berghmans, H., Thermoreversible gelation of solutions of vinyl polymers, *Makromol. Chem., Macromol. Symp.*, 39, 59. 1990.

1990CAB Cabezas, H., Kabiri-Badr, M., and Szlag, F.C., Statistical thermodynamics of phase separation and ion partitioning in aqueous two-phase systems, *Bioseparation*, 1, 227, 1990.

1990CHE Chen, L.-W. and Young, T.H., EVAL membranes for blood dialysis, *Makromol. Chem., Macromol. Symp.*, 33, 183, 1990.

1990CHI Chiu, G. and Mandelkern, L., Effect of molecular weight on the phase diagram of linear polyethylene in 1-dodecanol, *Macromolecules*, 23, 5356, 1990.

1990COW Cowie, J.M.G. and Swinyard, B., Location of three critical phase boundaries in poly(acrylic acid)-dioxane solutions, *Polymer*, 31, 1507, 1990.

1990FOR Forciniti, D., Hall, C.K., and Kula, M.R., Interfacial tension of poly(ethylene glycol)-dextran-water systems: influence of temperature and polymer molecular weight, *J. Biotechnol.*, 16, 279, 1990.

1990FUJ Fujita, H., *Polymer Solutions*, Elsevier, Amsterdam, 1990.

1990GOE Gödel, W.A., Zielesny, A., Belkoura, L., Engels, T., and Woermann, D., Temperature dependence of viscosity of polystyrene/cyclohexane mixtures at critical composition, *Ber. Bunsenges. Phys. Chem.*, 94, 17, 1990.

1990HEI Heinrich, M. and Wolf, B.A., Kinetics of phase separation. Trapping of molecules in nonequilibrium phases, *Macromolecules*, 23, 590, 1990.

1990INO Inomata, H., Goto, S., and Saito, S., Phase transition of *N*-substituted acrylamide gels, *Macromolecules*, 23, 4887, 1990.

1990IWA Iwai, Y., Matsuyama, S., Shigematsu, Y., Arai, Y., Tamura, K., and Shiojima, T., Measurement and quantitative representation of liquid-liquid equilibria of polystyrene-cyclopentane systems (exp. data by Y. Iwai), *Sekiyu Gakkaishi*, 33, 117, 1990.

1990JAC Jackson, C.L. and Shaw, M.T., The phase behaviour and gelation of a rod-like polymer in solution and implications for microcellular foam morphology, *Polymer*, 31, 1070, 1990.

1990KAM Kamide, K., *Thermodynamics of Polymer Solutions*, Elsevier, Amsterdam, 1990.

1990KAR Karlström, G., Carlsson, A., and Lindman, B., Phase diagrams of nonionic polymer-water systems. Experimental and theoretical studies of the effects of surfactants and other cosolutes, *J. Phys. Chem.*, 94, 5005, 1990.

1990KEN Kennis, H.A.J., Loos, Th.W. de, DeSwaan Arons, J., Van der Haegen, R., and Kleintjens, L.A., The influence of nitrogen on the liquid-liquid phase behaviour of the system n-hexane-polyethylene: experimental results and predictions with the mean-field lattice-gas model, (exp. data by Th.W. de Loos from M.Sc. thesis by H.A.J. Kennis, TU Delft, 1987), *Chem. Eng. Sci.*, 45, 1875, 1990.

1990KRE [Van] Krevelen, D.W., *Properties of Polymers*, 3rd ed., Elsevier, Amsterdam, 1990.

1990KYO Kyoumen, M., Baba, Y., Kagemoto, A., and Beatty, C.L., Determination of the consolute temperature of poly[styrene-*co*-(butyl methacrylate)] solutions by simultaneous measurement of differential thermal analysis and laser transmittance, *Macromolecules*, 23, 1085, 1990.

1990MAA Maasen, H.-P., Yang, J.L., and Wegner, G., The structure of poly(ethylene oxide)-poly(dimethyl siloxane) triblock copolymers in solution, *Makromol. Chem., Macromol. Symp.*, 39, 215, 1990.

1990MAR Marchetti, M., Prager, S., and Cussler, E.L., Thermodynamic predictions of volume changes in temperature-sensitive gels. I. Theory, II. Experiments, *Macromolecules*, 23, 1760, 3445, 1990.

1990OTA Otake, K., Inomata, H., Konno, M., and Saito, S., Thermal analysis of the volume phase transition with *N*-isopropylacrylamide gels, *Macromolecules*, 23, 283, 1990.

1990QIN Qin, J. and Jiang, C., Investigation of phase diagrams of polysulfone membrane forming systems (Chin.), *Membrane Sci. Technol., Lanzhou*, 10, (3), 13, 1990.

1990RA1 Rätzsch, M.T. and Wohlfarth, C., Continuous thermodynamics of copolymer systems, *Adv. Polym. Sci.*, 98, 49, 1990.

1990RA2 Rätzsch, M.T., Krüger, B., and Kehlen, H., Cloud-point curves and coexistence curves of several polydisperse polystyrenes in cyclohexane, *J. Macromol. Sci.-Chem. A*, 27, 683, 1990.

1990RAG Raghava Rao, K.S.M.S., Stewart, R.M., and Todd, P., Electrokinetic demixing of two-phase aqueous polymer systems, *Sep. Sci. Technol.*, 25, 985, 1990.

1990RUI Ruiz-Garcia, J. and Greer, S.C., Symmetric tricritical point in a living polymer solution (Erratum at page 3204), *Phys. Rev. Lett.*, 64, 1983, 1990.

1990SAM Samii, A.A., Lindman, B., and Karström, G., Phase behavior of some nonionic polymers in non-aqueous solvents, *Prog. Coll. Polym. Sci.*, 82, 280, 1990.

1990SCH Schild, H.G. and Tirrell, D.A., Microcalorimetric detection of lower critical solution temperatures in aqueous polymer solutions, *J. Phys. Chem.*, 94, 4352, 1990.

1990SHE Shen, W., Smith, G.R., Knobler, C.M., and Scott, R.L., Tricritical phenomena in bimodal polymer solutions. Three-phase coexistence curves for the system polystyrene (1) + polystyrene (2) + methylcyclohexane, *J. Phys. Chem.*, 94, 7943, 1990.

1990SIM Simek, L., Petrik, S., Hadobas, F., and Bohdanecky, M., Solubility in water of triblock (PEP) copolymers of ethylene and propylene oxides, *Eur. Polym. J.*, 26, 375, 1990.

1990STA Stafford, S.G., Ploplis, A.C., and Jacobs, D.T., Turbidity of polystyrene in diethyl malonate in the one-phase region near the critical solution point, *Macromolecules*, 23, 470, 1990.

1990STR Stroeks, A. and Nies, E., A modified hole theory of polymeric fluids. 2. Miscibility behavior and pressure dependence of the system polystyrene/cyclohexane, *Macromolecules*, 23, 4092, 1990.

1990TAG Tager, A.A., Safronov, A.P., Sharina, S.V., and Galaev, I.Yu., Thermodynamics of aqueous solutions of polyvinylcaprolactam (Russ.), *Vysokomol. Soedin., Ser. A*, 32, 529, 1990.

1990TON Tong, Z., Phase equilibria in ternary systems of two chemically dissimilar polymers and solvent-solvent effects, *Huanan Lioong Daxue Xuebao, Ziran Kexueban*, 18, (3), 1, 1990.

1990TS1 Tsai, F.-J. and Torkelson, J.M., Roles of phase separation mechanism and coarsening in the formation of poly(methyl methacrylate) asymmetric membranes, *Macromolecules*, 23, 775, 1990.

1990TS2 Tsai, F.-J. and Torkelson, J.M., Microporous poly(methyl methacrylate) membranes: Effect of a low-viscosity solvent on the formation mechanism, *Macromolecules*, 23, 4983, 1990.

1990VAN Van der Haegen, R. and Van Opstal, L., Thermodynamic investigation of the phase separation behaviour of ethylene-vinyl acetate copolymers in diphenyl ether, *Makromol. Chem.*, 191, 1871, 1990.

1990WIN Winnik, F.M., Fluorescence studies of aqueous solutions of poly(*N*-isopropylacylamides) below and above their LCST, *Macromolecules*, 23, 233, 1990.

1991AKH Akhmadeev, I.R., Gimerov, F.M., Sopin, V.F., and Marchenko, G.N., Study of phase equilibria in the cellulose nitrate-acetone system by the spin probe method (Russ.), *Vysokomol. Soedin., Ser. B*, 33, 543, 1991.

1991BAE Bae, Y.C., Lambert, S.M., Soane, D.S., and Prausnitz, J.M., Cloud-point curves of polymer solutions from thermooptical measurements, *Macromolecules*, 24, 4403, 1991.

1991BOT Bottino, A., Camera-Roda, G., Capanelli, G., and Munari, S., The formation of microporous polyvinylidene fluoride membranes by phase separation, *J. Membrane Sci.*, 57, 1, 1991.

1991CHU Chu, B., Linliu, K., Xie, P., Ying, Q., Wang, Z., and Shook, J.W., A modified centrifugal apparatus for coexistence curve measurements: polystyrene in methyl-cyclohexane, *Rev. Sci. Instr.*, 62, 2252, 1991.

1991CON Connemann, M., Gaube, J., Leffrang, U., Müller, S., and Pfennig, A., Phase equilibria in the system poly(ethylene glycol) + dextran + water, *J. Chem. Eng. Data*, 36, 446, 1991.

1991DEE Dee, G.T., The application of equation-of-state theories to polar-nonpolar liquid mixtures, *J. Supercrit. Fluids*, 4, 152, 1991.

1991EAG Eagland, D. and Crowther, N.J., Influence of composition and segment distribution upon lower critical demixing of aqueous poly(vinyl alcohol-*co*-vinyl acetate) solutions, *Eur. Polym. J.*, 27, 299, 1991.

1991FO1 Forciniti, D., Hall, C.K., and Kula, M.R., Influence of polymer molecular weight and temperature on phase composition in aqueous two-phase systems, *Fluid Phase Equil.*, 61, 243, 1991.

1991FO2 Forciniti, D., Hall, C.K., and Kula, M.R., Analysis of polymer molecular weight distributions in aqueous two-phase systems, *J. Biotechnol.*, 20, 151, 1991.

1991HAR Hartounian, H. and Sandler, S.I., Polymer fractionation in aqueous two-phase polymer systems, *Biotechnol. Progr.*, 7, 279, 1991.

1991HEL Hellmann, E.H., Hellmann, G.P., and Rennie, A.R., Solvent-induced phase separation in polycarbonate blends PC/TMPC, *Colloid Polym. Sci.*, 269, 343, 1991.

1991HOU Hourdet, D., Muller, G., Vincent, J.C., Avrillon, R., and Robert, E., Characterization of polyimides in solution. Relation with the preparation of asymmetric membranes, in *Polyimides and other High-Temperature Polymers*, Elsevier Sci. Publ., Amsterdam, 507, 1991.

1991KAW Kawate, K., Imagawa, I., and Nakata, M., Cloud-point curves of ternary system nitroethane + cyclohexane + polystyrene determined by a novel method, *Polym. J.*, 23, 233, 1991.

1991KIE Kiepen, F., Brinkmann, D., Koningsveld, R., and Borchard, W., Phase diagrams in temperature, pressure and concentration-space of polystyrenes in n-pentane and methylcyclohexane near the critical solution point, *Integr. Fundam. Polym. Sci. Technol.*, 5, 25, 1991.

1991LA1 Lau, W.W.Y., Guiver, M.D., and Matsuura, T., Phase separation in polysulfone/ solvent/water and polyethersulfone/solvent/water systems, *J. Membrane Sci.*, 59, 219, 1991.

1991LA2 Lau, W.W.Y., Guiver, M.D., and Matsuura, T., Phase separation in carboxylated polysulfone/solvent/water systems, *J. Appl. Polym. Sci.*, 42, 3215, 1991.

1991LEE Lee, K.D. and Lee, D.C., Phase equilibrium in multicomponent poly(α-methyl-styrene)-cyclohexane system, *Pollimo*, 15, 274, 1991.

1991LIA Liang, K., Wu, L, Grebowicz, J., Karasz, F.E., and MacKnight, W.J., Miscibility behavior in polyethersulfone/polyimde blends with and without solvents, *Progr. Pacific Polym. Sci.*, 213, 1991.

1991LLO Lloyd, D.R., Kim, S.S., and Kinzer, K.E., Microporous membrane formation via thermally induced phase separation. II. Liquid-liquid phase separation, *J. Membrane Sci.*, 64, 1, 1991.

1991LOU Louai, A., Sarazin, D., Pollet, G., Francois, J., and Moreaux, F., Properties of ethylene oxide-propylene oxide statistical copolymers in aqueous solution, *Polymer*, 32, 703, 1991.

1991MIK Mikhailov, Yu.M., Ganina, L.V., Chalykh, A.E., and Kulichikhin, S.G., Mutual diffusion and phase equilibrium in polysulfone-solvents systems (Russ.), *Vysokomol. Soedin., Ser. A*, 33, 1081, 1991.

1991OPS Opstal, L. van, Koningsveld, R., and Kleintjens, L.A., Description of partial miscibility in chain molecule mixtures, *Macromolecules*, 24, 161, 1991.

1991RAG Raghava Rao, K.S.M.S., Stewart, R., and Todd, P., Electrokinetic demixing of two-phase aqueous polymer systems, *Sep. Sci. Technol.*, 26, 257, 1991.

1991ROB Robitaille, L., Turcotte, N., Fortin, S., and Charlet, G., Calorimetric study of aqueous solutions of (hydroxypropyl)cellulose, *Macromolecules*, 24, 2413, 1991.

1991SAF Safronov, A.P. and Tager, A.A., Thermodynamic criterion of upper consolute thermodynamic temperature for glassy polymers (Russ.), *Vysokomol. Soedin., Ser. A*, 33, 2198, 1991.

1991SAK Sakkelariou, P. and Rowe, R.C., Phase separation and morphology in ethylcellulose/cellulose acetate phthalate blends, *J. Appl. Polym. Sci.*, 43, 845, 1991.

1991SA1 Samii, A.A., Karlström, G., and Lindman, B., Phase behavior of a nonionic cellulose ether in nonaqueous solution, *Langmuir*, 7, 653, 1991.

1991SA2 Samii, A.A., Karlström, G., and Lindman, B., Phase behavior of nonionic block copolymer in a mixed-solvent system, *J. Phys. Chem.*, 95, 7887, 1991.

1991SC1 Schild, H.G., Muthukumar, M., and Tirrell, D.A., Cononsolvency in mixed aqueous solutions of poly(*N*-isopropylacrylamide), *Macromolecules*, 24, 948, 1991.

1991SC2 Schild, H.G. and Tirrell, D.A., Interaction of poly(*N*-isopropylacrylamide) with sodium n-alkyl sulfates in aqueous solution, *Langmuir*, 7, 665, 1991.

1991SC3 Schubert, K.-V., Strey, R., and Kahlweit, M., A new purification technique for alkyl polyglycol ethers and miscibility gaps for water C_iE_j, *J. Colloid Interface Sci.*, 141, 21, 1991.

1991SHE Shen, W., Smith, G.R., Knobler, C.M., and Scott, R.L., Turbidity measurements of binary polystyrene solutions near critical solution points, *J. Phys. Chem.*, 95, 3376, 1991.

1991STO Stoks, W. and Berghmans, H., Phase behavior and gelation of solutions of poly(vinyl alcohol), *J. Polym. Sci.: Part B: Polym. Phys.*, 29, 609, 1991.

1991SZY Szydlowski, J. and van Hook, W.A., Isotope and pressure effects on liquid-liquid equilibria in polymer solutions. H/D solvent isotope effects in acetone-polystyrene solutions, *Macromolecules*, 24, 4883, 1991.

1991TA1 Tager, A. A., Yushkova, S. M., Adamova, L. V., Kovylin, S. V., Berezov, L. V., Mozzhukhin, V. B., and Guzeev, V. V., Thermodynamics of interaction of methyl methacrylate-methacrylic acid copolymers with plasticizers and their mixtures (Russ.), *Vysokomol. Soedin., Ser. A*, 33, 357, 1991.

1991TA2 Tager, A.A., Safronov, A.P., Berezyuk, E.A., and Galaev, I.Yu., Hydrophobic interactions and lower critical solution temperatures of aqueous polymer solutions (Russ.), *Vysokomol. Soedin., Ser. B*, 33, 572, 1991.

1991TOL Tolstoguzov, V.B., Functional properties of food proteins and role of protein-polysaccharide interaction, *Food Hydrocolloids*, 4, 429, 1991.

1991TSE Tseng, H.-S., Lloyd, D.R., and Ward, T.C., Phase behavior of polystyrene-polyisoprene-toluene systems in the temperature range 15-45°C, *J. Polym. Sci.: Part B: Polym. Phys.*, 29, 161, 1991.

1991VAN Vandeweerdt, P., Berghmans, H., and Tervoort, Y., Temperature-concentration behaviour of solutions of polydisperse, atactic poly(methyl methacrylate) and its influence on the formation of amorphous, microporous membranes, *Macromolecules*, 24, 3547, 1991.

1991VSH Vshivkov, S.A., Precipitants effect on the phase equilibrium of polystyrene solutions near upper and lower consolute temperatures (Russ.), *Vysokomol. Soedin., Ser. A*, 33, 2523, 1991.

1991WAK Wakker, A., Lower critical demixing of a polyketone in hexafluoroisopropanol. *Polymer*, 32, 279, 1991.

1991WOR Wormuth, K.R., Patterns of phase behavior in polymer and amphiphile mixtures, *Langmuir*, 7, 1622, 1991.

1991YOK Yokoyama, H., Takano, A., Okada, M., and Nose, T., Phase diagram of star-shaped polystyrene/cyclohexane system: location of critical point and profile of coexistence curve, *Polymer*, 32, 3218, 1991.

1992AHN Ahn, W., Kim, C.Y., Kim, H., and Kim, S.C., Phase behavior of polymer/liquid crystal blends, *Macromolecules*, 25, 5002, 1992.

1992AMB Ambrosino, S. and Sixou, P., Phase separation in mixtures of poly(vinylidene fluoride) and hydroxypropyl cellulose, *Polymer*, 33, 795, 1992.

1992BEN Benkhira, A., Franta, E., and Francois, J., Polydioxolane in aqueous solutions. 1. Phase diagram, *Macromolecules*, 25, 5697, 1992.

1992COO Cook, R.L., King, H.E., and Peiffer, D.G., Pressure-induced crossover from good to poor solvent behavior for polyethylene oxide in water, *Phys. Rev. Lett.*, 69, 3072, 1992.

1992HE1 Heinrich, M. and Wolf, B.A., Interfacial tension between solutions of polystyrenes: establishment of a useful master curve, *Polymer*, 33, 1926, 1992.

1992HE2 Heinrich, M. and Wolf, B.A., Interfacial tension between demixed solutions of molecularly nonuniform polymers, *Macromolecules*, 25, 3817, 1992.

1992HON Hong, J.-S. and Lee, D.C., Phase equilibrium in poly(α-methylstyrene)/*cis*-polyisoprene/cyclohexane system, *Pollimo*, 16, 470, 1992.

1992INO Inomata, H., Goto, S., Otake, K., and Saito, S., Effect of additives on phase transition of *N*-isopropylacrylamide gels, *Langmuir*, 8, 687, 1992.

1992JER	Jeremic, K., Karasz, F.E., and MacKnight, W.J., Influence of solvent and temperature on the phase behavior of polyarylsulfone/polyimide blends, *New Polym. Mater.*, 3, 163, 1992.
1992KAM	Kammer, H.W., Kummerlöwe, C., and Morgenstern, B., Structure formation in polymer blend solutions, *Makromol. Chem. Macromol. Symp.*, 58, 131, 1992.
1992KIM	Kim, S.S. and Lloyd, D.R., Thermodynamics of polymer/diluent systems for thermally induced phase separation: 3. Liquid-liquid phase separation systems, *Polymer*, 33, 1047, 1992.
1992KIR	Kiran, E. and Zhuang, W., Solubility of polyethylene in n-pentane at high pressures, *Polymer*, 33, 5259, 1992.
1992LEC	Lecointe, J.P., Pascault, J.P., Suspene, L., and Yang, Y.S., Cloud-point curves and interaction parameters of unsaturated polyester-styrene solutions, *Polymer*, 33, 3226, 1992.
1992LEE	Lee, H.K., Myerson, A.S., and Levon, K., Nonequilibrium liquid-liquid phase separation in crystallizable polymer solutions, *Macromolecules*, 25, 4002, 1992.
1992MAL	Malmsten, M. and Lindman, B., Self-assambly of aqueous block copolymer solutions, *Macromolecules*, 25, 5440, 1992.
1992MUE	Müller, K.F, Thermotropic aqueous gels and solutions of *N,N*-dimethylacrylamide-acrylate copolymers, *Polymer*, 33, 3470, 1992.
1992NAK	Nakamura, Y., Hirao, T., Einaga, Y., and Teramoto, A., Phase equilibrium of a three-component system consisting of polystyrene, *N,N*-dimethylformamide, and cyclohexane, *Polym. J.*, 24, 311, 1992.
1992SAK	Sakellariou, P., Eastmond, G.C., and Miles, I.S., Interfacial activity of graft copolymers in blends: effect of homopolymer molecular weight, *Polymer*, 33, 4493, 1992.
1992SAS	Sassen, C.L., Gonzales Casielles, A., Loos, Th.W. de, and DeSwaan Arons, J., The influence of pressure and temperature on the phase behaviour of the system H_2O + C_{12} + C_7E_5 and relevant binary subsystems, *Fluid Phase Equil.*, 72, 173, 1992.
1992SUV	Suvorova, A.I., Safronov, A.I., Mukhina, A.Yu., and Peshekhonova, A.L., Thermodynamics of interaction of cellulose diacetate with plasticizer mixtures (Russ.), *Vysokomol. Soedin., Ser. A*, 34, 92, 1992.
1992SZY	Szydlowski, J., Rebelo, L., and van Hook, W.A., A new apparatus for the detection of phase equilibria in polymer solvent systems by light scattering, *Rev. Sci. Instrum.*, 63, 1717, 1992.
1992VEN	Venugopal, G. and Krause, S., Development of phase morphologies of poly(methyl methacrylate)-polystyrene-toluene mixtures in electric fields, *Macromolecules*, 25, 4626, 1992.
1992VIA	Viallat, A., Cohen Addad, J.P., Bom, R.P., and Perez, S., Transient phase diagram of ternary polyetherimide solutions: effect of polymer association, *Polymer*, 33, 2784, 1992.
1992WEN	Wen, H., Elbro, H.S., and Alessi, P., *Polymer Solution Data Collection. III. Liquid-liquid equlibrium*, Chemistry Data Series, Vol. 15, DECHEMA, Frankfurt/Main, 1992.
1992YAM	Yamagishi, T.-A., Abe, J., Nakamoto, Y., and Ishida, S.-I., The role of hydrogen bonds on the fractionation efficiency of phenolic resins (Jap.), *J. Thermosetting Plastics Japan*, 13, 207, 1992.

1992YAN Yang, J., Wegner, G., and Koningsveld, R., Phase behavior of ethylene oxide-dimethylsiloxane PEO-PDMS-PEO triblock copolymers with water, *Colloid Polym. Sci.*, 270, 1080, 1992.

1992WAN Wang, J., Khokhlov, A., Peiffer, D.G., and Chu, B., Phase equilibria in the ternary system zinc sulfonated polystyrene/poly(ethyl acrylate-*co*-4-vinylpyridine)/tetrahydrofuran, *Macromolecules*, 25, 2566, 1992.

1992XIA Xia, K.-Q., Franck, C., and Widom, B., Interfacial tensions of phase-separated polymer solutions, *J. Chem. Phys.*, 97, 1446, 1992.

1992ZAS Zaslavskii, B.Yu., Borovskaya, A.A., Gulaeva, N.D., and Miheeva, L.M., Physicochemical features of solvent media in the phases of aqueous polymer two-phase systems, *Biotechnol. Bioeng.*, 40, 1, 1992.

1992ZHA Zhang, K.-W., Karlström, G., and Lindman, B., Phase behaviour of systems of a non-ionic surfactant and a non-ionic polymer in aqueous solution, *Colloids Surfaces*, 67, 147, 1992.

1993AER Aerts, L., Kunz, M., Berghmans, H., and Koningsveld, R., Relation between phase behaviour and morphology in polyethylene/diphenylether systems, *Makromol. Chem.*, 194, 2697, 1993.

1993ARN Arnauts, J., Berghmans, H., and Koningsveld, R., Structure formation in solutions of atactic polystyrene in *trans*-decalin, *Makromol. Chem.*, 194, 77, 1993.

1993BAE Bae, Y.C., Shim, J.J., Soane, D.S., and Prausnitz, J.M., Representation of vapor-liquid and liquid-liquid equilibria for binary systems containing polymers: applicability of an extended Flory-Huggins equation, *J. Appl. Polym. Sci.*, 47, 1193, 1993.

1993BE1 Beck, H.N., Mixtures of tetrabromobisphenol-A polycarbonate and bis(2-ethoxy ethyl)ether: Examples of a lower critical solution temperature, *J. Appl. Polym. Sci.*, 48, 21, 1993.

1993BE2 Beck, H.N., Phase behavior of mixtures of tetrabromobisphenol-A polycarbonate in solvents and nonsolvents, *J. Appl. Polym. Sci.*, 48, 13, 1993.

1993BOO Boom, R.M., van den Boomgard, Th., van den Berg, J.W.A., and Smolders, C.A., Linearized cloudpoint curve correlation for ternary systems consisting of one polymer, one solvent and one non-solvent, *Polymer*, 34, 2348, 1993.

1993DAN Danner, R.P. and High, M.S., *Handbook of Polymer Solution Thermodynamics*, American Institute of Chemical Engineers, New York, 1993.

1993DIS Dissing, U. and Mattiasson, B., Poly(ethyleneimine) as a phase-forming polymer in aqueous two-phase systems, *Biotechnol. Appl. Biochem.*, 17, 15, 1993.

1993DOB Dobashi, T. and Nakata, M., Coexistence curve of polystyrene in methylcyclohexane. VI. Two-phase behavior of ternary system near the critical point, *J. Chem. Phys.*, 99, 1419, 1993.

1993DUR Durrani, C.M., Prystupa, D.A., Donald, A.M., and Clark, A.H., Phase diagram of mixtures of polymers in aqueous solution using Fourier transform spectroscopy, *Macromolecules*, 26, 981, 1993.

1993FEI Feil, H., Bae, Y.H., Feifen, J., and Kim, S.W., Effect of comonomer hydrophobicity and ionization on the lower critical solution temperature of *N*-isopropylacrylamide copolymers, *Macromolecules*, 26, 2496, 1993.

1993GA1 Gaube, J., Höchemer, R., Keil, B., and Pfennig, A., Polydispersity effects in the system poly(ethylene glycol) + dextran + water, *J. Chem. Eng. Data*, 38, 207, 1993.

1993GA2 Gaube, J., Pfennig, A., and Stumpf, M., Thermodynamics of aqueous poly(ethylene glycol)-dextran two-phase systems using the consistent osmotic virial equation, *Fluid Phase Equil.*, 83, 365, 1993.

1993GE1 Geveke, D.J. and Danner, R.P., Ternary phase equilibria of polystyrene with a second polymer and a solvent, *J. Appl. Polym. Sci.*, 47, 565, 1993.

1993GE2 Geveke, D.J., Bernardin, F.E., and Danner, R.P., Ternary phase equilibria of tetrahydrofuran-polystyrene-polytetrahydrofuran, *J. Appl. Polym. Sci.*, 50, 251, 1993.

1993HAR Hartounian, H., Flöter, E., Kaler, E.W., and Sandler, S.I., Effect of temperature on the phase equilibrium of aqueous two-phase polymer systems, *AIChE-J.*, 39, 1976, 1993.

1993HAY Haynes, C.A., Benitez, F.J., Blanch, H.W., and Prausnitz, J.M., Application of integral-equation theory to aqueous two-phase partitioning systems, *AIChE-J.*, 39, 1593, 1993.

1993HEM He, M., Hill, R. M., Lin, Z., Scriven, L.E., and Davis, H.T., Phase behavior and microstructure of polyoxyethylene trisiloxane surfactants in aqueous solution, *J. Phys. Chem.*, 97, 8820, 1993.

1993HOS Hosokawa, H., Nakata, M., and Dobashi,T., Coexistence curve of polystyrene in methylcyclohexane. VII. Coexistence surface and critical double point of binary system in T-p-φ space (exp. data by M. Nakata and T. Dobashi), *J. Chem. Phys.*, 98, 10078, 1993.

1993IWA Iwai, Y., Shigematsu, Y., Furuya, T., Fukuda, H., and Arai, Y., Measurement and correlation of liquid-liquid equilibria for polystyrene-cyclopentane and polystyrene-cyclopentane-cyclohexane systems (exp. data by Y. Iwai), *Polym. Eng. Sci.*, 33, 480, 1993.

1993JOH Johansson, H.O., Karlström, G., and Tjerneld, F., Experimental and theoretical study of phase separation in aqueous solutions of clouding polymers and carboxylic acids, *Macromolecules*, 26, 4478, 1993.

1993KIR Kiran, E., Xiong, Y., and Zhunag, W., Modeling polyethylene solutions in near- and supercritical fluids using the Sanchez-Lacombe model, *J. Supercrit. Fluids*, 6, 193, 1993.

1993MAL Malmsten, M., Linse, P., and Zhang, K.W., Phase behavior of aqueous poly(ethylene oxide)/poly(propylene oxide) solutions (exp. data by P. Linse), *Macromolecules*, 26, 2905, 1993.

1993MAR Marsano, E., Tamagno, M., Bianchi, E., Terbojevich, M., and Cosani, A., Cellulose-polyacrylonitrile blends 1. Thermodynamic interaction parameters and phase diagram in dimethylacetamide-LiCl, *Polym. Adv. Technol.*, 4, 25, 1993.

1993MAT Matsuo, M., Kawase, M., Sugiura, Y., Takematsu, S., and Hara, C., Phase separation behavior of poly(vinyl alcohol) solutions in relation to the drawability of films prepared from the solutions, *Macromolecules*, 26, 4461, 1993.

1993MED Medin, A.S. and Janson, J.-C., Studies on aqueous polymer two-phase systems containing agarose, *Carbohydr. Polym.*, 22, 127, 1993.

1993MIS Mishima, K., Mori, S., Nagayusa, S., Eya, H., and Nagatani, M., Partition coefficient of benzene derivatives in aqueous two-phase systems containing polyethylene glycol, *Kagaku Kogaku Ronbunshu*, 19, 1171, 1993.

1993NEU Neuchl, C. and Mersmann, A., Phase equilibria of narrowly distributed dextranes in the ternary system dextran-water-ethanol, *Fluid Phase Equil.*, 90, 389, 1993.

1993OTA Otake, K., Karaki, R., Ebina, T., Yokoyama, C., and Takahashi, S., Pressure effects on the aggregation of poly(*N*-isopropylacrylamide) and poly(*N*-isopropylacrylamide-*co*-acrylic acid) in aqueous solutions, *Macromolecules*, 26, 2194, 1993.

1993PAS Pashkin, I.I., Kirsh, Yu.E., Zubov, V.P., Anisimova, T.V., Kuzkina, I.F., Voloshina, Ya.P., and Krylov, A.V., Synthesis of water-soluble *N*-vinylcaprolactam-based copolymers and physicochemical properties of their aqueous solutions (Russ.), *Vysokomol. Soedin., Ser. A*, 35, 481, 1993.

1993PO1 Podesva, J., Stejskal, J., and Kratochvil, P., Fractionation of a diblock copolymer in a demixing-solvent system, *J. Appl. Polym. Sci.*, 49, 1265, 1993.

1993PO2 Podesva, J., Stejskal, J., Prochazka, O., Spacek, P., and Enders, S., Fractionation of a statistical copolymer in a demixing-solvent system: Theory and experiment, *J. Appl. Polym. Sci.*, 48, 1127, 1993.

1993REB Rebelo, L.P. and van Hook, W.A., An unusual phase diagram: The polystyrene-acetone system in its hypercritical region: near tricritical behavior in a pseudo-binary solution, *J. Polym. Sci.: Part B: Polym. Phys.*, 31, 895, 1993.

1993SAK Sakellariou, P., Hassan, A., and Rowe, R.C., Phase separation and polymer interactions in aqueous poly(vinyl alcohol)/hydroxypropylmethylcellulose blend, *Polymer*, 34, 1240, 1993.

1993SAR Saraiva, A., Persson, O., and Fredenslund, A., An experimental investigation of cloud-point curves for the poly(ethylene glycol)/water system at varying molecular weight distributions, *Fluid Phase Equil.*, 91, 291, 1993.

1993TA1 Tager, A.A., Klyuzhin, E.S., Adamova, L.V., and Safronov, A.P., Thermodynamics of dissolution of acrylic acid-methyl acrylate copolymers in water (Russ.), *Vysokomol. Soedin., Ser. B*, 35, 1357, 1993.

1993TA2 Tager, A.A., Safronov, A.P., Sharina, S.V., and Galaev, I.Yu., Thermodynamic study of poly(*N*-vinylcaprolactam) hydration at temperatures close to lower critical solution temperature, *Colloid Polym. Sci.*, 271, 868, 1993.

1993TRE Treszczanowicz, T. and Cieslak, D., (Liquid + liquid) equilibria in a dimethyl ether of a polyethene glycol + an n-alkane, *J. Chem. Thermodyn.*, 25, 661, 1993.

1993VS1 Vshivkov, S.A., Rusinova, E.V., and Lemsh, O.S., New type of phase coexistence curves for a crystallizable polymer-solvent system (Russ.), *Vysokomol. Soedin., Ser. B*, 35, 159, 1993.

1993VS2 Vshivkov, S.A. and Rusinova, E.V., Phase equilibrium in solutions of crystalline polymers in poor solvents (Russ.), *Vysokomol. Soedin., Ser. B*, 35, 1353, 1993.

1993WAK Wakker, A., Van Dijk, F., and Van Dijk, M.A., Phase behavior of polystyrene in methyl acetate: A static light scattering study, *Macromolecules*, 26, 5088, 1993.

1993WEL Wells, P.A., Loos, Th.W. de, and Kleintjens, L.A., Pressure pulsed induced critical scattering: Spinodal and binodal curves for the system polystyrene + methylcyclohexane (exp. data by Th.W. de Loos), *Fluid Phase Equil.*, 83, 383, 1993.

1993ZHE Zheng, K.M., Greer, S.C., Corrales, L.R., and Ruiz-Garcia, J., Living poly(α-methylstyrene) near the polymerization line, *J. Chem. Phys.*, 98, 9873, 1993.

1994ALR Alred, P.A., Kozlowski, A., Harris, J.M., and Tjerneld, F., Application of temperature-induced phase partitioning at ambient temperature for enzyme purification, *J. Chromatogr. A*, 659, 289, 1994.

1994ARN Arnauts, J., De Cooman, R., Vandeweerdt, P., Koningsveld, R., and Berghmans, H., Calorimetric analysis of liquid-liquid phase separation, *Thermochim. Acta*, 238, 1, 1994.

1994ATK Atkinson, L. and Johns, M.R., Trypsin and α-chymotrypsin partitioning in polyethylene glycol/maltodextrin aqueous-two-phase systems, *Trans. Inst. Chem. Eng., Part C, Food Biopolym. Process.*, 72 (C2), 106, 1994.

1994BAR Barbarin-Castillo, J.-M. and McLure, I.A., Thermodynamics of the lower critical threshold line in solutions of cyclic and linear poly(dimethylsiloxane)s in tetramethyl-silane, *Polymer*, 35, 3075, 1994.

1994BO1 Boom, R.M., Van den Boomgard, Th., and Smolders, C.A., Mass transfer and thermodynamics during immersion precipitation for a two-polymer system. Evaluation with the system PES-PVP-NMP-water, *J. Membrane Sci.*, 90, 213, 1994.

1994BO2 Boom, R.M., Reinders, H.W., Rolevink, H.H.W., Van den Boomgard, Th., and Smolders, C.A., Equilibrium thermodynamics of a quaternary mambrane-forming system with two polymers. 2. Experiments, *Macromolecules*, 27, 2041, 1994.

1994BOR Borchard, W., Frahn, S., and Fischer, V., Determination of cloud-point, shadow and other coexistence curves in multicomponent systems from measurements of phase volume ratios, *Macromol. Chem. Phys.*, 195, 3311, 1994.

1994CHE Cheng, L.-P., Dwan, A.-H., and Gryte, C.C., Isothermal phase behavior of nylon-6, nylon-66, and nylon-610 polyamides in formic acid-water systems, *J. Polym. Sci.: Part B: Polym. Phys.*, 32, 1183, 1994.

1994DIS Dissing, U. and Mattiasson, B., Partition of proteins in polyelectrolyte-neutral polymer aqueous two-phase systems, *Bioseparation*, 4, 335, 1994.

1994HAR Hartounian, H., Sandler, S.I., and Kaler, E.W., Aqueous two-phase systems. 1. Salt partitioning, *Ind. Eng. Chem. Res.*, 33, 2288, 1994.

1994IMR Imre, A. and van Hook, W.A., Polymer-solvent demixing under tension. Isotope and pressure effects on liquid-liquid transitions. VII. Propionitrile-polystyrene solutions at negative pressure, *J. Polym. Sci.: Part B: Polym. Phys.*, 32, 2283, 1994.

1994KAW Kawai, T., Terajima, T., Ogaswara, T., Baba, H., and Teramachi, S., Phase separation phenomena in quasi-ternary systems of demixing solvents and poly(styrene-*stat*-methyl methacrylate) and composition fractionation, *Kogakuin Daigaku Kenkyu Hokoku*, 76, 23, 1994.

1994KI1 Kiran, E. and Zhuang, W., A new experimental method to study kinetics of phase separation in high-pressure polymer solutions. Multiple rapid pressure-drop technique – MRPD, *J. Supercrit. Fluids*, 7, 1, 1994.

1994KI2 Kiran, E., Xiong, Y., and Zhuang, W., Effect of polydispersity on the demixing pressures of polyethylene in near- or supercritical alkanes, *J. Supercrit. Fluids*, 7, 283, 1994.

1994KUR Kuramoto, N., Shishida, Y., and Nagai, K., Preparation of thermally responsive electroactive poly(*N*-acryloylpyrrolidine-*co*-vinylferrocene), *Macromol. Rapid Commun.*, 15, 441, 1994.

1994LAX Laxminarayan, A., McGuire, K.S., Kim, S.S., and Lloyd, D.R., Effect of initial composition, phase separation temperature and polymer crstallization on the formation of microcellular structures via thermally induced phase separation, *Polymer*, 35, 3060, 1994.

1994LOS LoStracco, M.A., Lee, S.-H., and McHugh, M.A., Comparison of the effect of density and hydrogen bonding on the cloud point behavior of poly(ethylene-*co*-methyl acrylate)-propane-cosolvent mixtures, *Polymer*, 35, 3272, 1994.

1994LUM Lu, M., Albertsson, P.-A., Johansson, G., and Tjerneld, F., Partitioning of proteins and membrane vesicles in aqueous two-phase systems with hydrophobically modified dextran, *J. Chromatogr. A*, 668, 215, 1994.

1994MCG McGuire, K.S., Laxminarayan, A., and Lloyd, D.R., A simple method of extrapolating the coexistence curve and predicting the melting point depression curve from cloud point data for polymer-diluent systems, *Polymer*, 35, 4404, 1994.

1994MIK Mikhailov, Yu.M., Ganina, L.V., and Chalykh, A.Ye., Phase equilibrium and mutual diffusion in the poly(sulfone)-dimethylsulfoxide system, *J. Polym. Sci.: Part B: Polym. Phys.*, 32, 1799, 1994.

1994MIY Miyashita, N., Okada, M., and Nose, T., Critical exponents and phase diagram of polymer blend solutions: Polystyrene/poly(methyl methacrylate)/d6-benzene system, *Polymer*, 35, 1038, 1994.

1994MOD Modlin, R.F., Alred, P. A., and Tjerneld, F., Utilization of temperature-induced phase separation for the purification of ecdysone and 20-hydroxyecdysone from spinach, *J. Chromatogr. A*, 668, 229, 1994.

1994MUM Mumick, P.S. and McCormick, C.L., Water soluble copolymers. 54: *N*-isopropylacrylamide-*co*-acrylamide copolymers in drag reduction: synthesis, characterization, and dilute solution behavior, *Polym. Eng. Sci.*, 34, 1419, 1994.

1994OKA Okano, K., Takada, M., Kurita, K., and Furusaka, M., Interaction parameters of poly(vinyl methyl ether) in aqueous solution determined by small-angle neutron scattering, *Polymer*, 35, 2284, 1994.

1994SAT Sato, H., Kuwahara, N., and Kubota, K., Phase separation in a dilute solution in a metastable region, *Phys. Rev. E*, 50, 1752, 1994.

1994SIQ Siqueira, D.F., Nunes, S.P., and Wolf, B.A., Solution properties of a diblock copolymer in a selective solvent of marginal quality 1. Phase diagram and rheological behavior, *Macromolecules*, 27, 1045, 1994.

1994SOE Soenen, H. and Berghmans, H., Phase behavior and gelation of solutions of poly(vinyl chloride), *Polym. Gels Networks*, 2, 159, 1994.

1994SON Song, S.-W. and Torkelson, J.M., Coarsening effects on microstructure formation in isopycnic polymer solutions and membranes produced via thermally induced phase separation, *Macromolecules*, 27, 6389, 1994.

1994TAG Tager, A.A., Safronov, A.P., Berezyuk, E.A., and Galaev, I.Yu., Lower critical solution temperature and hydrophobic hydration in aqueous polymer solutions, *Colloid Polym. Sci.*, 272, 1234, 1994.

1994TAK Takada, M., Okano, K., and Kurita, K., Coexistence curve of dilute polymer solution in a mixed solvent having critical demixing point, *Polym. J.*, 26, 113, 1994.

1994VA1 Vanhee, S., Kiepen, F., Brinkmann, D., Borchard, W., Koningsveld, R., and Berghmans, H., The system methylcyclohexane/polystyrene. Experimental critical curves, cloud-point and spinodal isopleths, and their description with a semi-phenomenological treatment, *Makromol. Chem. Phys.*, 195, 759, 1994.

1994VA2 Vandeweerdt, P., De Cooman, R., Berghmans, H., and Meijer, H., Gel spinning of porous poly(methyl methacrylate) fibers, *Polymer*, 35, 5141, 1994.

1994VSH Vshivkov, S.A., and Rusinova, E.V., Phase equilibrium in polystyrene solutions in mechanical field (Russ.), *Vysokomol. Soedin., Ser. A*, 36, 98, 1994.

1994ZH1 Zhang, K., Carlsson, M., Linse, P., and Lindman, B., Phase behavior of copolymer-homopolymer mixtures in aqueous solution, *J. Phys. Chem.*, 98, 2452, 1994.

1994ZH2 Zhang, K., Karlström, G., and Lindman, B., Ternary aqueous mixtures of a nonionic polymer with a surfactant or a second polymer, *J. Phys. Chem.*, 98, 4411, 1994.

1995ADA Adamova, L.V., Klyuzhin, E.S., and Nerush, N.T., Termodinamika vsaimodeistviya karboksilsoderzhashchikh akrilovykh sopolimerov s monokarbonovymi kislotami, *Khim. Volokna*, (3), 10, 1995.

1995AND Andrianova, G.P., Pakhomov, S.I., and Pustovoit, M.V., Study of phase equilibrium in the system polyethylene-xylene-dimethylformamide (Russ.), *Izv. Vyssh. Uchebn. Zav., Khim. Khim. Tekhnol.*, 38, 110, 1995.

1995BE1 Berggren, K., Johansson, H.-O., and Tjerneld, F., Effects of salts and the surface hydrophobicity of proteins on partitioning in aqueous two-phase systems containing thermoseparating ethylene oxide-propylene oxide copolymers, *J. Chromatogr. A*, 718, 67, 1995.

1995BE2 Berghmans, S., Mewis, J., Berghmans, H., and Meijer, H., Phase behavior and structure formation in solutions of poly(2,6-dimethyl-1,4-phenylene ether), *Polymer*, 36, 3085, 1995.

1995BOR Borrajo, J., Riccardi, C.C., Williams, R.J.J., Cao, Z.Q., and Pascault, J.P., Rubber-modified cyanate esters: Thermodynamic analysis of phase separation, *Polymer*, 36, 3541, 1995.

1995CHE Chen, G. and Hoffman, A.S., A new temperature- and pH-responsive copolymer for possible use in protein conjugation, *Macromol. Chem. Phys.*, 196, 1251, 1995.

1995CH1 Cheng, L.-P., Dwan, A.-H., and Gryte, C.C., Membrane formation by isothermal precipitation in polyamide-formic acid-water systems I. Description of membrane morphology, *J. Polym. Sci.: Part B: Polym. Phys.*, 33, 211, 1995.

1995CH2 Cheng, L.-P., Dwan, A.-H., and Gryte, C.C., Membrane formation by isothermal precipitation in polyamide-formic acid-water systems II. Precipitation dynamics, *J. Polym. Sci.: Part B: Polym. Phys.*, 33, 223, 1995.

1995DOE Döbert, F., Pfennig, A., and Stumpf, M., Derivation of the consistent osmotic virial equation and its application to aqueous poly(ethylene glycol)-dextran two-phase system, *Macromolecules*, 28, 7860, 1995.

1995FUR Furuya, T., Iwai, Y., Tanaka, Y., Uchida, H., Yamada, S., and Arai, Y., Measurement and correlation of liquid-liquid equilibria for dextran-poly(ethylene glycol)-water aqueous two-phase systems at 20°C, *Fluid Phase Equil.*, 103, 119, 1995.

1995GOM Gomez, C.M., Verdejo, E., Figueruelo, J.E., Campos, A., and Soria, V., On the thermodynamic treatment of poly(vinylidene fluoride)/polystyrene blend unter liquid-liquid phase separation conditions, *Polymer*, 36, 1487, 1995.

1995GRO Grossmann, C., Tintinger, R., Zhu, J., and Maurer, G., Aqueous two-phase systems of poly(ethylene glycol) and dextran – experimental results and modeling of thermodynamic properties, *Fluid Phase Equil.*, 106, 111, 1995.

1995GUI Guido, S., Phase behavior of aqueous solutions of hydroxypropylcellulose, *Macromolecules*, 28, 4530, 1995.

1995HAA Haas, C.K. and Torkelson, J.M., 2D coarsening in phase-separated polymer solutions: dependence on distance from criticality, *Phys. Rev. Lett.*, 75, 3134, 1995.

1995HAR Haraguchi, M., Nakagawa, T., and Nose, T., Miscibility of associated polymer blend solutions: 1. One-end-aminated polystyrene/one-end-carboxylated poly(ethylene glycol) blends in toluene, *Polymer*, 36, 2567, 1995.

1995IKI Ikier, C. and Klein, H., Slowing down of the diffusion process in polystyrene/cyclohexane mixtures approaching the coexistence curve and the critical point, *Macromolecules*, 28, 1003, 1995.

1995KIM Kim, Y.-H., Kwon, C., Bae, Y.H., and Kim, S.W., Saccharide effect on the lower critical solution temperature of thermosensitive polymers, *Macromolecules*, 28, 939, 1995.

1995LUM Lu, M., Johansson, G., Albertson, P.-A., and Tjerneld, F., Partitioning of proteins in dextran/hydrophobically modified dextran aqueous two-phase systems, *Bioseparation*, 5, 351, 1995.

1995LUS Luszczyk, M., Rebelo, L.P.N., and van Hook, W.A., Isotope and pressure dependence of liquid-liquid equilibria in polymer solutions. 5. Measurements of solute and solvent isotope effects. 6. A continuous polydisperse thermodynamic interpretation of demixing measurements in polystyrene-acetone and polystyrene-methylcyclopentane (exp. data by L.P.N. Rebelo), *Macromolecules*, 28, 745, 1995.

1995MCG McGuire, K.S., Laxminarayan, A., and Lloyd, D.R., Kinetics of droplet growth in liquid-liquid phase separation of polymer-diluent systems: Experimental results, *Polymer*, 36, 4951, 1995.

1995NEU Neuchl, C. and Mersmann, A., Fractionation of polydisperse dextran using ethanol, *Chem. Eng. Sci.*, 50, 951, 1995.

1995PAN Pandit, N.K., Kanjia, J., Patel, K., and Pontikes, D.G., Phase behavior of aqueous solutions containing nonionic surfactant-polyethylene glycol mixtures, *Int. J. Pharm.*, 122, 27, 1995.

1995PET Petri, H.-M., StammerR, A., and Wolf, B.A., Continuous polymer fractionation of poly(methyl vinyl ether) and a new Kuhn-Mark-Houwink relation, *Macromol. Chem. Phys.*, 196, 1453, 1995.

1995PFO Pfohl, O., Hino, T., and Prausnitz, J.M., Solubilities of styrene-based polymers and copolymers in common solvents, *Polymer*, 36, 2065, 1995.

1995SOH Soh, Y.S., Kim, J.H., and Gryte, C.C., Phase behavior of polymer/solvent/non-solvent systems, *Polymer*, 36, 3711, 1995.

1995SON Song, S.-W. and Torkelson, J.M., Coarsening effects on the formation of microporous membranes produced via thermally induced phase separation of polystyrene-cyclohexanol solutions, *J. Membrane Sci.*, 98, 209, 1995.

1995SVE Svensson, M., Linse, P., and Tjerneld, F., Phase behavior in aqueous two-phase systems containing micelle-forming block copolymers, *Macromolecules*, 28, 3597, 1995.

1995THU Thuresson, K., Karlström, G., and Lindman, B., Phase diagrams of mixtures of a nonionic polymer, hexanol, and water. An experimental and theoretical study of the effect of hydrophobic modification, *J. Phys. Chem.*, 99, 3823, 1995.

1995TIN Tintinger, R., Thermodynamische Eigenschaften ausgewählter wäßriger Zwei-Phasen Systeme, *Dissertation*, Universität Kaiserslautern, 1995.

1995TOK To, K., Chan, C.K., and Choi, H.J., Phase equilibrium and conformation of a polymer in a binary solvent, *Physica A*, 221, 223, 1995.

1995TON Tong, Z., Meissner, K., and Wolf, B.A., Phase equilibria and interfacial tension between coexisting phases for the system water/2-propanol/poly(acrylic acid), *Macromol. Chem. Phys.*, 196, 521, 1995.

1995TUM Tuminello, W.H., Brill, D.J., Walsh, D.J., and Paulaitis, M.E., Dissolving poly(tetrafluoroethylene) in low boiling halocarbons, *J. Appl. Polym. Sci.*, 56, 495, 1995.

1995VSH Vshivkov, S.A. and Safronov, A.P., Coil-globule transition of polystyrene in solution (Russ.), *Vysokomol. Soedin., Ser. B*, 37, 1779, 1995.

1995ZAS Zaslavsky, B.Y., *Aqueous Two-Phase Partitioning. Physical Chemistry and Bioanalytical Applications*, Marcel Dekker Inc., New York, 1995.

1995ZHA Zhang, K. and Kahn, A., Phase behavior of poly(ethylene oxide)-poly(propylene oxide)-poly(ethylene oxide) triblock copolymers in water, *Macromolecules*, 28, 3807, 1995.

1995ZRY Zryd, J.L. and Burghardt, W.R., Phase separation, crystallization, and structure formation in immiscible polymer solutions, *J. Appl. Polym. Sci.*, 57, 1525, 1995.

1996ALL Allcock, H.R. and Dudley, G.K., Lower critical solubility temperature study of alkyl ether based polyphosphazenes, *Macromolecules*, 29, 1313, 1996.

1996AND Andrianova, G.P., Pakhomov, S.I., and Pustovit, M.V., Phase equilibria in the polyethylene-xylene-dimethylformide system (Russ.), *Vysokomol. Soedin., Ser. B*, 38, 1761, 1996.

1996BE1 Bergfeldt, K. and Piculell, L., Phase behavior of weakly charged polymer/surfactant/water mixtures, *J. Phys. Chem.*, 100, 5935, 1996.

1996BE2 Berghmans, S., Berghmans, H., and Meijer, H.E.H., Spinning of hollow porous fibres via the TIPS mechanism, *J. Membrane Sci.*, 116, 171, 1996.

1996BUL Bulte, A.M.W., Naafs, E.M., van Eeten, F., Mulder, M.H.V., Smolders, C.A., and Strathmann, H., Equilibrium thermodynamics of the ternary membrane-forming system nylon, formic acid, and water, *Polymer*, 37, 1647, 1996.

1996CAM Campos, A., Gomez, C.M., Garcia, R., Figueruelo, J.E., and Soria, V., Extension of the Flory-Huggins theory to study incompatible polymer blends in solution from phase separation data, *Polymer*, 37, 3361, 1996.

1996CES Cesi, V., Katzbauer, B., Narodoslawsky, M., and Moser, A., Thermophysical properties of polymers in aqueous two-phase systems, *Int. J. Thermophys.*, 17, 127, 1996.

1996CHA Chalykh, A.E., Prokopov, N.I., Gritskova, I.A., and Gerasimov, V.K., Phase equilibrium diagrams for polystyrene-poly(methyl methacrylate)-styrene systems (Russ.), *Vysokomol. Soedin., Ser. A*, 38, 1888, 1996.

1996COO Cooman, R. de, Vandeweerdt, P., Berghmans, H., and Koningsveld, R., Solution spinning of fibers with oriented porosity, *J. Appl. Polym. Sci.*, 60, 1127, 1996.

1996DOB Dobashi, T., Koshiba, T., and Nakata, M., Coexistence curve of polystyrene in methylcyclohexane. IX. Pressure dependence of tricritical point, *J. Chem. Phys.*, 105, 2906, 1996.

1996ELE El-Ejmi, A.A.S. and Huglin, M.B., Characterization of *N,N*-dimethylacrylamide/2-methoxyethylacrylate copolymers and phase behaviour of their thermotropic aqueous solutions, *Polym. Int.*, 39, 113, 1996.

1996FIS Fischer, V., Borchard, W., and Karas, M., Thermodynamic properties of poly (ethylene glycol)/water systems. 1. A polymer sample with a narrow molar mass distribution, *J. Phys. Chem.*, 100, 15992, 1996.

1996FRA Francois, J., Maitre, S., Rawiso, M., Sarazin, D., Beinert, G., and Isel, F., Neutron and X-ray scattering studies of model hydrophobically end-capped poly(ethylene oxide) aqueous solutions at rest and under shear, *Coll. Surfaces A*, 112, 251, 1996.

1996FUR Furuya, T., Yamada, S., Zhu, J., Yamaguchi, Y., Iwai, Y., and Arai, Y., Measurement and correlation of liquid-liquid equilibria and partition coefficients of hydrolytic enzymes for DEX T500 + PEG20000 + water aqueous two-phase system at 20°C, *Fluid Phase Equil.*, 125, 89, 1996.

1996GAL Galera-Gomez, P.A. and Gu, T., Cloud point of mixtures of polypropylene glycol and Triton X-100 in aqueous solutions, *Langmuir*, 12, 2602, 1996.

1996GIR Girard-Reydet, E., *Ph.D. Thesis*, INSA de Lyon (quoted in 2004ZUC), France, 1996.

1996GOR Gorbunova, I.Yu., Kerber, M.L., Avdeev, N.N., Stepanova, A.V., and Vladimirova, S.I., Rheological properties and compatibility between polyethylene and some oligoesters (Russ.), *Vysokomol. Soedin., Ser. B*, 38, 1052, 1996.

1996IM1 Imre, A. and van Hook, W.A., Demixing in polystyrene/methylcyclohexane solutions, *J. Polym. Sci.: Part B: Polym. Sci.*, 34, 751, 1996.

1996IM2 Imre, A. and van Hook, W.A., Liquid-liquid demixing from solutions of polystyrene. 1. A review. 2. Improved correlation with solvent properties, *J. Phys. Chem. Ref. Data*, 25, 637, 1996.

1996KIR Kirsh, Yu.E., Krylov, A.V., Belova, T.A., Abdelsadek, G.G., and Pashkin, I.I., Phase transition of poly(*N*-vinylcaprolactam) in water-organic solvent systems (Russ.), *Zh. Fiz. Khim.*, 70, 1403, 1996.

1996KRA Krause, C., Schereinflüsse auf die Entmischung von Lösungsmittel/Polymer A/ Polymer B-Systemen, *Dissertation*, Johannes Gutenberg Universität Mainz, 1996.

1996KUM Kumaki, J., Hashimoto, T., and Granick, S., Temperature gradients induce phase separation in a miscible polymer solution, *Phys. Rev. Lett.*, 77, 1990, 1996.

1996LIM Li, M., Zhu, Z. Q., and Mei, L. H., Liquid-liquid equilibria for hydroxypropyl starch + poly(ethylene glycol) + water at 25°C, *J. Chem. Eng. Data*, 41, 500, 1996.

1996LOO Loos, Th.W.de, Graaf, L.J. de, and DeSwaan Arons, J., Liquid-liquid phase separation in linear low density polyethylene-solvent systems, *Fluid Phase Equil.*, 117, 40, 1996.

1996LUM Lu, M., Albertson, P.-A., Johansson, G., and Tjerneld, F., Ucon-benzoyl dextran aqueous two-phase systems: protein purification with phase component recycling, *J. Chromatogr. B*, 680, 65, 1996.

1996LUS Luszczyk, M. and van Hook, W.A., Isotope and pressure dependence of liquid-liquid equilibria in polymer solutions. 7. Solute and solvent H/D isotope effects in polystyrene-propionitrile solutions, *Macromolecules*, 29, 6612, 1996.

1996MAL Malkin, A.Ya., Kulichikhin, S.G., Karpov, E.E., and Lotmentsev, Yu.M., The effect of mechanical field on phase transitions in the poly(dimethylslioxane)-toluene-ethanol system (Russ.), *Vysokomol. Soedin., Ser. B*, 38, 728, 1996.

1996MAS Mas, A., Sledz, J., and Schue, F., Membranes de microfiltration en polyhydroxy-butyrate et poly(hydroxybutyrate-*co*-hydroxyvalerate): influence du pourcentage d'unites hydroxyvalerate, *Bull. Soc. Chim. Belg.*, 105, 223, 1996.

1996MIS Mishima, K., Wada, N., Sigematsu, Y., Oka, S., and Nagatani, M., Measurement and correlation of liquid-liquid phase equilibrium compositions of aqueous solutions containing methoxypolyethylene glycol and dextran, *Fukuoka Daigaku Kogaku Shuho*, 57, 141, 1996.

1996MIY Miyazaki, H. and Kataoka, K., Preparation of polyacrylamide derivatives showing thermo-reversible coacervate formation and their potential application to two-phase separation processes, *Polymer*, 37, 681, 1996.

1996PAN Pandit, N.K. and Kanjia, J., Phase behavior of nonionic surfactant solutions in the presence of polyvinylpyrrolidone, *Int. J. Pharm.*, 141, 197, 1996.

1996PET Petri, H.-M., Horst, R., and Wolf, B.A., Determination of interaction parameters for highly incompatible polymers, *Polymer*, 37, 2709, 1996.

1996PIC Piculell, L., Bergfeldt, K., and Gerdes, S., Segregation in aqueous mixtures of nonionic polymers and surfactant micelles. Effects of micelle size and surfactant headgroup/polymer interactions, *J. Phys. Chem.*, 100, 3675, 1996.

1996RON Rong, Z., Wang, H., Ying, X., and Hu, Y., Liquid-liquid equilibria for the systems of polystyrene in cyclohexane and methylcyclohexane (Chin.), *J. East China Univ. Sci. Technol.*, 22, 754, 1996.

1996SAF Safronov, A.P., Tager, A.A., and Koroleva, E.V., Thermodynamics of dissolution of polyacrylic acid in donor and acceptor solvents (Russ.), *Vysokomol. Soedin., Ser. B*, 38, 900, 1996.

1996SOE Soenen, H. and Berghmans, H., Gelation and structure formation in solutions of poly(vinyl chloride), *J. Polym. Sci.: Part B: Polym. Phys.*, 34, 241, 1996.

1996THU Thuresson, K., Nilsson, S., and Lindman, B., Influence of cosolutes on phase behavior and viscosity of a nonionic cellulose ether. The effect of hydrophobic modification, *Langmuir*, 12, 2412, 1996.

1996VS1 Vshivkov, S.A., Rusinova, E.V., Dubchak, V.N., and Zarubin, G.B., Stress-induced phase transitions in polydimethylsiloxane-methyl ethyl ketone system (Russ.), *Vysokomol. Soedin., Ser. A*, 38, 844, 1996.

1996VS2 Vshivkov, S.A., Rusinova, E.V., Dubchak, V.N., and Zarubin, G.B., Thermodynamics and structure of polydimethylsiloxane-methyl ethyl ketone system (Russ.), *Vysokomol. Soedin., Ser. A*, 38, 868, 1996.

1996VS3 Vshivkov, S.A. and Rusinova, E.V., Phase transitions in poly(methyl methacrylate) solutions (Russ.), *Vysokomol. Soedin., Ser. A*, 38, 1746, 1996.

1996WI1 Witte, P. van de, Boorsma, A., Esselburgge, H., Dijkstra, P.J., Berg, J.W.A. van den, and Feijen, J., Differential scanning calorimetry study of phase transition in poly(lactide)-chloroform-methanol systems, *Macromolecules*, 29, 212, 1996.

1996WI2 Witte, P. van de, Esselburgge, H., Dijkstra, P.J., Berg, J.W.A. van den, and Feijen, J., Phase transitions during membrane formation of polylactides. I. A morphological study of membranes obtained from the system polylactide-chloroform-methanol, *J. Membrane Sci.*, 113, 223, 1996.

1996WI3 Witte, P. van de, Dijkstra, P.J., Berg, J.W.A. van den, and Feijen, J., Phase behavior of polylactides in solvent-nonsolvent mixtures, *J. Polym. Sci.: Part B: Polym. Phys.*, 34, 2553, 1996.

1996XIA Xia, K.-Q., An, X.-Q., and Shen, W.-G., Measured coexistence curves of phase-separated polymer solutions, *J. Chem. Phys.*, 105, 6018, 1996.

1997BOU Boutris, C., Chatzi, E.G., and Kiparissides, C., Characterization of the LCST behaviour of aqueous poly(*N*-isopropylacrylamide) solutions by thermal and cloud-point techniques, *Polymer*, 38, 2567, 1997.

1997CHA Chalykh, A.E., Gerasimov, V.K., Vishnevskaya, I.A., and Morozova, N.I., Phase equilibria in segmented polyurethane solutions (Russ.), *Vysokomol. Soedin., Ser. A*, 39, 1485, 11997.

1997CHE Chevillard, C. and Axelos, M.A.V., Phase separation of aqueous solution of methylcellulose, *Colloid Polym. Sci.*, 275, 537, 1997.

1997CHO Cho, S.H., Jhon, M.S., Yuk, S.H., and Lee, H.B., Temperature-induced phase transition of poly(*N,N*-dimethylaminoethyl methacrylate-*co*-acrylamide), *J. Polym. Sci.: Part B: Polym. Phys.*, 35, 595, 1997.

1997CUN Cunha, M.T., Cabral, J.M.S., and Aires-Barros, M.R., Quantification of phase composition in aqueous two-phase systems of Breox/phosphate and Breox/Reppal PES 100 by isocratic HPLC, *Biotechnol. Techn.*, 11, 351, 1997.

1997DES DeSousa, H.C., Phase equilibria in polymer + solvent systems: experimental results and modeling (Portug.), *Ph.D. Thesis*, New University of Lisbon, Portugal, 1997.

1997ELE El-Ejmi, A.A.S. and Huglin, M.B., Behaviour of poly(*N,N*-dimethylacrylamide-*co*-2-methoxyethylacrylate) in non-aqueous solution and LCST behaviour in water, *Eur. Polym. J.*, 33, 1281, 1997.

1997END Enders, S. and Loos, Th.W. de, Pressure dependence of the phase behaviour of polystyrene in methylcyclohexane (exp. data by S. Enders), *Fluid Phase Equil.*, 139, 335, 1997.

1997FRA Franco, T.T., Galaev, I.Yu., Hatti-Kaul, R., Holmberg, N., Bülow, L., and Mattiasson, B., Aqueous two-phase system formed by thermoreactive vinyl imidazole/vinyl caprolactam copolymer and dextran for partitioning of a protein with a polyhistidine tail, *Biotechnol. Techn.*, 11, 231, 1997.

1997HO1 Holmberg, C., Nilsson, S., and Sundelöf, L.-O., Thermodynamic properties of surfactant/polymer/water systems with respect to clustering adsorption and intermolecular interactions as a function of temperature and polymer concentration, *Langmuir*, 13, 1392, 1997.

1997HO2 Holmqvist, P., Alexandridis, P., and Lindman, B., Phase behavior and structure of ternary amphiphilic block copolymer-alkanol-water systems: comparison of poly(ethylene oxide)/poly(propylene oxide) to poly(ethylene oxide)/poly(tetrahydrofuran) copolymers, *Langmuir*, 13, 2471, 1997.

1997HWA Hwang, J.H., Lee, K.H., and Lee, D.C., Study of phase behavior in SBS triblock copolymer/polybutadiene/methyl isobutyl ketone system, *Pollimo*, 21, 745, 1997.

1997IMR Imre, A. and van Hook, W.A., Continuity of solvent quality in polymer solutions. Poor-solvent to theta-solvent continuity in some polystyrene solutions, *J. Polym. Sci.: Part B: Polym. Phys.,* 35, 1251, 1997.

1997JOH Johansson, H.-O., Karlström, G., and Tjerneld, F., Effect of solute hydrophobicity on phase behaviour in solutions of thermoseparating polymers, *Colloid Polym. Sci.*, 275, 458, 1997.

1997KIM Kim, J.Y., Lee, H.K., Baik, K.J., and Kim, S.C., Liquid-liquid phase separation in polysulfone/solvent/water systems, *J. Appl. Polym. Sci.*, 65, 2643, 1997.

1997KIT Kita, R., Dobashi, T., Yamamoto, T., Nakata, M., and Kamide, K., Coexistence curve of a polydisperse polymer solution near the critical point, *Phys. Rev. E*, 55, 3159, 1997.

1997KIZ Kizhnyaev, V.N., Gorkovenko, O.P., Bazhenov, D.N., and Smirnov, A.I., The solubilities and enthalpies of a solution of polyvinyltetrazoles in organic solvents (Russ.), *Vysokomol. Soedin., Ser. A*, 39, 856, 1997.

1997KR1 Krause, C. and Wolf, B.A., Shear effects on the phase diagrams of solutions of highly incompatible polymers in a common solvent. 1. Equilibrium behavior and rheological properties, *Macromolecules*, 30, 885, 1997.

1997KR2 Krause, C., Horst, R., and Wolf, B.A., Shear effects on the phase diagrams of solutions of highly incompatible polymers in a common solvent. 2. Experiment and theory, *Macromolecules*, 30, 890, 1997.

1997KUN Kunugi, S., Takano, K., Tanaka, N., Suwa, K., and Akashi, M., Effects of pressure on the behavior of the thermoresponsive polymer poly(*N*-vinylisobutyramide) (PNVIBA), *Macromolecules*, 30, 4499, 1997.

1997KUR Kuramoto, N., Shishido, Y., and Nagai, K., Thermosensitive and redox-active polymers: preparation and properties of poly(*N*-ethylacrylamide-*co*-vinylferrocene) and poly(*N,N*-diethylacrylamide-*co*-vinylferrocene), *J. Polym. Sci.: Part A: Polym. Chem.*, 35, 1967, 1997.

1997LIM Li, M., Zhu, Z.-Q., and Mei, L.-H., Partitioning of amino acids by aqueous two-phase systems combined with temperature-induced phase formation, *Biotechnol. Progr.*, 13, 105, 1997.

1997MCL McLure, I.A., Mokhtari, A., and Bowers, J., Thermodynamics of linear dimethylsiloxane-perfluoroalkane mixtures. Part 1. Liquid-liquid coexistence curves for hexamethyldisiloxane-, octamethyltrisiloxane- or decamethyltetrasiloxane-tetradecafluorohexane near the upper critical endpoint and upper coexistence temperatures for 21 other dimethylsiloxane-perfluoroalkane mixtures, *J. Chem. Soc., Faraday Trans.*, 93, 249, 1997.

1997PAE Pae, B.J., Moon, T.J., Lee, C.H., Ko, M.B., Park, M., Lim S., Kim, J., and Choe, C.R., Phase behavior in PVA/water solution, *Korean Polym. J.*, 5, 126, 1997.

1997REB Rebelo, L.P.N., DeSousa, H.C., and van Hook, W.A., Hypercritically enhanced distortion of a phase diagram: The (polystyrene + acetaldehyde) system (experimental data by H.C. De Sousa), *J. Polym. Sci.: Part B: Polym. Phys.*, 35, 631, 1997.

1997SAR Sargantanis, I.G. and Karim, M.N., Prediction of aqueous two-phase equilibrium using the Flory-Huggins model, *Ind. Eng. Chem. Res.*, 36, 204, 1997.

1997SC1 Schäfer-Sönen, H., Mörkerke, R., Berghmans, H., Koningsveld, R., and Solc, K., Zero and off-zero critical concentrations in systems containing polydisperse polymers with very high molar masses. 2., *Macromolecules*, 30, 410, 1997.

1997SC2 Schneider, A. and Wolf, B.A., Interfacial tension of demixed polymer solutions: Augmentation by polymer additives, *Macromol. Rapid Commun.*, 19, 561, 1997.

1997SU1 Suwa, K., Wada, Y., Kikunaga, Y., Morishita, K., Kishida, A., and Akashi, M., Synthesis and functionalities of poly(*N*-vinylalkylamide). IV., *J. Polym. Sci.: Part A: Polym. Chem.*, 35, 1763, 1997.

1997SU2 Suwa, K., Morishita, K., Kishida, A., and Akashi, M., Synthesis and functionalities of poly(*N*-vinylalkylamide). V., *J. Polym. Sci.: Part A: Polym. Chem.*, 35, 3087, 1997.

1997TOP Topp, M.D.C., Dijkstra, P.J., Talsma, H., and Feijen, J., Thermosensitive micelle-forming block copolymers of poly(ethylene glycol) and poly(*N*-isopropyl-acrylamide), *Macromolecules*, 30, 8518, 1997.

1997VIN Vinches, C., Parker, A., and Reed, W.F., Phase behavior of aqueous gelatine/oligosaccharide mixtures, *Biopolymers*, 41, 607, 1997.

1997VSH Vshivkov, S.A., Rusinova, E.V., and Zarudko, I.V., Thermodynamics of the polystyrene-*tert*-butyl acetate system, *Vysokomol. Soedin., Ser. A*, 39, 1043, 1997.

1997WOL Wolf, B.A., Improvment of polymer solubility: influence of shear and pressure, *Pure Appl. Chem.*, 69, 929, 1997.

1997XIO Xiong, Y. and Kiran, E., Miscibility, density and viscosity of polystyrene in n-hexane at high pressures, *Polymer*, 38, 5185, 1997.

1997YOO Yoo, M.K., Sung, Y.K., Cho, C.S., and Lee, Y.M., Effect of polymer complex formation on the cloud-point of poly(*N*-isopropylacrylamide) (PNIPAAm) in the poly(NIPAAm-*co*-acrylic acid): polyelectrolyte complex between poly(acrylic acid) and poly(allylamine), *Polymer*, 38, 2759, 1997

1997YOU Young, T.-H., Lai, J.-Y., You, W.-M., and Cheng, L.-P., Equilibrium phase behavior of the membrane forming water-DMSO-EVAL copolymer system, *J. Membrane Sci.*, 128, 55, 1997.

1998ALM Almeida, M.C., Venancio, A., Teixeira, J.A., and Aires-Barros, M.R., Cutinase purification on poly(ethylene glycol)-hydroxypropyl starch aqueous two-phase systems, *J. Chromatogr. B*, 711, 151, 1998.

1998BE1 Berghmans, H., DeCooman, R., DeRudder, J., and Koningsveld, R., Structure formation in polymer solutions, *Polymer*, 39, 4621, 1998.

1998CHA Chalykh, A.E., Dement'eva, O.V., and Gerasimov, V.K., Phase diagrams of poly(methyl methacrylate)-poly(ethylene glycol) system (Russ.), *Vysokomol. Soedin., Ser. A*, 40, 815, 1998.

1998GOM Gomez, C.M., Figueruelo, J.E., and Campos, A., Evaluation of thermodynamic parameters for blends of polyethersulfone and poly(methyl methacrylate) or polystyrene in dimethylformamide, *Polymer*, 39, 4023, 1998.

1998GOS Gosselet, N.M., Borie, C., Amiel, C., and Sebille, B., Aqueous two-phase systems from cyclodextrin polymers and hydrophobically modified acrylic polymers, *J. Dispersion Sci.Technol.*, 19, 805, 1998.

1998IMR Imre, A. and van Hook, A.W., Liquid-liquid equilibria in polymer solutions at negative pressure, *Chem. Soc. Rev.*, 27, 117, 1998.

1998INO Inomata, K., Ohara, N., Shimizu, H., and Nose, T., Phase behaviour of rod with flexible side chains/coil/solvent systems: Poly(α,L-glutamate) with tri(ethylene glycol) side chains, poly(ethylene glycol), and dimethylformamide, *Polymer*, 39, 3379, 1998.

1998KI2 Kiran, E. and Xiong, Y., Miscibility of isotactic polypropylene in n-pentane and n-pentane + CO_2 mixtures at high pressures, *J. Supercrit. Fluids*, 11, 173, 1998.

1998KIS Kishida, A., Nakano, S., Kikunaga, Y., and Akashi, M., Synthesis and functionalities of poly(*N*-vinylalkylamide). VII. A novel aqueous two-phase system based on poly(*N*-vinyl acetamide) and dextran, *J. Appl. Polym. Sci.*, 67, 255, 1998.

1998KOA Koak, N., Loos, Th.W.de, and Heidemann, R.A., Upper-critical-solution-temperature behavior of the system polystyrene + methylcyclohexane. Influence of CO_2 on the liquid-liquid equilibria, *Fluid Phase Equil.*, 145, 311, 1998.

1998KUB Kubota, K., Kita, R., and Dobashi, T., Renormalized Ising behavior of critical nonionic surfactant solution, *J. Chem.,Phys.*, 109, 711, 1998.

1998KUR Kuramoto, N. and Shishido, Y., Property of thermo-sensitive and redox-active poly(*N*-cyclopropylacrylamide-*co*-vinylferrocene) and poly(*N*-isopropylacrylamide-*co*-vinylferrocene), *Polymer*, 39, 669, 1998.

1998LAI Lai, J.-Y., Lin, S.-F., Lin, F.-C., and Wang, D.-M., Construction of ternary phase diagrams in nonsolvent/solvent/PMMA systems, *J. Polym. Sci.: Part B: Polym. Phys.*, 36, 607, 1998.

1998LEE Lee, H.K., Kim, S.C., and Levon, K., Liquid-liquid phase separation and crystallization of polydisperse isotactic polypropylene solutions, *J. Appl. Polym. Sci.*, 70, 849, 1998.

1998LIM Li, M., Zhu, Z.-Q., Wu, Y.-T., and Lin, D.-Q., Measurement of phase diagrams for new aqueous two-phase systems and prediction by a generalized multicomponent osmotic virial equation, *Chem. Eng. Sci.*, 53, 2755, 1998.

1998MAG Maggioni, J.F., Nunes, S.P., Pires, A.T.N., Eich, A., Horst, R., and Wolf, B.A., Phase diagrams of the system tetrahydrofuran/γ-butyrolactone/polyetherimide and determination of interaction parameters, *Polymer*, 39, 5133, 1998.

1998MIY Miyashita, N. and Nose, T., Critical behavior of asymmetric polymer blend solutions: Poly(methyl methacrylate)/poly(dimethylsiloxane)/solvent, *J. Chem. Phys.*, 108, 4282, 1998.

1998MOE Mörkerke, R., Meeussen, F., Koningsveld, R., Berghmans, H., Mondelaers, W., Schacht, E., Dusek, K., and Solc, K., Phase transitions in swollen networks. 3. Swelling behavior of radiation cross-linked poly(vinyl methyl ether) in water, *Macromolecules*, 31, 2223, 1998.

1998NIL Nilsson, S., Blokhus, A.M., and Saure, A., Influence of hydrophobic cosolutes on aqueous two-phase polymer-surfactant systems, *Langmuir*, 14, 6082, 1998.

1998PLA Planas, J., Varelas, V., Tjerneld, F., and Hahn-Hägerdal, B., Amine-based aqueous polymers for the simultaneous titration and extraction of lactic acid in aqueous two-phase systems, *J. Chromatogr. B*, 711, 256, 1998.

1998RIC Riccardi, C.C., Borrajo, J., Williams, R.J.J., Siddiqi, H.M., Dumon, M., and Pascault, J.P., Multiple phase equilibria in polydisperse polymer/liquid crystal blends, *Macromolecules*, 31, 1124, 1998.

1998SCH Schneider, A., Homopolymer- und Copolymerlösungen im Vergleich: Wechselwirkungsparameter und Grenzflächenspannung, *Dissertation*, Johannes Gutenberg Universität Mainz, 1998.

1998STO Stöhr, T., Petzold, K., Wolf, B.A., and Klemm, D.O., Continuous polymer fractionation of polysaccharides using highly substituted trimethylsilylcellulose, *Macromol. Chem. Phys.*, 199, 1895, 1998.

1998SUW Suwa, K., Yamamoto, K., Akhashi, M., Takano, K., Tanaka, N., and Kunugi, S., Effects of salt on the temperature and pressure responsive properties of poly(*N*-vinylisobutyramide) aqueous solutions, *Colloid Polym. Sci.*, 276, 529, 1998.

1998SZY Szydlowski, J. and van Hook, W.A., Liquid-liquid demixing from polystyrene solutions. Studies on temperature and pressure dependeces using dynamic light scattering and neutron scattering, *Fluid Phase Equil.*, 150-151, 687, 1998.

1998TER Terao, K., Okumoto, M., Nakamura, Y., Norisuye, T., and Teramoto, A., Light-scattering and phase separation studies on cyclohexane solutions of four-arm star polystyrene, *Macromolecules*, 31, 6885, 1998.

1998YA1 Yamagishi, T.-A., Ozawa, M., Nakamoto, Y., and Ishida, S.-I., Phase separation of phenolic resins 2. The fractionation of phenol-formaldehyde resin based on molecular weight and structure, *Polym. Bull.*, 40, 69, 1998.

1998YA2 Yamagishi, T.-A., Nomoto, M., Yamashita, S., Yamazaki, T., Nakamoto, Y., and Ishida, S.-I., Characterization of high molecular weight novolak, *Macromol. Chem. Phys.*, 199, 423, 1998.

1998YOU Young, T.-H., Cheng, L.-P., Hsieh, C.-C., and Chen, L.-W., Phase behavior of EVAL polymers in water-2-propanol cosolvent, *Macromolecules*, 31, 1229, 1998.

1998ZHE Zheng, X., Tong, Z., Xie, X., and Zeng, F., Phase separation in poly(*N*-isopropyl acrylamide)/water solutions. I. Cloud point curves and microgelation, *Polym. J.*, 30, 284, 1998.

1998ZHU Zhuang, W. and Kiran, E., Kinetics of pressure-induced phase separation (PIPS) from polymer solutions by time resolved light scattering. Polyethylene + n-pentane, *Polymer*, 39, 2903, 1998.

1999AKI Akiba, I. and Akiyama, S., Phase diagram of poly(4-vinylphenol)-*N,N*-dimethyl-octadecylamine mixture, *Macromolecules*, 32, 3741, 1999.

1999BAI Baik, K.-J., Kim, J. Y., Lee, H. K., and Kim, S. C., Liquid-liquid phase separation in polysulfone/polyethersulfone/*N*-methyl-2-pyrrolidone/water quaternary system, *J. Appl. Polym. Sci.*, 74, 2113, 1999.

1999BAR Barth, C., Untersuchungen zum thermodynamischen Phasenverhalten membranbildender Systeme, *Dissertation* Johannes Gutenberg Universität Mainz, 1999.

1999BAT Baton, B.F. and McHugh, A.J., Kinetics of thermally induced phase separation in ternary polymer solutions. I., *J. Polym. Sci.: Part B: Polym. Phys.*, 37, 1449, 1999.

1999BER Berggren, K., Veide, A., Nygren, P.-A., and Tjerneld, F., Genetic engineering of protein-peptide fusions for control of protein partitioning in thermoseparating aqueous two-phase systems, *Biotechnol. Bioeng.*, 62, 135, 1999.

1999BEY Beyer, C., Oellrich, L.R., and McHugh, M.A., Effect of copolymer composition and solvent polarity on the phase behavior of mixtures of poly(ethylene-*co*-vinyl acetate) with cyclopentane and cyclopentene (exp. data by C. Beyer), *Chem.-Ing. Techn.*, 71, 1306, 1999.

1999BRA Brandrup, J., Immergut, E.H., and Grulke, E.A., Eds., *Polymer Handbook*, 4th ed., J. Wiley & Sons, New York, 1999.

1999BUN Bungert, B., Komplexe Phasengleichgewichte von Polymerlösungen, *Dissertation*, TU Berlin, Shaker Vlg., Aachen, 1999.

1999CHE1 Cheng, L.-P., Effect of temperature on the formation of microporous PVDF membranes by precipitation from 1-octanol/DMF/PVDF and water/DMF/PVDF systems, *Macromolecules*, 32, 6668, 1999.

1999CHE2 Cheng, L.-P., Lin, D.-J., Shih, C.-H., Dwan, A.-H., and Gryte, C.C., PVDF membrane formation by diffusion-induced phase separation, *J. Polym. Sci.: Part B: Polym. Phys.*, 37, 2079, 1999.

1999CHO1 Choi, J.J. and Bae, Y.C., Liquid-liquid equilibria of polydisperse polymer systems: Applicability of continuous thermodynamics, *Fluid Phase Equil.*, 157, 213, 1999.

1999CHO2 Choi, J.J. and Bae, Y.C., Liquid-liquid equilibria of polydisperse systems, *Eur. Polym. J.*, 35, 1703, 1999.

1999CHO3 Choi, J.J., Yi, S., and Bae, Y.C., Liquid-liquid equilibria of polydisperse polymer systems showing both UCST and LCST phase beahviors, *Macromol. Chem. Phys.*, 200, 1889, 1999.

1999ERB Erbil, C., Akpinar, F.D., and Uyanik, N., Investigation of the thermal aggregations in aqueous poly(*N*-isopropylacrylamide-*co*-itaconic acid) solutions, *Macromol. Chem. Phys.*, 200, 2448, 1999.

1999GOM Gomez, C.M., Figueruelo, J.E., and Campos, A., Thermodynamics of a polymer blend solution system studied by gel permeation chromatography and viscosity, *Macromol. Chem. Phys.*, 200, 246, 1999.

1999GRA Graham, P.D., Barton, B.F., and McHugh, A.J., Kinetics of thermally induced phase separation in ternary polymer solutions. II. Comparison of theory and experiment, *J. Polym. Sci.: Part B: Polym. Phys.*, 37, 1461, 1999.

1999HOO Hook, W.A. van, Wilczura, H., and Rebelo, L.P.N., Dynamic light scattering of polymer/solvent solutions under pressure. Near critical demixing ($0.1<P$/MPa<200) for polystyrene/cyclohexane and polystyrene/methylcyclohexane, *Macromolecules*, 32, 7299, 1999.

1999IM1 Imre, A.R., Melnichenko, G., and van Hook, W.A., Liquid-liquid equilibria in polystyrene solutions: The general pressure dependence, *Phys. Chem. Chem. Phys.*, 1, 4287, 1999.

1999IM2 Imre, A.R., Melnichenko, G., and van Hook, W.A., A polymer-solvent system with two homogeneous double critical points: Polystyrene (PS)/(n-heptane + methylcyclohexane), *J. Polym. Sci.: Part B: Polym. Phys.*, 37, 2747, 1999.

1999JAN Jang, J.G. and Bae, Y.C., Phase behaviors of hyperbranched polymer solutions, *Polymer*, 40, 6761, 1999.

1999JEO Jeong, B., Lee, D.S., Shon, J.-I., Bae, Y.H., and Kim, S.W., Thermoreversible gelation of poly(ethylene oxide) biodegradable polyester block copolymers, *J. Polym. Sci.: Part B: Polym. Phys.*, 37, 751, 1999.

1999JOH Johansson, H.-O., Persson, J., and Tjerneld, F., Thermoseparating water/polymer system: a novel one-polymer aqueous two-phase system for protein purification, *Biotechnol. Bioeng.*, 66, 247, 1999.

1999JON Jones, M.S., Effect of pH on the lower critical solution temperature of random copolymers of *N*-isopropylacrylamide and acrylic acid, *Eur. Polym. J.*, 35, 795, 1999.

1999KAN Kano, Y., Sato, H., Okamoto, M., Kotaka, T., and Akiyama, S., Phase separation process during solution casting of acrylate-copolymer/fluoro-copolymer blends, *J. Adhesion Sci. Technol.*, 13, 1243, 1999.

1999KIM Kim, Y.D., Kim, J.Y., Lee, H.K., and Kim, S.C., Formation of polyurethane membranes by immersion precipitation. I. Liquid-liquid phase separation in a polyurethane/ DMF/water system, *J. Appl. Polym. Sci.*, 73, 2377, 1999.

1999KIS Kishida, A., Kikunaga, Y., and Akashi, M., Synthesis and functionality of poly(*N*-vinylalkylamide). X. A novel aqueous two-phase system based on thermosensitive polymers and dextran, *J. Appl. Polym. Sci.*,73, 2545, 1999.

1999KIT Kita, R., Kaku, T., Kubota, K., and Dobashi, T., Pinning of phase separation of aqueous solution of hydroxypropylmethylcellulose by gelation, *Phys. Lett. A*, 259, 302, 1999.

1999KLE Klenin, V.J., *Thermodynamics of Systems Containing Flexible-Chain Polymers*, Elsevier, Amsterdam, 1999.

1999KUN Kunugi, S., Yamazaki, Y., Takano, K., and Tanaka, N., Effects of ionic additives and ionic comonomers on the temperature and pressure responsive behavior of thermoresponsive polymers in aqueous solutions, *Langmuir*, 15, 4056, 1999.

1999LAU Lau, A.C.W. and Wu, C., Thermally sensitive and biocompatible poly(*N*-vinylcaprolactam): synthesis and characterization of high molar mass linear chains, *Macromolecules*, 32, 581, 1999.

1999LIU Liu, H.Y. and Zhu, X.X., Lower critical solution temperatures of *N*-substituted acrylamide copolymers in aqueous solutions, *Polymer*, 40, 6985, 1999.

1999MAT Matsuyama, H., Teramoto, M., Nakatani, R., and Maki, T., Membrane formation via phase separation induced by penetration of nonsolvent from vapor phase. I. Phase diagram and mass transfer process, *J. Appl. Polym. Sci.*, 74, 159, 1999.

1999MIK Mikhailov, Yu.M., Ganina, L.V., Makhonina, L.I., Smirnov, V.S., and Shapayeva, N.V., Mutual diffusion and phase equilibrium in copolymers of nonyl acrylate and acrylic acid-diglycidyl ether of bisphenol-A, *J. Appl. Polym. Sci.*, 74, 2353, 1999.

1999NAK Nakata, M., Dobashi, T., Inakuma, Y.-I., and Yamamura, K., Coexistence curve of polystyrene in methylcyclohexane. X. Two-phase coexistence curves for ternary solutions near the tricritical compositions (exp. data by M. Nakata and T. Dobashi), *J. Chem. Phys.*, 111, 6617, 1999.

1999OCH Ochi, K., Saito, T., and Kojima, K., Determination of solubilities of polymers in solvents by a laser scattering technique, *Fluid Phase Equil.*, 158-160, 847, 1999.

1999PER Persson, J., Johansson, H.-O., and Tjerneld, F., Purification of protein and recycling of polymers in a new aqueous two-phase system using two thermoseparating polymers, *J. Chromatogr. A*, 864, 31, 1999.

1999PRA Prausnitz, J.M., Lichtenthaler, R.N., and de Azevedo, E.G., *Molecular Thermodynamics of Fluid Phase Equilibria*, 3rd ed., Prentice Hall, Upper Saddle River, NJ, 1999.

1999PRU Pruessner, M.D., Retzer, M.E., and Greer, S.C., Phase separation curves of poly(α-methylstyrene) in methylcyclohexane, *J. Chem. Eng. Data*, 44, 1419, 1999.

1999RUD Rudder, J. de, Berghmans, H., and Arnauts, J., Phase behaviour and structure formation in the system syndiotactic polystyrene/cyclohexanol, *Polymer*, 40, 5919, 1999.

1999SHI Shimofure, S., Kubota, K., Kita, R., and Dobashi, T., Coexistence curve and turbidity of aqueous solutions of oligooxyethylene alkyl ether with the addition of urea near the critical point, *J. Chem. Phys.*, 111, 4199, 1999.

1999SVE Svensson, M., Berggren, K., Veide, A., and Tjerneld, F., Aqueous two-phase systems containing self-associating block copolymers. Partitioning of hydrophilic and hydrophobic biomolecules, *J. Chromatogr. A*, 839, 71, 1999.

1999WAN Wang, D., Li, K., and Teo, W.K., Phase separation in polyetherimide/solvent/nonsolvent systems and membrane formation, *J. Appl. Polym. Sci.*, 71, 1789, 1999.

1999YEL Yelash, L.V. and Kraska, T., The global phase behaviour of binary mixtures of chain molecules. Theory and application, *Phys. Chem. Chem. Phys.*, 1, 4315, 1999.

1999YOU Young, T.-H., Cheng, L.-P., You, W.-M., and Chen, L.-Y., Prediction of EVAL membrane morphologies using the phase diagram of water-DMSO-EVAL at different temperatures, *Polymer*, 40, 2189, 1999.

2000AFR Afroze, F., Nies, E., and Berghmans, H., Phase transitions in the system poly(*N*-isopropylacrylamide)/water and swelling behaviour of the corresponding networks (exp. data by H. Berghmans), *J. Mol. Struct.*, 554, 55, 2000.

2000AZU Azuma, T., Tyagi, O.S., and Nose, T., Static and dynamic properties of block-copolymer solution in poor solvent, *Polym. J.*, 32, 151, 2000.

2000BAD Badiger, M.V., Lutz, A., and Wolf, B.A., Interrelation between the thermodynamic and viscometric behaviour of aqueous solutions of hydrophobically modified ethyl hydroxyethyl cellulose, *Polymer*, 41, 1377, 2000.

2000BA1 Barth, C. and Wolf, B.A., Quick and reliable routes to phase diagrams for polyethersulfone and polysulfone membrane formation, *Macromol. Chem. Phys.*, 201, 365, 2000.

2000BA2 Barth, C., Goncalves, M.C., Pires, A.T.N., Röder, J., and Wolf, B.A., Asymmetric polysulfone and polyethersulfone membranes: effects of thermodynamic conditions during formation on their performance, *J. Membrane Sci.*, 169, 287, 2000.

2000BEH Behme, S., Thermodynamik von Polymersystemen bei hohen Drucken, *Dissertation*, TU Berlin, 2000.

2000BEY Beyer, C., Oellrich, L.R., and McHugh, M.A., Effect of copolymer composition and solvent polarity on the phase behavior of mixtures of poly(ethylene-*co*-vinyl acetate) with cyclopentane and cyclopentene, *Chem. Eng. Technol.*, 23, 592, 2000.

2000BIG Bignotti, F., Penco, M., Sartore, L., Peroni, I. Mendichi, R., and Casolaro, M., Synthesis, characterisation and solution behaviour of thermo- and pH-responsive polymers bearing L-leucine residues in the side chains, *Polymer*, 41, 8247, 2000.

2000BO1 Bokias, G., Staikos, G., and Iliopoulos, I., Solution properties and phase behaviour of copolymers of acrylic acid with *N*-isopropylacrylamide, *Polymer*, 41, 7399, 2000.

2000BO2 Bokias, G., Vasilevskaya, V.V., Iliopoulos, I., Hourdet, D., and Khokhlov, A.R., Influence of migrating ionic groups on the solubility of polyelectrolytes: phase behavior of ionic poly(*N*-isopropylacrylamide) copolymers in water, *Macromolecules*, 33, 9757, 2000.

2000CHE Cheng, L.-P. and Shaw, H.-Y., Phase behavior of a water/2-propanol/poly(methyl methacrylate) cosolvent system, *J. Polym. Sci.: Part B: Polym. Phys.*, 38, 747, 2000.

2000CHU Chun, K.-Y., Jang, S.-H., Kim, H.-S., Kim, Y.-W., and Joe, Y.-I., Effects of solvent on the pore formation in asymmetric 6FDA-4,4'ODA polyimide membrane: terms of thermodynamics, precipitation kinetics, and physical factors, *J. Membrane Sci.*, 169, 197, 2000.

2000CLA Clausi, D.T. and Koros, W.J., Formation of defect-free polyimide hollow fiber membranes for gas separations, *J. Membrane Sci.*, 167, 79, 2000.

2000DE1 DeSousa, H.C. and Rebelo, L.P.N., (Liquid + liquid) equilibria of (polystyrene + nitroethane). Molecular weight, pressure, and isotope effects, *J. Chem. Thermodyn.*, 32, 355, 2000.

2000DE2 DeSousa, H.C. and Rebelo, L.P.N., A continuous polydisperse thermodynamic algorithm for a modified Flory-Huggins model: The (polystyrene + nitroethane) example, *J. Polym. Sci.: Part B: Polym. Phys.*, 38, 632, 2000.

2000FIS Fischer, V. and Borchard, W., Thermodynamic properties of poly(ethylene glycol)/ water systems. 2. Critical point data, *J. Phys. Chem. B*, 104, 4463, 2000.

2000FRI Friberg, S.E., Yin, Q., Barber, J.L., and Aikens, P.A., Vapor pressures of phenethyl alcohol in the system water-phenethyl alcohol and the triblock copolymer $EO_{4.5}PO_{59}EO_{4.5}$, *J. Dispersion Sci. Technol.*, 21, 65, 2000.

2000HON Hong, P.-D. and Chou, C.-M., Phase separation and gelation behaviours in poly(vinylidene fluoride)/tetra(ethylene glycol) dimethyl ether solutions, *Polymer*, 41, 8311, 2000.

2000IMR Imre, A. and van Hook, W.A., End group effects on liquid-liquid demixing of polystyrene/oligomethylene solutions. Polystyrene/dodecyl acetate solubility, *Macromolecules*, 33, 5308, 2000.

2000JIM Jimenez-Regalado, E., Selb, J., and Candau, F., Phase behavior and rheological properties of aqueous solutions containing mixtures of associating polymers, *Macromolecules*, 33, 8720, 2000.

2000JOA Joabsson, F., Nyden, M., and Thuresson, K., Temperature-induced fractionation of a quasi-binary self-associating polymer solution. A phase behavior and polymer self-diffusion investigation, *Macromolecules*, 33, 6772, 2000.

2000KAW Kawai, T., Teramachi, S., Tanaka, S., and Maeda, S., Comparison of chemical composition distributions of poly(methyl methacrylate)-graft-polydimethylsiloxane by high-performance liquid chromatography and demixing solvent fractionation, *Int. J. Polym. Anal. Char.*, 5, 381, 2000.

2000KIM Kim, C., Lee, S.C., Kang, S.W., Kwon, I.C., and Jeong, S.Y., Phase-transition characteristics of amphiphilic poly(2-ethyl-2-oxazoline)/poly(ε-caprolactone) block copolymers in aqueous solutions, *J. Polym. Sci., Part B: Polym. Phys.*, 38, 2400, 2000.

2000KOI Koizumi, J., Kawashima, Y., Kita, R., Dobashi, T., Hosokawa, H., and Nakata, M., Coexistence curves of polystyrene in cyclohexane near the critical double point in composition-pressure space, *J. Phys. Soc. Japan*, 69, 2543, 2000.

2000KUC Kuckling, D., Adler, H.-J.P., Arndt, K.F., Ling, L., and Habicher, W.D., Temperature and pH dependent solubility of novel poly(*N*-isopropylacrylamide) copolymers, *Macromol. Chem. Phys.*, 201, 273, 2000.

2000KUN Kunugi, S., Tada, T., Yamazaki, Y., Yamamoto, K., and Akashi, M., Thermodynamic studies on coil-globule transitions of poly(*N*-vinylisobutyramide-*co*-vinylamine) in aqueous solutions, *Langmuir*, 16, 2042, 2000.

2000LAM LaMesa, C., Phase equilibria in a water-block copolymer system, *J. Therm. Anal. Calorim.*, 61, 493, 2000.

2000LIW Li, W., Zhu, Z.-Q., and Li, M., Measurement and calculation of liquid-liquid equilibria of binary aqueous polymer solutions, *Chem. Eng. J.*, 78, 179, 2000.

2000LUC Luccio, M.D., Nobrega, R., and Borges, C.P., Microporous anisotropic phase inversion membranes from bisphenol-A polycarbonate: Study of a ternary system, *Polymer*, 41, 4309, 2000.

2000MAG Maggioni, J.F., Eich, A., Wolf, B.A., and Nunes, S.P., On the viscosity of moderately concentrated solutions of poly(etherimide) in a mixed solvent of marginal quality, *Polymer*, 41, 4743, 2000.

2000MA1 Matsuyama, H., Kudari, S., Kiyofuji, H., and Kitamura, Y., Kinetic studies of thermally induced phase separation in polymer-diluent system, *J. Appl. Polym. Sci.*, 76, 1028, 2000.

2000MA2 Matsuyama, H., Nishiguchi, H., and Kitamura, Y., Phase separation mechanism during membrane formation by dry-cast process, *J. Appl. Polym. Sci.*, 77, 776, 2000.

2000ME1 Meeussen, F., Bauwens, Y., Mörkerke, R., Nies, E., and Berghmans, H., Molecular complex formation in the system poly(vinyl methyl ether)/water, *Polymer*, 41, 3737, 2000.

2000ME2 Meeusen, F., Nies, E., Berghmans, H., Verbrugghe, S., Göthals, E., and DuPrez, F., Phase behaviour of poly(*N*-vinyl caprolactam) in water, *Polymer*, 41, 8597, 2000.

2000MOE Moerkerke, R., Berghmans, H., Vandeweerdt, P., Adriaensen, P., Ercken, M., Vandrzande, D., and Gelan, J., Phase behaviour and solvent diffusion in the system poly(methyl methacrylate)/methanol, *Macromol. Chem. Phys.*, 201, 308, 2000.

2000OLI Oliveira, J.V., Dariva, C., and Pinto, J.C., High-pressure phase equilibria for polypropylene-hydrocarbon systems, *Ind. Eng. Chem. Res.*, 39, 4627, 2000.

2000PEK Pekar, M., On the miscibility of liquid polybutadienes, *J. Appl. Polym. Sci.*, 78, 1628, 2000.

2000PE1 Persson, J., Johansson, H.-O., Galaev, I., Mattiasson, B., and Tjerneld, F., Aqueous polymer two-phase systems formed by new thermoseparating polymers, *Bioseparation*, 9, 105, 2000.

2000PE2 Persson, J., Kaul, A., and Tjerneld, F., Polymer recycling in aqueous two-phase extractions using thermoseparating ethylene oxide-propylene oxide copolymers, *J. Chromatogr. B*, 743, 115, 2000.

2000PIE Pietruszka, N., Galaev, I.Yu., Kumar, A., Brzozowski, Z.K., and Mattiasson, B., New polymers forming aqueous two-phase polymer systems, *Biotechnol. Progr.*, 16, 408, 2000.

2000PRI Principi, T., Goh, C.C.E., Liu, R.C.W., and Winnik, F.M., Solution properties of hydrophobically modified copolymers of *N*-isopropylacrylamide and *N*-glycineacryl-amide: a study by microcalorimetry and fluorescence spectroscopy, *Macromolecules*, 33, 2958, 2000.

2000SAH Sahakaro, K., Chaibundit, C., Kaligradaki, Z., Mai, S.-M., Heatley, F., Booth, C., Padget, J. C., and Shirley, I.M., Clouding of aqueous solutions of difunctional tapered statistical copolymers of ethylene oxide and 1,2-butylene oxide, *Eur. Polym. J.*, 36, 1835, 2000.

2000SCH Schneider, A. and Wolf, B.A., Specific features of the interfacial tension in the case of phase separated solutions of random copolymers, *Polymer*, 41, 4089, 2000.

2000SHI Shi, X., Li, J., Sun, C., and Wu, S., The aggregation and phase separation behavior of a hydrophobically modified poly(*N*-isopropylacrylamide), *Coll. Surfaces A*, 175, 41, 2000.

2000SIL Silva, L.H.M. da and Meirelles, A.J.A., Phase equilibrium in polyethylene glycol/ maltodextrin aqueous two-phase systems, *Carbohydr. Polym.*, 42, 273, 2000.

2000SIV Sivars, U. and Tjerneld, F., Mechanism of phase behaviour and protein partitioning in detergent/polymer aqueous two-phase systems for purification of integral membrane proteins, *Biochim. Biophys. Acta*, 1474, 133, 2000.

2000SPI Spitzer, M., Silva, L.H.M. da, and Loh, W., Liquid biphase systems formed in ternary miytures of two organic solvents and ethylene oxide oligomers or polymers (exp. data by W. Loh), *J. Braz. Chem. Soc.*, 11, 375, 2000.

2000SRI Srinivas, N.D., Barbate, R.S., Raghavarao, K.S.M.S., and Todd, P., Acoustic demixing of aqueous two-phase systems, *Appl. Microbiol. Biotechnol.*, 53, 650, 2000.

2000STR Strack, A., Polymakromonomere als anisotropes Modellsystem zur Untersuchung kolloidaler Mehrkomponentensysteme, *Dissertation*, Johannes Gutenberg Universität Mainz, 2000.

2000SUZ Suzuki, M., Dobashi, T., Mikawa, Y., Yamamura, K., and Nakata, M., Reentrant three-phase equilibrium of homologous polystyrene solution, *J. Phys. Soc. Japan*, 69, 1741, 2000.

2000TAM Tamura, T., Yamaoka, T., Kunugi, S., Panitch, A., and Tirrell, D.A., Effects of temperature and pressure on the aggregation properties of an engineered elastin model polypeptide in aqueous solution, *Biomacromolecules*, 1, 552, 2000.

2000XIO Xiong, Y. and Kiran, E., Kinetics of pressure-induced phase separation (PIPS) in polystyrene+methylcyclohexane solutions at high pressure, *Polymer*, 41, 3759, 2000.

2000YAM Yamazaki, Y., Tada, T., and Kunugi, S., Effect of acrylic acid incorporation on the pressure-temperature behavior and the calorimetric properties of poly(*N*-isopropyl-acrylamide) in aqueous solutions, *Colloid Polym. Sci.*, 278, 80, 2000.

2001ALB Albrecht, W., Weigel, Th., Schossig-Tiedemann, M., Kneifel, K., Peinemann, K.-V., and Paul, D., Formation of hollow fiber membranes from poly(etherimide) at wet phase inversion using binary mixtures of solvents for the preparation of the dope (exp. data by W. Albrecht), *J. Membrane Sci.*, 192, 217, 2001.

2001AND Anderson, V.J. and Jones, R.A.L., The influence of gelation on the mechanism of phase separation of a biopolymer mixture, *Polymer*, 42, 9601, 2001.

2001BOU Bouchaour, T., Benmouna, F., Roussel, F., Buisine, J.-M., Coqueret, X., Benmouna, M., and Maschke, U., Equilibrium phase diagram of poly(2-phenoxyethyl acrylate) and 5CB, *Polymer*, 42, 1663, 2001.

2001BUT Butler, M.F. and Heppenstall-Butler, M., Phase separation in gelatine/maltodextrin and gelatine/maltodextrin/gum Arabic mixtures studied using small-angle light scattering, turbidity, and microscopy, *Biomacromolecules*, 2, 812, 2001.

2001CAI Cai, W.S., Gan, L.H., and Tam, K.C., Phase transition of aqueous solutions of poly(*N,N*-diethylacrylamide-*co*-acrylic acid) by differential scanning calorimetric and spectrophotometric methods, *Colloid Polym. Sci.*, 279, 793, 2001.

2001CHE Cheng, L.-P., Young, T.-H., Chuang, W.-Y., Chen, L.-Y., and Chen, L.-W., The formation mechanism of membranes prepared from the nonsolvent-solvent-crystalline polymer systems, *Polymer*, 42, 443, 2001.

2001DES Desai, P.R., Jain, N.J., Sharma, R.K., and Bahadur, P., Effect of additives on the micellization of PEO/PPO/PEO block copolymer F127 in aqueous solution, *Coll. Surfaces A*, 178, 57, 2001.

2001DJO Djokpe, E. and Vogt, W., *N*-Isopropylacrylamide and *N*-isopropylmethacrylamide: cloud points of mixtures and copolymers, *Macromol. Chem. Phys.*, 202, 750, 2001.

2001EDE Edelman, M.W., Linden, E. van der, Hoog, E. de, and Tromp, R.H., Compatibility of gelatine and dextran in aqueous solution, *Biomacromolecules*, 2, 1148, 2001.

2001FER Fernandes, G.R., Pinto, J.C., and Nobrega, R., Modeling and simulation of the phase-inversion process during membrane preparation, *J. Appl. Polym. Sci.*, 82, 3036, 2001.

2001FIS Fischer, V., Bestimmung und Beschreibung der thermodynamischen Eigenschaften entmischender, wäßriger Polymerlösungen am Beispiel des Systems Polyethylenglykol/Wasser in einem weiten Konzentrations- und Temperaturbereich, *Dissertation*, Gerhard-Mercator-Universität-GH Duisburg 2001.

2001FUJ Fujii, S., Sasaki, N., and Nakata, M., Rheological studies on the phase separation of hydroxypropylcellulose solution systems, *J. Polym. Sci.: Part B: Polym. Phys.*, 39, 1976, 2001.

2001GAN Gan, L.-H., Cai, W., and Tam, K.C., Studies of phase transition of aqueous solutions of poly(*N,N*-diethylacrylamide-*co*-acrylic acid) by differential scanning calorimetry and spectrophotometry, *Eur. Polym. J.*, 37, 1773, 2001.

2001GAR Garcia, R., Gomez, C.M., Figueruelo, J.E., and Campos, A., Thermodynamic interpretation of the SEC behavior of polymers in a polystyrene gel matrix, *Macromol. Chem. Phys.*, 202, 1889, 2001.

2001GER Gerasimov, V.K., Chalykh, A.E., Aliev, A.D., Trankina, E.S., and Gritskova, I.A., Phase equilibrium and the morphology of the polystyrene-polydimethylsiloxane-styrene system (Russ.), *Vysokomol. Soedin., Ser. A*, 43, 1941, 2001.

2001GOG Gogibus, N., Maschke, U., Benmouna, F., Ewen, B., Coqueret, X., and Benmouna, M., Phase diagrams of poly(dimethylsiloxane) and 5CB blends, *J. Polym. Sci.: Part B: Polym. Phys.*, 39, 581, 2001.

2001GOM Gomes de Azevedo, R., Rebelo, L.P.N., Ramos, A.M., Szydlowski, J., DeSousa, H.C., and Klein, J., Phase behavior of (polyacrylamides + water) solutions: concentration, pressure and isotope effects, *Fluid Phase Equil.*, 185, 189, 2001.

2001HAM Hamley, I.W., Mai, S.-M., Ryan, A.J., Fairclough, J.P.A., and Booth, C., Aqueous mesophases of block copolymers of ethylene oxide and 1,2-butylene oxide, *Phys. Chem. Chem. Phys.*, 3, 2972, 2001.

2001HE1 Heijden, P.C. van der, A DSC-study on the demixing of binary polymer solutions (exp. data by P.C. van der Heijden), *Proefschrift*, Univ. Twente, 2001.

2001HE2 Heijden, P.C. van der, Mulder, M.H.V., and Wessling, M., Phase behavior of polymer-diluent systems characterized by temperature modulated differential scanning calorimetry (exp. data by P.C. van der Heijden), *Thermochim. Acta*, 378, 27, 2001.

2001HUA Hua, F.J., DoNam, J., and Lee, D.S., Preparation of a macroporous poly(L-lactide) scaffold by liquid-liquid phase separation of a PLLA/1,4-dioxane/water ternary system in the presence of NaCl, *Macromol. Rapid Commun.*, 22, 1053, 2001.

2001IM1 Imre, A.R. and van Hook, W.A., The effect of branching of alkanes on the liquid-liquid equilibrium of oligostyrene/alkane systems, *Fluid Phase Equil.*, 187-188, 363, 2001.

2001IM2 Imre, A.R., Melnichenko, G., van Hook, W.A., and Wolf, B.A., On the effect of pressure on the phase transition of polymer blends and polymer solutions. Oligostyrene-n-alkane systems, *Phys. Chem. Chem. Phys.*, 3, 1063, 2004.

2001JAN Jang, J.G. and Bae, Y.C., Phase behavior of hyperbranched polymer solutions with specific interactions, *J. Chem. Phys.*, 114, 5034, 2001.

2001KI1 Kim, J.H., Min, B.R., Park, H.C., Won, J., and Kang, Y.S., Phase behavior and morphological studies of polyimide/PVP/solvent/water systems by phase inversion, *J. Appl. Polym. Sci.*, 81, 3481, 2001.

2001KI2 Kim, J.H., Min, B.R., Won, J., Park, H.C., and Kang, Y.S., Phase behavior and mechanism of membrane formation for polyimide/DMSO/water system, *J. Membrane Sci.*, 187, 47, 2001.

2001KI3 Kim, S. and Willett, J.L., Phase separation in potato starch solutions, *Polym. Mater. Sci. Eng.*, 85, 528, 2001.

2001KIN King, A.D., The solubility of gases in aqueous solutions of poly(propylene glycol), *J. Colloid Interface Sci.*, 243, 457, 2001.

2001KON Koningsveld, R., Stockmayer, W.H., and Nies, E., *Polymer Phase Diagrams*, Oxford University Press, Oxford, 2001.

2001KUN Kunugi, S., Yoshida, D., and Kiminami, H., Effects of pressure on the behavior of (hydroxypropyl)cellulose in aqueous solution, *Colloid Polym. Sci.*, 279, 1139, 2001.

2001KUJ Kujawa, P. and Winnik, F. M., Volumetric studies of aqueous polymer solutions using pressure perturbation calorimetry: a new look at the temperature-induced phase transition of poly(*N*-isopropylacrylamide) in water and D_2O, *Macromolecules*, 34, 4130, 2001.

2001KWO Kwon, K.W., Park, M.J., Hwang, J., and Char, K., Effects of alcohol addition on gelation in aqueous solution of poly(ethylene oxide)-poly(propylene oxide)-poly(ethylene oxide) triblock copolymer, *Polym. J.*, 33, 404, 2001.

2001LEB Leblanc, N., LeCerf, D., Chappey, C., Langevin, D., Metayer, M., and Muller, G., Influence of solvent and non-solvent on polyimide asymmetric membranes formation in relation to gas permeation, *Separation Purification Technol.*, 22-23, 277, 2001.

2001LEE Lee, H.K., Kim, J.Y., Kim, Y.D., Shin, J.Y., and Kim, S.C., Liquid-liquid phase separation in a ternary system of segmented polyetherurethane/dimethylformamide/ water: effect of hard segment content, *Polymer*, 42, 3893, 2001.

2001LI1 Liu, F., Frere, Y., and Francois, J., Association properties of poly(ethylene oxide) modified by pendant aliphatic groups, *Polymer*, 42, 2969, 2001.

2001LI2 Liu, K. and Kiran, E., Pressure-induced phase separation in polymer solutions: kinetics of phase separation and crossover from nucleation and growth to spinodal decomposition in solutions of polyethylene in n-pentane, *Macromolecules*, 34, 3060, 2001.

2001LOR Loren, N., Altskaer, A., and Hermansson, A.-M., Structure evolution during gelation at later stages of spinodal decomposition in gelatine/maltodextrin mixtures, *Macromolecules*, 34, 8117, 2001.

2001MA1 Maeda, Y., Nakamura, T., and Ikeda, I., Change in the hydration states of poly(*N*-alkylacrylamide)s during their phase transitions in water observed by FTIR spectroscopy, *Macromolecules*, 34, 1391, 2001.

2001MA2 Maeda, Y., Nakamura, T., and Ikeda, I., Change in the hydration states of poly(*N*-n-propylmethacrylamide) and poly(*N*-isopropylmethacrylamide) during their phase transitions in water observed by FTIR spectroscopy, *Macromolecules*, 34, 8246, 2001.

2001MA3 Maeda, Y., IR spectroscopic study on the hydration and the phase transition of poly(vinyl methyl ether) in water, *Langmuir*, 17, 1737, 2001.

2001NER Nerli, B.B., Espariz, M., and Pico, G.G., Thermodynamic study of forces involved in bovine serum albumin and ovalbumin partitioning in aqueous two-phase systems, *Biotechnol. Bioeng.*, 72, 468, 2001.

2001PEN Pendyala, K.S., Greer, S.C., and Jacobs, D.T., Poly(α-methylstyrene) in methylcyclohexane: Densities and viscosities near the liquid-liquid critical point, *J. Chem. Phys.*, 115, 9995, 2001.

2001POS Poshamova, N., Schneider, A., Wünsch, M., Kuleznew, V., and Wolf, B.A., Polymer-polymer interaction parameters for homopolymers and copolymers from light scattering and phase separation experiments in a common solvent (exp. data by A. Schneider), *J. Chem. Phys.*, 115, 9536, 2001.

2001SCH Schuhmacher, E., Soldi, V., and Nunes Pires, A.T., PMMA or PEO in THF/H_2O mixture: phase diagram, separation mechanism and application, *J. Membrane Sci.*, 184, 187, 2001.

2001SIL Silva, L.H.M. da and Meirelles, A.J.A., Phase equilibrium and protein partitioning in aqueous mixtures of maltodextrin with polypropylene glycol, *Carbohydr. Polym.*, 46, 267, 2001.

2001TAK Takahashi, M., Shimazaki, M., and Yamamoto, J., Thermoreversible gelation and phase separation in aqueous methylcellulose solutions, *J. Polym. Sci.: Part B: Polym. Phys.*, 39, 91, 2001.

2001TOK To, K., Coexistence curve exponent of a binary mixture with a high molecular weight polymer, *Phys. Rev. E*, 63, 026108/1-4, 2001.

2001TOR Tork, T., Measurement and calculation of phase equilibria in polyolefin/solvent systems, *Dissertation*, TU Berlin, 2001.

2001WUE Wünsch, M., Ternäre polymerhaltige Lösungen. Phasenverhalten und Grenzflächenspannung, *Dissertation*, Johannes Gutenberg Universität Mainz, 2001.

2001WOH Wohlfarth, C., *CRC Handbook of Thermodynamic Data of Copolymer Solutions*, CRC Press, Boca Raton, 2001.

2001YA1 Yang, Y., Zeng, F., Xie, X., Tong, Z., and Liu, X., Phase separation and network formation in poly(vinyl methyl ether)/water solutions, *Polym. J.*, 33, 399, 2001.

2001YA2 Yang, Y., Zeng, F., Tong, Z., Liu, X., and Wu, S., Phase separation in poly(*N*-isopropylacrylamide)/water solutions. II. Salt effects on cloud-point curves and gelation, *J. Polym. Sci.: Part B: Polym. Phys.*, 39, 901, 2001.

2002ALM Al-Muallem, H.A., Wazeer, M.I.M., and Ali, Sk.A., Synthesis and solution properties of a new ionic polymer and its behavior in aqueous two-phase polymer systems, *Polymer*, 43, 1041, 2002.

2002BEH Behravesh, E., Shung, A.K., Jo, S., and Mikos, A.G., Synthesis and characterization of triblock copolymers of methoxy poly(ethylene glycol) and poly(propylene fumarate), *Biomacromolecules*, 3, 153, 2002.

2002BER Berlinova, I.V., Nedelcheva, A.N., Samchikov, V., and Ivanov, Ya., Thermally induced hydrogel formation in aqueous solutions of poly(*N*-isopropylacryalamide) and fluorocarbon-modified poly(oxyethylene)s, *Polymer*, 43, 7243, 2002.

2002CH1 Chaibundit, C., Ricardo, N.M.P.S., Crothers, M., and Booth, C., Micellization of diblock (oxyethylene/oxybutylene) copolymer $E_{11}B_8$ in aqueous solution. Micelle size and shape. Drug solubilization, *Langmuir*, 18, 4277, 2002.

2002CH2 Chang, Y., Bender, J.D., Phelps, M.V.B., and Allcock, H.R., Synthesis and self-association behavior of biodegradable amphiphilic poly[bis(ethyl glycinat-*N*-yl)phosphazene]-poly (ethylene oxide) block copolymers, *Biomacromolecules*, 3, 1364, 2002.

2002CH3 Chatterjee, J. and Alamo, R.G., Phase behavior of low molecular weight polyethylenes from homogeneous and heterogeneous solutions (exp. data by R.G. Alamo), *J. Polym. Sci.: Part B: Polym. Phys.*, 40, 878, 2002.

2002COS Costa, R.O.R. and Freitas, R.F.S., Phase behavior of poly(*N*-isopropylacrylamide) in binary aqueous solutions, *Polymer*, 43, 5879, 2002.

2002DER Derawi, S.O., Kontogeorgis, G.M., Stenby, E.H., Haugum, T., and Fredheim, A.O., Liquid-liquid equilibria for glycols + hydrocarbons: Data and correlation, *J. Chem. Eng. Data*, 47, 169, 2002.

2002DES Desai, P.R., Jain, N.J., and Bahadur, P., Anomalous clouding behavior of an ethylene oxide-propylene oxide block copolymer in aqueous solution, *Coll. Surfaces A*, 197, 19, 2002.

2002EDG Edgar, C.D. and Gray, D.G., Influence of dextran on the phase behavior of suspensions of cellulose nanocrystals, *Macromolecules*, 35, 7400, 2002.

2002FRE Freitag, R. and Garret-Flaudy, F., Salt effects on the thermoprecipitation of poly(*N*-isopropylacrylamide) oligomers from aqueous solution, *Langmuir*, 18, 3434, 2002.

2002FRI Frielinghaus, H., Schwahn, D., Willner, L., and Freed, K.F., Small angle neutron scattering studies of a polybutadiene/polystyrene blend with small additions of ortho-dichlorobenzene for varying temperatures and pressures. II. Phase boundaries and Flory-Huggins parameter (exp. data by H. Frielinghaus), *J. Chem. Phys.*, 116, 2241, 2002.

2002FUJ Fujii, S., Sasaki, N., and Nakata, M., Elongational flow studies on the phase separation of hydroxypropylcellulose solution, *J. Appl. Polym. Sci.*, 86, 2984, 2002.

2002GUP Gupta, V., Nath, S., and Chand, S., Role of water structure on phase separation in polyelectrolyte-polyethylene glycol based aqueous two-phase systems, *Polymer*, 43, 3387, 2002.

2002HAM Hammouda, B., Ho, D., and Kline, S., SANS from poly(ethylene oxide)/water systems, *Macromolecules*, 35, 8578, 2002.

2002HAN Han, M.-J. and Nam, S.-T., Thermodynamic and rheological variation in polysulfone solution by PVP and its effect in the preparation of phase inversion membrane, *J. Membrane Sci.*, 202, 55, 2002.

2002HOR Horst, M.H. ter, Behme, S., Sadowski, G., and Loos, Th.W. de, The influence of supercritical gases on the phase behavior of polystyrene-cyclohexane and polyethylene-cyclohexane systems: experimental results and modeling with the SAFT-equation of state, *J. Supercrit. Fluids*, 23, 181, 2002.

2002IMR Imre, A.R., van Hook, W., and Wolf, B.A., Liquid-liquid phase equilibria in polymer solutions and polymer mixtures, *Macromol. Symp.*, 181, 363, 2002.

2002KIR Kiran, E. and Liu, K., The miscibility and phase behavior of polyethylene with poly(dimethylsiloxane) in near critical pentane, *Korean J. Chem. Eng.*, 19, 153, 2002.

2002KUL Kuleznev, V.N., Wolf, B.A., and Pozharova, N.A., On intermolecular interactions in solutions of polymer blends (Russ.), *Vysokomol. Soedin., Ser. B*, 44, 512, 2002.

2002KUN Kunugi, S., Tada, T., Tanaka, N., Yamamoto, K., and Akashi, M., Microcalorimetric study of aqueous solution of a thermoresponsive polymer, poly(*N*-vinylisobutyramide) (PNVIBA), *Polym. J.*, 34, 383, 2002.

2002LAA Laatikainen, M., Markkanen, I., Tiihonen, J., and Paatero, E., Liquid-liquid equilibria in ternary systems of linear and cross-linked water-soluble polymers (exp. data by M. Laatikainen), *Fluid Phase Equil.*, 201, 381, 2002.

2002LEE Lee, D., Gong, Y., and Teraoka, I., High osmotic pressure chromatography of poly(ε-caprolactone) in near-theta solvent, *Macromolecules*, 35, 7093, 2002.

2002LIL Li, L., Shan, H., Yue, C.Y., Lam, Y.C., Tam, K.C., and Hu, X., Thermally induced association and dissociation of methylcellulose in aqueous solutions, *Langmuir*, 18, 7291, 2002.

2002LIN Lin, K.-Y., Wang, D.-M., and Lai, J.-Y., Nonsolvent-induced gelation and its effect on membrane morphology, *Macromolecules*, 35, 6697, 2002.

2002LIX1 Li, X.D. and Goh, S.H., Specific interactions and phase behavior of ternary poly(2-vinylpyridine)/poly(*N*-vinyl-2-pyrrolidone)/bisphenol-A blends, *J. Polym. Sci.: Part B: Polym. Phys.*, 40, 1125, 2002.

2002LIX2 Li, X.D. and Goh, S.H., Specific interactions and miscibility of ternary blends of poly(2-vinylpyridine), poly(*N*-vinyl-2-pyrrolidone) and aliphatic dicarboxylic acid: effect of spacer length of acid, *Polymer*, 43, 6853, 2002.

2002LOD Lodge, T.P., Pudil, B., and Hanley, K.J., The full phase behavior for block copolymers in solvents of varying selectivity, *Macromolecules*, 35, 4707, 2002.

2002LOS Loske, S., Fraktionierung und Lösungseigenschaften von Polymeren mit komplexer Struktur: Celluloseacetat, PMMA (Sterne, verzweigt) und SAN, *Dissertation*, Johannes Gutenberg Universität Mainz, 2002.

2002LUS Lu, S., Hu, Z., and Schwartz, J., Phase transition behavior of hydroxypropylcellulose under interpolymer complexation with poly(acrylic acid), *Macromolecules*, 35, 9164, 2002.

2002MAD Madbouly, S.A. and Wolf, B.A., Equilibrium phase behavior of polyethylene oxide and of its mixtures with tetrahydronaphthalene or/and poly(ethylene oxide-*b*-dimethylsiloxane) (exp. data by S.A. Madbouly), *J. Chem. Phys.*, 117, 7357, 2002.

2002MA1 Maeda, Y., Nakamura, T., and Ikeda, I., Hydration and phase behavior of poly(*N*-vinylcaprolactam) and poly(*N*-vinylpyrrolidone) in water (exp. data by Y. Maeda), *Macromolecules*, 35, 217, 2002.

2002MA2 Maeda, Y., Nakamura, T., and Ikeda, I., Change in solvation of poly(*N,N*-diethylacryl amide) during phase transition in aqueous solutions as observed by IR spectroscopy (exp. data by Y. Maeda), *Macromolecules*, 35, 10172, 2002.

2002MA3 Matsuo, M., Hashida, T., Tashiro, K., and Agari, Y., Phase separation of ultrahigh molecular weight isotactic polypropylene solutions in the gelation process estimated in relation to the morphology and mechanical properties of the resultant dry gel films, *Macromolecules*, 35, 3030, 2002.

2002MA4 Matsuyama, H., Takida, Y., Maki, T., and Teramoto, M., Preparation of porous membrane by combined use of thermally induced phase separation and immersion precipitation, *Polymer*, 43, 5243, 2002.

2002MA5 Matsuyama, H., Okafuji, H., Maki, T., Teramoto, M., and Tsujioka, N., Membrane formation via thermally induced phase separation in polypropylene/polybutene/diluent system, *J. Appl. Polym. Sci.*, 84, 1701, 2002.

2002MAK Makhaeva, E.E., Tenhu, H., and Khokhlov, A.R., Behavior of poly(*N*-vinyl-caprolactam-*co*-methacrylic acid) macromolecules in aqueous solution: interplay between coulombic and hydrophobic interaction, *Macromolecules*, 35, 1870, 2002.

2002MIK Mikhaliov, Yu.M., Ganina, L.V., Kurmaz, S.V., Smirnov, V.S., and Roshchupkin, V.P., Diffusion mobility of reactants, phase equilibrium, and specific features of radical copolymerization kinetics in the nonyl acrylate/2-methyl-5-vinyltetrazole system, *J. Polym. Sci.: Part B: Polym. Phys.*, 40, 1383, 2002.

2002MOR Morita, S., Tsunomori, F., and Ushiki, H., Polymer chain conformation in the phase separation process of a binary liquid mixture (exp. data by S. Morita), *Eur. Polym. J.*, 38, 1863, 2002.

2002NOW Nowicka, G. and Nowicki, W., On the role of electrolytes in precipitation of polyacrylamide from aqueous solution by addition of methanol (Russ.), *Vysokomol. Soedin., Ser. A*, 44, 2030, 2002.

2002PLA Plaza, M., Pons, R., Tadros, Th. F., and Solans, C., Phase behavior and formation of microemulsions in water/A-B-A block copolymer (polyhydroxystearic acid-polyethylene oxide-polyhydroxystearic acid)/1,2-alkanediol/isopropyl myristate systems, *Langmuir*, 18, 1077, 2002.

2002RAC Rackaitis, M., Strawhecker, K., and Manias, E., Water-soluble polymers with tunable temperature sensitivity. Solution behavior, *J. Polym. Sci.: Part B: Polym. Phys.*, 40, 2339, 2002.

2002REA Real, J. N., Iglesias, T. P., Pereira, S. M., and Rivas, M. A., Analysis of the temperature dependence of some physical properties of the binary system (n-nonane + tetraethylene glycol dimethyl ether), *J. Chem. Thermodyn.*, 34, 1, 2002.

2002RE1 Rebelo, L.P.N., Visak, Z.P., Sousa, H.C. de, Szydlowski, J., Gomes de Azevedo, R., Ramos, A.M., Najdanovic-Visak, V., Nunes da Ponte, M., and Klein, J., Double critical phenomena in (water + polyacrylamides) solutions, *Macromolecules*, 35, 1887, 2002.

2002RE2 Rebelo, L.P.N., Visak, Z.P., and Szydlowski, J., Metastable critical lines in (acetone + polystyrene) solutions and the continuity of solvent-quality states, *Phys. Chem. Chem. Phys.*, 4, 1046, 2002.

2002SAF Safronov, A.P. and Somova, T.V., Thermodynamics of poly(vinyl chloride) mixing with phthalate plasticizers (Russ.), *Vysokomol. Soedin., Ser. A*, 44, 2014, 2002.

2002SAX Saxena, R. and Caneba, G.T., Studies of spinodal decomposition in a ternary polymer-solvent-nonsolvent system, *Polym. Eng. Sci.*, 42, 1019, 2002.

2002SCH Schneider, A., Wünsch, M., and Wolf, B.A., An apparatus for automated turbidity titrations and its application to copolymer analysis and to the determination of phase diagrams (exp. data by A. Schneider), *Macromol. Chem. Phys.*, 203, 705, 2002.

2002SEI Seiler, M., Arlt, W., Kautz, H., and Frey, H., Experimental data and theoretical considerations on vapor-liquid and liquid-liquid equilibria of hyperbranched polyglycerol and PVA solutions, *Fluid Phase Equil.*, 201, 359, 2002.

2002SHI Shimizu, H., Kawakami, H., and Nagaoka, S., Membrane formation mechanism and permeation properties of a novel porous polyimide membrane, *Polym. Adv. Techn.*, 13, 370, 2002.

2002SHR Shresth, R.S., McDonald, R.C., and Greer, S.C., Molecular weight distributions of polydisperse polymers in coexisting liquid phases, *J. Chem. Phys.*, 117, 9037, 2002.

2002SON Soni, S.S., Sastry, N.V., Patra, A.K., Joshi, J.V., and Goyal, P.S., Surface activity, SANS, and viscosity studies in aqueous solutions of oxyethylene and oxybutylene di- and triblock copolymers, *J. Phys. Chem. B*, 106, 13069, 2002.

2002SPI Spitzer, M., Sabadini, E., and Loh, W., Entropically driven partitioning of ethylene oxide oligomers and polymers in aqueous/organic biphasic systems, *J. Phys. Chem. B*, 106, 12448, 2002.

2002VSH Vshvikov, S.A., Rusinova, E.V., and Gur'ev, A.A., Phase transitions in solutions of nitrile and methylstyrene rubbers (Russ.), *Vysokomol. Soedin., Ser. B*, 44, 504, 2002.

2002WUE Wünsch, M. and Wolf, B.A., Interfacial tension between coexisting polymer solutions in mixed solvents and its correlation with bulk thermodynamics: phase equilibria (liquid/gas and liquid/liquid) for the system toluene/ethanol/PDMS, *Polymer*, 43, 5027, 2002.

2002WOL Wolf, B., Kuleznev, V.N., and Pozhamova, N.A., Critical phenomena in solutions of the polystyrene-polyacrylonitrile random copolymer and its blends with polystyrene (Russ.), *Vysokomol. Soedin., Ser. A*, 44, 1212, 2002.

2002YEO Yeo, S.-D., Kang, I.-S., and Kiran, E., Critical polymer concentrations of polyethylene solutions in pentane, *J. Chem. Eng. Data*, 47, 571, 2002.

2002YIN Yin, X. and Stoever, H.D.H., Thermosensitive and pH-sensitive polymers based on maleic anhydride copolymers, *Macromolecules*, 35, 100178, 2002.

2002YOU Young, T.-H. and Chuang, W.-Y., Thermodynamic analysis on the cononsolvency of poly(vinyl alcohol) in water-DMSO mixtures through the ternary interaction parameter (exp. data by T.-H. Young), *J. Membrane Sci.*, 210, 349, 2002.

2002ZAV Zavarzina, A.G., Demin, V.V., Nifant'eva, T.I., Shkinev, V.M., Danilova, T.V., and Spivakov, B.Ya., Extraction of humic acids and their fractions in poly(ethylene glycol)-based aqueous biphasic systems, *Anal. Chim. Acta*, 452, 95, 2002.

2002ZHE Zheng, L., Suzuki, M., Inoue, T., and Lindman, B., Aqueous phase behavior of hexa(ethylene glycol) dodecyl ether studied by differential scanning calorimetry, Fourier transform infrared spectroscopy, and ^{13}C NMR spectroscopy, *Langmuir*, 18, 9204, 2002.

2002ZHO Zhou, C.-S., An, X.-Q., Xia, K.-Q., Yin, X.-L., and Shen, W.-G., Turbidity measurements and amplitude scaling of critical solutions of polystyrene in methylcyclohexane, *J. Chem. Phys.*, 117, 4557, 2002.

2003ABU Abu-Sharkh, B.F., Hamad, E.Z., and Ali, S.A., Influence of hydrophobe content on phase coexistence curves of aqueous two-phase solutions of associative polyacrylamide copolymers and poly(ethylene glycol), *J. Appl. Polym. Sci.*, 89, 1351, 2003.

2003BAD Badiger, M. V. and Wolf, B.A., Shear induced demixing and rheological behavior of aqueous solutions of poly(*N*-isopropylacrylamide), *Macromol. Chem. Phys.*, 204, 600, 2003.

2003BAL Balcan, M., Anghel, D.-F., and Raicu, V., Phase behavior of water/oil/nonionic surfactants with normal distribution of the poly(ethylene oxide) chain length, *Colloid Polym. Sci.*, 281, 143, 2003.

2003BAR Barker, I.C., Cowie, J.M.G., Huckeby, T.N., Shaw, D.A., Soutar, I., and Swanson, L., Studies of the 'smart' thermoresponsive behavior of copolymers of *N*-isopropylacrylamide and *N,N*-dimethylacrylamide in dilute aqueous solution, *Macromolecules*, 36, 7765, 2003.

2003CAM Campese, G.M., Rodriguez, E.M.G., Tambourgi, E.B., and Pessoa, Jr, A., Determination of cloud-point temperatures for different copolymers, *Brazil. J. Chem. Eng.*, 20, 335, 2003.

2003CHA Chang, Y., Powell, E.S., Allcock, H.R., Park, S.M., and Kim, C., Thermosensitive behavior of poly(ethylene oxide)-poly[bis(methoxyethoxyethoxy)-phosphazene] block copolymers, *Macromolecules*, 36, 2568, 2003.

2003CH1 Cheng, S.-K. and Chen, C.-Y., Study on the phase behavior of ethylene-vinyl acetate copolymer and poly(methyl methacrylate) blends by in situ polymerization, *J. Appl. Polym. Sci.*, 90, 1001, 2003.

2003CH2 Cheng, S.-K., Wang, and C.-C., Chen, C.-Y., Study of the phase behavior of EVA/PS blends during *in situ* polymerization, *Polym. Eng. Sci.*, 43, 1221, 2003.

2003CHR Christova, D., Velichkova, R., Loos, W., Goethals, E.J., and DuPrez, F., New thermo-responsive polymer materials based on poly(2-ethyl-2-oxazoline) segments, *Polymer*, 44, 2255, 2003.

2003DAV David, G., Alupei, V., Simionescu, B. C., Dincer, S., and Piskin, E., Poly(*N*-isopropylacryl amide)/poly[(*N*-acetylimino)ethylene] thermosensitive block and graft copolymers, *Eur. Polym. J.*, 39, 1209, 2003.

2003EDE Edelman, M.W., van der Linden, E., and Tromp, R.H., Phase separation of aqueous mixtures of poly(ethylene oxide) and dextran, *Macromolecules*, 36, 7783, 2003.

2003GOG Gogibus, N., Benmouna, F., Ewen, B., Pakula, T., Coqueret, X., Benmouna, M., and Maschke, U., Phase diagrams of poly(siloxane)/liquid crystal blends, *J. Polym. Sci.: Part B: Polym. Phys.*, 41, 39, 2003.

2003JI1 Jiang, S., An, L., Jiang, B., and Wolf, B. A., Pressure effects on the thermodynamics of *trans*-decahydronaphthalene/polystyrene polymer solutions: application of the Sanchez-Lacombe lattice fluid theory (exp. data by S. Jiang), *Macromol. Chem. Phys.*, 204, 692, 2003.

2003JI2 Jiang, S., An, L., Jiang, B., and Wolf, B.A., Liquid-liquid phase behavior of toluene/ polyethylene oxide/poly(ethylene oxide-b-dimethylsiloxane) polymer-containing ternary mixtures (exp. data by S. Jiang), *Phys. Chem. Chem. Phys.*, 5, 2066, 2003.

2003KOE Könderink, G.H., Aarts, D.G.A.L., Villeneuve, V.W.A. de, Philipse, A.P., Tuinier, R., and Lekkerkerker, H.N.W., Morphology and kinetics of phase separating transparent xanthan-colloid mixtures, *Biomacromolecules*, 4, 129, 2003.

2003KON Konishi, Y., Okubo, M., and Minami, H., Phase separation in the formation of hollow particles by suspension polymerization for divinylbenzene/toluene droplets dissolving polystyrene, *Colloid Polym. Sci.*, 281, 123, 2003.

2003KOT Koetz, J., Günther, C., Kosmella, S., Kleinpeter, E., and Wolf, G., Polyelectrolyte-induced structural changes in the isotropic phase of sulfobetaine/pentanol/toluene/ water system, *Progr. Colloid Polym. Sci.*, 122, 27, 2003.

2003LEE Lee, J.S., Lee, H.K., Kim, J.Y., Hyon, S.-H., and Kim, S.C., Thermally induced phase separation in poly(lactid acid)/dialkyl phthalate systems, *J. Appl. Polym. Sci.*, 88, 2224, 2003.

2003LES Lessard, D.G., Ousalem, M., Zhu, X.X., Eisenberg, A., and Carreau, P.J., Study of the phase transition of poly(*N,N*-diethylacrylamide) in water by rheology and dynamic light scattering, *J. Polym. Sci.: Part B: Polym. Phys.*, 41, 1627, 2003.

2003LIU Liu, S. and Liu, M., Synthesis and characterization of temperature- and pH-sensitive poly(*N,N*-diethylacrylamide-*co*-methacrylic acid), *J. Appl. Polym. Sci.*, 90, 3563, 2003.

2003LOS Loske, S., Schneider, A., and Wolf, B.A., Basis for the preparative fractionation of a statistical copolymer (SAN) with respect to either chain length or chemical composition, *Macromolecules*, 36, 5008, 2003.

2003LUT Lutz, A., Vergleich des Assoziationsverhaltens von Ethyl-Hydroxyethyl-Cellulose mit dem seines hydrophob modifizierten Analogons, *Dissertation*, Johannes Gutenberg Universität Mainz, 2003.

2003MAD Madbouly, S.A. and Wolf, B.A., Shear-induced crystallization and shear-induced dissolution of poly(ethylene oxide) in mixtures with tetrahydronaphthalene and oligo(dimethylsiloxane-*b*-ethylene oxide), *Macromol. Chem. Phys.*, 204, 417, 2003.

2003MA1 Maeda, Y., Tsubota, M., and Ikeda, I., Fourier transform IR spectroscopic study on phase transitions of copolymers of *N*-isopropylacrylamide and alkyl acrylates in water, *Colloid Polym. Sci.*, 281, 79, 2003.

2003MA2 Maeda, Y., Yamamoto, H., and Ikeda, I., Phase separation of aqueous solutions of poly(*N*-isopropylacrylamide) investigated by confocal Raman microscopy, *Macromolecules*, 36, 5055, 2003.

2003MA3 Matsuyama, H., Ohga, K., Maki, T., Teramoto, M., and Nakatsuka, S., Porous cellulose acetate membrane prepared by thermally induced phase separation, *J. Appl. Polym. Sci.*, 89, 3951, 2003.

2003MA4 Matsuyama, H., Nakagawa, K., Maki, T., and Teramoto, M., Studies on phase separation rate in porous polyimide membrane formation by immersion precipitation, *J. Appl. Polym. Sci.*, 90, 292, 2003.

2003MAO Mao, H., Li, C., Zhang, Y., Bergbreiter, D.E., and Cremer, P.S., Measuring LCSTs by novel temperature gradient methods: Evidence for intermolecular interactions in mixed polymer solutions, *J. Amer. Chem. Soc.*, 125, 2850, 2003.

2003MIL Milewska, A., Szydlowski, J., and Rebelo, L.P.N., Viscosity and ultrasonic studies of poly(*N*-isopropylacrylamide)-water solutions, *J. Polym. Sci.: Part B: Polym. Phys.*, 41, 1219, 2003.

2003OKH Okhapkin, I.M., Nasimova, I.R., Makhaeva, E.E., and Khokhlov, A.R., Effect of complexation of monomer units on pH- and temperature-sensitive properties of poly(*N*-vinylcaprolactam-*co*-methacrylic acid), *Macromolecules*, 36, 8130, 2003.

2003SCH Schmidt, J., Burchard, W., and Richtering, W., Shear-induced mixing and demixing in aqueous methyl hydroxypropyl cellulose solutions, *Biomacromolecules*, 4, 453, 2003.

2003SE1 Seiler, M., Rolker, J., and Arlt, W., Phase behavior and thermodynamic phenomena of hyperbranched polymer solutions, *Macromolecules*, 36, 2085, 2003.

2003SE2 Seiler, M., Köhler, D., and Arlt, W., Hyperbranched polymers: new selective solvents for extractive distillation and solvent extraction, *Separation Purification Technol.*, 30, 179, 2003.

2003SET Seto, Y., Kameyama, K., Tanaka, N., Kunugi, S., Yamamoto, K., and Akashi, M., High-pressure studies on the coacervation of copoly(*N*-vinylformamide-vinylacetate) and copoly(*N*-vinylacetylamide-vinylacetate), *Colloid Polym. Sci.*, 281, 690, 2003.

2003SH1 Shang, M., Matsuyama, H., Maki, T., Teramoto, M., and Lloyd, D.R., Effect of crystallization and liquid-liquid phase separation on phase-separation kinetics in poly(ethylene-*co*-vinyl alcohol)/glycerol solution, *J. Polym. Sci.: Part B: Polym. Phys.*, 41, 184, 2003.

2003SH2 Shang, M., Matsuyama, H., Maki, T., Teramoto, M., and Lloyd, D.R., Preparation and characterization of poly(ethylene-*co*-vinyl alcohol) membranes via thermally induced liquid-liquid phase separation, *J. Appl. Polym. Sci.*, 87, 853, 2003.

2003SH3 Shang, M., Matsuyama, H., Teramoto, M., Lloyd, D.R., and Kubota, N., Preparation and membrane performance of poly(ethylene-*co*-vinyl alcohol) hollow fiber membrane via thermally induced phase separation, *Polymer*, 44, 7441, 2003.

2003SHE Shen, L.-Q., Xu, Z.-K., and Xu, Y.-Y., Phase separation behavior of poly(ether imide)/*N,N*-dimethylacetamide/nonsolvent systems, *J. Appl. Polym. Sci.*, 89, 875, 2003.

2003SHT Shtanko, N.I., Lequieu, W., Goethals, E.J., and DuPrez, F.E., pH- and thermo-responsive properties of poly(*N*-vinylcaprolactam-*co*-acrylic acid) copolymers, *Polym. Int.*. 52, 1605, 2003.

2003SIL Silva, G.A., Eckelt, J., Goncalves, M.C., and Wolf, B.A., Thermodynamics of pseudo-ternary systems as a tool to predict the morphologies of cellulose acetate/polystyrene blends cast from tetrahydrofuran solutions, *Polymer*, 44, 1075, 2003.

2003SIP Siporska, A., Szydlowski, J., and Rebelo, L.P.N., Solvent H/D isotope effects on miscibility and theta-temperature in the polystyrene-cyclohexane system (exp. data by J. Szydlowski), *Phys. Chem. Chem. Phys.*, 5, 2996, 2003.

2003SLI Slimane, S.K., Roussel, F., Benmouna, F., Buisine, M., Coqueret, X., Benmouna, M., and Maschke, U., Effects of the method of preparation on the molar mass of polymer and phase diagrams of poly(2-ethylhexylacrylate)/5CB systems, *Macromolecules*, 36, 3443, 2003.

2003STI Stieger, M. and Richtering, W., Shear-induced phase separation in aqueous polymer solutions: temperature-sensitive microgels and linear polymer chains, *Macromolecules*, 36, 8811, 2003.

2003SWI Swier, S., Van Durme, K., Van Mele, B., Modulated-temperature differential scanning calorimetry study of temperature-induced mixing and demixing in poly(vinyl methyl ether)/water, *J. Polym. Sci.: Part B: Polym. Phys.*, 41, 1824, 2003.

2003TAK Takahashi, N., Kanaya, T., Nishida, K., and Kaji, K., Effects of cononsolvency on gelation of poly(vinyl alcohol) in mixed solvents of dimethyl sulfoxide and water, *Polymer*, 44, 4075, 2003.

2003TSY Tsypina, N.A., Kizhnyaev, V.N., and Adamova, L.V., Triazole-containing polymers: Solubility and thermodynamic behavior in solutions (Russ.), *Vysokomol. Soedin., Ser. A*, 45, 1718, 2003.

2003XUE Xue, W., Huglin, M.B., and Jones, T.G.J., Parameters affecting the lower critical solution temperature of linear and crosslinked poly(*N*-ethylacrylamide) in aqueous media, *Macromol. Chem. Phys.*, 204, 1956, 2003.

2003YA1 Yamamoto, K., Serizawa, T., and Akashi, M., Synthesis and thermosensitive properties of poly[(*N*-vinylamide)-*co*-(vinyl acetate)]s and their hydrogels, *Macromol. Chem. Phys.*, 204, 1027, 2003.

2003YA2 Yamaoka, T., Tamura, T., Seto, Y., Tada, T., Kunugi, S., and Tirrell, D.A., Mechanism for the phase transition of a genetically engineered elastin model peptide (VPGIG)40 in aqueous solution, *Biomacromolecules*, 4, 1680, 2003.

2003YAN Yang, Z., Crothers, M., Attwood, D., Collett, J.H., Ricardo, N.M.P., Martini, L.G.A., and Booth, C., Association properties of ethylene oxide/styrene oxide diblock copolymer E17S8 in aqueous solution, *J. Colloid Interface Sci.*, 263, 312, 2003.

2003VAR Variankaval, N.E., Rezac, M.E., and Abhiraman, A.S., The 'cold-solutioning' phenomenon in cellulose triacetate-acetone mixtures, *J. Appl. Polym. Sci.*, 90, 1697, 2003.

2003YEO Yeow, M.L., Liu, Y.T., and Li, K., Isothermal phase diagrams and phase-inversion behavior of poly(vinylidene fluoride)/solvents/additives/water systems, *J. Appl. Polym. Sci.*, 90, 2150, 2003.

2003YIN Yin, X. and Stoever, H.D.H., Hydrogel microspheres by thermally induced coacervation of poly(*N,N*-dimethylacrylamide-*co*-glycidyl methacrylate) aqueous solutions, *Macromolecules*, 36, 9817, 2003.

2003YOU Young, T.-H., Tao, C.-T., and Lai, P.-S., Phase behavior of poly(etherimide) in mixtures of *N*-methyl-2-pyrrolidinone and methylene chloride, *Polymer*, 44, 1689, 2003.

2003ZH1 Zhang, W. and Kiran, E., (p, V, T) Behaviour and miscibility of (polysulfone + THF + carbon dioxide) at high pressures, *J. Chem. Thermodyn.*, 35, 605, 2003.

2003ZH2 Zhang, W., Dindar, C., Bayraktar, Z., and Kiran, E., Phase behavior, density, and crystallization of polyethylene in n-pentane and in n-pentane/CO_2 at high pressures, *J. Appl. Polym. Sci.*, 89, 2201, 2003.

2004ALI Ali, M.M. and Stoever, H.D.H., Well-defined amphiphilic thermosensitive copolymers based on poly(ethylene glycol monomethacrylate) and methyl methacrylate prepared by atom transfer radical polymerization, *Macromolecules*, 37, 5219, 2004.

2004AGA Agarwal, R., Prasad, D., Maity, S., Gayen, K., and Ganguly, S., Experimental measurements and model based inferencing of solubility of polyethylene in xylene, *J. Chem. Eng. Japan*, 37, 1427, 2004.

2004ANT Antonov, Y.A., Van Puyvelde, P., Moldenaers, P., and Leuven, K.U., Effect of shear flow on the phase behavior of an aqueous gelatine-dextran emulsion, *Biomacromolecules*, 5, 276, 2004.

2004BER Berge, B., Koningsveld, R., and Berghmans, H., Influence of added components on the miscibility behavior of the (quasi-)binary system water/poly(vinyl methyl ether) and on the swelling behavior of the corresponding hydrogels. 1. Tetrahydrofuran, *Macromolecules*, 37, 8082, 2004.

2004CHE Chen, X., Yasuda, K., Sato, Y., Takishima, S., and Masuoka, H., Measurement and correlation of phase equilibria of ethylene + n-hexane + metallocene polyethylene at temperatures between 373 and 473 K and at pressures up to 20 MPa, *Fluid Phase Equil.*, 215, 105, 2004.

2004DER D'Errico, G., Paduano, L., and Khan, A.: Temperature and concentration effects on supramolecular aggregation and phase behavior for poly(propylene oxide)-*b*-poly(ethylene oxide)-*b*-poly(propylene oxide) copolymers of different composition in aqueous mixtures, *J. Colloid Interface Sci.*, 279, 379, 2004.

2004DIA Diab, C., Akiyama, Y., Kataoka, K., and Winnik, F.M., Microcalorimetric study of the temperature-induced phase separation in aqueous solutions of poly(2-isopropyl-2-oxazolines), *Macromolecules*, 37, 2556, 2004.

2004DOR Dormidontova, E.E., Influence of end groups on phase behavior and properties of PEO in aqueous solutions, *Macromolecules*, 37, 7747, 2004.

2004DU1 Durme, K. van, Verbrugghe, S., DuPrez, F.E., and Mele, B. van, Influence of poly(ethylene oxide) grafts on kinetics of LCST behavior in aqueous poly(*N*-vinylcaprolactam) solutions and networks studied by modulated temperature DSC, *Macromolecules*, 37, 1054, 2004.

2004DU2 Durme, K., van, Assche, G. van, and Mele, B. van, Kinetics of demixing and remixing in poly(*N*-isopropylacrylamide)/water studied by modulated temperature DSC, *Macromolecules*, 37, 9596, 2004.

2004ECK Eckelt, J., Fraktionierung und Membranbildung von Polysacchariden, *Dissertation*, Johannes Gutenberg Universität Mainz, 2004.

2004GAR Garay, M.T., Rodriguez, M., Vilas, J.L., and Leon, L.M., Study of polymer-polymer complexes of poly(*N*-isopropylacrylamide) with hydroxyl-containing polymers, *J. Macromol. Sci. Part B-Phys.*, 43, 437, 2004.

2004GIA Giannotti, M.I., Foresti, M.L., Mondragon, I., Galante, M.J., and Oyanguren, P.A., Reaction-induced phase separation in epoxy/polysulfone/poly(etherimide) systems. I. Phase diagrams, *J. Polym. Sci.: Part B: Polym. Phys.*, 42, 3953, 2004.

2004HAS Hasan, E., Zhang, M., Müller, A.H.E., and Tsvetanov, Ch.B., Thermoassociative block copolymers of poly(*N*-isopropylacrylamide) and poly(propylene oxide), *J. Macromol. Sci. Part A-Pure Appl. Chem.*, 41, 467, 2004.

2004HOP Hopkinson, I., Myatt, M., and Tajbakhsh, A., Static and dynamic studies of phase composition in a polydisperse system (exp. data by I. Hopkinson), *Polymer*, 45, 4307, 2004.

2004HUE Hüther, A. and Maurer, G., Swelling of *N*-isopropyl acrylamide hydrogels in aqueous solutions of poly(ethylene glycol), *Fluid Phase Equil.*, 226, 321, 2004.

2004INA Inamura, I., Jinbo, Y., Kittaka, M., and Asano, A., Solution properties of water-poly(ethylene glycol)-poly(*N*-vinylpyrrolidone) ternary system, *Polym. J.*, 36, 108, 2004.

2004JIA Jiang, S., An, L., Jiang, B., and Wolf, B. A., Temperature and pressure dependence of phase separation of *trans*-decahydronaphthalene/polystyrene solution, *Chem. Phys.*, 298, 37, 2004.

2004KAV Kavlak, S., Can, H.K., and Guener, A., Miscibility studies on poly(ethylene glycol)/dextran blends in aqueous solutions by dilute solution viscometry, *J. Appl. Polym. Sci.*, 94, 453, 2004.

2004KEL Kelarakis, A., Ming, X.-T., Yuan, X.-F., and Booth, C., Aqueous micellar solutions of mixed triblock and diblock copolymers studied using oscillatory shear, *Langmuir*, 20, 2036, 2004.

2004KIT Kitano, H., Hirabayashi, T., Gemmei-Ide, M., and Kyogoku, M., Effect of macrocycles on the temperature-responsiveness of poly[(methoxy diethylene glycol methacrylate)-graft-PEG], *Macromol. Chem. Phys.*, 205, 1651, 2004.

2004KUN Kunieda, H., Kaneko, M., Lopez-Quintela, M.A., and Tsukahara, M., Phase behavior of a mixture of poly(isoprene)-poly(oxyethylene) diblock copolymer and poly(oxyethylene) surfactant in water, *Langmuir*, 20, 2164, 2004.

2004LE1 Lee, J.S., Lee, H.K., and Kim, S.C., Thermodynamic parameters of poly(lactic acid) solutions in dialkyl phthalate, *Polymer*, 45, 4491, 2004.

2004LE2 Lee, H.J., Jung, B., Kang, Y.S., and Lee, H., Phase separation of polymer casting solution by nonsolvent vapor, *J. Membrane Sci.*, 245, 103, 2004.

2004LIS Liskova, A. and Berghmans, H., Phase-separation phenomena in the polymerization of styrene in the presence of polyethylene wax, *J. Appl. Polym. Sci.*, 91, 2234, 2004.

2004MAE Maeda, Y., Yamamoto, H., and Ikeda, I., Effects of ionization on the phase behavior of poly(*N*-isopropylacrylamide-*co*-acrylic acid) and poly(*N,N*-diethylacrylamide-*co*-acrylic acid) in water, *Colloid Polym. Sci.*, 282, 1268, 2004.

2004MAH Mahajan, R.K., Chawla, J., and Bakshi, M.S., Depression in the cloud point of Tween in the presence of glycol additives and triblock polymers, *Colloid Polym. Sci.*, 282, 1165, 2004.

2004MAN Mandala, I., Michon, C., and Launay, B., Phase and rheological behaviors of xanthan/amylose and xanthan/starch mixed systems, *Carbohydr. Polym.*, 58, 285, 2004.

2004MAO Mao, H., Li, C., Zhang, Y. Furyk, S., Cremer, P.S., and Bergbreiter, D.E., High-throughput studies of the effects of polymer structure and solution components on the phase separation of thermoresponsive polymers, *Macromolecules*, 37, 1031, 2004.

2004MAT Matsuyama, H., Hayashi, K., Maki, T., Teramoto, M., and Kubota, N., Effect of polymer density on polyethylene hollow fiber membrane formation via thermally induced phase separation, *J. Appl. Polym. Sci.*, 93, 471, 2004.

2004MEY Meyer, D.E. and Chilkoti, A., Quantification of the effects of chain length and concentration on the thermal behavior of elastin-like polypeptides, *Biomacromolecules*, 5, 846, 2004.

2004MIK Mikhailov, Yu.M., Ganina, L.V., Roshchupkin, V.P., Shapaeva, N.V., and Suchkova, L.I., Mutual dissolution of poly(nonyl acrylate-*co*-acrylic acid) in low-molecular-mass solvents and oligomers (Russ.), *Vysokomol. Soedin., Ser. A*, 46, 1583, 2004.

2004MOR Mori, T., Fukuda, Y., Okamura, H., Minagawa, K., Masuda, S., and Tanaka, M., Thermosensitive copolymers having soluble and insoluble monomer units, poly(*N*-vinylacetamide-*co*-methyl acrylate)s: effect of additives on their lower critical solution temperatures, *J. Polym. Sci.: Part A: Polym. Chem.*, 42, 2651, 2004.

2004NED Nedelcheva, A.N., Vladimirov, N.G., Novakov, C.P., and Berlinova, I.V., Associative block copolymers comprising poly(*N*-isopropylacrylamide) and poly(ethylene oxide) end-functionalized with a fluorophilic or hydrophilic group, *J. Polym. Sci.: Part B: Polym. Phys.*, 42, 5736, 2004.

2004NIE Nie, H., Li, M., Bansil, R., Konak, C., Helmstedt, M., and Lal, J., Structure and dynamics of a pentablock copolymer of polystyrene-polybutadiene in a butadiene-selective solvent, *Polymer*, 45, 8791, 2004.

2004NON Nonaka, T., Hanada, Y., Watanabe, T., Ogata, T., and Kurihara, S., Formation of thermosensitive water-soluble copolymers with phosphinic acid groups and the thermosensitivity of the copolymers and copolymer/metal complexes, *J. Appl. Polym. Sci.*, 92, 116, 2004.

2004OKA Okada, M., Inoue, G., Ikegami, T., Kimura, K., and Furukawa, H., Composition dependence of polymerization-induced phase separation of 2-chlorostyrene/ polystyrene mixtures, *Polymer*, 45, 4315, 2004.

2004OLS Olsson, M., Joabsson, F., and Piculell, L., Particle-induced phase separation in quasi-binary polymer solutions, *Langmuir*, 20, 1605, 2004.

2004PAG Pagonis, K. and Bokias, G., Upper critical solution temperature-type cononsolvency of poly(*N,N*-dimethylacrylamide) in water-organic solvent mixtures, *Polymer*, 45, 2149, 2004.

2004PAN Panayiotou, M., Garret-Flaudy, F., and Freitag, R., Co-nonsolvency effects in the thermoprecipitation of oligomeric polyacrylamides from hydroorganic solutions, *Polymer*, 45, 3055, 2004.

2004PA1 Park, M.J. and Char, K., Gelation of PEO-PLGA-PEO triblock copolymers induced by macroscopic phase separation, *Langmuir*, 20, 2456, 2004.

2004PA2 Park, J.-S., Akiyama, Y., Winnik, F.M., and Kataoka, K., Versatile synthesis of end-functionalized thermosensitive poly(2-isopropyl-2-oxazolines), *Macromolecules*, 37, 6786, 2004.

2004PER Pereira, M., Wu, Y.T., Madeira, P., Venancio, A., Macedo, E., and Teixeira, J., Liquid-liquid equilibrium phase diagrams of new aqueous two-phase systems: Ucon 50-HB5100 + ammonium sulfate + water, Ucon 50-HB5100 + poly(vinyl alcohol) + water, Ucon 50-HB5100 + hydroxypropyl starch + water, and poly(ethylene glycol) 8000 + poly(vinyl alcohol) + water, *J. Chem. Eng. Data* 49, 43,2004.

2004RAN Rangel-Yagui, C.O., Pessoa Jr, A. and Blankschtein, D., Two-phase aqueous micellar systems: an alternative method for protein purification, *Brazil. J. Chem. Eng.*, 21, 531, 2004.

2004RIC Riccardi, C.C., Borrajo, J., Meynie, L., Fenouillot, F., and Pascault, J.-P., Thermodynamic analysis of the phase separation during the polymerization of a thermoset system into a thermoplastic matrix. I. Effect of the composition on the cloud-point curves, *J. Polym. Sci.: Part B: Polym. Phys.*, 42, 1351, 2004.

2004SAL Salgado-Rodriguez, R., Licea-Claverie, A., and Arndt, K.F., Random copolymers of *N*-isopropylacrylamide and methacrylic acid monomers with hydrophobic spacers: pH-tunable temperature sensitive materials, *Eur. Polym. J.*, 40, 1931, 2004.

2004SAN Sanchez, M.S., Hanykova, L., Ilavsky, M., and Pradas, M.M., Thermal transitions of poly(*N*-isopropylmethacrylamide) in aqueous solutions, *Polymer*, 45, 4087, 2004.

2004SCH Schnell, M., Stryuk, S., and Wolf, B.A., Liquid/liquid demixing in the system n-hexane/narrowly distributed linear polyethylene (exp. data by M. Schnell and B.A. Wolf), *Ind. Eng. Chem. Res.*, 43, 2852, 2004.

2004SHI Shibayama, M., Isono, K., Okabe, S., Karino, T., and Nagao, M., SANS study on pressure-induced phase separation of poly(*N*-isopropylacrylamide) aqueous solutions and gels, *Macromolecules*, 37, 2909, 2004.

2004SOG Soga, O., Nostrum, C.F. van, and Hennik, W.E., Poly[*N*-(2-hydroxypropyl)methacrylamide mono/dilactate]: a new class of biodegradable polymers with tuneable thermosensitivity, *Biomacromolecules*, 5, 818, 2004.

2004SUG Sugihara, S., Kanaoka, S., and Aoshima, S., Thermosensitive random copolymers of hydrophilic and hydrophobic monomers obtained by living cationic copolymerization, *Macromolecules*, 37, 1711, 2004.

2004SUT Sutton, D., Stanford, J.L., and Ryan, A.J., Reaction-induced phase separation in polyoxyethylene/polystyrene blends. I. Ternary phase diagram, *J. Macromol. Sci.-Phys. B*, 43, 219, 2004.

2004TAD Tada, E. dos S., Loh, W., and Pessoa-Filho, P. de A., Phase equilibrium in aqueous two-phase systems containing ethylene oxide-propylene oxide block copolymers and dextran, *Fluid Phase Equil.*, 218, 221, 2004.

2004TAN Tanaka, T. and Lloyd, D.R., Formation of poly(L-lactid acid) microfiltration membranes via thermally induced phase separation, *J. Membrane Sci.*, 238, 65, 2004.

2004VAR Varade, D., Sharma, R., Aswal, V.K., Goyal, P.S., and Bahadur, P., Effect of hydrotopes on the solution behavior of PEO/PPO/PEO block copolymer L62 in aqueous solutions, *Eur. Polym. J.*, 40, 2457, 2004.

2004VOL Volkova, I.F., Gorshkova, M.Yu., Izumrudov, V.A., and Stotskaya, L.L., Interaction of a polycation with divinyl ether-maleic anhydride copolymer in aqueous solutions (Russ.), *Vysokomol. Soedin., Ser. A*, 46, 1388, 2004.

2004VSH Vshivkov, S.A. and Rusinova, E.V., Thermodynamics of solutions of blends of diene rubbers under deformation (Russ.), *Vysokomol. Soedin., Ser. B*, 46, 912, 2004.

2004WAZ Waziri, S.M., Abu-Sharkh, B.F., and Ali, S.A., Protein partitioning in aqueous two-phase systems composed of a pH-responsive copolymer and poly(ethylene glycol), *Biotechnol. Progr.*, 20, 526, 2004.

2004WEI Weiss-Malik, R.A., Solis, F.J., and Vernon, B.L.. Independent control of lower critical solution temperature and swelling behavior with pH for poly(*N*-isopropylacrylamide-*co*-maleic acid), *J. Appl. Polym. Sci.*, 94, 2110, 2004.

2004WIL Wilczura-Wachnik, H. and van Hook, W.A., Liquid-liquid phase equilibria for some polystyrene-methylcyclohexane mixtures, *Eur. Polym. J.*, 40, 251, 2004.

2004WOH Wohlfarth, C., *CRC Handbook of Thermodynamic Data of Aqueous Polymer Solutions*, CRC Press, Boca Raton, 2004.

2004ZUC Zucchi, I.A., Galante, M.J. Borrajo, J., and Williams, R.J.J., A model system for the thermodynamic analysis of reaction-induced phase separation: solutions of polystyrene in bifunctional epoxy/amine monomers (exp. data by I.A. Zucchi), *Macromol. Chem. Phys.*, 205, 676, 2004.

2005BAE Bae, S.J., Suh, J.M., Sohn, Y.S., Bae, Y.H., Kim, S.W., and Jeong, B., Thermogelling of poly(caprolactone-*b*-ethylene glycol-*b*-caprolactone) aqueous solutions, *Macromolecules*, 38, 5260, 2005.

2005BIS Bisht, H.S., Wan, L., Mao, G., and Oupicky, D., pH-Controlled association of PEG-containing terpolymers of *N*-isopropylacrylamide and 1-vinylimidazole, *Polymer*, 46, 7945, 2005.

2005BLA Blasig, A. and Thies, M.C., Rapid expansion of cellulose triacetate from ethyl acetate solutions, *J. Appl. Polym. Sci.*, 95, 290, 2005.

2005BOL Bolognese, B., Nerli, B., and Pico, G., Application of the aqueous two-phase systems of ethylene and propylene oxide copolymer-maltodextrin for protein purification, *J. Chromatogr. B*, 814, 347, 2005.

2005BUM Bumbu, G.-G., Vasile, C., Chitanu, G.C., and Staikos, G., Interpolymer complexes between hydroxypropylcellulose and copolymers of maleic acid: A comparative study, *Macromol. Chem. Phys.*, 206, 540, 2005.

2005CAO Cao, Z., Liu, W., Gao, P., Yao, K., Li, H., and Wang, G., Toward an understanding of thermoresponsive transition behavior of hydrophobically modified *N*-isopropylacrylamide copolymer solution, *Polymer*, 46, 5268, 2005.

2005CAR Carter, S., Hunt, B., and Rimmer, S., Highly branched poly(*N*-isopropylacrylamide)s with imidazole end groups prepared by radical polymerization in the presence of a styryl monomer containing a dithioester group, *Macromolecules*, 38, 4595, 2005.

2005CHA Chaibundit, C., Sumanatrakool, P., Chinchew, S., Kanatharana, P., Tattershall, C.E., Booth, C., and Yuan, X.-F., Association properties of diblock copolymer of ethylene oxide and 1,2-butylene oxide: $E_{17}B_{12}$ in aqueous solution, *J. Colloid Interface Sci.*, 283, 544, 2005.

2005CH1 Chen, X., Ding, X., Zheng, Z., and Peng, Y., Thermosensitive polymeric vesicles self-assembled by PNIPAAm-*b*-PPG-*b*-PNIPAAm triblock copolymers, *Colloid Polym. Sci.*, 283, 452, 2005.

2005CH2 Chen, H., Liang, Y., and Wang, C.-G., Solubility of highly isotactic polyacrylonitrile in dimethyl sulphoxide, *J. Polym. Res.*, 12, 325, 2005.

2005CIM Cimen, E.K., Rzaev, Z.M.O., and Piskin, E., Bioengineering functional copolymers: V. Synthesis, LCST,and thermal behavior of poly(*N*-isopropylacrylamide-*co*-p-vinyl-phenylboronic acid), *J. Appl. Polym. Sci.*, 95, 573, 2005.

2005DUB Dubovik, A.S., Makhaeva, E.E., Grinberg, V.Ya., and Khokhlov, A.R., Energetics of cooperative transitions of *N*-vinylcaprolactam polymers in aqueous solutions, *Macromol. Chem. Phys.*, 206, 915, 2005.

2005DUR Durme, K. van, Delellio, L.. Kudryashov, E., Buckin, V., and Mele, B. van: Exploration of high-resolution ultrasonic spectroscopy as an analytical tool to study demixing and remixing in poly(*N*-isopropyl acrylamide)/water solutions, *J. Polym. Sci.: Part B: Polym. Phys.*, 43, 1283, 2005.

2005ECK Eckelt, J. and Wolf, B.A., Membranes directly prepared from solutions of unsubstituted cellulose, *Macromol. Chem. Phys.*, 206, 227, 2005.

2005GAO Gaoa, C., Möhwald, H., and Shen, J., Thermosensitive poly(allylamine)-*g*-poly(*N*-isopropylacrylamide): synthesis, phase separation and particle formation, *Polymer*, 46, 4088, 2005.

2005GRA Gratson, G.M. and Lewis, J.A., Phase behavior and rheological properties of polyelectrolyte inks for direct-write assembly, *Langmuir*, 21, 457, 2005.

2005GUP Gupta, A., Mohanty, B., and Bohidar, H.B., Flory temperature and upper critical solution temperature of gelatine solutions, *Biomacromolecules*, 6, 1623, 2005.

2005HAM Hamada, N. and Einaga, Y., Effects of hydrophilic chain length on the characteristics of the micelles of octaoxyethylene tetradecyl $C_{14}E_8$, hexadecyl $C_{16}E_8$, and octadecyl $C_{18}E_8$ ethers, *J. Phys. Chem. B*, 109, 6990, 2005.

2005HEI Heijkants, R.G.J., VanCalck, R.V., DeGroot, J.H., Pennings, A.J., and Schouten, A.J., Phase transitions in segmented polyesterurethane-DMSO-water systems, *J. Polym. Sci.: Part B: Polym. Phys.*, 43, 716, 2005.

2005HO1 Holyst, R., Staniszewski, K., and Demyanchuk, I., Ordering in surfactant mixtures induced by polymers, *J. Phys. Chem. B*, 109, 4881, 2005.

2005HO2 Holyst, R., Staniszewski, K., Patkowski, A., and Gapinski, J., Hidden minima of the Gibbs free energy revealed in a phase separation in polymer/surfactant/water mixture, *J. Phys. Chem. B*, 109, 8533, 2005.

2005IHA Ihata, O., Kayaki, Y., and Ikariya, T., Aliphatic poly(urethane-amine)s synthesized by copolymerization of aziridines and supercritical carbon dioxide, *Macromolecules*, 38, 6429, 2005.

2005IMA Imanishi, K. and Einaga, Y., Effects of hydrophilic chain length on the characteristics of the micelles of pentaoxyethylene n-decyl $C_{10}E_5$ and hexaoxyethylene n-decyl $C_{10}E_6$ ethers, *J. Phys. Chem. B*, 109, 7574, 2005.

2005IZU Izumrudov, V.A., Gorshkova, M.Yu., and Volkova, I.F., Controlled phase separations in solutions of soluble polyelectrolyte complex of DIVEMA (copolymer of divinyl ether and maleic anhydride), *Eur. Polym. J.*, 41, 1251, 2005.

2005JON Jones, J.A., Novo, N., Flagler, K., Pagnucco, C.D., Carew, S., Cheong, C., Kong, Z., Burke, N.A.D., and Stoever, H.D.H., Thermoresponsive copolymers of methacrylic acid and poly(ethylene glycol) methyl ether methacrylate, *J. Polym. Sci.: Part A: Polym. Chem.*, 43, 6095, 2005.

2005KJO Kjoniksen, A.-L., Laukkanen, A., Galant, C., Knudsen, K.D., Tenhu, H., and Nyström, B., Association in aqueous solutions of a thermoresponsive PVCL-*g*-$C_{11}EO_{42}$ copolymer, *Macromolecules*, 38, 948, 2005.

2005KOZ Kozlowska, M.K., Domanska, U., Dudeka, D., and Rogalski, M., Surface tension, (solid + liquid) equilibria and (liquid + liquid) equilibria for (iPBu-1 + hydrocarbon, or alcohol) systems, *Fluid Phase Equil.*, 236, 184, 2005.

2005LAU Laukkanen, A., Valtola, L., Winnik, F.M., and Tenhu, H., Thermosensitive graft copolymers of an amphiphilic macromonomer and *N*-vinylcaprolactam: synthesis and solution properties in dilute aqueous solutions below and above the LCST, *Polymer*, 46, 7055, 2005.

2005LEE Lee, B.H. and Vernon, B., Copolymers of *N*-isopropylacrylamide, HEMA-lactate and acrylic acid with time-dependent lower critical solution temperature as a bioresorbable carrier, *Polym. Int.*, 54, 418, 2005.

2005LIC Li, C., Meng, L.-Z., Lu, X.-J., Wu, Z.-Q., Zhang, L.-F., and He, Y.-B., Thermo- and pH-sensitivities of thiosemicarbazone-incorporated, fluorescent and amphiphilic poly(*N*-isopropylacrylamide), *Macromol. Chem. Phys.*, 206, 1870, 2005.

2005LUN Lundell, C., Hoog, E.H.A. de, Tromp, R.H., and Hermansson, A.-M., Effects of confined geometry on phase-separated dextran/gelatine mixtures exposed to shear, *J. Colloid Interface Sci.*, 288, 222, 2005.

2005MA1 Matsuo, M., Miyoshi, S., Azuma, M., Bin, Y., Agari, Y., Sato, Y., and Kondo, A., Phase separation of several kinds of polyethylene solution under the gelation/crystallization process, *Macromolecules*, 38, 6688, 2005.

2005MA2 Matsuda, Y., Miyazaki, Y., Sugihara, S., Aoshima, S., Saito, K., and Sato, T., Phase separation behavior of aqueous solutions of a thermoresponsive polymer (exp. data by Y. Matsuda), *J. Polym. Sci.: Part B: Polym. Phys.*, 43, 2937, 2005.

2005MOO Moody, M.L., Willauer, H.D., Griffin, S.T., Huddleston, J.G., and Rogers, R.D., Solvent property characterization of poly(ethylene glycol)/dextran aqueous biphasic systems using the free energy of transfer of a methylene group and a linear solvation energy relationship, *Ind. Eng. Chem. Res.*, 44, 3749, 2005.

2005MOT Motokawa, R., Morishita, K., Koizumi, S., Nakahira, T., and Annaka, M., Thermosensitive diblock copolymer of poly(*N*-isopropylacrylamide) and poly(ethylene glycol) in water: polymer preparation and solution behavior, *Macromolecules*, 38, 5748, 2005.

2005NIE Nies, E., Ramzi, A., Berghmans, H., Li, T., Heenan, R.K., and King, S.M., Composition fluctuations, phase behavior, and complex formation in poly(vinyl methyl ether)/D_2O investigated by small-angle neutron scattering, *Macromolecules*, 38, 915, 2005.

2005OL1 Olsson, M., Joabsson, F., and Piculell, L., Particle-induced phase separation in mixed polymer solutions, *Langmuir*, 21, 1560, 2005.

2005OL2 Olsson, M., Boström, G., Karlson, L., and Piculell, L., Added surfactant can change the phase behavior of aqueous polymer-particle mixtures, *Langmuir*, 21, 2743, 2005.

2005RAY Ray, B., Okamoto, Y., Kamigito, M., Sawamoto, M., Seno, K., Kanaoka, S., and Aoshima, S., Effect of tacticity of poly(*N*-isopropylacrylamide) on the phase separation temperature of its aqueous solutions, *Polym. J.*, 37, 234, 2005.

2005RIC Ricoa, M., Borrajo, J., Abad, M.J., Barral, L., and Lopez, J., Thermodynamic analysis of phase separation in an epoxy/polystyrene mixture, *Polymer*, 46, 6114, 2005.

2005SET Seto, Y., Aoki, T., and Kunugi, S., Temperature- and pressure-responsive properties of L- and DL-forms of poly(*N*-(1-hydroxymethyl)propylmethacrylamide) in aqueous solutions, *Colloid Polym. Sci.*, 283, 1137, 2005.

2005SHA Shang, M., Matsuyama, H., Teramoto, M., Okuno, J., Lloyd, D.R., and Kubota, N., Effect of diluent on poly(ethylene-*co*-vinyl alcohol) hollow-fiber membrane formation via thermally induced phase separation, *J. Appl. Polym. Sci.*, 95, 219, 2005.

2005SP1 Spelzini, D., Rigatusso, R., Farruggia B., and Pico, G., Thermal aggregation of methyl cellulose in aqueous solution: a thermodynamic study and protein partitioning behaviour, *Cellulose*, 12, 293, 2005.

2005SP2 Spevacek, J., Phase separation in aqueous polymer solutions as studied by NMR methods, *Macromol. Symp.* 222, 1, 2005.

2005STA Starovoytova, L., Spevacek, J., and Ilavsky, M., ^{1}H-NMR study of temperature-induced phase transitions in D_2O solutions of poly(*N*-isopropylmethacrylamide)/poly(*N*-isopropylacrylamide) mixtures and random copolymers, *Polymer*, 46, 677, 2005.

2005TAD Tada, E. dos S., Loh, W., and Pessoa-Filho, P. de A., Erratum to [*Fluid Phase Equilibr.* 218 (2004) 221–228], *Fluid Phase Equil.*, 231, 250, 2005.

2005TAO Tao, C.-T. and Young, T.-H., Phase behavior of poly(*N*-isopropylacrylamide) in water-methanol cononsolvent mixtures and its relevance to membrane formation, *Polymer*, 46, 10077, 2005.

2005TIE Tiera, M.J., Santos, G.R. dos, Oliveira Tiera, V.A. de, Vieira, N.A.B., Frolini, E., Silva, R.C. da, and Loh, W., Aqueous solution behavior of thermosensitive (*N*-isopropylacrylamide-acrylic acid-ethyl methacrylate) terpolymers, *Colloid Polym. Sci.*, 283, 662, 2005.

2005UGU Uguzdogan, E., Camh, T., Kabasakal, O.S., Patir, S., Öztürk, E., Denkbas, E.B., and Tuncel, A., A new temperature-sensitive polymer: poly(ethoxypropylacrylamide), *Eur. Polym. J.*, 41, 2142, 2005.

2005VA1 Van Durme, K., Rahier, H., and Van Mele, B., Influence of additives on the thermoresponsive behavior of polymers in aqueous solution, *Macromolecules*, 38, 10155, 2005.

2005VA2 Van Durme, K., Loozen, E., Nies, E., and Van Mele, B., Phase behavior of poly(vinyl methyl ether) in deuterium oxide, *Macromolecules*, 38, 10234, 2005.

2005VEN Venkatesu, P., Effect of polymer chain in coexisting liquid phases by refractive index measurements, *J. Chem. Phys.*, 123, 024902, 2005.

2005VER Verdonck, B., Gohy, J.-F., Khousakoun, E., Jerome, R., and DuPrez, F., Association behavior of thermo-responsive block copolymers based on poly(vinyl ethers), *Polymer*, 46, 9899, 2005.

2005WOH Wohlfarth, C., *CRC Handbook of Thermodynamic Data of Polymer Solutions at Elevated Pressures*, Taylor & Francis, CRC Press, Boca Raton, 2005.

2005XIA Xia, Y., Yin, X., Burke, N.A.D., and Stoever, H.D., Thermal response of narrow-disperse poly(*N*-isopropylacrylamide) prepared by atom transfer radical polymerization, *Macromolecules*, 38, 5937, 2005.

2005YAM Yamazaki, R., Iizuka, K., Hiraoka, K., and Nose, T., Phase behavior and mechanical properties of mixed poly(ethylene glycol) monododecyl ether aqueous solutions, *Macromol. Chem. Phys.*, 206, 439, 2005.

2005YUE Yue, Z., Eccleston, M.E., and Slater, N.K.H., PEGylation and aqueous solution behaviour of pH responsive poly(L-lysine iso-phthalamide), *Polymer*, 46, 2497, 2005.

2005YI1 Yin, X. and Stoever, H.D.H., Probing the influence of polymer architecture on liquid-liquid phase transitions of aqueous poly(*N,N*-dimethylacrylamide) copolymer solutions, *Macromolecules*, 38, 2109, 2005.

2005YI2 Yin, X. and Stoever, H.D.H., Temperature-sensitive hydrogel microspheres formed by liquid-liquid phase transitions of aqueous solutions of poly(*N,N*-dimethylacrylamide-*co*-allyl methacrylate), *J. Polym. Sci.: Part A: Polym. Chem.*, 43, 1641, 2005.

2005ZH1 Zhang, D., Macias, C., and Ortiz, C., Synthesis and solubility of (mono-)end-functionalized poly(2-hydroxyethyl ethacrylate-*g*-ethylene glycol) graft copolymers with varying macromolecular architectures, *Macromolecules*, 38, 2530, 2005.

2005ZH2 Zhang, R., Liu, J., Han, B., Wang, B., Sun, D., and He, J., Effect of PEO–PPO–PEO structure on the compressed ethylene-induced reverse micelle formation and water solubilization, *Polymer*, 46, 3936, 2005.

2005ZH3 Zhang, W., Shi, L., Wu, K., and An, Y., Thermoresponsive micellization of poly(ethylene glycol)-*b*-poly(*N*-isopropylacrylamide) in water, *Macromolecules*, 38, 5743, 2005.

2005ZH4 Zhang, W., Shi, L., Ma, R., An, Y., Xu, Y., and Wu, K., Micellization of thermo- and pH-responsive triblock copolymer of poly(ethylene glycol)-*b*-poly(4-vinylpyridine)-*b*-poly(*N*-isopropylacrylamide), *Macromolecules*, 38, 8850, 2005.

2005ZHO Zhou, J., Yin, J., Lv, R., Du, Q., and Zhong, W., Preparation and properties of MPEG-grafted EAA membranes via thermally induced phase separation, *J. Membrane Sci.*, 267, 90, 2005.

2006ANT Antonov, Yu.A. and Wolf, B.A., Phase behavior of aqueous solutions of bovine serum albumin in the presence of dextran, at rest, and under shear, *Biomacromolecules*, 7, 1562, 2006.

2006CAL Calciu, D., Eckelt, J., Haase, T., and Wolf, B.A.: Inverse spin fractionation: a tool to fractionate sodium hyaluronate, *Biomacromolecules*, 7, 3544, 2006.

2006CAO Cao, Z., Liu, W., Ye, G., Zhao, X., Lin, X., Gao, P., and Yao, K., *N*-isopropylacrylamide/2-hydroxyethyl methacrylate star diblock copolymers: synthesis and thermoresponsive behavior, *Macromol. Chem. Phys.*, 207, 2329, 2006.

2006CAU Causse, J., Lagerge, S., Menorval, L.C. de, and Faure, S., Micellar solubilization of tributylphosphate in aqueous solutions of Pluronic block copolymers Part I. Effect of the copolymer structure and temperature on the phase behavior, *J. Colloid Interface Sci.*, 300, 713, 2006.

2006CHA Chalykh, A.E., Gerasimov, V.K., Rusanova, S.N., Stoyanov, O.V., Petukhova, O.G., Kulagina, G.S., and Pisarev, S.A., Phase structure of silanol-modified ethylene-vinyl acetate copolymers, *Polym. Sci., Ser. A*, 48, 1058, 2006.

2006CHE Chethana, S., Rastogi, N.K., and Raghavarao, K.S.M.S., New aqueous two phase system comprising polyethylene glycol and xanthan, *Biotechnol. Lett.*, 28, 25, 2006.

2006CRA Crawford, N. and Dadmun, M.D., The effect of polymer chain length on the thermodynamics of acrylate/cyanobiphenyl mixtures, *Liq. Cryst.*, 33, 195, 2006.

2006DAL Dalkas, G., Pagonis, K., and Bokias, G., Control of the lower critical solution temperature-type cononsolvency properties of poly(*N*-isopropylacrylamide) in water-dioxane mixtures through copolymerisation with acrylamide, *Polymer*, 47, 243, 2006.

2006DEM Demyanchuk, I., Wieczorek, S.A., and Holyst, A., Phase separation in binary polymer/liquid crystal mixtures: network breaking and domain growth by coalescence-induced coalescence, *J. Phys. Chem. B*, 110, 9869, 2006.

2006DIN Ding, H., Wu, F., Huang, Y., Zhang, Z., and Nie, Y., Synthesis and characterization of temperature-responsive copolymer of PELGA modified poly(*N*-isopropylacrylamide), *Polymer*, 47, 1575, 2006.

2006DUA Duan, Q., Narumi, A., Miura, Y., Shen, X., Sato, S.-I., Satoh, T., and Kakuchi, T., Thermoresponsive property controlled by end-functionalization of poly(*N*-isopropylacrylamide) with phenyl, biphenyl, triphenyl groups, *Polym. J.*, 38, 306, 2006.

2006EDA Edahiro, J., Sumaru, K., Takagi, T., Shinbo, T., and Kanamori, T., Photoresponse of an aqueous two-phase system composed of photochromic dextran, *Langmuir*, 22, 5224, 2006.

2006FAN Fang, J. and Kiran, E., Kinetics of pressure-induced phase separation in polystyrene + acetone solutions at high pressures, *Polymer*, 47, 7943, 2006.

2006GAR Garcia Sakai, V., Higgins, J.S., and Trusler, J.P.M., Cloud curves of polystyrene or poly(methyl methacrylate) or poly(styrene-*co*-methyl methacrylate) in cyclohexanol determined with a thermo-optical apparatus, *J. Chem. Eng. Data*, 51, 743, 2006.

2006GEE Geever, L.M., Devine, D.M., Nugent, M.J.D., Kennedy, J.E., Lyons,J.G., Hanley, A., and Higginbotham, C.L., Lower critical solution temperature control and swelling behaviour of physically crosslinked thermosensitive copolymers based on *N*-isopropylacrylamide, *Eur. Polym. J.*, 42, 2540, 2006.

2006HAL Halacheva, S., Rangelov, S., and Tsvetanov, C., Poly(glycidol)-based analogues to Pluronic block copolymers. Synthesis and aqueous solution properties, *Macromolecules*, 39, 6845, 2006.

2006HOD Ho, D.L., Hammouda, B., Kline, S.R., and Chen, W.R., Unusual phase behavior in mixtures of poly(ethylene oxide) and ethyl alcohol, *J. Polym. Sci.: Part B: Polym. Phys.*, 44, 557, 2006.

2006HU1 Hua, F., Jiang, X., and Zhao, B., Well-defined thermosensitive, water-soluble polyacrylates and polystyrenics with short pendant oligo(ethylene glycol) groups synthesized by nitroxide-mediated radical polymerization, *J. Polym. Sci.: Part A: Polym. Chem.*, 44, 2454, 2006.

2006HU2 Hua, F., Jiang, X., and Zhao, B., Temperature-induced self-association of doubly thermosensitive diblock copolymers with pendant methoxytris(oxyethylene) groups in dilute aqueous solutions, *Macromolecules*, 39, 3476, 2006.

2006KI1 Kim, Y.-C., Kil, D.-S., and Kim, J.C., Synthesis and phase separation of poly(*N*-isopropylacrylamide-*co*-methoxy polyethyleneglycol monomethacrylate), *J. Appl. Polym. Sci.*, 101, 1833, 2006.

2006KI2 Kim, Y.C., Bang, M.-S., and Kim, J.-C., Synthesis and characterization of poly(*N*-isopropyl acrylamide) copolymer with methoxy polyethyleneglycol monomethacrylate, *J. Ind. Eng. Chem.*, 12, 446, 2006.

2006KIR Kirpach, A. and Adolf, D., High pressure induced coil-globule transitions of smart polymers, *Macromol. Symp.*, 237, 7, 2006.

2006KUJ Kujawa, P., Segui, F., Shaban, S., Diab, C., Okada, Y., Tanaka, F., and Winnik, F.M., Impact of end-group association and main-chain hydration on the thermosensitive properties of hydrophobically modified telechelic poly(*N*-isopropylacrylamides) in water, *Macromolecules*, 39, 341, 2006.

2006LAZ Lazzara, G., Milioto, S., and Gradzielski, M., The solubilisation behaviour of some dichloroalkanes in aqueous solutions of PEO–PPO–PEO triblock copolymers, *Phys. Chem. Chem. Phys.*, 8, 2299, 2006.

2006LIC Li, C., Lee, D.H., Kim, J.K., Ryu, D.Y., and Russel, T.P., Closed-loop phase behavior for weakly interacting block copolymers, *Macromolecules*, 39, 5926, 2006.

2006LID Lide, D.R. (ed.): *CRC Handbook of Chemistry and Physics, Section 13: Polymer Properties*, 87th ed., Taylor & Francis, CRC Press, Boca Raton, 2006.

2006LIN Lin, D.-J., Chang, H.-H., Chen, T.-C., Lee, Y.-C., and Cheng, L.-P., Formation of porous poly(vinylidene fluoride) membranes with symmetric or asymmetric morphology by immersion precipitation in the water/TEP/PVDF system, *Eur. Polym. J.*, 42, 1581, 2006.

2006LOO Loozen, E., Nies, E., Heremans, K., and Berghmans, H., The influence of pressure on the lower critical solution temperature miscibility behavior of aqueous solutions of poly(vinyl methyl ether) and the relation to the compositional curvature of the volume of mixing, *J. Phys. Chem. B*, 110, 7793, 2006.

2006LUC Lu, C., Guo, S.-R., Zhang, Y., and Yin, M., Synthesis and aggregation behavior of four different shaped PCL-PEG block copolymers, *Polym. Int.*, 55, 694, 2006.

2006MAE Maeda, T., Kanda, T., Yonekura, Y., Yamamoto, Y., and Aoyagi, T., Hydroxylated poly(*N*-isopropylacrylamide) as functional thermoresponsive materials, *Biomacromolecules*, 7, 545, 2006.

2006MOR Mori, T., Nakashima, M., Fukuda, Y., Minagawa, K., Tanaka, M., and Maeda, Y., Soluble-insoluble-soluble transitions of aqueous poly(*N*-vinylacetamide-*co*-acrylic acid) solutions, *Langmuir*, 22, 4336, 2006.

2006MUN Mun, G.A., Nurkeeva, Z.S., Akhmetkalieva, G.T., Shmakov, S.N., Khutoryanskiy, V.V., Lee, S.C., and Park, K., Novel temperature-responsive water-soluble copolymers based on 2-hydroxyethylacrylate and vinyl butyl ether and their interactions with poly(carboxylic acids), *J. Polym. Sci.: Part B: Polym. Phys.*, 44, 195, 2006.

2006MYL Mylonas, Y., Bokias, G., Iliopoulos, I., and Staikos, G., Interpolymer association between hydrophobically modified poly(sodium acrylate) and poly(*N*-isopropyl-acrylamide) in water, *Eur. Polym. J.*, 42, 849, 2006.

2006NAG Nagy, I., Loos, Th.W. de, Krenz, R.A., and Heidemann, R.A., High pressure phase equilibria in the systems linear low density polyethylene + n-hexane and linear low density polyethylene + n-hexane + ethylene, *J. Supercrit. Fluids*, 37, 115, 2006.

2006NIE Nies, E., Li, T., Berghmans, H., Heenan, R.K., and King, S.M., Upper critical solution temperature phase behavior, composition fluctuations, and complex formation in poly(vinyl methyl ether)/D_2O solutions: small-angle neutron-scattering experiments and Wertheim lattice thermodynamic perturbation theory predictions, *J. Phys. Chem. B*, 110, 5321, 2006.

2006OKA Okada, Y., Tanaka, F., Kujawa, P., and Winnik, F.M., Unified model of association-induced lower critical solution temperature phase separation and its application to solutions of telechelic poly(ethylene oxide) and of telechelic poly(*N*-isopropylacrylamide) in water, *J. Chem. Phys.*, 125, 244902, 2006.

2006OSA Osaka, N., Okabe, S., Karino, T., Hirabaru, Y., Aoshima, S., and Shibayama, M., Micro- and macrophase separations of hydrophobically solvated block copolymer aqueous solutions induced by pressure and temperature, *Macromolecules*, 39, 5875, 2006.

2006PAG Pagonis, K. and Bokias, G., Simultaneous lower and upper critical solution temperature-type co-nonsolvency behaviour exhibited in water-dioxane mixtures by linear copolymers and hydrogels containing *N*-isopropylacrylamide and *N,N*-dimethylacrylamide, *Polym. Int.*, 55, 1254, 2006.

2006PLU Plummer, R., Hill, D.J.T., and Whittaker, A.K., Solution properties of star and linear poly(*N*-isopropylacrylamide), *Macromolecules*, 39, 8379, 2006.

2006QIU Qiu, G.-M., Zhu, B.-K., Xu, Y.-Y., and Geckeler, K.E., Synthesis of ultrahigh molecular weight poly(styrene-*alt*-maleic anhydride) in supercritical carbon dioxide, *Macromolecules*, 39, 3231, 2006.

2006RIC Rico, M., Ramirez, C., Montero, B., Diez, J., and Lopez, J., Phase diagram for a system of polydisperse components consisting of the precursor of an epoxy/diamine thermoset and a thermoplastic: analysis based on a lattice theory model, *Macromol. Theor. Simul.*, 15, 487, 2006.

2006RUS Rusinova, E.V., Adamova, L.V., and Vshivkov, S.A., Phase transitions in solutions of polystyrene with poly(methyl methacrylate) and polybutadiene under deformation, *Polym. Sci., Ser. A*, 48, 159, 2006.

2006SAR Saravanan, S., Reena, J.A., Rao, J.R., Murugesan, T., and Nair, B.U., Phase equilibrium compositions, densities, and viscosities of aqueous two-phase poly(ethylene glycol) + poly(acrylic acid) system at various temperatures, *J. Chem. Eng. Data*, 51, 1246, 2006.

2006SCH Schagerlöf, H., Johansson, M., Richardson, S., Brinkmalm, G., Wittgren, B., and Tjerneld, F., Substituent distribution and clouding behavior of hydroxypropyl methyl cellulose analyzed using enzymatic degradation, *Biomacromolecules*, 7, 3474, 2006.

2006SHA Sharma, S.C., Acharya, D.P., Garcia-Roman, M., Itami, Y., and Kunieda, H., Phase behavior and surface tensions of amphiphilic fluorinated random copolymer aqueous solutions, *Colloids Surfaces A*, 280, 140, 2006.

2006SHE Shen, Z., Terao, K., Maki, Y., Dobashi, T., Ma, G., and Yamamoto, T., Synthesis and phase behavior of aqueous poly(*N*-isopropylacrylamide-*co*-acrylamide), poly(*N*-iso-propylacrylamide-*co-N,N*-dimethylacrylamide) and poly(*N*-isopropylacrylamide-*co*-2-hydroxyethyl methacrylate), *Colloid Polym. Sci.*, 284, 1001, 2006.

2006SOU Soule, E.R., Jaffrennou, B., Mechin, F., Pascault, J.P., Borrajo, J., and Williams, R.J., Thermodynamic analysis of the reaction-induced phase separation of solutions of random copolymers of methyl methacrylate and *N,N*-dimethylacrylamide in the precursors of a polythiourethane network, *J. Polym. Sci.: Part B: Polym. Phys.*, 44, 2821, 2006.

2006SUG Sugi, R., Ohishi, T., Yokoyama, A., and Yokozawa, T., Novel water-soluble poly(m-benzamide)s: Precision synthesis and thermosensitivity in aqueous solution, *Macromol. Rapid Commun.*, 27, 716, 2006.

2006TAO Tao, C.-T. and Young, T.-H., Polyetherimide membrane formation by the co-nosolvent system and its biocompatibility of MG63 cell line, *J. Membrane Sci.*, 269, 66, 2006.

2006THO Tho, I., Kjoniksen, A.-L., Knudsen, K.-D., and Nyström, B., Effect of solvent composition on the association behavior of pectin in methanol-water mixtures, *Eur. Polym. J.*, 42, 1164, 2006.

2006TO1 Torrens, F., Soria, V:, Codoner, A., Abad, C., and Campos, A., Compatibility between polystyrene copolymers and polymers in solution via hydrogen bonding, *Eur. Polym. J.*, 42, 2807, 2006.

2006TO2 Torrens, F., Soria, V., Monzo, I.S., Abad, C., and Campos, A., Treatment of poly(styrene-*co*-methacrylic acid)/poly(4-vinylpyridine) blends in solution under liquid–liquid phase-separation conditions. A new method for phase-separation data attainment from viscosity measurements, *J. Appl. Polym. Sci.*, 102, 5039, 2006.

2006VAN Van Durme, K., Van Mele, B., Bernaerts, K.V., Verdonck, B., and DuPrez, F.E., End-group modified poly(methyl vinyl ether): Characterization and LCST demixing behavior in water, *J. Polym. Sci.: Part B: Polym. Phys.*, 44, 461, 2006.

2006VER Verezhnikov, V.N., Plaksitskaya, T.V., and Poyarkova, T.N., pH-Thermosensitive behavior of *N,N*-dimethylaminoethyl methacrylate (co)polymers with *N*-vinyl-caprolactam, *Polym. Sci., Ser. A*, 48, 870, 2006.

2006WEN Weng, Y., Ding, Y., and Zhang, G., Microcalorimetric investigation on the lower critical solution temperature behavior of *N*-isopropylacrylamide-*co*-acrylic acid copolymer in aqueous solution, *J. Phys. Chem. B*, 110, 11813, 2006.

2006WIN Winoto, W., Adidharma, H., Shen, Y., and Radosz, M., Micellization temperature and pressure for polystyrene-*b*-polyisoprene in subcritical and supercritical propane, *Macromolecules*, 39, 8140, 2006.

2006WO1 Wohlfarth, C., *CRC Handbook of Enthalpy Data of Polymer-Sovent Systems*, Taylor & Francis, CRC Press, Boca Raton, 2006.

2006WO2 Wohlfarth, C., Upper critical (UCST) and lower critical (LCST) solution temperatures of binary polymer solutions, in *CRC Handbook of Chemistry and Physics*, Lide, D.R. (ed.), 87th ed., Taylor & Francis, CRC Press, Boca Raton, 13-19, 2006.

2006XIA Xia, Y., Burke, N.A.D., and Stoever, H.D.H., End group effect on the thermal response of narrow-disperse poly(*N*-isopropylacrylamide) prepared by atom transfer radical polymerization, *Macromolecules*, 39, 2275, 2006.

2006YA1 Yang, Z., Li, P., Chang, H., and Wang, S., Effect of diluent on the morphology and performance of IPP hollow fiber microporous membrane via thermally induced phase separation, *Chin. J. Chem. Eng.*, 14, 394, 2006.

2006YA2 Yankov, D.S., Stateva, R.P., Trusler, J.P.M., and Cholakov, G.St., Liquid-liquid equilibria in aqueous two-phase systems of poly(ethylene glycol) and poly(ethyleneimine): experimental measurements and correlation, *J. Chem. Eng. Data*, 51, 1056, 2006.

2006YIN Yin, X., Hoffman, A.S., and Stayton, P.S., Poly(*N*-isopropylacrylamide-*co*-propylacrylic acid) copolymers that respond sharply to temperature and pH, *Biomacromolecules*, 7, 1381, 2006.

2006ZHA Zhao, G. and Chen, S.B., Clouding and phase behavior of nonionic surfactants in hydrophobically modified hydroxyethyl cellulose solutions, *Langmuir*, 22, 9129, 2006.

2006ZHO Zhou, J., Lin Y., Du, Q., Zhong, W., and Wang, H., Effect of MPEG on MPEG-grafted EAA membrane formation via thermally induced phase separation, *J. Membrane Sci.*, 283, 310, 2006.

2006ZUO Zuo, D.-Y., Zhu, B.-K., Cao, J.-H., and Xu, Y.-Y., Influence of alcohol-based nonsolvents on the formation and morphology in phase inversion process, *Chin. J. Polym. Sci.*, 24, 281, 2006.

APPENDICES

Appendix 1 List of polymers in alphabetical order

Appendix 2 List of solvents in alphabetical order

Name	Formula	CAS-RN	Page(s)
acetaldehyde	C_2H_4O	75-07-0	110-114, 259, 317
acetic acid	$C_2H_4O_2$	64-19-7	303, 517, 523, 561
acetonitrile	C_2H_3N	75-05-8	92, 261, 312, 354, 527, 529, 540
2-aminoethanol	C_2H_7NO	141-43-5	540
acetophenone	C_8H_8O	98-86-2	515
aniline	C_6H_7N	62-53-3	300
anisole	C_7H_8O	100-66-3	23, 28, 238, 242, 244, 254, 290, 295-296, 447
benzene	C_6H_6	71-43-2	239, 254, 266, 292, 304, 317, 360, 392-393, 447-449, 515-516, 522, 526-527, 529, 532, 537
benzonitrile	C_7H_5N	100-47-0	449
benzyl acetate	$C_9H_{10}O_2$	140-11-4	244, 296, 315
benzylamine	C_7H_9N	100-46-9	360
benzyl alcohol	C_7H_8O	100-51-6	233-234, 285, 287, 389
benzyl phenyl ether	$C_{13}H_{12}O$	946-80-5	244, 265, 296, 315
benzyl propionate	$C_{10}H_{12}O_2$	122-63-4	244, 265, 296, 315
bis(2-ethoxyethyl) ether	$C_8H_{18}O_3$	112-36-7	17
bis(2-ethylhexyl) phthalate	$C_{24}H_{38}O_4$	117-81-7	115, 267, 317, 450, 514, 532
bisphenol-A diglycidyl ether	$C_{21}H_{24}O_4$	1675-54-3	116, 288, 295, 317, 360-361, 517-518, 532, 561
biphenyl	$C_{12}H_{10}$	92-52-4	244, 262, 265, 296, 313
bromobenzene	C_6H_5Br	108-86-1	327, 450, 532
1-bromobutane	C_4H_9Br	109-65-9	117, 267, 532
2-bromobutane	C_4H_9Br	78-76-2	532
1-bromodecane	$C_{10}H_{21}Br$	112-29-8	117, 267
bromoethane	C_2H_5Br	74-96-4	532
1-bromooctane	$C_8H_{17}Br$	111-83-1	117, 267
2-bromooctane	$C_8H_{17}Br$	557-35-7	117, 267
1-bromopropane	C_3H_7Br	106-94-5	532
4-bromotoluene	C_7H_7Br	106-38-7	532
n-butane	C_4H_{10}	106-97-8	23, 63-64, 242, 254, 262, 295, 296, 304, 358

Name	Formula	CAS-RN	Page(s)
1-chlorononane	$C_9H_{19}Cl$	2473-01-0	121, 268
1-chlorooctadecane	$C_{18}H_{37}Cl$	3386-33-2	121, 268, 318
1-chlorooctane	$C_8H_{17}Cl$	111-85-3	121, 268
2-chlorostyrene	C_8H_7Cl	2039-87-4	515, 533
1-chlorotetradecane	$C_{14}H_{29}Cl$	2425-54-9	121, 268, 318
1-chloroundecane	$C_{11}H_{23}Cl$	2473-03-2	121, 268
cyclodecane	$C_{10}H_{20}$	293-96-9	268, 318
cycloheptane	C_7H_{14}	291-64-5	254, 268, 318, 365-366
cyclohexane	C_6H_{12}	110-82-7	31, 103-104, 121-149, 221, 245, 248-250, 254, 262, 265, 269-271, 281-282, 299, 314, 318-319, 323-324, 352, 365-369, 451-457, 506, 511, 514, 520, 525-527, 529-531, 533, 537, 539, 542, 551-555
cyclohexanol	$C_6H_{12}O$	108-93-0	93, 149-150, 220, 271, 282, 285, 294, 312, 319-320, 327, 529
cyclohexanone	$C_6H_{10}O$	108-94-1	245, 296, 390-392, 515, 533
cyclooctane	C_8H_{16}	292-64-8	254, 271, 320
cyclopentane	C_5H_{10}	287-92-3	31, 48, 105, 150-152, 221, 223, 238, 245, 248-250, 254, 262, 264-265, 271-272, 282, 293, 296, 299, 314, 320, 325, 530, 533
cyclopentene	C_5H_8	142-29-0	31, 48, 223, 296
decafluorobutane	C_4F_{10}	355-25-9	234-236
decahydronaphthalene	$C_{10}H_{18}$	91-17-8	152, 272, 294, 315, 320, 519, 531, 533, 537
cis-decahydronaphthalene	$C_{10}H_{18}$	493-01-6	108
trans-decahydronaphthalene	$C_{10}H_{18}$	493-02-7	105, 152-153, 212, 262, 272, 282, 320
n-decane	$C_{10}H_{22}$	124-18-5	153, 232, 238, 242, 245, 254, 272, 219, 315-316, 320, 533
1-decanol	$C_{10}H_{22}O$	112-30-1	32, 153-154, 245, 272, 296, 303, 312, 524, 529
1-decene	$C_{10}H_{20}$	872-05-9	315
decyl acetate	$C_{12}H_{24}O_2$	112-17-4	273

Name	Formula	CAS-RN	Page(s)
diphenylmethane	$C_{13}H_{12}$	101-81-5	246, 262, 265, 290, 297, 313, 316
dodecadeuterocyclohexane	C_6D_{12}	1735-17-7	157-158, 274, 321
dodecadeuteromethylcyclopentane	C_6D_{12}	144120-51-4	274, 370
dodecafluoropentane	C_5F_{12}	678-26-2	234-236
n-dodecane	$C_{12}H_{26}$	112-40-3	14, 159, 232, 242, 246, 254, 274, 316, 321, 511, 515
1-dodecanol	$C_{12}H_{26}O$	112-53-8	35, 159, 246, 274, 297, 321
n-eicosane	$C_{20}H_{42}$	112-95-8	534
ethane	C_2H_6	74-84-0	25, 242, 295
1,2-ethanediol	$C_2H_6O_2$	107-21-1	312, 326, 345, 379, 506, 517, 521
ethanol	C_2H_6O	64-17-5	17, 25, 238, 243, 291, 296, 301, 325, 329-337, 339, 353, 372, 374-375, 379, 382, 508-509, 513, 516-518, 524-527, 529, 532, 534, 538-541, 558, 562
ethoxybenzene	$C_8H_{10}O$	103-73-1	25, 242, 295
2-ethoxyethanol	$C_4H_{10}O_2$	110-80-5	94, 237, 260, 288, 515, 519
2-(ethoxyethoxy)ethanol	$C_6H_{14}O_3$	111-90-0	239, 292, 506, 519
ethyl acetate	$C_4H_8O_2$	141-78-6	160, 237, 239, 260, 274, 281, 285, 289-290, 312, 321, 323, 463, 503, 513-514, 516, 527, 529, 534, 539, 560, 562
ethylbenzene	C_8H_{10}	100-41-4	239, 254, 292, 301, 321, 464
2-ethylbutanal	$C_6H_{12}O$	97-96-1	94-95, 260, 312
ethyl butyrate	$C_6H_{12}O_2$	105-54-4	263, 274
ethyl chloroacetate	$C_4H_7ClO_2$	105-39-5	239, 292
ethylcyclohexane	C_8H_{16}	1678-91-7	160-161, 274-275, 321
ethylcyclopentane	C_7H_{14}	1640-89-7	254
ethyl formate	$C_3H_6O_2$	109-94-4	275, 321
ethyl heptanoate	$C_9H_{18}O_2$	106-30-9	254
2-ethyl-1,3-hexanediol	$C_8H_{18}O_2$	94-96-2	506
ethyl hexanoate	$C_8H_{16}O_2$	123-66-0	254
2-ethyl-1-hexanol	$C_8H_{18}O$	104-76-7	519
3-ethylpentane	C_7H_{16}	617-78-7	67, 238, 246, 248-250, 254, 262, 264-265

Name	Formula	CAS-RN	Page(s)
4-ethylphenol	$C_8H_{10}O$	123-07-9	265, 316
ethyl propionate	$C_5H_{10}O_2$	105-37-3	263
fluorotrichloromethane	CCl_3F	75-69-4	297
formamide	CH_3NO	75-12-7	286, 302-303, 314, 326, 529, 531, 540
formic acid	CH_2O_2	64-18-6	511
n-heptane	C_7H_{16}	142-82-5	35-36, 64, 68, 161, 220, 231, 238, 242, 246, 248-250, 254, 256, 262, 264-265-266, 275, 291-292, 297-299, 316, 321, 343-345-349, 380, 522, 526, 534, 561
1-heptanol	$C_7H_{16}O$	111-70-6	246, 297, 508
2-heptanone	$C_7H_{14}O$	110-43-0	519
3-heptanone	$C_7H_{14}O$	106-35-4	260, 312, 325, 516
4-heptanone	$C_7H_{14}O$	123-19-3	95, 260-261, 312, 529
hexadecafluoroheptane	C_7F_{16}	335-57-9	234-236
n-hexadecane	$C_{16}H_{34}$	544-76-3	161, 232, 242, 275, 291, 321, 324, 515, 531, 534
1-hexadecanol	$C_{16}H_{34}O$	36653-82-4	161, 275
hexadeuterobenzene	C_6D_6	1076-43-3	464
1,1,1,3,3,3-hexadeutero-2-propanone	C_3D_6O	666-52-4	162-165, 275, 321, 370-372, 555-558
1,1,1,3,3,3-hexafluoro-2-propanol	C_3D_8O	920-66-1	238
n-hexane	C_6H_{14}	110-54-3	36-45, 64, 68, 109, 165, 238, 242, 246-250, 255-256, 262, 264-266, 275, 290, 292, 297, 299, 304, 316, 321-322, 343, 511, 513, 519, 522, 529, 532-534, 560-561
1,2-hexanediol	$C_6H_{14}O_2$	6920-22-5	562
hexanoic acid	$C_6H_{12}O_2$	142-62-1	165, 275, 322
5,8,11,13,16,19-hexaoxatricosane	$C_{17}H_{36}O_6$	143-29-3	287, 289
1-hexanol	$C_6H_{14}O$	111-27-3	45, 166, 247, 275, 290, 322, 508, 533
3-hexanol	$C_6H_{14}O$	623-37-0	166, 275, 322
3-hexanone	$C_6H_{12}O$	589-38-8	261, 312
hexyl acetate	$C_8H_{16}O_2$	142-92-7	106, 166-167, 222, 262, 276, 282
iodomethane	CH_3I	74-88-4	534
isobutyric acid	$C_4H_8O_2$	79-31-2	522
isopropylbenzene	C_9H_{12}	98-82-8	239, 292, 312, 534

Name	Formula	CAS-RN	Page(s)
2-methylpropyl acetate	$C_6H_{12}O_2$	110-19-0	263, 279, 292
1-methyl-2-pyrrolidinone	C_5H_9NO	872-50-4	63, 295, 339, 342, 515, 517-518, 526, 528-530, 538-539, 541, 545, 558, 561-562
nitrobenzene	$C_6H_5NO_2$	98-95-3	297, 327, 470
nitroethane	$C_2H_5NO_2$	79-24-3	192-197, 279, 322, 368-369, 522, 537
nitroethane-d5	$C_2D_5NO_2$	57817-88-6	195-197
nitromethane	CH_3NO_2	75-52-5	285, 322, 534
1-nitropropane	$C_3H_7NO_2$	108-03-2	291, 322
n-nonane	C_9H_{20}	111-84-2	68, 232, 238, 247-250, 256, 262, 266, 519
1-nonanol	$C_9H_{20}O$	143-08-8	247, 298
5-nonanone	$C_9H_{18}O$	502-56-7	285
4-nonylphenol	$C_{15}H_{24}O$	104-40-5	247, 298
octadecafluorooctane	C_8F_{18}	307-34-6	234-236
n-octadecane	$C_{18}H_{38}$	593-45-3	198, 279, 323
1-octadecanol	$C_{18}H_{38}O$	112-92-5	198, 279, 323
octamethylcyclotetrasiloxane	$C_8H_{24}O_4Si_4$	556-67-2	45-46, 298
n-octane	C_8H_{18}	111-65-9	46, 69, 198, 238, 242, 247-250, 255-256, 262, 264, 266, 279, 291, 298, 304, 316, 323, 519, 531, 535, 560
1,2-octanediol	$C_8H_{18}O_2$	1117-86-8	562
1-octanol	$C_8H_{18}O$	111-87-5	17, 46, 198-199, 247, 279, 298, 541, 561
2-octanone	$C_8H_{16}O$	111-13-7	96-97, 260, 312
3-octanone	$C_8H_{16}O$	106-68-3	97-100, 260-261, 313
1-octene	C_8H_{16}	111-66-0	46-47, 199, 279, 323, 343
4-octylphenol	$C_{14}H_{22}O$	1806-26-4	266, 298, 316
n-pentadecane	$C_{15}H_{32}$	629-62-9	199, 279
n-pentane	C_5H_{12}	109-66-0	47-48, 65, 199, 236, 238, 242, 247-250, 255, 262, 264-266, 279, 292, 298-299, 304, 316, 323, 516, 519
1-pentanol	$C_5H_{12}O$	71-41-0	18, 200, 238-239, 247, 290-292, 298, 323, 362-364, 372-373, 509, 560
2-pentanone	$C_5H_{10}O$	107-87-9	236, 287
3-pentanone	$C_5H_{10}O$	96-22-0	261, 313

Appendix 3 List of solvents in order of their molecular formulas

Formula	Name	CAS-RN	Page(s)
CCl_2F_2	dichlorodifluoromethane	75-71-8	297
CCl_3F		75-69-4	297
CCl_4	tetrachloromethane	56-23-5	239, 292, 325, 394, 472-473, 515, 518, 534, 538
CD_4O	methanol-d4	811-98-3	379
$CHCl_3$	trichloromethane	67-66-3	54, 101-102, 261, 289, 295, 313, 347-349, 443, 445, 496-502, 513-516, 522-525, 528-530, 534-539, 542-543, 560-561
CH_2Cl_2	dichloromethane	75-09-2	53, 285, 295, 328, 343-345, 513, 515, 517, 522, 529, 534, 543, 560
CH_2O_2	formic acid	64-18-6	511
CH_3I		4-88-4	534
CH_3NO	formamide	75-12-7	286, 302-303, 314, 326, 529, 531, 540
CH_3NO_2	nitromethane	75-52-5	285, 322, 534
CH_4O	methanol	67-56-1	53-54, 223, 243, 296, 325, 335, 346-347, 349, 352-362, 379-380, 382, 508-509, 511-514, 516-518, 520, 525-529, 532-534, 536, 538-542, 559-560-561
CS_2	carbon disulfide	75-15-0	318
$C_2Cl_2F_2$	dichlorodifluoroethene	79-35-6	297
$C_2Cl_3F_3$	1,1,2-trichloro-1,2,2-trifluoroethane	76-13-1	282, 298, 324
C_2Cl_4	tetrachloroethene	127-18-4	239, 292, 530, 537
$C_2D_5NO_2$	nitroethane-d5	57817-88-6	195-197
C_2HCl_3	trichloroethene	79-01-6	536, 542, 561
$C_2H_2Cl_2O_2$	dichloroacetic acid	79-43-6	289, 514
$C_2H_2Cl_4$	1,1,2,2-tetrachloroethane	79-34-5	511, 513-514, 530, 536, 560-561
C_2H_3Cl	vinyl chloride	75-01-4	514, 541
C_2H_3N	acetonitrile	75-05-8	92, 261, 312, 354, 527, 529, 540

Formula	Name	CAS-RN	Page(s)
C_3H_7NO	*N,N*-dimethylformamide	68-12-2	63, 233, 282, 285, 289, 324, 339-342, 374-378, 394-397, 427-428, 460-462, 502-503, 505, 514, 516-519, 525, 527, 529, 533, 537-539, 541, 545, 561-562
C_3H_7NO	*N*-methylacetamide	79-16-3	349, 524
$C_3H_7NO_2$	1-nitropropane	108-03-2	291, 322
C_3H_8	propane	74-98-6	26, 66, 242, 255, 295, 304, 324
$C_3H_8N_2O$	1,3-dimethylurea	598-94-7	531, 540
C_3H_8O	1-propanol	71-23-8	62-63, 101, 239, 243, 253, 261, 296, 312, 353, 364-365, 509, 513, 516-518, 525, 527, 530, 532-533, 538-540
C_3H_8O	2-propanol	67-63-0	238-239, 291-292, 299, 324-325, 339-341, 350, 359, 375, 383, 506, 514, 516-517, 520, 525-527, 530, 532, 539-541
$C_3H_8O_2$	dimethoxymethane	109-87-5	273, 321
$C_3H_8O_2$	1,3-propanediol	504-63-2	520
$C_3H_8O_3$	1,2,3-propanetriol	56-81-5	61-62, 253, 299, 347, 353, 510, 520, 522, 527, 541
C_4F_{10}	decafluorobutane	355-25-9	234-236
$C_4H_6O_2$	γ-butyrolactone	96-48-0	63, 327, 517
$C_4H_6O_2$	vinyl acetate	108-05-4	219, 281, 323
$C_4H_6O_3$	propylene carbonate	108-32-7	327, 513
$C_4H_6O_4$	dimethyl oxalate	553-90-2	157, 273, 321
$C_4H_7ClO_2$	ethyl chloroacetate	105-39-5	239, 292
C_4H_8	1-butene	106-98-9	358-359
$C_4H_8Br_2$	1,3-dibromobutane	107-80-2	533
C_4H_8O	2-butanone	78-93-3	233, 252, 260-261, 267, 285, 291, 295, 304, 312-313, 317, 323-324, 351, 361-365, 506, 516, 526, 530, 532, 538-539
C_4H_8O	tetrahydrofuran	109-99-9	15, 237-238, 264, 282, 285, 324, 337-338, 350-351, 379-382, 384-385, 473-480, 506, 513-514,

Formula	Name	CAS-RN	Page(s)
$C_6H_5NO_2$	nitrobenzene	98-95-3	297, 327, 470
C_6H_6	benzene	71-43-2	239, 254, 266, 292, 304, 317, 360, 392-393, 447-449, 515-516, 522, 526-527, 529, 532, 537
C_6H_6O	phenol	108-95-2	316, 509, 524, 540, 561
C_6H_7N	aniline	62-53-3	300
$C_6H_{10}O$	cyclohexanone	108-94-1	245, 296, 390-392, 515, 533
$C_6H_{10}O_4$	butanedioic acid dimethyl ester	106-65-0	117, 267, 317
$C_6H_{10}O_4$	diethyl oxalate	95-92-1	155-156, 273, 321
$C_6H_{11}NO$	ε-caprolactam	105-60-2	287, 294, 299
C_6H_{12}	cyclohexane	110-82-7	31, 103-104, 121-149, 221, 245, 248-250, 254, 262, 265, 269-271, 281-282, 299, 314, 318-319, 323-324, 352, 365-369, 451-457, 506, 511, 514, 520, 525-527, 529-531, 533, 537, 539, 542, 551-555
C_6H_{12}	methylcyclopentane	96-37-7	189-191, 247-250, 255, 266, 278-279, 299, 322, 370, 535
$C_6H_{12}N_2O$	dimethylpropyleneurea	89607-25-0	518, 538
$C_6H_{12}O$	cyclohexanol	108-93-0	93, 149-150, 220, 271, 282, 285, 294, 312, 319-320, 327, 529
$C_6H_{12}O$	2-ethylbutanal	97-96-1	94-95, 260, 312
$C_6H_{12}O$	3-hexanone	589-38-8	261, 312
$C_6H_{12}O$	4-methyl-2-pentanone	108-10-1	191, 263, 279, 314, 537
$C_6H_{12}O_2$	butyl acetate	123-86-4	102, 118, 221, 239, 245, 262-263, 267-268, 282, 290, 296, 314-315, 527, 529
$C_6H_{12}O_2$	2-butyl acetate	105-46-4	268
$C_6H_{12}O_2$	*tert*-butyl acetate	540-88-5	119, 239, 251, 268, 292, 300, 318, 533
$C_6H_{12}O_2$	ethyl butyrate	105-54-4	263, 274
$C_6H_{12}O_2$	hexanoic acid	142-62-1	165, 275, 322
$C_6H_{12}O_2$	2-methylpropyl acetate	110-19-0	263, 279, 292
$C_6H_{12}O_2$	propyl propionate	106-36-5	264
C_6H_{14}	2,2-dimethylbutane	75-83-2	156, 238, 248-250, 254, 262, 264-265

Formula	Name	CAS-RN	Page(s)
C_6H_{14}	2,3-dimethylbutane	79-29-8	156, 248-250, 254, 265
C_6H_{14}	n-hexane	110-54-3	36-45, 64, 68, 109, 165, 238, 242, 246-250, 255-256, 262, 264-266, 275, 290, 292, 297, 299, 304, 316, 321-322, 343, 511, 513, 519, 522, 529, 532-534, 560-561
C_6H_{14}	2-methylpentane	107-83-5	191, 248, 255, 299
C_6H_{14}	3-methylpentane	96-14-0	191, 255-256, 299, 522
$C_6H_{14}O$	1-hexanol	111-27-3	45, 166, 247, 275, 290, 322, 508, 533
$C_6H_{14}O$	3-hexanol	623-37-0	166, 275, 322
$C_6H_{14}O_2$	2-butoxyethanol	111-76-2	519-520, 522
$C_6H_{14}O_2$	1,2-hexanediol	6920-22-5	562
$C_6H_{14}O_2$	2-methyl-2,4-pentanediol	107-41-5	505-506
$C_6H_{14}O_3$	2-(ethoxyethoxy)ethanol	111-90-0	239, 292, 506, 519
$C_6H_{14}O_4$	tri(ethylene glycol)	112-27-6	102, 261, 313, 515, 521
$C_6H_{15}O_4P$	triethyl phosphate	78-40-0	541
C_7F_{16}	hexadecafluoroheptane	335-57-9	234-236
C_7H_5N	benzonitrile	100-47-0	449
C_7H_7Br	4-bromotoluene	106-38-7	532
C_7H_8	toluene	108-88-3	101, 219-220, 238-239, 255, 261, 281, 290, 292-294, 301, 304, 313, 323, 328, 335-337, 351, 358-359, 430, 443-444, 481-496, 514-516, 518, 526-527, 529-531, 534-537, 539, 542, 544, 560, 562
C_7H_8O	anisole	100-66-3	23, 28, 238, 242, 244, 254, 290, 295-296, 447
C_7H_8O	benzyl alcohol	100-51-6	233-234, 285, 287, 389
C_7H_8O	4-methylphenol	106-44-5	266, 316
C_7H_9N	benzylamine	100-46-9	360
C_7H_9N	2,6-dimethylpyridine	108-48-5	512, 522
$C_7H_{12}O_2$	butyl acrylate	141-32-2	514, 541
$C_7H_{12}O_4$	diethyl malonate	105-53-3	154, 273, 320-321
C_7H_{14}	cycloheptane	291-64-5	254, 268, 318, 365-366
C_7H_{14}	ethylcyclopentane	1640-89-7	254
C_7H_{14}	methylcyclohexane	108-87-2	106-107, 169-189, 247-250, 255, 262, 265-266, 277-278, 314, 322, 367-368, 464-470, 530, 534-535, 537

Formula	Name	CAS-RN	Page(s)
C_8H_{10}	ethylbenzene	100-41-4	239, 254, 292, 301, 321, 464
$C_8H_{10}O$	ethoxybenzene	103-73-1	25, 242, 295
$C_8H_{10}O$	4-ethylphenol	123-07-9	265, 316
$C_8H_{10}O$	phenetole	103-73-1	227, 248, 255, 298, 316, 516, 526
$C_8H_{10}O$	2-phenylethanol	60-12-8	524
$C_8H_{11}N$	*N,N*-dimethylaniline	121-69-7	459
C_8H_{16}	cyclooctane	292-64-8	254, 271, 320
C_8H_{16}	1,4-dimethylcyclohexane	589-90-2	273, 321, 534
C_8H_{16}	ethylcyclohexane	1678-91-7	160-161, 274-275, 321
C_8H_{16}	1-octene	111-66-0	46-47, 199, 279, 323, 343
C_8H_{16}	propylcyclopentane	2040-96-2	255
$C_8H_{16}O$	2-octanone	111-13-7	96-97, 260, 312
$C_8H_{16}O$	3-octanone	106-68-3	97-100, 260-261, 313
$C_8H_{16}O_2$	ethyl hexanoate	123-66-0	254
$C_8H_{16}O_2$	hexyl acetate	142-92-7	106, 166-167, 222, 262, 276, 282
$C_8H_{17}Br$	1-bromooctane	111-83-1	117, 267
$C_8H_{17}Br$	2-bromooctane	557-35-7	117, 267
$C_8H_{17}Cl$	1-chlorooctane	111-85-3	121, 268
C_8H_{18}	2,2-dimethylhexane	590-73-8	254
C_8H_{18}	2,4-dimethylhexane	589-43-5	254
C_8H_{18}	2,5-dimethylhexane	592-13-2	66, 238, 250, 256
C_8H_{18}	3,4-dimethylhexane	583-48-2	66, 238, 245, 248-250, 254, 256, 264-265
C_8H_{18}	2-methylheptane	592-27-8	255
C_8H_{18}	3-methylheptane	589-81-1	255
C_8H_{18}	n-octane	111-65-9	46, 69, 198, 238, 242, 247-250, 255-256, 262, 264, 266, 279, 291, 298, 304, 316, 323, 519, 531, 535, 560
C_8H_{18}	2,2,4-trimethylpentane	540-84-1	70, 219, 238, 248-250, 255-256, 264, 266, 290-291, 293
$C_8H_{18}O$	dibutyl ether	142-96-1	526
$C_8H_{18}O$	2-ethyl-1-hexanol	104-76-7	519
$C_8H_{18}O$	1-octanol	111-87-5	17, 46, 198-199, 247, 279, 298, 541, 561
$C_8H_{18}O_2$	2-ethyl-1,3-hexanediol	94-96-2	506
$C_8H_{18}O_2$	1,2-octanediol	1117-86-8	562
$C_8H_{18}O_3$	2-(2-butoxyethoxy)ethanol	112-34-5	239, 292, 531

Formula	Name	CAS-RN	Page(s)
$C_{12}H_{10}O$	diphenyl ether	101-84-8	32-35, 49, 59, 244-246, 250, 252, 254, 262, 265, 290, 297, 299, 313, 315-316, 519, 526, 531
$C_{12}H_{18}$	4-*tert*-butyl-ethylbenzene	7364-19-4	120, 268
$C_{12}H_{18}$	1-phenylhexane	1077-16-3	531
$C_{12}H_{24}O_2$	decyl acetate	112-17-4	273
$C_{12}H_{25}Cl$	1-chlorododecane	112-52-7	120, 268, 318
$C_{12}H_{26}$	n-dodecane	112-40-3	14, 159, 232, 242, 246, 254, 274, 316, 321, 511, 515
$C_{12}H_{26}O$	1-dodecanol	112-53-8	35, 159, 246, 274, 297, 321
$C_{12}H_{27}O_4P$	tributyl phosphate	126-73-8	524
$C_{13}H_{12}$	diphenylmethane	101-81-5	246, 262, 265, 290, 297, 313, 316
$C_{13}H_{12}O$	benzyl phenyl ether	946-80-5	244, 265, 296, 315
$C_{13}H_{20}$	1-phenylheptane	1078-71-3	531
$C_{13}H_{28}$	n-tridecane	629-50-5	248, 536
$C_{14}H_{14}O$	dibenzyl ether	103-50-4	245, 265, 297, 315
$C_{14}H_{22}$	1-phenyloctane	2189-60-8	202, 531
$C_{14}H_{22}O$	4-octylphenol	1806-26-4	266, 298, 316
$C_{14}H_{29}Cl$	1-chlorotetradecane	2425-54-9	121, 268, 318
$C_{14}H_{30}O$	1-tetradecanol	112-72-1	219, 280
$C_{15}H_{24}O$	4-nonylphenol	104-40-5	247, 298
$C_{15}H_{32}$	n-pentadecane	629-62-9	199, 279
$C_{16}H_{22}O_4$	dibutyl phthalate	84-74-2	259, 283, 285, 287, 289-290, 292, 310, 313, 315, 324, 327, 530-531, 533
$C_{16}H_{26}$	1-phenyldecane	104-72-3	201, 280, 323
$C_{16}H_{33}Cl$	1-chlorohexadecane	4860-03-1	120, 268, 318
$C_{16}H_{34}$	n-hexadecane	544-76-3	161, 232, 242, 275, 291, 321, 324, 515, 531, 534
$C_{16}H_{34}O$	1-hexadecanol	36653-82-4	161, 275
$C_{17}H_{34}O_2$	isopropyl myristate	110-27-0	525, 562
$C_{17}H_{36}O_6$	5,8,11,13,16,19-hexaoxatricosane	143-29-3	287, 289
$C_{18}H_{26}O_4$	dipentyl phthalate	131-18-0	259, 310
$C_{18}H_{30}$	1-phenyldodecane	123-01-3	201-202, 280
$C_{18}H_{37}Cl$	1-chlorooctadecane	3386-33-2	121, 268, 318
$C_{18}H_{38}$	n-octadecane	593-45-3	198, 279, 323
$C_{18}H_{38}O$	1-octadecanol	112-92-5	198, 279, 323
$C_{20}H_{30}O_4$	dihexyl phthalate	84-75-3	259, 310, 315
$C_{20}H_{42}$	n-eicosane	112-95-8	534

INDEX

Milton Keynes UK
Ingram Content Group UK Ltd.
UKHW051928141024
449569UK00027B/1393